Modern
Structural
Analysis

Modern Structural Analysis

The Matrix Method Approach

Anthony E. Armenàkas

Professor of Aerospace and Mechanical Engineering
Polytechnic University
Brooklyn, New York

McGraw-Hill, Inc.

New York   St. Louis   San Francisco   Auckland   Bogotá
Caracas   Hamburg   Lisbon   London   Madrid
Mexico   Milan   Montreal   New Delhi   Paris
San Juan   São Paulo   Singapore
Sydney   Tokyo   Toronto

Library of Congress Cataloging-in-Publication Data

Armenàkas, Anthony E., date
    Modern structural analysis: the matrix method approach /
Anthony E. Armenàkas.

        p.   cm.
    Includes index.
    ISBN 0-07-002348-4
    1. Structural analysis (Engineering)—Matrix methods.   I. Title.
TA642.A75   1991
624.1'71—dc20   90-46020

1 2 3 4 5 6 7 8 9 0   DOC/DOC   9 6 5 4 3 2 1 0

ISBN   0-07-002348-4

*The sponsoring editor for this book was Joel Stein, the editing
supervisor was Stephen M. Smith, the designer was Naomi Auerbach,
and the production supervisor was Thomas G. Kowalczyk. It was set in
Century Schoolbook by The Universities Press (Belfast) Ltd.*

Printed and bound by R. R. Donnelley & Sons Company.

To the memory of my mother,
Euterpe Sakis-Armenàka,
for her immense kindness and generosity

Contents

Part 1   Preliminary Considerations

Part 3   The Direct Stiffness Method

Part 4   The Modern Flexibility Method

Part 5   Derivation and Applications of the Principle of Virtual Work

Preface

This is the second of a series of two books which cover in a unified way the most important methods for analyzing framed structures. The first book is entitled *Classical Structural Analysis: A Modern Approach,* and as its title indicates it is devoted entirely to the classical methods. These methods are best suited for analyzing relatively simple structures by hand calculations. This second book is devoted to the modern (matrix) methods for analyzing framed structures. These methods have been developed in the last 30 years and are best suited for writing programs for analyzing framed structures by computer.

These two books have been written as reference texts for practicing engineers and could be used as texts for undergraduate and graduate courses in structural analysis. The knowledge required for studying them is contained in basic books in statics, strength of materials, calculus, and elements of matrix algebra. In each section of these books, the presentation of the pertinent theory is followed by a number of solved examples which contribute to a better understanding of the theory and illustrate its application. The structures analyzed in these examples have only a few unknown quantities and thus lengthy calculations are avoided. However, these structures can be readily analyzed employing one of the more physically obvious classical methods and, consequently, the advantage of the matrix methods may not always become apparent. A number of photographs of interesting structures with a brief description and sketches of important details are presented in the two books. They bring to the attention of the reader some of the interesting structural or aesthetic features of these structures, as well as ingenious aspects of their construction.

This book is divided into six parts. Part 1 consists of Chaps. 1 to 3. In Chap. 1 certain preliminary concepts are presented, including definitions of terminology, sign conventions, and a discussion of the idealizations and assumptions made in the analysis of framed structures. In Chap. 2 the fundamental relations of the mechanics of materials theories which are pertinent to the analysis of framed structures are established, and the strong forms of the boundary-value problems for computing the components of displacement and the internal forces and moments of the elements of framed structures are formulated. In Chap. 3 the modern methods (direct stiffness and modern flexibility) for analyzing framed structures are introduced by applying them to a simple example. Our aim in this chapter is to demonstrate the salient features of these methods and to describe the important steps which must be followed when applying them, without delving into details. Moreover, we indicate some of the advantages and disadvantages of these methods.

In modern structural analysis a framed structure is considered to be an assemblage of one-dimensional (line) elements whose ends are connected to a number of points called *nodes*. The response of a structure is determined from the response of its elements. Thus in Part 2 (Chaps. 4 to 6) the matrices and the equations which are used to describe the response of an element are established, while in Parts 3 and 4 methods for assembling and solving the equations which describe the response of a structure are presented.

In modern structural analysis the response of an element is regarded as the sum of its response when subjected to the given loads acting along its length with its ends fixed and its response when subjected only to the displacements of its ends. In the first case the response of an element is described by its fixed-end actions (see Chap. 4). In the second case the response of an element is described either by its stiffness equations (see Chap. 4) or by its flexibility equations (see Chap. 5). In Chap. 6 the transformation matrices are formed for transforming the components of internal actions and of displacements of an element from local to global and vice versa.

Part 3 consists of Chaps. 7 to 14 and is devoted to the direct stiffness or direct displacement method. This method is used in practice almost exclusively when writing programs for analyzing framed structures by computer. In this method the analysis of a structure (statically determinate or indeterminate) is formulated in terms of the components of displacements of its nodes.

In modern structural analysis the response of a structure is regarded as the sum of its response when subjected to the given loads with its nodes fixed and its response when subjected to equivalent actions on its nodes, that is, to the concentrated forces and moments which, when applied to the nodes of a structure, displace each one of them by an amount equal to the displacement of the corresponding node of the structure subjected to the given loads. Chapter 7 is devoted to the analysis of structures subjected to given loads with their nodes fixed and to the computation of the equivalent actions. The response of a structure subjected to equivalent actions on its nodes is expressed either by its stiffness equations or by its flexibility equations. In Chap. 8 methods for assembling the stiffness equations for a structure directly from the stiffness equations for its elements are presented. In Chap. 9 the boundary conditions of the structure are introduced into its stiffness equations, which are then solved to give the components of displacements of the nodes of the structure and its reactions. Subsequently, the components of displacements of the nodes of the structure are used to compute the internal actions acting on the ends of its elements. In Chap. 10 the direct stiffness method is extended to structures having skew supports or other special constraints. In Chap. 11 an effective procedure for programming the analysis of framed structures using the direct stiffness method is described.

Chapters 7 to 9, due to their introductory nature, cover only the basic procedures used in the direct stiffness method. Chapters 10 and 12 to 14 cover some of the special procedures which are essential for the efficient solution of the out-of-the-ordinary problems encountered in practice. In Chap. 12 a method is presented for condensing the stiffness equations for a structure by eliminating a number of specific unknown components of displacements of its nodes. Moreover, a method is presented for computing the approximate response of elements of complex geometry and loading. Chapter 13 is devoted to the method of substructures. In this method a structure is subdivided into parts, referred to as substructures, each substructure is analyzed, and the results are combined to obtain the components of displacements of the nodes of the structure. In Chap. 14 procedures for analyzing redesigned structures are presented, using the results of the analysis of the original structure.

Part 4 consists of Chaps. 15 to 17 and is devoted to the modern flexibility or modern force method. In this method the procedures used to analyze statically determinate structures differ from those used to analyze statically indeterminate structures. In Chap. 15 systematic procedures are presented for writing (1) the equations of equilibrium for the nodes of a structure and (2) the equations of compatibility of the components of displacements of the ends of each element of a structure with the components of displacements of the nodes, to which the element is connected. In Chap. 16 the components of displacement of statically determinate structures are computed by using the flexibility and stiffness methods. The procedures described in Chaps. 15 and 16 can be used very effectively to write programs for analyzing statically determinate structures by computer. However, such programs are of rather limited scope and, consequently, practical application. In Chap. 17 the modern flexibility method for analyzing statically indeterminate framed structures is presented. In this method the analysis of a structure is formulated in terms of some of its reactions and/or internal actions (redundants). Up to now this method has found little practical use in writing general computer programs for analyzing statically indeterminate structures. A major reason for this is that the choice of the unknown internal actions and/or reactions (redundants) is not unique and it affects the stability of the matrices which must be inverted in order to analyze a statically indeterminate structure. Thus it is possible that the modern flexibility method could find more practical applications when effective procedures are established for obtaining the optimum choice of redundants by computer.

In Part 2 we establish the response of an element by solving the differential equations of equilibrium for the segments of the element and by satisfying its boundary conditions. In Parts 3 and 4 we establish the response of a framed structure by directly satisfying the requirements for equilibrium of its nodes and for compatibility of the components of displacements of the ends of its elements with the components of displacements of its nodes. In Part 5 we establish the response of an element and of a structure by using the principle of virtual work.

The principle of virtual work for a body represents an integral form of the boundary-value problem for computing the components of displacement strain and stress of this body. Integral forms of boundary-value problems have been used extensively in constructing approximate solutions for them (for example, weighted residual and finite-element methods). In Chap. 18 the principle of virtual work for a body is derived and specialized to framed structures. In Chap. 19 the finite-element method is described and applied in conjunction with the principle of virtual work to establish approximate stiffness equations and matrices of fixed-end actions for elements of framed structures. Moreover, the principle of virtual work is employed to establish exact formulas for the flexibility coefficients for certain types of elements, including tapered and curved. In Chap. 20 the principle of virtual work is employed to (1) compute a component of displacement of a point of a framed structure (method of virtual work or dummy load method) and (2) obtain the stiffness equations for a structure.

The last part of this book contains two appendices. Appendix A is a brief introduction to the concept and applications of the functions of discontinuity. Appendix B is an outline of the elements of vector analysis used in this book.

The author is deeply appreciative of and forever grateful to his valued colleague, his late wife Stella, who provided inestimable assistance and indefatigable support during the preparation of this book. Moreover, the author wishes to express his appreciation and thanks to Dr. Theodore Balderes of Grumman Aerospace Corp. for reading the entire manuscript and making many valuable suggestions, and to Professor H. H. Pan of Polytechnic University and Professors John T. Katsikadelis, Vlasis C. Koumousis, and M. Papadrakakis of the National Technical University of Athens for reading parts of the manuscript and making helpful comments.

This book was written during the time the author was professor and director of the Institute of Structural Analysis of the National Technical University of Athens. Miss Dia Troullinou typed the first draft of the manuscript and Mrs. Evgenia Kapou typed the revised manuscript. Their superior ability and patience is greatly appreciated.

Anthony E. Armenàkas

Partial List of Symbols

A	Area.
A_e	Area of element e.
$\{A\}$ or $\{\bar{A}\}$	Local or global matrix of the nodal actions, respectively, of an element of a structure subjected to given loads. For an element of a planar truss in the $x_1 x_2$ plane $\{A\}^T = [F_1^j \quad F_1^k]$, while $\{\bar{A}\}^T = [\bar{F}_1^j \quad \bar{F}_2^j \quad \bar{F}_1^k \quad \bar{F}_2^k]$; for an element of a planar beam or frame in the $x_1 x_2$ plane $\{A\}^T = [F_1^j \quad F_2^j \quad M_3^j \quad F_1^k \quad F_2^k \quad M_3^k]$; for an element of a space beam or frame $\{A\}^T = [F_1^j \quad F_2^j \quad F_3^j \quad M_1^j \quad M_2^j \quad M_3^j \quad F_1^k \quad F_2^k \quad F_3^k \quad M_1^k \quad M_2^k \quad M_3^k]$.
$\{A^e\}$ or $\{\bar{A}^e\}$	Local or global matrix of nodal actions, respectively, of element e of a structure ($e = 1, 2, \ldots, NE$).
$\{A^E\}$	Local matrix of nodal actions of an element of a structure subjected to the equivalent actions on its nodes.
$\{A^{Ee}\}$	Local matrix of nodal actions of element e of a structure subjected to the *equivalent actions on its nodes*.
$\{A^R\}$ or $\{\bar{A}^R\}$	Local or global matrix of nodal actions, respectively, of an element *of the restrained structure,* that is, the structure subjected to the given loads with its nodes fixed against translation and rotation.
$\{A^{Re}\}$ or $\{\bar{A}^{Re}\}$	Local or global matrix of nodal actions, respectively, of element e of the restrained structure.
$\{\hat{A}^R\}$	Global matrix of nodal actions of all the elements of the restrained structure. For a structure with NE elements $\{\hat{A}^R\}^T = [\{\bar{A}^{R1}\}^T \quad \{\bar{A}^{R2}\}^T \quad \ldots \quad \{\bar{A}^{RNE}\}^T]$.
$\{a\}$	Matrix of basic nodal actions of an element of a structure subjected to the given loads. For an element of a truss we choose $\{a\} = F_1^k$; for an element of a planar beam or frame in the $x_1 x_2$ plane we choose $\{a\}^T = [F_1^k \quad F_2^k \quad M_3^k]$. For an

	element of a space beam or frame we choose $\{a\}^T = [F_1^k \quad F_2^k \quad F_3^k \quad M_1^k \quad M_2^k \quad M_3^k]$.
$\{a^E\}$	Matrix of basic nodal actions of an element of a structure subjected to equivalent actions on its nodes.
$\{a^R\}$	Matrix of basic nodal actions of an element of the restrained structure, that is, of the structure subjected to the given loads with its nodes fixed against translation and rotation.
$\{a^{Re}\}$	Matrix of basic nodal actions of element e of the restrained structure.
$\{\hat{a}^E\}$	Matrix of basic nodal actions of all the elements of a structure subjected to equivalent actions on its nodes. For a structure with NE elements $\{\hat{a}^E\}^T = [\{a^{E1}\}^T \quad \{a^{E2}\}^T \quad \ldots \quad \{a^{ENE}\}^T]$.
$\{\hat{a}^{EM}\}$	Matrix of the basic nodal actions of all the elements of the model of a structure subjected to equivalent actions on its nodes.
$\{a_x^{EM}\}$	Matrix of the chosen redundant basic nodal actions of the model of a structure subjected to equivalent actions on its nodes.
$\{a_0^{EM}\}$	Matrix of the basic nodal actions of the model for a structure subjected to equivalent actions on its nodes which are not included in the matrix $\{a_x^{EM}\}$.
$[B]$	Equilibrium matrix for a structure subjected to equivalent actions on its nodes. It is defined by $\{P^{EF}\} = [B]\{\hat{a}^E\}$.
$[\hat{B}]$	Equilibrium matrix for a structure subjected to equivalent actions on its nodes. It is defined by $\{\hat{P}^E\} = [\hat{B}] = \left\{ \begin{array}{c} \{\hat{a}^E\} \\ -\{R\} \end{array} \right\}$.
$[b] = [B]^{-1}$	This matrix exists only for statically determinate structures.
$[C]$	Compatibility matrix for a structure subjected to equivalent actions on its nodes. It is defined by $[\hat{d}] = [C]\{\Delta^F\}$.
$[\hat{C}]$	Compatibility matrix for a structure subjected to equivalent actions on its nodes. It is defined as $\left\{ \begin{array}{c} \{\hat{d}\} \\ \{\Delta^S\} \end{array} \right\} = [\hat{C}][\hat{\Delta}]$.
$[c] = [C]^{-1}$	This matrix exists only for statically determinate structures.

$\{D\}$ or $\{\bar{D}\}$	Local or global matrix of nodal displacements, respectively, of an element of a structure subjected to given loads. It is identical to that of an element of the structure subjected to equivalent actions on its nodes. For an element of a truss $\{D\}^T = [u_1^j \quad u_1^k]$, while $\{\bar{D}\}^T = [\bar{u}_1^j \quad \bar{u}_2^j \quad \bar{u}_1^k \quad \bar{u}_2^k]$; for an element of a planar beam or frame $\{D\}^T = [u_1^j \quad u_2^j \quad \theta_3^j \quad u_1^k \quad u_2^k \quad \theta_3^k]$; for an element of a space beam or frame $\{D\}^T = [u_1^j \quad u_2^j \quad u_3^j \quad \theta_1^j \quad \theta_2^j \quad \theta_3^j \quad u_1^j \quad u_2^j \quad u_3^j \quad \theta_1^j \quad \theta_2^j \quad \theta_3^j]$.
$\{D^e\}$ or $\{\bar{D}^e\}$	Local or global matrix of nodal displacements, respectively, of element e of a structure subjected to given loads or to the equivalent actions on its nodes.
$\{d\}$	Matrix of the basic deformation parameters of an element of a structure.
$\{d^e\}$	Matrix of the basic deformation parameters of element e of a structure.
$\{\hat{d}\}$	Matrix of the basic deformation parameters of all the elements of a structure. For a structure with NE elements $\{\hat{d}\}^T = [\{d^1\}^T \quad \{d^2\}^T \quad \dots \quad \{d^{NE}\}^T]$.
E	Modulus of elasticity.
E_e	Modulus of elasticity of element e.
F_i^q or $\bar{F}_i^q$ ($i = 1, 2, 3; q = j$ or k)	Local or global component in the x_i or $\bar{x}_i$ direction, respectively, of the internal force acting at the end q ($q = j$ or k) of an element of a structure subjected to given loads.
F_i^{eq} or $\bar{F}_i^{eq}$ ($i = 1, 2, 3; q = j$ or k)	Local or global component in the x_i or $\bar{x}_i$ direction, respectively, of the internal force acting at the end q ($q = j$ or k) of element e of a structure subjected to given loads.
F_i^{Eq} or $\bar{F}_i^{Eq}$ ($i = 1, 2, 3; q = j$ or k)	Local or global component in the x_i or $\bar{x}_i$ direction, respectively, of the internal force acting at the end q ($q = j$ or k) of an element of a structure subjected to equivalent actions on its nodes.
F_i^{Rq} or $\bar{F}_i^{Rq}$ ($i = 1, 2, 3; q = j$ or k)	Local or global component in the x_i or $\bar{x}_i$ direction of the internal force, respectively, acting at the end q ($q = j$ or k) of an element of the restrained structure, that is, the structure subjected to the given loads with its nodes fixed against translation and rotation.
F_i^{Req} or $\bar{F}_i^{Req}$ ($i = 1, 2, 3; q = j$ or k)	Local or global component in the x_i or $\bar{x}_i$ direction of the internal force, respectively, acting at the end q ($q = j$ or k) of an element e

	of the restrained structure, that is, the structure subjected to the given loads with its nodes fixed against translation and rotation.
$[f]$	Flexibility matrix for an element of a structure.
$[f^e]$	Flexibility matrix for element e of a structure.
$[\hat{f}]$	Flexibility matrix of all the elements of a structure defined by relation (16.9).
G	Shear modulus.
I_i $(i = 2, 3)$	Moment of inertia of the cross section of an element about its x_i $(i = 2, 3)$ principal centroidal axis.
$I_i^{(e)}$ $(i = 2, 3)$	Moment of inertia of the cross section of element e about its x_i $(i = 2, 3)$ principal centroidal axis.
J	Polar moment of inertia of the cross section of an element.
K	Torsional constant of the cross section of an element.
$[K]$ or $[\bar{K}]$	Local or global stiffness matrix for an element of a structure, respectively.
$[K^e]$ or $[\bar{K}^e]$	Local or global stiffness matrix for element e of a structure, respectively.
$[k]$	Basic local stiffness matrix for an element of a structure. It relates its basic nodal actions to its basic deformation parameters $\{a\} = \{k\}\{d\}$.
$[k^e]$	Basic local stiffness matrix for element e of a structure defined by relations (16.12).
L	Length.
L_e	Length of element e.
$\mathcal{M}_i^{(A)}$ or $\mathcal{M}_i^{(3)}$	Local component in the x_i $(i = 1, 2, 3)$ direction of the concentrated external moment acting on point A or 3 of an element of a structure, respectively.
$\mathcal{M}^{(A)}$ or $\mathcal{M}^{(3)}$	Concentrated external moment vector acting on point A or 3 of a structure, respectively.
$M_i(x_i)$	Local component in the x_i $(i = 1, 2, 3)$ direction of the internal moment of an element.
M_i^q or $\bar{M}_i^q$ $(i = 1, 2, 3; q = j$ or $k)$	Local or global component in the x_i or $\bar{x}_i$ direction, respectively, of the internal moment acting at the end q $(q = j$ or $k)$ of an element.
M_i^{eq} or $\bar{M}_i^{eq}$ $(i = 1, 2, 3; q = j$ or $k)$	Local or global component in the x_i or $\bar{x}_i$ direction, respectively, of the internal moment acting at the end q $(q = j$ or $k)$ of element e.

$m_i(x_1)$ $(i = 1, 2, 3)$	Component in the x_i direction of the external distributed moment acting on an element. It is given in units of moment per unit length of the element.
$\mathbf{m}(x_i)$	External distributed moment vector acting on an element.
N or $N(x_1)$	Axial component of internal force in an element.
$P_i^{(A)}$ or $P_i^{(3)}$ $(i = 1, 2, 3)$	Local components in the x_i direction of the concentrated external force acting on point A or 3 of an element of a structure, respectively.
$\mathbf{P}^{(A)}$	Concentrated external force vector acting on point A of a structure.
$\{\hat{P}^E\}$	Matrix of the global components of the equivalent actions acting on all the nodes of a structure.
$\{P^{EF}\}$	Matrix of the global components of the equivalent actions acting on the nodes of a structure, which are not directly absorbed by its supports.
$\{P^{ES}\}$	Matrix of the global components of the equivalent actions acting on the nodes of a structure, which are directly absorbed by its supports.
$\{\hat{P}^G\}$	Matrix of the global components of the given actions acting on the nodes of a structure.
$\mathbf{p}$	Distributed external forces acting on an element. It is given in units of force per unit of length of the element.
p_i or $p_i(x_1)$ $(i = 1, 2, 3)$	Local component in the x_i direction of the distributed external forces acting on an element.
$R_i^{(n)}$	Global component in the $\bar{x}_i$ direction of the reaction at support n of a structure.
$\bar{S}_i^{(n)}$	Global component in the $\bar{x}_i$ direction of the restraining action acting on node n of a structure.
$\mathbf{u}$ or $\mathbf{u}(x_1)$	Translation vector of the points of an element.
$u_i(x_1)$ $(i = 1, 2, 3)$	Local component in the x_1 direction of the translation vector of the points of an element.
$u_i^e(x_1)$ $(i = 1, 2, 3)$	Local component in the x_1 direction of the translation vector of the points of element e of a structure.
$u_i^{(A)}$ or $u_i^{(3)}$ $(i = 1, 2, 3)$	Local component in the x_i direction of the translation vector of point A or point 3 of an element of a structure, respectively.

$u_i^E(x_i)$ $(i = 1, 2, 3)$ Local component in the x_i direction of the translation vector of the points of an element of a structure subjected to equivalent actions.

$u_i^R(x_i)$ $(i = 1, 2, 3)$ Local component in the x_i direction of the translation vector of the points of an element of the restrained structure.

u_i^q $(q = j$ or $k; i = 1, 2, 3)$ Local component in the x_i direction of the translation of the end q $(q = j$ or $k)$ of an element of a structure subjected either to the given loads or to the equivalent actions.

u_i^{eq} $(q = j$ or $k; i = 1, 2, 3)$ Local component in the x_i direction of the translation of the end q $(q = j$ or $k)$ of element e of a structure subjected either to the given loads or to the equivalent actions.

$\{u\}$ Matrix of the basic components of displacement of an element. For an axial deformation element $\{u\} = u_1(x_1)$. For a general planar element $\{u\}^T = [u_1(x_1) \quad u_2(x_1)]$. For a general space element $\{u\}^T = [u_1(x_1) \quad u_2(x_1) \quad u_3(x_1) \quad \theta_1(x_1)]$.

$\hat{u}_i(x_1, x_2, x_3)$ $(i = 1, 2, 3)$ Component of displacement in the x_i direction of the particles of a deformable body.

$T_k^{(+)}, T_k^{(-)}$ Temperature of the points of a cross section of an element of the structure where the positive and negative x_k $(k = 2, 3)$ axis, respectively, intersects the perimeter of its cross section.

T_0 Uniform temperature at the stress-free state of a structure, that is, the temperature at which the structure was constructed.

ΔT_c Change of temperature at the centroid of the cross sections of an element of a structure. It could be a function of x_1.

$[T]$ Matrix which transforms the matrix of basic nodal actions to the matrix of nodal actions of an element of a structure subjected to equivalent actions on its nodes $\{A^E\} = [T]\{a^E\}$.

x_i $(i = 1, 2, 3)$ Local cartesian coordinate of a point of an element. The coordinate x_i is measured along the axis of the element from its end j.

x_i^e $(i = 1, 2, 3)$ Local cartesian coordinate of a point of element e.

$\bar{x}_i$ $(i = 1, 2, 3)$ Global cartesian coordinate of the points of a structure.

α Coefficient of linear thermal expansion.

θ_i^q $(i = 1, 2, 3)$ Component of rotation about the x_i axis of the end q $(q = j$ or $k)$ of an element of a structure.

θ_i^{eq} $(i = 1, 2, 3)$ — Component of rotation, about the x_i axis, of the end q $(q = j$ or k) of element e of a structure.

$\boldsymbol{\theta} = \boldsymbol{\theta}(x_1)$ — Rotation vector of the points of an element.

$\theta_i(x_i)$ $(i = 1, 2, 3)$ — Component about the x_i axis $(i = 1, 2, 3)$ of the rotation vector of the points of an element.

$\theta_i^e(x_i)$ $(i = 1, 2, 3)$ — Component about the x_i axis of the rotation vector of the points of element e of a structure.

$\theta_i^{(A)}$ or $\theta_i^{(3)}$ — Component of rotation about the x_i axis of point A or point 3 of a structure, respectively.

Δ_i $(i = 1, 2, \ldots, N)$ — Global components of the displacements (translations and rotations) of the nodes of a structure. They are numbered consecutively.

$\Delta_i^{(n)}$ — Global component of displacement of node n of a structure.

$[\hat{\Delta}]$ — Matrix of the global components of displacements of all the nodes of a structure subjected to the given loads or to the equivalent actions.

$\Delta^{(S)}$ — Matrix of the given global components of displacements (translations and rotations) of the supports of a structure.

$\{\Delta^F\}$ — Matrix of the global components of displacements of the nodes of a structure which are not inhibited by the supports of the structure.

$\phi_{qi}(x_1)$ $(i = 1, 2, 3; q = j$ or k) — Shape function associated with the end q $(q = j$ or k) of an element. It is used in the solution of the boundary-value problem for computing the component of translation $u_i(x_1)$ $(i = 1, 2, 3)$.

$\phi_q(x_1)$ $(q = j$ or k) — Shape function associated with the end q $(q = j$ or k) of an element. It is used in the solution of the boundary-value problems for computing the axial or the torsional component of displacement of an element.

$[\phi]$ — Matrix of shape functions for an element. It is used in the solution of the boundary-value problems for computing the basic components of displacement of an element [see relations (4.57) and (19.1)].

$[\phi^e]$ — Matrix of shape functions for element e.

$\phi_q^u(x_1)$ $(q = j$ or k) — Shape function associated with the end q $(q = j$ or k) of an element. It is used in the solution of the boundary-value problem for computing the transverse component of translation $u_3(x_1)$ or $u_2(x_1)$ of an element. It is equal to the component of translation $u_3(x_1)$ or $u_2(x_1)$ of the element when it is subjected at its end q to the

	actions required to displace it by a unit translation u_3^q or u_2^q and zero rotation, while its other end is fixed.
$\phi_q^\theta(x_1)$ $(q = j$ or k $)$	Shape function associated with the end q $(q = j$ or k $)$ of an element. It is used in the solution of the boundary-value problem for computing the transverse component of translation $u_3(x_1)$ or $u_2(x_1)$. It is equal to the translation $u_3(x_1)$ or $u_2(x_1)$ of an element when subjected at its end q to the actions required to displace it by a unit rotation θ_2^q or $-\theta_3^q$, respectively, and zero translation, while its other end is fixed.
$(\bar{\ })$	A bar at the top of a component of an action $\bar{F}_i^{ej}$ or a displacement $\bar{u}_i^k$ indicates that it refers to global axes.

Photographs and a Brief Description of Drilling Rigs and Platforms for Offshore Oil Exploration and Production

When geophysical and geological data indicate the possibility of the existance of oil or natural gas under the ocean floor, an exploratory well is drilled. If the results from the exploratory well are encouraging, appraisal wells are drilled to provide information as to the capacity of the reservoir and its

Figure a Jack-up rig. (*Courtesy Exxon Corp.*)

anticipated production rates. Exploration and appraisal drilling is done by one of the following three types of mobile drilling rigs:

1. *The jack-up rig.* This is the preferred rig for water depths up to 91.5 m (300 ft). The work deck of this rig is supported on telescopic legs (usually three). It is mounted on a barge with its legs up and is towed to the site. Its legs are then lowered to the sea bed, and the deck is jacked up to the desired elevation above the sea (see Fig. a).

2. *The semisubmersible rig (see Fig. b).* The work deck of this rig is mounted on steel columns supported on large

Figure b   Semisubmersible rig. (*Courtesy American Bureau of Shipping.*)

pontoons which can be filled with water or air to raise or lower the deck as required. It is held on location by eight or more heavy anchors. Some semisubmersible rigs can move under their own power.

3. *The drill ship.* This is more mobile than the semisubmersible rig but tends to move considerably more when anchored.

Some drill ships and semisubmersibles have been adapted for drilling in water depths of up to 1830 m (6000 ft). In this case they are equipped with computer-controlled, motor-driven propellers on their sides which keep them on the desired

Figure c The Cognac platform located 19 km (12 mi) south of the mouth of the Mississippi River. (*Courtesy Shell Oil Co.*)

position. The design of mobile drilling rigs presents interesting structural analysis problems.

Once exploration and appraisal drilling is completed and oil and gas are assessed as being recoverable in commercial quantities, the mobile drilling rigs are replaced by fixed production offshore platforms. These platforms provide a large, stable work area above the sea where drilling and production equipment, accommodation facilities for over 120 workers, a power station, and a helicopter landing pad can be installed. The two basic types of conventional offshore platforms are as follows.

1. *Pile-supported, template-type platforms (see Figs. c to f).* They are suitable when the sea floor soil is soft. Many of the existing template-type platforms are at sea depths of less than 91 m (300 ft). However, a few have been placed at much greater sea depths. For example, the Hobo platform (see Fig. d) has been built for Exxon Corporation in the Santa Barbara

Figure d The Hobo platform located in the Santa Barbara Channel off the coast of California.

Figure e The base section of the substructure of the Cognac platform during its 360-km (225-mi) trip to the launch site. (*Courtesy Shell Oil Co.*)

Figure f Substructure of the Hobo platform. (*Courtesy Exxon Corp.*)

Channel off the shore of California in 259 m (850 ft) of water, while the Cognac platform (see Fig. c) has been built for the Shell Oil Co. off the shore of Louisiana in over 305 m (1000 ft) of water. Template-type platforms consist of a prefabricated steel deck and a prefabricated tubular steel space frame substructure, which extends from the sea bed to above the water level. The substructure usually has eight legs; however, ice-resistant offshore platforms in the North Sea have been designed with either one or four hollow cylindrical steel legs of large diameter. The deck of template-type platforms is supported on steel piles driven 60 to 104 m (200 to 350 ft) into the sea floor through the tubular legs of the substructure. The piles also resist the lateral loading of the wind and the waves. The substructure of platforms placed in not very deep water [less than 140 m (450 ft)] is fabricated on land and transported as a unit to the site on a barge. It is then skidded off the barge, kept in the upright position, and lowered to the sea bed with the help of a derrick barge. The piles are then driven to their design penetration, and the prefabricated steel deck is usually welded on them. The substructure of the Hobo platform was fabricated and transported in two pieces, each about 137 m (450 ft) long (see Fig. f). The substructure of the Cognac platform was built and transported to location in three pieces (see Fig. e).

2. *Gravity-type platforms* (*see Figs. g to i*). They are made of steel or concrete or a combination of steel and concrete. They are used when the soil of the sea floor is hard because they rest directly on the ocean floor without pile support; that is, they rely on their own weight to stay in place against the forces of wind and waves. Concrete platforms are larger and heavier than steel platforms. The base of concrete platforms consists of huge subsea concrete tanks (see Figs. h and i) used for storage of crude oil. For example, the base of a gravity-type platform (the Beryl A) built for Mobil Oil in the North Sea consists of 19 hollow concrete cylinders of 0.61-m (2-ft) wall thickness, 20-m (66-ft) diameter, and 50-m (164-ft) height. Concrete platforms are built as monoliths from the base up. First a shallow dry dock is excavated at water's edge behind a retaining wall facing the sea. The cylindrical concrete storage tanks of the base are then built to such a height that when the dock is flooded and the retaining wall removed (see Fig. i), the

Figure g   The Brent B gravity-type platform is towed to its site in the North Sea. (*Courtesy Exxon Corp.*)

Figure h Sketch of the Brent B platform. (*Courtesy Exxon Corp.*)

base has sufficient freeboard to float on its own bouyancy. The base is then towed out into a sheltered deep-water site where it is moored. There the storage tanks are completed. On top of some tanks (usually three) the concrete walls of the tapered columns are built to the designed height. The columns of the Beryl A platform described previously rise 90 m (300 ft) above the top of the storage tanks. Finally the steel superstructure is placed on top of the completed concrete substructure, and the platform is towed to the location (see Fig. h) where it is slowly subsided in the water by adding water in the storage tanks. The Beryl A platform had three steel dowels 5 m (16 ft) long extending down below its bottom. These dowels penetrate into the sea bed and help arrest lateral movement of the structure. Each cylindrical tank was fitted around the edge of its base by a steel skirt which reached down 3.5 m (11 ft). As the 19 tanks were filled with water, the steel skirts were forced down into the sea bed until the concrete ribs projecting below the bottom

Figure i Concrete storage tanks at the base of a gravity-type platform during construction. (*Courtesy Mobil Oil Corp.*)

edge of each cell touched the sea bed. Finally grout was pumped beneath the bottom of the tanks to provide a firmer foundation. The Beryl A platform is heavy enough to remain stable without piles in 120 m (394 ft) of rough water. The concrete tanks are filled with oil or water at all times.

Fixed-leg platforms are limited to a maximum water-depth range of about 183 m (600 ft) due to construction constraints. For this reason several concepts have been proposed for expanding the water-depth range of platforms. Two of the most promising ones are the guyed tower and the tension-leg platforms. The guyed tower platform is a trussed shaft held upright by several guy lines which run to clump weights on the sea floor. The tension-leg platform consists of a structure with a large amount of buoyancy and a vertical member anchored to the sea floor. The vertical member is in tension under the expected sea conditions.

The steel substructure of the Cognac platform was built on shore as three sections. First the 54-m (178-ft) base section was transported on site on a large barge and was lowered on the ocean floor (see Fig. e). Then steel piles of 2.1-m (6.9-ft) diameter were inserted through sleeves in the base section and driven 135 m (443 ft) into the sea bed. Then the 98-m- (321-ft) high midsection was transported on site and was mated with the base section. Finally the 160-m- (525-ft) high top section was transported on site and was mated with the other two sections.

In Fig. f the substructure of the Hobo platform is shown. A barge can be seen preparing to load its lower half which it transported to a protected cove and left there floating at anchor. Then the second half of the substructure was transported and the two halves were joined together, while floating on their sides, using a number of hydraulic jacks. The superstructure was then towed to the site 64 km (40 mi) away where it was upended and submerged. Eight hollow steel piles of 1.22-m (4-ft) outside diameter and 366-m (1200-ft) length were driven 104 m (340 ft) into the sea bed through the eight hollow legs of the substructure. The outside diameter of the legs is 1.37 m (4.5 ft). Twelve additional piles were driven through sleeves extending 30 m (98 ft) above the ocean floor.

Preliminary Considerations

Framed Structures

1.1 Introduction

Many structures of engineering interest may be considered as an assemblage of *line (one-dimensional) members,* that is, parts whose lengths are large compared with their other dimensions. The locus of the centroids of the cross sections of a line member is called the *axis* of the line member. This axis can be a straight line or a curve. Moreover the dimensions and orientation of the cross sections of a line member can change along its length. A line member is referred to as *prismatic* or *cylindrical* when the dimensions and orientation of its cross sections do not change along its length.

Structures made up of line members joined together are called *framed structures.* In this text *we limit our attention to framed structures whose members either have a constant cross section or cross sections whose geometry changes in such a way that the direction of their principal*† *centroidal axis remains constant throughout their length.* Moreover, we consider primarily structures with straight-line members. Curved-line members of framed structures can be approximated by a model consisting of a series of straight-line elements (see Fig. 1.1). The directions of the principal centroidal axes of the cross sections of each straight-line element are considered constant and equal to those of the cross section at its one end. It is evident that

---

† A set of rectangular axes x_2 and x_3 are principal axes of a plane surface of area A if the product of inertia of the surface with respect to this axis vanishes. That is,

$$I_{23} = \int\int_A x_2 x_3 \, dA = 0$$

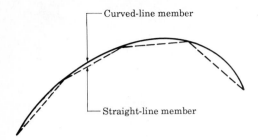

Curved-line member

Straight-line member

Figure 1.1 Approximation of a curved-line member by a series of straight-line elements.

as the number of straight-line elements of this model increases, the model's properties approach those of the actual curved member.

In general, framed structures have a three-dimensional configuration. Often, however, for purposes of analysis and design, a framed structure may be broken down into planar parts whose response can be considered as two-dimensional. They are called *planar* framed structures. The axes of their members lie in one plane, and, moreover, they are subjected to external disturbances (such as forces and moments) which do not induce movement of their particles in the direction normal to their plane. Furthermore, one principal centroidal axis of the cross section of a planar structure is normal to its plane. Thus, framed structures may be classified as

1. Planar

2. Space

1.2   Idealization of Framed Structures

As a result of the geometry of line members, it is possible to make certain assumptions as to their deformed configuration and as to the distribution of the components of stress. These assumptions form the basis of the mechanics of materials theories which are employed in the analysis of framed structures. These theories are presented in Chap. 2. In this section we will only mention their following attributes:

1. In the mechanics of materials theories the components of stress at any point of a cross section of a line member are expressed in terms of the components of the resultant force and moment acting on this cross section by simple relations.

2. In the mechanics of materials theories it is assumed that plane

sections normal to the axis of a member remain plane after deformation. Consequently, the movement of a cross section of a member of a space framed structure, due to deformation, is specified by the three components of the displacement vector of its centroid, referred to a set of rectangular axes, and by the three components of its rotation about the same axes. Moreover, the movement of a cross section of a member of a planar framed structure, due to its deformation, is specified by the two components (in the plane of the structure) of the displacement vector of its centroid and by its component of rotation about the axis normal to the plane of the structure (see Fig. 1.2). The components of the displacement vector of the centroid of a cross section of a member of a structure are referred to as the *components of translation of this cross section.*

3. In the mechanics of materials theories the weight of a member is considered as acting along its axis. Moreover, the surface forces acting on a member are converted to equivalent forces or moments acting on its axis. When surface forces of very high intensity are applied over a very small portion of the length of a member, these forces are replaced, depending upon the nature of their distribution by an equivalent concentrated force and/or moment. Thus, line members may be subjected to concentrated external forces and moments and to distributed external forces and moments acting on their axis. The distributed external forces and moments are given in units of force and moment, respectively, per unit length of the axis of the member (see Fig. 1.3a) or per unit length along some other direction as, for instance, the horizontal direction (see Fig. 1.3b). The external forces and moments are called the *external actions.*

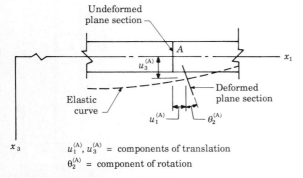

$u_1^{(A)}, u_3^{(A)}$ = components of translation

$\theta_2^{(A)}$ = component of rotation

Figure 1.2 Components of displacement of a cross section of a planar structure.

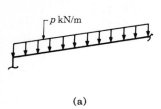
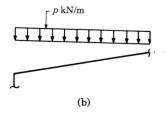

(a)   (b)

Figure 1.3 Uniformly distributed forces.

In the analysis of framed structures we are interested in establishing

1. The components of internal forces and moments acting on their cross sections

2. The components of translational and rotation of certain of their cross sections

The components of internal forces and moments acting on the cross sections of a member of a framed structure and the components of translation and rotation of its cross sections are functions only of its axial coordinate. Thus, on the basis of the previously described attributes of the mechanics of materials theories the configuration of framed structures is conveniently described by a line diagram (see Fig. 1.4). Therein a line member is represented by a line, and a cross section by a point. Prismatic members are represented by their axis, while tapered members are represented by a convenient line. For example, as shown in Fig. 1.5*b* the line diagram for the beam of Fig. 1.5*a* is taken as a straight line. That is, the effect of the haunch at the

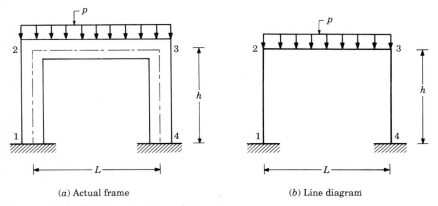

(*a*) Actual frame   (*b*) Line diagram

Figure 1.4 Idealization of a framed structure.

(a) Actual structure   (b) Line diagram

Figure 1.5 Idealization of a haunched beam.

middle support on the line diagram of the beam is disregarded. However, the effect of the haunch on the area and moments of inertia of the cross sections of the beam cannot be disregarded.

The components of the internal force and moment acting on a cross section are referred to as the components of the *internal action* acting on this cross section. Moreover, the components of translation and rotation of a cross section are called its *components of displacement*.

1.2.1 Idealization of the joints of framed structures

The joints of a planar framed structure are usually idealized as *rigid joints* or *pinned joints*. Rigid joints develop full continuity of the connected members. That is, the ends of the members connected to a rigid joint undergo the same translation and rotation. Moreover, a member connected to a rigid joint can transfer a force and a moment to it. Pinned joints permit rotation of the connected ends of the members about the axis of the connecting pin, and, consequently, the components of translation of the ends of the members of the structure connected to a pinned joint are the same, while their components of rotation are not. Moreover, a member cannot transfer a moment to a pinned joint. That is, generally, on a cross section of a member adjacent to a pinned joint, it is possible to have only an internal force which lies in the plane of the structure.

The joints of a space structure are idealized as *rigid, ball-and-socket,* or *pinned* joints. Rigid joints develop full continuity of the connected members. In general, on a cross section of a member adjacent to a rigid joint, it is possible to have an internal force and an internal moment, each of which could have three components with respect to a given system of axes. Ball-and-socket joints permit rotation of the ends of the connected members about any axis. Thus, a member does not transfer a moment to a ball-and-socket joint. Pinned joints permit rotation of the ends of the connected members only about the axis of the pin. Thus, a member does not transfer to a pinned joint component of moment in the direction of the axis of the pin.

1.2.2   Internal action release mechanism

In certain cases, a mechanism is introduced at a point of a member of a beam or frame which renders one or more of the internal actions at this point equal to zero. We refer to this mechanism as an *internal action release mechanism*. For example, the internal action release mechanism (rollers) in the beam of Fig. 1.6*a* renders the internal axial force and the bending moment equal to zero.

Pinned joints of planar or space beams and frames and ball-and-socket joints of space beams and frames are also internal action release mechanisms. For example, the pin at the apex of the frame of Fig. 1.6*b* is an internal action release mechanism. It renders the bending moment at that point equal to zero.

1.2.3   Idealization of the supports of framed structures

The supports of a planar framed structure are idealized as

1. *Roller support.*   This support permits the supported ends of the member of the structure to rotate about an axis normal to the plane of the structure and to move only in one direction, referred to as the direction of rolling. It can exert a reacting force on the structure acting in the direction normal to the direction of rolling and of magnitude equal to that required to counteract the applied loads.

2. *Hinged support.*   This support restrains the supported end of the members of the structure from translating. However, it permits

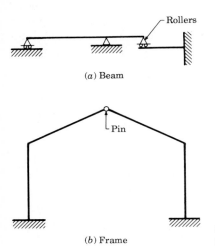

(*a*) Beam

(*b*) Frame

Figure   1.6 Internal   action release mechanisms.

them to rotate about an axis normal to the plane of the structure. A hinged support can exert a reacting force **R** on the structure, passing through the center of a hinge and having the magnitude and direction required to counteract the applied loads.

3. *Fixed support.* This support restrains the supported end of the members of the structure from translating and rotating. It can exert a reacting force on the structure acting in any required direction in the plane of the structure and a moment whose vector is normal to the plane of the structure.

4. *Helical spring support.* This support partially restrains the supported end of the members of the structure from moving in the direction of the axis of the spring. However, it permits it to translate freely in the direction normal to the axis of the spring and to rotate about an axis normal to the plane of the structure. This support can exert a reacting force on the structure in the direction of the axis of the spring whose magnitude is a known function (usually a linear function) of the deformation of the spring.

5. *Spiral spring support.* This support restrains the supported end of the members of the structure from translating and partially from rotating. It can exert a reacting moment on the structure whose magnitude is a known function (usually linear) of the rotation of the connected end of the members.

The supports of space framed structures restrain one or more of the components of translation and/or rotation of the supported end of the members of the structures. Supports which permit rotation of the supported end of the members of a structure about any axis are referred to as *ball-and-socket supports*. Supports which permit rotation of the supported end of the members of a structure about only one axis are referred to as *cylindrical* or *pin supports*. Supports which do not permit rotation of the supported end of the members of a structure are referred to as *fixed-against-rotation supports*. Each of the aforementioned types of supports can be either nontranslating or translating in one or two directions. Usually, however supports fixed against rotation are nontranslating and are referred to as *fixed supports*. These supports can exert a reacting force and a reacting moment on the structure, both acting in any direction required to counteract the applied loads.

1.3 Loads on Framed Structures

We refer to the disturbances which cause internal forces and moments in the members of a structure and/or displacements of its points as the *loads* on the structure. These disturbances may be classified as

1. External actions
2. Displacements of the supports
3. Change of environmental conditions (usually change of temperature)
4. Lack of fit of members

In certain instances a member of a structure is manufactured with its length or curvature slightly different than that required to fit. In order for such a member to be connected to the structure it must be subjected to the external actions required to make it fit. When these actions are subsequently removed, the member may or may not assume its undeformed geometry, depending on whether or not the other members of the structure can restrain it. If the member assumes its undeformed geometry, the members of the structure will neither be subjected to internal actions nor deform but will move as rigid bodies. If, however, the member is restrained from assuming its undeformed geometry, the members of the structure will be subjected to a distribution of internal forces and moments and they will deform. These initial forces and moments could be added to those induced by other loads, and, consequently, the capacity of the member of the structure to carry the other loads would decrease.

When analyzing framed structures, we consider their members, which are manufactured with lengths or curvatures slightly different than that required in order to fit, to be *in a state of initial strain*. The initial components of strain of an element are equal and opposite to those induced by the initial actions which must be applied to it in order to make it fit in the geometry of the structure.

1.4   Types of Framed Structures

We distinguish the following types of framed structures:

1. Planar trusses
2. Planar beams
3. Planar frames
4. Space trusses
5. Space beams
6. Space frames
7. Grids

Trusses are framed structures whose members are straight and assumed connected by frictionless pins; moreover, the axes of their members which are connected to the same joint are assumed to intersect at a point. Trusses are loaded by concentrated forces acting on their joints (see Fig. 1.7a). The weight of truss members is usually neglected or considered as acting on their joints. Thus, it is assumed that the truss members are not subjected to end moments or to intermediate external actions. Consequently, they are subject only to internal axial forces, inducing a uniform state of axial tension or compression.

Beams are framed structures whose line diagram is a straight line. They are subjected to external disturbances which induce internal forces and moments on their cross sections. Planar beams are subjected to external forces lying on a plane which passes through the shear center[1] of their cross sections; this plane is parallel to a

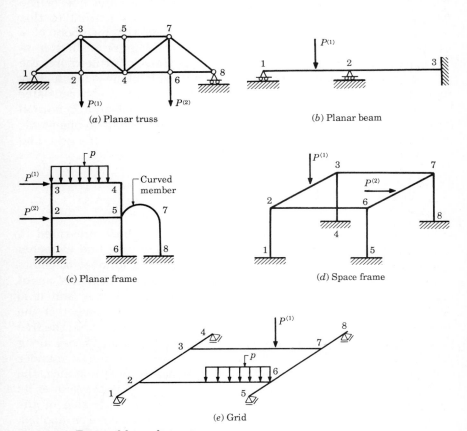

(a) Planar truss

(b) Planar beam

(c) Planar frame

(d) Space frame

(e) Grid

Figure 1.7 Types of framed structures.

plane which contains their axis and a principal centroidal axis of their cross sections. Moreover the vector of the external moments acting on planar beams is normal to the plane of the external forces. Consequently, every cross section of a planar beam rotates only about the axis normal to the plane of the external forces. It does not twist, and it does not translate in the direction normal to the plane of the external forces. When the external forces and moments acting on a beam do not meet one or more of the requirements described previously, the beam is called a *space beam*.

Frames are the most general type of framed structures. Their members can be subjected to axial and shearing forces, bending moments, and torsional moments. They can have both rigid and nonrigid joints and can be loaded in any way. Usually, frames are space structures. Frequently, however, they can be analyzed by being broken down into planar frames and/or grids (see Fig. 1.7c and 1.7e). The members of a planar frame lie in one plane, and one of the principal centroidal axes of their cross sections is normal to this plane. Moreover, the external forces acting on the members of a frame lie on a plane which contains the shear centers of their cross sections and is parallel to the plane of the frame. Furthermore, the vector of the external moments is normal to this plane. Thus a cross section of a member of a planar frame does not twist—it translates only in the plane of the frame and rotates only about an axis normal to the plane of the frame. The members of a grid also lie in one plane. However, the external forces are normal to the plane of the grid, and the vector of the external moments lies in this plane.

1.5 Elements and Nodes of Framed Structures

When analyzing framed structures using the methods presented in this text, the framed structures are subdivided into line elements whose ends are imagined as being connected to a number of points called *nodes*. The ends of an element are referred to as its *nodal points*. Thus a line element extends between two nodes, and it is either a member of the structure or a portion of a member of the structure. The nodes of a structure are its joints, its supports, the free ends of its members, and any other points that we may choose along the length of its members. As a rule, we choose the smallest number of nodes required for the analysis of a structure. For instance, the smallest number of nodes for the beam of Fig. 1.8b is the sum of its support points (1, 3, and 6), the two points on each side of the internal rollers, and the point where the external force is applied. The latter has been chosen because we want the external forces to act at nodes of the beam.

The nodes and the elements of a structure are numbered consecutively, and the number of each element is placed in a circle, as shown in Fig. 1.8. Moreover, the ends of each element are denoted by j and k (j being the end of the element connected to the node having the smallest number). It is preferable that the nodes of framed structures are numbered so that the difference between the numbers of the nodes at the ends of each element is as small as possible (see Sec. 8.4).

After the nodes and the elements of a structure are numbered, their connectivity can be expressed, for example, as shown in Table 1.1 for the structure of Fig. 1.8c.

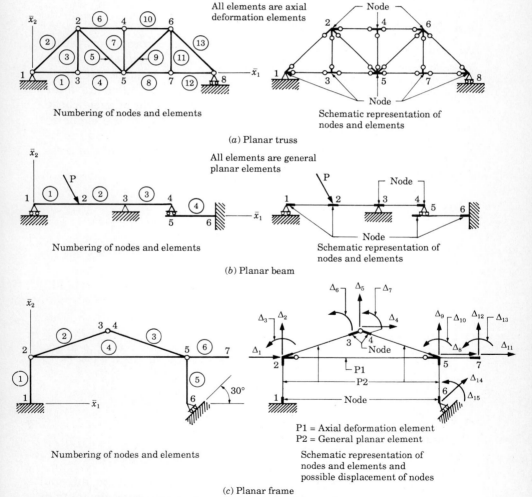

(a) Planar truss

(b) Planar beam

P1 = Axial deformation element
P2 = General planar element

(c) Planar frame

Figure 1.8 Numbering the nodes and elements of framed structures.

TABLE 1.1 Connectivity of the
Elements and Nodes of the
Structure of Fig. 1.8c

Element number	Node number	
	End j	End k
1	1	2
2	2	3
3	4	5
4	2	5
5	5	6
6	5	7

1.6 Global and Local Axes of Reference

We refer each framed structure to a right-handed rectangular system of axes (cartesian axes) $\bar{x}_1$, $\bar{x}_2$, $\bar{x}_3$ called the *global axes* of the structure. Moreover, we refer each element of a structure to a right-handed rectangular system of axes x_1, x_2, x_3 called its *local axes*. As the local axes of an element, we choose the set of axes whose origin is the centroid of the cross section at the end j of the element; its x_1 axis is directed along the axis of the element from its end j to its end k; its x_2 and x_3 axes are the principal centroidal axes of the cross section at the end j of the element (see Fig. 1.9).

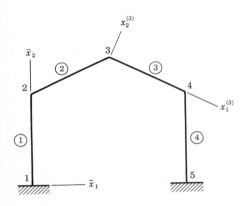

Figure 1.9 Global axes of a planar frame and local axes of element 3.

1.7 Classification of the Elements of Framed Structures

We classify the elements of planar framed structures as follows:

1. Elements whose cross sections are subjected only to internal axial force. Each end of these elements is pinned to other elements or

to a support. We consider the pins at the end of these elements as being part of them. We refer to these elements as *axial deformation elements*.

2. Elements whose cross sections are subjected to an internal force in the plane of the structure and to an internal moment whose vector is normal to the plane of the structure. One end of an element of this type is rigidly connected to other elements or to a support, while the other end is either free or connected in some way (rigidly, with pins, rollers, etc.) to other elements or to a support. These elements can be subjected to external actions along their lengths. We consider the ends of these elements as being rigidly connected to nodes. That is, if one end of an element of this type is connected to other elements or to a support by a connection which is not rigid (pin, rollers), we consider this connection as part of the node. (See nodes 1, 4, and 5 of the beam of Fig. 1.8*b* and nodes 3, 4, and 6 of the frame of Fig. 1.8*c*.) We refer to these elements as *general planar elements*.

We classify the elements of a space structure as follows:

1. Elements whose cross sections are subjected only to an axial force. Each end of these elements is connected by a ball-and-socket connection to other elements or to a support. We consider the ball-and-socket connections at the ends of these elements as being part of them. We refer to these elements as *axial deformation elements*.

2. Elements whose cross sections are subjected to an internal force and moment acting in any direction. One end of an element of this type is rigidly connected to other elements or to a support, while the other end is either free or connected in some way (rigidly, with ball-and-socket, pin, rollers, etc.) to other elements or to a support. These elements can be subjected to external actions along their lengths. We consider the ends of these elements as being rigidly connected to nodes. That is, if one end of an element of this type is connected to other elements or to a support by a connection which is not rigid (ball-and-socket, pin, rollers, etc.), we consider this connection as part of the node. We refer to these elements as *general space elements*.

With the previous classification the elements of a truss (planar or space) are axial deformation elements (see Fig. 1.8*a*); the elements of a planar beam are general planar elements (see Fig. 1.8*b*); while the elements of a planar frame can be either general planar elements or axial deformation elements (see Fig. 1.8*c*). Moreover, the elements of a space beam are general space elements, while the elements of a space frame are either general space elements or axial deformation elements. Furthermore, when a general planar or a general space element is adjacent to an internal action release mechanism, we assume that there is a node between the internal action release mechanism and the end of this element. Thus we have two or more nodes adjacent to an action release mechanism which we call *connected nodes*. For example, there is a node on each side of the hinge at the apex of the frame of Fig. 1.8c and on each side of the internal rollers of the beam of Fig. 1.8b. However, a node does not exist between the internal action release mechanism and the end of an axial force element. That is, in this case, the internal action release mechanism is considered as part of the element (see element 4 of the frame of Fig. 1.8c). Furthermore, when an end of a general planar or general space element is connected to a nonrigid support, we assume that there is a node between this end of the element and the nonrigid support. (See element 1 of the beam of Fig. 1.8b and element 5 of the frame of Fig. 1.8c.)

Depending on the type of the internal action release mechanism, one or more components of the relative motion of the connected nodes vanish. For instance, nodes 4 and 5 of the beam of Fig. 1.8b can rotate and translate in any direction. However, their components of translation along the $\bar{x}_2$ axis are equal. Moreover, nodes 3 and 4 of the frame of Fig. 1.8c can rotate and translate in any direction. However, their components of translation are equal.

When analyzing planar structures by hand calculations using a classical displacement method (the slope deflection or the displacement method with moment distribution), it is preferable to classify their elements into the following three types:[2]

Type M1. These are axial deformation elements. Each end of these elements is pinned to other elements or to a support.

Type M2. One end of an element of this type is rigidly connected to other elements or to a support, while the other end is either free or is rigidly connected to other elements or to a support.

Type M3. One end of an element of this type is rigidly connected to other elements or to a support, while the other end is pinned to other elements or to a support.

With this classification an action release mechanism or the pin at a support is regarded as being part of the adjacent elements, while with the previous classification it is regarded as being part of the node. For instance, the pin at support 5 of the frame of Fig. 1.10 is regarded as being part of element 5. Thus, it is assumed that node 5 can only translate in the direction of rolling. Moreover, the hinge at the apex of the frame of Fig. 1.10 is considered part of the two adjacent elements (2 and 3). Thus, these elements do not transfer

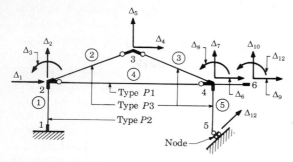

Figure 1.10 Schematic representation of the nodes and elements of the frame and possible displacements of its nodes.

moment to node 3 which consequently does not rotate (see Fig. 1.10). Referring to Figs. 1.8c and 1.10 it can be seen that more independent components of displacements of the nodes are introduced in the analysis of a beam or a frame when its action release mechanisms and the pin or ball-and-socket connections of its supports are regarded as being part of its nodes than when they are regarded as being part of the adjacent members. When analyzing a structure by a displacement method, the components of displacement of its nodes are computed first. The internal actions in the elements of the structure are established from the displacements of its nodes. The number of simultaneous equations which must be solved is equal to the number of unknown components of displacement of its nodes. Consequently, fewer simultaneous equations must be solved when the action release mechanisms of a structure and the pins or ball-and-socket connections of its supports are considered part of its elements than when they are considered part of its nodes. This is a great advantage when analyzing a structure by hand calculations using a displacement method. However, when writing a program for analyzing a group of structures with the aid of a computer using a displacement method, it is important to have as much uniformity as possible, and thus the disadvantages accrued by the introduction of additional types of elements outweigh the advantage of having to manipulate matrices of smaller order.

1.8 Degree of Freedom of a Framed Structure

The deformed configuration of a framed structure is described relative to an initial configuration wherein the structure is not subjected to external loads. In the initial configuration the length or the curvature of some elements of the structure may be different than that required in order for these elements to fit in the geometry of the structure (see Secs. 1.3 and 2.3).

When a structure is subjected to loads, some of its nodes undergo translations and/or rotations which are not known, while others undergo translations and rotations which are known. For instance, the components of translation and rotation of a fixed support are zero. We refer to the components of translation and rotation of a node as its *components of displacement*. The degree of freedom of a body is equal to the smallest number of independent components of displacement of its particles required for the specification of its configuration. However, in this text we call the number of unknown components of displacement of the nodes of a framed structure its *degree of freedom* or *its degree of kinematic indeterminacy*.

In general, the displacement of an unrestrained node of a space frame has three components of translation and three components of rotation with respect to a rectangular system of axes. Moreover, the displacement of an unrestrained node of a planar frame has two components of translation with respect to a set of two rectangular axes lying in the plane of the frame, and one component of rotation whose vector is normal to the plane of the frame. Furthermore, generally, the nodes of a truss do not rotate since the truss elements are assumed to be connected to the nodes by pins, and, consequently, they cannot transfer a moment to the nodes. Thus, the displacement of an unrestrained (free) node of a space truss has three components of translation, while the displacement of an unrestrained (free) node or a planar truss has two components of translation. For example, the translation of each of nodes 2, 3, 4, and 5 of the simple planar truss shown in Fig. 1.11 has a horizontal and a vertical component, while the translation of node 6 has only a horizontal component because

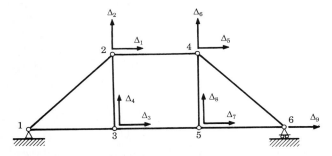

Figure 1.11   Degree of freedom of a planar truss.

this node cannot move in the vertical direction. Thus, the simple planar truss of Fig. 1.11 is kinematically indeterminate to the ninth degree.

As discussed in Sec. 1.7 one or more components of the relative motion of nodes connected by an internal action release mechanism vanish. Thus, the degree of freedom of a number of connected nodes is less than that of the same number of unrestrained nodes. For example, the degree of freedom of nodes 3 and 4 of the planar frame of Fig. 1.8c is 4. The degree of freedom of this frame is 15.

1.9 Components of Internal Actions—
Sign Convention

When a structure is subjected to external loads, a distribution of normal and shearing components of stress could exist on any cross section of its elements. This distribution of stress on a cross section is statically equivalent to a force acting at its centroid and to a moment about its centroid. As mentioned in Sec. 1.2 we refer to the components of this force and moment as the *components of internal action* acting on the cross section of the element. The components of the internal action acting on a cross section of an element of a structure are referred either to the local axes of the element or to the global axes of the structure. The local component of the internal force acting on a cross section of an element in the direction of its axis is called *axial*, while the local components of the internal force acting along the other two local axes of the element are called *shearing*. The local component of internal moment acting in the direction of the axis of an element is called *torsional*, while the local components of internal moment acting along the other two local axes of the element are called *bending*. We represent a moment either by a vector with two arrowheads (see Fig. 1.12) or by a curl.

We denote the axial component of internal force by N, the shearing components of internal force acting in the direction of the x_i axis by Q_i ($i = 2, 3$), and the component (torsional and bending) of the internal moment about the x_i axis by M_i ($i = 1, 2, 3$).

We consider as positive the components of force and moment acting on a positive cross section if their sense coincides with the positive sense of the local axes x_1, x_2, x_3 (see Fig. 1.12a). Furthermore, we consider as positive the components of force and moment acting on a negative cross section† if their sense coincides with the negative

---

† We call a cross section of an element positive or negative if the unit vector normal to it is directed along the positive or negative x_1 axis, respectively.

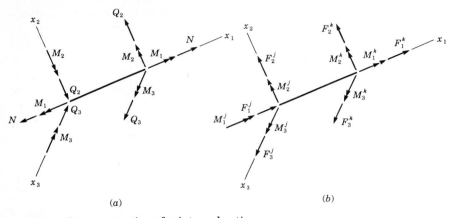

Figure 1.12 Sign conventions for internal actions.

sense of the local axes x_1, x_2, x_3 (see Fig. 1.12a). Thus, a tensile axial force is considered positive, while a compressive axial force is considered negative. On the basis of the sign convention described previously the components of the internal action on a cross section are related to the components of stress acting on this cross section by the following relations:

$$N = \iint_A \tau_{11} \, dA \tag{1.1a}$$

$$Q_2 = \iint_A \tau_{12} \, dA \tag{1.1b}$$

$$Q_3 = \iint_A \tau_{13} \, dA \tag{1.1c}$$

$$M_1 = \iint_A (\tau_{13}x_2 - \tau_{12}x_3) \, dA \tag{1.1d}$$

$$M_2 = \iint_A \tau_{11}x_3 \, dA \tag{1.1e}$$

$$M_3 = -\iint_A \tau_{11}x_2 \, dA \tag{1.1f}$$

When we analyze structures using the modern methods of structural analysis presented in this book, we denote the components of the internal force and moment acting on the end q ($q = j$ or k) of an element by F_m^q and M_m^q ($m = 1, 2, 3$), respectively, when referred to local axes and by $\bar{F}_m^q$ and $\bar{M}_m^q$, respectively, when referred to global axes. Moreover, we consider as positive the components of force and moment acting on the ends of an element if their sense coincides with the positive sense of the corresponding local axes x_1, x_2, x_3 (see Fig. 1.12b). Thus,

$$N(0) = -F_1^j \tag{1.2a}$$

$$Q_2(0) = -F_2^j \tag{1.2b}$$

$$Q_3(0) = -F_3^j \tag{1.2c}$$

$$M_1(0) = -M_1^j \tag{1.2d}$$

$$M_2(0) = -M_2^j \tag{1.2e}$$

$$M_3(0) = -M_3^j \tag{1.2f}$$

$$N(L) = F_1^k \tag{1.2g}$$

$$Q_2(L) = F_2^k \tag{1.2h}$$

$$Q_3(L) = F_3^k \tag{1.2i}$$

$$M_1(L) = M_1^k \tag{1.2j}$$

$$M_2(L) = M_2^k \tag{1.2k}$$

$$M_3(L) = M_3^k \tag{1.2l}$$

The component of the internal forces and moments acting at the ends of an element are called the *components of its nodal actions*. If it is necessary to specify the element on which the components of a force or a moment act, we add the number of the element as a superscript to their symbol. For example, we denote the components of the force and of the moment acting at the end j of element e by F_i^{ej} and M_i^{ej} ($i = 1, 2, 3$), respectively. Moreover, we denote the shearing components of the internal force and moment of element e by $N^{(e)}(x_1)$, $Q_i^{(e)}(x_1)$ ($i = 2, 3$), and $M_i^{(e)}(x_1)$ ($i = 1, 2, 3$), respectively.

1.10   Components of Displacement

When framed structures are subjected to external loads, their elements translate and rotate as rigid bodies and they deform. In structural analysis the warping of the cross sections of the elements of a structure is disregarded when their deformed configuration is specified. Consequently, the deformed configuration of a cross section of an element of a space framed structure is specified by the three components of the displacement vector of its centroid, referred to a set of rectangular axes, and by the three components of its rotation vector,† referred to the same set of axes. Moreover, the deformed configuration of a cross section of an element of a planar framed structure, due to its deformation, is specified by the two components (in the plane of the structure) of the displacement vector of its centroid and by the component of its rotation vector about the axis normal to the plane of the structure. The components of the displacement vector of the centroid of a cross section of an element of a structure are referred to as the *components of translation* of this cross section. Inasmuch as a cross section of a structure is represented by a point on the line diagram of the structure, we refer to the components of translation and rotation of the cross section represented by point A on the line diagram of an element of a structure as the *components of translation and rotation of point A of the element.*

We denote the components of translation and rotation of a cross section of an element by u_i and θ_i $(i = 1, 2, 3)$, respectively, when referred to local axes, and by $\bar{u}_i$ and θ_i $(i = 1, 2, 3)$, respectively, when referred to global axes. Moreover, if we want to specify the point with which a component of translation or rotation is associated, we add the letter specifying the point as a superscript to its symbol. Thus, u_i^j is the component of translation of the end j of an element in the direction of its local axis x_i (see Fig. 1.13a), while u_i^A is the local component of translation of point A of an element in the direction of its local axis x_i. If it is necessary to specify the element with which a component of translation or rotation is associated, we add the number of the element as a superscript to its symbol. For example, we denote the components of translation and rotation of the end j of element e by u_i^{ej} and θ_i^{ej} $(i = 1, 2, 3)$, respectively. Moreover we denote the components of translation and rotation of element e by $u_i^{(e)}(x_1)$ and $\theta_i^{(e)}(x_1)$ respectively.

As mentioned in Sec. 1.1 the components of translation u_1, u_2, u_3

---

† A rotation about an axis is represented by a vector acting along this axis and pointing in the direction in which a right-hand screw moves when subjected to this rotation. Small rotations are vector quantities, while large rotations are not (see Sec. B.1). In this text we consider structures whose deformation involves only small rotations.

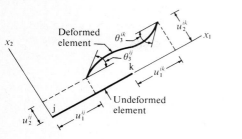 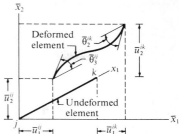

(a) Local components of displacement   (b) Global components of displacement

Figure 1.13 Positive local and global components of nodal displacements of an element of a planar structure.

and the components of rotation θ_1, θ_2, θ_3 of a cross section of an element of a structure are called its *components of displacement*. The components of translations and rotations of the ends of an element are called *components of its nodal displacements*. Moreover, the deformed axis of an element is called its *elastic curve*.

Throughout this text *the local or global components of displacement (translation and rotation) of a point of an element are assumed positive when their sense coincides with the positive sense of the corresponding local or global axes*. The positive components of displacement of the ends of an element of a planar structure are shown in Fig. 1.13.

1.11 Theory of Small Deformation

In this text we consider framed structures which are subjected to external disturbances of such magnitudes that the deformation of their elements is within the range of validity of the theory of small deformations. As a result of this assumption the following approximations can be made.

1. The change of length, area, or volume of a segment of an element due to its deformation is negligible compared to its undeformed length, area, or volume, respectively.

2. The deformation of a particle is completely specified by its six components of strain which are related to its components of displacement $\hat{u}_1(x_1, x_2, x_3)$, $\hat{u}_2(x_1, x_2, x_3)$, and $\hat{u}_3(x_1, x_2, x_3)$ by the

following linear relations:

$$e_{11} = \frac{\partial \hat{u}_1}{\partial x_1} \qquad e_{22} = \frac{\partial \hat{u}_2}{\partial x_2} \qquad e_{33} = \frac{\partial \hat{u}_3}{\partial x_3}$$

$$e_{21} = e_{12} = \frac{1}{2}\left(\frac{\partial \hat{u}_2}{\partial x_1} + \frac{\partial \hat{u}_1}{\partial x_2}\right)$$

$$\text{(1.3)}$$

$$e_{31} = e_{13} = \frac{1}{2}\left(\frac{\partial \hat{u}_3}{\partial x_1} + \frac{\partial \hat{u}_1}{\partial x_3}\right)$$

$$e_{32} = e_{23} = \frac{1}{2}\left(\frac{\partial \hat{u}_2}{\partial x_3} + \frac{\partial \hat{u}_3}{\partial x_2}\right)$$

3. The effect of the change of geometry of a structure due to its deformation on the magnitude of its internal forces and moments is negligible. Consequently, when we consider the equilibrium of a part of a structure, we use its undeformed configuration. For example, when we consider the equilibrium of joint 2 of the truss of Fig. 1.14a, we use the free-body diagram of Fig. 1.14b. In this case we have

$$\sum \bar{F}_1 = 0 \qquad N^{(1)} = N^{(2)}$$

$$\text{(1.4)}$$

$$\sum \bar{F}_2 = 0 \qquad N^{(1)} = N^{(2)} = \frac{P}{2\cos(\alpha/2)}$$

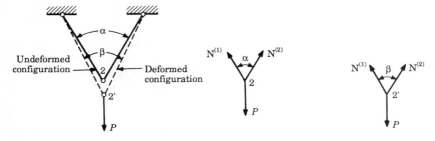

(a) Geometry and loading of the truss

(b) Free-body diagram of node 2 disregarding the change of the angle α due to the deformation

(c) Free-body diagram of node 2 including the change of the angle α due to the deformation

Figure 1.14 Two-element truss subjected to an external force.

It is apparent that the relation between the internal forces in the members of the truss $N^{(1)} = N^{(2)}$ and the external force P is linear. This is not true if the effect of the change of the geometry of the truss due to its deformation on the magnitude of its internal forces is taken into account. In this case referring to Fig. 1.14c we have

$$\sum \bar{F}_1 = 0 \qquad N^{(1)} = N^{(2)}$$

$$\sum \bar{F}_2 = 0 \qquad N^{(1)} = N^{(2)} = \frac{P}{2 \cos (\beta/2)}$$

(1.5)

The magnitude of the angle β depends on the magnitude of the external force P. Consequently, the relation between the internal forces in the members of the truss and the external force P is not linear.

As a second example we consider the equilibrium of a segment of a beam. If the effect of the change, due to the deformation, of the geometry of the beam on the magnitude of its internal moment is not taken into account, we use the free-body diagram of Fig. 1.15b. In

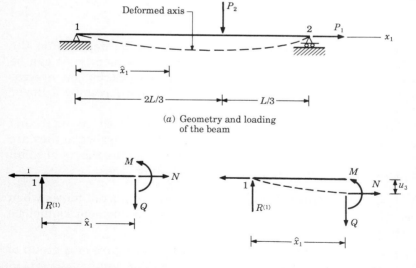

(a) Geometry and loading
of the beam

(b) Free-body diagram of a segment
of the beam using its undeformed
configuration

(c) Free-body diagram of a segment of the
beam using its deformed configuration

Figure 1.15 Beam subjected to transverse and axial forces.

this case we have

$$\sum \bar{F}_1 = 0 \qquad N = P_1 \tag{1.6a}$$

$$\sum \bar{F}_2 = 0 \qquad Q = R^{(1)} = \frac{P_2}{3} \tag{1.6b}$$

$$\sum \bar{M}_3^{(1)} = 0 \qquad M = R^{(1)}x_1 = \frac{P_2 x_1}{3} \tag{1.6c}$$

Thus, the internal force and moment acting on a cross section of the beam are related to the external forces by a linear relation. This is not true if the effect of the change of the geometry of the beam, due to its deformation, on the magnitude of its internal moment is taken into account. In this case referring to Fig. 1.15c we have

$$\sum \bar{F}_1 = 0 \qquad N = P_1 \tag{1.7a}$$

$$\sum \bar{F}_2 = 0 \qquad Q = R^{(1)} = \frac{P_2}{3} \tag{1.7b}$$

$$\sum \bar{M}^{(1)} = 0 \qquad M = R^{(1)}x_1 - P_1 u_3 = \frac{P_2 x_1}{3} - P_1 u_3 \tag{1.7c}$$

The magnitude of the component of translation u_3 depends on the magnitude of the external forces P_1 and P_2. Consequently, as can be seen from relation (1.7c), the internal moment acting on a cross section of the beam is not related to the external forces by a linear relation.

The theory of small deformation cannot be used in analyzing certain structures of interest to the structural designer when they are subjected to certain types of loading. For example, the theory of small deformation cannot be employed in the following cases:

1. In analyzing beams subjected to transverse and axial forces when the effect of the axial forces on their bending moment cannot be neglected

2. In establishing the loading under which a structure or a group of its members reaches a state of unstable equilibrium

3. In analyzing long cables subjected to transverse forces

4. In analyzing a kinematically unstable structure.[3]

For certain types of loading a number of elements of a kinematically unstable structure can move without deforming. Moreover, the internal actions acting on some elements of kinematically unstable structures cannot be computed if the change of their geometry due to their deformation is not taken into account. That is, the internal actions in the elements of kinematically unstable structures cannot be established using the theory of small deformation. Furthermore, when the geometry of a structure approaches a kinematically unstable configuration, certain of its elements are subjected to very large internal actions even for relatively small values of the external actions. Thus, kinematically, unstable structures should be avoided. Therefore, the structural designer is not concerned with the analysis of kinematically unstable structures but rather with the detection of kinematically unstable structures or of structures whose geometry approaches a kinematically unstable configuration. When analyzing structures with the aid of a computer, kinematically unstable structures are detected automatically because their basic stiffness matrix is singular.

If a structure or a group of its elements can move as a rigid body when subjected to external actions or if a structure is kinematically unstable, we say that the *structure is a mechanism*.

1.12   Linear Response of Structures

Consider a structure, subjected to external disturbances, which has the following attributes.

1. Its elements are made of linearly elastic materials. Consequently, the relations between the components of stress and strain of each of its particles are linear.

2. The magnitude of the external disturbances is such that the deformation of the elements of the structure is within the range of validity of the theory of small deformation. Thus, the relations between the components of strain and the components of displacement of each of its particles are linear.

Referring to Sec. 1.11 we may conclude that the internal forces and moments and consequently the components of stress acting on a particle of the structure under consideration are related to the external disturbances acting on the structure by linear relations. Moreover, the components of stress acting on a particle of this structure are related to the components of strain of this particle by linear relations. Furthermore, the components of strain of a particle of this structure are related to its components of displacement by linear relations. Consequently, the relations between the external loads acting on the structure under consideration and the internal forces, moments, components of stress, components of strain, or components of displacement are linear. That is, the effects are linearly related to the cause. In this case we say that the *response of the structure is linear.*

A direct consequence of the linear response of a structure is that *the principle of superposition is valid* for this structure. That is, its response due to a number of simultaneously applied disturbances is equal to the sum of its responses due to the application of each of these disturbances separately.

References

1. A. E. Armenàkas, *Classical Structural Analysis: A Modern Approach,* McGraw-Hill Book Co., New York, 1988, p. 679.
2. Ibid., p. 505.
3. Ibid., p. 51.

Mechanics of Materials Theories

2.1 Introduction

In this chapter we derive the fundamental relations of the mechanics of materials theories, which are pertinent to the analysis of framed structures, and we use them to formulate the boundary-value problems for computing the components of displacement and the components of the internal action of an element. More precisely, we do the following:

1. We derive the action equations of equilibrium.
2. We establish the internal action-displacement relations, for elements made from an isotropic linearly elastic material.
3. We establish the relations between the components of stress acting on a cross section of an element and the internal actions acting on this cross section.
4. We derive the displacement equations of equilibrium, for elements made from an isotropic linearly elastic material.
5. We formulate the strong form of the boundary-value problems for computing the components of displacement and the components of internal action of an element.

2.2 Action Equations of Equilibrium

Consider an element of a framed structure, subjected to external concentrated forces $\mathbf{P}^{(i)}$ ($i = 1, 2, \ldots, n$), concentrated moments $\mathbf{M}^{(i)}$ ($i = 1, 2, \ldots, m$), and external distributed forces $\mathbf{p}(x_1)$ and distributed moments $\mathbf{m}(x_1)$ given in units of force or moment per unit length. Moreover, consider a segment of this element of length Δx_1,

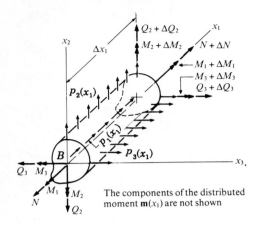

The components of the distributed moment $\mathbf{m}(x_1)$ are not shown

Figure 2.1 Segment of an element with positive internal actions.

at point B (see Fig. 2.1). This segment is loaded only by distributed forces $\mathbf{p}(x_1)$ and moments $\mathbf{m}(x_1)$ which are continuous functions of x_1 at point B. The latter are not shown in Fig. 2.1 to avoid cluttering it. The internal actions at the end faces of the segment under consideration are assumed positive, as shown in Fig. 2.1. Since the segment is in equilibrium, we have

$$\sum F_1 = 0 = -N + p_1\,\Delta x_1 + N + \Delta N$$

$$\sum F_2 = 0 = -Q_2 + p_2\,\Delta x_1 + Q_2 + \Delta Q_2$$

$$\sum F_3 = 0 = -Q_3 + p_3\,\Delta x_1 + Q_3 + \Delta Q_3$$

$$\sum M_1 = 0 = -M_1 + m_1\,\Delta x_1 + M_1 + \Delta M_1$$
$$\tag{2.1}$$

$$\sum M_2 = 0 = -M_2 + m_2\,\Delta x_1 + M_2 + \Delta M_2 + \frac{p_3(\Delta x_1)^2}{2} - Q_3\,\Delta x_1$$

$$\sum M_3 = 0 = -M_3 + m_3\,\Delta x_1 + M_3 + \Delta M_3 - \frac{p_2(\Delta x_1)^2}{2} + Q_2\,\Delta x_1$$

In the limit as $\Delta x_1 \to 0$, the above relations reduce to

$$p_1 = -\frac{dN}{dx_1} \tag{2.2}$$

$$p_2 = -\frac{dQ_2}{dx_1} \tag{2.3}$$

$$p_3 = -\frac{dQ_3}{dx_1} \tag{2.4}$$

$$m_1 = -\frac{dM_1}{dx_1} \tag{2.5}$$

$$Q_2 = -\frac{dM_3}{dx_1} - m_3 \tag{2.6}$$

$$Q_3 = \frac{dM_2}{dx_1} + m_2 \tag{2.7}$$

These relations are the action equations of equilibrium at point B. Each one of them must be satisfied at every point of an element in equilibrium provided that there are no corresponding concentrated external actions acting at this point. For example, referring to Fig. 2.2 the force $p_3(x_1)$ has a simple discontinuity (jump) at point E and a Dirac delta–type (see App. A) discontinuity at point F. The internal shearing force $Q_3(x_1)$ is continuous at point E, but it has a simple discontinuity (jump) at point F. Consequently, its derivative does not exist at this point, and relation (2.4) is not valid at this point. Moreover the moment m_2 has a Dirac delta–type discontinuity at point G. The internal bending moment $M_2(x_1)$ has a simple discontinuity at point G. Consequently its derivative does not exist at this point, and relation (2.7) is not valid at point G.

Notice that the satisfaction of relations (2.2) to (2.7) by a distribution of internal actions in an element ensures that every segment of infinitesimal length of this element which is not subjected to concentrated external actions is in equilibrium. However, it does not ensure that every particle of the element is in equilibrium. In order to accomplish this the components of stress τ_{ij} $(i, j = 1, 2, 3)$ must satisfy the equations of equilibrium[1] at every particle inside the volume of the body. In the mechanics of materials theories we do not attempt to satisfy these equations. In fact the components of stress obtained on the basis of the mechanics of materials theories often do not satisfy these equations at all particles of a line element.

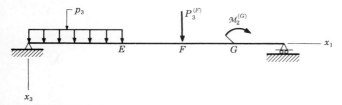

Figure 2.2 Planar beam subjected to external actions.

2.3   Stress-Strain Relations for a Particle of a Framed Structure Made from an Isotropic Linearly Elastic Material

The dimensions of the cross sections of line elements are small compared to their lengths. For this reason, at the particles of line elements the normal components of stress τ_{22} and τ_{33} acting on the planes normal to the x_2 and x_3 axes, respectively, and the shearing component of stress $\tau_{23} = \tau_{32}$ are negligible compared to the other components of stress (see Fig. 2.3). That is,

$$\tau_{22} \simeq \tau_{33} \simeq \tau_{23} \simeq 0 \tag{2.8}$$

The relations between the components of stress and strain for a particle of a body made from an isotropic linearly elastic material are

$$e_{11} = \frac{1}{E}[\tau_{11} - \nu(\tau_{22} + \tau_{33})] + \alpha(T - T_0) + e_{11}^I$$

$$e_{22} = \frac{1}{E}[\tau_{22} - \nu(\tau_{11} + \tau_{33})] + \alpha(T - T_0) + e_{22}^I$$

$$e_{33} = \frac{1}{E}[\tau_{33} - \nu(\tau_{11} + \tau_{22})] + \alpha(T - T_0) + e_{33}^I \tag{2.9}$$

$$e_{12} = \frac{\tau_{12}}{2G}$$

$$e_{13} = \frac{\tau_{13}}{2G}$$

$$e_{23} = \frac{\tau_{23}}{2G}$$

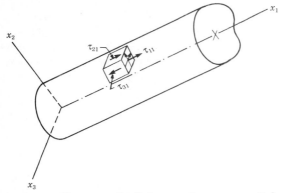

Figure 2.3 Components of stress acting on a particle of an element of a framed structure.

where E, G, v, and α are the modulus of elasticity, the shear modulus, Poisson's ratio, and the coefficient of linear thermal expansion, respectively, of the material from which the structure is made. $T(x_1, x_2, x_3)$ is the temperature of the particle at the present state; T_0 is the uniform temperature of the structure at the stress-free state, that is, the temperature during construction of the structure; e_{11}^I is the axial component of the initial strain which was present in the particle at the stress-free state.

Substituting relations (2.8) into (2.9) and assuming that the components of initial strain e_{22}^I and e_{33}^I vanish, the relations among the components of stress and strain for a particle of a framed structure, whose elements are made from isotropic linearly elastic materials, are

$$e_{11} = \frac{\tau_{11}}{E} + \alpha(T - T_0) + e_{11}^I$$

$$e_{22} = -\frac{v\tau_{11}}{E} + \alpha(T - T_0)$$

$$e_{33} = -\frac{v\tau_{11}}{E} + \alpha(T - T_0) \qquad (2.10)$$

$$e_{12} = \frac{\tau_{12}}{2G}$$

$$e_{13} = \frac{\tau_{13}}{2G}$$

$$e_{23} = 0$$

In general it is not possible to measure the temperature T of the particles of an element which are not located on its surface. For this reason in what follows we express the temperature of a particle of an element in terms of the temperature of certain particles located on the surface of this element. For this purpose we introduce the following notation:

1. $T_2^{(+)}(x_1)$ and $T_2^{(-)}(x_1)$ are the temperatures at the points of the two lines of intersection of the plane $x_1 x_2$ and the lateral surface of the element (see Fig. 2.4a). We define the difference in temperature $\delta T_2(x_1)$ as

$$\delta T_2 = T_2^{(+)} - T_2^{(-)} \qquad (2.11)$$

2. $T_3^{(+)}(x_1)$ and $T_3^{(-)}(x_1)$ are the temperatures at the points of the two lines of intersection of the plane $x_1 x_3$ and the lateral surface of the element (see Fig. 2.4a). We define the difference in temperature

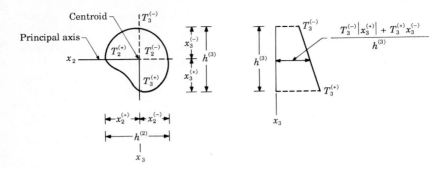

(a) Temperature at the points of intersection of the x_1x_2 or x_1x_3 plane with the lateral surface of an element

(b) Distribution of temperature along the x_3 axis.

Figure 2.4   Distribution of temperature on a cross section of a member.

$\delta T_3(x_1)$ as

$$\delta T_3 = T_3^{(+)} - T_3^{(-)} \tag{2.12}$$

3. $T_c(x_1)$ is the temperature of the points of the centroidal axis of an element. We define the change of temperature $\Delta T_c(x_1)$ at the points of the centroidal axis of an element as

$$\Delta T_c = T_c - T_0 \tag{2.13}$$

4. α is the coefficient of linear thermal expansion of the material from which the element is made. Thus, the change in length of a fiber of an element of length dx_1 due to an increase of temperature ΔT is equal to

$$\Delta L = \alpha \, \Delta T \, dx_1 \tag{2.14}$$

We assume that the variation of the temperature $T(x_1, x_2, x_3)$ is such that plane sections normal to the axis of an element before the temperature changes remain plane after the temperature changes. This implies that the change of temperature is a linear function of the coordinates x_2 and x_3, and consequently it, can be expressed as

$$T - T_0 = A_1 + A_2 x_2 + A_3 x_3 \tag{2.15}$$

where A_1, A_2, A_3 are functions of x_1 only. Referring to relation (2.13) the change of temperature ΔT_c at the points of the centroidal axis

$(x_2 = x_3 = 0)$ of an element is equal to

$$\Delta T_c = T_c - T_0 = A_1 \tag{2.16}$$

Notice that

$$T = \begin{cases} T_2^{(+)} & \text{at } x_3 = 0, \ x_2 = x_2^{(+)} \\ T_2^{(-)} & \text{at } x_3 = 0, \ x_2 = -x_2^{(-)} \end{cases} \tag{2.17}$$

Substituting conditions (2.17) into relation (2.15) and using (2.16) we obtain

$$\begin{aligned} T_2^{(+)} - T_0 &= A_2 x_2^{(+)} + \Delta T_c \\ T_2^{(-)} - T_0 &= -A_2 x_2^{(-)} + \Delta T_c \end{aligned} \tag{2.18}$$

Moreover, notice that

$$T = \begin{cases} T_3^{(+)} & \text{at } x_2 = 0, \ x_3 = x_3^{(+)} \\ T_3^{(-)} & \text{at } x_2 = 0, \ x_3 = -x_3^{(-)} \end{cases} \tag{2.19}$$

Substituting conditions (2.19) into relation (2.15) and using (2.16) we get

$$\begin{aligned} T_3^{(+)} - T_0 &= A_3 x_3^{(+)} + \Delta T_c \\ T_3^{(-)} - T_0 &= -A_3 x_3^{(-)} + \Delta T_c \end{aligned} \tag{2.20}$$

Multiplying the first of relations (2.18) by $|x_2^{(-)}|$ and the second by $|x_2^{(+)}|$ and adding, we have

$$T_c = \frac{T_2^{(+)} |x_2^{(-)}| + T_2^{(-)} |x_2^{(+)}|}{h^{(2)}} \tag{2.21}$$

Multiplying the first of relations (2.20) by $|x_3^{(-)}|$ and the second by $|x_3^{(+)}|$ and adding, we obtain

$$T_c = \frac{T_3^{(+)} |x_3^{(-)}| + T_3^{(-)} |x_3^{(+)}|}{h^{(3)}} \tag{2.22}$$

Notice that as expected when three of the temperatures $T_2^{(+)}$, $T_2^{(-)}$, $T_3^{(+)}$, $T_3^{(-)}$ are given, the fourth may be established from relations (2.21) and (2.22). Moreover, notice that $\Delta T_c = T_c - T_0$ can be established if $T_2^{(+)}$ and $T_2^{(-)}$ or $T_3^{(+)}$ and $T_3^{(-)}$ are given. Subtracting the second of relation (2.18) from the first and using relation (2.11) we get

$$A_2 = \frac{\delta T_2}{h^{(2)}} \tag{2.23a}$$

Subtracting the second of relations (2.20) from the first and using relation (2.12) we obtain

$$A_3 = \frac{\delta T_3}{h^{(3)}} \qquad (2.23b)$$

Substituting relations (2.16), (2.23a), and (2.23b) into relation (2.15) we have

$$T - T_0 = \Delta T_c + \frac{\delta T_2 x_2}{h^{(2)}} + \frac{\delta T_3 x_3}{h^{(3)}} \qquad (2.24)$$

where ΔT_c, δT_2, and δT_3 are functions of the axial coordinate of the element.

In certain instances an element of a framed structure is manufactured with its length or curvature slightly different than that required to fit in the geometry of the structure. In order that such an element is connected to the structure it must be subjected to the external actions required to change its length or curvature and make it fit. When these actions are subsequently removed, the structure assumes a deformed configuration. The elements of statically indeterminate structures restrain their nonfitting elements from assuming their undeformed geometry, while the elements of statically determinate structures do not. For this reason the elements of statically indeterminate structures deform and are subjected to internal actions, while the elements of statically determinate structures neither deform nor are subjected to internal actions; they only move as rigid bodies.

In order to establish the effect of nonfitting elements of a structure on the components of displacement of its points and on the internal actions in its elements we may regard its nonfitting elements as being subjected to a state of initial strain. Thus, an element which is manufactured with a length which is either shorter or longer than that required in order to fit may be regarded as being subjected to a uniform axial component of initial strain. That is,

$$e_{11}^I = \frac{\text{manufactured length} - \text{required length}}{\text{required length}} \qquad (2.25)$$

Moreover, an element which is manufactured with a different curvature than that required in order to fit may be regarded as being subjected to an initial, axial component of strain. For example, consider an element which has been manufactured with a curvature in the $x_1 x_3$ plane although it is required to be straight. We denote the initial curvature of its axis by $k_3^I(x_1)$. That is,

$$k_3^I(x_1) = \frac{1}{\rho_3^I(x_1)} \qquad (2.26)$$

where $\rho_3^I(x_1)$ is the radius of curvature of the axis of the element. The element may be regarded as being subjected to an axial component of initial strain which varies linearly with x_3. On the basis of this assumption referring to Fig. 2.5b we have

$$dx_1 = \rho_3^I \, d\theta_2^I \tag{2.27}$$

$$AD = (\rho_3^I + x_3) \, d\theta_2^I \tag{2.28a}$$

$$AB + CD = e_{11}^I\big|_{x_3} \, dx_1 \tag{2.28b}$$

$$AD = dx_1 + AB + CD \tag{2.29}$$

Substituting relations (2.27) and (2.28) into (2.29) and using (2.26), we obtain

$$e_{11}^I\big|_{x_3} = \frac{x_3}{\rho_3^I} = x_3 k_3^I \tag{2.30}$$

Similarly if we assume that the component of initial strain e_{11}^I varies linearly with x_2, we can show that

$$e_{11}^I\big|_{x_2} = -\frac{x_2}{\rho_2^I} = -x_2 k_2^I \tag{2.31}$$

Thus in general if we assume that the initial strain varies linearly

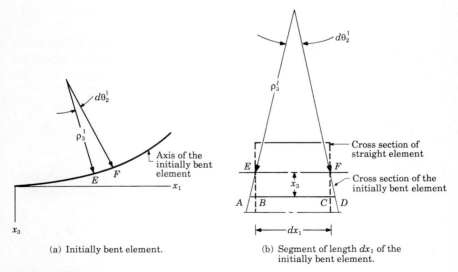

(a) Initially bent element.

(b) Segment of length dx_1 of the initially bent element.

Figure 2.5 Element with initial curvature in the $x_1 x_3$ plane.

with the x_2 and x_3 axes, we have

$$e_{11}^I = e_{11}^{Ic} + \frac{\delta e_{11}^{I2} x_2}{h^{(2)}} + \frac{\delta e_{11}^{I3} x_3}{h^{(3)}} = e_{11}^{Ic} - k_2^I x_2 + k_3^I x_3 \qquad (2.32)$$

where

$e_{11}^{Ic}(x_1) =$ initial strain at centroidal axis of cross section of element

$\delta e_{11}^{Ij}(x_1)$ $(j = 2, 3) =$ difference of values of initial strain at points of cross sections of element which are located on intersections of $x_1 x_j$ $(j = 2, 3)$ plane with lateral surface of element

$k_2^I, k_3^I =$ initial curvature of projection of deformed axis (elastic curve) of element on $x_1 x_2$ and $x_1 x_3$ planes, respectively

Substituting relations (2.24) and (2.32) into (2.9) we have

$$e_{11} = \frac{\tau_{11}}{E} + \alpha \, \Delta T_c + \frac{\alpha \, \delta T_2 x_2}{h^{(2)}} + \frac{\alpha \, \delta T_3 x_3}{h^{(3)}} + e_{11}^{Ic} - k_2^I x_2 + k_3^I x_3$$

$$e_{22} = e_{33} = -\frac{\nu \tau_{11}}{E} + \alpha \, \Delta T_c + \frac{\alpha \, \delta T_2 x_2}{h^{(2)}} + \frac{\alpha \, \delta T_3 x_3}{h^{(3)}}$$

$$e_{12} = \frac{\tau_{12}}{2G}$$

$$e_{13} = \frac{\tau_{13}}{2G} \qquad (2.33)$$

$$e_{23} = 0$$

Relations (2.33) represent a convenient form of the stress-strain relations for a particle of an element of a framed structure made of an isotropic linearly elastic material. ΔT_c, δT_2, δT_3, e_{11}^{Ic}, k_2^I, and k_3^I are functions only of the axial coordinate of the element.

2.4 Internal Action-Displacement Relations—Displacement Equations of Equilibrium

In this section we derive the internal action-displacement relations and the displacement equations of equilibrium for elements made of isotropic linearly elastic materials which deform in one of the following modes:

1. Axial deformation only (axial deformation elements)

2. Bending only

3. Twisting only

Moreover, we use the displacement equations of equilibrium to compute the components of translation of elements which are fixed at both ends and of cantilever elements.

2.4.1 Elements undergoing axial deformation only

When an element undergoes axial deformation only, its axis does not bend $(u_2 = u_3 = \theta_2 = \theta_3 = 0)$. Moreover, plane sections normal to its axis before deformation do not rotate about its axis $(\theta_1 = 0)$ and can be considered plane and normal to its axis $[\hat{u}_1 = u_1(x_1)\dagger]$ after deformation. An element undergoes axial deformation only when it is subjected to one or more of the following disturbances:

1. External axial centroidal forces. These could be distributed forces $p_1(x_1)$ or concentrated forces $P_1^{(i)}$ $(i = 1, 2, \ldots, n_1)$.

2. Change of temperature which does not vary on the cross sections of the element $[\delta T_2 = \delta T_3 = 0, \Delta T_c(x_1) \neq 0]$ (see Sec. 2.3).

3. Initial strain which does not vary on the cross sections of the element $[k_2^I = k_3^I = 0, e_{11}^{Ic}(x_1) \neq 0]$ (see Sec. 2.3).

Referring to the first of relations (1.3) the normal component of strain e_{11} of the particles of an element undergoing axial deformation only is given as

$$e_{11} = \frac{\partial \hat{u}_1}{\partial x_1} = \frac{du_1}{dx_1} \tag{2.34}$$

Moreover, referring to relations (2.33) and using relation (2.34) the normal component of stress τ_{11} at a particle of an element made from an isotropic linearly elastic material and undergoing only axial deformation is equal to

$$\tau_{11} = E\frac{du_1}{dx_1} - E\alpha\,\Delta T_c - Ee_{11}^{Ic} \tag{2.35}$$

Substituting relation (2.35) into the first of relations (1.1) we get the following internal action-displacement relation for an element of area $A(x_1)$ made of an isotropic linearly elastic material:

$$N(x_1) = EA\left(\frac{du_1}{dx_1} - H_1\right) \tag{2.36}$$

where

$$H_1(x_1) = \alpha\,\Delta T_c + e_{11}^{Ic} \tag{2.37}$$

---

† $\hat{u}_1(x_1, x_2, x_3)$ is the component of displacement in the direction of the x_1 axis of a particle of an element. $u_1(x_1)$ is the component of displacement in the direction of the x_1 axis of a particle of the centroidal axis of an element.

Notice that by eliminating du_1/dx_1 from relations (2.35) and (2.36) as expected we obtain the following stress-action relation for an element undergoing only axial deformation:

$$\tau_{11} = \frac{N}{A} \tag{2.38}$$

The equilibrium of a segment of infinitesimal length of the element requires that the axial component of the internal and external forces acting on the element satisfy relation (2.2) at all its points except those at which the distribution of the axial component of the external forces acting on the element is discontinuous. For the element of Fig. 2.6, relation (2.2) is not valid at point $x_1 = c_1$ where the axial component of the external force has a simple discontinuity and at points $x_1 = a_{1i}$ $(i = 1, 2, \ldots, n_1)$ where the axial component of the external force has a Dirac delta–type discontinuity (see App. A). In order to treat relation (2.2) as if the distribution of the axial component of the external forces acting on the element of Fig. 2.6 is continuous, we express this distribution using functions of discontinuity (see App. A). That is, referring to relations (A.39) and (A.45) the action equation of equilibrium (2.2) for the element of Fig. 2.6 can be written as

$$\frac{dN}{dx_1} = -p_1(x_1)\,\Delta(x_1 - c_1) - \sum_{i=1}^{n_1} P_1^{(i)}\,\delta(x_1 - a_{1i}) \tag{2.39}$$

Substituting relation (2.36) into (2.39) for an element subjected to the external disturbances shown in Fig. 2.6, we obtain

$$\frac{d}{dx_1}\left(EA\frac{du_1}{dx_1}\right) = -p_1(x_1)\,\Delta(x_1 - c_1) - \sum_{i=1}^{n_1} P_1^{(i)}\,\delta(x_1 - a_{1i}) + \frac{d}{dx_1}(EAH_1) \tag{2.40}$$

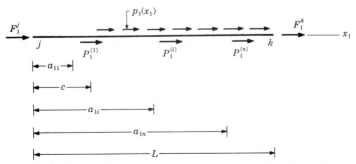

Figure 2.6 Free-body diagram of an element of a framed structure undergoing axial deformation only.

This is the displacement equation of equilibrium for an element made of an isotropic linearly elastic material when subjected to disturbances producing only axial deformation. It is a linear differential equation of the second order. Consequently its solution involves two arbitrary constants which can be evaluated if either u_1 or N is specified at each end of the element.

2.4.2 Elements subjected to bending only

When an element is subjected to bending only, its axis does not elongate ($u_1 = 0$) and its cross sections do not twist ($\theta_1 = 0$). This occurs when an element is subjected to one or more of the following disturbances:

1. External forces whose plane contains the shear center of the cross sections of the element. These could be distributed forces $\mathbf{p}(x_1) = p_2\mathbf{i}_2 + p_3\mathbf{i}_3$ or concentrated forces $\mathbf{P}^{(i)} = P_2^{(i)}\mathbf{i}_2 + P_3^{(i)}\mathbf{i}_3$ ($i = 1, 2, \ldots, n$).

2. External moments which do not have a torsional component. These could include distributed moments $\mathbf{m}(x_1) = m_2\mathbf{i}_2 + m_3\mathbf{i}_3$ or concentrated moments $\mathcal{M}^{(i)} = \mathcal{M}_2^{(i)}\mathbf{i}_2 + \mathcal{M}_3^{(i)}\mathbf{i}_3$ ($i = 1, 2, \ldots, m$).

3. A change of temperature which is a linear function of x_2 and x_3 and moreover vanishes on the axis of the element [$\Delta T_c = 0$, $\delta T_2(x_1) \neq 0$, $\delta T_3(x_1) \neq 0$].

4. An initial strain which is a linear function of x_2 and x_3 and, moreover, vanishes on the axis of the element [$e_{11}^{Ic} = 0$, $k_2^I(x_1) \neq 0$, $k_3^I(x_1) \neq 0$] (see Sec. 2.3).

Notice that for deformation within the range of validity of the theory of small deformation the effect of the axial force acting on the cross sections of an element on the magnitude of its bending moment is negligible. Consequently, the effect of the axial force on the transverse components of translation is negligible. That is, axial forces produce only axial components of translation, while transverse forces and bending moments do not produce axial components of translation. We say that the effect of the axial forces is not coupled to that of the transverse forces and bending moments.

The following fundamental assumptions are made as to the geometry of the deformed configuration of the cross sections of an element subjected to bending only:

1. Plane sections normal to the axis of an element, prior to deformation, can be considered plane subsequent to deformation; that is, the warping of its cross sections is assumed negligible.

2. Plane sections normal to the axis of an element, before deformation, can be considered normal to its deformed axis subsequent to deformation. This implies that the effect of the shearing components of strain e_{12} and e_{13} on the deformed configuration of an element is negligible.

When an element is subjected only to end bending moments, plane sections normal to its axis do remain plane subsequent to deformation and normal to its deformed axis. However, when an element is subjected to transverse external forces, its cross sections warp and do not remain normal to its deformed axis. Nevertheless, when the length of an element is considerably larger than its other dimensions, the warping of its cross sections and the change of the angle between its axis and its cross sections, due to its deformation, do not affect appreciably the components of displacement of its points, and they can be disregarded.

The theory based only on the first of the above assumptions is referred to as the *Timoshenko theory of beams,* while the theory based on both assumptions is referred to as the *classical theory of beams.* In the Timoshenko theory of beams the effect of shear deformation of an element on its deformed configuration is taken into account, while in the classical theory of beams it is disregarded. In this book unless we state otherwise we use the classical theory of beams.

On the basis of the first fundamental assumption stated previously the component of displacement $\hat{u}_1$ of any point (x_1, x_2, x_3) of a cross section of an element may be expressed as

$$\hat{u}_1(x_1, x_2, x_3) = -x_2\theta_3(x_1) + x_3\theta_2(x_1) \tag{2.41}$$

where $\theta_2(x_1)$ and $\theta_3(x_1)$ are the components of rotation of the cross section about the x_2 and x_3 axes, respectively. Substituting relation (2.41) into (1.3) we obtain

$$e_{11} = \frac{\partial \hat{u}_1}{\partial x_1} = -x_2\frac{d\theta_3}{dx_1} + x_3\frac{d\theta_2}{dx_1} \tag{2.42}$$

Moreover, substituting relation (2.41) into (2.33) the normal component of stress τ_{11} at a particle of an element made from an isotropic linearly elastic material is equal to

$$\tau_{11} = E\left(e_{11} - \frac{\alpha\,\delta T_2\,x_2}{h^{(2)}} - \frac{\alpha\,\delta T_3\,x_3}{h^{(3)}} + k_2^I x_2 - k_3^I x_3\right)$$

$$= E\left(x_3\frac{d\theta_2}{dx_1} - x_2\frac{d\theta_3}{dx_1} - \frac{\alpha\,\delta T_2\,x_2}{h^{(2)}} - \frac{\alpha\,\delta T_3\,x_3}{h^{(3)}} + k_2^I x_2 - k_3^I x_3\right) \tag{2.43}$$

Substituting relations (2.43) into relation (1.1e) and (1.1f) and integrating, we get the following internal action-displacement relations for an element made of an isotropic linearly elastic material:

$$M_2(x_1) = EI_2\left(\frac{d\theta_2}{dx_1} - H_2\right) \qquad (2.44a)$$

$$M_3(x_1) = EI_3\left(\frac{d\theta_3}{dx_1} - H_3\right) \qquad (2.44b)$$

where $I_2(x_1)$ and $I_3(x_1)$ are the moments of inertia of the cross sections of the element about their principal centroidal axes x_2 and x_3, respectively, and

$$H_3(x_1) = -\frac{\alpha\,\delta T_2}{h^{(2)}} + k_2^I \qquad (2.45a)$$

$$H_2(x_1) = \frac{\alpha\,\delta T_3}{h^{(3)}} + k_3^I \qquad (2.45b)$$

Substituting relations (2.44) into (2.43) we obtain the following expression for the normal component of stress at a particle of an element made of an isotropic linearly elastic material and subjected only to bending.

$$\tau_{11} = \frac{M_2 x_3}{I_2} - \frac{M_3 x_2}{I_3} \qquad (2.46)$$

The second fundamental assumption indicates that the angle between a deformed cross section and the tangent to the elastic curve may be approximated by a right angle. Hence, the components of rotation of a cross section are related to the slopes of the projections of the elastic curve on the $x_1 x_3$ and $x_1 x_2$ planes, respectively, as follows:

$$\theta_2 = -\frac{du_3}{dx_1} \qquad (2.47a)$$

$$\theta_3 = \frac{du_2}{dx_1} \qquad (2.47b)$$

Thus, on the basis of the classical theory of beams, the components of rotation θ_2 and θ_3 can be established if the components of translation u_2 and u_3 are known functions of x_1. This implies that on the basis of the classical theory of beams, the deformed configuration of an element of a space framed structure subjected to bending without twisting is completely specified if the components of translation u_2 and u_3 of its axis are known functions of the axial coordinate x_1. Moreover, choosing the x_3 axis in the plane of the structure, the deformed configuration of an element of a planar structure subjected to bending without twisting is completely specified if the component of translation u_3 of its axis is a known function of x_1.

Substituting relations (2.47) into (2.42) we get the following strain displacement relation for the classical theory of beams.

$$e_{11} = \frac{\partial \hat{u}_1}{\partial x_1} = -x_3 \frac{d^2 u_3}{dx_1^2} - x_2 \frac{d^2 u_2}{dx_1^2} \qquad (2.48)$$

Moreover substituting relations (2.47) into (2.44) we get the following internal action-displacement relations for an element made of an isotropic linearly elastic material:

$$M_2(x_1) = -EI_2\left(\frac{d^2 u_3}{dx_1^2} + H_2\right) \qquad (2.49a)$$

$$M_3(x_1) = EI_3\left(\frac{d^2 u_2}{dx_1^2} - H_3\right) \qquad (2.49b)$$

Substituting relations (2.49) into the action equations of equilibrium (2.6) and (2.7) we obtain the following internal action-displacement relations for an element made of an isotropic linearly elastic material:

$$Q_2 = -\frac{d}{dx_1}\left[EI_3\left(\frac{d^2 u_2}{dx_1^2} - H_3\right)\right] - m_3 \qquad (2.50a)$$

$$Q_3 = -\frac{d}{dx_1}\left[EI_2\left(\frac{d^2 u_3}{dx_1^2} + H_2\right)\right] + m_2 \qquad (2.50b)$$

The equilibrium of a segment of infinitesimal length of an element subjected to bending only requires that the components of the internal and external actions acting on it satisfy relations (2.3), (2.4), (2.6), and (2.7). Combining relations (2.3) with (2.6) and (2.4) with (2.7) we get

$$\frac{d^2 M_3}{dx_1^2} + \frac{dm_3}{dx_1} = p_2 \qquad (2.51a)$$

$$\frac{d^2 M_2}{dx_1^2} + \frac{dm_2}{dx_1} = -p_3 \qquad (2.51b)$$

These are the action equations of equilibrium for an element subjected to bending only. As mentioned in Sec. 2.2 these equations are not valid at the points of discontinuity of the distribution of the external actions acting on the element. That is, for the element of Fig. 2.7, relation (2.51b) is not valid at the following points:

1. $x_1 = c_3$ where the distribution of external forces $p_3(x_1)$ has a simple discontinuity (jump)

2. $x_1 = a_{3i}$ $(i = 1, 2, \ldots, n_3)$ where the distribution of external forces has Dirac delta–type discontinuities (see App. A)

3. $x_1 = d_2$ where the distribution of external moments $m_2(x_1)$ has a simple discontinuity (jump)

4. $x_1 = b_{2i}$ $(i = 1, 2, \ldots, m_2)$ where the distribution of external moments has Dirac delta–type discontinuities

In order to treat relations (2.51) as if the distribution of external actions on an element is continuous, we express this distribution using functions of discontinuity. That is, referring to Fig. 2.7 and to relations (A.39), (A.45), and (A.51) the transverse components of the external forces and moments in relations (2.51) may be replaced by the following expressions:

External force in the x_2 direction

$$= p_2(x_1)\, \Delta(x_1 - c_2) + \sum_{i=1}^{n_2} P_2^{(i)}\, \delta(x_1 - a_{2i})$$

External force in the x_3 direction

$$= p_3(x_1)\, \Delta(x_1 - c_3) + \sum_{i=1}^{n_3} P_3^{(i)}\, \delta(x_1 - a_{3i}) \qquad (2.52)$$

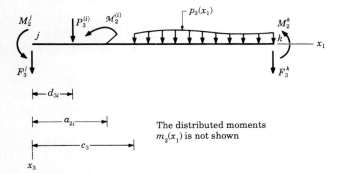

Figure 2.7 Free-body diagram of an element of a framed structure undergoing bending about the x_2 axis.

External moment about the x_2 axis

$$= m_2(x_1) \, \Delta(x_1 - d_2) + \sum_{i=1}^{m_2} \mathcal{M}_2^{(i)} \, \delta(x_1 - b_{2i})$$

External moment about the x_3 axis

$$= m_3(x_1) \, \Delta(x_1 - d_3) + \sum_{i=1}^{m_3} \mathcal{M}_3^{(i)} \, \delta(x_1 - b_{3i})$$

Substituting relations (2.49) into (2.51) and using (2.52) we obtain

$$\frac{d^2}{dx_1^2} \left[EI_3 \left(\frac{d^2 u_2}{dx_1^2} - H_3 \right) \right] = p_2(x_1) \, \Delta(x_1 - c_2) - \frac{d}{dx_1} [m_3 \, \Delta(x_1 - d_3)]$$

$$\times \sum_{i-1}^{n_2} P_2^{(i)} \, \delta(x_1 - a_{2i}) - \sum_{i=1}^{m_3} \mathcal{M}_3^{(i)} \, \delta^1(x_1 - b_{3i}) \quad (2.53a)$$

$$\frac{d^2}{dx_1^2} \left[EI_2 \left(\frac{d^2 u_3}{dx_1^2} + H_2 \right) \right] = p_3(x_1) \, \Delta(x_1 - c_3) + \frac{d}{dx_1} [m_2 \, \Delta(x_1 - d_2)]$$

$$\times \sum_{i-1}^{n_3} P_3^{(i)} \, \delta(x_1 - a_{3i}) + \sum_{i=1}^{m_2} \mathcal{M}_2^{(i)} \, \delta^1(x_1 - b_{2i}) \quad (2.53b)$$

where $p_2(x_1)$ and $p_3(x_1)$ are the transverse components of the external forces distributed along the length of the element; $\delta^1(x_1 - b_{3i})$ is the doublet function defined by relation (A.14); $H_2(x_1)$ and $H_3(x_1)$ are defined by relations (2.45).

Relations (2.53a) and (2.53b) are the displacement equations of equilibrium for an element made from an isotropic linearly elastic material, subjected to disturbances inducing only bending about the x_3 and x_2 axes, respectively. They are linear differential equations of the fourth order. Consequently their solution involves four arbitrary constants which can be evaluated if four conditions at the ends of the element are specified. The transverse component of translation $u_2(x_1)$ of an element can be established to within a rigid-body motion of the element from relation (2.53a) if p_2 and m_3 are given functions of the axial coordinate of the element, the values and points of application of $p_2^{(i)}$ ($i = 1, 2, \ldots, n_2$) and $\mathcal{M}_3^{(i)}$ ($i = 1, 2, \ldots, m_3$) are given, and one quantity of the following pairs is specified at each end of the element:

$$\begin{array}{cc} u_2 \quad \text{or} \quad Q_2 \\ \theta_3 \quad \text{or} \quad \mathcal{M}_3 \end{array} \quad (2.54a)$$

Similarly the transverse component of translation $u_3(x_1)$ of an element can be established to within a rigid-body motion of the element from relation (2.53b) if p_3 and m_2 are given functions of the axial coordinate of the element, the values and points of application of $p_3^{(i)}$ $(i = 1, 2, \ldots, n_3)$ and $\mathcal{M}_2^{(i)}$ $(i = 1, 2, \ldots, m_2)$ are given, and one quantity of the following pairs is specified at each end of the element:

$$u_3 \quad \text{or} \quad Q_3$$
$$\theta_2 \quad \text{or} \quad \mathcal{M}_2 \tag{2.54b}$$

2.4.3 Shearing components of stress in an element subjected to transverse forces

Consider an element of a framed structure made from an isotropic linearly elastic material subjected to transverse forces which pass through the shear center of its cross sections. Moreover, referring to Fig. 2.8 consider a line AB on a cross section of this element and denote the unit vectors in the direction of this line and normal to this direction by $\mathbf{s}$ and $\mathbf{n}$, respectively. In general on the basis of the mechanics of materials theories we can compute the average value of the shearing component of stress τ_{1n} acting on the particles of line AB in the direction of the unit vector $\mathbf{n}$. It is equal[2] to

$$(\tau_{1n})_Q = \frac{Q_2 Z_3}{I_3 b_s} + \frac{Q_3 Z_2}{I_2 b_s} \tag{2.55}$$

where $\qquad Z_2 = \iint_{A_n} x_3 \, dA = \bar{x}_{3n} A_n \qquad Z_3 = \iint_{A_n} x_2 \, dA = \bar{x}_{2n} A_n \qquad (2.56)$

and $\qquad Q_j \ (j = 2, 3) =$ shearing components of the force acting on the cross section in the direction of axis x_j $(j = 2, 3)$

$\qquad\quad I_j \ (j = 2, 3) =$ moment of inertia of the cross section about its principal centroidal axis x_j $(j = 2, 3)$

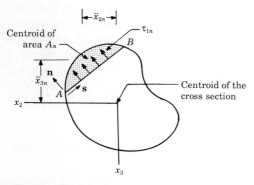

Figure 2.8 Cross section of an element subjected to transverse forces.

$b_s = $ length of line AB

$A_n = $ area of any one of the two parts into which the cross section is divided by AB†

$\mathbf{n} = $ unit vector directed toward A_n‡

$$\iint_{A_n} x_j \, dA \ (j = 2, 3) = \text{first moment of } A_n \text{ about the axis } x_j \ (j = 2, 3)$$

$\bar{x}_{jn} \ (j = 2, 3) = $ distance of the centroid of A_n from axis x_i $(i = 3, 2, i \neq j)$

Notice that if the average shearing component of stress established on the basis of relation (2.55) is positive, its sense is that of the unit vector $\mathbf{n}$.

For certain elements we know a priori that the variation of the shearing component of stress τ_{1n} acting on the particles of a line AB of its cross section is negligible. For those elements, the average value of the shearing component of stress τ_{1n} acting on the particles of line AB can be considered equal to its actual value. This occurs in the following cases of practical interest:

1. In elements having thin-walled, open cross sections when subjected to external actions acting in any plane (see Fig. 2.9a and b). In this case if we choose the unit vector $\mathbf{s}$ normal to the lateral boundary of the element, τ_{1n} does not vary much in the direction of $\mathbf{s}$ and, moreover, τ_{1s} is negligible. This becomes apparent by noting that the shearing stress acting on a cross section at the particles of its boundary must be tangent to the boundary.

2. In elements having thin-walled, closed cross sections, which are symmetric with respect to an axis, when subjected to external actions acting in a plane specified by their axis and the axis of symmetry (see Fig. 2.9c). In this case if we choose the unit vector $\mathbf{s}$ normal to the lateral boundary of the element, τ_{1n} does not vary in the direction of $\mathbf{s}$ and, moreover, τ_{1s} is negligible.

3. In elements having solid cross sections, which are symmetric with respect to an axis, when subjected to external actions acting in a plane specified by their axis and the axis of symmetry of their cross sections (see Fig. 2.9d). In this case if we choose the unit vector $\mathbf{s}$ normal to the axis of symmetry of the cross section of the element, τ_{1n} does not vary much in the direction of $\mathbf{s}$. Moreover, τ_{1s} is negligible at points close to the axis of symmetry of the cross section. However, as

---

† Referring to Fig. 2.8, A_n is either the shaded part of the cross section or the remaining part. However, it is more convenient to choose the part which does not include the centroid of the cross section, that is, the shaded area of Fig. 2.8.

‡ Unit vector $\mathbf{n}$ shown in Fig. 2.8 corresponds to the shaded area.

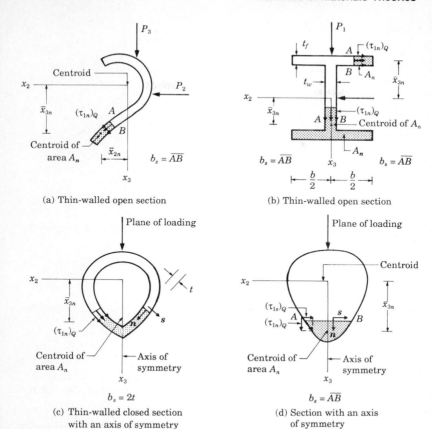

(a) Thin-walled open section

(b) Thin-walled open section

(c) Thin-walled closed section
with an axis of symmetry

(d) Section with an axis
of symmetry

Figure 2.9 Cross sections of members of a structure.

shown in Fig. 2.9d, τ_{1s} may not be negligible at points close to the boundary of the element.

In what follows we illustrate the use of relation (2.55) by an example.

Example. Compute the components of shearing stress at particles A and B of the beam loaded as shown in Fig. a.

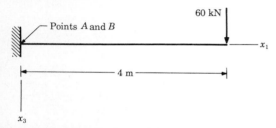

Figure a Geometry and loading of the beam.

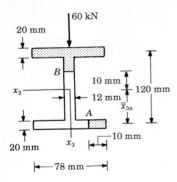

Figure b Cross section of the beam.

solution Referring to Fig. b the moment of inertia of the cross section of the beam about the x_2 axis is equal to

$$I_2 = 2\left[\frac{(20)^3(78)}{12} + 20(78)(60)^2\right] + \frac{(100)^3(12)}{12} = 12.336 \times 10^6 \text{ mm}^4$$

Referring to relation (2.55) the shearing component of stress $\tau_{13}^{(B)}$ is equal to

$$\tau_{13}^{(B)} = \frac{Q_3 \iint_{A_n} x_3 \, dA}{I_2 b_s}$$

Referring to Fig. b the integral $\iint_{A_n} x_3 \, dA$ is the first moment of the shaded area above point B about the axis x_2. That is,

$$\iint_{A_n} x_3 \, dA = 78(20)(-60) + 12(40)(-30) = -108,000 \text{ mm}^3$$

Moreover, $b_s = 12 \text{ mm}$

Thus, $\tau_{13}^{(B)} = -\dfrac{60(108,000)}{12.336 \times 10^6(12)} = -43,770 \text{ kN/m}^2$

The minus sign indicates that the shearing stress $\tau_{13}^{(B)}$ is acting in the direction opposite to that of the unit vector **n** (see Fig. c). Referring to

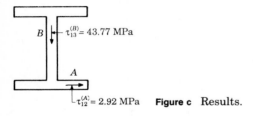

Figure c Results.

relation (2.55) the shearing component of stress $\tau_{12}^{(A)}$ is equal to

$$\tau_{12}^{(A)} = \frac{Q_3 \displaystyle\iint_{A_n} x_3 \, dA}{I_2 b_s}$$

where referring to Fig. b the integral $\iint_{A_n} x_3 \, dA$ is the first moment of the shaded area to the right of point A about the axis x_2. That is,

$$\iint_A x_3 \, dA = A_n \bar{x}_{3n} = 20(10)60 = 12{,}000 \text{ mm}^2$$

Moreover, $$b_s = 20 \text{ mm}$$

Thus, $$\tau_{12}^{(A)} = \frac{60(12{,}000)}{12\,336 \times 10^6 (20)} = 2920 \text{ kN/m}^2$$

The results are shown in Fig. c.

2.4.4 Elements subjected to twisting only

When a prismatic straight element is subjected to two equal and opposite external torsional moments at its ends, its axis does not elongate or bend. However, a centroidal, straight line normal to the axis of the element, prior to deformation, does not remain always straight subsequent to deformation. It can become a curve whose projection on a plane normal to the axis of the element is a centroidal straight line which is rotated by an angle θ_1 about the axis of the element. The projections of the deformed configuration of all centroidal lines of a cross section of an element rotate by the same angle θ_1 about the axis of the element. The angle θ_1 is called the *rotation of this cross section* or the *angle of twist of this cross section*. Moreover the length of the projection of a deformed centroidal line on a plane normal to the axis of the element is equal to its undeformed length. That is, the radial component of displacement of the particles of an element vanishes.

On the basis of the foregoing discussion plane sections normal to the axis of a prismatic straight element subjected to two equal and opposite external torsional moments at its ends do not remain plane after deformation; they rotate about the axis of the element, and they warp. Only plane sections normal to the axis of prismatic straight elements of circular cross section remain plane subsequent to deformation; that is, they do not warp.

Consider a prismatic straight element of noncircular cross section subjected to two equal and opposite torsional moments at its ends. When all the cross sections of this element are free to warp, the length of its longitudinal fibers does not change due to deformation, and, consequently, the normal component of stress acting on the cross sections of the element vanishes. This implies that the axial component of translation of the particles of the element is not a function of x_1. In practice, however, one or more cross sections of an element are usually restrained from warping. In this case, the change of length of the longitudinal fibers of an element having thin, open cross sections may not be negligible. Consequently, the normal component of stress acting at particles located at or near the cross sections which are restrained from warping and away from the axis of the element may be large and must be taken into account. Moreover, the effect of restraining the warping of a cross section of an element having thin, open cross sections on the angle of twist may not be negligible. For elements of other cross sections (hollow, closed, or full) the effect of restraining the warping of a cross section on the values of the normal component of stress and on the angle of twist is small and is neglected.

In structural analysis the effect of warping of an element subjected to torsional moments is taken into account only in computing the torsional constant [see Eq. (2.64)].

Consider an element of constant cross section subjected to equal and opposite torsional moments at its ends. Assume that the torsional moments are applied in a way that the cross sections of the element are free to warp. A segment of length dx_1 of the element under consideration rotates as a rigid body and deforms. As shown in Fig. 2.10 because of the deformation of the segment, the radial plane $ABCD$ deforms to $ABC'D$ and line BC rotates relative to line AD by an angle $d\theta_1$. Although in general the cross sections of the segment warp the length of its fibers, fiber AB, for instance, does not change because of the deformation. Moreover, the radial component of displacement $\hat{u}_r$ of the particles of the element is zero. Thus, the components of displacement of the particles of the element have the following form:

$$\hat{u}_1 = F_1(x_2, x_3)$$

$$\hat{u}_r = 0 \qquad\qquad (2.57)$$

$$\hat{u}_t = r\theta_1$$

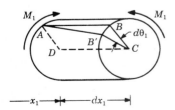

Figure 2.10 Segment of an element subjected to end torsional moments.

where r is the distance of the particle from the axis of the element. The function $F_1(x_2, x_3)$ is called the *warping function* and depends on the geometry of the cross section of the element. Elements of circular or hollow circular cross section do not warp, and thus the warping function for these elements vanishes. Notice that the rate of change of θ_1 along the axis of an element of constant cross section which is subjected to end torsional moments is constant. That is,

$$\frac{d\theta_1}{dx_1} = \gamma \tag{2.58}$$

The constant γ is referred to as the *twist per unit length*. Assuming that the element is supported in a way that $\theta_1 = 0$ at $x_1 = 0$, we have

$$\theta_1 = \gamma x_1 \tag{2.59}$$

Consider a centroidal line CP which prior to deformation lies on the cross section of the element under consideration. In Fig. 2.11 line CP' represents the projection of the deformed line CP on a plane normal to the axis of the element. Inasmuch as the radial component of displacement vanishes, line $P'P$ is normal to line CP. Consequently, the triangles PGP' and CPE are similar. From the similarity of these triangles and using relation (2.59) we have

$$\hat{u}_2 = -\frac{x_3 \hat{u}_t}{r} = -x_3 \theta_1 = -x_3 x_1 \gamma$$

$$\tag{2.60}$$

$$\hat{u}_3 = \frac{x_2 \hat{u}_t}{r} = x_2 \theta_1 = x_2 x_1 \gamma$$

Moreover, it is convenient to express the component of displacement $\hat{u}_1$ [see Eq. (2.57)] as

$$\hat{u}_1 = \gamma f_1(x_2, x_3) \tag{2.61}$$

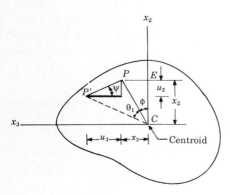

Figure 2.11 Cross section of an element subjected to end torsional moments.

The component of strain of a particle of the line element under consideration may be obtained by substituting relations (2.60) and (2.61) into the strain displacement relations (1.3). Thus,

$$e_{11} = e_{22} = e_{33} = e_{23} = 0$$

$$e_{13} = \frac{\gamma}{2}\left(\frac{\partial f_1}{\partial x_3} + x_2\right)$$

$$e_{12} = \frac{\gamma}{2}\left(\frac{\partial f_1}{\partial x_2} - x_3\right)$$

(2.62)

Using the above relations in (1.1d) we obtain the following expression for the internal torsional moment M_1 in an element made of an isotropic linearly elastic material:

$$M_1 = \iint_A (\tau_{13}x_2 - \tau_{12}x_3)\, dA = 2G \iint_A (e_{13}x_2 - e_{12}x_3)\, dA$$

$$= G\gamma\left[\iint_A \left(\frac{\partial f_1}{\partial x_3}x_2 - \frac{\partial f_1}{\partial x_2}x_3\right)dA + \iint_A (x_2^2 + x_3^2)\, dA\right]$$

$$= G\gamma\left[\iint_A \left(\frac{\partial f_1}{\partial x_3}x_2 - \frac{\partial f_1}{\partial x_2}x_3\right)dA + J\right]$$

(2.63)

where J is the polar moment of inertia of the cross section of the element. Referring to relation (2.58), relation (2.63) can be rewritten as

$$\frac{d\theta_1}{dx_1} = \gamma = \frac{M_1}{KG}$$

(2.64)

where K is referred to as the *torsional constant* and is dependent upon the geometry of the cross section of the element. In order to compute the value of the torsional constant for a cylindrical element of noncircular cross section, the warping $\hat{u}_1(x_2, x_3) = \gamma f_1(x_2, x_3)$ corresponding to this cross section must be computed.[†] The torsional constant of prismatic elements of circular or hollow circular cross section is equal to their polar moment of inertia.[‡]

[†] For a detailed analysis of the torsion of prismatic elements see S. Timoshenko and J. N. Goodier, *Theory of Elasticity*, McGraw-Hill Book Co., New York, 1970, p. 291.

[‡] The torsional constants of prismatic elements of rectangular cross section are given in A. E. Armenàkas, *Classical Structural Analysis: A Modern Approach*, McGraw-Hill Book Co., New York, 1988, p. 686.

Relation (2.64) is the internal action displacement relation for an element subjected only to twisting. When an element is subjected only to end torsional moments (M_1 = constant), relation (2.64) may be integrated to give

$$\theta_1 = \gamma x_1 = \frac{M_1 x_1}{KG} \tag{2.65}$$

When prismatic elements are subjected to a distribution of external torsional moments along their length, the warping of their cross sections varies along their length. That is the twist per unit length γ is a function of x_1. This implies that the cross sections of the elements cannot warp freely because they are partially restrained from warping by the neighboring cross sections which warp by a different amount. In this case, the normal component of stress in elements having thin, open cross sections may not be negligible.

Consider a prismatic element made of an isotropic linearly elastic material subjected to external distributed torsional moments $m_1(x_1)$ applied at $d_1 \leq x_1 \leq L$ and to external concentrated torsional moments $\mathcal{M}_1^{(i)}$ applied at $x_1 = b_{1i}$ ($i = 1, 2, 3, \ldots, m_1$). The equilibrium of a segment of infinitesimal length of this element requires that the internal and external torsional moments satisfy relation (2.5) at all points of the element except those at which the distribution of the external torsional moments is discontinuous. Thus for the element under consideration relation (2.5) is not valid at point $x_1 = d_1$ where the external torsional moment has a simple discontinuity (jump) and at points $x_1 = b_{1i}$ ($i = 1, 2, \ldots, m_1$) where the external torsional moment has a Dirac delta–type discontinuity (see App. A). In order to treat relation (2.5) as if the distribution of the external torsional moments acting on the element is continuous, we express this distribution using functions of discontinuity (see App. A). That is, referring to relations (A.39) and (A.45) the external torsional moment in relation (2.5) may be replaced by the following expression:

External torsional moment

$$= m_1(x_1)\,\Delta(x_1 - d_1) + \sum_{i=1}^{m} \mathcal{M}_1^{(i)}\,\delta(x_1 - b_{1i}) \tag{2.66}$$

Differentiating relations (2.64) and using relations (2.66) and (2.5) we obtain

$$\frac{d}{dx_1}\left(KG\frac{d\theta_1}{dx_1}\right) = \frac{dM_1}{dx_1} = -m_1\,\Delta(x_1 - d_1) - \sum_{i=1}^{m_1} \mathcal{M}_1^{(i)}\,\delta(x_1 - b_{1i}) \tag{2.67}$$

This is the displacement equation of equilibrium for an element made of an isotropic linearly elastic material undergoing twisting only. It is a linear differential equation of the second order. Consequently, its solution involves two arbitrary constants which can be evaluated if either θ_1 or M_1 is specified at each end of the element.

2.5   The Boundary-Value Problems for Computing the Components of Displacement and of Internal Action of an Element of a Framed Structure

The deformed configuration of a framed structure is described relative to a reference configuration wherein the structure is not subjected to external loads. However, in the reference configuration the length and/or the curvature of some elements of the structure may be different than that required in order that these elements fit in the geometry of the structure. These elements may be regarded as being in a state of initial strain (see Sec. 2.3).

As discussed in Sec. 1.10 the deformed configuration of a cross section of an element of a space framed structure is completely specified if its components of translation $u_1, u_2,\ u_3$ and its components of rotation $\theta_1,\ \theta_2,\ \theta_3$ are known. However, in the classical theory of beams employed in this text the components of rotation θ_2 and θ_3 of the cross sections of an element can be established if the components of translation u_3 and u_2, respectively, are known functions of its axial coordinate [see relations (2.47)]. Thus the deformed configuration of an element of a space framed structure is completely specified if the components of translation u_1, u_2, u_3 and the component of rotation θ_1 are known functions of the axial coordinate of the element. Moreover, the deformed configuration of an element of a planar framed structure is completely specified if the components of translation in its plane are known functions of the axial coordinate of the element. We call the components of displacement which are needed in order to specify the deformed configuration of an element *its independent components of displacement.*

The internal forces acting on an element are not affected by its rigid-body motions. Consequently, if a component of displacement represents only rigid-body motion of an element, it does not affect its internal actions. As shown in Sec. 5.3 the transverse components of translation u_2 and u_3 and the components of rotation θ_2 and θ_3 of an axial deformation element represent only rigid-body motion. Hence, they do not contribute to its deformation, and they do not affect its internal actions.

We call the independent components of displacement of an element which contribute to its deformation and thus affect its internal actions its *basic components of displacement*. The basic components of displacement of the types of elements which we are considering are

Axial deformation element	u_1
General planar element in the $x_1 x_2$ plane	$u_1 \, u_2$
General space element	u_1, u_2, u_3, θ_1

The basic components of displacement of an element are not coupled. Each one of them can be established as a function of the axial coordinate of the element by solving one of the boundary-value problems described in the following subsections.

2.5.1 The strong or classical form of the boundary-value problem for establishing the axial component of translation of an element of a framed structure

Consider an element of a framed structure initially in a reference stress-free state of mechanical[†] and thermal[‡] equilibrium at a uniform temperature T_0. In this state the element may have a different length than the one required to fit in the geometry of the structure. Thus as discussed in Sec. 2.3 it may be regarded as being in a state of uniform initial strain. It is assumed that the axial component of initial strain $e_{11}^{Ic}(x_1)$ of the element is specified. The element reaches a second state of mechanical equilibrium due to the application on it of one or more of the following disturbances:

1. Specified distributed axial centroidal forces $p_1(x_1)$ acting at $c_1 \leq x_1 \leq L$ (see Fig. 2.6 or 2.12)
2. Specified concentrated axial centroidal forces $P_1^{(i)}$ acting at $x_1 = a_{1i}$ $(i = 1, 2, \ldots, n_1)$ (see Fig. 2.6 or 2.12)
3. Specified change of temperature $[\Delta T_c(x_1) \neq 0, \; \delta T_2 = \delta T_3 = 0]$ (see Sec. 2.3)

Because of the application of these disturbances the element translates, rotates, and deforms. However, as shown in Sec. 5.3 the transverse components of translation $u_2(x_1)$, $u_3(x_1)$, and the components of rotation represent only rigid-body motion of the element,

---

† When a body is in a state of mechanical equilibrium, its particles do not accelerate. That is, the sum of the forces acting on any portion of the body and the sum of the moments of these forces about a point vanish.

‡ When a body is in a state of thermal equilibrium, heat does not flow in or out of it. That is, all its particles have the same temperature.

while its component of translation $u_1(x_1)$ represents both deformation and rigid-body motion.

Notice that if in the second state of mechanical equilibrium the temperature of the element is a function of its axial coordinate, heat will flow into or out of it and consequently it will not be in a state of thermal equilibrium.

We are interested to establish the axial component of translation $u_1(x_1)$ and of the internal force $N(x_1)$ of the element under consideration. This is a two-point boundary-value problem. Its domain is the length of the element $0 \le x_1 \le L$. Its boundary consists of the two points $x_1 = 0$ and $x_1 = L$.

When this boundary-value problem is formulated in its so-called *strong or classical form*, we are looking for the function $u_1(x_1)$ which has the following attributes:

1. When substituted into the appropriate internal action-displacement relation for the material from which the element is made, it gives an axial component of internal force $N(x_1)$ which satisfies the action equation of equilibrium (2.39). For an element made of an isotropic linearly elastic material $N(x_1)$ is related to the axial component of translation on the basis of relation (2.36).

2. It satisfies the specified boundary conditions at the ends of the element.

In the second state of mechanical equilibrium either the axial component of translation u_1 or the axial component of force N must be specified at each end of the element. The specification of u_1 is called an *essential boundary condition*, while the specification of N is called a *natural boundary condition*. For a unique solution it is essential that the element is inhibited from translating in the x_1 direction as a rigid body. Thus, the axial component of translation u_1 must be specified at least at one end of the element.

In modern structural analysis we are interested in establishing the axial component of translation of an element of a framed structure as a function of its axial coordinate involving its axial components of nodal translations u_1^j and u_1^k. In this case we consider both boundary conditions of the element as essential. That is,

$$u_1(0) = u_1^j \qquad (2.68a)$$

$$u_1(L) = u_1^k \qquad (2.68b)$$

Moreover, in modern structural analysis we are interested in establishing the axial component of translation $u_1(x_1)$ and the internal force of one-element structures subjected to given loads as, for example, those shown in Fig. 2.12. The boundary conditions for the

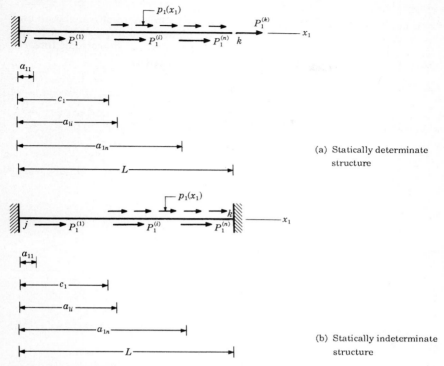

Figure 2.12 One-element structures subjected to external axial forces.

one-element structure of Fig. 2.12a are

$$u_1(0) = 0 \qquad (2.69a)$$

$$N(L) = P_1^k \qquad (2.69b)$$

where P_1^k is the specified axial component of external force at the end k of the element. The boundary condition (2.69a) is *essential*, while the boundary condition (2.69b) is *natural*. The boundary conditions for the one-element structure of Fig. 2.12b are

$$u_1(0) = 0 \qquad (2.70a)$$

$$u_1(L) = 0 \qquad (2.70b)$$

Thus for the one-element structure of Fig. 2.12b both boundary conditions are *essential*.

In what follows we illustrate the solution of the boundary-value problem described in this subsection by finding the exact axial component of translation of a tapered bar.

Example. Establish the expression for the axial component of translation and for the internal force of the tapered bar of constant width b subjected to the external axial forces shown in Fig. a. The bar is made of an isotropic linearly elastic material of modulus of elasticity E.

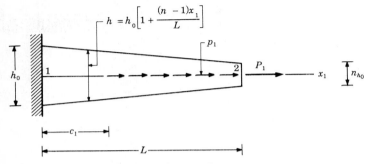

Figure a Geometry and loading of the bar.

solution Referring to Fig. a the area of the bar is equal to

$$A(x_1) = A_0\left[1 + \frac{(n-1)x_1}{L}\right] \tag{a}$$

where
$$A_0 = h_0 b \tag{b}$$

The displacement equation of equilibrium (2.40) for this bar reduces to

$$\frac{d}{dx_1}\left(EA\frac{du_1}{dx_1}\right) = -p_1\,\Delta(x_1 - c_1) \tag{c}$$

Moreover the boundary conditions for the bar are

$$u_1(0) = 0$$
$$\tag{d}$$
$$N(L) = EA\frac{du_1}{dx_1}\bigg|_{x_1=L} = P_1^k$$

Integrating relation (c) and referring to relation (A.3) we obtain

$$N(x_1) = EA\frac{du_1}{dx_1} = -p_1(x_1 - c_1)\,\Delta(x_1 - c_1) + C_1 \tag{e}$$

Using relation (a) we get

$$\frac{du_1}{dx_1} = -\frac{P_1(x_1 - c_1)\,\Delta(x - c_1)}{EA_0[1 + (n-1)x_1/L]} + \frac{C_1}{EA_0[1 + (n-1)x_1/2]} \tag{f}$$

Integrating relation (f) we get

$$u_1(x_1) = -\frac{P_1}{EA_0}\left\{\frac{L(x_1-c_1)}{n-1} - \frac{L^2}{(n-1)^2}\left[1+\frac{c_1(n-1)}{L}\right]\ln\left[\frac{L+(n-1)x_1}{L+(n-1)c_1}\right]\right\}$$

$$\times \Delta(x_1-c_1) + \frac{LC_1}{EA_0(n-1)}\ln\left[1+\frac{(n-1)x_1}{L}\right] + C_2 \quad (g)$$

The constants C_1 and C_2 are obtained by requiring that the solution (g) satisfies the boundary conditions (d). Thus

$$N(L) = \left(EA\frac{du_1}{dx_1}\right)_{x_1=L} = P_1^k = -p_1(L-c_1) + C_1$$

$$u(0) = 0 = C_2 \quad (h)$$

or
$$C_1 = P_1^k + p_1(L-c_1) \qquad C_2 = 0 \quad (i)$$

Substituting relations (i) into (e) and (g) we get

$$N(x_1) = -p_1(x_1-c_1)\,\Delta(x_1-c_1) + P_1^k + p_1(L-c_1) \quad (j)$$

$$u_1(x_1) = \frac{p_1L^2}{EA_0(n-1)^2}\left(\left\{-\frac{(x_1-c_1)(n-1)}{L} + \left[1+\frac{c_1(n-1)}{L}\right]\right.\right.$$

$$\times\ln\left.\left[\frac{L+(n-1)x_1}{L+(n-1)c_1}\right]\right\}\Delta(x_1-c_1)$$

$$+\left(1-\frac{c_1}{L}\right)(n-1)\ln\left.\left[1+\frac{(n-1)x_1}{L}\right]\right)$$

$$+\frac{P_1^kL}{EA_0(n-1)}\ln\left[1+\frac{(n-1)x_1}{L}\right] \quad (k)$$

For $c_1 = L/3$ and $n = 0.5$ relation (k) reduces to

$$u_1(x_1) = \frac{4p_1L^2}{EA_0}\left(\left\{\frac{x_1-L/3}{2L} + \frac{5}{6}\ln\left[\frac{3(2L-x_1)}{5L}\right]\right\}\Delta\left(x_1-\frac{L}{3}\right)\right.$$

$$\left.-\frac{1}{3}\ln\left(1-\frac{x_1}{2L}\right)\right) - \frac{2P_1^kL}{EA_0}\ln\left(1-\frac{x_1}{2L}\right) \quad (l)$$

TABLE a   Values of $u_1(x_1)$ for $c_1 = L/3$ and $n = 0.5$

x_1	$EA_0 u_1(x_1)/L$
$L/6$	$0.11602p_1L + 0.17402P_1$
$L/3$	$0.24376p_1L + 0.36464P_1$
$L/2$	$0.36571p_1L + 0.57536P_1$
$2L/3$	$0.46347p_1L + 0.81093P_1$
$5L/6$	$0.52957p_1L + 1.07799P_1$
L	$0.55478p_1L + 1.38629P_1$

The values of $u_1(x_1)$ at $x_1 = L/6$, $L/3$, $L/2$, $2L/3$, $5L/6$, and L are given in Table a.

2.5.2 The strong or classical form of the boundary-value problem for establishing the transverse component of translation $u_3(x_1)$ of an element of a framed structure

Consider an element of a framed structure initially in a reference stress-free state of mechanical and thermal equilibrium at a uniform temperature T_0. In this state the element may have been manufactured with a different curvature than that required to fit in the geometry of the structure. We refer to the difference of the manufactured from the required curvature as the initial curvature of the element. We limit our discussion to elements having an initial curvature $k_3^I(x_1)$ in the $x_1 x_3$ plane. As discussed in Sec. 2.3, the element may be regarded as being subjected to a distribution of an axial component of initial strain which is not a function of x_2, is a specified function of x_1, and is a linear function of x_3 which vanishes at $x_3 = 0$. The element reaches a second state of mechanical equilibrium because of the application on it of one or more of the following disturbances:

1. Specified distributed transverse forces $p_3(x_1)$ acting at $c_3 \leq x_1 \leq L$ through the shear center of the cross sections of the element (see Fig. 2.7 or 2.13)

2. Specified concentrated transverse force $P_3^{(i)}$ acting at $x_1 = a_{3i}$ ($i = 1, 2, \ldots, n_3$) through the shear center of the cross sections of the element (see Fig. 2.7 or 2.13)

3. Specified distributed moments $m_2(x_1)$ acting at $d_2 \leq x_1 \leq L$

4. Specified concentrated moments $\mathcal{M}_2^{(i)}$ acting at $x_1 = b_{2i}$ ($i = 1, 2, \ldots, m_2$) (see Fig. 2.7 or 2.13)

5. Change of temperature δT_3 ($\Delta T_c = 0$, $\delta T_2 = 0$) [see relation (2.12)]

It is assumed that because of the application of these disturbances the cross sections of the element do not warp. They only translate in the x_3 direction and rotate about the x_2 axis. Moreover, it is assumed that the cross sections of the element remain normal to its deformed axis. That is, their deformed configuration is specified by the transverse component of translation $u_3(x_1)$. The component of rotation $\theta_2(x_1)$ is obtained from the component of translation $u_3(x_1)$ on the basis of relation (2.47a).

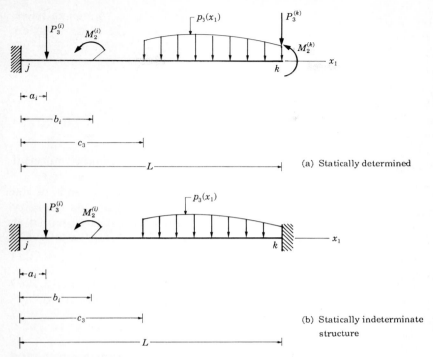

(a) Statically determined

(b) Statically indeterminate
structure

Figure 2.13 One-element structure subjected to external actions producing only a transverse component of translation $u_3(x_1)$.

Notice that if in the second state the temperature of the surface of the element is maintained constant but nonuniform, the temperature inside the element will be constant but nonuniform. Thus, heat will flow into or out of the element, and consequently the element will not be in a state of thermal equilibrium. It is assumed that the distribution of the temperature inside the element is not a function of x_2 and is a linear function of x_3 which vanishes at $x_3 = 0$.

We are interested in establishing the transverse component of translation $u_3(x_1)$ and the internal actions $Q_3(x_1)$ and $M_2(x_1)$ of the element under consideration. This is a two-point boundary-value problem. Its domain is the length of the element $0 \le x_1 \le L$. Its boundary consists of the two points $x_1 = 0$ and $x_1 = L$. When this boundary-value problem is formulated in its so-called strong or classical form, the function $u_3(x_1)$ has the following attributes:

1. When substituted into the appropriate internal action-displacement relation for the material from which the element is

made, it gives a component of internal moment $M_2(x_1)$ which satisfies the action equation of equilibrium $(2.51b)$. That is, using relation (2.52) we have

$$\frac{d^2 M_2}{dx_1^2} + \frac{d}{dx_1}[m_2 \Delta(x_1 - d_2)] + p_3(x_1)\Delta(x_1 - c_3)$$

$$+ \sum_{i-1}^{n_3} P_3^{(i)} \delta(x_1 - a_{3i}) + \sum_{i-1}^{m_2} \mathcal{M}_2^{(i)} \delta^1(x_1 - b_{2i}) = 0 \quad (2.71)$$

For an element made from an isotropic linearly elastic material $M_2(x_1)$ is given by relation $(2.49a)$.

2. It satisfies the specified boundary conditions at the ends of the element. In the second state of mechanical equilibrium one quantity from each of the following two sets of quantities must be specified at each end of the element:

$$u_3 \quad \text{or} \quad P_3 \qquad\qquad (2.72a)$$

$$\theta_2 \quad \text{or} \quad \mathcal{M}_2 \qquad\qquad (2.72b)$$

The specification of u_3 or θ_2 is an *essential* boundary condition, while the specification of P_3 or $\mathcal{M}_2$ is a *natural* boundary condition. For a unique solution the rigid-body motion of the element must be specified. Consequently, either the component of translation u_3 and the component of rotation θ_2 must be specified at least at one end of the element, or the component of translation u_3 must be specified at least at both ends of the element.

In modern structural analysis we are interested in establishing the transverse component of translation $u_3(x_1)$ of an element of a framed structure as a function of its axial coordinate involving its nodal displacements u_3^j, u_3^k, θ_2^j, and θ_2^k. In this case we consider all boundary conditions of the element as essential. That is,

$$u_3(0) = u_3^j \qquad\qquad (2.73a)$$

$$u_3(L) = u_3^k \qquad\qquad (2.73b)$$

$$\theta_2(0) = \theta_2^j \qquad\qquad (2.73c)$$

$$\theta_2(L) = \theta_2^k \qquad\qquad (2.73d)$$

Moreover, in structural analysis we are interested in establishing the transverse component of translation $u_3(x_1)$ and the internal actions $Q_3(x_1)$ and $M_2(x_1)$ of one-element structures subjected to given loads as, for example, those shown in Fig. 2.13. The boundary conditions for the one-element structure of Fig. 2.13a are

$$u_3(0) = 0 \qquad (2.74a)$$

$$\theta_2(0) = -\frac{du_3}{dx_1}\bigg|_{x_1=0} = 0 \qquad (2.74b)$$

$$Q_3(L) = -\left\{\frac{d}{dx_1}\left[EI_2\left(-\frac{d^2u_3}{dx_1^2}+H_3\right)\right]\right\}_{x_1=L} = P_3^k \qquad (2.74c)$$

$$M_2(L) = -\left[EI_2\left(\frac{d^2u_3}{dx_1^2}+H_3\right)\right]_{x_1=L} = \mathcal{M}_2^k \qquad (2.74d)$$

where H_3 is defined by relation (2.45b), and P_3^k and $\mathcal{M}_2^k$ are the specified components of external force and moment at the end k of the element. The boundary conditions (2.74a) and (2.74b) are *essential*, while the boundary conditions (2.74c) and (2.74d) are *natural*. The boundary conditions for the one-element structure of Fig. 2.13b are

$$\begin{aligned} u_3(0) = 0 \qquad \theta_2(0) = 0 \\ u_3(L) = 0 \qquad \theta_2(L) = 0 \end{aligned} \qquad (2.75)$$

2.5.3 The strong or classical form of the boundary-value problem for establishing the angle of twist $\theta_1(x_1)$ of an element of a framed structure

Consider an element of a framed structure initially in a reference stress-free, strain-free state of equilibrium at a uniform temperature T_0. Because of the application of specified torsional moments $m_1(x_1)$ at $d_1 \le x_1 \le L$ and $\mathcal{M}_1^{(i)}$ at $x_1 = b_{1i}$ ($i = 1, 2, \ldots, m_1$) (see Fig. 2.14), the element reaches a second state of equilibrium. In this state either the torsional moment M_1 or the angle of rotation (twist) θ_1 must be specified at each end of the element. For a unique solution the rigid-body motion of the element must be specified. Consequently the angle of twist θ_1 must be specified at least at one end of the element.

We are interested in establishing the angle of twist $\theta_1(x_1)$ and the internal torsional moment $M_1(x_1)$ of the element. When this boundary-value problem is formulated in its strong or classical form, the function $\theta_1(x_1)$ has the following attributes:

1. When substituted into the appropriate internal action-displacement relation for the material from which the element is made, it gives an internal torsional moment which satisfies the action equation of equilibrium (2.5). For an element made of an isotropic

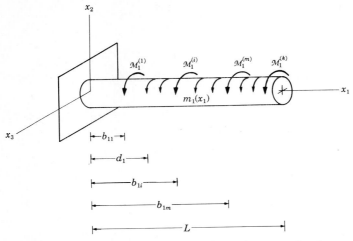

Figure 2.14 One-element structure subjected to torsional moments.

linearly elastic material, $M_1(x_1)$ is related to the angle of twist on the basis of relation (2.64).

2. It satisfies the specified boundary conditions at the ends of the element. In modern structural analysis we are interested in establishing the angle of twist $\theta_1(x_1)$ of an element of a framed structure as a function of its axial coordinate involving its nodal angles of twist θ_1^j and θ_1^k. In this case we consider all boundary conditions of the element as essential. That is,

$$\theta_1(0) = \theta_1^j$$
$$\theta_1(L) = \theta_1^k \tag{2.76}$$

Moreover, in structural analysis we are interested in establishing the angle of twist $\theta_1(x_1)$ and the internal torsional moment $M_1(x_1)$ of one-element structures, subjected to given torsional moments as, for example, the one shown in Fig. 2.14. The boundary conditions for this structure are

$$\theta_1(0) = 0 \tag{2.77a}$$

$$M_1(L) = \mathcal{M}_1^k \tag{2.77b}$$

where $\mathcal{M}_1^k$ is the specified torsional moment at the end k of the structure. The boundary condition (2.77a) is essential, while the boundary condition (2.77b) is natural.

2.6   Problems

1 and 2.   Compute the axial component of translation $u_1(x_1)$ and of the internal force $N(x_1)$ of the bar subjected to the axial forces shown in Fig. 2.P1. The bar has a constant cross section, and it is made from an isotropic linearly elastic material. Repeat with the bar of Fig. 2.P2.

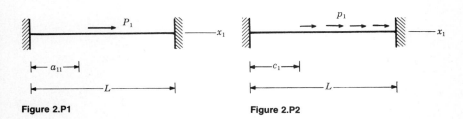

Figure 2.P1   **Figure 2.P2**

3 to 10.   Compute the transverse component of translation $u_3(x_1)$ and the internal actions $Q_3(x_1)$ and $M_2(x_1)$ of the beam subjected to the external actions shown in Fig. 2.P3. The beam has a constant cross section, and it is made from an isotropic linearly elastic material. Repeat with the beams of Figs. 2.P4 to 2P.10.

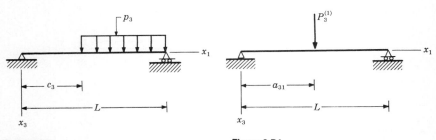

Figure 2.P3   **Figure 2.P4**

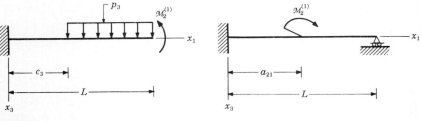

Figure 2.P5   **Figure 2.P6**

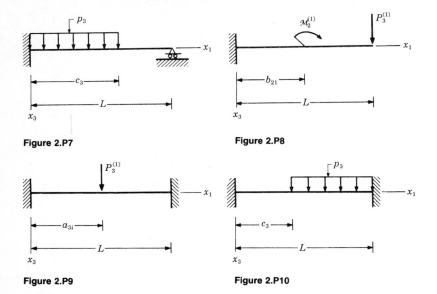

Figure 2.P7

Figure 2.P8

Figure 2.P9

Figure 2.P10

11 and 12.   Compute the axial component of the angle of rotation $\theta_1(x_1)$ and of the internal moment $M_1(x_1)$ of the cylindrical bar of Fig 2.P11. The bar has a constant cross section, and it is made from an isotropic linearly elastic material. Repeat with the bar of Fig 2.P12.

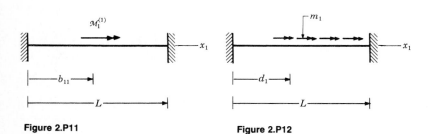

Figure 2.P11

Figure 2.P12

References

1. A. E. Armenàkas, *Classical Structural Analysis: A Modern Approach*, McGraw-Hill Book Co., New York, 1988, p. 634.
2. Ibid., p. 670.

3

Modern Methods for Analyzing Framed Structures

3.1 Introduction

When we analyze a framed structure, we are interested in establishing the internal actions in each of its elements as functions of its axial coordinate as well as one or more components of displacement of some of its points.

The analysis of a framed structure is formulated either in terms of the components of displacement of its nodes or in terms of some of its reactions and/or internal actions of its elements. The methods which employ the first formulation are called *displacement* or *stiffness methods,* while those which use the second formulation are called *force* or *flexibility methods*.

The objective of this chapter is to give a brief description of the modern methods for analyzing framed structures by applying them to the truss of Fig. 3.1. These methods are the *direct stiffness* or *direct displacement method* and the *modern flexibility* or *modern force method*. Our purpose is to demonstrate the salient features of these methods and to describe the important steps which must be followed when applying them, without delving into details. Moreover we indicate some of the advantages and disadvantages of these methods.

In the modern methods of structural analysis the element approach is used. That is, a structure is subdivided into a number of elements, and its response is determined from that of its elements. Thus the analysis of a structure is subdivided into two parts. In the first part the response of its elements is established. In the second part the response of the structure is established from that of its elements.

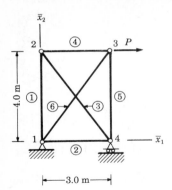

Figure 3.1 Geometry and loading of a truss.

In the modern methods of structural analysis the response of an element is established as the sum of its responses when the element is subjected to the following two loading cases.

Case 1. The given loads acting along the length of the element and the restraining actions (fixed-end actions) which must be applied to its ends in order to restrain them from moving (translating or rotating). In this case the response of the element is described by its fixed-end actions.

Case 2. The unknown components of its nodal displacements. In this case the response of the element is described either by its stiffness equations or by its flexibility equations. When the elements of a structure are not subjected to external loads along their length, as are, for example, the elements of the truss of Fig. 3.1, their response is described only by their response when subjected to this loading case. That is, their response is described only by their stiffness equations or by their flexibility equations.

In the modern methods of structural analysis the response of a structure is established as the sum of its responses when subjected to the following two loading cases.

Case 1. Both the given loads acting along the length of its elements and the unknown restraining actions which must be applied to the nodes of the structure in order to restrain them from moving (translating or rotating). The movement of the supports of the structure may or may not be included in this loading.

Case 2. Both the equivalent actions on its nodes and the given translations and rotations of its supports if they have not been included in loading case 1. The components of equivalent actions which must be applied to a node of a structure are equal to the sum of the corresponding components of the following actions:

1. The given actions acting on this node of the structure.
2. Actions equal and opposite to the restraining actions acting on this node of the structure, when subjected to loading case 1.

When a structure is subjected only to given actions on its nodes, as is the truss of Fig. 3.1, its response is described only by its response when subjected to loading case 2. The equivalent actions acting on each of its nodes are equal to the given external actions acting on each node.

3.2   Element Response

The relations which specify the response of an element of a structure are established using the following facts:

1. The components of stress and strain at a particle of an element are related by the stress-strain relations for the material from which this element is made. These relations are used to obtain its internal action-displacement relations for the element [see relations (2.36), (2.49), (2.50), and (2.64)].
2. The element is in equilibrium. This fact is satisfied by requiring that either the components of internal action satisfy the action equations of equilibrium (2.2) to (2.7) or the components of displacement satisfy the displacement equations of equilibrium. For an element made of an isotropic linearly elastic material these equations have been derived in Sec. 2.4 [see Eqs. (2.40), (2.53), and (2.67)].

3.2.1   The stiffness equations for an element of the truss

In the direct stiffness method the response of an element of a structure subjected to equivalent actions on its nodes is specified by either its local stiffness or global stiffness equations. As discussed in Sec. 4.3 these equations express the local (global) components of nodal actions for an element as a linear combination of the local (global) components of its nodal displacements.

In what follows we establish the local stiffness equations for an element of the truss of Fig. 3.1 by adhering to the following steps.

Step 1. We establish the axial component of translation of the element as a function of its axial coordinate involving its nodal translation. In order to accomplish this we solve the displacement equation of equilibrium (2.40) for the element. That is,

$$\frac{d}{dx_1}\left(EA\frac{du_1}{dx_1}\right) = 0 \tag{3.1}$$

Integrating Eq. (3.1) twice we obtain

$$EA\frac{du_1}{dx_1} = C_1 \tag{3.2}$$

and

$$EAu_1(x_1) = C_1x_1 + C_2 \tag{3.3}$$

The constants C_1 and C_2 are evaluated from the conditions at the ends (boundary conditions) of the element. That is,

$$u_1(0) = u_1^j \qquad u_1(L) = u_1^k \tag{3.4}$$

Substituting relation (3.3) into (3.4) we get

$$C_2 = EAu_1^j \tag{3.5}$$

$$C_1 = EA\frac{u_1^k - u_1^j}{L} \tag{3.6}$$

Substituting relations (3.5) and (3.6) into (3.3) we have

$$EAu_1(x_1) = EA\left[\left(1 - \frac{x_1}{L}\right)u_1^j + \left(\frac{x_1}{L}\right)u_1^k\right] \tag{3.7}$$

This relation may be rewritten as

$$u_1(x_1) = [\phi]\{D\} \tag{3.8}$$

where

$$\{D\} = \begin{Bmatrix} u_1^j \\ u_1^k \end{Bmatrix} = \text{local matrix of nodal displacements of element}$$
$$[\phi] = [\phi_{1j} \quad \phi_{1k}] = \text{matrix of shape functions for element} \tag{3.9}$$

$$\left.\begin{aligned} \phi_{1j} &= 1 - \frac{x_1}{L} \\ \phi_{1k} &= \frac{x_1}{L} \end{aligned}\right\} \text{shape functions for element} \tag{3.10}$$

Step 2. We establish the local stiffness equations for the element. To accomplish this we substitute relation (3.8) into the internal force-displacement relation (2.36) for the element. That is,

$$N(x_1) = EA\frac{du_1}{dx_1} = EA\frac{d[\phi]}{dx_1}\{D\} = EA\left[-\frac{1}{L}\quad\frac{1}{L}\right]\{D\} \qquad (3.11)$$

Referring to relations (1.2a) and (1.2g) and using relation (3.11) we obtain

$$\{A\} = \begin{Bmatrix} F_1^j \\ F_1^k \end{Bmatrix} = \begin{Bmatrix} -N(0) \\ N(L) \end{Bmatrix} = \frac{EA}{L}\begin{bmatrix} 1 & -1 \\ -1 & 1 \end{bmatrix}\begin{Bmatrix} u_1^j \\ u_1^k \end{Bmatrix} = \begin{bmatrix} K_{11} & K_{12} \\ K_{21} & K_{22} \end{bmatrix}\begin{Bmatrix} u_1^j \\ u_1^k \end{Bmatrix}$$

$$(3.12)$$

or
$$\{A\} = [K]\{D\} \qquad (3.13)$$

where
$$[K] = \frac{EA}{L}\begin{bmatrix} 1 & -1 \\ -1 & 1 \end{bmatrix} \qquad (3.14)$$

is the local stiffness matrix for the element. Its terms are the local stiffness coefficients for the element. $\{A\}$ is the local matrix of nodal actions.

Referring to relation (3.14) it is apparent that the stiffness matrix for the element is symmetric. That is,

$$[K] = [K]^T \qquad (3.15)$$

Equations (3.13) describe the response of an element of the truss. They are called the *stiffness equations* for the element and give the nodal forces of the element as a linear combination of its nodal translations. Notice that the second row of the stiffness matrix in Eq. (3.14) is a multiple of the first. Consequently the stiffness matrix for the element is singular. This implies that the components of nodal translations of the element cannot be expressed as a linear combination of its nodal forces. Actually when the components of nodal actions of an element are known, its components of nodal displacement are specified only to within a rigid-body movement of the element. That is, the magnitude of the nodal actions of an element is the same for all sets of its nodal displacements which differ by a rigid-body motion.

In order to establish the physical significance of the stiffness coefficients K_{11} and K_{21} we consider the element subjected to the nodal forces required to produce the following nodal translations

$$u_1^j = 1 \qquad u_1^k = 0$$

In this case relation (3.12) gives

$$F_1^j = K_{11} = \frac{EA}{L} \qquad F_1^k = K_{21} = -\frac{EA}{L} \qquad (3.16)$$

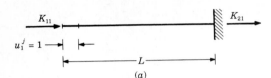

(a)

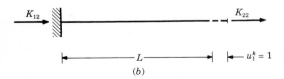

(b)

Figure 3.2 Physical significance of the stiffness coefficients for an element of a truss.

That is, as shown in Fig. 3.2a the stiffness coefficients K_{11} and K_{21} are the axial forces which must be placed at the ends j and k of the element, respectively, in order to shorten it by a unit. The physical significance of the stiffness coefficients K_{12} and K_{22} is shown in Fig. 3.2b.

3.2.2 The flexibility equations for an element of the truss

In the modern flexibility method the response of an element of a structure subjected to equivalent actions on its nodes is specified by *its flexibility equations*. These equations express the quantities which characterize the deformation (not the rigid-body motion) of the element as a linear combination of its basic (independent) components of nodal actions. In what follows we establish the flexibility equations for an element of a truss.

The nodal actions of an element are not independent. They are related by the equations of equilibrium for the element. That is, referring to Fig. 3.3 we have

$$\sum F_1 = 0 \qquad F_1^j = -F_1^k \tag{3.17}$$

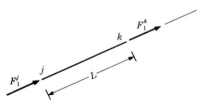

Figure 3.3 Free-body diagram of an element of a truss.

It is apparent that if either F_1^j or F_1^k is known, the internal force $N(x_1)$ in the element can be established. We choose F_1^k as the basic (independent) nodal action of the element, and using relation (3.17) we write the matrix of nodal actions of the element as

$$\{A\} = \begin{Bmatrix} F_1^j \\ F_1^k \end{Bmatrix} = \begin{Bmatrix} -1 \\ 1 \end{Bmatrix} F_1^k \qquad (3.18)$$

Consider a cross section of the element under consideration located at a distance x_1 from its end j when the truss is at its reference configuration at the uniform temperature T_0. When the truss is subjected to loads, its elements translate and rotate as rigid bodies and deform. The position of the cross section under consideration is specified by its components of translation $u_1(x_1)$ and $u_2(x_1)$ and by its component of rotation $\theta_3(x_1)$.

Referring to Fig. 3.4 these components of displacement can be expressed as

$$u_1(x_1) = u_1^j + [x_1 + \delta_1(x_1)] \cos \theta_3^j - x_1 \qquad (3.19a)$$

$$u_2(x_1) = u_2^j + [x_1 + \delta_1(x_1)] \sin \theta_3^j \qquad (3.19b)$$

$$\theta_3 = \theta_3^j \qquad (3.19c)$$

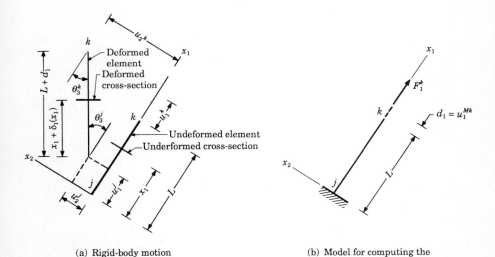

(a) Rigid-body motion and deformation

(b) Model for computing the relation between the basic nodal action and the basic deformation parameter

Figure 3.4 Element subjected only to its nodal displacements.

For very small rotations θ_3^j, relations (3.19) reduce to the following

$$u_1(x_1) = u_1^j + \delta_1(x_1) \tag{3.20a}$$

$$u_2(x_1) = u_2^j + x_1\theta_3^j \tag{3.20b}$$

$$\theta_3 = \theta_3^j \tag{3.20c}$$

The components of nodal displacement u_1^j, u_2^j, and θ_3^j specify the rigid-body motion of the element; $\delta_1(x_1)$ represents the change in length of the portion of the element from its end j to the cross section under consideration. That is, $\delta_1(x_1)$ is the axial component of translation of the cross section under consideration when the end j of the element is fixed. Hence, $\delta(x_1)$ is equal to the component of translation $u_1^M(x_1)$ of the model of Fig. 3.4b.

It is apparent that the value of $\delta_1(x_1)$ at $x_1 = L$ specifies the deformation of the element. This quantity is the *basic deformation parameter* of the element corresponding to our choice of basic nodal action. We denote it by d_1. Referring to relation (3.20a) we have

$$d_1 = \delta(L) = u_1^k - u_1^j = [-1 \quad 1]\begin{Bmatrix} u_1^j \\ u_1^k \end{Bmatrix} \tag{3.21}$$

The internal force $N(x_1)$ of the element must satisfy its action equation of equilibrium (2.2). That is, for an element subjected only to its nodal displacements, we have

$$\frac{dN}{dx_1} = 0 \tag{3.22}$$

Integrating this relation we obtain

$$N(x_1) = C \tag{3.23}$$

At $x_1 = L$ we get

$$F_1^k = N(L) = C \tag{3.24}$$

Substituting relation (3.24) into (3.23) as expected we get

$$N(x_1) = F_1^k \tag{3.25}$$

Substituting relation (3.25) into the internal action-displacement relation (2.36) for the element we have

$$EA\frac{du_1}{dx_1} = N(x_1) = F_1^k \tag{3.26}$$

Integrating relations (3.26) from $x_1 = 0$ to $x_1 = L$ we obtain

$$d_1 = u_1^k - u_1^j = \frac{LF_1^k}{EA} = f_{11}F_1^k \tag{3.27}$$

where

$$f_{11} = \frac{L}{EA} = \text{flexibility coefficient for element} \tag{3.28}$$

For elements of beams and frames we have more than one basic nodal action and more than one basic deformation parameter. That is, in general, relations (3.18), (3.21), and (3.27) can be written as

$$\{A\} = [T]\{a\} \tag{3.29}$$

$$\{d\} = [T]^T\{D\} \tag{3.30}$$

$$\{d\} = [f]\{a\} \tag{3.31}$$

where $\{a\}$ = matrix of basic nodal actions of element
   $\{d\}$ = matrix of basic deformation parameters of element
   $[T]$ = matrix which transforms matrix of basic nodal actions of an element to its matrix of nodal actions
   $[f]$ = flexibility matrix for element, whose terms are called the flexibility coefficients for the element

The flexibility matrix for an element is a square, nonsingular matrix. Thus relations (3.31) can be written as

$$\{a\} = [k]\{d\} \tag{3.32}$$

$$[k] = [f]^{-1} \tag{3.33}$$

where $[k]$ is called the *basic stiffness matrix for the element*. Equations (3.31) are the *flexibility equations* for an element, while Eqs. (3.32) are the *basic stiffness equations* for the element.

3.3 Transformation Matrix for an Element of the Truss

In this section we establish the transformation matrix which transforms the local matrix of nodal actions and of nodal displacements of an element of a truss to global and vice versa. Moreover, we use this transformation matrix to obtain the global stiffness equations for the elements of the truss. In the direct stiffness method the global stiffness equations for the elements of a structure are used to assemble the stiffness equations for the structure.

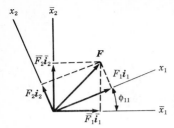

Figure 3.5 Transformation of the components of a planar vector.

Referring to Fig. 3.5 the components F_1 and F_2 of a planar vector **F** with respect to a set of orthogonal axes x_1 and x_2 are related to its components $\bar{F}_1$ and $\bar{F}_2$ with respect to another set of orthogonal axes $\bar{x}_1$ and $\bar{x}_2$ by the following relations

$$\begin{Bmatrix} F_1 \\ F_2 \end{Bmatrix} = \begin{bmatrix} \cos\phi_{11} & \sin\phi_{11} \\ -\sin\phi_{11} & \cos\phi_{11} \end{bmatrix} \begin{Bmatrix} \bar{F}_1 \\ \bar{F}_2 \end{Bmatrix} \tag{3.34}$$

For an element of a planar truss the local components of nodal actions F_2^j and F_2^k vanish; thus relation (3.34) can be written as

$$F_1^j = \begin{bmatrix} \cos\phi_{11} & \sin\phi_{11} \end{bmatrix} \begin{Bmatrix} \bar{F}_1^j \\ \bar{F}_2^j \end{Bmatrix} \tag{3.35}$$

and

$$F_1^k = \begin{bmatrix} \cos\phi_{11} & \sin\phi_{11} \end{bmatrix} \begin{Bmatrix} \bar{F}_1^k \\ \bar{F}_2^k \end{Bmatrix} \tag{3.36}$$

Combining relations (3.35) and (3.36) we have

$$\begin{Bmatrix} F_1^j \\ F_1^k \end{Bmatrix} = \begin{bmatrix} \cos\phi_{11} & \sin\phi_{11} & 0 & 0 \\ 0 & 0 & \cos\phi_{11} & \sin\phi_{11} \end{bmatrix} \begin{Bmatrix} \bar{F}_1^j \\ \bar{F}_2^j \\ \bar{F}_1^k \\ \bar{F}_2^k \end{Bmatrix} \tag{3.37}$$

or in general

$$\{A\} = [\Lambda]\{\bar{A}\} \tag{3.38}$$

where $\{A\}$ = local matrix of nodal actions of element
$\{\bar{A}\}$ = global matrix of nodal actions of element
$[\Lambda]$ = transformation matrix for element

Referring to relation (3.37), for an element of a planar truss we have

$$[\Lambda] = \begin{bmatrix} \cos\phi_{11} & \sin\phi_{11} & 0 & 0 \\ 0 & 0 & \cos\phi_{11} & \sin\phi_{11} \end{bmatrix} \tag{3.39}$$

Similarly it can be shown that

$$\begin{Bmatrix} \bar{F}_1^j \\ \bar{F}_2^j \\ \bar{F}_1^k \\ \bar{F}_2^k \end{Bmatrix} = \begin{bmatrix} \cos \phi_{11} & 0 \\ \sin \phi_{11} & 0 \\ 0 & \cos \phi_{11} \\ 0 & \sin \phi_{11} \end{bmatrix} \begin{Bmatrix} F_1^j \\ F_1^k \end{Bmatrix} \tag{3.40}$$

or
$$\{\bar{A}\} = [\Lambda]^T \{A\} \tag{3.41}$$

Referring to relations (3.38) and (3.41) it is apparent that

$$[\Lambda]^T [\Lambda] = [I] \tag{3.42}$$

where $[I]$ is the unit matrix.

Inasmuch as the nodal translations of an element of a truss are vector quantities corresponding to the nodal forces, their transformation relations are

$$\{D\} = [\Lambda]\{\bar{D}\} \tag{3.43}$$

$$\{\bar{D}\} = [\Lambda]^T \{D\} \tag{3.44}$$

where $\{D\}$ is defined by (3.9). Substituting relations (3.41) and (3.43) into (3.13) we obtain

$$[\Lambda]\{\bar{A}\} = [K][\Lambda]\{\bar{D}\} \tag{3.45}$$

Multiplying both sides of relation (3.45) by $[\Lambda]^T$ and taking into account (3.42) we get

$$\{\bar{A}\} = [\bar{K}]\{\bar{D}\} \tag{3.46}$$

where $[\bar{K}]$ is the *global stiffness* matrix for the element and is defined as

$$[\bar{K}] = [\Lambda]^T [K][\Lambda] \tag{3.47}$$

Equations (3.46) are the global stiffness equations for an element of a truss.

3.4 Structure Response

The response of the truss can be established by taking into account the following facts:

1. The response of the elements of the truss is specified either by their stiffness equations or by their flexibility equations.
2. The nodal components of displacement of the elements of the truss

must be compatible to the components of displacement of its nodes.

3. The nodal forces of the elements of the truss must satisfy the equations of equilibrium for its nodes.

4. The components of displacement of the nodes of the truss must satisfy the given conditions at its supports (boundary conditions).

3.4.1 The direct stiffness or displacement method

In order to analyze the truss of Fig. 3.1 by the direct stiffness or displacement method we adhere to the following steps.

Step 1. We form the matrix of the given forces acting on the nodes of the truss $\{\hat{P}^G\}$. Referring to Fig. 3.1 we have

$$\{\hat{P}^G\} = \begin{Bmatrix} 0 \\ 0 \\ 0 \\ 0 \\ P \\ 0 \\ 0 \\ 0 \end{Bmatrix} \begin{matrix} \left.\vphantom{\begin{matrix}0\\0\end{matrix}}\right\}\text{components of force on node 1} \\ \left.\vphantom{\begin{matrix}0\\0\end{matrix}}\right\}\text{components of force on node 2} \\ \left.\vphantom{\begin{matrix}P\\0\end{matrix}}\right\}\text{components of force on node 3} \\ \left.\vphantom{\begin{matrix}0\\0\end{matrix}}\right\}\text{components of force on node 4} \end{matrix} \qquad (3.48)$$

Step 2. We establish the local stiffness matrix for each element of the truss. Referring to relation (3.14), we have

$$[K^e] = \frac{E_e A_e}{L_e} \begin{bmatrix} 1 & -1 \\ -1 & 1 \end{bmatrix} \qquad e = 1, 2, \ldots, 6 \qquad (3.49)$$

Step 3. We compute the global stiffness matrix for each element of the truss and write its stiffness equations. Referring to relation (3.39), the transformation matrices for the elements of the truss of Fig. 3.1 are

$$[\Lambda^1] = \begin{bmatrix} 0 & 1 & 0 & 0 \\ 0 & 0 & 0 & 1 \end{bmatrix} \qquad\qquad [\Lambda^4] = [\Lambda^2]$$

$$[\Lambda^2] = \begin{bmatrix} 1 & 0 & 0 & 0 \\ 0 & 0 & 1 & 0 \end{bmatrix} \qquad\qquad [\Lambda^5] = \begin{bmatrix} 0 & -1 & 0 & 0 \\ 0 & 0 & 0 & -1 \end{bmatrix}$$

$$[\Lambda^3] = \begin{bmatrix} 0.6 & -0.8 & 0 & 0 \\ 0 & 0 & 0.6 & -0.8 \end{bmatrix} \qquad [\Lambda^6] = \begin{bmatrix} 0.6 & 0.8 & 0 & 0 \\ 0 & 0 & 0.6 & 0.8 \end{bmatrix}$$

$$(3.50)$$

Substituting relations (3.50) and (3.49) into (3.47), we transform the local stiffness matrix for each element to global and use it to write the global stiffness equations for the element. That is,

$$\{\bar{A}^1\} = \begin{Bmatrix} \bar{F}_1^{1j} \\ \bar{F}_2^{1j} \\ \bar{F}_1^{1k} \\ \bar{F}_2^{1k} \end{Bmatrix} = \frac{EA}{4} \begin{bmatrix} 0 & 0 & 0 & 0 \\ 0 & 1 & 0 & -1 \\ 0 & 0 & 0 & 0 \\ 0 & -1 & 0 & 1 \end{bmatrix} \begin{Bmatrix} \bar{u}_1^{1j} \\ \bar{u}_2^{1j} \\ \bar{u}_1^{1k} \\ \bar{u}_2^{1k} \end{Bmatrix} = [\bar{K}^1][\bar{D}^1]$$

$$\{\bar{A}^2\} = \begin{Bmatrix} \bar{F}_1^{2j} \\ \bar{F}_2^{2j} \\ \bar{F}_1^{2k} \\ \bar{F}_2^{2k} \end{Bmatrix} = \frac{EA}{3} \begin{bmatrix} 1 & 0 & -1 & 0 \\ 0 & 0 & 0 & 0 \\ -1 & 0 & 1 & 0 \\ 0 & 0 & 0 & 0 \end{bmatrix} \begin{Bmatrix} \bar{u}_1^{2j} \\ \bar{u}_2^{2j} \\ \bar{u}_1^{2k} \\ \bar{u}_2^{2k} \end{Bmatrix} = [\bar{K}^2][\bar{D}^2]$$

$$\{\bar{A}^3\} = \begin{Bmatrix} \bar{F}_1^{3j} \\ \bar{F}_2^{3j} \\ \bar{F}_1^{3k} \\ \bar{F}_2^{3k} \end{Bmatrix} = \frac{EA}{5} \begin{bmatrix} 0.36 & -0.48 & -0.36 & 0.48 \\ -0.48 & 0.64 & 0.48 & -0.64 \\ -0.36 & 0.48 & 0.36 & -0.48 \\ 0.48 & -0.64 & -0.48 & 0.64 \end{bmatrix} \begin{Bmatrix} \bar{u}_1^{3j} \\ \bar{u}_2^{3j} \\ \bar{u}_1^{3k} \\ \bar{u}_2^{3k} \end{Bmatrix}$$

$$= [\bar{K}^3]\{\bar{D}^3\} \tag{3.51}$$

$$\{\bar{A}^4\} = \begin{Bmatrix} \bar{F}_1^{4j} \\ \bar{F}_2^{4j} \\ \bar{F}_1^{4k} \\ \bar{F}_2^{4k} \end{Bmatrix} = \frac{EA}{3} \begin{bmatrix} 1 & 0 & -1 & 0 \\ 0 & 0 & 0 & 0 \\ -1 & 0 & 1 & 0 \\ 0 & 0 & 0 & 0 \end{bmatrix} \begin{Bmatrix} \bar{u}_1^{4j} \\ \bar{u}_2^{4j} \\ \bar{u}_1^{4k} \\ \bar{u}_2^{4k} \end{Bmatrix} = [\bar{K}^4]\{\bar{D}^4\}$$

$$\{\bar{A}^5\} = \begin{Bmatrix} \bar{F}_1^{5j} \\ \bar{F}_2^{5j} \\ \bar{F}_1^{5k} \\ \bar{F}_2^{5k} \end{Bmatrix} = \frac{EA}{4} \begin{bmatrix} 0 & 0 & 0 & 0 \\ 0 & 1 & 0 & -1 \\ 0 & 0 & 0 & 0 \\ 0 & -1 & 0 & 1 \end{bmatrix} \begin{Bmatrix} \bar{u}_1^{5j} \\ \bar{u}_2^{5j} \\ \bar{u}_1^{5k} \\ \bar{u}_2^{5k} \end{Bmatrix} = [\bar{K}^5]\{\bar{D}^5\}$$

$$\{\bar{A}^6\} = \begin{Bmatrix} \bar{F}_1^{6j} \\ \bar{F}_2^{6j} \\ \bar{F}_1^{6k} \\ \bar{F}_2^{6k} \end{Bmatrix} = \frac{EA}{5} \begin{bmatrix} 0.36 & 0.48 & -0.36 & -0.48 \\ 0.48 & 0.64 & -0.48 & -0.64 \\ -0.36 & -0.48 & 0.36 & 0.48 \\ -0.48 & -0.64 & 0.48 & 0.64 \end{bmatrix} \begin{Bmatrix} \bar{u}_1^{6j} \\ \bar{u}_2^{6j} \\ \bar{u}_1^{6k} \\ \bar{u}_2^{6k} \end{Bmatrix}$$

$$= [\bar{K}^6]\{\bar{D}^6\}$$

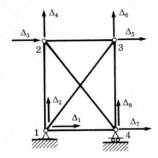

Figure 3.6 Numbering of the components of translations of the nodes of the truss.

Step 4. Referring to Fig. 3.6, we write the compatibility relations for the truss:

$$
\begin{array}{lll}
\bar{u}_1^{1j} = \Delta_1 & \bar{u}_1^{3j} = \Delta_3 & \bar{u}_1^{5j} = \Delta_5 \\[4pt]
\bar{u}_2^{1j} = \Delta_2 & \bar{u}_2^{3j} = \Delta_4 & \bar{u}_2^{5j} = \Delta_6 \\[4pt]
\bar{u}_1^{1k} = \Delta_3 & \bar{u}_1^{3k} = \Delta_7 & \bar{u}_1^{5k} = \Delta_7 \\[4pt]
\bar{u}_2^{1k} = \Delta_4 & \bar{u}_2^{3k} = \Delta_8 & \bar{u}_2^{5k} = \Delta_8 \\[4pt]
\bar{u}_1^{2j} = \Delta_1 & \bar{u}_1^{4j} = \Delta_3 & \bar{u}_1^{6j} = \Delta_1 \\[4pt]
\bar{u}_2^{2j} = \Delta_2 & \bar{u}_2^{4j} = \Delta_4 & \bar{u}_2^{6j} = \Delta_2 \\[4pt]
\bar{u}_1^{2k} = \Delta_7 & \bar{u}_1^{4k} = \Delta_5 & \bar{u}_1^{6k} = \Delta_5 \\[4pt]
\bar{u}_2^{2k} = \Delta_8 & \bar{u}_2^{4k} = \Delta_6 & \bar{u}_2^{6k} = \Delta_6
\end{array}
\tag{3.52}
$$

We use the compatibility relations (3.52) to replace the global components of nodal translations in the global stiffness equations for the elements of the truss (3.51) with the appropriate components of translations of the nodes of the truss. That is,

$$
\{\bar{A}^1\} = \left\{ \begin{array}{c} \bar{F}_1^{1j} \\ \bar{F}_2^{1j} \\ \hdashline \bar{F}_1^{1k} \\ \bar{F}_2^{1k} \end{array} \right\} = EA \left[\begin{array}{cc:cc} 0 & 0 & 0 & 0 \\ 0 & 0.25 & 0 & -0.25 \\ \hdashline 0 & 0 & 0 & 0 \\ 0 & -0.25 & 0 & 0.25 \end{array} \right] \left\{ \begin{array}{c} \Delta_1 \\ \Delta_2 \\ \Delta_3 \\ \Delta_4 \end{array} \right\}
$$

$$
\{\bar{A}^2\} = \left\{ \begin{array}{c} \bar{F}_1^{2j} \\ \bar{F}_2^{2j} \\ \hdashline \bar{F}_1^{2k} \\ \bar{F}_2^{2k} \end{array} \right\} = EA \left[\begin{array}{cc:cc} 0.333 & 0 & -0.333 & 0 \\ 0 & 0 & 0 & 0 \\ \hdashline -0.333 & 0 & 0.333 & 0 \\ 0 & 0 & 0 & 0 \end{array} \right] \left\{ \begin{array}{c} \Delta_1 \\ \Delta_2 \\ \Delta_7 \\ \Delta_8 \end{array} \right\}
$$

$$\{\bar{A}^3\} = \begin{Bmatrix} \bar{F}_1^{3j} \\ \bar{F}_2^{3j} \\ \bar{F}_1^{3k} \\ \bar{F}_2^{3k} \end{Bmatrix} = EA \begin{bmatrix} 0.072 & -0.096 & -0.072 & 0.096 \\ -0.096 & 0.128 & 0.096 & -0.128 \\ \hline -0.072 & 0.096 & 0.072 & -0.096 \\ 0.096 & -0.128 & -0.096 & 0.128 \end{bmatrix} \begin{Bmatrix} \Delta_3 \\ \Delta_4 \\ \Delta_7 \\ \Delta_8 \end{Bmatrix}$$

$$\{\bar{A}^4\} = \begin{Bmatrix} \bar{F}_1^{4j} \\ \bar{F}_2^{4j} \\ \bar{F}_1^{4k} \\ \bar{F}_2^{4k} \end{Bmatrix} = EA \begin{bmatrix} 0.333 & 0 & -0.333 & 0 \\ 0 & 0 & 0 & 0 \\ \hline -0.333 & 0 & 0.333 & 0 \\ 0 & 0 & 0 & 0 \end{bmatrix} \begin{Bmatrix} \Delta_3 \\ \Delta_4 \\ \Delta_5 \\ \Delta_6 \end{Bmatrix}$$

$$\{\bar{A}^5\} = \begin{Bmatrix} \bar{F}_1^{5j} \\ \bar{F}_2^{5j} \\ \bar{F}_1^{5k} \\ \bar{F}_2^{5k} \end{Bmatrix} = EA \begin{bmatrix} 0 & 0 & 0 & 0 \\ 0 & 0.25 & 0 & -0.25 \\ \hline 0 & 0 & 0 & 0 \\ 0 & -0.25 & 0 & 0.25 \end{bmatrix} \begin{Bmatrix} \Delta_5 \\ \Delta_6 \\ \Delta_7 \\ \Delta_8 \end{Bmatrix}$$

$$\{\bar{A}^6\} = \begin{Bmatrix} \bar{F}_1^{6j} \\ \bar{F}_2^{6j} \\ \bar{F}_1^{6k} \\ \bar{F}_2^{6k} \end{Bmatrix} = EA \begin{bmatrix} 0.072 & 0.096 & -0.072 & -0.096 \\ 0.096 & 0.128 & -0.096 & -0.128 \\ \hline -0.072 & -0.096 & 0.072 & 0.096 \\ -0.096 & -0.128 & 0.096 & 0.128 \end{bmatrix} \begin{Bmatrix} \Delta_1 \\ \Delta_2 \\ \Delta_5 \\ \Delta_6 \end{Bmatrix}$$

$$(3.53)$$

Step 5.   We assemble the stiffness equations for the truss from the global stiffness equations for its elements. In order to illustrate how this is done we consider the equilibrium of the nodes of the truss. That is,

Node 1 $\begin{cases} \sum \bar{F}_1 = 0 \\ \sum \bar{F}_2 = 0 \end{cases}$ $\begin{Bmatrix} P_1^G \\ P_2^G \end{Bmatrix} + \begin{Bmatrix} R_1^{(1)} \\ R_2^{(1)} \end{Bmatrix} = \begin{Bmatrix} \bar{F}_1^{1j} \\ \bar{F}_2^{1j} \end{Bmatrix} + \begin{Bmatrix} \bar{F}_1^{2j} \\ \bar{F}_2^{2j} \end{Bmatrix} + \begin{Bmatrix} \bar{F}_1^{6j} \\ \bar{F}_2^{6j} \end{Bmatrix}$

Node 2 $\begin{cases} \sum \bar{F}_1 = 0 \\ \sum \bar{F}_2 = 0 \end{cases}$ $\begin{Bmatrix} P_3^G \\ P_4^G \end{Bmatrix} = \begin{Bmatrix} \bar{F}_1^{1k} \\ \bar{F}_2^{1k} \end{Bmatrix} + \begin{Bmatrix} \bar{F}_1^{3j} \\ \bar{F}_2^{3j} \end{Bmatrix} + \begin{Bmatrix} \bar{F}_1^{4j} \\ \bar{F}_2^{4j} \end{Bmatrix}$

Node 3 $\begin{cases} \sum \bar{F}_1 = 0 \\ \sum \bar{F}_2 = 0 \end{cases}$ $\begin{Bmatrix} P_5^G \\ P_6^G \end{Bmatrix} = \begin{Bmatrix} \bar{F}_1^{4k} \\ \bar{F}_2^{4k} \end{Bmatrix} + \begin{Bmatrix} \bar{F}_1^{5j} \\ \bar{F}_2^{5j} \end{Bmatrix} + \begin{Bmatrix} \bar{F}_1^{6k} \\ \bar{F}_2^{6k} \end{Bmatrix}$

$$(3.54)$$

Node 4 $\begin{cases} \sum \bar{F}_1 = 0 \\ \sum \bar{F}_2 = 0 \end{cases}$ $\begin{Bmatrix} P_7^G \\ P_8^G \end{Bmatrix} + \begin{Bmatrix} 0 \\ R_2^{(4)} \end{Bmatrix} = \begin{Bmatrix} \bar{F}_1^{2k} \\ \bar{F}_2^{2k} \end{Bmatrix} + \begin{Bmatrix} \bar{F}_1^{3k} \\ \bar{F}_2^{3k} \end{Bmatrix} + \begin{Bmatrix} \bar{F}_1^{5k} \\ \bar{F}_2^{5k} \end{Bmatrix}$

where P_i^G $(i = 1, 2, \ldots, 8)$ are the global components of the given external forces acting on the nodes of the truss. Their numbering corresponds to that of the components of displacement of the nodes of the truss shown in Fig. 3.6.

In relations (3.53) the global components of nodal forces of the elements of the truss are expressed as a linear combination of the global components of translation of its nodes. We use relations (3.53) to eliminate the global components of the nodal forces from relations (3.54), and we combine the resulting relations to obtain

$$
\begin{Bmatrix} 0 \\ 0 \\ 0 \\ 0 \\ P \\ 0 \\ 0 \\ 0 \end{Bmatrix} + \begin{Bmatrix} R_1^{(1)} \\ R_2^{(1)} \\ 0 \\ 0 \\ 0 \\ 0 \\ 0 \\ R_2^{(4)} \end{Bmatrix}
$$

$$
= EA \begin{bmatrix} 0.405 & 0.096 & 0 & 0 & -0.072 & -0.096 & -0.333 & 0 \\ 0.096 & 0.378 & 0 & -0.250 & -0.096 & -0.128 & 0 & 0 \\ 0 & 0 & 0.405 & -0.096 & -0.333 & 0 & -0.072 & 0.096 \\ 0 & -0.250 & -0.096 & 0.378 & 0 & 0 & 0.096 & -0.128 \\ -0.072 & -0.096 & -0.333 & 0 & 0.405 & 0.096 & 0 & 0 \\ -0.096 & -0.128 & 0 & 0 & 0.096 & 0.378 & 0 & -0.250 \\ -0.333 & 0 & -0.072 & 0.096 & 0 & 0 & 0.405 & -0.096 \\ 0 & 0 & 0.096 & -0.128 & 0 & -0.250 & -0.096 & 0.378 \end{bmatrix} \begin{Bmatrix} \Delta_1 \\ \Delta_2 \\ \Delta_3 \\ \Delta_4 \\ \Delta_5 \\ \Delta_6 \\ \Delta_7 \\ \Delta_8 \end{Bmatrix}
$$

(3.55)

or

$$\{\hat{P}^G\} + \{\hat{R}\} = [\hat{S}]\{\hat{\Delta}\} \tag{3.56}$$

where $\{\hat{P}^G\}$ is a matrix of global components of the given actions acting on the nodes of the truss including those directly absorbed by its supports, the nonvanishing terms of matrix $\{\hat{R}\}$ are the reactions of the truss and the terms of matrix $\{\hat{\Delta}\}$ are the components of translations of the nodes of the truss including those which are inhibited by its supports. Equations (3.55) are called the *stiffness equations for the truss*. $[\hat{S}]$ is called the *stiffness matrix for the truss*. Its terms are called the stiffness coefficients for the truss. On the basis of their derivation it is apparent that the stiffness equations (3.55) for the truss represent equations of equilibrium for its nodes. Moreover, referring to relations (3.55) it can be seen that the stiffness matrix for the truss is symmetric. That is,

$$[\hat{S}] = [\hat{S}]^T \tag{3.57}$$

The stiffness matrix for the truss is singular. Thus, the stiffness equations (3.55) cannot be solved directly to yield the components of translations of the nodes of the truss.

Step 6. We incorporate the conditions imposed by the supports of the truss into its stiffness equations (3.55), and we solve them. For this purpose we rearrange the rows and columns of the stiffness equations (3.55) and partition them as follows:

$$
\begin{Bmatrix} 0 \\ 0 \\ P \\ 0 \\ 0 \\ \hdashline 0 \\ 0 \\ 0 \end{Bmatrix} + \begin{Bmatrix} 0 \\ 0 \\ 0 \\ 0 \\ 0 \\ \hdashline R_1^{(1)} \\ R_2^{(1)} \\ R_2^{(4)} \end{Bmatrix}
$$

$$
= EA \begin{bmatrix}
0.405 & -0.096 & -0.333 & 0 & -0.072 & 0 & 0 & 0.096 \\
-0.096 & 0.378 & 0 & 0 & 0.096 & 0 & -0.250 & -0.128 \\
-0.333 & 0 & 0.405 & 0.096 & 0 & -0.072 & -0.096 & 0 \\
0 & 0 & 0.096 & 0.378 & 0 & -0.096 & -0.128 & -0.250 \\
-0.072 & 0.096 & 0 & 0 & 0.405 & -0.333 & 0 & -0.096 \\
\hdashline
0 & 0 & -0.072 & -0.096 & -0.333 & 0.405 & 0.096 & 0 \\
0 & -0.250 & -0.096 & -0.128 & 0 & 0.096 & 0.378 & 0 \\
0.096 & -0.128 & 0 & -0.250 & -0.096 & 0 & 0 & 0.378
\end{bmatrix}
\begin{Bmatrix} \Delta_3 \\ \Delta_4 \\ \Delta_5 \\ \Delta_6 \\ \Delta_7 \\ \hdashline \Delta_1 \\ \Delta_2 \\ \Delta_8 \end{Bmatrix}
$$

$$\text{(3.58)}$$

or

$$
\begin{Bmatrix} \{P^{GF}\} \\ \{P^{GS}\} \end{Bmatrix} + \begin{Bmatrix} \{0\} \\ \{R\} \end{Bmatrix} = \begin{bmatrix} [S^{FF}] & [S^{FS}] \\ [S^{SF}] & [S^{SS}] \end{bmatrix} \begin{Bmatrix} \{\Delta^F\} \\ \{\Delta^S\} \end{Bmatrix} \qquad \text{(3.59)}
$$

where the terms of matrix $\{P^{GF}\}$ are the known components of the external actions acting on the nodes of the structure which are not absorbed directly by its supports, the terms of matrix $\{P^{GS}\}$ are the known components of external actions which are directly absorbed by its supports (for the truss of Fig. 3.1 they are equal to zero), the terms of matrix $\{R\}$ are the reactions of the truss, the terms of matrix $\{\Delta^F\}$ are the unknown components of displacements of the nodes of the truss, the terms of matrix $\{\Delta^S\}$ are the known (given) components of displacements of the supported nodes of the truss (for the structure under consideration they are equal to zero), and $[S^{FF}]$ is the *basic stiffness matrix for the truss*.

Relation (3.59) can be expanded to yield

$$
\begin{aligned}
\{P^{GF}\} &= [S^{FF}]\{\Delta^F\} + [S^{FS}]\{\Delta^S\} \\
\{P^{GS}\} + \{R\} &= [S^{SF}]\{\Delta^F\} + [S^{SS}]\{\Delta^S\}
\end{aligned} \qquad \text{(3.60)}
$$

Inasmuch as the known components of displacement $\{\Delta^S\}$ are zero, relations (3.60) reduce to

$$\{P^{GF}\} = [S^{FF}]\{\Delta^F\} \tag{3.61}$$

$$\{P^{GS}\} + \{R\} = [S^{SF}]\{\Delta^F\} \tag{3.62}$$

If a structure is not a mechanism, its basic stiffness matrix $[S^{FF}]$ is not a singular matrix. Thus, referring to Eqs. (3.58) and (3.59) for the truss of Fig. 3.1, Eqs. (3.61) and (3.62) give

$$\{\Delta^F\} = [S^{FF}]^{-1}\{P^{GF}\}$$

$$= \frac{1}{EA}
\begin{bmatrix}
0.405 & -0.096 & -0.333 & 0 & -0.072 \\
-0.096 & 0.378 & 0 & 0 & 0.096 \\
-0.333 & 0 & 0.405 & 0.096 & 0 \\
0 & 0 & 0.096 & 0.378 & 0 \\
-0.072 & 0.096 & 0 & 0 & 0.405
\end{bmatrix}^{-1}
\begin{Bmatrix}
0 \\
0 \\
P \\
0 \\
0
\end{Bmatrix}$$

$$= \frac{P}{AE}
\begin{Bmatrix}
10.50000 \\
2.33333 \\
11.81261 \\
-3.00003 \\
1.31261
\end{Bmatrix} \tag{3.63}$$

and

$$\{R\} = [S^{SF}]\{\Delta^F\}$$

$$= P
\begin{bmatrix}
0 & 0 & -0.072 & -0.096 & -0.333 \\
0 & -0.250 & -0.096 & -0.128 & 0 \\
0.096 & -0.128 & 0 & -0.250 & -0.096
\end{bmatrix}
\begin{Bmatrix}
10.50000 \\
2.33333 \\
11.81261 \\
-3.00003 \\
1.31261
\end{Bmatrix}$$

$$= P
\begin{Bmatrix}
-1.000 \\
-1.333 \\
1.333
\end{Bmatrix} \tag{3.64}$$

Step 7. We compute the local components of nodal forces of the elements of the truss from the components of translations of its nodes. In order to accomplish this we use the compatibility relations (3.52) to establish the global components of nodal translations of each element of the truss from the components of displacements of its nodes. Thus, referring to relations (3.52) and (3.63) we have

$$\{\bar{D}^1\} = \begin{Bmatrix} \bar{u}_1^{1j} \\ \bar{u}_2^{1j} \\ \bar{u}_1^{1k} \\ \bar{u}_2^{1k} \end{Bmatrix} = \frac{P}{AE} \begin{Bmatrix} 0 \\ 0 \\ 10.50000 \\ 2.33333 \end{Bmatrix} \qquad \{\bar{D}^2\} = \begin{Bmatrix} \bar{u}_1^{2j} \\ \bar{u}_2^{2j} \\ \bar{u}_1^{2k} \\ \bar{u}_2^{2k} \end{Bmatrix} = \frac{P}{AE} \begin{Bmatrix} 0 \\ 0 \\ 1.31261 \\ 0 \end{Bmatrix}$$

$$\{\bar{D}^3\} = \begin{Bmatrix} \bar{u}_1^{3j} \\ \bar{u}_2^{3j} \\ \bar{u}_1^{3k} \\ \bar{u}_2^{3k} \end{Bmatrix} = \frac{P}{AE} \begin{Bmatrix} 10.50000 \\ 2.33333 \\ 1.31261 \\ 0 \end{Bmatrix} \qquad \{\bar{D}^4\} = \begin{Bmatrix} \bar{u}_1^{4j} \\ \bar{u}_2^{4j} \\ \bar{u}_1^{4k} \\ \bar{u}_2^{4k} \end{Bmatrix} = \frac{P}{AE} \begin{Bmatrix} 10.50000 \\ 2.33333 \\ 11.81261 \\ -3.00003 \end{Bmatrix}$$

$$\{\bar{D}^5\} = \begin{Bmatrix} \bar{u}_1^{5j} \\ \bar{u}_2^{5j} \\ \bar{u}_1^{5k} \\ \bar{u}_2^{5k} \end{Bmatrix} = \frac{P}{AE} \begin{Bmatrix} 11.81261 \\ -3.00003 \\ 1.31261 \\ 0 \end{Bmatrix} \qquad \{\bar{D}^6\} = \begin{Bmatrix} \bar{u}_1^{6j} \\ \bar{u}_2^{6j} \\ \bar{u}_1^{6k} \\ \bar{u}_2^{6k} \end{Bmatrix} = \frac{P}{AE} \begin{Bmatrix} 0 \\ 0 \\ 11.81261 \\ -3.00003 \end{Bmatrix}$$

$$(3.65)$$

From the global components of nodal translations of each element of the truss we compute the local components of its nodal actions. That is,

$$\{A\} = [\check{K}]\{\bar{D}\} \tag{3.66}$$

where $[\check{K}]$ is called the *hybrid stiffness matrix* for the element and is defined as

$$[\check{K}] = [K][\Lambda] \tag{3.67}$$

We obtain the hybrid matrices of the elements of the truss of Fig. 3.1 by substituting in relation (3.67) the transformation matrix for each element of the truss given by relations (3.50) and its local stiffness matrix given by relation (3.49). Thus,

$$[\check{K}^1] = \frac{E_1 A_1}{L_1} \begin{bmatrix} 1 & -1 \\ -1 & 1 \end{bmatrix} \begin{bmatrix} 0 & 1 & 0 & 0 \\ 0 & 0 & 0 & 1 \end{bmatrix} = \frac{EA}{4} \begin{bmatrix} 0 & 1 & 0 & -1 \\ 0 & -1 & 0 & 1 \end{bmatrix}$$

$$[\check{K}^2] = \frac{E_2 A_2}{L_2} \begin{bmatrix} 1 & -1 \\ -1 & 1 \end{bmatrix} \begin{bmatrix} 1 & 0 & 0 & 0 \\ 0 & 0 & 1 & 0 \end{bmatrix} = \frac{EA}{3} \begin{bmatrix} 1 & 0 & -1 & 0 \\ -1 & 0 & 1 & 0 \end{bmatrix}$$

$$[\check{K}^3] = \frac{E_3 A_3}{L_3} \begin{bmatrix} 1 & -1 \\ -1 & 1 \end{bmatrix} \begin{bmatrix} 0.6 & -0.8 & 0 & 0 \\ 0 & 0 & 0.6 & -0.8 \end{bmatrix}$$

$$= \frac{EA}{5} \begin{bmatrix} 0.6 & -0.8 & -0.6 & 0.8 \\ -0.6 & 0.8 & 0.6 & -0.8 \end{bmatrix}$$

$$[K^4] = \frac{E_4 A_4}{L_4}\begin{bmatrix} 1 & -1 \\ -1 & 1 \end{bmatrix}\begin{bmatrix} 1 & 0 & 0 & 0 \\ 0 & 0 & 1 & 0 \end{bmatrix} = \frac{EA}{3}\begin{bmatrix} 1 & 0 & -1 & 0 \\ -1 & 0 & 1 & 0 \end{bmatrix}$$

$$[K^5] = \frac{E_5 A_5}{L_5}\begin{bmatrix} 1 & -1 \\ -1 & 1 \end{bmatrix}\begin{bmatrix} 0 & -1 & 0 & 0 \\ 0 & 0 & 0 & -1 \end{bmatrix} = \frac{EA}{4}\begin{bmatrix} 0 & -1 & 0 & 1 \\ 0 & 1 & 0 & -1 \end{bmatrix}$$

$$[K^6] = \frac{E_6 A_6}{L_6}\begin{bmatrix} 1 & -1 \\ -1 & 1 \end{bmatrix}\begin{bmatrix} 0.6 & 0.8 & 0 & 0 \\ 0 & 0 & 0.6 & 0.8 \end{bmatrix}$$

$$= \frac{EA}{5}\begin{bmatrix} 0.6 & 0.8 & -0.6 & -0.8 \\ -0.6 & -0.8 & 0.6 & 0.8 \end{bmatrix} \tag{3.68}$$

We substitute relations (3.68) and (3.65) into (3.66) to obtain the local nodal actions for each element of the truss. That is,

$$\{A^1\} = \frac{P}{4}\begin{bmatrix} 0 & 1 & 0 & -1 \\ 0 & -1 & 0 & 1 \end{bmatrix}\begin{Bmatrix} 0 \\ 0 \\ 10.50000 \\ 2.33333 \end{Bmatrix} = P\begin{Bmatrix} -0.5833 \\ 0.5833 \end{Bmatrix}$$

$$\{A^2\} = \frac{P}{3}\begin{bmatrix} 1 & 0 & -1 & 0 \\ -1 & 0 & 1 & 0 \end{bmatrix}\begin{Bmatrix} 0 \\ 0 \\ 1.31261 \\ 0 \end{Bmatrix} = P\begin{Bmatrix} -0.4375 \\ 0.4375 \end{Bmatrix}$$

$$\{A^3\} = \frac{P}{5}\begin{bmatrix} 0.6 & -0.8 & -0.6 & 0.8 \\ -0.6 & 0.8 & 0.6 & -0.8 \end{bmatrix}\begin{Bmatrix} 10.50000 \\ 2.33333 \\ 1.31261 \\ 0 \end{Bmatrix} = P\begin{Bmatrix} 0.72916 \\ -0.72916 \end{Bmatrix}$$

$$\tag{3.69}$$

$$\{A^4\} = \frac{P}{3}\begin{bmatrix} 1 & 0 & -1 & 0 \\ -1 & 0 & 1 & 0 \end{bmatrix}\begin{Bmatrix} 10.50000 \\ 2.33333 \\ 11.81261 \\ -3.00003 \end{Bmatrix} = P\begin{Bmatrix} -0.4375 \\ 0.4375 \end{Bmatrix}$$

$$\{A^5\} = \frac{P}{4}\begin{bmatrix} 0 & -1 & 0 & 1 \\ 0 & 1 & 0 & -1 \end{bmatrix}\begin{Bmatrix} 1.81261 \\ -3.00003 \\ 1.31261 \\ 0 \end{Bmatrix} = P\begin{Bmatrix} 0.7500 \\ -0.7500 \end{Bmatrix}$$

$$\{A^6\} = \frac{P}{5}\begin{bmatrix} 0.6 & 0.8 & -0.6 & -0.8 \\ -0.6 & -0.8 & 0.6 & 0.8 \end{bmatrix}\begin{Bmatrix} 0 \\ 0 \\ 11.81261 \\ -3.00003 \end{Bmatrix} = P\begin{Bmatrix} -0.9375 \\ 0.9375 \end{Bmatrix}$$

3.4.2 The modern flexibility or force method

The components of the reactions of a statically determinate structure and the components of nodal actions of its elements can be established from the applied loads by writing and solving the independent equations of equilibrium for its elements and nodes. However, a structure which is statically indeterminate to the Nth degree has N components of reactions and/or nodal actions in excess of the number of independent equations of equilibrium which can be written for its elements and nodes. Consequently, the components of the reactions of a statically indeterminate structure and the components of nodal actions of its elements cannot be computed from the independent equations of equilibrium which can be written for its elements and nodes.

When we use a force or flexibility method to establish the components of nodal actions in the elements of a statically indeterminate structure to the Nth degree, we choose N of its reactions and/or internal actions of its elements as the unknown quantities which we call the *redundants of the structure*. The statically determinate structure which results by removing the constraints which induce the redundants from the actual structure is called the *primary structure* (see Fig. 3.7b to d). When the primary structure is subjected to the given loads and to the chosen redundants, the internal actions in its elements and the components of displacement of its nodes are equal to the corresponding quantities of the actual structure subjected to the given loads.

In the modern flexibility or force method it is convenient to consider a model of the structure made from the actual structure by replacing its supports with elements referred to as *reaction-elements*. Each reaction-element is connected with its end j to the supporting body (ground) and with its end k to a supported node of the structure. A roller support of a planar structure is replaced by an axial deformation reaction-element whose axis is normal to the plane of rolling. A hinged support of a planar structure is replaced by two mutually perpendicular axial deformation reaction-elements. The axis of each of these reaction-elements is parallel to a global axis of the structure. The model is subjected to the known equivalent actions acting on the nodes of the structure. The model for the truss of Fig. 3.1 is shown in Fig. 3.8. Referring to Fig. 3.8 it is apparent that the basic nodal action of a reaction-element of the model is equal and opposite to the corresponding component of reaction of the truss. In order to analyze the truss of Fig. 3.1 by the modern flexibility method we adhere to the following steps.

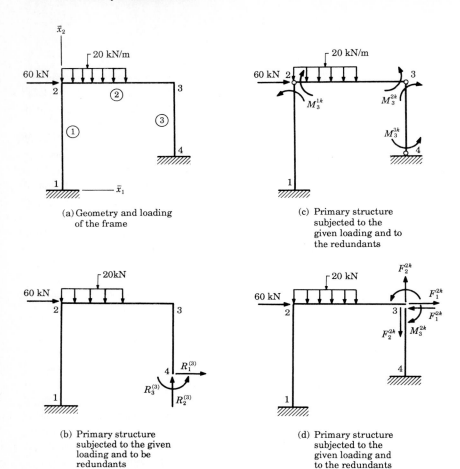

(a) Geometry and loading
of the frame

(c) Primary structure
subjected to the
given loading and to
the redundants

(b) Primary structure
subjected to the given
loading and to be
redundants

(d) Primary structure
subjected to the
given loading and
to the redundants

Figure 3.7   Redundants of a statically indeterminate frame.

Step 1. We form the matrix of the given actions $\{\hat{P}^G\}$ acting on the nodes of the truss [see relation (3.48)].

Step 2. We form a model for the truss by replacing its supports by reaction-elements (see Fig. 3.8).

Step 3. We establish the flexibility equations for each element of the truss. Referring to relation (3.27) we have

$$d_1^e = u_1^{ek} - u_1^{ej} = f_{11}^e F_1^{ek} \tag{3.70}$$

where
$$f_{11}^{eM} = \frac{L_e}{E_e A_e} \qquad e = 1, 2, \ldots, 6$$

$$f_{11}^{eM} = 0 \qquad e = 7, 8, 9 \tag{3.71}$$

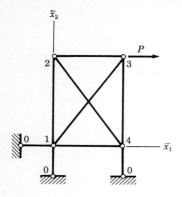

Figure 3.8 Model for the truss of Fig. 3.1.

Step 4. We combine the flexibility equations for all the elements of the model as follows:

$$
\begin{Bmatrix} d_1^1 \\ d_1^2 \\ d_1^3 \\ d_1^4 \\ d_1^5 \\ d_1^7 \\ d_1^8 \\ d_1^9 \\ \hline d_1^6 \end{Bmatrix} = \frac{1}{EA} \begin{bmatrix} 4 & 0 & 0 & 0 & 0 & 0 & 0 & 0 & \vdots & 0 \\ 0 & 3 & 0 & 0 & 0 & 0 & 0 & 0 & \vdots & 0 \\ 0 & 0 & 5 & 0 & 0 & 0 & 0 & 0 & \vdots & 0 \\ 0 & 0 & 0 & 3 & 0 & 0 & 0 & 0 & \vdots & 0 \\ 0 & 0 & 0 & 0 & 4 & 0 & 0 & 0 & \vdots & 0 \\ 0 & 0 & 0 & 0 & 0 & 0 & 0 & 0 & \vdots & 0 \\ 0 & 0 & 0 & 0 & 0 & 0 & 0 & 0 & \vdots & 0 \\ 0 & 0 & 0 & 0 & 0 & 0 & 0 & 0 & \vdots & 0 \\ \hline 0 & 0 & 0 & 0 & 0 & 0 & 0 & 0 & \vdots & 5 \end{bmatrix} \begin{Bmatrix} F_1^{1k} \\ F_1^{2k} \\ F_1^{3k} \\ F_1^{4k} \\ F_1^{5k} \\ F_1^{7k} \\ F_1^{8k} \\ F_1^{9k} \\ \hline F_1^{6k} \end{Bmatrix}
\tag{3.72}
$$

or

$$
\left\{ \frac{\{d_0^M\}}{\{d_x^M\}} \right\} = \left[\frac{[f_{00}^M] \;\vdots\; [f_{0x}^M]}{[f_{x0}^M] \;\vdots\; [f_{xx}^M]} \right] \left\{ \frac{\{a_0^M\}}{\{a_x^M\}} \right\}
\tag{3.73}
$$

In relations (3.72) the basic deformation parameters of the elements of the model of the truss are expressed as a linear combination of their basic nodal forces. These relations depend on the geometry of the elements of the truss and on properties of the materials from which they are made.

Step 5. We write the equations of equilibrium for the un-restrained nodes of the model in terms of the basic nodal forces of its elements. In Fig. 3.9 we express the nodal forces of each element of

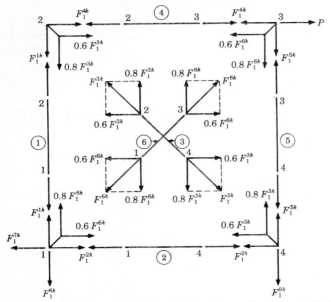

Figure 3.9 Free-body diagrams of the nodes and elements of the truss.

the model in terms of its basic nodal force by considering the equilibrium of the element [see Eq. (3.17)]. Moreover, we transform the local components of the basic nodal force of each element to global. Referring to Fig. 3.9 the equations of equilibrium for the unrestrained nodes of the model can be written as

$$
\begin{array}{c}
\text{Element no.} \\
\begin{array}{l}
\text{Node 1} \begin{cases} \sum \bar{F}_1 = 0 \\ \sum \bar{F}_2 = 0 \end{cases} \\
\text{Node 2} \begin{cases} \sum \bar{F}_1 = 0 \\ \sum F_2 = 0 \end{cases} \\
\text{Node 3} \begin{cases} \sum \bar{F}_1 = 0 \\ \sum F_2 = 0 \end{cases} \\
\text{Node 4} \begin{cases} \sum F_1 = 0 \\ \sum F_2 = 0 \end{cases}
\end{array}
\left\{ \begin{array}{c} 0 \\ 0 \\ 0 \\ 0 \\ P \\ 0 \\ 0 \\ 0 \end{array} \right\}
-
\begin{array}{cccccccccc}
& 1 & 2 & 3 & 4 & 5 & 6 & 7 & 8 & 9 \\
\left[\begin{array}{ccccccccc}
0 & -1 & 0 & 0 & 0 & -0.6 & 1 & 0 & 0 \\
-1 & 0 & 0 & 0 & 0 & -0.8 & 0 & 1 & 0 \\
0 & 0 & -0.6 & -1 & 0 & 0 & 0 & 0 & 0 \\
1 & 0 & 0.8 & 0 & 0 & 0 & 0 & 0 & 0 \\
0 & 0 & 0 & 1 & 0 & 0.6 & 0 & 0 & 0 \\
0 & 0 & 0 & 0 & 1 & 0.8 & 0 & 0 & 0 \\
0 & 1 & 0.6 & 0 & 0 & 0 & 0 & 0 & 0 \\
0 & 0 & -0.8 & 0 & -1 & 0 & 0 & 0 & 1
\end{array} \right]
\end{array}
\left\{ \begin{array}{c} F_1^{1k} \\ F_1^{2k} \\ F_1^{3k} \\ F_1^{4k} \\ F_1^{5k} \\ F_1^{6k} \\ F_1^{7k} \\ F_1^{8k} \\ F_1^{9k} \end{array} \right\} = 0
$$

$$(3.74)$$

or
$$\{\hat{P}^G\} = [B^M]\{a^M\} \qquad (3.75)$$

where $\{\hat{P}^G\}$ is the matrix of the global components of the given external forces acting on all the nodes of the truss, and $\{a^M\}$ is the matrix of the basic nodal actions of all the elements of the truss and of the reaction elements.

Step 6. We choose the redundant and express the basic nodal forces of the elements of the model in terms of the known external forces and the chosen redundant.

The model of Fig. 3.8 is statically indeterminate to the first degree. That is, the number of the unknown basic nodal forces of its elements exceeds by one the number of the equations of equilibrium for its nodes. Thus the model has one redundant force. We choose the basic nodal force of element 6 as the redundant for the model. We could have chosen the basic nodal force of any element of the truss as the redundant for the model provided that the structure which results by removing the constraint which causes the chosen redundant is not a mechanism (see Sec. 1.11).

We express the basic nodal forces of the elements of the model of Fig. 3.8 in terms of the given external force and the chosen redundant. To accomplish this, we modify the equations of equilibrium (3.74) by interchanging the columns of the matrix $[B^M]$ and the rows of the matrix $\{a^M\}$ so that the chosen redundant is in the last row of the matrix $\{a^M\}$. That is,

$$
\begin{Bmatrix} 0 \\ 0 \\ 0 \\ 0 \\ P \\ 0 \\ 0 \\ 0 \end{Bmatrix} =
\begin{bmatrix}
0 & -1 & 0 & 0 & 0 & 1 & 0 & 0 & \vdots & -0.6 \\
-1 & 0 & 0 & 0 & 0 & 0 & 1 & 0 & \vdots & -0.8 \\
0 & 0 & -0.6 & -1 & 0 & 0 & 0 & 0 & \vdots & 0 \\
1 & 0 & 0.8 & 0 & 0 & 0 & 0 & 0 & \vdots & 0 \\
0 & 0 & 0 & 1 & 0 & 0 & 0 & 0 & \vdots & 0.6 \\
0 & 0 & 0 & 0 & 1 & 0 & 0 & 0 & \vdots & 0.8 \\
0 & 1 & 0.6 & 0 & 0 & 0 & 0 & 0 & \vdots & 0 \\
0 & 0 & -0.8 & 0 & -1 & 0 & 0 & 1 & \vdots & 0
\end{bmatrix}
\begin{Bmatrix} F_1^{1k} \\ F_1^{2k} \\ F_1^{3k} \\ F_1^{4k} \\ F_1^{5k} \\ F_1^{7k} \\ F_1^{8k} \\ F_1^{9k} \\ \text{-} \\ F_1^{6k} \end{Bmatrix} \qquad (3.76)
$$

Denoting the matrix of the redundant basic nodal actions of the model by $\{a_x^M\}$ and the matrix of the remaining basic nodal actions of the model by $\{a_0^M\}$, the equations of equilibrium (3.76) can be written as

$$\{\hat{P}^G\} = [[B_0^M] \vdots [B_x^M]]\begin{Bmatrix} \{a_a^M\} \\ \{a_x^M\} \end{Bmatrix} \qquad (3.77)$$

or
$$\{\hat{P}^G\} = [B_0^M]\{a_0^M\} + [B_x^M]\{a_x^M\} \qquad (3.78)$$

where $[B_0^M]$ is a square nonsingular matrix. Thus, relation (3.78) can be solved for $\{a_0^M\}$ to give

$$\{a_0^M\} = [F_0]\{\hat{P}^G\} + [F_x]\{a_x^M\} \tag{3.79}$$

where
$$[F_0] = [B_0^M]^{-1} \qquad [F_x] = -[B_0^M]^{-1}[B_x^M] \tag{3.80}$$

For the truss of Fig. 3.1 referring to relations (3.76) and (3.77) we have

$$[F_0] = [B_0^M]^{-1} = \begin{bmatrix} 0 & 0 & 1.333 & 1 & 1.333 & 0 & 0 & 0 \\ 0 & 0 & 1 & 0 & 1 & 0 & 1 & 0 \\ 0 & 0 & -1.667 & 0 & -1.667 & 0 & 0 & 0 \\ 0 & 0 & 0 & 0 & 1 & 0 & 0 & 0 \\ 0 & 0 & 0 & 0 & 0 & 1 & 0 & 0 \\ 1 & 0 & 1 & 0 & 1 & 0 & 1 & 0 \\ 0 & 1 & 1.333 & 1 & 1.333 & 0 & 0 & 0 \\ 0 & 0 & -1.333 & 0 & -1.333 & 1 & 0 & 1 \end{bmatrix} \tag{3.81}$$

$$[F_x] = -[B_0^M]^{-1}[B_x^M] = \begin{Bmatrix} -0.8 \\ -0.6 \\ 1.0 \\ -0.6 \\ -0.8 \\ 0 \\ 0 \\ 0 \end{Bmatrix} \tag{3.82}$$

Substituting relations (3.81), (3.82), and (3.48) into (3.79) we obtain

$$\{a_0^M\} = \begin{Bmatrix} F_1^{1k} \\ F_1^{2k} \\ F_1^{3k} \\ F_1^{4k} \\ F_1^{5k} \\ F_1^{7k} \\ F_1^{8k} \\ F_1^{9k} \end{Bmatrix} = \begin{Bmatrix} 1.333 \\ 1 \\ -1.667 \\ 1 \\ 0 \\ 1 \\ 1.333 \\ -1.333 \end{Bmatrix} P + \begin{Bmatrix} -0.8 \\ -0.6 \\ 1.0 \\ -0.6 \\ -0.8 \\ 0 \\ 0 \\ 0 \end{Bmatrix} F_1^{6k} \tag{3.83}$$

Relations (3.83) give the basic nodal forces of the elements of the model of Fig. 3.8 in terms of the given external force and the chosen redundant. The terms of the first matrix on the right-hand side of relation (3.83) represent the basic nodal forces of the elements of the statically determined truss shown in Fig. 3.10b subjected to the given force. This truss is called the *primary structure*. It is obtained from the model of Fig. 3.8 by disconnecting element 6 from node 3. The terms of the second matrix on the right-hand side of relation (3.83) represent the internal forces in the elements of the primary structure when subjected to the redundant as shown in Fig. 3.10c, that is, when subjected to a pair of opposite forces equal to F_1^{6k} applied at the points of the primary structure where element 6 has been disconnected from node 3. Thus, we may conclude that in relation (3.79) or (3.83) the basic nodal actions of the elements of the model for the truss are expressed as the sum of the corresponding basic nodal actions of the elements of the primary structure for the model subjected to the following two loading cases:

1. The given loads

2. The chosen redundants

Step 7. Referring to Figs. 3.6 and 3.8 we write the equations of compatibility for the model of the truss under consideration.

$$u_1^{1j} = \Delta_2 \qquad u_1^{4j} = \Delta_3 \qquad u_1^{7j} = 0$$

$$u_1^{1k} = \Delta_4 \qquad u_1^{4k} = \Delta_5 \qquad u_1^{7k} = \Delta_1$$

$$u_1^{2j} = \Delta_1 \qquad u_1^{5j} = -\Delta_6 \qquad u_1^{8j} = 0$$

$$u_1^{2k} = \Delta_7 \qquad u_1^{5k} = -\Delta_8 \qquad u_1^{8k} = \Delta_2 \tag{3.84}$$

$$u_1^{3j} = \tfrac{3}{5}\Delta_3 - \tfrac{4}{5}\Delta_4 \qquad u_1^{6j} = \tfrac{3}{5}\Delta_1 + \tfrac{4}{5}\Delta_2 \qquad u_1^{9j} = 0$$

$$u_1^{3k} = \tfrac{3}{5}\Delta_7 - \tfrac{4}{5}\Delta_8 \qquad u_1^{6k} = \tfrac{3}{5}\Delta_5 + \tfrac{4}{5}\Delta_6 \qquad u_1^{9k} = \Delta_8$$

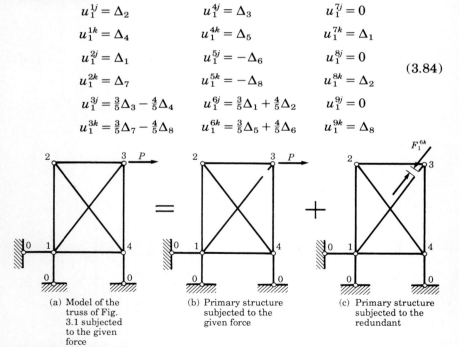

(a) Model of the truss of Fig. 3.1 subjected to the given force

(b) Primary structure subjected to the given force

(c) Primary structure subjected to the redundant

Figure 3.10 Superposition of the primary structure for the model of the truss of Fig. 3.1 subjected to the given force and to the redundant.

Substituting relation (3.84) into (3.21) we obtain

$$d_1^1 = \Delta_4 - \Delta_2$$

$$d_1^2 = \Delta_7 - \Delta_1$$

$$d_1^3 = \tfrac{3}{5}\Delta_7 - \tfrac{4}{5}\Delta_8 + \tfrac{4}{5}\Delta_4 - \frac{3\Delta_3}{5}$$

$$d_1^4 = \Delta_5 - \Delta_3$$

$$d_1^5 = -\Delta_8 + \Delta_6 \qquad\qquad (3.85)$$

$$d_1^6 = \tfrac{3}{5}\Delta_5 + \tfrac{4}{5}\Delta_6 - \tfrac{3}{5}\Delta_1 - \tfrac{4}{5}\Delta_2$$

$$d_1^7 = \Delta_1$$

$$d_1^8 = \Delta_2$$

$$d_1^9 = \Delta_8$$

Relation (3.85) can be rewritten as

$$\Delta_1 = d_1^7$$

$$\Delta_2 = d_1^8$$

$$\Delta_4 = d_1^1 + d_1^8$$

$$\Delta_7 = d_1^2 + d_1^7$$

$$\Delta_3 = \tfrac{4}{3}d_1^1 + d_1^2 - \tfrac{5}{3}d_1^3 + d_1^7 + \tfrac{4}{3}d_1^8 - \tfrac{4}{3}d_1^9 \qquad\qquad (3.86)$$

$$\Delta_5 = \tfrac{4}{3}d_1^1 + d_1^2 - \tfrac{5}{3}d_1^3 + d_1^4 + d_1^7 + \tfrac{4}{3}d_1^8 - \tfrac{4}{3}d_1^9$$

$$\Delta_6 = d_1^5 + d_1^9$$

$$\Delta_8 = d_1^9$$

$$d_1^6 = \tfrac{4}{3}d_1^1 + \tfrac{3}{5}d_1^2 - d_1^3 + \tfrac{3}{5}d_1^4 + \tfrac{4}{5}d_1^5$$

or

$$
\begin{Bmatrix}
\Delta_1 \\ \Delta_2 \\ \Delta_3 \\ \Delta_4 \\ \Delta_5 \\ \Delta_6 \\ \Delta_7 \\ \Delta_8
\end{Bmatrix}
=
\begin{bmatrix}
0 & 0 & 0 & 0 & 0 & 1 & 0 & 0 \\
0 & 0 & 0 & 0 & 0 & 0 & 1 & 0 \\
\tfrac{4}{3} & 1 & -\tfrac{5}{3} & 0 & 0 & 1 & \tfrac{4}{3} & -\tfrac{4}{3} \\
1 & 0 & 0 & 0 & 0 & 0 & 1 & 0 \\
\tfrac{4}{3} & 1 & -\tfrac{5}{3} & 1 & 0 & 1 & \tfrac{4}{3} & -\tfrac{4}{3} \\
0 & 0 & 0 & 0 & 1 & 0 & 0 & 1 \\
0 & 1 & 0 & 0 & 0 & 1 & 0 & 0 \\
0 & 0 & 0 & 0 & 0 & 0 & 0 & 1
\end{bmatrix}
\begin{Bmatrix}
d_1^1 \\ d_1^2 \\ d_1^3 \\ d_1^4 \\ d_1^5 \\ d_1^7 \\ d_1^8 \\ d_1^9
\end{Bmatrix}
\qquad (3.87)
$$

and

$$d_1^6 = [0.8 \quad 0.6 \quad -1.0 \quad 0.6 \quad 0.8 \quad 0 \quad 0 \quad 0] \begin{Bmatrix} d_1^1 \\ d_1^2 \\ d_1^3 \\ d_1^4 \\ d_1^5 \\ d_1^7 \\ d_1^8 \\ d_1^9 \end{Bmatrix} \tag{3.88}$$

Referring to relations (3.81) and (3.82), relations (3.87) and (3.88) can be rewritten as

$$\{\hat{\Delta}\} = [F_0]^T \{d_0^M\} \tag{3.89}$$

$$\{d_x^M\} = -[F_x]^T \{d_0^M\} \tag{3.90}$$

where $\{d_x^M\}$ is the matrix of basic deformation parameters corresponding to the basic nodal actions which have been chosen as the redundants, and $\{d_0^M\}$ is the matrix of basic deformation parameters not included in $\{d_x^M\}$.

Combining relations (3.89) and (3.90) we obtain

$$\begin{Bmatrix} \{\hat{\Delta}\} \\ \{0\} \end{Bmatrix} = \begin{bmatrix} [F_0]^T & [0] \\ [F_x] & [I] \end{bmatrix} \begin{Bmatrix} \{d_0^M\} \\ \{d_x^M\} \end{Bmatrix} \tag{3.91}$$

Relations (3.91) represent a convenient form of the equations of compatibility for the structure. They ensure that the displacements of the nodes of a structure are compatible to the deformation of its elements.

Step 8. We compute the redundants of the model and, if required, the components of displacement of its nodes. To accomplish this we substitute relations (3.73) into (3.91) and use relation (3.79) to eliminate $\{a_0^M\}$. That is,

$$\{\hat{\Delta}\} = [F_{11}]\{\hat{P}^G\} + [F_{12}]\{a_x^M\} \tag{3.92}$$

$$\{0\} = [F_{21}]\{\hat{P}^G\} + [F_{22}]\{a_x^M\} \tag{3.93}$$

where

$$[F_{11}] = [F_0]^T [f_{00}^M][F_0]$$

$$[F_{12}] = [F_{21}]^T = [F_0]^T [f_{00}^M][F_x] + [F_0]^T [f_{0x}^M] \tag{3.94}$$

$$[F_{22}] = [F_x]^T [f_{00}^M][F_x] + [F_x]^T [f_{0x}^M] + [f_{x0}^M][F_x] + [f_{xx}^M]$$

Relation (3.93) can be solved for the redundants $\{a_x^M\}$ to give

$$\{a_x^M\} = -[F_{22}]^{-1}[F_{21}]\{\hat{P}^G\} \tag{3.95}$$

For the truss of Fig. 3.1 referring to relations (3.81), (3.82), (3.72), and (3.73) we have

$$[F_{11}] = \begin{bmatrix} 0 & 0 & 0 & 0 & 0 & 0 & 0 & 0 \\ 0 & 0 & 0 & 0 & 0 & 0 & 0 & 0 \\ 5.3335 & 3 & -8.3335 & 0 & 0 & 0 & 0 & 0 \\ 4.0000 & 0 & 0 & 0 & 0 & 0 & 0 & 0 \\ 5.3335 & 3 & -8.3335 & 3 & 0 & 0 & 0 & 0 \\ 0 & 0 & 0 & 0 & 4 & 0 & 0 & 0 \\ 0 & 0 & 0 & 0 & 0 & 0 & 0 & 0 \\ 0 & 0 & 0 & 0 & 0 & 0 & 0 & 0 \end{bmatrix} \tag{3.96}$$

$$[F_{12}] = [F_{21}]^T = \begin{Bmatrix} 0 \\ 0 \\ -14.40014 \\ -3.20000 \\ -16.20018 \\ -3.20000 \\ -1.80000 \\ 0 \end{Bmatrix} \tag{3.97}$$

$$[F_{22}] = 17.28 \tag{3.98}$$

Substituting relations (3.98), (3.97), and (3.48) into relation (3.95) we obtain

$$\{a_x^M\} = F_1^{6k} = 0.9375P \tag{3.99}$$

Substituting relations (3.99), (3.48), (3.81), and (3.82) into (3.79) we get

$$\{a_0^M\} = \begin{Bmatrix} F_1^{1k} \\ F_1^{2k} \\ F_1^{3k} \\ F_1^{4k} \\ F_1^{5k} \\ F_1^{7k} \\ F_1^{8k} \\ F_1^{9k} \end{Bmatrix} = \begin{Bmatrix} 0.5833 \\ 0.4375 \\ -0.72916 \\ 0.4375 \\ -0.7500 \\ 1.0000 \\ 1.3333 \\ -1.3333 \end{Bmatrix} P \tag{3.100}$$

Referring to results (3.99) and (3.100) we have

$$\{a\} = \begin{Bmatrix} F_1^{1k} \\ F_1^{2k} \\ F_1^{3k} \\ F_1^{4k} \\ F_1^{5k} \\ F_1^{6k} \end{Bmatrix} = \begin{Bmatrix} 0.5833 \\ 0.4375 \\ -0.7292 \\ 0.4375 \\ -0.7500 \\ 0.9375 \end{Bmatrix} P \tag{3.101}$$

$$[R] = \begin{Bmatrix} -1.000 \\ -1.333 \\ 1.333 \end{Bmatrix} P \tag{3.102}$$

Substituting relations (3.96), (3.97), (3.48), and (3.99) into relation (3.92) we obtain

$$\{\hat{\Delta}\} = \begin{Bmatrix} \Delta_1 \\ \Delta_2 \\ \Delta_3 \\ \Delta_4 \\ \Delta_5 \\ \Delta_6 \\ \Delta_7 \\ \Delta_8 \end{Bmatrix} = \frac{P}{EA} \begin{Bmatrix} 0 \\ 0 \\ 10.500 \\ 2.333 \\ 11.812 \\ -3.000 \\ 1.312 \\ 0 \end{Bmatrix} \tag{3.103}$$

3.5   The Direct Stiffness and the Modern Flexibility Methods

In the direct stiffness method, the analysis of a structure (statically determinate or indeterminate) is formulated in terms of the displacements of its nodes. That is, the sum of the component of the known external actions acting on the nodes of a structure and its unknown reactions are expressed as a linear combination of the components of displacements of its nodes. This is accomplished by considering the equilibrium of the nodes of the structure and using

1. The stiffness equations of its elements

2. The compatibility of the components of nodal displacements of its elements and the components of displacements of its nodes.

Once the components of displacements of the nodes of a structure are established, the global components of the nodal displacements of its elements can be computed from the components of displacements of the nodes to which they are connected. Moreover, the local components of nodal actions of the elements of the structure are established from the global components of their nodal displacements using their hybrid stiffness equations (3.66). In the direct stiffness method all structures (statically determinate or indeterminate) are treated the same way. The direct stiffness method may be readily programmed on an electronic computer. Practically all major general programs for analyzing framed structures have been written using the direct stiffness method because of its simplicity, generality, and cost-effectiveness.

The modern flexibility method is used for analyzing statically indeterminate structures. The analysis of these structures is formulated in terms of some of their reactions and/or nodal actions which are called the *redundants*. The number of redundants of a structure is equal to the degree of its static indeterminacy. The equations from which the redundants of a structure are established represent conditions of continuity of the components of displacement at certain points of the structure. These equations are obtained after a formidable sequence of matrix operations. An advantage of the flexibility method is that the reactions and nodal actions of the elements of a structure are established without first computing the components of displacements of its nodes. When we analyze a structure, we are primarily interested in establishing the magnitudes of the internal actions of its elements on which its design usually depends. The major shortcoming of the modern flexibility method is that the choice of the redundants is not unique, and it affects the stability of the matrices which must be inverted in order to analyze a structure. Thus, a procedure must be programmed for the optimum choice of the redundants by the computer. Up to now the flexibility method has found very little practical use in writing general computer programs for analyzing groups of framed structures. However, it is possible that this may change when more effective procedures are established for overcoming its shortcomings.

3.6 Problems

1 and 2. Using the direct stiffness and the modern flexibility methods compute the internal forces in the elements of the truss subjected to the loads shown in Fig. 3.P1. The elements of the truss have the same constant cross section and are made from the same isotropic linearly elastic material. Repeat with the truss of Fig. 3.P2.

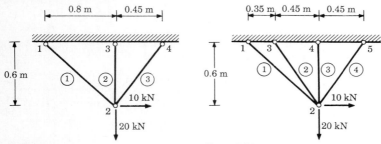

Figure 3.P1   **Figure 3.P2**

3. Using the direct stiffness and the modern flexibility methods compute the internal actions in the elements of the structure subjected to the loads shown in Fig. 3.P3. The area of the cross section of the steel cable is $A = 800 \text{ mm}^2$. The moment of inertia of the steel beam is $I = 369.7 \times 10 \text{ mm}^4$. Disregard the effect of axial deformation of beam 1, 3, 4. *Hint*: The component of translation $u_3(x_1)$ of an element of a planar beam or frame which is not subjected to external disturbances along its length has the following form:

$$u_3(x_1) = c_{30} + c_{31}x_1 + c_{32}x_1^2 + c_{33}x_1^3$$

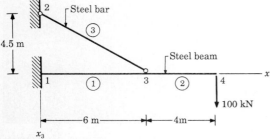

Figure 3.P3

Photographs and a Brief Description of Timber Dome Roofs

What makes timber attractive as a construction material is its low cost, its light weight, and its ease of construction. Recent technological developments have greatly increased the use of timber for long, deep, structurally reliable, moisture-resistant, load-carrying members. The glue-laminated Glulam timber and the laminated veneer lumber (LVL) have a proven long-term reliability in both wet and dry use. In what follows we present photographs and a brief description of two impressive timber dome roof structures made of Glulam.

The framework of the dome shown in Fig. 3.11 is made of Glulam ribs which are connected by patented steel hubs to form triangles. Each triangle contains parallel purlins. The framework is covered with a wood deck at the top of which is insulation and roofing.

This structure defers from the conventional timber dome roof structures in that the dome is not supported on a cylindrical wall or on columns with a compression ring at the top but on 36 reinforced concrete buttress supports which are 1.53 m (5 ft) high. The horizontal forces are resisted by a cast-in-place posttensioned concrete ring beam which links the buttresses together. The floor of the stadium is 9.1 m (30 ft) below the ground level. Its dome has a diameter of 153 m (502 ft). An important aspect of this structure is its very low cost.

Figure 3.11 The timber dome roof stadium of Northern Arizona University, Flagstaff, Arizona. (*Courtesy Western Wood Structures, Inc., Portland, Oregon.*)

Figure 3.12 Inside view of the timber roof of the Tacoma Dome, the sports and convention center of the city of Tacoma, Washington. (*Courtesy Western Wood Structures, Inc., Portland, Oregon.*)

Figure 3.13 View of the timber roof of the Tacoma Dome during construction. (*Courtesy Western Wood Structures, Inc., Portland, Oregon.*)

Figure 3.14 The timber dome roof Tacoma Dome. (*Courtesy Western Wood Structures, Inc., Portland, Oregon.*)

The beautiful timber dome roof shown in Figs. 3.12 to 3.14 has a diameter of 161.6 m (530 ft) and is the largest framework of Glulam ribs in the world. The Glulam ribs form triangles connected by patented steel hubs. Each triangle contains parallel purlins. This framework is covered with a 51 mm- (82-in-) thick tongue-and-groove wood deck, topped with 38.1-mm- ($1\frac{1}{2}$-in-) thick urethane foam insulation and 30 mm (1.18 in) of urethane roofing. The dome is supported on 36 cast-in-place concrete columns of 660 mm (26 in) diameter. The horizontal forces are resisted by a cast-in-place post-tensioned concrete tension ring beam located at the top of the columns. In the spaces between the columns a concrete and concrete masonry unit shear wall has been constructed.

Element Response

The Stiffness Equations for an Element

4.1 Introduction

In Part 1 of this book (Chaps. 1 to 3) we presented certain preliminary concepts of structural analysis. In Part 2 (Chaps. 4 to 6) we describe the quantities which we use to specify the response of an element of a framed structure and present methods for computing them. In Parts 3 and 4 (Chaps. 7 to 16) we present in detail the modern methods for analyzing framed structures.

In general we express the response of an element as the sum of its response when it is subjected only to its nodal displacements and its response when it is subjected to the given loading with its ends fixed (see Fig. 4.1). This approach is analogous to that adhered to in the slope deflection method of classical structural analysis.[1]

In this chapter we describe the response of an element subjected only to its nodal displacements by its local stiffness equations. In these equations the local components of the nodal actions of an element subjected only to its nodal displacements are expressed as a linear combination of the local components of its nodal displacements. The components of nodal actions of an element are not independent. They are related by the equations of equilibrium for the element. Moreover, the components of nodal displacements of an element represent not only its deformation but also its rigid-body motion. For this reason the stiffness equations for an element cannot be solved to yield the local components of its nodal displacements as a linear combination of the local components of its nodal actions (see Sec. 4.3).

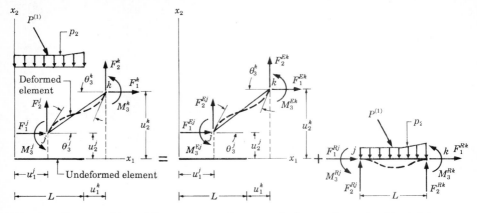

Figure 4.1 Superposition of the response of a general planar element subjected to the given loads acting along its length with its ends fixed and the element subjected only to its nodal displacements.

4.2 Matrices of Nodal Actions and Nodal Displacements for an Element

We store the local or the global components of the internal force and moment acting at the end q ($q = j$ or k) of an element of a structure in a matrix which we denote by $\{A^q\}$ or $\{\bar{A}^q\}$, respectively. Moreover, we store the local or global components of the translation and of the rotation of the end q ($q = j$ or k) of an element of a structure in the matrix $\{D^q\}$ or $\{\bar{D}^q\}$, respectively. If we want to specify the element to which the matrix $\{A^q\}$, $\{\bar{A}^q\}$, $\{D^q\}$, $\{\bar{D}^q\}$ belongs, we denote it as $\{A^{eq}\}$, $\{\bar{A}^{eq}\}$, $\{D^{eq}\}$, or $\{\bar{D}^{eq}\}$, respectively, where e stands for the number of the element.

The elements of a truss do not transfer a moment to its nodes. Consequently, the nodes of the truss do not rotate. Moreover, as discussed in Sec. 2.5 the transverse components of translation and the components of rotation of an element of a truss represent only rigid-body motion of the element. Consequently, we are not interested in computing them. Furthermore, an element of a truss is not subjected to a transverse component of force. For these reasons we do not include the components of moment at the end q ($q = j$ or k) of an element of a truss in the matrices $\{A^q\}$ and $\{\bar{A}^q\}$ and the components of rotation of its end q ($q = j$ or k) in its matrices $\{D^q\}$ and $\{\bar{D}^q\}$. Moreover, we do not include the transverse components of force at the end q ($q = j$ or k) of an element of a truss in the matrix $\{A^q\}$ or the transverse components of translation of its end q ($q = j$ or k) in the matrix $\{D^q\}$. Thus, the matrices $\{A^q\}$, $\{\bar{A}^q\}$, $\{D^q\}$, and $\{\bar{D}^q\}$ $(q = j$ or $k)$ of an element of a planar truss in the $x_1 x_2$ plane are

$$\{A^q\} = F^q_1 \tag{4.1a}$$

$$\{\bar{A}^q\} = \left\{ \begin{matrix} \bar{F}^q_1 \\ \bar{F}^q_2 \end{matrix} \right\} \tag{4.1b}$$

$$\{D^q\} = u^q_1 \tag{4.1c}$$

$$\{\bar{D}^q\} = \left\{ \begin{matrix} \bar{u}^q_1 \\ \bar{u}^q_2 \end{matrix} \right\} \tag{4.1d}$$

The matrices $\{A^q\}$, $\{\bar{A}^q\}$, $\{D^q\}$, and $\{\bar{D}^q\}$ $(q = j$ or $k)$ of a general planar element of a planar beam or frame in the $x_1 x_2$ plane are

$$\{A^q\} = \left\{ \begin{matrix} F^q_1 \\ F^q_2 \\ M^q_3 \end{matrix} \right\} \tag{4.2a}$$

$$\{\bar{A}^q\} = \left\{ \begin{matrix} \bar{F}^q_1 \\ \bar{F}^q_2 \\ \bar{M}^q_3 = M^q_3 \end{matrix} \right\} \tag{4.2b}$$

$$\{D^q\} = \left\{ \begin{matrix} u^q_1 \\ u^q_2 \\ \theta^q_3 \end{matrix} \right\} \tag{4.2c}$$

$$\{\bar{D}^q\} = \left\{ \begin{matrix} \bar{u}^q_1 \\ \bar{u}^q_2 \\ \bar{\theta}^q_3 = \theta^q_3 \end{matrix} \right\} \tag{4.2d}$$

Axial deformation elements of planar or space frames do not carry moments to the nodes to which they are connected. However, other elements of a frame may carry moments to these nodes (see nodes 2 and 4 of the frame of Fig. 4.2). Consequently, in order to ensure the equilibrium of these nodes the sum of the moments acting on them must be set equal to zero. For this reason we must include the end moments in the matrices $\{A^q\}$ and $\{\bar{A}^q\}$ of axial deformation elements of planar frames or of space frames.

Thus, the matrices $\{A^q\}$, $\{\bar{A}^q\}$, $\{D^q\}$, and $\{\bar{D}^q\}$ of an axial

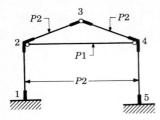

P1 = Axial deformation element
P2 = General planar element

Figure 4.2 Planar frame having two types of elements.

deformation element of a planar frame in the $x_1 x_2$ plane are

$$\{A^q\} = \begin{Bmatrix} F_1^q \\ 0 \\ 0 \end{Bmatrix} \tag{4.3a}$$

$$\{\bar{A}^q\} = \begin{Bmatrix} \bar{F}_1^q \\ \bar{F}_2^q \\ 0 \end{Bmatrix} \tag{4.3b}$$

$$\{D^q\} = \begin{Bmatrix} u_1^q \\ u_2^q \\ \theta_3^q \end{Bmatrix} \tag{4.3c}$$

$$\{\bar{D}^q\} = \begin{Bmatrix} \bar{u}_1^q \\ \bar{u}_2^q \\ \bar{\theta}_3^q \end{Bmatrix} \tag{4.3d}$$

We store the local or the global components of the actions at both ends of an element in the *matrix of nodal actions of the element*. That is,

$$\{A\} = \begin{Bmatrix} \{A^j\} \\ \{A^k\} \end{Bmatrix} \qquad \{\bar{A}\} = \begin{Bmatrix} \{\bar{A}^j\} \\ \{\bar{A}^k\} \end{Bmatrix} \tag{4.4}$$

We store the local or the global components of displacement of both ends of an element in the *matrix of nodal displacements of the element*. That is,

$$\{D\} = \begin{Bmatrix} \{D^j\} \\ \{D^k\} \end{Bmatrix} \qquad \{\bar{D}\} = \begin{Bmatrix} \{\bar{D}^j\} \\ \{\bar{D}^k\} \end{Bmatrix} \tag{4.5}$$

If it is necessary to specify the element with which a matrix of nodal actions or of nodal displacements is associated, we denote it as $\{A^e\}$ or $\{\bar{A}^e\}$ and $\{D^e\}$ or $\{\bar{D}^e\}$, respectively, where e is the number of the element.

The term of the nth row of the matrix of nodal displacements of an element is the component of displacement over which the component of nodal action represented by the term of the nth row of the matrix of nodal actions of this element performs work during the process of deformation of the structure. Two matrices having this property are termed *conjugate*. The work of the nodal actions of an element during the process of deformation of the structure is equal to

$$W = \tfrac{1}{2}\{A\}^T\{D\} \tag{4.6}$$

4.3 The Stiffness Equations for an Element

As discussed in Sec. 1.12 for the structures which we are considering in this book, the principle of superposition is valid. Thus we can express the matrix of nodal actions of an element as

$$\{A\} = \{A^E\} + \{A^R\} \tag{4.7}$$

where $\{A^E\}$ is the matrix of nodal actions of the element when subjected only to its nodal displacements, and $\{A^R\}$ is the matrix of nodal actions of the element when subjected to the given external disturbances with its ends fixed. $\{A^R\}$ is called *the matrix of fixed-end actions* of the element. The relations between the nodal actions and the nodal displacements of the elements of the structures which we are considering in this text are linear (see Sec. 1.12). Thus we express the local components of the nodal actions of an element of a framed structure subjected only to its nodal displacements as a linear combination of its nodal displacements. That is,

$$\{A^E\} = [K]\{D\} \tag{4.8}$$

Relations (4.8) are called the stiffness equations for the element.

The matrix $[K]$ is called the *local stiffness matrix for the element*. Its terms are called the local stiffness coefficients for the element. The local stiffness coefficient K_{mn} represents the nodal action A^E_m (the action in the mth row of the matrix $\{A^E\}$) of the element when it is subjected only to a nodal displacement $D_n = 1$, while all other nodal displacements vanish (D_n is the displacement in the nth row of the matrix $\{D\}$).

The internal actions of an element do not change when the element moves as a rigid body. That is, the magnitude of the nodal actions of an element is the same for all sets of values of its nodal displacements which differ by a rigid-body motion. Consequently, the components of nodal displacements of an element are specified only to within a rigid-body movement of the element when the components of its nodal actions are known. Thus, the components of nodal displacements of an element cannot be expressed as a linear combination of its nodal actions. Consequently, Eqs. (4.8) cannot be solved to give the nodal displacements of an element as a linear combination of its nodal actions. This implies that the local stiffness matrix $[K]$ of an element is singular.

Notice that the third row of the stiffness equations (4.8) for an element of a planar beam or frame is the slope deflection equation of classical structural analysis.

As it is shown subsequently, a number of terms of the local stiffness matrix of an element vanish. Moreover, its terms are not independent. They must satisfy certain relations resulting from the fact that the nodal actions of an element are in equilibrium.

In what follows we specialize the stiffness equations (4.8) for the elements of the various types of structures which we are considering.

Local stiffness matrix for an element of a truss. For an element of a planar or space truss the stiffness equations (4.8) reduce to

$$\{A\} = \{A^E\} = \begin{Bmatrix} F^j_1 \\ F^k_1 \end{Bmatrix} = \begin{bmatrix} K_{11} & K_{12} \\ K_{21} & K_{22} \end{bmatrix} \begin{Bmatrix} u^j_1 \\ u^k_1 \end{Bmatrix} = [K]\{D\} \qquad (4.9)$$

When the ends of an element of a truss are displaced by $u^j_1 = 1$ and $u^k_1 = 0$, relation (4.9) gives

$$F^j_1 = K_{11} \qquad F^k_1 = K_{21} \qquad (4.10)$$

That is, as shown in Fig. 4.3a, the stiffness coefficients K_{11} and K_{21} are the axial forces which must be placed at the ends j and k of the element, respectively, in order to shorten it by one unit. Inasmuch as the element is in equilibrium, referring to Fig. 4.3a, we have

$$K_{21} = -K_{11} \qquad (4.11)$$

Similarly, as shown in Fig. 4.3b, the stiffness coefficients K_{22} and K_{12} are the axial forces which must be placed at the ends k and j of the element, respectively, in order to elongate it by one unit. Moreover, from the equilibrium of the element we get

$$K_{12} = -K_{22} \qquad (4.12)$$

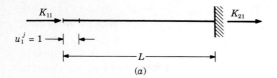

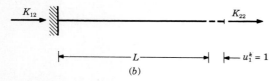

Figure 4.3 Physical significance of the local stiffness coefficients K_{mn} $(m, n = 1, 2)$ for an element of a truss.

Finally; referring to Fig. 4.3 and employing the Betti-Maxwell reciprocal theorem,[2] we obtain

$$K_{12} = K_{21} \tag{4.13}$$

That is, the local stiffness matrix of an element of a truss is symmetric.

Using relations (4.10) to (4.13) and referring to Eqs. (4.9), the local stiffness matrix for an element of a truss can be written as

$$[K] = K_{11}\begin{bmatrix} 1 & -1 \\ -1 & 1 \end{bmatrix} \tag{4.14}$$

It is apparent that the local stiffness matrix for an element of a truss is completely specified by one quantity—its independent stiffness coefficient.

Local stiffness matrix for an element of a planar beam or a planar frame. For a general planar element the stiffness equations (4.8) reduce to

$$\{A^E\} = \begin{Bmatrix} F_1^j \\ F_2^j \\ M_3^j \\ F_1^k \\ F_2^k \\ M_3^k \end{Bmatrix} = \begin{bmatrix} K_{11} & K_{12} & K_{13} & K_{14} & K_{15} & K_{16} \\ K_{21} & K_{22} & K_{23} & K_{24} & K_{25} & K_{26} \\ K_{31} & K_{32} & K_{33} & K_{34} & K_{35} & K_{36} \\ K_{41} & K_{42} & K_{43} & K_{44} & K_{45} & K_{46} \\ K_{51} & K_{52} & K_{53} & K_{54} & K_{55} & K_{56} \\ K_{61} & K_{62} & K_{63} & K_{64} & K_{65} & K_{66} \end{bmatrix} \begin{Bmatrix} u_1^j \\ u_2^j \\ \theta_3^j \\ u_1^k \\ u_2^k \\ \theta_3^k \end{Bmatrix} \tag{4.15}$$

The effect of axial deformation of general planar elements of beams and frames is in general small, and it is disregarded when analyzing these structures by hand calculations. In this case, we do not include the axial components of the nodal forces of an element in its matrix $\{A\}$ or the axial components of its nodal displacements in its matrix $\{D\}$. Hence, in this case the stiffness equations (4.8) for a general planar element subjected only to its nodal displacements [$\{A^R\} = 0$] reduce to

$$\{A^E\} = \begin{Bmatrix} F_2^j \\ M_3^j \\ F_2^k \\ M_3^k \end{Bmatrix} = \begin{bmatrix} K_{22} & K_{23} & K_{25} & K_{26} \\ K_{32} & K_{33} & K_{35} & K_{36} \\ K_{52} & K_{53} & K_{55} & K_{56} \\ K_{62} & K_{63} & K_{65} & K_{66} \end{bmatrix} \begin{Bmatrix} u_2^j \\ \theta_3^j \\ u_2^k \\ \theta_3^k \end{Bmatrix} \qquad (4.16)$$

The physical significance of the stiffness coefficients of the first column of the stiffness matrix of a general planar element can be established by considering such an element subjected only to the axial components of nodal actions of the magnitude required to induce the following nodal displacements:

$$u_1^j = 1 \qquad u_2^j = \theta_3^j = u_1^k = u_2^k = \theta_3^k = 0 \qquad (4.17)$$

In this case the stiffness equations (4.15) reduce to

$$\{A^E\} = \begin{Bmatrix} F_1^j \\ 0 \\ 0 \\ F_1^k \\ 0 \\ 0 \end{Bmatrix} = \begin{Bmatrix} K_{11} \\ K_{21} \\ K_{31} \\ K_{41} \\ K_{51} \\ K_{61} \end{Bmatrix} \qquad (4.18)$$

Thus,

$$K_{21} = K_{31} = K_{51} = K_{61} = 0 \qquad (4.19)$$

Moreover, as shown in Fig. 4.4a the stiffness coefficients K_{11} and K_{41} represent the axial forces which must be placed at the ends j and k of the element, respectively, in order to shorten it by one unit.

The physical significance of the stiffness coefficients of the second column of the stiffness matrix for a general planar element can be established by considering such an element subjected only to the components of nodal actions which are required to induce the

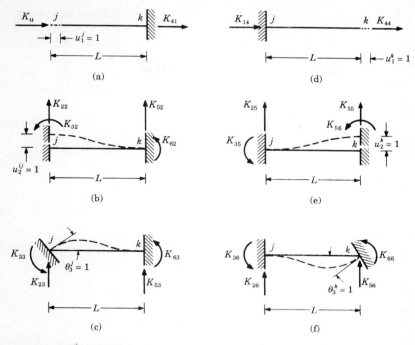

Figure 4.4 Physical significance of the local stiffness coefficients K_{mn} $(m, n = 1, 2, \ldots, 6)$ of a general planar element.

following nodal displacements.

$$u_2^j = 1 \qquad u_1^j = \theta_3^j = u_1^k = u_2^k = \theta_3^k = 0 \tag{4.20}$$

In this case the stiffness equations (4.15) reduce to

$$\{A^E\} = \begin{Bmatrix} 0 \\ F_2^j \\ M_3^{ij} \\ 0 \\ F_2^{ik} \\ M_3^{ik} \end{Bmatrix} = \begin{Bmatrix} K_{12} \\ K_{22} \\ K_{32} \\ K_{42} \\ K_{52} \\ K_{62} \end{Bmatrix} \tag{4.21}$$

Thus,

$$K_{12} = K_{42} = 0 \tag{4.22}$$

Moreover, as shown in Fig. 4.4b the stiffness coefficients K_{22}, K_{32},

K_{52}, and K_{62} represent the nodal actions which are required to induce the nodal displacements specified by relations (4.20).

The physical significance of the stiffness coefficients for the third column of the stiffness matrix of a general planar element can be established by considering an element subjected only to the components of nodal actions which are required to induce the following nodal displacements.

$$\theta_3^j = 1 \qquad u_1^j = u_2^j = u_1^k = u_2^k = \theta_3^k = 0 \qquad (4.23)$$

In this case the stiffness equations (4.15) reduce to

$$\{A^E\} = \begin{Bmatrix} 0 \\ F_2^j \\ M_3^j \\ 0 \\ F_2^k \\ M_3^k \end{Bmatrix} = \begin{Bmatrix} K_{13} \\ K_{23} \\ K_{33} \\ K_{43} \\ K_{53} \\ K_{63} \end{Bmatrix} \qquad (4.24)$$

Thus,

$$K_{13} = K_{43} = 0 \qquad (4.25)$$

Moreover, as shown in Fig. 4.4c the stiffness coefficients K_{23}, K_{33}, K_{53}, and K_{63} represent the nodal actions which are required to induce the nodal displacements specified by relations (4.23).

The physical significance of the remaining nonvanishing stiffness coefficients for a general planar element is shown in Fig. 4.4d to f. Thus, for a general planar element the stiffness equations (4.15) reduce to

$$\{A^E\} = \begin{Bmatrix} F_1^j \\ F_2^j \\ M_3^j \\ F_1^k \\ F_2^k \\ M_3^k \end{Bmatrix} = \begin{bmatrix} K_{11} & 0 & 0 & K_{14} & 0 & 0 \\ 0 & K_{22} & K_{23} & 0 & K_{25} & K_{26} \\ 0 & K_{32} & K_{33} & 0 & K_{35} & K_{36} \\ K_{41} & 0 & 0 & K_{44} & 0 & 0 \\ 0 & K_{52} & K_{53} & 0 & K_{55} & K_{56} \\ 0 & K_{62} & K_{63} & 0 & K_{65} & K_{66} \end{bmatrix} \begin{Bmatrix} u_1^j \\ u_2^j \\ \theta_3^j \\ u_1^k \\ u_2^k \\ \theta_3^k \end{Bmatrix} \qquad (4.26)$$

In what follows we prove that the stiffness matrix for an element of a planar beam or frame is symmetric. Moreover, we establish the relations between the stiffness coefficients for an element of a planar beam or frame by considering its equilibrium.

Applying the Betti-Maxwell reciprocal theorem[2] to an element subjected to the nodal actions shown in Fig. 4.4a and d, we obtain

$$K_{41} = K_{14} \tag{4.27}$$

Moreover, applying the Betti-Maxwell reciprocal theorem to an element subjected to the nodal actions shown in Fig. 4.4b and e, we get

$$K_{25} = K_{52} \tag{4.28}$$

Furthermore, applying the Betti-Maxwell reciprocal theorem to an element subjected to the nodal actions shown in Fig. 4.4b and c, we have

$$K_{23} = K_{32} \tag{4.29}$$

Thus, it is apparent that the local stiffness matrix of a planar beam or a planar frame is symmetric. That is,

$$[K] = [K]^T \tag{4.30}$$

Consider an element of a planar beam or frame subjected to the set of nodal actions shown in Fig. 4.4. From the equilibrium of this element we have

From Fig. 4.4a: $\qquad\qquad K_{41} = -K_{11}$

From Fig. 4.4b: $\quad K_{22} = -K_{52} \qquad K_{32} = -K_{62} - LK_{52}$

From Fig. 4.4c: $\quad K_{23} = -K_{53} \qquad K_{33} = -K_{63} - LK_{53}$

From Fig. 4.4d: $\qquad\qquad K_{14} = -K_{44}$

From Fig. 4.4e: $\quad K_{25} = -K_{55} \qquad K_{35} = -K_{65} - LK_{55}$

From Fig. 4.4f: $\quad K_{26} = -K_{56} \qquad K_{36} = -K_{66} - LK_{56}$

$$\tag{4.31}$$

Using relations (4.30) and (4.31) the nonvanishing stiffness coefficients for a general planar element can be expressed in terms of four independent stiffness coefficients. We choose K_{44}, K_{55}, K_{56}, and K_{66} as the independent stiffness coefficients for a general planar element. In this case the local stiffness matrix for a general planar element can be written as

$$[K] = \begin{bmatrix} K_{44} & 0 & 0 & -K_{44} & 0 & 0 \\ 0 & K_{55} & (K_{56} + LK_{55}) & 0 & -K_{55} & -K_{56} \\ 0 & (K_{56} + LK_{55}) & (K_{66} + 2LK_{56} + L^2 K_{55}) & 0 & (-K_{56} - LK_{55}) & (-K_{66} - LK_{56}) \\ -K_{44} & 0 & 0 & K_{44} & 0 & 0 \\ 0 & -K_{55} & (-K_{56} - LK_{55}) & 0 & K_{55} & K_{56} \\ 0 & -K_{56} & (-K_{66} - LK_{56}) & 0 & K_{56} & K_{66} \end{bmatrix}$$

$$\tag{4.32}$$

Moreover, the stiffness matrix for an axial deformation element of a planar frame is given as

$$[K] = K_{44} \begin{bmatrix} 1 & 0 & 0 & -1 & 0 & 0 \\ 0 & 0 & 0 & 0 & 0 & 0 \\ 0 & 0 & 0 & 0 & 0 & 0 \\ -1 & 0 & 0 & 1 & 0 & 0 \\ 0 & 0 & 0 & 0 & 0 & 0 \\ 0 & 0 & 0 & 0 & 0 & 0 \end{bmatrix} \qquad (4.33)$$

Local stiffness matrix for an element of a space beam or frame. The axial components of the nodal forces of an element induce only axial translation of its cross sections. Thus, the axial components of nodal forces of an element are related only to the axial components of its nodal translations. Moreover, the torsional components of nodal moments of an element induce only twisting of its cross sections. Thus, the torsional components of nodal moments of an element are related only to the twisting components of its nodal rotations. Furthermore, the transverse components of nodal forces F_2^j and F_2^k and the components of nodal moments M_3^j and M_3^k of an element do not induce components of translation u_1 and u_3 and components of rotation θ_1 and θ_2 of its cross sections. Thus, the transverse components of nodal forces F_2^j and F_2^k and the components of nodal moments M_3^j and M_3^k of an element are not related to the components of its nodal displacements u_1^j, u_3^j, θ_1^j, θ_2^j, u_1^k, u_3^k, θ_1^k, and θ_2^k. Similarly, the transverse components of the nodal forces F_3^j and F_3^k and the components of nodal moments M_2^j and M_2^k of an element are not related to the components of its nodal displacements u_1^j, u_2^j, θ_1^j, θ_3^j, u_1^k, u_2^k, θ_1^k, and θ_3^k. Consequently, a number of the stiffness coefficients of an element of a space beam or a space frame vanish. The physical significance of the nonvanishing stiffness coefficients of a general space element is shown in Fig. 4.5. Referring to this figure and using the Betti-Maxwell reciprocal theorem,[2] it can be shown that the local stiffness matrix of a general space element is symmetric. Moreover, by considering the equilibrium of a general space element subjected to the sets of nodal actions shown in Fig. 4.5 we can obtain 18 independent relations among the stiffness coefficients of a general space element. Using these relations and the symmetry of its stiffness matrix we can express the nonvanishing stiffness coefficients for a general space element in terms of eight independent stiffness coefficients. If we choose K_{77}, K_{88}, $K_{12,8}$, K_{99}, $K_{11,9}$, $K_{10,10}$, $K_{11,11}$, and $K_{12,12}$ as its independent stiffness coefficients, the stiffness matrix for a general space element is given by relation (4.34):

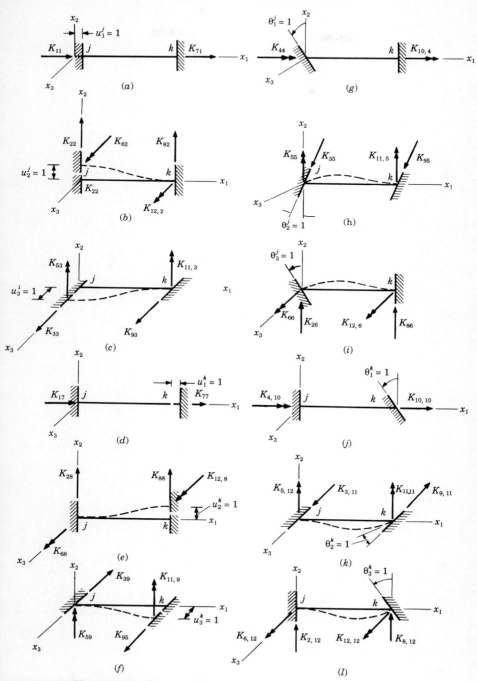

Figure 4.5 Physical significance of the local stiffness coefficients of a general space element.

$$
[K] =
\begin{bmatrix}
K_{77} & 0 & 0 & 0 & 0 & 0 & -K_{77} & 0 & 0 & 0 & 0 & 0 \\
0 & K_{88} & 0 & 0 & 0 & K_{12,8}+LK_{88} & 0 & -K_{88} & 0 & 0 & 0 & -K_{12,8} \\
0 & 0 & K_{99} & 0 & K_{11,9}-LK_{99} & 0 & 0 & 0 & -K_{99} & 0 & -K_{11,9} & 0 \\
0 & 0 & 0 & K_{10,10} & 0 & 0 & 0 & 0 & 0 & -K_{10,10} & 0 & 0 \\
0 & 0 & K_{11,9}-LK_{99} & 0 & k_{11,11}-2LK_{11,9}+L^2K_{99} & 0 & 0 & 0 & -K_{11,9}+LK_{99} & 0 & -K_{11,11}+LK_{11,9} & 0 \\
0 & K_{12,8}+LK_{88} & 0 & 0 & 0 & K_{12,12}+2LK_{12,8}+L^2K_{88} & 0 & -K_{12,8}-LK_{88} & 0 & 0 & 0 & -K_{12,12}-LK_{12,8} \\
-K_{77} & 0 & 0 & 0 & 0 & 0 & K_{77} & 0 & 0 & 0 & 0 & 0 \\
0 & -K_{88} & 0 & 0 & 0 & -K_{12,8}-LK_{88} & 0 & K_{88} & 0 & 0 & 0 & K_{12,8} \\
0 & 0 & -K_{99} & 0 & -K_{11,9}+LK_{99} & 0 & 0 & 0 & K_{99} & 0 & K_{11,9} & 0 \\
0 & 0 & 0 & -K_{10,10} & 0 & 0 & 0 & 0 & 0 & K_{10,10} & 0 & 0 \\
0 & 0 & -K_{11,9} & 0 & -K_{11,11}+LK_{11,9} & 0 & 0 & 0 & K_{11,9} & 0 & K_{11,11} & 0 \\
0 & -K_{12,8} & 0 & 0 & 0 & -K_{12,12}-LK_{12,8} & 0 & K_{12,8} & 0 & 0 & 0 & K_{12,12}
\end{bmatrix}
\qquad (4.34)
$$

Moreover, the stiffness matrix for an axial deformation element of a space frame is equal to

$$[K] = K_{77} \begin{bmatrix} 1 & 0 & 0 & 0 & 0 & 0 & -1 & 0 & 0 & 0 & 0 & 0 \\ 0 & 0 & 0 & 0 & 0 & 0 & 0 & 0 & 0 & 0 & 0 & 0 \\ 0 & 0 & 0 & 0 & 0 & 0 & 0 & 0 & 0 & 0 & 0 & 0 \\ 0 & 0 & 0 & 0 & 0 & 0 & 0 & 0 & 0 & 0 & 0 & 0 \\ 0 & 0 & 0 & 0 & 0 & 0 & 0 & 0 & 0 & 0 & 0 & 0 \\ 0 & 0 & 0 & 0 & 0 & 0 & 0 & 0 & 0 & 0 & 0 & 0 \\ -1 & 0 & 0 & 0 & 0 & 0 & 1 & 0 & 0 & 0 & 0 & 0 \\ 0 & 0 & 0 & 0 & 0 & 0 & 0 & 0 & 0 & 0 & 0 & 0 \\ 0 & 0 & 0 & 0 & 0 & 0 & 0 & 0 & 0 & 0 & 0 & 0 \\ 0 & 0 & 0 & 0 & 0 & 0 & 0 & 0 & 0 & 0 & 0 & 0 \\ 0 & 0 & 0 & 0 & 0 & 0 & 0 & 0 & 0 & 0 & 0 & 0 \\ 0 & 0 & 0 & 0 & 0 & 0 & 0 & 0 & 0 & 0 & 0 & 0 \end{bmatrix} \qquad (4.35)$$

Local stiffness matrix for an element of a grid. Consider a grid in the $\bar{x}_1\bar{x}_2$ plane subjected at its nodes to external forces acting in the direction normal to its plane and to external moments whose vector is in the plane of the grid (see Fig. 4.6). An element of this grid is subjected only to end actions F_3^j, F_3^k, M_1^j, M_1^k, M_2^j, M_2^k. Thus, its stiffness equations (4.8) reduce to

$$[A^E] = \begin{bmatrix} F_3^j \\ M_1^j \\ M_2^j \\ F_3^k \\ M_1^k \\ M_2^k \end{bmatrix} = \begin{bmatrix} K_{11} & K_{12} & K_{13} & K_{14} & K_{15} & K_{16} \\ K_{21} & K_{22} & K_{23} & K_{24} & K_{25} & K_{26} \\ K_{31} & K_{32} & K_{33} & K_{34} & K_{35} & K_{36} \\ K_{41} & K_{42} & K_{43} & K_{44} & K_{45} & K_{46} \\ K_{51} & K_{52} & K_{53} & K_{54} & K_{55} & K_{56} \\ K_{61} & K_{62} & K_{63} & K_{64} & K_{65} & K_{66} \end{bmatrix} \begin{bmatrix} u_3^j \\ \theta_1^j \\ \theta_2^j \\ u_3^k \\ \theta_1^k \\ \theta_2^k \end{bmatrix} \qquad (4.36)$$

Referring to relation (4.34) and noting that for the equilibrium of an element $M_1^k = -M_1^j$, the local stiffness matrix $[K]$ for an element of a

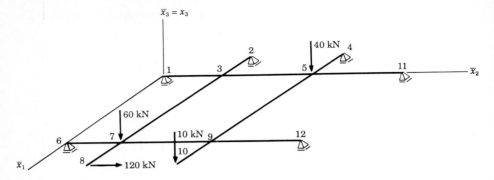

Figure 4.6 Grid.

grid is

$$[K] = \begin{bmatrix} K_{44} & 0 & (K_{46}+LK_{44}) & -K_{44} & 0 & -K_{46} \\ 0 & K_{55} & 0 & 0 & -K_{55} & 0 \\ (K_{46}+LK_{44}) & 0 & (K_{66}+2LK_{46}+L^2K_{44}) & -(K_{46}+LK_{44}) & 0 & (-K_{66}-LK_{46}) \\ -K_{44}^i & 0 & (-K_{46}-LK_{44}) & K_{44} & 0 & K_{46} \\ 0 & -K_{55} & 0 & 0 & K_{55} & 0 \\ -K_{46} & 0 & (-K_{66}-LK_{46}) & K_{46} & 0 & K_{66} \end{bmatrix}$$

$$(4.37)$$

4.4 Computation of the Exact Response of an Element

In this section we establish the exact response of an element. That is, we compute the following:

1. The exact stiffness matrix for an element
2. The exact matrix of fixed-end actions $\{A^R\}$ of an element. This is the matrix of nodal actions of the element subjected to the given loads acting along its length with its ends fixed.

We accomplish this by adhering to the following steps:

Step 1. We solve the boundary-value problems described in Sec. 2.5, and we find the basic components of displacements for the element as functions of its axial coordinate involving the corresponding components of its nodal displacements and the given loads acting along its length. For example, for a general space element we get

$$u_1(x_1) = u_1^E(x_1, u_1^j, u_1^k) + u_1^R(x_1, L_1)$$

$$u_2(x_1) = u_2^E(x_1, u_2^j, u_2^k, \theta_3^j, \theta_3^k) + u_2^R(x_1, L_2)$$

$$u_3(x_1) = u_3^E(x_1, u_3^j, u_3^k, \theta_2^j, \theta_2^k) + u_3^R(x_1, L_3)$$

$$\theta_1(x_1) = \theta_1^E(x_1, \theta_1^j, \theta_1^k) + \theta_1^R(x_1, L_4)$$

(4.38)

where L_1, L_2, L_3, or L_4 stand for the components of the external loads which produce the components of displacement u_1, u_2, u_3, or θ_1, respectively.

In relations (4.38) the effect of the nodal displacements of the element and the effect of the loads acting along its length are separated. The first term on the right-hand side of each of these relations is the solution of the homogeneous part of the differential equation (2.40), (2.53a), (2.53b), or (2.67), respectively, which when evaluated at the ends of the element gives the corresponding component of its nodal displacement. It represents the component of displacement of the element subjected only to its corresponding components of nodal displacements and thus it vanishes when the ends of the element are fixed. The second term on the right-hand side of each of relations (4.38) is a particular solution of the differential equation (2.40), (2.53a), (2.53b), or (2.67), respectively, which when evaluated at the ends of the element is equal to zero. It represents the displacement of the element subjected to the given loads with its ends fixed and thus it vanishes when the element is not subjected to given loads along its length. Moreover, the derivatives of u_2^R and u_3^R are equal to zero at the ends of the element. For example, as shown in Fig. 4.7, $u_3^E(x_1)$ is the component of translation of the element when subjected only to its nodal displacements u_3^j, θ_2^j, u_3^k, and θ_2^k, while $u_3^R(x_1)$ is the component of translation of the same element when subjected to the given loads with its ends fixed.

In general, relations (4.38) can be written as

$$\{u\} = [\phi]\{D\} + \{u^R\}$$

(4.39)

where $\{u\}$ = matrix of basic components of displacements of element; for an axial deformation element $\{u\} = u_1(x_1)$, for a general planar element $\{u\}^T = [u_1(x_1), u_2(x_1)]$, for a general space element $\{u\}^T = [u_1(x_1), u_2(x_1), u_3(x_1), \theta_1(x_1)]$ (see Sec. 2.5)

$[\phi]$ = matrix of shape functions for element

$\{D\}$ = local matrix of nodal displacements [see relations (4.1c), (4.2c), and (4.3c)]

$\{u^R\}$ = matrix of basic components of displacements of element subjected to given loads with its ends fixed

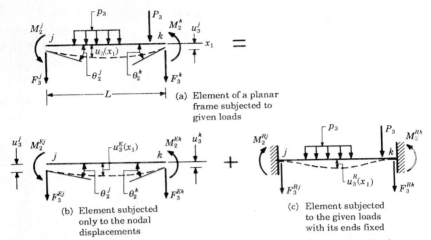

(a) Element of a planar frame subjected to given loads

(b) Element subjected only to the nodal displacements

(c) Element subjected to the given loads with its ends fixed

Figure 4.7 Superposition of the element subjected only to its nodal displacements and the element subjected to the given loads with its ends fixed.

Step 2. We establish the exact response of the element. In order to accomplish this, we substitute each one of relations (4.38) into the corresponding action-displacement relation and evaluate the resulting equations at $x_1 = 0$ and $x_1 = L$. Thus for a general space element made from an isotropic linearly elastic material, referring to relation (1.2), we obtain

$$F_1^j = -N(0) = F_1^{Ej}(u_1^j, u_1^k) + F_1^{Rj}(L_1)$$

$$F_2^j = -Q_2(0) = F_2^{Ej}(u_2^j, u_2^k, \theta_3^j, \theta_3^k) + F_2^{Rj}(L_2)$$

$$F_3^j = -Q_3(0) = F_3^{Ej}(u_3^j, u_3^k, \theta_2^j, \theta_2^k) + F_3^{Rj}(L_3)$$

$$M_1^j = -M_1(0) = M_1^{Ej}(\theta_1^j, \theta_1^k) + M_1^{Rj}(L_4)$$

$$M_2^j = -M_2(0) = M_2^{Ej}(u_3^j, u_3^k, \theta_2^j, \theta_2^k) + M_2^{Rj}(L_3)$$

$$M_3^j = -M_3(0) = M_3^{Ej}(u_2^j, u_2^k, \theta_3^j, \theta_3^k) + M_3^{Rj}(L_2)$$

$$F_1^k = N(L) = F_1^{Ek}(u_1^j, u_1^k) + F_1^{Rk}(L_1)$$

$$F_2^k = Q_2(L) = F_2^{Ek}(u_2^j, u_2^k, \theta_3^j, \theta_3^k) + F_2^{Rk}(L_2)$$

$$F_3^k = Q_3(L) = F_3^{Ek}(u_3^j, u_3^k, \theta_2^j, \theta_2^k) + F_3^{Rk}(L_3)$$

$$M_1^k = M_1(L) = M_1^{Ek}(\theta_1^j, \theta_1^k) + M_1^{Rk}(L_4)$$

$$M_2^k = M_2(L) = M_2^{Ek}(u_3^j, u_3^k, \theta_2^j, \theta_2^k) + M_2^{Rk}(L_3)$$

$$M_3^k = M_3(L) = M_3^{Ek}(u_2^j, u_2^k, \theta_3^j, \theta_3^k) + M_3^{Rk}(L_2) \qquad (4.40)$$

or $$\{A\} = \{A^E\} + \{A^R\} = [K]\{D\} + \{A^R\} \qquad (4.40a)$$

where L_1, L_2, L_3, and L_4 are the components of the given loads which produce the components of displacement u_1, u_2, u_3, and θ_1, respectively. $[K]$ is the stiffness matrix for the element. Relations (4.40) express each component of nodal action of a general space element as the sum of

1. The corresponding component of the nodal action of the element when subjected only to its nodal displacements. (This is the term with the superscript E.) (See Fig. 4.7b.)

2. The corresponding component of the fixed-end action of the element, that is, the nodal action of the element when subjected to the given loads with its ends fixed. (This is the term with the superscript R.) (See Fig. 4.7c.)

The components of nodal actions of an element obtained by substituting into Eqs. (4.40) the given loads acting on it and a set of nodal displacements have the following attributes:

1. They are in equilibrium with the given loads.

2. They are based on the stress-strain relations for the material from which the element is made.

The approach presented in this section can be used to establish the response of elements for which the exact solution of the boundary-value problems described in Sec. 2.5 can be easily found (for example, elements of constant cross section subjected to concentrated actions and to uniformly distributed or linearly varying loads along their length, or elements of varying cross section of simple geometry which are not subjected to loads along their length).

In most commercially available programs for analyzing groups of framed structures the exact stiffness matrix for elements of variable cross section is computed from its flexibility matrix using relation (5.93). The flexibility coefficients for an element of variable cross section are established using exact formulas derived either by solving the strong form of the boundary-value problems described in Sec. 2.5 (see Sec. 5.6) or by employing a procedure based on the principle of virtual work (see Sec. 19.3).

If it is difficult to find the exact stiffness equations for an element and/or its exact matrix of fixed-end actions, $\{A^R\}$ approximate stiffness equations and/or approximate matrices of fixed-end actions are computed using one of the following methods:

1. The element is approximated by a model of simplified geometry

and loading and its exact stiffness matrix and its exact matrix of fixed-end actions are computed. They represent approximations to the stiffness matrix and to the matrix of fixed-end actions of the element, respectively (see Secs. 12.3 and 12.4).

2. The finite-element method is used. This method can also be used to find the approximate response of two- and three-dimensional elements. In Sec. 19.2 we use the finite-element method in conjunction with the principle of virtual work to establish the approximate response (approximate stiffness matrix and matrix of fixed-end actions) of elements of framed structures.

4.4.1 Computation of the stiffness matrix for elements of constant cross section

Element of a planar or space truss. Consider an element of a planar or space truss of length L and constant cross section of area A. The element is made of an isotropic linearly elastic material of modulus of elasticity E and is subjected only to its basic nodal displacements u_1^j and u_1^k. We establish its stiffness matrix by adhering to the following steps.

Step 1. We solve the boundary-value problem described in Sec. 2.5.1 and find the axial component of translation $u_1^E(x_1)$ of the element as a function of its axial coordinate involving the axial components of its nodal displacements u_1^j and u_1^k. For the element under consideration the displacement equation of equilibrium (2.40) reduces to

$$EA\frac{d^2 u_1^E}{dx_1^2} = 0 \tag{4.41}$$

Moreover, the boundary conditions for the problem are

$$u_1^E(0) = u_1^j \qquad u_1^E(L) = u_1^k \tag{4.42}$$

The solution of Eq. (4.41) is

$$u_1^E(x_1) = c_{10} + c_{11}x_1 = [1 \quad x_1]\begin{Bmatrix} c_{10} \\ c_{11} \end{Bmatrix} \tag{4.43}$$

The constants c_{10} and c_{11} are evaluated by requiring that $u_1^E(x_1)$ satisfies the boundary conditions (4.42). That is,

$$\{D\} = \begin{Bmatrix} u_1^E(0) \\ u_1^E(L) \end{Bmatrix} = \begin{Bmatrix} u_1^j \\ u_1^k \end{Bmatrix} = \begin{bmatrix} 1 & 0 \\ 1 & L \end{bmatrix}\begin{Bmatrix} c_{10} \\ c_{11} \end{Bmatrix} \tag{4.44}$$

From relation (4.44) we get

$$\begin{Bmatrix} c_{10} \\ c_{11} \end{Bmatrix} = \begin{bmatrix} 1 & 0 \\ 1 & L \end{bmatrix}^{-1} \begin{Bmatrix} u_1^j \\ u_1^k \end{Bmatrix} = \frac{1}{L} \begin{bmatrix} L & 0 \\ -1 & 1 \end{bmatrix} \begin{Bmatrix} u_1^j \\ u_1^k \end{Bmatrix} \tag{4.45}$$

Substituting relations (4.45) into (4.43) we get

$$u_1^E(x_1) = \frac{1}{L} \begin{bmatrix} 1 & x_1 \end{bmatrix} \begin{bmatrix} L & 0 \\ -1 & 1 \end{bmatrix} \begin{Bmatrix} u_1^j \\ u_1^k \end{Bmatrix} = \begin{bmatrix} \phi_j & \phi_k \end{bmatrix} \begin{Bmatrix} u_1^j \\ u_1^k \end{Bmatrix}$$

or
$$u_1^E(x_1) = [\phi]\{D\} \tag{4.46}$$

$[\phi]$ is called the *matrix of shape functions for the element*; it is defined as follows:

$$[\phi] = [\phi_j \quad \phi_k] \tag{4.47a}$$

where ϕ_j and ϕ_k are called the *shape functions for the element*. They are equal to

$$\phi_j = 1 - \frac{x_1}{L} \tag{4.47b}$$

$$\phi_k = \frac{x_1}{L} \tag{4.47c}$$

We have shown that the exact solution of the boundary-value problem for computing the axial component of translation $u_1(x_1)$ for an element of constant cross section subjected only to nodal displacements u_1^j and u_1^k is given by relation (4.46).

Step 2. We establish the stiffness matrix for the element under consideration. Substituting relation (4.46) into the action-displacement relation (2.36) we get

$$\begin{Bmatrix} F_1^j \\ F_1^k \end{Bmatrix} = \begin{Bmatrix} -N(0) \\ N(L) \end{Bmatrix} = \begin{Bmatrix} -\left[EA \dfrac{du_1^E}{dx_1} \right]_{x_1=0} \\ \left[EA \dfrac{du_1^E}{dx_1} \right]_{x_1=L} \end{Bmatrix}$$

$$= EA \frac{d}{dx_1} \begin{bmatrix} -\phi_j & -\phi_k \\ \phi_j & \phi_k \end{bmatrix} \begin{Bmatrix} u_1^j \\ u_1^k \end{Bmatrix}$$

$$= \frac{EA}{L} \begin{bmatrix} 1 & -1 \\ -1 & 1 \end{bmatrix} \begin{Bmatrix} u_1^j \\ u_1^k \end{Bmatrix} \tag{4.48}$$

or
$$[A^E] = [K]\{D\} \tag{4.49}$$

where
$$[K] = \frac{EA}{L} \begin{bmatrix} 1 & -1 \\ -1 & 1 \end{bmatrix} \tag{4.50}$$

Relation (4.50) is the stiffness matrix for an element of a truss of length L and constant cross section of area A made from an isotropic linearly elastic material of modulus of elasticity E.

Element of a planar beam or frame.   Consider a general planar element (in the $x_1 x_2$ plane) of length L and constant cross section of area A and moment of inertia I_3 about its x_3 axis. The element is made of an isotropic linearly elastic material of modulus of elasticity E and is subjected only to its nodal displacements u_1^j, u_2^j, θ_3^j, u_1^k, u_2^k, and θ_3^k. We establish its stiffness matrix by adhering to the following steps.

Step 1.   We solve the boundary-value problems described in Secs. 2.5.1 and 2.5.2 and find the basic components of displacement $[u_1^E(x_1)$ and $u_2^E(x_1)]$ of the element as functions of its axial coordinate involving the corresponding components of its nodal displacements. For the element under consideration the displacement equations of equilibrium (2.40) and (2.53a) reduce to

$$EA \frac{d^2 u_1^E}{dx_1^2} = 0 \tag{4.51a}$$

and

$$EI_3 \frac{d^4 u_2^E}{dx_1^4} = 0 \tag{4.51b}$$

Moreover the boundary conditions for the problem are

$$
\begin{aligned}
u_1^E(0) &= u_1^j \\
u_2^E(0) &= u_2^j \\
\theta_3^E(0) &= \theta_3^j \\
u_1^E(L) &= u_1^k \\
u_2^E(L) &= u_2^k \\
\theta_3^E(L) &= \theta_3^k
\end{aligned}
\tag{4.52}
$$

The solutions of Eqs. (4.51) are

$$u_1^E(x_1) = [1 \quad x_1] \begin{Bmatrix} c_{10} \\ c_{11} \end{Bmatrix} \tag{4.53a}$$

$$u_2^E(x_1) = [1 \quad x_1 \quad x_1^2 \quad x_1^3] \begin{Bmatrix} c_{20} \\ c_{21} \\ c_{22} \\ c_{23} \end{Bmatrix} \tag{4.53b}$$

Moreover substituting relation (4.53b) into (2.47) we have

$$\theta_3^E(x_1) = \frac{du_2^E}{dx_1} = [0 \quad 1 \quad 2x_1 \quad 3x_1^2] \begin{Bmatrix} c_{20} \\ c_{21} \\ c_{22} \\ c_{23} \end{Bmatrix} \tag{4.54}$$

The constants c_{20}, c_{21}, c_{22}, and c_{23} are evaluated by requiring that the solution (4.53) satisfies the boundary conditions (4.52) for the problem. That is,

$$\left\{ \begin{array}{c} u_1^E(0) \\ u_1^E(L) \end{array} \right\} = \left\{ \begin{array}{c} u_1^j \\ u_1^k \end{array} \right\} = \begin{bmatrix} 1 & 0 \\ 1 & L \end{bmatrix} \left\{ \begin{array}{c} c_{10} \\ c_{11} \end{array} \right\} \tag{4.55a}$$

$$\left\{ \begin{array}{c} u_2^E(0) \\ \theta_3^E(0) \\ u_2^E(L) \\ \theta_3^E(L) \end{array} \right\} = \left\{ \begin{array}{c} u_2^j \\ \theta_3^j \\ u_2^k \\ \theta_3^k \end{array} \right\} = \begin{bmatrix} 1 & 0 & 0 & 0 \\ 0 & 1 & 0 & 0 \\ 1 & L & L^2 & L^3 \\ 0 & 1 & 2L & 3L^2 \end{bmatrix} \left\{ \begin{array}{c} c_{20} \\ c_{21} \\ c_{22} \\ c_{23} \end{array} \right\} \tag{4.55b}$$

From relations (4.54) we get

$$\left\{ \begin{array}{c} c_{10} \\ c_{11} \end{array} \right\} = \frac{1}{L} \begin{bmatrix} L & 0 \\ -1 & 1 \end{bmatrix} \left\{ \begin{array}{c} u_1^j \\ u_1^k \end{array} \right\} \tag{4.56a}$$

$$\left\{ \begin{array}{c} c_{20} \\ c_{21} \\ c_{22} \\ c_{23} \end{array} \right\} = \begin{bmatrix} 1 & 0 & 0 & 0 \\ 0 & 1 & 0 & 0 \\ 1 & L & L^2 & L^3 \\ 0 & 1 & 2L & 3L^3 \end{bmatrix}^{-1} \left\{ \begin{array}{c} u_2^j \\ \theta_3^j \\ u_2^k \\ \theta_3^k \end{array} \right\}$$

$$= \frac{1}{L^3} \begin{bmatrix} L^3 & 0 & 0 & 0 \\ 0 & L^3 & 0 & 0 \\ -3L & -2L^2 & 3L & -2L^2 \\ 2 & L & -2 & L \end{bmatrix} \left\{ \begin{array}{c} u_2^j \\ \theta_3^j \\ u_2^k \\ \theta_3^k \end{array} \right\} \tag{4.56b}$$

Substituting relations (4.56) into (4.53) we get

$$u_1^E(x_1) = \begin{bmatrix} 1 & x_1 \end{bmatrix} \frac{1}{L} \begin{bmatrix} L & 0 \\ -1 & 1 \end{bmatrix} \left\{ \begin{array}{c} u_1^j \\ u_1^k \end{array} \right\} = \begin{bmatrix} \phi_j & \phi_k \end{bmatrix} \left\{ \begin{array}{c} u_1^j \\ u_1^k \end{array} \right\} \tag{4.57a}$$

$$u_2^E(x_1) = \begin{bmatrix} 1, x_1 & x_1^2 & x_1^3 \end{bmatrix} \begin{bmatrix} L^3 & 0 & 0 & 0 \\ 0 & L^3 & 0 & 0 \\ -3L & -2L^2 & 3L & -2L^2 \\ 2 & L & -2 & L \end{bmatrix} \left\{ \begin{array}{c} u_2^j \\ \theta_3^j \\ u_2^k \\ \theta_3^k \end{array} \right\}$$

$$= \begin{bmatrix} \phi_{2j}^u & \phi_{2j}^\theta & \phi_{2k}^u & \phi_{2k}^\theta \end{bmatrix} \left\{ \begin{array}{c} u_2^j \\ \theta_3^j \\ u_2^k \\ \theta_3^k \end{array} \right\} \tag{4.57b}$$

where
$$\phi_j = 1 - \frac{x_1}{L}$$

$$\phi_k = \frac{x_1}{L}$$

$$\phi_{2j}^u = 1 - 3\left(\frac{x_1}{L}\right)^2 + 2\left(\frac{x_1}{L}\right)^3$$

$$\phi_{2j}^\theta = x_1\left(1 - \frac{x_1}{L}\right)^2 \qquad (4.58)$$

$$\phi_{2k}^u = 3\left(\frac{x_1}{L}\right)^2 - 2\left(\frac{x_1}{L}\right)^3$$

$$\phi_{2k}^\theta = x_1\left[-\frac{x_1}{L} + \left(\frac{x_1}{L}\right)^2\right]$$

are called the shape functions for the element. Combining relations (4.57a) and (4.57b) we get

$$\{u^E\} = \left\{\begin{matrix} u_1^E(x_1) \\ u_2^E(x_1) \end{matrix}\right\} = \begin{bmatrix} \phi_j & 0 & 0 & \phi_k & 0 & 0 \\ 0 & \phi_{2j}^u & \phi_{2j}^\theta & 0 & \phi_{2k}^u & \phi_{2k}^\theta \end{bmatrix} \left\{\begin{matrix} u_1^j \\ u_2^j \\ \theta_3^j \\ u_1^k \\ u_2^k \\ \theta_3^k \end{matrix}\right\}$$

or
$$\{u^E\} = [\phi]\{D\} \qquad (4.59)$$

where
$$[\phi] = \begin{bmatrix} \phi_j & 0 & 0 & \phi_k & 0 & 0 \\ 0 & \phi_{2j}^u & \phi_{2j}^\theta & 0 & \phi_{2k}^u & \phi_{2k}^\theta \end{bmatrix} \qquad (4.59a)$$

is called the matrix of shape functions for the element. We have shown that the exact solution of the boundary-value problem for computing the components of translation $u_1^E(x_1)$ and $u_2^E(x_1)$ for an element of constant cross section subjected only to nodal displacements u_1^j, u_2^j, θ_3^j, u_1^k, u_2^k, and θ_3^k is given by relation (4.59).

The matrix $[\phi]$ for an axial deformation element of a frame in the x_1x_2 plane is obtained from that for a general planar element (4.59a) by setting the shape functions ϕ_{2j}^u, ϕ_{2j}^θ, ϕ_{2k}^u, and ϕ_{2k}^θ equal to zero.

Step 2. We establish the stiffness matrix for the element under consideration. Referring to relations (1.2) and to the action-displacement relations (2.36), (2.50a), and (2.49b), we have

$$\{A^E\} = \begin{Bmatrix} F_1^j \\ F_2^j \\ M_3^j \\ F_1^k \\ F_2^k \\ M_3^k \end{Bmatrix} = \begin{Bmatrix} -N(0) \\ -Q_2(0) \\ -M_3(0) \\ N(L) \\ Q_2(L) \\ M_3(L) \end{Bmatrix} = \begin{Bmatrix} \left(-EA\dfrac{du_1^E}{dx_1}\right)_{x_1=0} \\[2mm] \left(EI_3\dfrac{d^3 u_2^E}{dx_1^3}\right)_{x_1=0} \\[2mm] \left(-EI_3\dfrac{d^2 u_2^E}{dx_1^2}\right)_{x_1=0} \\[2mm] \left(EA\dfrac{du_1^E}{dx_1}\right)_{x_1=L} \\[2mm] \left(-EI_3\dfrac{d^3 u_2^E}{dx_1^3}\right)_{x_1=L} \\[2mm] \left(EI_3\dfrac{d^2 u_2^E}{dx_1^2}\right)_{x_1=L} \end{Bmatrix}$$

$$= \begin{bmatrix} \left(-EA\dfrac{d}{dx_1}\right)_{x_1=0} & 0 \\[2mm] 0 & \left(EI_3\dfrac{d^3}{dx_1^3}\right)_{x_1=0} \\[2mm] 0 & \left(-EI_3\dfrac{d^2}{dx_1^2}\right)_{x_1=0} \\[2mm] \left(EA\dfrac{d}{dx_1}\right)_{x_1=L} & 0 \\[2mm] 0 & \left(-EI_3\dfrac{d^3}{dx_1^3}\right)_{x_1=L} \\[2mm] 0 & \left(EI_3\dfrac{d^2}{dx_1^2}\right)_{x_1=L} \end{bmatrix} \begin{Bmatrix} u_1^E(x_1) \\ u_2^E(x_1) \end{Bmatrix}$$

$$= [\mathscr{D}]\{u^E\} \tag{4.60}$$

Substituting relation (4.59) into (4.60) and using (4.58) we get

$$\{A^E\} = [K]\{D\} \tag{4.61}$$

where the stiffness matrix $[K]$ for a general planar element of constant cross section is

$$[K] = [\mathscr{D}]\{\phi\} = \begin{bmatrix} \dfrac{AE}{L} & 0 & 0 & -\dfrac{AE}{L} & 0 & 0 \\[2ex] 0 & \dfrac{12EI_3}{L^3} & \dfrac{6EI_3}{L^2} & 0 & -\dfrac{12EI_3}{L^3} & \dfrac{6EI_3}{L^2} \\[2ex] 0 & \dfrac{6EI_3}{L^2} & \dfrac{4EI_3}{L} & 0 & -\dfrac{6EI_3}{L^2} & \dfrac{2EI_3}{L} \\[2ex] -\dfrac{AE}{L} & 0 & 0 & \dfrac{AE}{L} & 0 & 0 \\[2ex] 0 & -\dfrac{12EI_3}{L^3} & -\dfrac{6EI_3}{L^2} & 0 & \dfrac{12EI_3}{L^3} & -\dfrac{6EI_3}{L^2} \\[2ex] 0 & \dfrac{6EI_3}{L^2} & \dfrac{2EI_3}{L} & 0 & -\dfrac{6EI_3}{L^2} & \dfrac{4EI_3}{L} \end{bmatrix}$$

$$(4.62)$$

where L = length of element

A = area of cross section of element

I_3 = moment of inertia of cross section of element about its x_3 local axis

E = modulus of elasticity of material from which element is made

Element of a space beam or frame.   Referring to Secs. 2.5.1 and 2.5.3, we see that the boundary-value problem for computing the twisting component of rotation $\theta_1(x_1)$ of an element of a framed structure is analogous to the boundary-value problem for computing its axial component of translation $u_1(x_1)$. Thus referring to relation (4.46) we may conclude that the solution of the boundary-value problem for computing the torsional component of rotation of an element of constant cross section subjected only to nodal components of rotation θ_1^j and θ_1^k is given as

$$\theta_1^E(x_1) = [\phi]\begin{Bmatrix} \theta_1^j \\ \theta_1^k \end{Bmatrix} \qquad (4.63)$$

where the matrix $[\phi]$ is defined by relations (4.47). Moreover, following a procedure analogous to the one adhered to in the solution of the boundary-value problem for computing the component of translation $u_2(x_1)$, the solution of the boundary-value problem for computing the component of translation $u_3(x_1)$ is

$$u_3(x_1) = [\phi]\begin{Bmatrix} u_3^j \\ \theta_2^j \\ u_3^k \\ \theta_2^k \end{Bmatrix} \qquad (4.64)$$

where
$$[\phi] = [\phi^u_{3j} \quad \phi^\theta_{3j} \quad \phi^u_{3k} \quad \phi^\theta_{3k}] \tag{4.65}$$

and
$$\begin{aligned}
\phi^u_{3j} &= \phi^u_{2j} = \phi^u_j \\
\phi^\theta_{3j} &= -\phi^\theta_{2j} = \phi^\theta_j \\
\phi^u_{3k} &= \phi^u_{2k} = \phi^u_j \\
\phi^\theta_{3k} &= -\phi^\theta_{2k} = \phi^\theta_j
\end{aligned} \tag{4.66}$$

The shape functions ϕ^u_{2j}, ϕ^θ_{2j}, ϕ^u_{2k}, and ϕ^θ_{2k} are given by relations (4.58).

Referring to relations (4.46), (4.59), and (4.63) to (4.66) the solution of the boundary-value problem for computing the basic components of displacement of a general space element of constant cross section subjected only to nodal displacements u^j_1, u^j_2, u^j_3, θ^j_1, θ^j_2, θ^j_3, u^k_1, u^k_2, u^k_3, θ^k_1, θ^k_2, and θ^k_3 is given as

$$\{u^E\} = \begin{Bmatrix} u^E_1(x_1) \\ u^E_2(x_1) \\ u^E_3(x_1) \\ \theta^E_1(x_1) \end{Bmatrix} = [\phi] \begin{Bmatrix} u^j_1 \\ u^j_2 \\ u^j_3 \\ \theta^j_1 \\ \theta^j_2 \\ \theta^j_3 \\ u^k_1 \\ u^k_2 \\ u^k_3 \\ \theta^k_1 \\ \theta^k_2 \\ \theta^k_3 \end{Bmatrix} \tag{4.67}$$

where

$$[\phi] = \begin{bmatrix}
\phi_j & 0 & 0 & 0 & 0 & 0 & \phi_k & 0 & 0 & 0 & 0 & 0 \\
0 & \phi^u_j & 0 & 0 & 0 & -\phi^\theta_j & 0 & \phi^u_k & 0 & 0 & 0 & -\phi^\theta_k \\
0 & 0 & \phi^u_j & 0 & \phi^\theta_j & 0 & 0 & 0 & \phi^u_k & 0 & \phi^\theta_k & 0 \\
0 & 0 & 0 & \phi_j & 0 & 0 & 0 & 0 & 0 & \phi_k & 0 & 0
\end{bmatrix} \tag{4.68}$$

The shape functions ϕ_j and ϕ_k are given by relations (4.47), while the shape functions ϕ^u_j, ϕ^θ_j, ϕ^u_k, and ϕ^θ_k are given by relations (4.58).

The stiffness matrix for a general space element of constant cross section is obtained following a procedure analogous to that employed

$$[K] = \begin{bmatrix}
\dfrac{EA}{L} & 0 & 0 & 0 & 0 & 0 & -\dfrac{EA}{L} & 0 & 0 & 0 & 0 & 0 \\[6pt]
0 & \dfrac{12EI_3}{L^3} & 0 & 0 & 0 & \dfrac{6EI_3}{L^2} & 0 & -\dfrac{12EI_3}{L^3} & 0 & 0 & 0 & \dfrac{6EI_3}{L^2} \\[6pt]
0 & 0 & \dfrac{12EI_2}{L^3} & 0 & -\dfrac{6EI_2}{L^2} & 0 & 0 & 0 & -\dfrac{12EI_2}{L^3} & 0 & -\dfrac{6EI_2}{L^2} & 0 \\[6pt]
0 & 0 & 0 & \dfrac{GK}{L} & 0 & 0 & 0 & 0 & 0 & -\dfrac{GK}{L} & 0 & 0 \\[6pt]
0 & 0 & -\dfrac{6EI_2}{L^2} & 0 & \dfrac{4EI_2}{L} & 0 & 0 & 0 & \dfrac{6EI_2}{L^2} & 0 & \dfrac{2EI_2}{L} & 0 \\[6pt]
0 & \dfrac{6EI_3}{L^2} & 0 & 0 & 0 & \dfrac{4EI_3}{L} & 0 & -\dfrac{6EI_3}{L^2} & 0 & 0 & 0 & \dfrac{2EI_3}{L} \\[6pt]
-\dfrac{EA}{L} & 0 & 0 & 0 & 0 & 0 & \dfrac{EA}{L} & 0 & 0 & 0 & 0 & 0 \\[6pt]
0 & -\dfrac{12EI_3}{L^3} & 0 & 0 & 0 & -\dfrac{6EI_3}{L^2} & 0 & \dfrac{12EI_3}{L^3} & 0 & 0 & 0 & -\dfrac{6EI_3}{L^2} \\[6pt]
0 & 0 & -\dfrac{12EI_2}{L^3} & 0 & \dfrac{6EI_2}{L^2} & 0 & 0 & 0 & \dfrac{12EI_2}{L^3} & 0 & \dfrac{6EI_2}{L^2} & 0 \\[6pt]
0 & 0 & 0 & -\dfrac{GK}{L} & 0 & 0 & 0 & 0 & 0 & \dfrac{GK}{L} & 0 & 0 \\[6pt]
0 & 0 & -\dfrac{6EI_2}{L^2} & 0 & \dfrac{2EI_2}{L} & 0 & 0 & 0 & \dfrac{6EI_2}{L^2} & 0 & \dfrac{4EI_2}{L} & 0 \\[6pt]
0 & \dfrac{6EI_3}{L^2} & 0 & 0 & 0 & \dfrac{2EI_3}{L} & 0 & -\dfrac{6EI_3}{L^2} & 0 & 0 & 0 & \dfrac{4EI_3}{L}
\end{bmatrix} \tag{4.69}$$

for a general planar element. It is given by relation (4.69) (see p. 134). In this relation L is the length of the element, A is the cross-sectional area of the element, and I_2 and I_3 are the moments of inertia of the cross section of the element about its x_2 and x_3 local axes, respectively.

The stiffness matrix for an axial deformation element of a space frame is obtained from that of a general space element by setting its bending (EI_2 and EI_3) and torsional (KG) rigidities equal to zero.

4.4.2 Computation of the matrix of fixed-end actions for an element of constant cross section

Axial deformation element. Consider an axial deformation element of a framed structure of constant cross section of area A made of an isotropic linearly elastic material of modulus of elasticity E and subjected with its ends fixed to the following loads:

1. Uniformly distributed axial force $p_1(x_1)$ applied from $x_1 = c_1$ to $x_1 = L$.
2. Concentrated axial forces $P_1^{(i)}$ applied at $x_1 = a_{1i}$ $(i = 1, 2, \ldots, n_1)$
3. Uniform change (increase) of temperature ΔT_c
4. Uniform initial strain e_{11}^{IC}

The displacement equation of equilibrium for this element is given by relation (2.40). That is,

$$EA \frac{d^2 u_1}{dx_1^2} = -p_1 \Delta(x_1 - c_1) - \sum_{i=1}^{n_1} P_1^{(i)} \delta(x_1 - a_{1i}) \qquad (4.70)$$

where $\Delta(x_1 - c_1)$ is the unit step function and $\delta(x_1 - a_{1i})$ is the Dirac δ-function defined in App. A. Moreover, the boundary conditions of the element are

$$u_1(0) = 0 \qquad u_1(L) = 0 \qquad (4.71)$$

Integrating Eq. (4.70) twice we obtain

$$EA \frac{du_1}{dx_1} = -p_1(x_1 - c_1) \Delta(x_1 - c_1) - \sum_{i=1}^{n_1} P_1^{(i)} \Delta(x_1 - a_{1i}) + C_1 \qquad (4.72)$$

$$EAu_1 = -\frac{p_1(x_1 - c_1)^2}{2}\Delta(x_1 - c_1)$$

$$- \sum_{i=1}^{n_1} [P_1^{(i)}(x_1 - a_{1i})\,\Delta(x_1 - a_{1i})] + C_1 x_1 + C_2 \qquad (4.73)$$

The constants C_1 and C_2 are evaluated by requiring that the solution satisfies the boundary conditions (4.71). Taking into account that $\Delta(x_1 - c_1)$ and $\Delta(x_1 - a_{1i})$ vanish at $x_1 = 0$ and are equal to unity at $x_1 = L$, and substituting relations (4.72) and (4.73) into (4.71), we get

$$C_2 = 0$$

$$C_1 = \frac{p_1(L - c_1)^2}{2L} + \sum_{i=1}^{n_1} \frac{P_1^{(i)}(L - a_{1i})}{L} \qquad (4.74)$$

Substituting relations (4.74) into (4.73) we have

$$u_1(x_1) = -\frac{p_1(x_1 - c_1)^2}{2EA}\Delta(x_1 - c_1) - \sum_{i=1}^{n_1}\left[\frac{P_1^{(i)}(x_1 - a_{1i})}{EA}\Delta(x_1 - a_{1i})\right]$$

$$+ \frac{p_1(L - c_1)^2 x_1}{2EAL} + \sum_{i=1}^{n_1} \frac{P_1^{(i)}(L - a_{1i})x_1}{EAL} \qquad (4.75)$$

Substituting relations (4.74) into (4.72) and the resulting expression into (2.36), we obtain

$$N(x_1) = -p_1(x_1 - c_1)\Delta(x_1 - c) - \sum_{i=1}^{n_1} P_1^{(i)}\Delta(x_1 - a_{1i}) + \frac{p_1(L - c_1)^2}{2L}$$

$$+ \sum_{i=1}^{n_1} \frac{P_1^{(i)}(L - a_{1i})}{L} - EAH_1 \qquad (4.76)$$

where H_1 is specified by relation (2.37).

From relation (4.74) we get

$$\{A^R\} = \begin{Bmatrix} F_1^{Rj} \\ F_1^{Rk} \end{Bmatrix} = \begin{Bmatrix} -N(0) \\ N(L) \end{Bmatrix} = \begin{Bmatrix} -\dfrac{p_1(L - c_1)^2}{2L} - \displaystyle\sum_{i=1}^{n_1} \dfrac{P_1^{(i)}(L - a_{1i})}{L} + EAH_1 \\[4mm] -\dfrac{p_1(L - c_1)(L + c_1)}{2L} - \displaystyle\sum_{i=1}^{n_1} \dfrac{P_1^{(i)}a_{1i}}{L} - EAH_1 \end{Bmatrix}$$

$$(4.77)$$

Element of a beam or frame subjected to bending about its x_3 axis. Consider an element of constant cross section of a beam or frame made of an isotropic linearly elastic material and subjected with its ends fixed to the following loading:

1. Uniformly distributed transverse forces $p_2(x_1)$ applied from $x_1 = c_2$ to $x_1 = d_2$ (see Fig. 4.8)

2. Concentrated transverse forces $P_2^{(i)}$ applied at $x_1 = a_{2i}$ ($i = 1, 2, \ldots, n_2$)

3. Concentrated bending moments $\mathcal{M}_3^{(i)}$ applied at $x_1 = b_{3i}$ ($i = 1, 2, \ldots, m_3$)

4. Temperature difference $\delta T_2 = \text{constant}$

5. Initial strain specified by $k_3^I = \text{constant}$

The displacement equation of equilibrium (2.53a) for this element reduces to

$$EI_3 \frac{d^4 u_2}{dx_1^4} = p_2 [\Delta(x_1 - c_2) - \Delta(x_1 - d_2)]$$

$$+ \sum_{i=1}^{n_2} P_2^{(i)} \delta(x_1 - a_{2i}) + \sum_{i=1}^{m_2} \mathcal{M}_3^{(i)} \delta^1(x_1 - b_{3i}) \qquad (4.78)$$

where $\Delta(x_1 - c_2)$ is the unit step function defined by relation (A.1); $\delta(x_1 - a_{2i})$ is the Dirac delta function satisfying relation (A.11); and $\delta^1(x_1 - b_{3i})$ is the doublet function defined by relation (A.14).

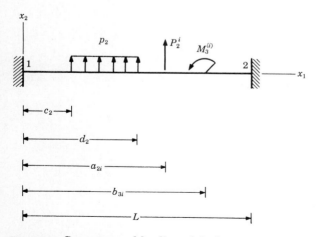

Figure 4.8 Geometry and loading of the beam.

Moreover, the boundary conditions of the problem are

$$u_2(0) = 0 \tag{4.79a}$$

$$u_2(L) = 0 \tag{4.79b}$$

$$\theta_3(0) = \frac{du_2}{dx_1}\bigg|_{x_1=0} = 0 \tag{4.79c}$$

$$\theta_3(L) = \frac{du_2}{dx_1}\bigg|_{x_1=L} = 0 \tag{4.79d}$$

Integrating relation (4.78) four times we have

$$EI_3 \frac{d^3 u_2}{dx_1^3} = p_2[(x_1 - c_2)\,\Delta(x_1 - c_2) - (x_1 - d_2)\,\Delta(x_1 - d_2)]$$

$$+ \sum_{i=1}^{n_2} P_2^{(i)}\,\Delta(x_1 - a_{2i}) + \sum_{i=1}^{m_3} M_3^{(i)}\,\delta(x_1 - b_{3i}) + C_1 \tag{4.80}$$

$$EI_3 \frac{d^2 u_2}{dx_1^2} = p_2\left[\frac{(x_1 - c_2)^2}{2}\,\Delta(x_1 - c_2) - \frac{(x_1 - d_2)^2}{2}\,\Delta(x_1 - d_2)\right]$$

$$+ \sum_{i=1}^{n_2} P_2^{(i)}(x_1 - a_{2i})\,\Delta(x_1 - a_{2i})$$

$$+ \sum_{i=1}^{m_3} M_3^{(i)}\,\Delta(x_1 - b_{3i}) + C_1 x_1 + C_2 \tag{4.81}$$

$$EI_3 \frac{du_2}{dx_1} = p_2\left[\frac{(x_1 - c_2)^3}{6}\,\Delta(x_1 - c_2) - \frac{(x_1 - d_2)^3}{6}\,\Delta(x_1 - d_2)\right]$$

$$+ \sum_{i=1}^{n_2} \frac{P_2^{(i)}(x_1 - a_{2i})^2}{2}\,\Delta(x_1 - a_{2i}) + \sum_{i=1}^{m_3} M_3^{(i)}(x_1 - b_{3i})\,\Delta(x_1 - b_{3i})$$

$$+ \frac{C_1 x_1^2}{2} + C_2 x_1 + C_3 \tag{4.82}$$

$$EI_3 u_2 = p_2\left[\frac{(x_1 - c_2)^4}{24}\,\Delta(x_1 - c_2) - \frac{(x_1 - d_2)^4}{24}\,\Delta(x_1 - d_2)\right]$$

$$+ \sum_{i=1}^{n_2} \frac{P_2^{(i)}(x_1 - a_{2i})^3}{6}\,\Delta(x_1 - a_{2i})$$

$$+ \sum_{i=1}^{m_3} \frac{M_3^{(i)}(x_1 - b_{3i})^2}{2}\,\Delta(x_1 - b_{3i}) + \frac{C_1 x_1^3}{6}$$

$$+ \frac{C_2 x_1^2}{2} + C_3 x_1 + C_4 \tag{4.83}$$

The constants C_i $(i = 1, 2, 3, 4)$ are evaluated by requiring that relations (4.82) and (4.83) satisfy the boundary conditions of the problem (4.79). Substituting relation (4.83) into (4.79a) and (4.82) into (4.79c) we get

$$C_4 = 0 \qquad C_3 = 0 \tag{4.84}$$

Substituting relation (4.83) into (4.79b) and (4.82) into (4.79d) we obtain

$$EI_3 u_2(L) = 0 = \frac{p_2}{24}[(L - c_2)^4 - (L - d_2)^4] + \sum_{i=1}^{n_2} \frac{P_2^{(i)}(L - a_{2i})^3}{6}$$

$$+ \sum_{i=1}^{m_2} \frac{M_3^{(i)}(L - b_{3i})^2}{2} + \frac{C_1 L^3}{6} + \frac{C_2 L^2}{2} \tag{4.85}$$

$$EI_3 \frac{du_2}{dx_1}\bigg|_{x_1=L} = 0 = \frac{p_2}{6}[(L - c_2)^3 - (L - d_2)^3] + \sum_{i=1}^{n_3} \frac{P_2^{(i)}(L - a_{2i})^2}{2}$$

$$+ \sum_{i=1}^{m_3} M_3^{(i)}(L - b_{3i}) + \frac{C_1 L^2}{2} + C_2 L \tag{4.86}$$

Relations (4.85) and (4.86) give

$$C_1 = \frac{p_2}{2L^3}[-(L - c_2)^3(L + c_2) + (L - d_2)^3(L + d_2)]$$

$$- \sum_{i=1}^{n_2} \frac{P_2^{(i)}(L - a_{2i})^2(L + 2a_{2i})}{L^3} - \sum_{i=1}^{m_3} \frac{6M_3^{(i)}(L - b_{3i})b_{3i}}{L^3}$$

$$C_2 = \frac{p_2}{12L^2}[(L - c_2)^3(L + 3c_2) - (L - d_2)^3(L + 3d_2)]$$

$$+ \sum_{i=1}^{n_2} \frac{P_2^{(i)}(L - a_{2i})^2 a_{2i}}{L^2} - \sum_{i=1}^{m_3} \frac{M_3^{(i)}(L - b_{3i})(L - 3b_{3i})}{L^2} \tag{4.87}$$

Substituting relation (4.83) into the action-displacement relations (2.49a) and (2.50b) the fixed-end actions of the element are

$$F_2^{Rj} = -Q_2(0) = EI_3 \frac{d^3 u_2}{dx_1^3}\bigg|_{x_1=0} = C_1$$

$$M_3^{Rj} = -M_3(0) = -EI_3 \frac{d^2 u_2}{dx_1^2}\bigg|_{x_1=0} - H_2 = -C_2 - H_2$$

$$F_2^{Rk} = Q_2(L) = -EI_3 \frac{d^3 u_2}{dx_1^3}\bigg|_{x_1=L} = -p_2(d - c_2) - \sum_{i=1}^{n_3} P_2^{(i)} - C_1 \tag{4.88}$$

$$M_3^{Rk} = M_3(L) = EI_3 \frac{d^2 u_2}{dx_1^2}\bigg|_{x_1=L} + H_2 = \frac{P_2}{2}[(L - c_2)^2$$

$$- (L - d_2)^2] + \sum_{i=1}^{n_2} P_2^{(i)}(L - a_{2i})$$

$$+ \sum_{i=1}^{m_3} M_3^{(i)} + C_1 L + C_2 + H_2$$

$$\{A^R\} = \begin{Bmatrix} F_2^{Rj} \\ M_3^{Rj} \\ F_2^{Rk} \\ M_3^{Rk} \end{Bmatrix} = \begin{Bmatrix} C_1 \\ -C_2 - H_2 \\ -p_2(d - c_2) - P_2 - C_1 \\ \left[\frac{p_2}{2}[(L - c_2)^2 - (L - d_2)^2] \\ + \sum_{i=1}^{n_2} P_2^{(i)}(L - a_{21}) + \sum_{i=1}^{m_3} \mathcal{M}_3^{(i)} + C_1 L + C_2 + H_2 \right] \end{Bmatrix}$$

$$(4.89)$$

4.5 Problems

1. Compute the stiffness matrix for a general planar element of constant cross section made of two isotropic linearly elastic materials as shown in Fig. 4.P1.

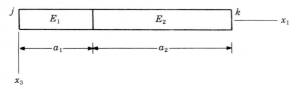

Figure 4.P1

2. Compute the stiffness matrix for a general planar element made of two parts as shown in Fig. 4.P2. The one part has a constant cross section $A_1 I_2^{(1)}$, while the other part has a constant cross section $A_2 I_2^{(2)}$. Both parts are made of the same isotropic linearly elastic material.

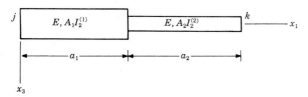

Figure 4.P2

3. Compute the stiffness matrix for the general planar element of constant width b shown in Fig. 4.P3. The element is made from an isotropic linearly elastic material.

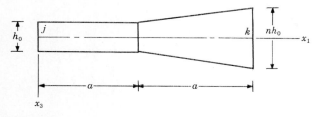

Figure 4.P3

4 and 5. Compute the fixed-end actions of the axial deformation element of constant cross section subjected to the forces shown in Fig. 4.P4 and to an increase in temperature ΔT_c. The element is made of two isotropic linearly elastic materials. Repeat with the element of Fig. 4.P5.

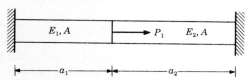

Figure 4.P4

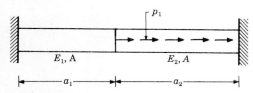

Figure 4.P5

6. Compute the fixed-end actions of the axial deformation element subjected to the forces shown in Fig. 4.P6 and to an increase in temperature ΔT_c. The element is made of an isotropic linearly elastic material.

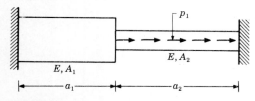

Figure 4.P6

7 and 8.   Compute the fixed-end actions of the element of a planar beam subjected to the forces shown in Fig. 4.P7. The element is made of two isotropic linearly elastic materials. Repeat with the element of Fig. 4.P8.

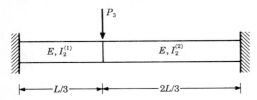

Figure 4.P7

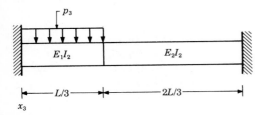

Figure 4.P8

9.   Compute the fixed-end actions of the element of a planar beam subjected to the forces shown in Fig. 4.P9. The element is made of an isotropic linearly elastic material.

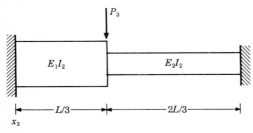

Figure 4.P9

References

1. A. E. Armenàkas, *Classical Structural Analysis: A Modern Approach,* McGraw-Hill Book Co., New York, 1988, sec. 8.6.
2. Ibid., p. 330.

Flexibility Equations
for an Element

5.1 Introduction

In Sec. 4.3 we expressed the response of an element subjected only to nodal displacements by its stiffness equations. In these equations the local components of the nodal actions of an element are expressed as linear combinations of the local components of its nodal displacements. That is, for an element subjected only to nodal actions we have

$$\{A^E\} = [K]\{D\} \qquad (5.1)$$

where $\{A^E\}$ = matrix of local components of nodal actions of element subjected only to nodal displacements

$\{D\}$ = matrix of local components of nodal displacements of element

$[K]$ = stiffness matrix for element; its terms depend on geometry of element and on properties of material from which element is made

As mentioned in Sec. 4.1 the components of nodal actions of an element included in the matrix $\{A\}$, $\{A^E\}$, or $\{A^R\}$ are not independent. They are related by the equations of equilibrium for the element. Moreover, the components of the nodal displacements of an element included in the matrix $\{D\}$ do not represent only the deformation of the element but also its rigid-body motion. Inasmuch as the internal actions of an element of a structure do not change when the element moves as a rigid body, the components of the nodal displacements of an element are specified only to within a rigid-body movement of the element when the components of its nodal actions are known. Thus, the components of the nodal displacements of an element cannot be expressed as a linear combination of its nodal actions.

In Sec. 5.2 we choose a set of independent components of nodal actions (basic nodal actions) for each type of element which we are considering and store them in a matrix which we denote by $\{a\}$ and call the *matrix of basic nodal actions*. For an element of a structure subjected to given loads this matrix can be expressed as

$$\{a\} = \{a^E\} + \{a^R\} \tag{5.2}$$

where $\{a^E\}$ is the matrix of basic nodal actions of the element subjected only to its nodal displacements, and $\{a^R\}$ is the matrix of basic nodal actions of the element subjected to the given loads acting along its length with its ends fixed. Moreover, in Sec. 5.2 we establish the relation between the matrices $\{a^E\}$ and $\{A^E\}$ of an element by considering its equilibrium.

In Sec. 5.3 we choose a set of independent quantities which specify completely the deformation of an element and are not affected by its rigid-body motion. We call these quantities the *basic deformation parameters of the element*. They are the relative displacements of the nodal points of an element which correspond to the set of independent components of nodal actions which we have chosen as the basic nodal actions. We store the basic deformation parameters of an element in a matrix which we denote by $\{d\}$.

In Sec. 5.4 we express the response of an element subjected only to its nodal displacements by its flexibility equations. In these equations the basic deformation parameters for an element are expressed as a linear combination of its basic nodal actions. In Sec. 5.5 we express the response of an element subjected only to its nodal displacements by its basic stiffness equations. In these equations the basic nodal actions of an element are expressed as a linear combination of its basic deformation parameters.

It is apparent that a change of the basic nodal actions for an element results in a change of its basic deformation parameters. Moreover, a change of the basic deformation parameters for an element requires a change of its basic nodal actions. Thus, the relation between the basic nodal actions and the basic deformation parameters for an element can be written in either flexibility or basic stiffness form. That is,

$$\{d\} = [f]\{a^E\} \tag{5.3}$$

or

$$\{a^E\} = [k]\{d\} \tag{5.4}$$

where $[f]$ and $[k]$ are the flexibility and the basic stiffness matrices for the element, respectively. They are both square, nonsingular matrices. Thus

$$[k] = [f]^{-1} \tag{5.5}$$

The flexibility equations (5.3) for an element are required in the modern flexibility method of structural analysis presented in Chap. 17. However, as discussed in Sec. 3.5 this method has found only very limited practical use in writing general computer programs for analyzing groups of structures. Nevertheless, the formulation of the flexibility equations for an element of a framed structure is of considerable practical interest because they can be converted to its stiffness equations (see Sec. 5.8). There are elements for which the direct calculation of their stiffness coefficients is considerably more difficult than the direct calculation of their flexibility coefficients. The stiffness matrix for such elements is obtained from their flexibility matrix. Examples of such elements are

1. Tapered elements
2. Elements of beams or frames when the effect of shear deformation is taken into account
3. Elements with curved or in general not straight axes

In all general computer programs known to the author the stiffness matrix of tapered elements is established from the elements' flexibility matrix.

In Sec. 5.6 we establish the flexibility equations for elements of constant cross section made from an isotropic linearly elastic material. We accomplish this by solving the boundary-value problems for computing the components of displacement of a model consisting of the element with its end j fixed and subjected to the chosen basic nodal actions on its end k.

In Sec. 5.7 we establish the flexibility equations for tapered elements made from an isotropic linearly elastic material by solving boundary-value problems analogous to the ones solved in Sec. 5.6. In Sec. 5.8 we establish the stiffness matrix $[K]$ for an element from its flexibility matrix $[f]$.

5.2 Matrix of Basic Nodal Actions for an Element

In this section we choose the matrix of basic nodal actions for each type of element which we are considering. Moreover, we establish the

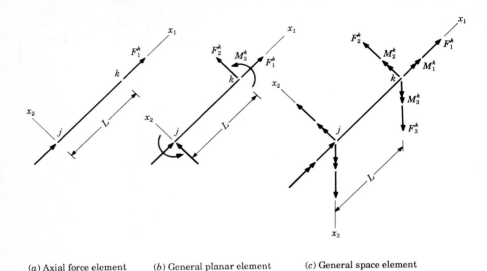

(*a*) Axial force element   (*b*) General planar element   (*c*) General space element

Figure 5.1   Basic nodal actions of elements of framed structures.

relation between the matrices of basic nodal actions $\{a^E\}$ and of nodal actions $\{A^E\}$ for the elements of the various types of framed structures which we are considering, subjected only to their nodal displacement. Referring to Fig. 5.1 we choose

Axial force element.

$$\{a\} = F^k_1 \tag{5.6}$$

General planar element.

$$\{a\} = \begin{Bmatrix} F^k_1 \\ F^k_2 \\ M^k_3 \end{Bmatrix} \tag{5.7}$$

General space element.

$$\{a\} = \begin{Bmatrix} F^k_1 \\ F^k_2 \\ F^k_3 \\ M^k_1 \\ M^k_2 \\ M^k_3 \end{Bmatrix} \tag{5.8}$$

Notice that the choice of the basic nodal actions is not unique. For instance, we could have chosen the axial component of force F_1^k and the moments M_3^j and M_3^k as the basic nodal actions for a general planar element.

Consider an element subjected only to its nodal displacements. We can establish a direct relation between the local components of its nodal actions $\{A^E\}$ and the basic components of its nodal actions $\{a^E\}$ by considering the element's equilibrium. This relation can be written as

$$\{A^E\} = [T]\{a^E\} \tag{5.9}$$

The matrix $[T]$ transforms the basic components of the nodal actions of an element into the local components of its nodal actions.

In what follows we give the matrix $[T]$ for the elements of the different types of structures which we are considering.

Element of a planar or space truss.

$$\{A^E\} = \begin{Bmatrix} F_1^{Ej} \\ F_1^{Ek} \end{Bmatrix} = \begin{Bmatrix} -1 \\ 1 \end{Bmatrix} F_1^{Ek} = \{T\}\{a^E\} \tag{5.10}$$

Element of a planar beam or a planar frame. For a general planar element we have

$$\{A^E\} = \begin{Bmatrix} F_1^{Ej} \\ F_2^{Ej} \\ M_3^{Ej} \\ \hdashline F_1^{Ek} \\ F_2^{Ek} \\ M_3^{Ek} \end{Bmatrix} = \begin{bmatrix} -1 & 0 & 0 \\ 0 & -1 & 0 \\ 0 & -L & -1 \\ \hdashline 1 & 0 & 0 \\ 0 & 1 & 0 \\ 0 & 0 & 1 \end{bmatrix} \begin{Bmatrix} F_1^{Ek} \\ F_2^{Ek} \\ M_3^{Ek} \end{Bmatrix} = [T]\{a^E\} \tag{5.11}$$

Moreover, for an axial deformation element of a planar frame we have

$$\{A^E\} = \begin{Bmatrix} F_1^{Ej} \\ F_2^{Ej} \\ M_3^{Ej} \\ \hdashline F_1^{Ek} \\ F_2^{Ek} \\ M_3^{Ek} \end{Bmatrix} = \begin{Bmatrix} -1 \\ 0 \\ 0 \\ \hdashline 1 \\ 0 \\ 0 \end{Bmatrix} F_1^{Ek} = [T]\{a^E\} \tag{5.12}$$

Element of a space beam or a space frame.　For a general space element we have

$$\{A^E\} = \begin{Bmatrix} F_1^{Ej} \\ F_2^{Ej} \\ F_3^{Ej} \\ M_1^{Ej} \\ M_2^{Ej} \\ M_3^{Ej} \\ F_1^{Ek} \\ F_2^{Ek} \\ F_3^{Ek} \\ M_1^{Ek} \\ M_2^{Ek} \\ M_3^{Ek} \end{Bmatrix} = \left[\begin{array}{cccccc} -1 & 0 & 0 & 0 & 0 & 0 \\ 0 & -1 & 0 & 0 & 0 & 0 \\ 0 & 0 & -1 & 0 & 0 & 0 \\ 0 & 0 & 0 & -1 & 0 & 0 \\ 0 & 0 & L & 0 & -1 & 0 \\ 0 & -L & 0 & 0 & 0 & -1 \\ \hline 1 & 0 & 0 & 0 & 0 & 0 \\ 0 & 1 & 0 & 0 & 0 & 0 \\ 0 & 0 & 1 & 0 & 0 & 0 \\ 0 & 0 & 0 & 1 & 0 & 0 \\ 0 & 0 & 0 & 0 & 1 & 0 \\ 0 & 0 & 0 & 0 & 0 & 1 \end{array} \right] \begin{Bmatrix} F_1^{Ek} \\ F_2^{Ek} \\ F_3^{Ek} \\ M_1^{Ek} \\ M_2^{Ek} \\ M_3^{Ek} \end{Bmatrix} = [T]\{a^E\}$$

$$(5.13)$$

Moreover, for an axial deformation element of a space frame we have

$$\{A^E\} = \begin{Bmatrix} F_1^{Ej} \\ F_2^{Ej} \\ F_3^{Ej} \\ M_1^{Ej} \\ M_2^{Ej} \\ M_3^{Ej} \\ \hline F_1^{Ek} \\ F_2^{Ek} \\ F_3^{Ek} \\ M_1^{Ek} \\ M_2^{Ek} \\ M_3^{Ek} \end{Bmatrix} = \begin{Bmatrix} -1 \\ 0 \\ 0 \\ 0 \\ 0 \\ 0 \\ \hline 1 \\ 0 \\ 0 \\ 0 \\ 0 \\ 0 \end{Bmatrix} F_1^{Ek} = [T]\{a^E\} \qquad (5.14)$$

5.3　Basic Deformation Parameters for an Element

When a structure deforms, its elements move as rigid bodies and deform. In this section we establish certain parameters which are sufficient to specify the deformation of an element and which are not affected by its rigid-body motion. We call these parameters the *basic deformation parameters for the element,* and we store them in a matrix which we denote by $\{d\}$. The choice of the basic deformation parameters for an element is not unique; we select those which correspond to our choice of basic nodal actions. This implies that the matrices $\{d\}$ and $\{a\}$ are conjugate.

Axial deformation element of a planar framed structure. Consider a cross section of an axial deformation element of a planar framed structure. In the reference configuration of the structure this cross section is located at a distance x_1 from the end j of the element. When the structure is subjected to external loads, its elements translate and rotate as rigid bodies and deform. The deformed position of the cross section under consideration is specified by the components of translation $u_1(x_1)$ and $u_2(x_1)$ and by the component of rotation θ_3. Referring to Fig. 5.2b these components of displacement can be expressed as

$$u_1(x_1) = u_1^j + [x_1 + \delta_1(x_1)] \cos \theta_3^j - x_1$$

$$u_2(x_1) = u_2^j + [x_1 + \delta_1(x_1)] \sin \theta_3^j \qquad (5.15)$$

$$\theta_3 = \theta_3^j$$

For very small rotation θ_3, relations (5.15) reduce to

$$u_1(x_1) = u_1^j + \delta_1(x_1)$$

$$u_2(x_1) = u_2^j + x_1 \theta_3^j \qquad (5.16)$$

$$\theta_3 = \theta_3^j$$

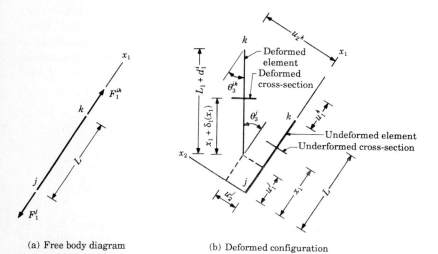

(a) Free body diagram   (b) Deformed configuration

Figure 5.2 Axial deformation element.

In relations (5.16) the components of nodal displacements u_1^j, u_2^j, and θ_3^j specify the rigid-body motion of the element; $\delta_1(x_1)$ represents the change of length of the portion of the element from its end j to the cross section under consideration. That is, $\delta_1(x_1)$ is equal to the component of translation of the cross section under consideration when the end j of the element is fixed. It is apparent that the component of translation $u_2(x_1)$ and the component of rotation θ_3 of a cross section of an axial deformation element of a planar structure represent only rigid-body motion of the element, while its component of translation $u_1(x_1)$ represents both rigid-body motion of the element and deformation of the portion of the element from its end j to the cross section under consideration.

Referring to relation (5.16) the components of displacement of the end k of the element are

$$u_1(L) = u_1^k = u_1^j + d_1$$
$$u_2(L) = u_2^k = u_2^j + L\theta_3^j \qquad (5.17)$$
$$\theta_3(L) = \theta_3^k = \theta_3^j$$

where
$$d_1 = \delta_1(L) = u_1^k - u_1^j \qquad (5.18)$$

is the change of length of the element.

Referring to Fig. 5.2b it is apparent that the deformed configuration of an axial deformation element of a planar structure may be obtained by a rigid-body translation of the element, specified by the vector $u_1^j \mathbf{i}_1 + u_2^j \mathbf{i}_2$; a rigid-body rotation specified by the angle θ_3^j; and a deformation characterized by the change in length d_1 of the element, which is its basic deformation parameter. It is apparent that the relation between the basic nodal action F_1^k and the basic deformation parameter d_1 of an axial deformation element of a planar framed structure may be established by considering the model of Fig. 5.3. This model consists of the element fixed at its end j and

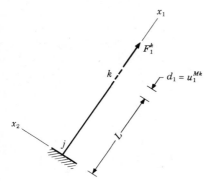

Figure 5.3 Model for computing the relation between the basic nodal action and the basic deformation parameter for an axial deformation element.

subjected to an axial force F_1^k at its end k. The translation u_1^{Mk} of the end k of the model is the deformation parameter d_1 of the element.

General planar element. Consider a cross section of a general planar element. In the reference configuration of the structure this cross section is located at a distance x_1 from the end j of the element. Its deformed position is specified by the components of translation $u_1(x_1)$ and $u_2(x_1)$ and by the component of rotation $\theta_3(x_1)$. Referring to Fig. 5.4b these components of displacement can be expressed as

$$u_1(x_1) = u_1^j + [x_1 + \delta_1(x_1)] \cos \theta_3^j - x_1$$

$$u_2(x_1) = u_2^j + [x_1 + \delta_1(x_1)] \sin \theta_3^j + \delta_2(x_1) \cos \theta_3^j \qquad (5.19)$$

$$\theta_3(x_1) = \frac{du_2}{dx_1} = \theta_3^j + \frac{d\delta_2}{dx_1}$$

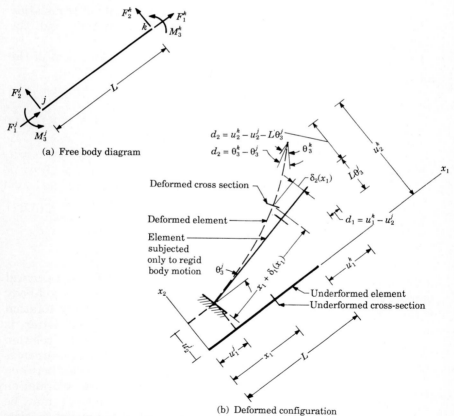

(a) Free body diagram

$d_2 = u_2^k - u_2^j - L\theta_3^j$

$d_2 = \theta_3^k - \theta_3^j$

Deformed cross section

Deformed element

Element subjected only to regid body motion

$d_1 = u_1^k - u_2^j$

Underformed element
Underformed cross-section

(b) Deformed configuration

Figure 5.4 General planar element.

For very small rotation θ_3, relations (5.19) reduce to

$$u_1(x_1) = u_1^j + \delta_1(x_1)$$

$$u_2(x_1) = u_2^j + x_1\theta_3^j + \delta_2(x_1) \tag{5.20}$$

$$\theta_3(x_1) = \theta_3^j + \frac{d\delta_2}{dx_1}$$

In relation (5.20) the components of nodal displacement u_1^j, u_2^j, and θ_3^j specify the rigid-body motion of a general planar element. $\delta_1(x_1)$ represents the change in length of the portion of the element from its end j to the cross section under consideration. That is, $\delta_1(x_1)$ is equal to the axial component of translation of the cross section under consideration when the element is subjected to all the loads acting on it with its end j fixed. $\delta_2(x_1)$ represents the transverse component of translation (deflection) of the cross section under consideration when the element is subjected to all the loads acting on it with its end j fixed (see Fig. 5.4). It is apparent that the components of translation $u_1(x_1)$ and $u_2(x_1)$ specify both the rigid-body motion and the deformation of a general planar element.

Referring to relation (5.20) the components of displacement of the end k of the element are

$$u_1(L) = u_1^k = u_1^j + d_1$$

$$u_2(L) = u_2^k = u_2^j + L\theta_3^j + d_2 \tag{5.21}$$

$$\theta_3^k = \theta_3^j + d_3$$

where

$$d_1 = \delta_1(L) = u_1^k - u_1^j$$

$$d_2 = \delta_2(L) = u_2^k - u_2^j - L\theta_3^j \tag{5.22}$$

$$d_3 = \frac{d\delta_2}{dx_1}\bigg|_{x_1=L} = \theta_3^k - \theta_3^j$$

Relations (5.21) indicate that the deformed configuration of a general planar element can be obtained by subjecting it to a rigid-body translation specified by the vector $u_1^j\mathbf{i}_1 + u_2^j\mathbf{i}_2$, a rigid-body rotation about its end j equal to θ_3^j, and a deformation. The latter is characterized by the quantities d_1, d_2, and d_3 defined by relation (5.22). We choose d_1, d_2, and d_3 as the basic deformation parameters of a general planar element. It is apparent that the relations between the basic nodal actions F_1^k, F_2^k, and M_3^k and the basic deformation parameters d_1, d_2, and d_3 of a general planar element may be established by considering the model of Fig. 5.5. This model consists of the element fixed at its end j and subjected to the basic nodal

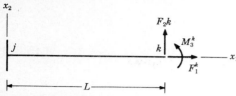

(a) Model subjected to the basic nodal actions

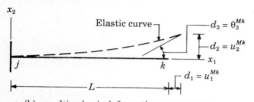

(b) resulting basic deformation parameters

Figure 5.5 Model for computing the relations between the basic nodal actions and the basic deformation parameters of a general planar element.

actions F_1^k, F_2^k, and M_3^k at its end k. The components of translation u_1^{Mk} and u_2^{Mk} of the end k of the model are equal to the basic deformation parameters d_1 and d_2 of the element. Moreover, the component of rotation θ_3^{Mk} of the model is equal to the basic deformation parameter d_3 of the element.

Notice that the choice of the parameters which specify the rigid-body motion of an element is not unique. For example, referring to Fig. 5.6 we can express the components of translation $u_1(x_1)$ and $u_2(x_1)$ and the component of rotation $\theta_3(x_1)$ of a cross section of a

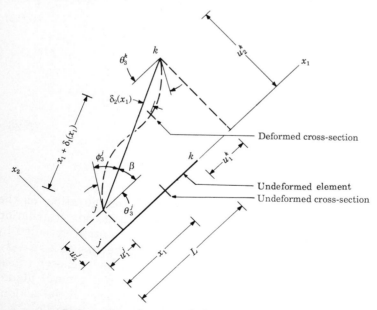

Figure 5.6 Deformation of a general planar element.

general planar element as

$$u_1(x_1) = u_1^j + [x_1 + \delta_1(x_1)]\cos\beta - x_1$$
$$u_2(x_1) = u_2^j + [x_1 + \delta_1(x_1)]\sin\beta + \delta_2(x_1)\cos\beta \qquad (5.23)$$
$$\theta_3(x_1) = \beta + \phi_3(x_1) \qquad \beta = \theta_3^j - \phi_3^j$$

For very small rotation θ_3, relations (5.23) reduce to

$$u_1(x_1) = u_1^j + \delta_1(x_1)$$
$$u_2(x_1) = u_2^j + x_1\beta + \delta_2(x_1) \qquad (5.24)$$
$$\theta_3(x_1) = \beta + \phi_3(x_1) \qquad \beta = \theta_3^j - \phi_3^j$$

In relation (5.24) the components of nodal translation u_1^j and u_2^j and the rotation β specify the rigid-body motion of the element. $\delta_1(x_1)$ represents the change in length of the portion of the element from its end j to the cross section under consideration. $\delta_2(x_1)$ represents the transverse component of translation (deflection) of the cross section under consideration when the element is subjected to all the loads acting on it with its end j pinned to a fixed support and its end k pinned to a sliding support.

Referring to relation (5.24) and noting that $\delta_2(L) = 0$, the components of displacement of the end k of the element are

$$u_1(L) = u_1^k = u_1^j + d_1$$
$$u_2(L) = u_2^k = u_2^j + L(\theta_3^j - d_2) \qquad (5.25)$$
$$\theta_3(L) = \theta_3^k = \theta_3^j - d_2 + d_3$$

where

$$d_1 = \delta_1(L) = u_1^k - u_1^j$$
$$d_2 = \phi_3(0) = \phi_3^j = \theta_3^j - \frac{u_2^k - u_2^j}{L} \qquad (5.26)$$
$$d_3 = \phi_3(L) = \phi_3^k = \theta_3^k - \frac{u_2^k - u_2^j}{L}$$

Relations (5.25) indicate that the deformed configuration of the element can be obtained by subjecting it to a rigid-body translation specified by the vector $u_1^j\mathbf{i}_1 + u_2^j\mathbf{i}_2$, a rigid-body rotation specified by the angle β, and a deformation characterized by the quantities d_1, d_2, and d_3 defined by relations (5.26). We could have chosen these quantities as the basic deformation parameters of a general planar element. In this case the relations between the basic nodal actions

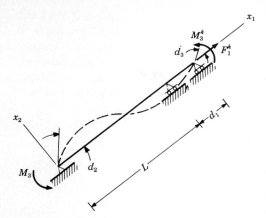

Figure 5.7 Model for computing the relation between the basic nodal actions and the basic deformation parameters of a general planar element.

F_1^k, M_3^j, and M_3^k and the basic deformation parameters d_1, d_2, and d_3 of a general planar element may be established by considering the model of Fig. 5.7. This model consists of the element with its end j pinned to a fixed support, while its end k is pinned to a sliding support; the element is then subjected to an axial force F_1^k at its end k and bending moments M_3^j and M_3^k at each of its ends. The component of translation u_1^{Mk} of the end k of the model is equal to the basic deformation parameter d_1 of the element. Moreover, the components of rotation ϕ_3^{Mj} and ϕ_3^{Mk} of the ends of the model are equal to the basic deformation parameters d_2 and d_3 of the element.

Axial deformation element of a space framed structure. Consider a cross section of an axial deformation element of a space framed structure. In the reference configuration of the structure this cross section is located at a distance x_1 from the end j of the element. Its deformed position is specified by the components of translation $u_1(x_1)$, $u_2(x_1)$, and $u_3(x_1)$ and by the components of rotation θ_2 and θ_3. For very small rotations θ_2 and θ_3, these components of displacement can be expressed as

$$u_1(x_1) = u_1^j + \delta_1(x_1)$$

$$u_2(x_1) = u_2^j + x_1\theta_3^j$$

$$u_3(x_1) = u_3^j - x_1\theta_2^j \qquad (5.27)$$

$$\theta_1 = 0 \qquad \theta_2 = \theta_2^j \qquad \theta_3 = \theta_3^j$$

In relations (5.27) the components of nodal displacements u_1^j, u_2^j, u_3^j, θ_2^j, and θ_3^j specify the rigid body motion of the element; $d_1(x_1)$ represents the change in length of the portion of the element from its end j to the cross section under consideration. It is apparent that the components of translation $u_2(x_1)$ and $u_3(x_1)$ and the components of rotation θ_2 and θ_3 of a cross section of an axial deformation element of a space structure represent only rigid-body motion of the element, while its component of translation u_1 represents both rigid-body motion of the element and deformation of the portion of the element from its end j to the cross section under consideration.

Referring to relations (5.27) the components of displacement of the end k of the element are

$$u_1(L) = u_1^k = u_1^j + d_1$$
$$u_2(L) = u_2^k = u_2^j + L\theta_3^j$$
$$u_3(L) = u_3^k = u_3^j - L\theta_2^j \qquad (5.28)$$
$$\theta_2(L) = \theta_2^k = \theta_2^j$$
$$\theta_3(L) = \theta_3^k = \theta_3^j$$

where
$$d_1 = \delta_1(L) = u_1^k - u_1^j \qquad (5.29)$$

is the basic deformation parameter of the element. It is apparent that the relation between the basic nodal action F_1^k and the basic deformation parameter d_1 of an axial deformation element of a space framed structure may be established by considering the model of Fig. 5.3. This model consists of the element fixed at its end j and subjected to an axial force F_1^k at its end k. The translation u_1^{Mk} of the end k of the model is the deformation parameter d_1 of the element.

General space element.   Consider a cross section of a general space element. In the reference configuration of the structure this cross section is located at a distance x_1 from the end j of the element. Its deformed configuration is specified by the components of translation $u_1(x_1)$, $u_2(x_1)$, and $u_3(x_1)$ and by the components of rotation $\theta_1(x_1)$, $\theta_2(x_1)$, and $\theta_3(x_1)$. For very small rotations θ_1^i, θ_2^i, and θ_3^i, these components of displacement can be expressed as

$$u_1(x_1) = u_1^j + \delta_1(x_1)$$
$$u_2(x_1) = u_2^j + x_1\theta_3^j + \delta_2(x_1)$$
$$u_3(x_1) = u_3^j - x_1\theta_2^j + \delta_3(x_1) \qquad (5.30)$$
$$\theta_1(x_1) = \theta_1^j + \hat{\theta}_1(x_1)$$
$$\theta_2(x_1) = \theta_2^j - \frac{d\delta_3}{dx_1} \qquad \theta_3 = \theta_3^j + \frac{d\delta_2}{dx_1}$$

In relations (5.30) the components of displacement u_1^j, u_2^j, and u_3^j and θ_1^j, θ_2^j, and θ_3^j specify the rigid-body motion of the element; $\delta_1(x_1)$ represents the change in length of the portion of the element from its end j to the cross section under consideration. $\delta_2(x_1)$ and $\delta_3(x_1)$ represent the transverse components of translation of the cross section under consideration when the element is subjected to all the loads acting on it with its end j fixed. $\hat{\theta}_1(x_1)$ represents the angle of twist of the cross section under consideration when the element is subjected to all the loads acting on it with its end j fixed.

Referring to relation (5.30) the components of displacement of the end k of the element are

$$u_1(L) = u_1^k = u_1^j + d_1$$
$$u_2(L) = u_2^k = u_2^j + L\theta_3^j + d_2$$
$$u_3(L) = u_3^k = u_3^k + L\theta_2^j + d_3$$
$$\theta_1(L) = \theta_1^k = \theta_1^j + d_4 \qquad\qquad (5.31)$$
$$\theta_2(L) = \theta_2^k = \theta_2^j + d_5$$
$$\theta_3(L) = \theta_3^k = \theta_3^j + d_6$$

where

$$d_1 = \delta_1(L) = u_1^k - u_1^j$$
$$d_2 = \delta_2(L) = u_2^k - u_2^j - L\theta_3^j$$
$$d_3 = \delta_3(L) = u_3^k - u_3^j - L\theta_2^j$$
$$d_4 = \hat{\theta}_1(L) = \theta_1^k - \theta_1^j \qquad\qquad (5.32)$$
$$d_5 = -\left.\frac{d\delta_3}{dx_1}\right|_{x_1=L} = \theta_2^k - \theta_2^j$$
$$d_6 = \left.\frac{d\delta_2}{dx_1}\right|_{x_1=L} = \theta_3^k - \theta_3^j$$

Relations (5.30) indicate that the deformed configuration of a general space element can be obtained by subjecting it to a rigid-body translation specified by the vector $u_1^j\mathbf{i}_1 + u_2^j\mathbf{i}_2 + u_3^j\mathbf{i}_3$, a rigid-body rotation about its end j equal to $\theta_1^j\mathbf{i}_1 + \theta_2^j\mathbf{i}_2 + \theta_3^j\mathbf{i}_3$, and a deformation. The latter is characterized by the quantities d_1, d_2, d_3, d_4, d_5, and d_6 defined by relation (5.32). We choose these quantities as the basic deformation parameters of a general space element. It is apparent that the relations between the basic nodal actions F_1^k, F_2^k, F_3^k, M_1^k, M_2^k, and M_3^k, and the basic deformation parameters d_1, d_2, d_3, d_4, d_5, and d_6 of a general space element may be established by considering the

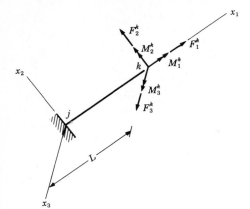

Figure 5.8 Model for computing the relation between the basic nodal actions and the basic deformation parameters of a general space element.

model of Fig. 5.8. This model consists of the element fixed at its end j and subjected to the basic nodal actions F_1^k, F_2^k, and F_3^k at its end k. The components of translation u_1^{Mk}, u_2^{Mk}, and u_3^{Mk} of the end k of the model are equal to the basic deformation parameters d_1, d_2, and d_3 of the element, respectively. Moreover, the components of rotation θ_1^{Mk}, θ_2^{Mk}, and θ_3^{Mk} of the model are equal to the basic deformation parameters d_4, d_5, and d_6 of the element, respectively.

5.3.1 Relations between the matrices of the basic deformation parameters and the matrices of the nodal displacements of an element

In this subsection we list the relations between the basic deformation parameters which we have chosen in the previous subsection and the nodal displacements for the elements of each type of structure which we are considering.

Element of a planar or space truss. Referring to relations (5.18) or (5.29) and (5.10) we have

$$\{d\} = u_1^k - u_1^j = [-1 \quad 1]\begin{Bmatrix} u_1^j \\ u_1^k \end{Bmatrix} = [T]^T\{D\} \qquad (5.33)$$

Element of a planar beam or frame. Referring to relation (5.11) and (5.22) for a general planar element of a planar beam or frame we

have

$$\{d\} = \begin{bmatrix} -1 & 0 & 0 & 1 & 0 & 0 \\ 0 & -1 & -L & 0 & 1 & 0 \\ 0 & 0 & -1 & 0 & 0 & 1 \end{bmatrix} \begin{Bmatrix} u_1^j \\ u_2^j \\ \theta_3^j \\ u_1^k \\ u_2^k \\ \theta_3^k \end{Bmatrix} = [T]^T \{D\} \qquad (5.34)$$

Moreover, referring to relations (5.12) and (5.18) for an axial deformation element of a planar beam or frame we get

$$\{d\} = u_1^k - u_1^j = \begin{bmatrix} -1 & 0 & 0 & 1 & 0 & 0 \end{bmatrix} \begin{Bmatrix} u_1^j \\ u_2^j \\ \theta_3^j \\ u_1^k \\ u_2^k \\ \theta_3^k \end{Bmatrix} = [T]^T \{D\}$$

$$(5.35)$$

Element of a space beam or frame.   Referring to relation (5.32) for a general space element of a space beam or frame we have

$$\{d\} = \begin{Bmatrix} d_1 \\ d_2 \\ d_3 \\ d_4 \\ d_5 \\ d_6 \end{Bmatrix} = \begin{Bmatrix} u_1^k - u_1^j \\ u_2^k - u_2^j - L\theta_3^j \\ u_3^k - u_3^j - L\theta_2^j \\ \theta_1^k - \theta_1^j \\ \theta_2^k - \theta_2^j \\ \theta_3^k - \theta_3^j \end{Bmatrix} = [T]^T \{D\} \qquad (5.36)$$

where the matrix $[T]$ is defined by relation (5.13). Moreover, referring to relation (5.29) for an axial deformation element of a space beam or frame we get

$$\{d\} = u_1^k - u_1^j = [T]^T \{D\} \qquad (5.37)$$

where the matrix $[T]$ is defined by relation (5.14).

By considering the geometry of the deformed element, in this section, we have shown that if we write the relation between the matrix of the local components of the nodal actions $\{A^E\}$ of an element subjected only to its nodal displacements and its matrix of its basic nodal actions $\{a^E\}$ in the form

$$\{A^E\} = [T]\{a^E\} \tag{5.38}$$

the relation between the matrix of the basic deformation parameters and the matrix of the local components of the nodal displacements of this element is

$$\{d\} = [T]^T\{D\} \tag{5.39}$$

We say that the action (5.38) and displacement (5.39) transformations are *contragradient*.

From relation (5.39) as well as from physical considerations, it is apparent that when the nodal displacements $\{D\}$ of an element are known, its basic deformation parameters can be established. However, when the basic deformation parameters of an element are known, its nodal displacements $\{D\}$ cannot be established uniquely. For any given set of basic deformation parameters of an element, an infinite number of sets of nodal displacements $\{D\}$ can be established. The differences of the corresponding components of nodal displacements of any two sets of nodal displacements obtained from the same set of basic deformation parameters represent rigid-body motion of the element.

5.4 Element Flexibility Equations

For the structures which we are considering the relations between the basic deformation parameters and the basic nodal actions of an element made of a linearly elastic material are linear. Consequently, they can be expressed as

$$\{d\} = [f]\{a^E\} \tag{5.40}$$

where the matrix $[f]$ is referred to as the *flexibility matrix for the element*. Inasmuch as the matrices $\{d\}$ and $\{a^E\}$ are of the same order, the matrix $[f]$ is a square matrix. The terms of the flexibility matrix $[f]$ are referred to as the *flexibility coefficients of the element*. They depend on the properties of the material from which the element is made, as well as on the geometry of its cross section and its length. On the basis of definition (5.40), the flexibility coefficient f_{mn} represents the basic deformation parameter d_m of the element, when its end k is subjected to a unit value of the basic nodal action a_n, while its other basic nodal actions vanish. Consequently, the flexibility coefficients for any element may be obtained by referring to the models described in Sec. 5.3. Moreover, from the Betti-Maxwell reciprocal theorem[1] we may conclude that

$$f_{mn} = f_{nm} \tag{5.41}$$

or
$$[f] = [f]^T \tag{5.42}$$

That is, the flexibility matrix of an element is symmetric.

In the sequel we specialize the flexibility equations (5.40) to the different types of elements which we are considering.

Axial deformation element. Referring to the model of Fig. 5.3 the flexibility equations (5.40) of an axial deformation element are

$$d_1 = u_1^{Mk} = f_{11} F_1^k \tag{5.43}$$

In this case, the flexibility coefficient f_{11} represents the elongation of the end k of the model of the element shown in Fig. 5.3 when it is subjected to an axial force equal to unity, at its end k.

General planar element. As discussed in Sec. 5.3 the relations between the basic nodal actions and the basic deformation parameters of a general planar element may be obtained by considering the model of Fig. 5.5a. This model consists of the element fixed at its end j, free at its end k, and subjected to the basic nodal actions F_1^k, F_2^k, and M_3^k. Thus, for a general planar element, the flexibility equations (5.40) reduce to

$$\begin{Bmatrix} d_1 \\ d_2 \\ d_3 \end{Bmatrix} = \begin{bmatrix} f_{11} & 0 & 0 \\ 0 & f_{22} & f_{23} \\ 0 & f_{32} & f_{33} \end{bmatrix} \begin{Bmatrix} F_1^k \\ F_2^k \\ M_3^k \end{Bmatrix} \tag{5.44}$$

The physical significance of the flexibility coefficients of the second column of the flexibility matrix can be established by subjecting the model of Fig. 5.5a to $F_2^k = 1$, $F_1^k = M_3^k = 0$. In this case relation (5.44) reduces to

$$d_1 = 0$$
$$d_2 = f_{22} \tag{5.45}$$
$$d_3 = f_{32}$$

Thus, as shown in Fig. 5.9b, the flexibility coefficients f_{22} and f_{23} are

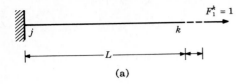

(a)

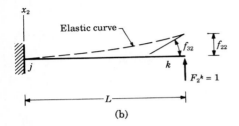

(b)

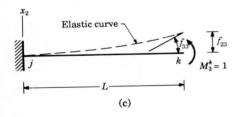

(c)

Figure 5.9 Flexibility coefficients of a general planar element.

the translation and the rotation, respectively, of the end k of the model of Fig. 5.5a when it is subjected to $F_2^k = 1$, $F_1^k = M_3^k = 0$.

The physical significance of the flexibility coefficients of the third column of the flexibility matrix can be established by subjecting the model of Fig. 5.5a to $M_3^k = 1$, $F_1^k = F_2^k = 0$. In this case relation (5.44) reduces to

$$d_1 = 0$$

$$d_2 = f_{23} \qquad (5.46)$$

$$d_3 = f_{33}$$

Thus, as shown in Fig. 5.9c the flexibility coefficients f_{23} and f_{33} represent the translation and the rotation, respectively, of the end k of the model of Fig. 5.5a when subjected to $M_3^k = 1$, $F_1^k = F_2^k = 0$.

In beams the effect of the axial forces is not coupled with the effect of the transverse forces and bending moments; thus, it can be computed separately. Moreover, in frames, the effect of axial deformation of their elements on the components of their internal actions is very small and is often disregarded. In these cases, we may use the following simplified flexibility equations for a general planar element:

$$\begin{Bmatrix} d_2 \\ d_3 \end{Bmatrix} = \begin{bmatrix} f_{22} & f_{23} \\ f_{32} & f_{33} \end{bmatrix} \begin{Bmatrix} F_2^k \\ M_3^k \end{Bmatrix} \tag{5.47}$$

General space element. The flexibility equations for a general space element are

$$\begin{Bmatrix} d_1 \\ d_2 \\ d_3 \\ d_4 \\ d_5 \\ d_6 \end{Bmatrix} = \begin{bmatrix} f_{11} & 0 & 0 & 0 & 0 & 0 \\ 0 & f_{22} & 0 & 0 & 0 & f_{26} \\ 0 & 0 & f_{33} & 0 & f_{35} & 0 \\ 0 & 0 & 0 & f_{44} & 0 & 0 \\ 0 & 0 & f_{53} & 0 & f_{55} & 0 \\ 0 & f_{62} & 0 & 0 & 0 & f_{66} \end{bmatrix} \begin{Bmatrix} F_1^k \\ F_2^k \\ F_3^k \\ M_1^k \\ M_2^k \\ M_3^k \end{Bmatrix} \tag{5.48}$$

As discussed in Sec. 5.3 the flexibility equations for a general space element may be obtained by considering the model of Fig. 5.8. This model consists of the element fixed at its end j and free at its end k. The flexibility coefficient f_{11} of the element is equal to the axial translation of end k of the model of Fig. 5.8 resulting from the application to this end of $F_1^k = 1$, $F_2^k = F_3^k = M_1^k = M_2^k = M_3^k = 0$. The flexibility coefficients f_{22} and f_{62} represent the translation along the x_2 axis and the rotation about the x_3 axis, respectively, of end k of the model of Fig. 5.8 resulting from the application to this end of $F_2^k = 1$, $F_1^k = F_3^k = M_1^k = M_2^k = M_3^k = 0$. The flexibility coefficients f_{33} and f_{53} represent the translation along the x_3 axis and the rotation about the x_2 axis, respectively, of the end k of the model of Fig. 5.8, resulting from the application to this end of $F_3^k = 1$, $F_1^k = F_2^k = M_1^k = M_2^k = M_3^k = 0$. The flexibility coefficient f_{44} represents the twist of the end k of the model of Fig. 5.8 resulting from the application to this end of $M_1^k = 1$, $F_1^k = F_2^k = F_3^k = M_2^k = M_3^k = 0$. The flexibility coefficients f_{35} and f_{55} represent the translation along the x_3 axis and the rotation about the x_2 axis, respectively, of the end k of the model of Fig. 5.8, resulting from the application to this end of $M_2^k = 1$, $F_1^k = F_2^k = F_3^k = M_1^k = M_3^k = 0$. The flexibility coefficients f_{26} and f_{66} represent the translation along the x_2 axis and the rotation about the x_3 axis, respectively, of the end k of the model of Fig. 5.8, resulting from the application to this end of $M_3^k = 1$, $F_1^k = F_2^k = F_3^k = M_1^k = M_2^k = 0$.

5.5 Element Basic Stiffness Equations

The relations between the matrix of the basic nodal actions and the matrix of the basic deformation parameters of an element can be expressed as

$$\{a^E\} = [k]\{d\} \tag{5.49}$$

where the matrix $[k]$ is referred to as the *basic stiffness matrix* for the element. The terms of this matrix are referred to as the *basic* or the *independent stiffness coefficients* for the element. Referring to relations (5.40) and (5.49) it is apparent that

$$[f] = [k]^{-1} \tag{5.50}$$

On the basis of its definition (5.49), the basic stiffness coefficient k_{mn} represents the nodal action a_m of the element when it is subjected to the basic deformation parameter $d_n = 1$, while all other basic deformation parameters vanish. Since the flexibility matrix for an element is symmetric, as can be seen from relation (5.50), its basic stiffness matrix is also symmetric.

The physical significance of the basic stiffness coefficients for a general planar element is illustrated in Fig. 5.10. Moreover, the physical significance of the basic stiffness coefficients for a general space element is illustrated in Fig. 5.11. Referring to these figures it is apparent that the basic stiffness coefficients for a general planar or a general space element are the independent stiffness coefficients in relations (4.32) and (4.34), respectively.

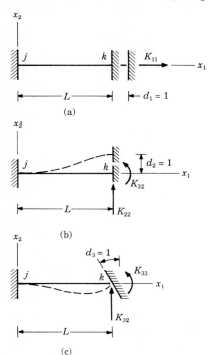

Figure 5.10 Physical significance of the basic stiffness coefficients of a general planar element.

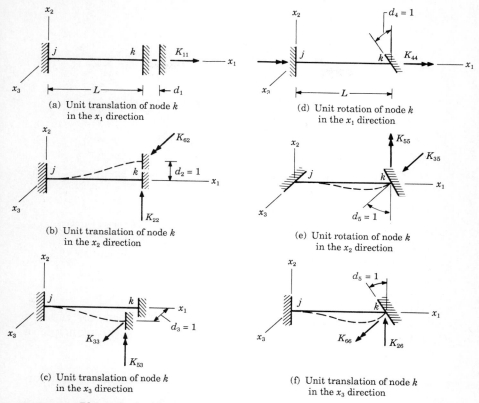

Figure 5.11 Physical significance of the basic stiffness coefficients of a general space element.

5.6 Computation of the Flexibility Matrix for an Element of Constant Cross Section by Solving the Boundary-Value Problems Described in Sec. 2.5

As discussed in Sec. 5.4 the flexibility equations for an element can be established by considering a model made from the element with its end j fixed and subjected at its end k to the basic nodal actions of the element (see Fig. 5.12). The components of displacement of the end k of the model are equal to the basic deformation parameters of the element.

In what follows we establish the flexibility equations for elements of constant cross section made of an isotropic linearly elastic material by solving the displacement equations of equilibrium (2.40), (2.53), or (2.67). This approach can also be used to establish the flexibility

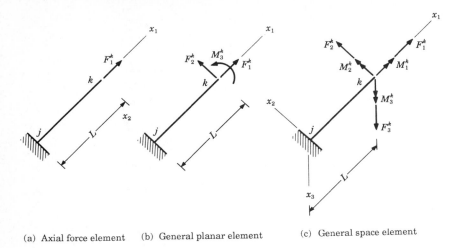

(a) Axial force element   (b) General planar element   (c) General space element

Figure 5.12 Models for establishing the flexibility matrix of the various types of elements being considered.

equations for tapered elements (see Sec. 5.7). However, for tapered elements it is preferable to use a method employing the principle of virtual work described in Sec. 19.3.

Axial deformation element. The flexibility equation for an axial deformation element is

$$d_1 = f_{11}F_1^k \tag{5.51}$$

The model for computing the flexibility coefficient f_{11} for this element is shown in Fig. 5.12a. We denote by u_1^M the axial component of displacement of this model. If the element is made from an isotropic linearly elastic material, the displacement equation of equilibrium (2.40) for the model becomes

$$EA\frac{d^2u_1^M}{dx_1^2} = 0 \tag{5.52}$$

Moreover referring to Fig. 5.12a and to relation (2.36), the boundary conditions for the model are

$$u_1^M(0) = 0 \qquad N^M(L) = EA\frac{du_1^M}{dx_1}\bigg|_{x_1=L} = F_1^k \tag{5.53}$$

By inspection it is apparent that the solution of the homogeneous equation (5.52) has the following form

$$u_1^M(x_1) = c_{10} + c_{11}x_1 = [1 \quad x_1]\begin{Bmatrix} c_{10} \\ c_{11} \end{Bmatrix} \tag{5.54}$$

Substituting relation (5.54) into (5.53), we obtain

$$\begin{Bmatrix} c_{10} \\ c_{11} \end{Bmatrix} = \frac{F_1^k}{EA}\begin{Bmatrix} 0 \\ 1 \end{Bmatrix} \tag{5.55}$$

Substituting results (5.55) into relation (5.54) we get

$$u_1^M(x_1) = \frac{x_1 F_1^k}{EA} \tag{5.56}$$

Thus
$$u_1^M(L) = d_1 = \left(\frac{L}{EA}\right)F_1^k \tag{5.57}$$

Consequently, the stiffness coefficient for an axial deformation element of constant cross section made of an isotropic linearly elastic material is given as

$$f_{11} = \frac{L}{EA} \tag{5.58}$$

General planar element. The flexibility equations (5.40) for a general planar element in the $x_1 x_2$ plane are

$$\begin{Bmatrix} d_1 \\ d_2 \\ d_3 \end{Bmatrix} = \begin{bmatrix} f_{11} & 0 & 0 \\ 0 & f_{22} & f_{23} \\ 0 & f_{32} & f_{33} \end{bmatrix}\begin{Bmatrix} F_1^k \\ F_2^k \\ M_3^k \end{Bmatrix} \tag{5.59}$$

The model for computing the flexibility coefficients for a general planar element is shown in Fig. 5.12b. It is apparent that the flexibility coefficient f_{11} for this element is equal to the flexibility coefficient for an axial deformation element. Thus, the flexibility coefficient f_{11} for an element made from an isotropic linearly elastic material is given by relation (5.58). The remaining flexibility coefficients for a general planar element made from an isotropic linearly elastic material can be obtained by solving the boundary-value problem described in Sec. 2.5.2. For the model of Fig. 5.12b the displacement equation of equilibrium (2.53a) reduces to

$$EI_3 \frac{d^4 u_2^M}{dx_1^4} = 0 \tag{5.60}$$

Moreover, referring to Fig. 5.12b and to relations (2.49b) and (2.50a) the boundary conditions for the model are

$$u_2^M(0) = 0 \qquad Q_2^M(L) = -EI_3 \frac{d^3 u_2^M}{dx_1^3} = F_2^k$$

$$\left. \frac{du_2^M}{dx_1} \right|_{x_1=0} = 0 \qquad M_3^M(L) = EI_3 \frac{d^2 u_2^M}{dx_1^2} = M_3^k \tag{5.61}$$

By inspection, it is apparent that the solution of the homogeneous Eq. (5.60) has the following form:

$$u_2^M(x_1) = c_{20} + c_{21}x_1 + c_{22}x_1^2 + c_{23}x_1^3 \tag{5.62}$$

Substituting relation (5.62) into (5.61) we get

$$\begin{Bmatrix} c_{20} \\ c_{21} \\ c_{22} \\ c_{23} \end{Bmatrix} = \begin{Bmatrix} 0 \\ 0 \\ \dfrac{M_3^k + LF_2^k}{2EI_3} \\ -\dfrac{F_2^k}{6EI_3} \end{Bmatrix} \tag{5.63}$$

Substituting results (5.63) into relation (5.62) we obtain

$$u_2^M(x_1) = \left(\frac{M_3^k + LF_2^k}{2EI_3} \right) x_1^2 - \left(\frac{F_2^k}{6EI_3} \right) x_1^3 \tag{5.64}$$

Moreover, referring to relation (2.47b) we have

$$\theta_3^M(x_1) = \frac{du_2^M}{dx_1} = \left(\frac{M_3^k + LF_2^k}{EI_3} \right) x_1 - \left(\frac{F_2^k}{2EI_3} \right) x_1^2 \tag{5.65}$$

From relations (5.64) and (5.65) we get

$$\begin{Bmatrix} u_2^M(L) \\ \theta_3^M(L) \end{Bmatrix} = \begin{Bmatrix} d_2 \\ d_3 \end{Bmatrix} = \begin{bmatrix} \dfrac{L^3}{3EI_3} & \dfrac{L^2}{2EI_3} \\ \dfrac{L^2}{2EI_3} & \dfrac{L}{EI_3} \end{bmatrix} \begin{Bmatrix} F_2^k \\ M_3^k \end{Bmatrix} \tag{5.66}$$

Consequently, the flexibility matrix of a general planar element of constant cross section made of an isotropic linearly elastic material is

$$[f] = \begin{bmatrix} \dfrac{L}{AE} & 0 & 0 \\[2ex] 0 & \dfrac{L^3}{3EI_3} & \dfrac{L^2}{2EI_3} \\[2ex] 0 & \dfrac{L^2}{2EI_3} & \dfrac{L}{EI_3} \end{bmatrix} \tag{5.67}$$

General space element. Following a procedure analogous to that above it can be shown that the flexibility matrix of a general space element of constant cross section made of an isotropic linearly elastic material is

$$[f] = \begin{bmatrix} \dfrac{L}{AE} & 0 & 0 & 0 & 0 & 0 \\[2ex] 0 & \dfrac{L^3}{3EI_3} & 0 & 0 & 0 & \dfrac{L^2}{2EI_3} \\[2ex] 0 & 0 & \dfrac{L^3}{3EI_2} & 0 & -\dfrac{L^2}{2EI_2} & 0 \\[2ex] 0 & 0 & 0 & \dfrac{L}{KG} & 0 & 0 \\[2ex] 0 & 0 & -\dfrac{L^2}{2EI_2} & 0 & \dfrac{L}{EI_2} & 0 \\[2ex] 0 & \dfrac{L^2}{2EI_3} & 0 & 0 & 0 & \dfrac{L}{EI_3} \end{bmatrix} \tag{5.68}$$

5.7 Computation of the Flexibility Matrix for Tapered Elements by Solving the Boundary-Value Problems Described in Sec. 2.5

In this section we establish the flexibility matrix for the tapered element of constant width made of an isotropic linearly elastic material shown in Fig. 5.13.

Axial deformation element. The flexibility equation for an axial deformation element is (5.51). The model for computing the flexibility coefficient f_{11} of this element is shown in Fig. 5.12a. The axial component of translation u_1^{Mk} of the end k of this model is the basic

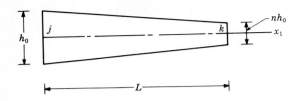

Figure 5.13 Geometry of a tapered element.

deformation parameter of the element. If the element is made from an isotropic linearly elastic material, the displacement equation of equilibrium (2.40) for the model becomes

$$\frac{d}{dx_1}\left(EA\frac{du_1^M}{dx_1}\right) = 0 \tag{5.69}$$

where

$$A = A_0\left(1 + \frac{n-1}{L}x_1\right) \tag{5.70}$$

Moreover, the boundary conditions of the problem are

$$u_1^M(0) = 0 \qquad N^M(L) = EA\frac{du_1^M}{dx_1}\bigg|_{x_1=L} = F_1^k \tag{5.71}$$

Integrating Eq. (5.69) twice we obtain

$$N^M(x_1) = EA\frac{du_1^M}{dx_1} = C_1 \tag{5.72}$$

and

$$u_1^M(x_1) = -\frac{L\ln\left[1 + (n-1)x_1/L\right]C_1}{EA_0(1-n)} + C_2 \tag{5.73}$$

The constants C_1 and C_2 are evaluated from the boundary conditions (5.71). Substituting relations (5.72) and (5.73) into relations (5.71) we get

$$C_2 = 0 \tag{5.74a}$$

$$C_1 = F_1^k \tag{5.74b}$$

Substituting relation (5.74) into (5.73) we obtain

$$u_1^M(L) = d_1 = -\frac{L \ln (n) F_1^k}{EA_0(1-n)} \tag{5.75}$$

Thus comparing Eqs. (5.51) and (5.75) the flexibility coefficient for the axial deformation tapered element shown in Fig. 5.13 made from an isotropic linearly elastic material is

$$f_{11} = -\frac{L \ln n}{EA_0(1-n)} \tag{5.76}$$

General planar element. The flexibility equations for a general planar element are (5.59). The model for computing the flexibility matrix for this element is shown in Fig. 5.12b. The axial component of translation u_1^{Mk}, the transverse component of translation u_2^{Mk}, and the component of rotation θ_3^{Mk} of the end k of the model are equal to the basic deformation parameters d_1, d_2, and d_3 of the element, respectively. It is apparent that the flexibility coefficient f_{11} for a general planar element is equal to that of an axial deformation element given by relation (5.76). The remaining flexibility coefficients for a general planar element can be obtained by solving the boundary-value problem described in Sec. 2.5.2. If the element is made from an isotropic linearly elastic material, the displacement equation of equilibrium (2.53a) for its model becomes

$$\frac{d^2}{dx_1^2}\left(EI_3 \frac{d^2 u_2^M}{dx_1^2}\right) = 0 \tag{5.77}$$

The moment of inertia of the cross section of the model is given as

$$I_3 = \frac{bh_0^3}{12}\left[1 + \frac{(n-1)x_1}{L}\right]^3 = I_0\left[1 + \frac{(n-1)x_1}{L}\right]^3 \tag{5.78}$$

where

$$I_0 = \frac{bh_0^3}{12} \tag{5.79}$$

Moreover, referring to Fig. 5.12b the boundary conditions of the problem are

$$u_2^M(0) = 0 \qquad Q_2^M(L) = -EI_3\frac{d^3 u_2^M}{dx_1^3} = F_2^k$$

$$\left.\frac{du_2^M}{dx_1}\right|_{x_1=0} = 0 \qquad M_3^M(L) = EI_3\frac{d^2 u_2^M}{dx_1^2} = M_3^k \tag{5.80}$$

Substituting relation (5.78) into (5.77) we get

$$\frac{d^2}{dx_1^2}\left\{\left[1+\frac{(n-1)x_1}{L}\right]^3\frac{d^2u_2^M}{dx_1^2}\right\}=0 \tag{5.81}$$

Integrating relation (5.81) we obtain

$$\frac{d}{dx_1}\left\{\left[1+\frac{(n-1)x_1}{L}\right]^3\frac{d^2u_2^M}{dx_1^2}\right\}=C_1 \tag{5.82}$$

$$\frac{d^2u_2^M}{dx_1^2}=\frac{C_1x_1}{[1+(n-1)x_1/L]^3}+\frac{C_2}{[1+(n-1)x_1/L]^3} \tag{5.83}$$

$$\frac{du_2^M}{dx_1}=\left\{\frac{1}{2[1+(n-1)x_1/L]^2}-\frac{1}{[1+(n-1)x_1/L]}\right\}\frac{L^2C_1}{(n-1)^2}$$

$$-\frac{LC_2}{2(n-1)[1+(n-1)x_1/L]^2}+C_3 \tag{5.84}$$

$$u_2^M=-\left\{\frac{1}{2[1+(n-1)x_1/L]}+\ln\left[1+(n-1)x_1/L\right]\right\}\frac{L^3C_1}{(n-1)^3}$$

$$+\frac{L^2C_2}{2(n-1)^2[1+(n-1)x_1/L]}+C_3x_1+C_4 \tag{5.85}$$

The constants C_1, C_2, C_3, and C_4 are evaluated from the boundary conditions (5.80) for the model. Substituting relations (5.82) to (5.85) into (5.80) we get

$$u_2^M(0)=0\rightarrow-\frac{L^3C_1}{2(n-1)^3}+\frac{L^2C_2}{2(n-1)^2}+C_4=0$$

$$\left.\frac{du_2^M}{dx_1}\right|_{x_1=0}=0\rightarrow-\frac{L^2C_1}{2(n-1)^2}-\frac{LC_2}{2(n-1)}+C_3=0$$

$$\tag{5.86}$$

$$M_3^{Mk}=EI_3\left.\frac{d^2u_2^M}{dx_1^2}\right|_{x_1=L}=\left[\frac{EI_2LC_1}{n^3}+\frac{EI_2C_2}{n^3}\right]_{x_1=L}=EI_0(LC_1+C_2)$$

$$F_2^{Mk}=-\frac{d}{dx_1}\left(EI_2\frac{d^2u_3^M}{dx_1^2}\right)\bigg|_{x_1=L}=-EI_0C_1$$

Solving relations (5.86) we get

$$C_1 = -\frac{F_2^{Mk}}{EI_0} \qquad C_2 = \frac{1}{EI_0}(F_2^{Mk}L - M_3^{Mk})$$

$$C_3 = \frac{L}{2EI_0(n-1)^2}[F_2^{Mk}L(n-2) - M_3^{Mk}(n-1)] \qquad (5.87)$$

$$C_4 = \frac{L^2}{2EI_0(n-1)^3}[-F_2^{Mk}Ln + M_3^{Mk}(n-1)]$$

Referring to relation (2.47a) and using relations (5.84) and (5.85) the rotation and the translation at the free end of the model are

$$d_2 = u_2^M(L) = \frac{L^2C_2}{2n(n-1)^2} - \left(\frac{1}{2n} + \ln n\right)\frac{L^3C_1}{(n-1)^3} + C_3L + C_4$$

$$d_3 = \theta_3^M(L) = \frac{du_2^M}{dx_1}\bigg|_{x_1=L} = -\frac{LC_2}{2n^2(n-1)} + \frac{L^2C_1(1-2n)}{2n^2(n-1)^2} + C_3$$

$$(5.88)$$

Relations (5.88) may be rewritten as

$$\begin{Bmatrix} d_2 \\ d_3 \end{Bmatrix} = \begin{bmatrix} f_{22} & f_{23} \\ f_{32} & f_{33} \end{bmatrix}\begin{Bmatrix} F_2^k \\ M_3^k \end{Bmatrix} = [f]\begin{Bmatrix} F_2^k \\ M_3^k \end{Bmatrix} \qquad (5.89)$$

where

$$f_{22} = \frac{L^3[2\ln n + n^2 - 4n + 3]}{2EI_0(n-1)^3}$$

$$f_{23} = \frac{L^2}{2nEI_0} = f_{32} \qquad (5.90)$$

$$f_{33} = \frac{L(n+1)}{2n^2EI_0}$$

For an element of constant cross section ($n = 1$) relations (5.90) reduce to

$$f_{22} = \frac{L^3}{3EI_0} \qquad f_{23} = \frac{L^2}{2EI_0} = f_{32} \qquad f_{33} = \frac{L}{EI_0} \qquad (5.91)$$

The first of relations (5.91) has been obtained by applying L'Hopital's rule.

5.8 Derivation of the Stiffness Matrix for an Element from its Flexibility Matrix

Consider the relation between the local components of nodal actions of an element and its basic nodal actions (5.9). Using relations (5.39), (5.49), and (5.50), relation (5.9) gives

$$\{A^E\} = [T]\{a^E\} = [T][k]\{d\} = [T][k][T]^T\{D\}$$

$$= [T][f]^{-1}[T]^T\{D\} \qquad (5.92)$$

Comparing relations (5.92) and (4.8) it is apparent that

$$[K] = [T][f]^{-1}[T]^T \qquad (5.93)$$

Relation (5.93) can be used to compute the stiffness matrix for an element from its flexibility matrix. In what follows we use relation (5.93) to compute the stiffness matrix of elements of constant cross section made of an isotropic linearly elastic material.

Element of constant cross section of a truss. Substituting relation (5.58) and the matrix $[T]$ given in relation (5.10) into relation (5.93) we have

$$[K] = \left\{ \begin{matrix} -1 \\ 1 \end{matrix} \right\} \left(\frac{L}{EA} \right)^{-1} [-1 \quad 1] = \frac{EA}{L} \begin{bmatrix} 1 & -1 \\ -1 & 1 \end{bmatrix} \qquad (5.94)$$

Element of constant cross section of a planar beam or frame. Substituting relation (5.67) and the matrix $[T]$ given in relation (5.11) into relation (5.93) we obtain

$$[K] = \begin{bmatrix} -1 & 0 & 0 \\ 0 & -1 & 0 \\ 0 & L & -1 \\ 1 & 0 & 0 \\ 0 & 1 & 0 \\ 0 & 0 & 1 \end{bmatrix} \begin{bmatrix} \dfrac{EA}{L} & 0 & 0 \\ 0 & \dfrac{12EI_3}{L^3} & -\dfrac{6EI_3}{L^2} \\ 0 & \dfrac{-6EI_3}{L^2} & \dfrac{4EI_3}{L} \end{bmatrix}$$

$$\times \begin{bmatrix} -1 & 0 & 0 & 1 & 0 & 0 \\ 0 & -1 & L & 0 & 1 & 0 \\ 0 & 0 & -1 & 0 & 0 & 1 \end{bmatrix}$$

$$
= \begin{bmatrix}
\dfrac{AE}{L} & 0 & 0 & -\dfrac{AE}{L} & 0 & 0 \\[2.2ex]
0 & \dfrac{12EI_3}{L^3} & \dfrac{6EI_3}{L^2} & 0 & -\dfrac{12EI_3}{L^3} & \dfrac{6EI_3}{L^2} \\[2.2ex]
0 & \dfrac{6EI_3}{L^2} & \dfrac{4EI_3}{L} & 0 & -\dfrac{6EI_3}{L^2} & \dfrac{2E_i I}{L} \\[2.2ex]
-\dfrac{AE}{L} & 0 & 0 & \dfrac{AE}{L} & 0 & 0 \\[2.2ex]
0 & -\dfrac{12EI_3}{L^3} & -\dfrac{6EI_3}{L^2} & 0 & \dfrac{12EI_3}{L^3} & -\dfrac{6EI_3}{L^2} \\[2.2ex]
0 & \dfrac{6EI_3}{L^2} & \dfrac{2EI_3}{L} & 0 & -\dfrac{6EI_3}{L^2} & \dfrac{4EI_3}{L}
\end{bmatrix}
$$

$$(5.95)$$

The stiffness matrix of an axial deformation element of a frame made of an isotropic linearly elastic material is obtained from relation (5.95) by setting equal to zero all the terms involving EI_3.

5.9 Problems

1. Compute the flexibility matrix for a general planar element of constant cross section made of two isotropic linearly elastic materials as shown in Fig. 5.P1.

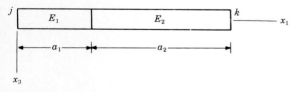

Figure 5.P1

2. Compute the flexibility matrix for a general planar element made of two parts as shown in Fig. 5.P2. The one part has a constant cross section $A_1 I_2^{(1)}$, while the other part has a constant cross section $A_2 I_2^{(2)}$. Both parts are made of the same isotropic linearly elastic material.

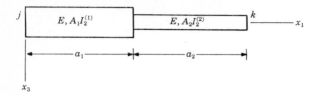

Figure 5.P2

3. Compute the flexibility matrix for the general planar element of constant width b shown in Fig. 5.P3. The element is made from an isotropic linearly elastic material.

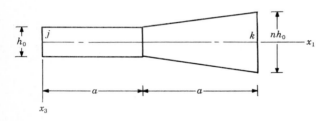

Figure 5.P3

Reference

1. A. E. Armenàkas, *Classical Structural Analysis: A Modern Approach,* McGraw-Hill Book Co., New York, 1988, p. 330.

6

The Transformation Matrix for an Element

6.1 Introduction

When analyzing a structure, it is necessary to transform the local components of nodal actions and of nodal displacements to global, and vice versa. The local and global components of a vector quantity of an element of a space framed structure are related by relations (B.15) or (B.16). For example, the components of the internal force acting on a cross section of an element of a space framed structure are related as follows:

$$\{F\} = \begin{Bmatrix} F_1 \\ F_2 \\ F_2 \end{Bmatrix} = [\Lambda_S] \begin{Bmatrix} \bar{F}_1 \\ \bar{F}_2 \\ \bar{F}_3 \end{Bmatrix} = [\Lambda_S]\{\bar{F}\} \tag{6.1}$$

or

$$\{\bar{F}\} = [\Lambda_S]^T \{F\} \tag{6.2}$$

The 3×3 matrix $[\Lambda_S]$ is the *transformation matrix for an element of a space structure*. Referring to relation (B.19) it is equal to

$$[\Lambda_S] = \begin{bmatrix} \lambda_{11} & \lambda_{12} & \lambda_{13} \\ \lambda_{21} & \lambda_{22} & \lambda_{23} \\ \lambda_{31} & \lambda_{32} & \lambda_{33} \end{bmatrix} \tag{6.3}$$

where λ_{nm} is the direction cosine of the local x_n axis relative to the global $\bar{x}_m$ axis.

The internal force acting on a cross section or the translation of a cross section of an element of a planar structure is a vector which lies in the plane of the structure. Moreover, one of the principal centroidal axes of a cross section of an element of a planar structure lies in the plane of the structure. We denote this axis by x_2. Furthermore as global axes $\bar{x}_1$ and $\bar{x}_2$ for a planar structure we choose two axes in its plane. Thus, referring to relation (B.29) or (B.30) the relation between the local and global components of an internal force acting on a cross section of an element of a planar structure can be written as

$$\{F\} = \begin{Bmatrix} F_1 \\ F_2 \end{Bmatrix} = [\Lambda_P] \begin{Bmatrix} \bar{F}_1 \\ \bar{F}_2 \end{Bmatrix} = [\Lambda_P]\{\bar{F}\} \tag{6.4}$$

or

$$\{\bar{F}\} = [\Lambda_P]^T \{F\} \tag{6.5}$$

The 2×2 matrix $[\Lambda_P]$ is the *transformation matrix for an element of a planar structure*. Referring to relation (B.31), we have

$$[\Lambda_P] = \begin{bmatrix} \lambda_{11} & \lambda_{12} \\ \lambda_{21} & \lambda_{22} \end{bmatrix} = \begin{bmatrix} \cos \phi_{11} & \sin \phi_{11} \\ -\sin \phi_{11} & \cos \phi_{11} \end{bmatrix} \tag{6.6}$$

where as shown in Fig. 6.1 ϕ_{11} is the angle between the local x_1 axis and the global $\bar{x}_1$ axis. From relations (6.1) and (6.2) or (6.4) and (6.5) it is apparent that the inverses of the matrices $[\Lambda_S]$ and $[\Lambda_P]$ are equal to their transposes.

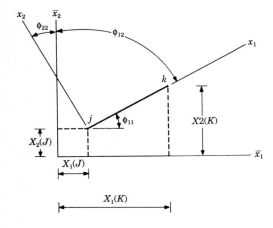

Figure 6.1 Element of a planar structure.

6.2 Computation of the Transformation Matrix for an Element of a Planar Structure

The length and the transformation matrix for each element of a planar structure are established from the global coordinates of the nodes to which the element is connected. This is accomplished by forming the following vector for each element of the structure.

$$\{L\}^T = [[X1(K) - X1(J)] \quad [X2(K) - X2(J)]] \tag{6.7}$$

where as shown in Fig. 6.1, $X1(K)$ is the $\bar{x}_1$ coordinate of the end k of the element.

The length of the element is equal to

$$L = [\{L\}^T \{L\}]^{1/2} \tag{6.8}$$

Referring to Fig. 6.1 or to relation (6.6) it is apparent that the orientation of the local x_1 and x_2 axes of an element is specified relative to the global $\bar{x}_1$ and $\bar{x}_2$ axes if the angle ϕ_{11} is known. The direction cosines λ_{11} and λ_{12} of the x_1 axis are equal to

$$\lambda_{11} = \cos \phi_{11} = \frac{X1(K) - X1(J)}{L}$$

$$\lambda_{12} = \cos \phi_{12} = \sin \phi_{11} = \frac{X2(K) - X2(J)}{L} \tag{6.9}$$

and

$$[\lambda] = [\lambda_{11} \quad \lambda_{12}] = \frac{\{L\}^T}{L} \tag{6.10}$$

Thus, the transformation matrix $[\Lambda_P]$ for an element of a planar structure [see relation (6.6)] can be formed if the coordinates of the nodes of the structure to which the element is connected are given.

6.3 Computation of the Transformation Matrix for an Element of a Space Structure

The length of an element of a space structure and the direction cosines of its axis with respect to the global axes of the structure are obtained from the global coordinates of the nodes to which the element is connected. To accomplish this we form the following vector for the element:

$$\{L\}^T = [[X1(K) - X1(J)] \quad [X2(K) - X2(J)] \quad [X3(K) - X3(J)]] \tag{6.11}$$

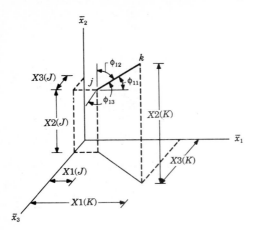

Figure 6.2 Element of a space structure.

Referring to Fig. 6.2, it can be seen that the length of an element of a space structure may be established by substituting relation (6.11) into (6.8). Moreover, the direction cosines of the axis of an element of a space structure are

$$\lambda_{11} = \cos \phi_{11} = \frac{X1(K) - X1(J)}{L}$$

$$\lambda_{12} = \cos \phi_{12} = \frac{X2(K) - X2(J)}{L} \qquad (6.12)$$

$$\lambda_{13} = \cos \phi_{13} = \frac{X3(K) - X3(J)}{L}$$

or

$$[\lambda] = [\lambda_{11} \quad \lambda_{12} \quad \lambda_{13}] = \frac{[L]^T}{L} \qquad (6.13)$$

It is apparent that the three direction cosines λ_{11}, λ_{12}, λ_{13} (6.13) of the axis of an element of a space structure do not specify the orientation of its local x_2 and x_3 axes with respect to the global $\bar{x}_1$, $\bar{x}_2$ and $\bar{x}_3$ axes of the structure. To accomplish this, additional information is required.

Referring to relations (B.23) it can be shown that the direction cosines of the local axes of an element of a space structure with respect to the global axes of the structure can be established if in addition to the three direction cosines of the axis of the element one direction cosine of its x_2 or x_3 axis is known. Often, however, the direction cosines of the local axes of an element of a space structure relative to the global axes of the structure are established from the following data:

1. The global coordinates of the nodes to which the element is connected.
2. The global coordinates of a point in the x_1x_2 or in the x_1x_3 plane which is not located on the x_1 axis. We call this point *node i*.

In the sequel, we present two procedures for establishing the transformation matrix $[\Lambda_S]$ of an element of a space structure from the data described above.

Procedure I. We assume that node i, whose coordinates are given, is located on the x_1x_2 plane.

Step 1. The direction cosines of the position vector of node i relative to end j of the element are computed using the following relation:

$$[\lambda^i] = [\lambda_{i1} \quad \lambda_{i2} \quad \lambda_{i3}]$$
$$= \frac{[[X1(I) - X1(J)] \quad [X2(I) - X2(J)] \quad [X3(I) - X3(J)]]}{\{[X1(I) - X1(J)]^2 + [X2(I) - X2(J)]^2 + [X3(I) - X3(J)]^2\}^{1/2}}$$

$$(6.14)$$

where $X1(I)$ is the $\bar{x}_1$ coordinate of node i.

Step 2. The direction cosines of the x_3 axis are computed from the following relation:

$$\mathbf{n}_3 = \lambda_{31}\bar{\mathbf{i}}_1 + \lambda_{32}\bar{\mathbf{i}}_2 + \lambda_{33}\bar{\mathbf{i}}_3 = \frac{\mathbf{n}_1 \times \mathbf{n}_i}{|\mathbf{n}_1 \times \mathbf{n}_i|}$$

$$(6.15)$$

where

$$\mathbf{n}_1 = \lambda_{11}\bar{\mathbf{i}}_1 + \lambda_{12}\bar{\mathbf{i}}_2 + \lambda_{13}\bar{\mathbf{i}}_3$$

$$(6.16a)$$

$$\mathbf{n}_i = \lambda_{i1}\bar{\mathbf{i}}_1 + \lambda_{i2}\bar{\mathbf{i}}_2 + \lambda_{i3}\bar{\mathbf{i}}_3$$

$$(6.16b)$$

$$|\mathbf{n}_1 \times \mathbf{n}_i| = \text{magnitude of vector } \mathbf{n}_1 \times \mathbf{n}_i$$

and $\bar{\mathbf{i}}_1$, $\bar{\mathbf{i}}_2$, and $\bar{\mathbf{i}}_3$ are the unit vectors in the direction of the global $\bar{x}_1$, $\bar{x}_2$, and $\bar{x}_3$ axes, respectively.

Step 3. The direction cosines of the x_2 axis of the element are computed from the following relation:

$$\mathbf{n}_2 = \lambda_{21}\bar{\mathbf{i}}_1 + \lambda_{22}\bar{\mathbf{i}}_2 + \lambda_{21}\bar{\mathbf{i}} = \frac{\mathbf{n}_3 \times \mathbf{n}_1}{|\mathbf{n}_3 \times \mathbf{n}_1|}$$

$$(6.17)$$

Procedure II. We assume that node i, whose coordinates are given, is located on the x_1x_2 plane. Moreover, we assume that the element is not parallel to the $\bar{x}_2$ axis. If an element is parallel to the $\bar{x}_2$ axis, the terms of its transformation matrix $[\Lambda_S]$ cannot be obtained directly from the formulas established in this section [see relation (6.35)]. However, in this case, the x_2 and x_3 local axes of the element specify a plane parallel to the $\bar{x}_1\bar{x}_3$ plane, and their direction cosines can be obtained by inspection.

Consider an element which is not parallel to the $\bar{x}_2$ axis and a set of axes $\bar{x}_1'$, $\bar{x}_2'$, $\bar{x}_3'$ which are parallel to the global axes of the structure and have their origin at the end j of the element. Moreover, consider a set of axes x_1^α, x_2^α, x_3^α obtained by rotating the set of axes $\bar{x}_1'$, $\bar{x}_2'$, $\bar{x}_3'$ about the $\bar{x}_2'$ axis by an angle α (see Fig. 6.3). The magnitude of the angle α is such that the x_1^α axis coincides with the line of intersection of the planes $\bar{x}_1'\bar{x}_3'$ and $\bar{x}_2'x_1$. Referring to Fig. 6.3, the direction cosines of the x_1^α, x_2^α, and x_3^α axes with respect to the $\bar{x}_1'$, $\bar{x}_2'$, and $\bar{x}_3'$ axes are

$$\lambda_{11}^\alpha = \cos\alpha \qquad \lambda_{21}^\alpha = 0 \qquad \lambda_{31}^\alpha = -\sin\alpha$$

$$\lambda_{12}^\alpha = 0 \qquad \lambda_{22}^\alpha = 1 \qquad \lambda_{32}^\alpha = 0 \qquad (6.18)$$

$$\lambda_{13}^\alpha = \sin\alpha \qquad \lambda_{23}^\alpha = 0 \qquad \lambda_{33}^\alpha = \cos\alpha$$

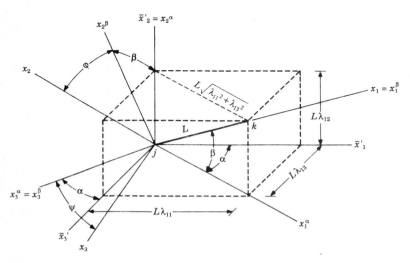

Figure 6.3 Rotation of the $\bar{x}_1'$, $\bar{x}_2'$, $\bar{x}_3'$ axes.

where

$$\sin \alpha = \frac{\lambda_{13}}{\sqrt{\lambda_{11}^2 + \lambda_{13}^2}} \qquad \cos \alpha = \frac{\lambda_{11}}{\sqrt{\lambda_{11}^2 + \lambda_{13}^2}} \tag{6.19}$$

Thus, referring to relations (6.1) and (6.3) the components of a vector **A** with respect to the x_1^α, x_2^α, x_3^α axes may be established from its global components using the following transformation relation:

$$\{A^\alpha\} = \begin{Bmatrix} A_1^\alpha \\ A_2^\alpha \\ A_3^\alpha \end{Bmatrix} = [\Lambda^\alpha] \begin{Bmatrix} \bar{A}_1 \\ \bar{A}_2 \\ \bar{A}_3 \end{Bmatrix} = [\Lambda^\alpha]\{\bar{A}\} \tag{6.20}$$

where
$$[\Lambda^\alpha] = \begin{bmatrix} \cos \alpha & 0 & \sin \alpha \\ 0 & 1 & 0 \\ -\sin \alpha & 0 & \cos \alpha \end{bmatrix} \tag{6.21}$$

Substituting relations (6.19) into (6.21), we obtain

$$[\Lambda^\alpha] = \begin{bmatrix} \dfrac{\lambda_{11}}{\sqrt{\lambda_{11}^2 + \lambda_{13}^2}} & 0 & \dfrac{\lambda_{13}}{\sqrt{\lambda_{11}^2 + \lambda_{13}^2}} \\ 0 & 1 & 0 \\ \dfrac{-\lambda_{13}}{\sqrt{\lambda_{11}^2 + \lambda_{13}^2}} & 0 & \dfrac{\lambda_{11}}{\sqrt{\lambda_{11}^2 + \lambda_{13}^2}} \end{bmatrix} \tag{6.22}$$

Consider another set of axes, x_1^β, x_2^β, x_3^β obtained by rotating the set of axes x_1^α, x_2^α, x_3^α about the x_3^α axis by an angle β. The magnitude of the angle β is such that the x_1^α axis coincides with the local x_1 axis of the element. It is apparent that the x_2^β and x_3^β axes are located in the plane perpendicular to the axis of the element. Referring to Fig. 6.3, the direction cosines of the x_1^β, x_2^β, x_3^β axes with respect to the x_1^α, x_2^α, and x_3^α axes are

$$
\begin{aligned}
&\lambda_{11}^\beta = \cos \beta & &\lambda_{21}^\beta = -\sin \beta & &\lambda_{31}^\beta = 0 \\
&\lambda_{12}^\beta = \sin \beta & &\lambda_{22}^\beta = \cos \beta & &\lambda_{32}^\beta = 0 \\
&\lambda_{13}^\beta = 0 & &\lambda_{23}^\beta = 0 & &\lambda_{33}^\beta = 1
\end{aligned} \tag{6.23}
$$

where

$$\sin \beta = \lambda_{12} \tag{6.24}$$

$$\cos \beta = \sqrt{\lambda_{11}^2 + \lambda_{13}^2} \tag{6.25}$$

Thus, referring to relations (6.1) and (6.3), the components of a vector **A** with respect to the x_1^β, x_2^β, x_3^β axes may be established from its components with respect to the x_1^α, x_2^α, x_3^α axes using the following transformation relation:

$$\{A^\beta\} = \begin{Bmatrix} A_1^\beta \\ A_2^\beta \\ A_3^\beta \end{Bmatrix} = [\Lambda^\beta] \begin{Bmatrix} A_1^\alpha \\ A_2^\alpha \\ A_3^\alpha \end{Bmatrix} = [\Lambda^\beta]\{A^\alpha\} \tag{6.26}$$

where

$$[\Lambda^\beta] = \begin{bmatrix} \cos\beta & \sin\beta & 0 \\ -\sin\beta & \cos\beta & 0 \\ 0 & 0 & 1 \end{bmatrix} \tag{6.27}$$

Substituting relations (6.24) and (6.25) into (6.27), we obtain

$$[\Lambda^\beta] = \begin{bmatrix} \sqrt{\lambda_{11}^2 + \lambda_{13}^2} & \lambda_{12} & 0 \\ -\lambda_{12} & \sqrt{\lambda_{11}^2 + \lambda_{13}^2} & 0 \\ 0 & 0 & 1 \end{bmatrix} \tag{6.28}$$

The local axes of the element are obtained by rotating the x_1^β, x_2^β, x_3^β axes about the $x_1^\beta = x_1$ axis by an angle ψ having the magnitude required for the x_2^β and x_3^β axes to coincide with the local x_2 and x_3 axes, respectively. The angle ψ is referred to as the *angle of the roll*. Referring to Fig. 6.3 the direction cosines of the local x_1, x_2, and x_3 axes of the element with respect to the x_1^β, x_2^β, and x_3^β axes are

$$\begin{array}{lll} \lambda_{11}^\psi = 1 & \lambda_{21}^\psi = 0 & \lambda_{31}^\psi = 0 \\ \lambda_{12}^\psi = 0 & \lambda_{22}^\psi = \cos\psi & \lambda_{32}^\psi = -\sin\psi \\ \lambda_{13}^\psi = 0 & \lambda_{23}^\psi = \sin\psi & \lambda_{33}^\psi = \cos\psi \end{array} \tag{6.29}$$

Thus, referring to relations (6.1) and (6.3) the local components of vector **A** may be obtained from its components with respect to the x_1^β, x_2^β, x_3^β axes by the following transformation relation:

$$\{A\} = [\Lambda^\psi]\{A^\beta\} \tag{6.30}$$

where

$$[\Lambda^\psi] = \begin{bmatrix} 1 & 0 & 0 \\ 0 & \cos\psi & \sin\psi \\ 0 & -\sin\psi & \cos\psi \end{bmatrix} \tag{6.31}$$

Substituting relation (6.20) into (6.26) and the resulting relation into

(6.30), we obtain

$$\{A\} = [\Lambda^\psi][\Lambda^\beta][\Lambda^\alpha]\{\bar{A}\} \tag{6.32}$$

Referring to relation (6.1), we have

$$\{A\} = [\Lambda_S]\{\bar{A}\} \tag{6.33}$$

where the transformation matrix $[\Lambda_S]$ is given by relation (6.3). Comparing relations (6.32) and (6.33), we have

$$[\Lambda_S] = [\Lambda^\psi][\Lambda^\beta][\Lambda^\alpha] \tag{6.34}$$

Substituting relations (6.22), (6.28), and (6.31) into relation (6.34) and performing the matrix multiplication, we obtain

$$[\Lambda_S] =$$

$$\begin{bmatrix} \lambda_{11} & \lambda_{12} & \lambda_{13} \\ -\dfrac{\lambda_{11}\lambda_{12}\cos\psi + \lambda_{13}\sin\psi}{\sqrt{\lambda_{11}^2 + \lambda_{13}^2}} & \sqrt{\lambda_{11}^2 + \lambda_{13}^2}\cos\psi & -\dfrac{\lambda_{12}\lambda_{13}\cos\psi - \lambda_{11}\sin\psi}{\sqrt{\lambda_{11}^2 + \lambda_{13}^2}} \\ \dfrac{\lambda_{11}\lambda_{12}\sin\psi - \lambda_{13}\cos\psi}{\sqrt{\lambda_{11}^2 + \lambda_{13}^2}} & -\sqrt{\lambda_{11}^2 + \lambda_{13}^2}\sin\psi & \dfrac{\lambda_{12}\lambda_{13}\sin\psi + \lambda_{11}\cos\psi}{\sqrt{\lambda_{11}^2 + \lambda_{13}^2}} \end{bmatrix}$$

$$\tag{6.35}$$

Using relation (6.12) the direction cosines λ_{11}, λ_{12}, and λ_{13} of the axis of an element can be established from the global coordinates of the nodes to which the element is connected. In what follows we establish formulas for the sine and cosine of the angle of the roll ψ of the local axes of an element involving the direction cosines of the axis of the element and the global coordinates $X1(I)$, $X2(I)$, and $X3(I)$ of a point (node i) on the $x_1 x_2$ plane which is not located on the x_1 axis of the element. Referring to relations (6.20) and (6.26) the coordinates $x_1^{i\beta}$, $x_2^{i\beta}$, and $x_3^{i\beta}$ of node i with respect to the x_1^β, x_2^β, and x_3^β axes are

$$\begin{Bmatrix} x_1^{i\beta} \\ x_2^{i\beta} \\ x_3^{i\beta} \end{Bmatrix} = [\Lambda^\beta][\Lambda^\alpha] \begin{Bmatrix} X1(I) - X1(J) \\ X2(I) - X2(J) \\ X3(I) - X3(J) \end{Bmatrix} \tag{6.36}$$

where $X1(J)$, $X2(J)$, and $X3(J)$ are the global coordinates of the end j of the element. That is, the term $[X1(I) - X1(J)]$ is the coordinate $\bar{x}_1'$ of node i. Substituting relations (6.22) and (6.27) in relation

(6.36) and performing the matrix multiplication, we obtain

$$x_1^{i\beta} = \lambda_{11}[X1(I) - X1(J)] + \lambda_{12}[X2(I) - X2(J)] + \lambda_{13}[X3(I) - X3(J)]$$

$$x_2^{i\beta} = -\frac{\lambda_{11}\lambda_{12}[X1(I) - X1(J)]}{\sqrt{\lambda_{11}^2 + \lambda_{13}^2}} + (\sqrt{\lambda_{11}^2 + \lambda_{13}^2})[X2(I) - X2(J)]$$

$$\qquad - \frac{\lambda_{12}\lambda_{13}[X3(I) - X3(J)]}{\sqrt{\lambda_{11}^2 + \lambda_{13}^2}} \qquad\qquad (6.37)$$

$$x_3^{i\beta} = -\frac{\lambda_{13}[X1(I) - X1(J)]}{\sqrt{\lambda_{11}^2 + \lambda_{13}^2}} + \frac{\lambda_{11}[X3(I) - X3(J)]}{\sqrt{\lambda_{11}^2 + \lambda_{13}^2}}$$

Referring to Fig. 6.4 and assuming that coordinate x_2 of node i is positive, it can be seen that

$$\sin \psi = \frac{x_3^{i\beta}}{\sqrt{(x_2^{i\beta})^2 + (x_3^{i\beta})^2}} \qquad \cos \psi = \frac{x_2^{i\beta}}{\sqrt{(x_2^{i\beta})^2 + (x_3^{i\beta})^2}} \qquad (6.38)$$

On the basis of the foregoing, it is apparent that the transformation matrix $[\Lambda_S]$ of an element of a space structure can be established from the global coordinates of the nodes to which the element is connected and from the global coordinates of a point (node i) on the $x_1 x_2$ plane which is not located on the x_1 axis of the element. This is accomplished, by adhering to the following steps:

Step 1. The direction cosines of the axis of the element are computed from the global coordinates of the nodes to which the element is connected using relation (6.13).

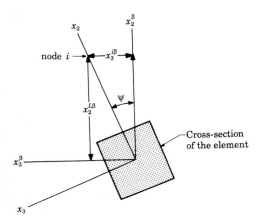

Figure 6.4 The angle of the roll ψ of the local axes of an element.

Step 2. The coordinates $x_1^{i\beta}$, $x_2^{i\beta}$, and $x_3^{i\beta}$ of node i are computed from its global coordinates $X1(I)$, $X2(I)$, and $X3(I)$ using relations (6.37).

Step 3. The sine and the cosine of the angle ψ are computed using relations (6.38).

Step 4. The transformation matrix $[\Lambda_S]$ is formed using relation (6.35). Notice that the matrix $[\Lambda_S]$ specifies the set of local axes x_1, x_2, and x_3 of the element under consideration with respect to which node i has a positive x_2 coordinate (see example).

Example. Using Procedure II, form the transformation matrix $[\Lambda_S]$ of each element of the space frame shown in Fig. a. The $\bar{x}_2$ axis is parallel to a principal axis is chosen as the local x_2 axis of these two elements. A principal centroidal axis of the cross section of element 3 is in the plane specified by points 3, 4, and node i, where node i is the projection of point 3 on the $\bar{x}_1\bar{x}_3$ plane.

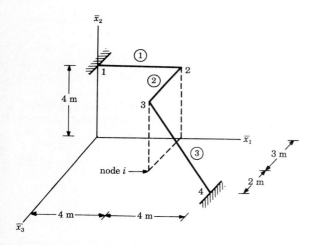

Figure a Geometry of the space frame.

---

solution The transformation matrices for elements 1 and 2 may be obtained by referring to Fig. a. They are

$$[\Lambda_S^1] = \begin{bmatrix} 1 & 0 & 0 \\ 0 & 1 & 0 \\ 0 & 0 & 1 \end{bmatrix} \qquad (a)$$

$$[\Lambda_S^2] = \begin{bmatrix} 0 & 0 & 1 \\ 0 & 1 & 0 \\ -1 & 0 & 0 \end{bmatrix} \qquad (b)$$

Formation of the transformation matrix $[\Lambda_S^3]$

STEP 1. We compute the direction cosines of the axis of element 3. The coordinates of nodes 3 and 4 of the structure of Fig. a are

$$\bar{x}_1^{(3)} = 4 \text{ m} \qquad \bar{x}_2^{(3)} = 4 \text{ m} \qquad \bar{x}_3^{(3)} = 3 \text{ m}$$
$$\bar{x}_1^{(4)} = 8 \text{ m} \qquad \bar{x}_2^{(4)} = 0 \qquad \bar{x}_3^{(4)} = 5 \text{ m} \qquad \text{(c)}$$

Substituting the values of these coordinates in relation (6.11) we get

$$\{L\}^T = [(8-4) \quad (0-4) \quad (5-3)] = [4 \quad -4 \quad 2] \qquad \text{(d)}$$

Substituting the matrix $\{L\}$ into relation (6.8), we obtain the length of element 3 as

$$L_3 = [\{L\}^T \{L\}]^{1/2} = \left[[4 \quad -4 \quad 2] \begin{Bmatrix} 4 \\ -4 \\ 2 \end{Bmatrix} \right]^{1/2} = (36)^{1/2} = 6 \qquad \text{(e)}$$

Substituting relations (d) and (e) into (6.13), the direction cosines of the axis of element 3 are

$$[\lambda] = [\lambda_{11} \quad \lambda_{12} \quad \lambda_{13}] = \frac{\{L\}^T}{L_3} = \begin{bmatrix} \dfrac{2}{3} & -\dfrac{2}{3} & \dfrac{1}{3} \end{bmatrix} \qquad \text{(f)}$$

STEP 2. We compute the coordinates $x_1^{i\beta}$, $x_2^{i\beta}$, and $x_3^{i\beta}$ of node i. The global coordinates of node i which is located in the plane specified by the x_1, x_2 axes of element 3 are

$$X1(I) = 4 \text{ m} \qquad X2(I) = 0 \qquad X3(I) = 3 \text{ m} \qquad \text{(g)}$$

The global coordinates of end j of the element (node 3) are

$$X1(J) = 4 \text{ m} \qquad X2(J) = 4 \text{ m} \qquad X3(J) = 3 \text{ m} \qquad \text{(h)}$$

Substituting relations (f) to (h) into (6.37) we obtain

$$x_1^{i\beta} = \tfrac{2}{3}(4-4) - \tfrac{2}{3}(0-4) + \tfrac{1}{3}(3-3) = 2.66667$$
$$x_2^{i\beta} = \tfrac{1}{3}(\sqrt{2^2+1})(0-4) = -\frac{4\sqrt{5}}{3} \qquad \text{(i)}$$
$$x_3^{i\beta} = 0$$

STEP 3. We compute the sine and cosine of the angle ψ. Substituting relation (i) into (6.38), we have

$$\sin \psi = 0 \qquad \cos \psi = -1 \qquad \text{(j)}$$

STEP 4. We form the transformation matrix $[\Lambda_S]$. Substituting relations (f) and (j) into (6.35) we obtain

$$[\Lambda_S^3] = \begin{bmatrix} 0.66667 & -0.66667 & 0.33333 \\ -0.59628 & -0.745356 & -0.29814 \\ 0.44721 & 0 & -0.89442 \end{bmatrix} \tag{k}$$

The direction of the x_2 axis of element 3 is shown in Fig. b. Notice that node i has a positive x_2 coordinate.

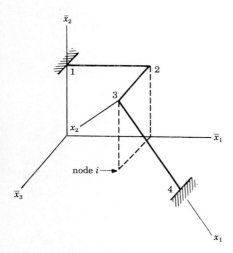

Figure b Direction of the x_2 axis of element 3.

6.4   Transformation of the Matrices of Nodal Actions of an Element

In this section, we give the transformation relations between the local and global matrices of nodal actions of the elements of the various types of structures that we are considering.

Element of a planar truss.   Consider an element of a planar truss and choose the x_1, x_2 and $\bar{x}_1, \bar{x}_2$ axes in the plane of the truss. Referring to Fig. 6.5 we have

$$\begin{aligned} F_1^j &= \bar{F}_1^j \cos \phi_{11} + \bar{F}_2^j \sin \phi_{11} \\ F_1^k &= \bar{F}_1^k \cos \phi_{11} + \bar{F}_2^k \sin \phi_{11} \end{aligned} \tag{6.39}$$

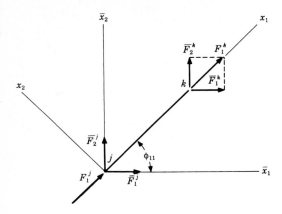

Figure 6.5 Free-body diagram of an element of a truss.

Hence,

$$\{A\} = \begin{Bmatrix} F_1^j \\ F_1^k \end{Bmatrix} = \begin{bmatrix} \cos\phi_{11} & \sin\phi_{11} & 0 & 0 \\ 0 & 0 & \cos\phi_{11} & \sin\phi_{11} \end{bmatrix} \begin{Bmatrix} \bar{F}_1^j \\ \bar{F}_2^j \\ \bar{F}_1^k \\ \bar{F}_2^k \end{Bmatrix} = [\Lambda_{PT}]\{\bar{A}\}$$

(6.40)

Moreover,

$$\{\bar{A}\} = \begin{Bmatrix} \bar{F}_1^j \\ \bar{F}_2^j \\ \bar{F}_1^k \\ \bar{F}_2^k \end{Bmatrix} = \begin{bmatrix} \cos\phi_{11} & 0 \\ \sin\phi_{11} & 0 \\ 0 & \cos\phi_{11} \\ 0 & \sin\phi_{11} \end{bmatrix} \begin{Bmatrix} F_1^j \\ F_1^k \end{Bmatrix} = [\Lambda_{PT}]^T\{A\}$$

(6.41)

where

$$[\Lambda_{PT}] = \begin{bmatrix} \lambda_{11} & \lambda_{12} & 0 & 0 \\ 0 & 0 & \lambda_{11} & \lambda_{12} \end{bmatrix} = \begin{bmatrix} \cos\phi_{11} & \sin\phi_{11} & 0 & 0 \\ 0 & 0 & \cos\phi_{11} & \sin\phi_{11} \end{bmatrix}$$

(6.42)

Element of a planar beam or a planar frame. Consider an element of a planar beam or a planar frame and choose the x_1, x_2 and the $\bar{x}_1$, $\bar{x}_2$ axes in the plane of the structure. At any cross section of this element, there could be two components of force in the $x_1 x_2$ plane and one component of moment, whose vector is normal to the $x_1 x_2$ plane. Noting that M_3^q is equal to $\bar{M}_3^q$ ($q = j$ or k), and referring to relations (6.4) and (6.5) the matrices $\{A^q\}$ and $\{\bar{A}^q\}$ ($q = j$ or k) of the element under consideration are related by the following transformation relations:

$$\{A^q\} = \begin{Bmatrix} F_1^q \\ F_2^q \\ M_3^q \end{Bmatrix} = [\hat{\Lambda}_{PF}] \begin{Bmatrix} \bar{F}_1^q \\ \bar{F}_2^q \\ \bar{M}_3^q \end{Bmatrix} = [\hat{\Lambda}_{PF}]\{\bar{A}^q\} \qquad q = j \text{ or } k \quad (6.43)$$

and
$$\{\bar{A}^q\} = [\hat{\Lambda}_{PF}]^T \{A^q\} \tag{6.44}$$

where
$$[\hat{\Lambda}_{PF}] = \begin{bmatrix} [\Lambda_P] & [0] \\ [0] & 1 \end{bmatrix} \tag{6.45}$$

The matrix $\{\Lambda_P\}$ is the transformation matrix for the element given by relations (6.6).

The matrices of nodal actions $\{A\}$ and $\{\bar{A}\}$ of an element of a planar beam or a planar frame are related by the following transformation relations:

$$\{A\} = \begin{Bmatrix} F_1^j \\ F_2^j \\ M_3^j \\ F_1^k \\ F_2^k \\ M_3^k \end{Bmatrix} = \begin{Bmatrix} \{A^j\} \\ \{A^k\} \end{Bmatrix} = [\Lambda_{PF}] \begin{Bmatrix} \{\bar{A}^j\} \\ \{\bar{A}^k\} \end{Bmatrix} = [\Lambda_{PF}]\{\bar{A}\} \tag{6.46}$$

and
$$\{\bar{A}\} = [\Lambda_{PF}]^T \{A\} \tag{6.47}$$

where
$$[\Lambda_{PF}] = \begin{bmatrix} [\hat{\Lambda}_{PF}] & [0] \\ [0] & [\hat{\Lambda}_{PF}] \end{bmatrix} = \begin{bmatrix} [\Lambda_P] & [0] & [0] & [0] \\ [0] & 1 & [0] & [0] \\ [0] & [0] & [\Lambda_P] & [0] \\ [0] & [0] & [0] & 1 \end{bmatrix} \tag{6.48}$$

The matrix $[\Lambda_P]$ is the transformation matrix for the element given by relation (6.6).

Element of a space truss. The matrices of nodal actions $\{A\}$ and $\{\bar{A}\}$ of an element of a space truss are related by the following transformation relations:

$$\{A\} = \begin{Bmatrix} F_1^j \\ F_1^k \end{Bmatrix} = [\Lambda_{ST}] \begin{Bmatrix} \bar{F}_1^j \\ \bar{F}_2^j \\ \bar{F}_3^j \\ \bar{F}_1^k \\ \bar{F}_2^k \\ \bar{F}_3^k \end{Bmatrix} = [\Lambda_{ST}]\{\bar{A}\} \tag{6.49}$$

and
$$\{\bar{A}\} = [\Lambda_{ST}]^T \{A\} \qquad (6.50)$$

where
$$[\Lambda_{ST}] = \begin{bmatrix} \lambda_{11} & \lambda_{12} & \lambda_{13} & 0 & 0 & 0 \\ 0 & 0 & 0 & \lambda_{11} & \lambda_{12} & \lambda_{13} \end{bmatrix} \qquad (6.51)$$

λ_{11}, λ_{12}, and λ_{13} are the direction cosines of the axis of the element.

Element of a space beam or a space frame. Consider an element of a space beam or a space frame. At any cross section of this element, there could be three components of force and three components of moment. Thus, the matrices $\{A^q\}$ and $\{\bar{A}^q\}$ of the local and global components of the internal actions, respectively, acting at the end q ($q = j$ or k) of this element are related by the following transformation relations:

$$\{A^q\} = \begin{Bmatrix} F_1^q \\ F_2^q \\ F_3^q \\ M_1^q \\ M_2^q \\ M_3^q \end{Bmatrix} = [\hat{\Lambda}_{SF}] \begin{Bmatrix} \bar{F}_1^q \\ \bar{F}_2^q \\ \bar{F}_3^q \\ \bar{M}_1^q \\ \bar{M}_2^q \\ \bar{M}_3^q \end{Bmatrix} = [\hat{\Lambda}_{SF}]\{\bar{A}^q\} \qquad q = j \text{ or } k \quad (6.52)$$

and
$$\{\bar{A}^q\} = [\hat{\Lambda}_{SF}]^T \{A^q\} \qquad (6.53)$$

where
$$[\hat{\Lambda}_{SF}] = \begin{bmatrix} [\Lambda_S] & [0] \\ [0] & [\Lambda_S] \end{bmatrix} \qquad (6.54)$$

The matrix $\{\Lambda_S\}$ is the transformation matrix of the element given by relation (6.3). The (6×6) matrix $[\hat{\Lambda}_{SF}]$ is referred to as a *superdiagonal matrix*.

The matrices of nodal actions $\{A\}$ and $\{\bar{A}\}$ of an element of a space beam or a space frame are related by the following transformation relations:

$$\{A\} = \begin{Bmatrix} \{A^{-j}\} \\ \{A^k\} \end{Bmatrix} = [\Lambda_{SF}] \begin{Bmatrix} \{\bar{A}^j\} \\ \{\bar{A}^k\} \end{Bmatrix} = [\Lambda_{SF}]\{\bar{A}\} \qquad (6.55)$$

and
$$\{\bar{A}\} = [\Lambda_{SF}]^T \{A\} \qquad (6.56)$$

where
$$[\Lambda_{SF}] = \begin{bmatrix} [\hat{\Lambda}_{SF}] & [0] \\ [0] & [\hat{\Lambda}_{SF}] \end{bmatrix} \qquad (6.57)$$

where the matrix $[\hat{\Lambda}_{SF}]$ is given by relation (6.54).

Element of a grid. Consider a grid in the x_1x_2 plane subjected at its nodes to external forces acting in the direction normal to its plane and to external moments whose vector is in the plane of the grid (see Fig. 6.6). The axis normal to the plane of the grid is a principal centroidal axis of its elements and is chosen as their local axis x_3. An element of this grid is subjected to nodal actions F_3^j, F_3^k, M_1^j, M_1^k, M_2^j, and M_2^k. Noting that F_3^q ($q = j$ or k) is identical to $\bar{F}_3^q$ and referring to relation (B.29) the matrices $\{A^q\}$ and $\{\bar{A}^q\}$ of the element under consideration are related by the following transformation relations:

$$\{A^q\} = \begin{Bmatrix} F_3^q \\ M_1^q \\ M_2^q \end{Bmatrix} = [\Lambda_G] \begin{Bmatrix} \bar{F}_3^q \\ \bar{M}_1^q \\ \bar{M}_2^q \end{Bmatrix} = [\Lambda_G]\{\bar{A}^q\} \qquad q = j \text{ or } k \quad (6.58)$$

and

$$\{\bar{A}^q\} = [\Lambda_G]^T \{A^q\} \tag{6.59}$$

where

$$[\Lambda_G] = \begin{bmatrix} 1 & 0 & 0 \\ 0 & \lambda_{11} & \lambda_{12} \\ 0 & \lambda_{21} & \lambda_{22} \end{bmatrix} \tag{6.60}$$

The matrices of nodal actions $[A]$ and $[\bar{A}]$ of an element of a grid are related by the following relations:

$$\{A\} = \begin{Bmatrix} \{A^j\} \\ \{A^k\} \end{Bmatrix} = [\Lambda_{GG}] \begin{Bmatrix} \{\bar{A}^j\} \\ \{\bar{A}^k\} \end{Bmatrix} = [\Lambda_{GG}]\{\bar{A}\} \tag{6.61}$$

$$\{\bar{A}\} = [\Lambda_{GG}]^T \{A\} \tag{6.62}$$

where

$$[\Lambda_{GG}] = \begin{bmatrix} [\Lambda_G] & [0] \\ [0] & [\Lambda_G] \end{bmatrix} \tag{6.63}$$

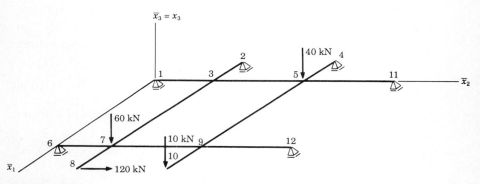

Figure 6.6 Grid.

Generalization of the results. On the basis of the foregoing, the relations between the matrices of nodal actions $\{A\}$ and $\{\bar{A}\}$ of an element of a structure may be written as

$$\{A\} = [\Lambda]\{\bar{A}\} \tag{6.64}$$

$$\{\bar{A}\} = [\Lambda]^T\{A\} \tag{6.65}$$

where depending on the type of the structure the matrix $[\Lambda]$ is one of the matrices $[\Lambda_{PT}]$, $[\Lambda_{PF}]$, $[\Lambda_{ST}]$, $[\Lambda_{SF}]$, or $[\Lambda_{GG}]$ given by relations (6.42), (6.48), (6.51), (6.57), and (6.63), respectively. From relations (6.64) and (6.65), it is apparent that

$$[\Lambda][\Lambda]^T = [I] \tag{6.66}$$

where $[I]$ is the unit matrix.

6.5 Transformation of the Matrices of Nodal Displacements of an Element

It is apparent that the transformation relations between the local and global matrices of nodal displacements are analogous to those between the nodal and global matrices of nodal actions. That is, in general, referring to relations (6.64) and (6.65) the relations between the local and global matrices of nodal displacements can be written as

$$\{D\} = [\Lambda]\{\bar{D}\} \tag{6.67}$$

and $$\{\bar{D}\} = [\Lambda]^T\{D\} \tag{6.68}$$

where, depending on the type of structure, the matrix $[\Lambda]$ is one of the matrices $[\Lambda_{PT}]$, $[\Lambda_{PF}]$, $[\Lambda_{ST}]$, $[\Lambda_{SF}]$, or $[\Lambda_{GG}]$, given by relations (6.42), (6.48), (6.51), (6.57), and (6.63), respectively.

6.6 Transformation of the Local Stiffness Matrix to Global

In this section we express the stiffness equations (4.8) for an element of a framed structure in global form. That is,

$$\{\bar{A}^E\} = [\bar{K}]\{\bar{D}\} \tag{6.69}$$

For this purpose consider relations (4.8) for an element of a framed structure. That is,

$$[A^E] = [K][D] \tag{6.70}$$

Substituting relation (6.67) into (6.70), we obtain

$$[A^E] = [\bar{K}][\bar{D}] \tag{6.71}$$

We call the matrix $[\overset{\square}{K}]$ the *hybrid stiffness matrix for the element*. It transforms the global components of nodal displacements of an element to the local components of its nodal actions. It is given as

$$[\overset{\square}{K}] = [K][\Lambda] \tag{6.72}$$

where depending on the type of the structure the matrix $[\Lambda]$ is one of the matrices $[\Lambda_{PT}]$, $[\Lambda_{PF}]$, $[\Lambda_{ST}]$, $[\Lambda_{SF}]$, or $[\Lambda_{GG}]$ given by relations (6.42), (6.48), (6.51), (6.57), and (6.63), respectively.

Premultiplying each side of relation (6.71) by $[\Lambda]^T$ and using relation (6.67) we obtain

$$\{\bar{A}^E\} = [\Lambda]^T\{A^E\} = [\Lambda]^T[\overset{\square}{K}]\{\bar{D}\} = [\Lambda]^T[K][\Lambda][\bar{D}] \tag{6.73}$$

Referring to relations (6.69) and (6.73) we may conclude that

$$[\bar{K}] = [\Lambda]^T[K][\Lambda] = [\Lambda]^T[\overset{\square}{K}] \tag{6.74}$$

Equation (6.74) is employed in obtaining the global stiffness matrix of an element from its local stiffness matrix.

The global matrix of nodal actions of an element of a framed structure subjected to given loads can be expressed as

$$\{\bar{A}\} = \{\bar{A}^E\} + \{\bar{A}^R\} = [\bar{K}]\{\bar{D}\} + \{\bar{A}^R\} \tag{6.75}$$

6.7 Problems

1. Construct the transformation matrix $[\Lambda_S]$ of each element of the space frame shown in Fig. 6.P1. The principal centroidal axis x_2 of the cross section of each element of the frame is in the plane specified by the axis of the element and the projection of node 2 on the $\bar{x}_1 x_3$ plane (point A).

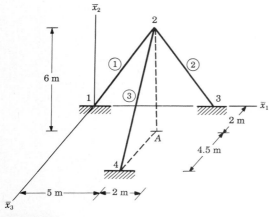

Figure 6.P1

2. Construct the transformation matrix $[\Lambda_S]$ of each element of the space frame shown in Fig. 6.P2. The $\bar{x}_3$ axis is a principal centroidal axis of the cross section of elements 1 and 4, and it is chosen as the local x_2 axis of these elements. The $\bar{x}_2$ axis is a principal centroidal axis of element 3, and it is chosen as its local x_2 axis. The principal centroidal axis x_2 of the cross section of element 2 is in the plane specified by points 2, 3, and A, where point A is the projection of point 2 on the $\bar{x}_2$ axis. The principal centroidal axis x_2 of element 5 is in the plane parallel to the $\bar{x}_2\bar{x}_3$ plane. The sense of this axis is such that its points have positive x_2 coordinates.

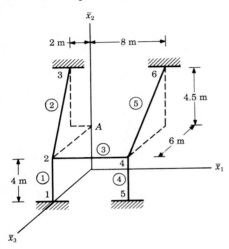

Figure 6.P2

3. Construct the transformation matrix $[\Lambda_S]$ of each element of the space frame shown in Fig. 6.P3. The $\bar{x}_2$ axis is a principal centroidal axis of

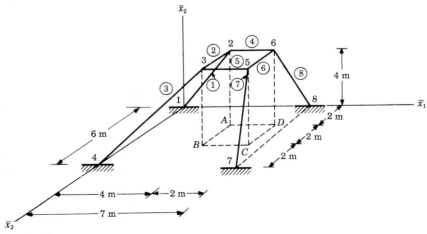

Figure 6.P3

elements 2, 4, 5, and 6, and it is taken as their local x_2 axis. The principal centroidal axis x_2 of elements 1, 3, 7, and 8 is in the planes specified by the axis of the element and points A, B, C, and D, respectively. Points A, B, C, and D are the projections of nodes 2, 3, 5, and 6, respectively, on the x_1x_3 plane.

4. Compute the global stiffness matrix for the element of constant cross section and length L of a truss shown in Fig. 6.P4. The element is made from an isotropic linearly elastic material.

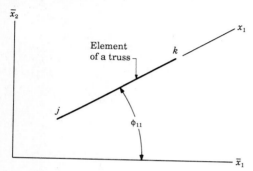

Figure 6.P4

5. Compute the global stiffness matrix for the general planar element of constant cross section and length L of a planar frame shown in Fig. 6.P5. The element is made from an isotropic linearly elastic material.

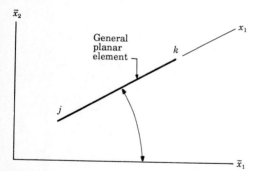

Figure 6.P5

Photographs and a Brief Description of an Airline Hangar

The Eastern Airlines hangar in Miami, Florida (see Fig. 6.7) is 85.35 × 143.12 m (280 × 440 ft) in plan. Its roof is a 6.10-m-(20-ft-) deep space truss whose top chord members are skewed 45° to the bottom chord members. This truss has simpler chord connections and shorter top chord members than the usual space trusses whose top and bottom chord members are parallel (see Fig. 6.8). Six members, lying in three planes, frame into each top chord joint (Fig. 6.8a), and eight members, lying in only two planes, frame into each bottom chord joint (Fig. 6.8b). Notice that the members of the top chord of a simply supported truss are subjected to axial compression. Consequently, their critical loads at buckling are an important design criterion. The critical load at buckling is inversely proportional to the square of the length of the member and directly proportional to the moment of inertia of its cross section ($P_{cr} = EI\pi^2/L^2$). Consequently, shorter members require a cross section with less moment of inertia and, thus, less area than longer members. For this reason, the space truss of the Eastern Airlines hanger is lighter and its cost is less than that of a truss with parallel top and bottom chord members covering the same area. The truss is supported on steel columns encased in concrete made of shapes specially rolled in the length required [maximum column height is 33.53 m (110 ft)]. The columns are located as shown in the plans of the top and bottom chords (see Fig. 6.9). The truss members and the columns are made of A572 steel [344.73 N/mm² (50 kips/in²)], while the wind bracing and the purlings are made of A36 steel [248.21 N/mm² (36 kips/in²)]. The weight of the truss is 0.862 kN/m² (18 lb/ft²). All connections are shop-welded or field-bolted. The top chord of the truss supports the roof skin consisting of a 76.2-mm-(3-in-) thick metal deck, which is covered by a 50.8-mm-(2-in-) thick insulation plank topped with two layers of bituminous cardboard. The bottom chord of the truss supports a fireproof hung ceiling. The columns are braced, and the spacing between them is filled by concrete blocks. The hangar was designed to withstand hurricane winds. The horizontal forces of the top of the 24.38-m- (80-ft-) high doors and walls are transmitted to the horizontal edge trusses and through the space truss acting as a horizontal diaphragm to the wind-bracing system.

(a) View of the hanger

(b) View of the roof truss during construction

Figure 6.7 The Eastern Airlines hangar in Miami, Florida. (*Courtesy Geiger Gossen Hamilton Liao Engineers P.C., New York, N.Y.*)

(a) Joint of the top chord

(b) Joint of the bottom chord

Figure 6.8 Details of the joints of the roof truss of the Eastern Airlines hangar. (*Courtesy Geiger Gossen Hamilton Liao Engineers P.C., New York, N.Y.*)

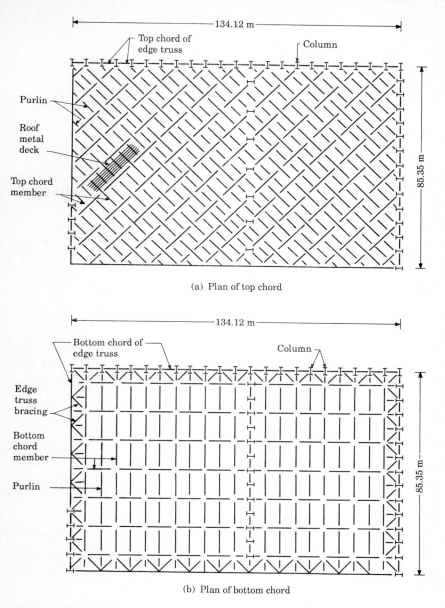

(a) Plan of top chord

(b) Plan of bottom chord

Figure 6.9 Plans of the top and bottom chords of the roof truss of the Eastern Airlines hangar. (*Courtesy Geiger Gossen Hamilton Liao Engineers P.C., New York, N.Y.*)

The Direct Stiffness Method

Analysis of the Restrained Structure and Computation of the Equivalent Actions

7.1 Restrained Structure—Structure Subjected to Equivalent Actions

Consider the frame subjected to the loading shown in Fig. 7.1a. The external actions acting on this frame are equal to the sum of the corresponding external actions acting on the frame subjected to the loads shown in Fig. 7.1b and c. Consequently, since the principle of superposition is valid for the structures which we are considering (see Sec. 1.12), the internal actions at any cross section of an element and the components of displacement of any point of the frame, loaded as shown in Fig. 7.1a, are equal to the sum of the corresponding quantities of the frame loaded as shown in Fig. 7.1b and c.

Notice that the values of the external actions $\bar{S}_1^{(2)}$, $\bar{S}_2^{(2)}$, $\bar{S}_3^{(2)}$, $\bar{S}_1^{(3)}$, $\bar{S}_2^{(3)}$, $\bar{S}_3^{(3)}$ can be chosen so that the components of translation and of rotation of nodes 2 and 3 of the frame loaded as shown in Fig. 7.1b vanish. For this choice of the external actions $\bar{S}_1^{(2)}$, $\bar{S}_2^{(2)}$, $\bar{S}_3^{(2)}$, $\bar{S}_1^{(3)}$, $\bar{S}_2^{(3)}$, and $\bar{S}_3^{(3)}$, the frame of Fig. 7.1b becomes kinematically determinate (see Sec. 1.8). In this case the structure of Fig. 7.1b is called the *restrained structure*.

The external actions $\bar{S}_1^{(2)}$, $\bar{S}_2^{(2)}$, $\bar{S}_3^{(2)}$, $\bar{S}_1^{(3)}$, $\bar{S}_2^{(3)}$, and $\bar{S}_2^{(3)}$ are called the *restraining actions*. Moreover, the external actions acting on the nodes of the structure of Fig. 7.1c are called the *equivalent actions*.

In the modern methods of structural analysis the *element approach is used*. That is, a structure is considered as an assemblage of

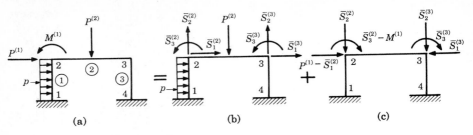

Figure 7.1 Principle of superposition.

elements, and its response is determined from the responses of its elements. This is accomplished by considering the response of a structure as the sum of its responses when it is subjected to the following loading cases.

Case 1. The structure subjected to the given loads acting along the length of its elements and to the unknown restraining actions which must be applied to its nodes in order to restrain them from moving (translating and rotating). As mentioned previously we call the structure subjected to these loads the *restrained structure* (see Fig. 7.2b). The translations and rotations of the supports of a structure may or may not be included in this loading.

Case 2. The structure subjected to the equivalent actions on its nodes (see Fig. 7.2c) and to the given translations and rotations of its supports, if they have not been included in the loading of the restrained structure. The components of the equivalent actions which must be applied to a node of a structure are equal to the sum of the

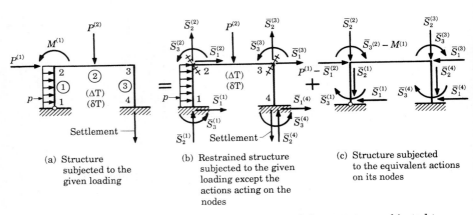

(a) Structure subjected to the given loading

(b) Restrained structure subjected to the given loading except the actions acting on the nodes

(c) Structure subjected to the equivalent actions on its nodes

Figure 7.2 Superposition of the restrained structure and the structure subjected to the equivalent actions.

corresponding components of the following actions:

1. The given actions acting on this node of the structure.
2. Actions equal and opposite to the restraining actions acting on this node of the restrained structure.

The restrained structure is comprised of elements which are either fixed at both ends or pinned at both ends. Thus, the external loads acting along the length of an element of the restrained structure affect only this element. These loads could include one or more of the following disturbances:

1. The given external action acting along the length of the corresponding element of the actual structure.
2. The given change of temperature of the corresponding element of the actual structure.
3. The effect of lack of fit of the corresponding element of the actual structure (see Secs. 1.3 and 2.3).
4. Possibly the loads due to the given translations and rotations of the supports of the actual structure. As becomes apparent in Example 3 of Sec. 9.3 it is preferable not to include the given translations and rotations of the supports of a structure in the loading of the restrained structure.

Notice that by considering the equilibrium of a node of the restrained structure, it can be shown that the global components of the restraining actions acting on this node are equal to the sum of the corresponding global components of the actions acting at the nodal points of the elements which are connected to this node. Moreover, notice that for uniformity of treatment we consider the reactions of the supports of the restrained structure as restraining actions (see Fig. 7.2).

Referring to Fig. 7.2, it is apparent that the superposition of the components of displacement of nodes 1, 2, and 3 of the restrained structure and those of the structure subjected to the equivalent actions must yield the corresponding components of displacement of nodes 1, 2, and 3, respectively, of the structure subjected to the given loading. Inasmuch as the components of displacement of nodes 1, 2, and 3 of the restrained structure are zero, the components of displacement of nodes 1, 2, and 3 of the structure, subjected to the equivalent actions on its nodes, must be equal to the corresponding components of displacement of the nodes of the structure subjected to the given loading. Moreover, the superposition of the reactions and the equivalent actions acting at the supports of the structure, subjected to the equivalent actions and the corresponding reactions of the restrained structure, yields the corresponding reactions of the actual structure subjected to the given loads. However, each component of reaction of a support of the restrained structure (see Fig. 7.2b) is equal and opposite to the corresponding component of equivalent action applied to the same support of the structure subjected to the equivalent actions (see Fig. 7.2c). Thus, it is apparent that the reactions of the structure subjected to the equivalent actions are equal to those of the structure subjected to the given loads. Finally, the nodal actions of the elements of the structure subjected to the given loads are equal to the sum of the corresponding nodal actions of the elements of the restrained structure and of the elements of the structure subjected to the equivalent actions.

7.2 Computation of the Equivalent Actions

The restrained structure consists of elements fixed at both ends or pinned at both ends and subjected to the given loads. The response of the elements of the restrained structure is specified by their matrix of fixed-end actions $\{A^R\}$ or $\{\bar{A}^R\}$ [see Eq. (4.8)]. The terms of the matrices $\{\bar{A}^R\}$ of the elements of a structure are used to establish the equivalent actions to be placed on the nodes of the structure. For elements of constant cross section subjected to certain loads which are frequently encountered in practice the matrix $\{A^R\}$ can be established by referring to the table on the inside of the back cover. For elements of constant cross section subjected to complex load distributions or for elements of variable cross section subjected to any loads, the matrix $\{A^R\}$ is established using approximate methods (see Secs. 12.3 and 19.1.2).

Our intent in this section is to familiarize the reader with the analysis of the restrained structure and the computation of the equivalent actions. For this reason we consider only structures whose elements have constant cross section and, moreover, are subjected to loads which are included in the table on the inside of the back cover of the book. Thus the matrices $\{A^R\}$ of the elements of the restrained structure can be established by referring to that table.

In order to compute the equivalent actions to be placed on the nodes of a structure we adhere to the following steps.

1. We compute the local matrix of fixed-end actions $\{A^R\}$ of each element of the structure subjected to the given loading. For elements of constant cross section subjected to certain loads which are frequently encountered in practice, the matrix $\{A^R\}$ is given in the table on the inside of the back cover.

2. We transform the local matrix of fixed-end actions of each

element to global. That is, referring to relation (6.65) we have

$$\{\bar{A}^R\} = \{A\}^T\{A^R\} \tag{7.1}$$

where, depending on the type of structure, the matrix $[\Lambda]$ is one of the matrices $[\Lambda_{PT}]$, $[\Lambda_{PF}]$, $[\Lambda_{ST}]$, $[\Lambda_{SF}]$, or $[\Lambda_{GG}]$ given by relations (6.42), (6.48), (6.51), (6.57), or (6.63), respectively.

3. We establish the matrix of the restraining actions $\{\widehat{RA}\}$ which must be placed on the nodes of the restrained structure by considering their equilibrium. In the matrix $\{\widehat{RA}\}$ we store the global components of the restraining actions acting on the nodes of the restrained structure including the reactions of its supports. The global components of the restraining action acting on a node are equal to the sum of the global components of the nodal actions acting at the ends of the elements of the restrained structure (fixed-end actions) connected to this node.

4. We store in a matrix $\{\hat{P}^G\}$ the global components of the given actions acting on the nodes of the structure including those directly absorbed by its supports.

5. We establish the matrix of the equivalent actions $\{\hat{P}^E\}$ to be applied to the nodes of the structure. That is,

$$\{\hat{P}^E\} = \{\hat{P}^G\} - \{\widehat{RA}\} \tag{7.2}$$

We illustrate the computation of the equivalent actions for planar structures whose elements have a constant cross section by the following six examples.

Example 1. Illustrates the computation of the equivalent actions for a frame subjected to external actions along the length of its elements.

Example 2. Illustrates the computation of the equivalent actions for a truss when some of its elements are subjected to a uniform temperature increase.

Example 3. Illustrates the computation of the equivalent actions for a frame subjected to (a) a uniform change of temperature ΔT and (b) a difference in temperature δT between its internal and external fibers.

Example 4. Illustrates the computation of the equivalent actions for a frame subjected to a settlement of one of its supports.

Example 5. Illustrates the computation of the equivalent actions for a frame subjected to a rotation of one of its supports.

Example 6. Illustrates the computation of the equivalent actions for a frame constructed with an element which did not fit.

Example 1. Establish the matrix of equivalent actions for the frame loaded as shown in Fig. a.

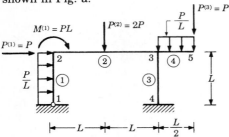

Figure a Geometry and loading of the frame.

solution Notice that we could analyze the structure shown in Fig. b instead of that of Fig. a since the distribution of the internal actions in the elements of the two structures is the same.

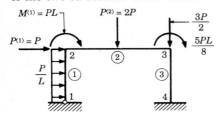

Figure b

As shown in Fig. c, the restrained structure is formed by restraining the nodes of the frame from moving by applying restraining actions to them.

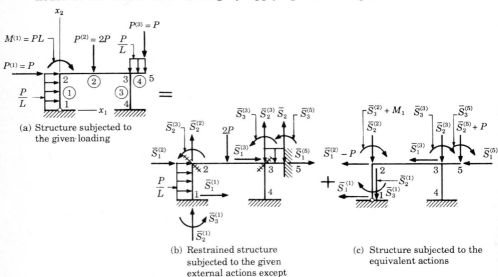

(a) Structure subjected to the given loading

(b) Restrained structure subjected to the given external actions except those acting on the nodes

(c) Structure subjected to the equivalent actions

Figure c Superposition of the restrained structure and the structure subjected to the equivalent actions.

Notice that the loading of the restrained structure does not include the moment $M^{(1)}$ and the force $P^{(1)}$ because they act on the nodes of the frame.

STEP 1. We establish the matrix of fixed-end actions of each element of the structure by referring to the table on the inside of the back cover. The free-body diagrams of the elements and the nodes of the restrained structure are shown in Fig. d. Referring to this figure we have

$$\{A^{R1}\} = \begin{Bmatrix} F_1^{1j} \\ F_2^{1j} \\ M_3^{1j} \\ \hline F_1^{1k} \\ F_2^{1k} \\ M_3^{1k} \end{Bmatrix} = \begin{Bmatrix} 0 \\ P/2 \\ PL/12 \\ \hline 0 \\ P/2 \\ -PL/12 \end{Bmatrix} \qquad \{A^{R2}\} = \begin{Bmatrix} 0 \\ P \\ PL/2 \\ \hline 0 \\ P \\ -PL/2 \end{Bmatrix}$$

$$\{A^{R3}\} = \{0\} \qquad\qquad\qquad \{A^{R4}\} = \begin{Bmatrix} 0 \\ P/4 \\ PL/48 \\ \hline 0 \\ P/4 \\ -PL/48 \end{Bmatrix} \qquad (a)$$

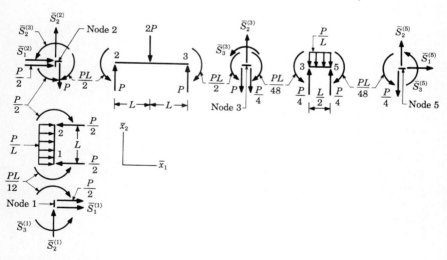

Figure d Free-body diagrams of the elements and nodes of the restrained structure.

STEP 2. We transform the local matrix of fixed-end actions of each element to global. The local axes of the elements of the frame are shown in Fig. e. Referring to this figure we form the transformation matrix for each element of the frame, [see relations (6.6) and (6.48)] and we substitute it in relation

(6.7) to obtain the global components of its fixed-end actions. That is,

$$\{\bar{A}^{R1}\} = \begin{Bmatrix} \{\bar{A}^{R1j}\} \\ \hline \{\bar{A}^{R1k}\} \end{Bmatrix} = \begin{bmatrix} 0 & -1 & 0 & 0 & 0 & 0 \\ 1 & 0 & 0 & 0 & 0 & 0 \\ 0 & 0 & 1 & 0 & 0 & 0 \\ 0 & 0 & 0 & 0 & -1 & 0 \\ 0 & 0 & 0 & 1 & 0 & 0 \\ 0 & 0 & 0 & 0 & 0 & 1 \end{bmatrix} \begin{Bmatrix} 0 \\ P/2 \\ PL/12 \\ 0 \\ P/2 \\ -PL/12 \end{Bmatrix} = \begin{Bmatrix} -P/2 \\ 0 \\ PL/12 \\ -P/2 \\ 0 \\ -PL/12 \end{Bmatrix}$$

(b)

$$\{\bar{A}^{R2}\} = \{A^{R2}\} \qquad \{\bar{A}^{R3}\} = \{0\} \qquad \{\bar{A}^{R4}\} = \{A^{R4}\}$$

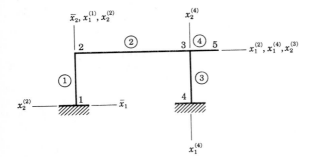

Figure e Global axes of the frame and local axes of its elements.

STEP 3.   We establish the matrix of the restraining actions $\{\widehat{RA}\}$ of the restrained structure. That is, referring to Fig. cb from the equilibrium of the nodes of the restrained structure we have

$$\{\widehat{RA}\} = \begin{Bmatrix} \{A^{R1j}\} \\ \{\bar{A}^{R1k}\} + \{\bar{A}^{R2j}\} \\ \{\bar{A}^{R2k}\} + \{\bar{A}^{R3j}\} + \{\bar{A}^{R4j}\} \\ \{\bar{A}^{R3k}\} \\ \{\bar{A}^{R4k}\} \end{Bmatrix} = \left\{ \begin{array}{l} \left. \begin{array}{c} -P/2 \\ 0 \\ PL/12 \end{array} \right\} \text{node 1} \\ \hline \left. \begin{array}{c} -P/2 \\ P \\ 5PL/12 \end{array} \right\} \text{node 2} \\ \hline \left. \begin{array}{c} 0 \\ 5P/4 \\ -23PL/48 \end{array} \right\} \text{node 3} \\ \hline \left. \begin{array}{c} 0 \\ 0 \\ 0 \end{array} \right\} \text{node 4} \\ \hline \left. \begin{array}{c} 0 \\ P/4 \\ -PL/48 \end{array} \right\} \text{node 5} \end{array} \right.$$

(c)

This result can be verified by referring to the free-body diagrams of the nodes of the restrained structure shown in Fig. d and considering their equilibrium.

STEP 4. We form the matrix of the given actions $\{\hat{P}^G\}$ acting on the nodes of the frame. Referring to Fig. a we have

$$\{\hat{P}^G\}^T = [0 \quad 0 \quad 0 \quad P \quad 0 \quad PL \quad 0 \quad 0 \quad 0 \quad 0 \quad 0 \quad 0 \quad 0 \quad -P \quad 0] \qquad \text{(d)}$$

STEP 5. We establish the matrix of the equivalent actions $[\hat{P}^E]$. Substituting relations (c) and (d) into (7.2) we obtain

$$\{\hat{P}^E\} = \{\hat{P}^G\} - \{\widehat{RA}\} = \left\{ \begin{array}{c} P/2 \\ 0 \\ -PL/12 \\ \hdashline 3P/2 \\ -P \\ -17PL/12 \\ \hdashline 0 \\ -5P/4 \\ 23PL/48 \\ \hdashline 0 \\ 0 \\ 0 \\ \hdashline 0 \\ -5P/4 \\ PL/48 \end{array} \right\} \qquad \text{(e)}$$

The results are shown in Fig. f.

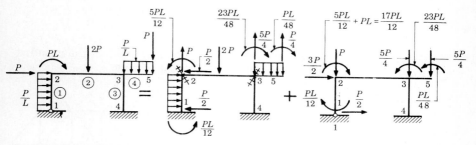

(a) Structure subjected to the given external actions

(b) Restrained structure subjected to the given actions except those acting at the nodes (the restraining actions are also shown)

(c) Structure subjected to the equivalent actions

Figure f Superposition of the restrained structure and the structure subjected to the equivalent actions.

Example 2. Consider the truss shown in Fig. a, whose elements 5 and 9 are subjected to a uniform temperature increase ΔT. Compute the equivalent forces which must be applied to the nodes of the truss in order to account for the effect of the change of temperature. The elements of the truss are all made of the same material and have the same constant cross section.

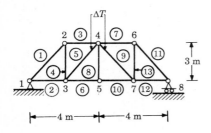

Figure a Geometry and loading of the truss.

solution   When elements 5 and 9 of the truss are subjected to a temperature increase ΔT, they elongate and, thus, the nodes of the truss move. The elongation of an element of length L due to an increase in temperature ΔT is equal to

$$\Delta L = \Delta T\, \alpha L \tag{a}$$

where α is the coefficient of linear thermal expansion.

The nodes of a truss do not rotate because its elements are connected to them by pins and, consequently, moments are not transferred to them. Thus, the restrained structure of a truss is made from the actual structure by restraining the translation of its nodes with the application of restraining forces. In this example, it is apparent that elements 5 and 9 of the restrained truss must be subjected to axial compressive nodal forces in order to prevent them from elongating. The magnitude of these forces is equal to

$$F_1^{R5j} = -F_1^{R5k} = F_1^{R9j} = -F_1^{R9k} = \frac{EA\,\Delta L}{L} = EA\,\Delta T\,\alpha = P \tag{b}$$

where A is the area of the cross section of the element and E is the modulus of elasticity of the material from which the element is made.

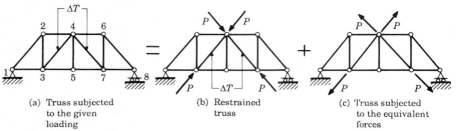

(a) Truss subjected to the given loading

(b) Restrained truss

(c) Truss subjected to the equivalent forces

Figure b Superposition of the restrained structure and the structure subjected to equivalent forces.

The equivalent forces acting on the nodes of the truss are shown in Fig. bc. They are equal and opposite to the restraining forces of Fig. bb.

Notice that the elements of statically determinate trusses are free to expand and, consequently, the change of temperature does not induce internal forces in them. That is, the sum of the internal forces in the elements of the truss loaded as shown in Fig. bb and c must vanish. However, as a result of the change of temperature, the nodes of the truss translate. The components of translation of the nodes of the truss subjected to the given temperature change (see Fig. ba) are equal to those of the truss subjected to the equivalent actions (see Fig. bc).

Example 3. The elements of the frame of Fig. a have constant cross sections and are subjected to the following two cases of loading:

Case 1

To a uniform temperature increase ΔT. This temperature increase represents the difference between the uniform temperature to which the frame is exposed in its present state and the temperature existing during its construction.

Case 2

To a temperature T_1 at the external fibers of its elements and to a temperature $T_2 > T_1$ at their internal fibers. Assume that in this case the temperature of the axis of the elements of the frame is the same as the temperature during construction. Thus, the axis of the elements of the frame does not elongate.

For each case of loading compute the equivalent actions to be applied to the nodes of the frame.

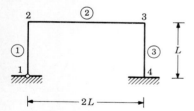

Figure a Geometry of the frame.

solution

Loading Case 1

Because of the increase of temperature ΔT, the elements of the frame elongate. Thus, nodes 2 and 3 translate.

STEP 1. The nodal actions required to prevent element e of the restrained structure from elongating are equal to

$$
\{A^{Re}\} = \left\{ \begin{array}{c} F_1^{Rej} \\ 0 \\ 0 \\ \hline F_1^{Rek} \\ 0 \\ 0 \end{array} \right\} = \left\{ \begin{array}{c} E_e A_e \alpha_e \, \Delta T_c \\ 0 \\ 0 \\ \hline -E_e A_e \alpha_e \, \Delta T \\ 0 \\ 0 \end{array} \right\} \qquad e = 1, 2, 3 \qquad \text{(a)}
$$

where E_e and α_e are the modulus of elasticity and the coefficient of linear thermal expansion of the material, respectively, from which element e of the structure is made. A_e is the area of the cross section of element e.

STEP 2. We transform the local matrix of the fixed-end actions of each element to global. The local axes of the elements of the frame are shown in Fig. b. Referring to this figure we form the transformation matrices for each element of the frame [see relations (6.6) and (6.48)] and substitute it and relation (a) in relation (6.47) to obtain the global components of the fixed-end actions of the element. That is,

$$
\{\bar{A}^{R1}\} = \left\{ \begin{array}{c} \{\bar{A}^{R1j}\} \\ \{\bar{A}^{R1k}\} \end{array} \right\} = \begin{bmatrix} 0 & -1 & 0 & 0 & 0 & 0 \\ 1 & 0 & 0 & 0 & 0 & 0 \\ 0 & 0 & 1 & 0 & 0 & 0 \\ 0 & 0 & 0 & 0 & -1 & 0 \\ 0 & 0 & 0 & 1 & 0 & 0 \\ 0 & 0 & 0 & 0 & 0 & 1 \end{bmatrix} \left\{ \begin{array}{c} E_1 A_1 \alpha_1 \, \Delta T \\ 0 \\ 0 \\ -E_1 A_1 \alpha_1 \, \Delta T \\ 0 \\ 0 \end{array} \right\}
$$

$$
= \left\{ \begin{array}{c} 0 \\ E_1 A_1 \alpha_1 \, \Delta T \\ 0 \\ 0 \\ -E_1 A_1 \alpha_1 \, \Delta T \\ 0 \end{array} \right\} \qquad \text{(b)}
$$

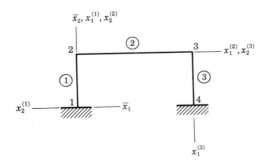

Figure b Global axes of the frame and local axes of its elements.

$$\{\bar{A}^{R2}\} = \{A^{R2}\}$$

$$\{\bar{A}^{R3}\} = \left\{\begin{matrix} \{\bar{A}^{R3j}\} \\ \hdashline \{\bar{A}^{R3k}\} \end{matrix}\right\} = \begin{bmatrix} 0 & 1 & 0 & 0 & 0 & 0 \\ -1 & 0 & 0 & 0 & 0 & 0 \\ 0 & 0 & 1 & 0 & 0 & 0 \\ 0 & 0 & 0 & 0 & 1 & 0 \\ 0 & 0 & 0 & -1 & 0 & 0 \\ 0 & 0 & 0 & 0 & 0 & 1 \end{bmatrix} \left\{\begin{matrix} E_3 A_3 \alpha_3 \, \Delta T \\ 0 \\ 0 \\ -E_3 A_3 \alpha_3 \, \Delta T \\ 0 \\ 0 \end{matrix}\right\}$$

$$= \left\{\begin{matrix} 0 \\ -E_3 A_3 \alpha_3 \, \Delta T \\ 0 \\ \hdashline 0 \\ E_3 A_3 \alpha_3 \, \Delta T \\ 0 \end{matrix}\right\}$$

STEP 3. We establish the matrix of the restraining actions $\widehat{\{RA\}}$ of the restrained structure. That is, from the equilibrium of the nodes of the restrained structure we have

$$\widehat{\{RA\}} = \left\{\begin{matrix} \{\bar{A}^{R1j}\} \\ \{\bar{A}^{R1k}\} + \{\bar{A}^{R2j}\} \\ \{\bar{A}^{R2k}\} + \{\bar{A}^{R3j}\} \\ \{\bar{A}^{R3k}\} \end{matrix}\right\} = \left\{\begin{matrix} 0 \\ E_1 A_1 \alpha_1 \\ 0 \\ \hdashline E_2 A_2 \alpha_2 \\ -E_1 A_1 \alpha_1 \\ 0 \\ \hdashline -E_2 A_2 \alpha_2 \\ -E_3 A_3 A_3 \\ 0 \\ \hdashline 0 \\ E_3 A_3 \alpha_3 \\ 0 \end{matrix}\right\} \Delta T \qquad (c)$$

The restraining actions are shown in Fig. c*b*.

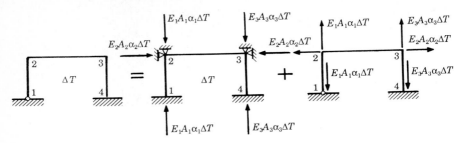

(a) Structure subjected
   to the given loading

(b) Restrained structure
   subjected to the given
   rotation of support

(c) Structure subjected
   to the equivalent
   actions on its nodes

Figure c Superposition of the restrained structure and the structure subjected to the equivalent actions.

STEP 4. We form the matrix of the given actions $\{\hat{P}^G\}$ acting on the nodes of the structure. Referring to Fig. a we see that

$$\{\hat{P}^G\} = \{0\} \tag{d}$$

STEP 5. We establish the matrix of equivalent actions $\{\hat{P}^E\}$. Substituting relations (c) and (d) into (7.2) we have

$$\{\hat{P}^E\} = \{\hat{P}^G\} - \{\widehat{RA}\} = \left\{ \begin{array}{c} 0 \\ -E_1A_1\alpha_1 \\ 0 \\ \hdashline -E_2A_2\alpha_2 \\ E_1A_1\alpha_1 \\ 0 \\ \hdashline E_2A_2\alpha_2 \\ E_3A_3\alpha_3 \\ 0 \\ \hdashline 0 \\ -E_3A_3\alpha_3 \\ 0 \end{array} \right\} \Delta T \tag{e}$$

The results are shown in Fig. cc.

Loading Case 2

Because of the given difference in temperature, the internal fibers of the elements of the frame elongate more than their external fibers and, consequently, the elements of the frame bend. However, the nodes of the frame do not translate.

STEP 1. The restrained structure is subjected to the given difference in temperature. The nodal actions required in order to prevent the nodal points of the elements of the restrained structure from rotating are established by referring to the table on the inside of the back cover and are given in Fig. d. Referring to this figure the matrix of fixed-end actions of element e is equal to

$$\{A^{Re}\} = (T_2 - T_1) \begin{Bmatrix} 0 \\ 0 \\ E_e I^{(e)} \alpha_e / h^{(e)} \\ \hline 0 \\ 0 \\ -E_e I^{(e)} \alpha_e / h^{(e)} \end{Bmatrix} \qquad e = 1, 2, 3 \qquad (f)$$

where $I^{(e)}$ and $h^{(e)}$ are the moment of inertia about the x_3 axis and the depth of the cross section, respectively, of element e; E_e and α_e are the modulus of elasticity and the coefficient of linear thermal expansion, respectively, of the material from which element e is made.

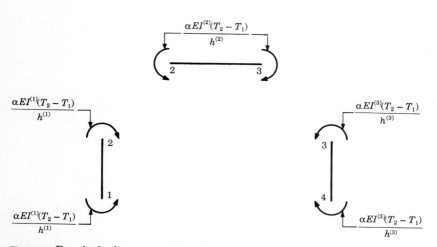

Figure d Free-body diagrams of the elements of the restrained structure.

STEP 2. We transform the local matrix of fixed-end actions of each element to global. However, in this case the global components of the fixed-end actions of the elements of the restrained structure are equal to their local components. That is,

$$\{\bar{A}^{Re}\} = \{A^{Re}\} \qquad e = 1, 2, 3 \qquad (g)$$

STEP 3. We establish the matrix of the restraining actions $\{\widehat{RA}\}$ of the restrained structure. That is, referring to Fig. eb from the equilibrium of the

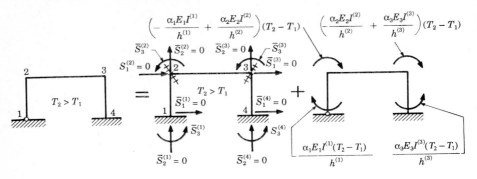

(a) Structure subjected to the given loading

(b) Restrained structure subjected to given temperature difference

(c) Structure subjected to the equivalent actions at its nodes

Figure e Superposition of the restrained structure and the structure subjected to the equivalent actions.

nodes of the restrained structure we have

$$\widehat{\{RA\}} = \begin{Bmatrix} \{\bar{A}^{R1j}\} \\ \{\bar{A}^{R1k}\} + \{\bar{A}^{R2j}\} \\ \{\bar{A}^{R2k}\} + \{\bar{A}^{R3j}\} \\ \{\bar{A}^{R3k}\} \end{Bmatrix}$$

$$= (T_2 - T_1) \left\{ \begin{matrix} 0 \\ 0 \\ 0 \\ \alpha_1 E_1 I^{(1)}/h^{(1)} \\ \hline 0 \\ 0 \\ -\alpha_1 E_1 I^{(1)}/h^{(1)} + \alpha_2 E_2 I^{(2)}/h^{(2)} \\ \hline 0 \\ 0 \\ -\alpha_2 E_2 I^{(2)}/h^{(2)} + \alpha_3 E_3 I^{(3)}/h^{(3)} \\ \hline 0 \\ 0 \\ -\alpha_3 E_3 I^{(3)}/h^{(3)} \end{matrix} \right\} \begin{matrix} \\ \Big\}\ \text{node 1} \\ \\ \\ \Big\}\ \text{node 2} \\ \\ \\ \Big\}\ \text{node 3} \\ \\ \\ \Big\}\ \text{node 4} \end{matrix} \qquad \text{(h)}$$

STEP 4. We form the matrix of the given external actions $\{\hat{P}^G\}$ acting on the nodes of the frame. That is, referring to Fig. a we have

$$\{\hat{P}^G\} = \{0\} \qquad \text{(i)}$$

STEP 5. We establish the matrix of equivalent actions $\{\hat{P}^E\}$. Substituting relations (h) and (i) into (7.2) we obtain

$$\{\hat{P}^E\} = \{\hat{P}^G\} - \{\widehat{RA}\} = -\{\widehat{RA}\} \tag{j}$$

The results are shown in Fig. ec.

The procedure for establishing the equivalent actions to be applied to the nodes of a structure in order to account for the lack of fit of some of its elements is analogous to that used for establishing the effect of a change of temperature (see Example 6).

Example 4. Support 4 of the structure of Fig. a settles by Δ^s. Compute the equivalent forces to be applied to the nodes of the structure.

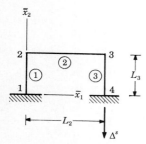

Figure a   Geometry of the frame.

---

solution

STEP 1. The restrained structure is subjected to the given settlement of support 4 (see Fig. b b). Under this loading, only element 3 of the restrained

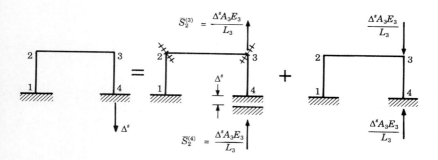

(a) Structure subjected to the given loading

(b) Restrained structure subjected to the given settlement of support

(c) Structure subjected to the equivalent actions on its nodes

Figure b   Superposition of the restraint structure and the structure subjected to the equivalent actions.

structure is stressed. It is subjected to an axial tensile force equal to $\Delta^s A_3 E_3/L_3$. Thus

$$\{A^{R1}\} = \{A^{R2}\} = 0 \tag{a}$$

$$\{A^{R3}\} = \left\{\begin{array}{c} -1 \\ 0 \\ 0 \\ \hline 1 \\ 0 \\ 0 \end{array}\right\} \frac{E_3 A_3 \, \Delta^s}{L_3}$$

STEP 2.   We transform the local matrix of fixed-end actions of each element to global using relations (6.6) and (6.48). Thus

$$\{\bar{A}^{R1}\} = \{\bar{A}^{R2}\} = 0$$

$$\{\bar{A}^{R3}\} = \begin{bmatrix} 0 & 1 & 0 & 0 & 0 & 0 & -1 \\ -1 & 0 & 0 & 0 & 0 & 0 & 0 \\ 0 & 0 & 1 & 0 & 0 & 0 & 0 \\ 0 & 0 & 0 & 0 & 1 & 0 & 1 \\ 0 & 0 & 0 & -1 & 0 & 0 & 0 \\ 0 & 0 & 0 & 0 & 0 & 1 & 0 \end{bmatrix} \frac{E_3 A_3 \, \Delta^s}{L_3} = \left\{\begin{array}{c} 0 \\ 1 \\ 0 \\ 0 \\ -1 \\ 0 \end{array}\right\} \frac{E_3 A_3 \, \Delta^s}{L_3} \tag{b}$$

STEP 3.   We establish the matrix of the restraining actions $\widehat{\{\text{RA}\}}$ of the restrained structure. That is, from the equilibrium of the nodes of the restrained structure we have

$$\widehat{\{\text{RA}\}} = \left\{\begin{array}{c} \{\bar{A}^{R1j}\} \\ \{\bar{A}^{R1k}\} + \{\bar{A}^{R2j}\} \\ \{\bar{A}^{R2k}\} + \{\bar{A}^{R3j}\} \\ \{\bar{A}^{R3k}\} \end{array}\right\} = \left\{\begin{array}{c} 0 \\ 0 \\ 0 \\ \hline 0 \\ 0 \\ 0 \\ \hline 0 \\ 1 \\ 0 \\ \hline 0 \\ -1 \\ 0 \end{array}\right\} \frac{E_3 A_3 \, \Delta^s}{L_3} \tag{c}$$

STEP 4.   We form the matrix of the given actions $\{\hat{P}^G\}$ acting on the nodes of

the frame. Referring to Fig. a we have

$$\{\hat{P}^G\} = \{0\} \tag{d}$$

STEP 5.   We establish the matrix of equivalent actions $\{\hat{P}^E\}$. Substituting relations (c) and (d) into (7.2) we obtain

$$\{\hat{P}^E\} = -\{\widehat{RA}\} \tag{e}$$

The results are shown in Fig. b*c*.

Example 5.   Support 4 of the structure of Fig. a rotates by Δ^s. Compute the equivalent actions to be applied to the nodes of the structure.

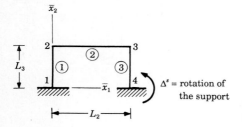

Δ^s = rotation of
the support

Figure a   Geometry and loading of the structure.

solution

STEP 1.   The restrained structure is subjected to the given rotation of support 4 (see Fig. b*b*). Under this loading, only element 3 of the restrained structure is stressed. The fixed-end actions of element 3 are obtained by

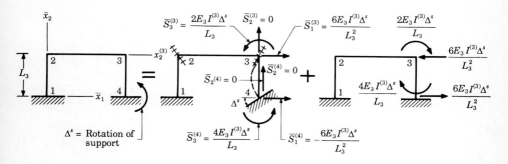

(a) Structure subjected to the given loading

(b) Restrained structure subjected to the given rotation of support

(c) Structure subjected to the equivalent actions on its nodes

Figure b   Superposition of the restrained structure and the structure subjected to the equivalent actions.

referring to the table on the inside of the back cover. Thus,

$$\{A^{R1}\} = \{A^{R2}\} = 0 \tag{a}$$

$$\{A^{R3}\} = \left\{ \begin{array}{c} 0 \\ 3 \\ L_3 \\ \hline 0 \\ -3 \\ 2L_3 \end{array} \right\} \frac{2E_3 I^{(3)} \Delta^s}{L_3^2}$$

STEP 2. We transform the local matrix of fixed-end actions of each element to global using relations (6.6) and (6.48). Thus

$$\{\bar{A}^{R1}\} = \{\bar{A}^{R2}\} = 0 \tag{b}$$

$$\{\bar{A}^{R3}\} = \begin{bmatrix} 0 & 1 & 0 & 0 & 0 & 0 \\ -1 & 0 & 0 & 0 & 0 & 0 \\ 0 & 0 & 1 & 0 & 0 & 0 \\ 0 & 0 & 0 & 0 & 1 & 0 \\ 0 & 0 & 0 & -1 & 0 & 0 \\ 0 & 0 & 0 & 0 & 0 & 1 \end{bmatrix} \left\{ \begin{array}{c} 0 \\ 3 \\ L_3 \\ 0 \\ -3 \\ 2L_3 \end{array} \right\} \frac{2E_3 I^{(3)} \Delta^s}{L_3^2} = \left\{ \begin{array}{c} 3 \\ 0 \\ L_3 \\ \hline -3 \\ 0 \\ 2L_3 \end{array} \right\} \frac{2E_3 I^{(3)} \Delta^s}{L_3^2}$$

STEP 3. We establish the matrix of the restraining actions $\{\widehat{RA}\}$ of the restrained structure. That is, from the equilibrium of the nodes of the restrained structure we have

$$\{\widehat{RA}\} = \left\{ \begin{array}{c} \{\bar{A}^{R1j}\} \\ \{\bar{A}^{R1k}\} + \{\bar{A}^{R2j}\} \\ \{\bar{A}^{R2k}\} + \{\bar{A}^{R3j}\} \\ \{\bar{A}^{R3k}\} \end{array} \right\} = \left\{ \begin{array}{c} 0 \\ 0 \\ 0 \\ \hline 0 \\ 0 \\ 0 \\ \hline 3 \\ 0 \\ L_3 \\ \hline -3 \\ 0 \\ 2L_3 \end{array} \right\} \frac{2E_3 I^{(3)} \Delta^s}{L_3^2} \tag{c}$$

STEP 4. We form the matrix of the given actions $\{\hat{P}^G\}$ acting on the nodes of the structure. Referring to Fig. a we have

$$\{\hat{P}^G\} = \{0\} \tag{d}$$

STEP 5.   We establish the matrix of equivalent action $\{\hat{P}^E\}$. Substituting relations (c) and (d) into (7.2) we obtain

$$\{\hat{P}^E\} = -\{\widehat{RA}\} \tag{e}$$

The results are shown in Fig. bc.

Example 6.   The elements of the frame of Fig. a have constant cross sections. Element 2 had an initial deformation (lack of fit) characterized by an elongation $d_1^{(2)}$, an upward transverse translation $d_2^{(2)}$ of its end k relative to its end j, and a rotation $d_3^{(2)}$ of its end k relative to its end j (see Fig. b). During construction of the frame, this element was subjected to the initial nodal actions required to render $d_1^{(2)}$, $d_2^{(2)}$, and $d_3^{(2)}$ equal to zero, and it was connected to nodes 2 and 5. Subsequently, the initial nodal actions applied to element 2 were removed, and, consequently, element 2 would have assumed the geometry it had prior to its connection to the frame if the remaining structure did not restrain it. This induced internal actions in the elements of the frame.

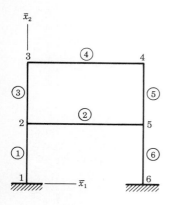

Figure a   Geometry of the structure.

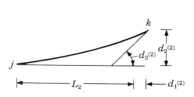

Figure b   Initial deformation of element 2.

Compute the equivalent actions to be applied to the nodes of the structure of Fig. a in order to account for the initial deformation of element 2 (see Fig. b).

---

solution

STEP 1.   Consider the initially deformed element 2, with its ends fixed. The external actions which must be applied to the ends of this element in order to

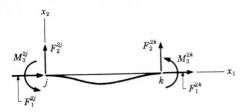

Figure c Initially deformed element subjected to the end actions required to render $d_1^{(2)}$, $d_2^{(2)}$, and $d_3^{(2)}$ equal to zero.

render its initial displacements $d_1^{(2)}$, $d_2^{(2)}$, and $d_3^{(2)}$ equal to zero may be established by referring to the table on the inside of the back cover (also see Fig. c). Thus

$$\{A^{R1}\} = \{A^{R3}\} = \{A^{R4}\} = \{A^{R5}\} = \{A^{R6}\} = 0 \tag{a}$$

$$\{A^{R2}\} = \left\{ \begin{array}{c} E_2 A_2 d_1^{(2)}/L_2 \\[2mm] 12E_2 I^{(2)} d_2^{(2)}/L_2^3 - 6E_2 I^{(2)} d_3^{(2)}/L_2^2 \\[2mm] 6E_2 I^{(2)} d_2^{(2)}/L_2^2 - 2E_2 I^{(2)} d_3^{(2)}/L_2 \\[1mm] \hline \\[-2mm] -E_2 A_2 d_1^{(2)}/L_2 \\[2mm] -12E_2 I^{(2)} d_2^{(2)}/L_2^3 + 6E_2 I^{(2)} d_3^{(2)}/L_2^2 \\[2mm] 6E_2 I^{(2)} d_2^{(2)}/L_2^2 - 4E_2 I^{(2)} d_3^{(2)}/L_2 \end{array} \right\}$$

STEP 2.   We transform the local matrix of fixed-end actions for each element to global using relations (6.8) and (6.48). Thus

$$\{\bar{A}^{R1}\} = \{\bar{A}^{R3}\} = \{\bar{A}^{R4}\} = \{\bar{A}^{R5}\} = \{\bar{A}^{R6}\} = 0 \qquad \{\bar{A}^{R2}\} = \{A^{R2}\} \tag{b}$$

STEP 3.   We establish the matrix of the restraining actions $\widehat{\{RA\}}$ of the restrained structure. That is, from the equilibrium of the nodes of the restrained structure we have

$$\widehat{\{RA\}} = \left\{ \begin{array}{c} \{\bar{A}^{R1j}\} \\[2mm] \{\bar{A}^{R1k}\} + \{\bar{A}^{R2j}\} + \{\bar{A}^{R3j}\} \\[2mm] \{\bar{A}^{R3k}\} + \{\bar{A}^{R4j}\} \\[2mm] \{\bar{A}^{R4k}\} + \{\bar{A}^{R5j}\} \\[2mm] \{A^{R2k}\} + \{\bar{A}^{R5k}\} + \{A^{R6j}\} \\[2mm] \{\bar{A}^{R6k}\} \end{array} \right\}$$

$$= \left\{ \begin{array}{c} 0 \\ 0 \\ 0 \\ \hline E_2 A_2 d_1^{(2)}/L_2 \\ 12E_2 I^{(2)} d_2^{(2)}/L_2^3 - 6E_2 I^{(2)} d_3^{(2)}/L_2^2 \\ 6E_2 I^{(2)} d_2^{(2)}/L_2^2 - 2E_2 I^{(2)} d_3^{(2)}/L_2 \\ \hline 0 \\ 0 \\ 0 \\ \hline 0 \\ 0 \\ 0 \\ \hline -E_2 A_2 d_1^{(2)}/L_2 \\ -12E_2 I^{(2)} d_2^{(2)}/L_2^3 + 6E_2 I^{(2)} d_3^{(2)}/L_2^2 \\ 6E_2 I^{(2)} d_2^{(2)}/L_2^2 - 4E_2 I^{(2)} d_3^{(2)}/L_2 \\ \hline 0 \\ 0 \\ 0 \end{array} \right\} \qquad \text{(c)}$$

STEP 4.   We form the matrix of the given external actions $\{\hat{P}^G\}$ acting on the nodes of the structure. That is, referring to Fig. a we have

$$\{\hat{P}^G\} = 0 \qquad \text{(d)}$$

STEP 5.   We establish the matrix of equivalent actions $\{\hat{P}^E\}$. Substituting

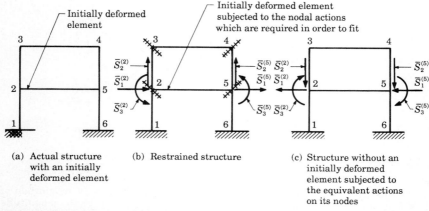

(a) Actual structure with an initially deformed element

(b) Restrained structure

(c) Structure without an initially deformed element subjected to the equivalent actions on its nodes

Figure d Superposition of the restrained structure and the structure without an initially deformed element subjected to the equivalent actions.

relations (c) and (d) into (7.2) we obtain

$$\{\hat{P}^E\} = \{\hat{P}^G\} - \{\widehat{RA}\} = -\{\widehat{RA}\} \tag{e}$$

The results are shown in Fig. dc.

7.3 Problems

1 to 16. Compute and show on a sketch the equivalent actions to be applied to the nodes of the structure shown in Fig. 7.P1. Repeat with Figs. 7.P2 to 7.P16.

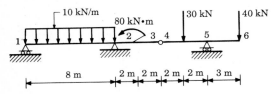

Figure 7.P1

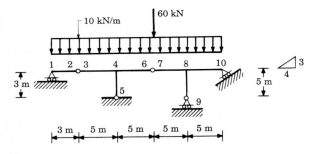

Figure 7.P2

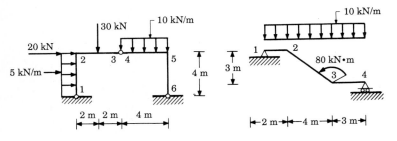

Figure 7.P3   **Figure 7.P4†**

---

† In this figure, the uniformly distributed vertical forces have a magnitude of 10 kN per meter of horizontal length. For the inclined element 2,3 the magnitude of these forces should be converted to kilonewtons per meter along the axis of the element and the perpendicular to it.

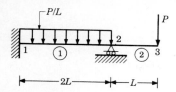

Figure 7.P5

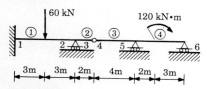

Figure 7.P7

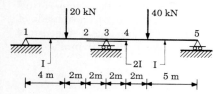

Figure 7.P9

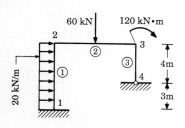

Figure 7.P11

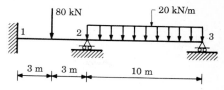

Figure 7.P6

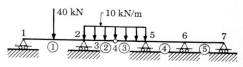

Figure 7.P8

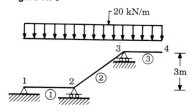

Figure 7.P10†

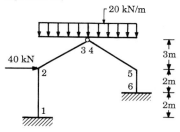

Figure 7.P12†

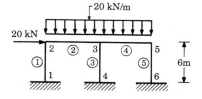

Figure 7.P13†

Figure 7.P14

† See footnote on p. 230.

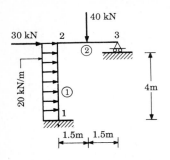

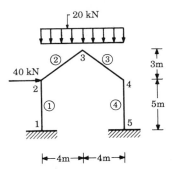

Figure 7.P15   Figure 7.P16†

17 to 20. Compute and show on a sketch the equivalent actions to be applied to the nodes of the structure shown in Fig. 7.P17. The temperature during construction was $T_0 = 10°C$. The elements of the structure are made of the same material $(E = 210\,\text{kN/mm}^2,\ \alpha = 10^{-5}/°C)$ and have the same constant cross section $(I = 117.7 \times 10^6\,\text{mm}^4,\ A = 6.26 \times 10^3\,\text{mm}^2,\ h = 330\,\text{mm})$. Repeat with Figs. 7.P18 to Fig. 7.P20.

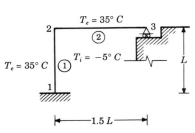

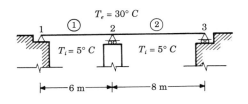

Figure 7.P17   Figure 7.P18

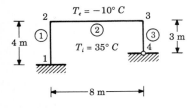

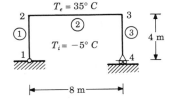

Figure 7.P19   Figure 7.P20

21 to 23. Compute and show on a sketch the equivalent actions to be applied to the nodes of the structure of Fig. 7.P6 for a settlement of support 1 of 20 mm. The elements of the structure are made of the same material $(E = 210\,\text{kN/mm}^2)$ and have the same constant cross section $(I = 117.7 \times 10^6\,\text{mm}^4,\ A = 6.62 \times 10^3\,\text{mm}^2,\ h = 330\,\text{mm})$. Repeat with Figs. 7.P14 and 7.P16.

24 to 29. Compute and show on a sketch the equivalent actions to be applied to the nodes of the structure of Fig. 7.P24 for each of the following loading cases:

(a) The external actions shown in the figure.

(b) A settlement of 20 mm of support 1.

(c) A temperature of the external fibers $T_e = 25°C$ and of the internal fibers $T_i = -5°C$. The temperature during construction was $T_0 = 10°C$ ($\Delta T_c = 0$).

The elements of the structure are made of the same material ($E = 210 \text{ kN/mm}^2$, $\alpha = 10^{-5}/°C$) and have the same constant cross section ($I = 369.7 \times 10^6 \text{ mm}^4$, $A = 13.2 \times 10^3 \text{ mm}^2$, $h = 425 \text{ mm}$). Repeat with Figs. 7.P25 to 7.P29.

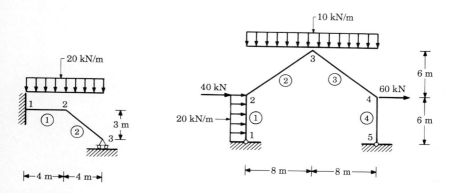

Figure 7.P24†

Figure 7.P25†

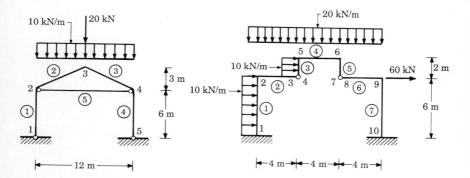

Figure 7.P26†

Figure 7.P27

---

† See footnote on p. 230.

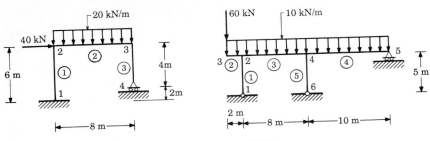

Figure 7.P28

Figure 7.P29

30. Compute and show on a sketch the equivalent actions to be applied to the nodes of the truss of Fig. 7.P30 when element 1 is initially longer than required by 20 mm. The elements of the truss are made of steel ($E = 210 \text{ kN/mm}^2$) and have the same constant cross section ($A = 4 \times 10^3 \text{ mm}^2$).

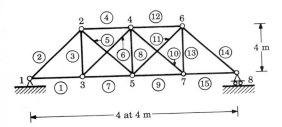

Figure 7.P30

31. Compute and show on a sketch the equivalent actions to be applied to the nodes of the frame of Fig. 7.P31a when element 2 is initially deformed, as shown in Fig. 7.P31b. The elements of the frame are made of the same material ($E = 210 \text{ kN/mm}^2$) and have the same constant cross section ($I = 83.6 \times 10^6 \text{ mm}^4$, $A = 5.38 \times 10^3 \text{ mm}^2$).

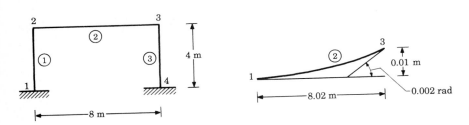

Figure 7.P31

The Stiffness Equations for Framed Structures

8.1   Introduction

In this part of the book we present the direct stiffness method for analyzing framed structures. In this method the response of a structure subjected to the equivalent actions on its nodes is described by its stiffness equations. In these equations the known global components of the equivalent actions corresponding to the given loads acting on a structure and the unknown global components of the reactions of its supports are expressed as a linear combination of the global components of displacements of its nodes. That is,

$$\{\hat{P}^E\} + \{\hat{R}\} = [\hat{S}]\{\hat{\Delta}\} \tag{8.1}$$

$\{\hat{P}^E\}$ is the matrix of equivalent actions of the structure defined by relation (7.2). $\{\hat{R}\}$ is the matrix of the unknown reactions of the structure. $\{\hat{\Delta}\}$ is the matrix of the components of displacements of the nodes of the structure; in this matrix we include all possible components of displacements of the nodes of the structure including those inhibited by its supports. $\{\hat{S}\}$ is the *stiffness matrix for the structure;* its terms are called the stiffness coefficients of the structure.

Equations (8.1) can be rewritten as

$$
\left\{ \begin{array}{c} P_1^E \\ P_2^E \\ \cdot \\ \cdot \\ \cdot \\ P_N^E \end{array} \right\} + \{\hat{R}\} =
\begin{bmatrix}
S_{11} & S_{12} \cdots S_{1N} \\
S_{21} & S_{22} \cdots S_{2N} \\
\cdot & \cdot & \cdot \\
\cdot & \cdot & \cdot \\
\cdot & \cdot & \cdot \\
S_{N1} & S_{N2} \cdots S_{NN}
\end{bmatrix}
\left\{ \begin{array}{c} \Delta_1 \\ \Delta_2 \\ \cdot \\ \cdot \\ \cdot \\ \Delta_N \end{array} \right\} \tag{8.2}
$$

where N is the number of all possible components of displacements of the nodes of the structure including those inhibited by its supports. Moreover, Eqs. (8.1) can be written as

$$
\begin{Bmatrix} \{P^E\}_1 \\ \{P^E\}_2 \\ \cdot \\ \cdot \\ \cdot \\ \{P^E\}_i \\ \cdot \\ \cdot \\ \cdot \\ \{P^E\}_Q \end{Bmatrix} + \begin{Bmatrix} \{R\}_1 \\ \{R\}_2 \\ \cdot \\ \cdot \\ \cdot \\ \{R\}_i \\ \cdot \\ \cdot \\ \cdot \\ \{R\}_Q \end{Bmatrix} = \begin{bmatrix} [S]_{11} & [S]_{12} & \cdots & [S]_{1i} & \cdots & [S]_{1Q} \\ [S]_{21} & [S]_{22} & \cdots & [S]_{2i} & \cdots & [S]_{2Q} \\ & & \cdot & & & \\ & & & \cdot & & \\ & & & & \cdot & \\ [S]_{i1} & [S]_{i2} & \cdots & [S]_{ii} & \cdots & [S]_{iQ} \\ & & \cdot & & & \\ & & & \cdot & & \\ & & & & \cdot & \\ [S]_{Q1} & [S]_{Q2} & \cdots & [S]_{Qi} & \cdots & [S]_{QQ} \end{bmatrix} \begin{Bmatrix} \{\Delta\}_1 \\ \{\Delta\}_2 \\ \cdot \\ \cdot \\ \cdot \\ \{\Delta\}_i \\ \cdot \\ \cdot \\ \cdot \\ \{\Delta\}_Q \end{Bmatrix}
$$

$$(8.3)$$

where Q = number of nodes of structure

$\{P^E\}_i$ = matrix of global components of equivalent actions acting on ith node of structure

$\{R\}_i$ = matrix of global components of reactions acting on ith node of structure (is a zero matrix if ith node is not supported)

$\{\Delta\}_i$ = matrix of global components of displacement of ith node of structure

As discussed in Chap. 7 the matrix of equivalent actions $\{\hat{P}^E\}$ for a structure is established from the matrices of nodal actions $\{\bar{A}^R\}$ of the elements of the restrained structure, that is, of the elements subjected to the given loads with their ends fixed. This is accomplished by considering the equilibrium of the nodes of the restrained structure.

In the direct stiffness method the stiffness matrix $[\hat{S}]$ for a structure is assembled directly from the global stiffness matrices of its elements. In Sec. 8.3 we present two procedures for assembling the stiffness matrix for a structure and apply them to a number of planar structures. The stiffness equations for a structure are the equations of equilibrium for the nodes of the structure subjected to the given loads. Moreover, the stiffness equations for a structure are the equations of equilibrium for the nodes of the structure subjected to the equivalent actions on its nodes.

When the matrix of equivalent actions $\{\hat{P}^E\}$ corresponding to the given loads acting on a structure and the stiffness matrix $[\hat{S}]$ for the structure are established, the structure's stiffness equations (8.1) can be solved to obtain the components of displacements of its nodes as well as the components of the reactions of its supports. To accomplish this the conditions at the supports (boundary conditions) of the structure must be introduced into its stiffness equations (8.1). The components of the reactions and the components of displacements of the nodes of the structure subjected to the given loads are equal to the corresponding components of the reactions and the components of displacements of the nodes of the structure subjected to the equivalent actions on its nodes. From the components of displacements of the nodes of a structure the components of nodal displacements of its elements can be established and employed to compute their local components of nodal actions.

In Sec. 8.2 we derive the stiffness equations for a framed structure using a different approach than the one used in the direct stiffness method (see Sec. 8.3). This approach is conceptually straightforward and can be easily programmed for establishing the stiffness equations of framed structures by computer. However, as explained in Sec. 8.2, this approach requires considerable computer time and storage, and for this reason it has found little practical application. We present this approach because it employs directly the relations which describe the facts on which the stiffness equations for a structure are based. Analogous relations describing the same facts are employed directly in the modern flexibility method in order to derive the compatibility equations for a structure. Furthermore, in the approach of Sec. 8.2 the stiffness equations for a structure are established without having to think of the restrained structure and the structure subjected to the equivalent actions.

In Sec. 9.1 we describe a procedure for solving the stiffness equations for a structure subjected to given loads. Moreover, in Sec. 9.3 we describe the step required to analyze a framed structure using the direct stiffness method, and we analyze a number of planar structures.

8.2 Derivation of the Stiffness Equations for a Framed Structure

The stiffness equations for a framed structure are based on the following facts.

Fact 1. The nodes of the structure are in equilibrium. This fact gives a set of relations between the given external actions and the unknown reactions acting on the nodes of a structure and the global components of the nodal actions of its elements which can be written in the following form:

$$\{\hat{P}^G\} + \{\hat{R}\} = [\hat{B}]\{\hat{A}\} \tag{8.4}$$

$\{\hat{P}^G\}$ is the matrix of the given concentrated external actions acting on the nodes of the structure. The number of rows in this matrix is equal to the number of all possible components of external actions which can be applied to the nodes of the structure including those directly absorbed by its supports. We place a 0 in the rows corresponding to vanishing components of external actions. $\{\hat{R}\}$ is the matrix of unknown reactions of the structure. $[\hat{B}]$ is called the equilibrium matrix for the structure. Its terms are either 0 or 1. For statically determinate structures which are not mechanisms the matrix $[\hat{B}]$ is a square nonsingular matrix. For statically indeterminate structures it is not a square matrix. $\{\hat{A}\}$ is the matrix of the global components of the nodal actions of all the elements of the structure. The matrix $\{\hat{A}\}$ for a structure with NE elements is

$$\{\hat{A}\} = \left\{ \begin{array}{c} \{\bar{A}^1\} \\ \{\bar{A}^2\} \\ \cdot \\ \cdot \\ \cdot \\ \{\bar{A}^{NE}\} \end{array} \right\} \tag{8.5}$$

where $\{\bar{A}^e\}$ $(e = 1, 2, \ldots, NE)$ is the matrix of the global components of the nodal actions of element e of the structure (see Sec. 4.2).

Fact 2. The nodal displacements of the elements of the structure are compatible with the displacements of its nodes. This requirement gives a relation between the global components of nodal displacements of the elements of a structure and the global components of displacements of its nodes which can be written in the following form:

$$[\hat{D}] = [\hat{C}][\hat{\Delta}] \tag{8.6}$$

$\{\hat{\Delta}\}$ is the matrix of all node displacements of the structure. In this matrix we include all possible global components of displacements of the nodes of the structure including those inhibited by its supports. $\{\hat{D}\}$ is the global matrix of the nodal displacements of all the elements of the structure. The matrix $[\hat{D}]$ for a structure with NE

elements is equal to

$$\{\hat{D}\} = \begin{Bmatrix} \{\bar{D}^1\} \\ \{\bar{D}^2\} \\ \cdot \\ \cdot \\ \cdot \\ \{\bar{D}^{NE}\} \end{Bmatrix} \tag{8.7}$$

where $\{\bar{D}^e\}$ $(e = 1, 2, \ldots, NE)$ is the matrix of the global components of the nodal displacements of element e of the structure (see Sec. 4.2). In general, the matrix $[\hat{C}]$ is not a square matrix; moreover, its terms are either 0 or 1. It can be shown (see Sec. 15.5) that the matrix $[\hat{B}]$ and $[\hat{C}]$ of a structure are related as follows:

$$[\hat{C}] = [\hat{B}]^T \tag{8.8}$$

Fact 3. The response of the elements of a structure is specified by their stiffness equations and their matrices of fixed-end actions. Referring to relations (6.65) for a structure with NE elements we have:

$$\{\bar{A}^1\} = [\bar{K}^1]\{\bar{D}^1\} + \{\bar{A}^{R1}\}$$
$$\{\bar{A}^2\} = [\bar{K}^2]\{\bar{D}^2\} + \{\bar{A}^{R2}\}$$

$$\cdot \qquad \cdot$$
$$\cdot \qquad \cdot$$
$$\cdot \qquad \cdot$$

$$\underline{\{\bar{A}^{NE}\} = [\bar{K}^{NE}]\{\bar{D}^{NE}\} + \{\bar{A}^{RNE}\}}$$

or $\qquad \{\hat{A}\} \quad = [\hat{K}] \quad \{\hat{D}\} \quad + \{\hat{A}^R\}$ $\qquad\qquad$ (8.9)

$\{\hat{A}^R\}$ is the global matrix of the nodal actions of all the elements of the structure subjected to the given loads acting along their length with their ends fixed. $[\hat{K}]$ is the global stiffness matrix of all the elements of the structure. It is equal to

$$[\hat{K}] = \begin{bmatrix} [\bar{K}^1] & & & & \\ & [\bar{K}^2] & & & \\ & & \cdot & & \\ & & & \cdot & \\ & & & & \cdot \\ & & & & & [\bar{K}^{NE}] \end{bmatrix} \tag{8.10}$$

In relation (8.9) the global components of the nodal actions of all the elements of a structure subjected to given loads are expressed as the sum of the global components of the nodal actions of the elements of the structure subjected to the equivalent actions on its nodes $[[\hat{K}]\{\hat{D}\}]$ and those of the restrained structure $\{\hat{A}^R\}$.

Traditionally, structures are classified as *statically determinate* or *statically indeterminate*. As mentioned previously the equilibrium matrix $[\hat{B}]$ for statically determinate structures is square and nonsingular. Consequently, the nodal actions in the elements of statically determinate structures can be established using relations (8.4). When the nodal actions in the elements of a statically determinate structure are established, relations (8.6) and (8.9) may be employed to establish the components of displacements of its nodes. However, in order to find the nodal actions of the elements of statically indeterminate structures, all three relations (8.4), (8.6), and (8.9) must be employed.

In what follows we derive the stiffness equations for a framed structure using relations (8.4), (8.6), and (8.9). To accomplish this we substitute relation (8.9) into (8.4) and use relation (8.6) to obtain

$$\{\hat{P}^G\} + \{\hat{R}\} = [\hat{B}]\{\hat{A}\} = [\hat{B}][\hat{K}]\{\hat{D}\} + [\hat{B}]\{\hat{A}^R\}$$
$$= [\hat{B}][\hat{K}][\hat{C}]\{\hat{A}\} + [\hat{B}]\{\hat{A}^B\} \qquad (8.11)$$

As discussed in Sec. 7.1 the restrained structure is obtained from the actual structure subjected to the given loads by applying to its nodes the external actions and reactions (restraining actions) required to restrain them from moving (translating or rotating). Thus, if we apply relation (8.4) to the restrained structure, it is apparent that the terms of the matrix $\{\hat{P}^G\} + \{\hat{R}\} = [\hat{B}]\{\hat{A}^R\}$ are the components of the restraining actions which must be applied to the nodes of the restrained structure in order to restrain them from moving. That is

$$[\hat{B}]\{\hat{A}^R\} = \{\widehat{RA}\} \qquad (8.12)$$

Consequently, substituting relation (8.12) into (7.2), the matrix of equivalent actions $\{\hat{P}^E\}$ to be applied to the nodes of the structure is equal to

$$\{\hat{P}^E\} = \{\hat{P}^G\} - [\hat{B}]\{\hat{A}^R\} \qquad (8.13)$$

Using relation (8.13), relation (8.11) can be rewritten as

$$\{\hat{P}^E\} + \{\hat{R}\} = [\hat{S}]\{\hat{A}\} \qquad (8.14)$$

where, using relation (8.8), we have

$$[\hat{S}] = [\hat{B}][\hat{K}][\hat{C}] = [\hat{B}][\hat{K}][\hat{B}]^T \tag{8.15}$$

Equations (8.14) are the stiffness equations for the structure. $[\hat{S}]$ is the *stiffness matrix* for the structure. Referring to relations (8.10) and (8.15) it is apparent that inasmuch as the terms of the matrices $[\hat{B}]$ and $[\hat{C}]$ are either 1 or 0, each term of the stiffness matrix $[\hat{S}]$ of a structure is the sum of certain terms of the global stiffness matrices of some elements of the structure.

An algorithm can be written for generating by computer the matrices $[\hat{K}]$ and $[\hat{B}]$ or $[\hat{C}]$ for any structure. These matrices can be employed in relation (8.15) to yield the stiffness matrix $[\hat{S}]$ of the structure. However, the matrices $[\hat{B}]$, $[\hat{C}]$, and $[\hat{K}]$ are sparsely populated and for structures having many nodes (many degrees of freedom) are very large. Thus, for such structures this approach consumes large computer storage and time, and consequently it has found little practical use. As shown in Sec. 8.3, in the direct stiffness method each term of the stiffness matrix for a structure is established by adding the appropriate terms of the global stiffness matrices for its elements. The storage used, and the time required for assembling the stiffness matrix $[\hat{S}]$ for a large structure by the computer, using the direct stiffness method, is considerably smaller than the storage used and the time required for generating the matrices $[\hat{K}]$, $[\hat{B}]$, or $[\hat{C}]$ by the computer and carrying out the multiplication indicated by relation (8.15).

In what follows, we demonstrate by an example the formation of relations (8.4), (8.6), and (8.9) and the derivation of the stiffness equations for a planar frame using relations (8.14) and (8.15).

Example 1. Establish the stiffness equations for the frame loaded as shown in Fig. a using the approach described in this section. The elements of the frame are made of the same isotropic linearly elastic material and have the same constant cross section.

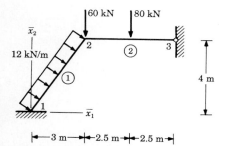

Figure a Geometry and loading of the frame.

solution

STEP 1.   We rewrite the equations of equilibrium for the nodes of the frame. Referring to Fig. a we have

$$\begin{Bmatrix} 0 \\ 0 \\ 0 \\ 0 \\ -60 \\ 0 \\ 0 \\ 0 \\ 0 \end{Bmatrix} + \begin{Bmatrix} \bar{R}_1^{(1)} \\ \bar{R}_2^{(1)} \\ \bar{R}_3^{(1)} \\ 0 \\ 0 \\ 0 \\ \bar{R}_1^{(3)} \\ \bar{R}_2^{(3)} \\ 0 \end{Bmatrix} = \begin{bmatrix} 1 & 0 & 0 & 0 & 0 & 0 & 0 & 0 & 0 & 0 & 0 & 0 \\ 0 & 1 & 0 & 0 & 0 & 0 & 0 & 0 & 0 & 0 & 0 & 0 \\ 0 & 0 & 1 & 0 & 0 & 0 & 0 & 0 & 0 & 0 & 0 & 0 \\ 0 & 0 & 0 & 1 & 0 & 0 & 1 & 0 & 0 & 0 & 0 & 0 \\ 0 & 0 & 0 & 0 & 1 & 0 & 0 & 1 & 0 & 0 & 0 & 0 \\ 0 & 0 & 0 & 0 & 0 & 1 & 0 & 0 & 1 & 0 & 0 & 0 \\ 0 & 0 & 0 & 0 & 0 & 0 & 0 & 0 & 0 & 1 & 0 & 0 \\ 0 & 0 & 0 & 0 & 0 & 0 & 0 & 0 & 0 & 0 & 1 & 0 \\ 0 & 0 & 0 & 0 & 0 & 0 & 0 & 0 & 0 & 0 & 0 & 1 \end{bmatrix} \begin{Bmatrix} \bar{F}_1^{1j} \\ \bar{F}_2^{1j} \\ \bar{M}_3^{1j} \\ \bar{F}_1^{1k} \\ \bar{F}_2^{1k} \\ \bar{M}_3^{1k} \\ \bar{F}_1^{2j} \\ \bar{F}_2^{2j} \\ \bar{M}_3^{2j} \\ \bar{F}_1^{2k} \\ \bar{F}_2^{2k} \\ \bar{M}_3^{2k} \end{Bmatrix} \qquad \text{(a)}$$

or

$$\{\hat{P}^G\} + \{\hat{R}\} = [\hat{B}]\{\hat{A}^T\}$$

STEP 2.   We write the equations of compatibility of the nodes of the frame. Referring to Fig. b we have

$$[\hat{D}] = \begin{Bmatrix} \bar{u}_1^{1j} \\ \bar{u}_2^{1j} \\ \bar{\theta}_3^{1j} \\ \bar{u}_1^{1k} \\ \bar{u}_2^{1k} \\ \bar{\theta}_3^{1k} \\ \bar{u}_1^{2j} \\ \bar{u}_2^{2j} \\ \bar{\theta}_3^{2j} \\ \bar{u}_1^{2k} \\ \bar{u}_2^{2k} \\ \bar{\theta}_3^{2k} \end{Bmatrix} = \begin{bmatrix} 1 & 0 & 0 & 0 & 0 & 0 & 0 & 0 & 0 \\ 0 & 1 & 0 & 0 & 0 & 0 & 0 & 0 & 0 \\ 0 & 0 & 1 & 0 & 0 & 0 & 0 & 0 & 0 \\ 0 & 0 & 0 & 1 & 0 & 0 & 0 & 0 & 0 \\ 0 & 0 & 0 & 0 & 1 & 0 & 0 & 0 & 0 \\ 0 & 0 & 0 & 0 & 0 & 1 & 0 & 0 & 0 \\ 0 & 0 & 0 & 1 & 0 & 0 & 0 & 0 & 0 \\ 0 & 0 & 0 & 0 & 1 & 0 & 0 & 0 & 0 \\ 0 & 0 & 0 & 0 & 0 & 1 & 0 & 0 & 0 \\ 0 & 0 & 0 & 0 & 0 & 0 & 1 & 0 & 0 \\ 0 & 0 & 0 & 0 & 0 & 0 & 0 & 1 & 0 \\ 0 & 0 & 0 & 0 & 0 & 0 & 0 & 0 & 1 \end{bmatrix} \begin{Bmatrix} \Delta_1 \\ \Delta_2 \\ \Delta_3 \\ \Delta_4 \\ \Delta_5 \\ \Delta_6 \\ \Delta_7 \\ \Delta_8 \\ \Delta_9 \end{Bmatrix} \qquad \text{(b)}$$

or

$$\{\hat{D}\} = [\hat{C}]\{\hat{\Delta}\}$$

Referring to relations (a) and (b) it can be seen that

$$[\hat{C}] = [\hat{B}]^T \qquad \text{(c)}$$

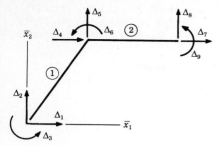

Figure b Numbering of the components of displacement of the nodes of the structure.

STEP 3.   We establish the matrix of equivalent actions. To accomplish this we do the following.

1. We compute the local components of fixed-end actions of the elements of the frame. This is done by referring to the table on the inside of the back cover. The results are shown in Fig. c. Referring to this figure we have

$$\{A^{R1}\} = \left\{ \begin{array}{c} 0 \\ 30 \\ 25 \\ \hline 0 \\ 30 \\ -25 \end{array} \right\} \quad \{A^{R2}\} = \left\{ \begin{array}{c} 0 \\ 40 \\ 50 \\ \hline 0 \\ 40 \\ -50 \end{array} \right\} \quad \text{(d)}$$

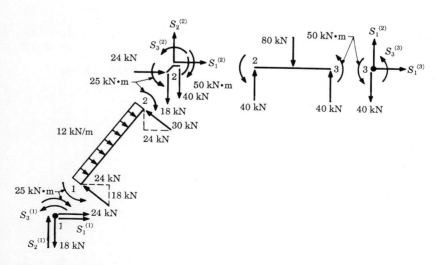

Figure c Free-body diagrams of the elements and nodes of the restrained structure.

2. We transform the local matrices of nodal actions $\{A^{Ri}\}$ of each element of the structure to global. The transformation matrix for the elements of the frame is given by relation (6.44). Referring to Fig. a the transformation matrix of element 1 is

$$[\Lambda^1] = \begin{bmatrix} 0.6 & 0.8 & 0 & 0 & 0 & 0 \\ -0.8 & 0.6 & 0 & 0 & 0 & 0 \\ 0 & 0 & 1 & 0 & 0 & 0 \\ 0 & 0 & 0 & 0.6 & 0.8 & 0 \\ 0 & 0 & 0 & -0.8 & 0.6 & 0 \\ 0 & 0 & 0 & 0 & 0 & 1 \end{bmatrix} \tag{e}$$

Substituting relation (e) into (6.65) we get

$$\{\bar{A}^{R1}\} = [\Lambda^1]^T \{A^{R1}\} = \begin{Bmatrix} -24 \\ 18 \\ 25 \\ \hline -24 \\ 18 \\ -25 \end{Bmatrix} \tag{f}$$

Moreover,
$$\{\bar{A}^{R2}\} = \{A^{R2}\} \tag{g}$$

3. We form the matrix $\{\hat{A}^R\}$ of the nodal actions of all the elements of the frame subjected to the given loads acting along their length with their ends fixed. That is, using relation (d), (f), and (g), we have

$$\{\hat{A}^R\}^T = [\{A^{R1}\}^T, \{A^{R2}\}^T]$$
$$= [-24, 18, 25, -24, 18, -25, 0, 40, 50, 0, 40, 50] \tag{h}$$

4. We compute the equivalent actions to be applied to the nodes of the frame. Substituting the matrices $\{\hat{A}^R\}$, $[\hat{B}]$, and $\{\hat{P}^G\}$ into relation (8.13) we obtain

$$\{\hat{P}^E\} = \begin{Bmatrix} 0 \\ 0 \\ 0 \\ \hline 0 \\ -60 \\ 0 \\ \hline 0 \\ 0 \\ 0 \end{Bmatrix} - \begin{bmatrix} 1 & 0 & 0 & 0 & 0 & 0 & 0 & 0 & 0 & 0 & 0 & 0 \\ 0 & 1 & 0 & 0 & 0 & 0 & 0 & 0 & 0 & 0 & 0 & 0 \\ 0 & 0 & 1 & 0 & 0 & 0 & 0 & 0 & 0 & 0 & 0 & 0 \\ 0 & 0 & 0 & 1 & 0 & 0 & 1 & 0 & 0 & 0 & 0 & 0 \\ 0 & 0 & 0 & 0 & 1 & 0 & 0 & 1 & 0 & 0 & 0 & 0 \\ 0 & 0 & 0 & 0 & 0 & 1 & 0 & 0 & 1 & 0 & 0 & 0 \\ 0 & 0 & 0 & 0 & 0 & 0 & 0 & 0 & 0 & 1 & 0 & 0 \\ 0 & 0 & 0 & 0 & 0 & 0 & 0 & 0 & 0 & 0 & 1 & 0 \\ 0 & 0 & 0 & 0 & 0 & 0 & 0 & 0 & 0 & 0 & 0 & 1 \end{bmatrix} \begin{Bmatrix} -24 \\ 18 \\ 25 \\ \hline -24 \\ 18 \\ -25 \\ \hline 0 \\ 40 \\ 50 \\ \hline 0 \\ 40 \\ -50 \end{Bmatrix}$$

$$
= \left\{ \begin{array}{c}
24 \\
-18 \\
-25 \\
\hline
24 \\
-118 \\
-25 \\
\hline
0 \\
-40 \\
50
\end{array} \right\} \tag{i}
$$

The equivalent actions acting on the nodes of the frame are shown in Fig. d. Notice that the second term after the first equation sign of relation (i) represents actions opposite to the restraining actions which must be applied to the nodes of the frame in order to restrain them from moving. These actions can be obtained directly by referring to the free-body diagram of the nodes of the restrained structure shown in Fig. c and considering their equilibrium.

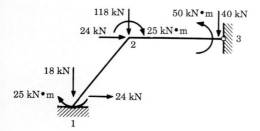

Figure d Structure subjected to the equivalent actions.

STEP 4.   We establish the stiffness matrix for the frame. To accomplish this we do the following.

1. We compute the local stiffness matrix of each element for the frame. That is, referring to relation (4.62) we have

$$
[K^1] = [K^2] = \frac{EI}{125} \begin{bmatrix}
1000 & 0 & 0 & -1000 & 0 & 0 \\
0 & 12 & 30 & 0 & -12 & 30 \\
0 & 30 & 100 & 0 & -30 & 50 \\
-1000 & 0 & 0 & 1000 & 0 & 0 \\
0 & -12 & -30 & 0 & 12 & -30 \\
0 & 30 & 50 & 0 & -30 & 100
\end{bmatrix} \tag{j}
$$

2. From the local stiffness matrix of each element of the structure, we compute its global stiffness matrix. Referring to relation (6.74), we have

$$[\bar{K}^e] = [\Lambda^e]^T [K^e][\Lambda^e] \tag{k}$$

Substituting into (k) the transformation matrix $[\Lambda^1]$ for element 1 given by relation (e), we obtain

$$[\bar{K}^1] = \frac{EI}{125} \begin{bmatrix} 367.68 & 474.24 & -24 & -367.68 & -474.24 & -24 \\ 474.24 & 644.32 & 18 & -474.24 & -644.32 & 18 \\ -24.00 & 18.00 & 100 & 24.00 & -18.00 & 50 \\ -367.68 & -474.24 & 24 & 367.68 & 474.24 & 24 \\ -474.24 & -644.32 & -18 & 474.24 & 644.32 & -18 \\ -24.00 & 18.00 & 50 & 24.00 & -18.00 & 100 \end{bmatrix} \tag{l}$$

Moreover, $$[\bar{K}^2] = [K^2] \tag{m}$$

3. The stiffness matrix for all the elements of the structure is

$$[\hat{K}] = \begin{bmatrix} [\bar{K}^1] & 0 \\ 0 & [\bar{K}^2] \end{bmatrix} \tag{n}$$

Referring to relations (j), (l), and (m) it is apparent that $[\hat{K}]$ is a 12×12 matrix.

4. We establish the stiffness matrix for the structure. Substituting the matrix $[\hat{B}]$, $[\hat{C}]$, and $[\hat{K}]$ from relations (a), (b), and (n), respectively, into relation (8.15) we obtain

$$[\hat{S}] = \frac{EI}{125} \begin{bmatrix} 367.68 & 474.24 & -24 & -367.68 & -474.24 & -24 & 0 & 0 & 0 \\ 474.24 & 644.32 & 18 & -474.24 & -644.32 & 18 & 0 & 0 & 0 \\ -24.00 & 18.00 & 100 & 24.00 & -18.00 & 50 & 0 & 0 & 0 \\ -367.68 & -474.24 & 24 & 1367.68 & 474.24 & 24 & -1000 & 0 & 0 \\ -474.24 & -644.32 & -18 & 474.24 & 656.32 & 12 & 0 & -12 & 30 \\ -24.00 & 18.00 & 50 & 24.00 & 12.00 & 200 & 0 & -30 & 50 \\ 0 & 0 & 0 & -1000.00 & 0 & 0 & 1000 & 0 & 0 \\ 0 & 0 & 0 & 0 & -12.00 & -30 & 0 & 12 & -30 \\ 0 & 0 & 0 & 0 & 30.00 & 50 & 0 & -30 & 100 \end{bmatrix} \tag{o}$$

STEP 5. We form the stiffness equations for the structure. Substituting relations (i) and (o) into (8.14) we get

$$
\begin{Bmatrix}
24 \\
-18 \\
-25 \\
\hline
24 \\
-118 \\
-25 \\
\hline
0 \\
-40 \\
50
\end{Bmatrix}
+
\begin{Bmatrix}
R_1^{(1)} \\
R_2^{(1)} \\
R_3^{(1)} \\
\hline
0 \\
0 \\
0 \\
\hline
R_1^{(3)} \\
R_2^{(3)} \\
0
\end{Bmatrix}
= [\hat{S}]
\begin{Bmatrix}
\Delta_1 \\
\Delta_2 \\
\Delta_3 \\
\Delta_4 \\
\Delta_5 \\
\Delta_6 \\
\Delta_7 \\
\Delta_8 \\
\Delta_9
\end{Bmatrix}
$$

8.3   Procedures for Assembling the Stiffness Matrix for a Structure

In this section, we describe two slightly different procedures for assembling the stiffness matrix for a structure from the global stiffness matrices for its elements. In one procedure, each stiffness coefficient for the structure is computed from the global stiffness coefficients for its elements and is placed in the appropriate slot of the stiffness matrix for the structure [see relation (8.2)]. In the other procedure, each block of stiffness coefficients $[S]_{ij}$ for a structure is established from blocks of stiffness coefficients for its elements and is placed in the appropriate slots of the stiffness matrix for the structure [see relation (8.3)]. Both procedures can be used in writing computer programs for assembling the stiffness matrix for structures. In Sec. 11.4 we present a more efficient procedure for writing a program for assembling directly the submatrices $[S^{FF}]$, $[S^{SF}]$, $[S^{FS}]$, and $[S^{SS}]$ [see Eqs. (9.1)] of the stiffness matrix $[S]$ of a structure by computer.

8.3.1   Procedure 1 for assembling the stiffness matrix for a structure

Consider node m of a planar structure, and as shown in Fig. 8.1 assume that three elements are connected to it. The components of displacements of the nodes into which the three elements under consideration are connected are shown in Fig. 8.1. Referring to this figure the relations between the global components of the nodal actions and the global components of the nodal displacements of

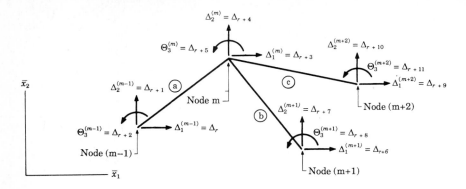

Figure 8.1 Elements connected to node m of a planar structure and components of displacements of the nodes to which these elements are connected.

these elements are expressed as follows:

$$
\begin{Bmatrix}
\bar{F}_1^{aj} \\
\bar{F}_2^{aj} \\
\bar{M}_3^{aj} \\
\bar{F}_1^{ak} \\
\bar{F}_2^{ak} \\
\bar{M}_3^{ak}
\end{Bmatrix}
=
\begin{bmatrix}
\bar{K}_{rr}^a & \bar{K}_{r(r+1)}^a & \bar{K}_{r(r+2)}^a \\
\bar{K}_{(r+1)r}^a & \bar{K}_{(r+1)(r+1)}^a & \bar{K}_{(r+1)(r+2)}^a \\
\bar{K}_{(r+2)r}^a & \bar{K}_{(r+2)(r+1)}^a & \bar{K}_{(r+2)(r+2)}^a \\
\bar{K}_{(r+3)r}^a & \bar{K}_{(r+3)(r+1)}^a & \bar{K}_{(r+3)(r+2)}^a \\
\bar{K}_{(r+4)r}^a & \bar{K}_{(r+4)(r+1)}^a & \bar{K}_{(r+4)(r+2)}^a \\
\bar{K}_{(r+5)r}^a & \bar{K}_{(r+5)(r+1)}^a & \bar{K}_{(r+5)(r+2)}^a
\end{bmatrix}
$$

$$
\begin{bmatrix}
\bar{K}_{r(r+3)}^a & \bar{K}_{r(r+4)}^a & \bar{K}_{r(r+5)}^a \\
\bar{K}_{(r+1)(r+3)}^a & \bar{K}_{(r+1)(r+4)}^a & \bar{K}_{(r+1)(r+5)}^a \\
\bar{K}_{(r+2)(r+3)}^a & \bar{K}_{(r+2)(r+4)}^a & \bar{K}_{(r+2)(r+5)}^a \\
\bar{K}_{(r+3)(r+3)}^a & \bar{K}_{(r+3)(r+4)}^a & \bar{K}_{(r+3)(r+5)}^a \\
\bar{K}_{(r+4)(r+3)}^a & \bar{K}_{(r+4)(r+4)}^a & \bar{K}_{(r+4)(r+5)}^a \\
\bar{K}_{(r+5)(r+3)}^a & \bar{K}_{(r+5)(r+4)}^a & \bar{K}_{(r+5)(r+5)}^a
\end{bmatrix}
\begin{Bmatrix}
\Delta_r \\
\Delta_{r+1} \\
\Delta_{r+2} \\
\Delta_{r+3} \\
\Delta_{r+4} \\
\Delta_{r+5}
\end{Bmatrix}
\qquad (8.16)
$$

$$
\begin{Bmatrix}
\bar{F}_1^{bj} \\
\bar{F}_2^{bj} \\
\bar{M}_3^{bj} \\
\bar{F}_1^{bk} \\
\bar{F}_2^{bk} \\
\bar{M}_3^{bk}
\end{Bmatrix}
=
\begin{bmatrix}
\bar{K}_{(r+3)(r+3)}^b & \bar{K}_{(r+3)(r+4)}^b & \bar{K}_{(r+3)(r+5)}^b \\
\bar{K}_{(r+4)(r+3)}^b & \bar{K}_{(r+4)(r+4)}^b & \bar{K}_{(r+4)(r+5)}^b \\
\bar{K}_{(r+5)(r+3)}^b & \bar{K}_{(r+5)(r+4)}^b & \bar{K}_{(r+5)(r+5)}^b \\
\bar{K}_{(r+6)(r+3)}^b & \bar{K}_{(r+6)(r+4)}^b & \bar{K}_{(r+6)(r+5)}^b \\
\bar{K}_{(r+7)(r+3)}^b & \bar{K}_{(r+7)(r+4)}^b & \bar{K}_{(r+7)(r+5)}^b \\
\bar{K}_{(r+8)(r+3)}^b & \bar{K}_{(r+8)(r+4)}^b & \bar{K}_{(r+8)(r+5)}^b
\end{bmatrix}
$$

$$
\begin{bmatrix}
\bar{K}^b_{(r+3)(r+6)} & \bar{K}^b_{(r+3)(r+7)} & \bar{K}^b_{(r+3)(r+8)} \\
\bar{K}^b_{(r+4)(r+6)} & \bar{K}^b_{(r+4)(r+7)} & \bar{K}^b_{(r+4)(r+8)} \\
\bar{K}^b_{(r+5)(r+6)} & \bar{K}^b_{(r+5)(r+7)} & \bar{K}^b_{(r+5)(r+8)} \\
\bar{K}^b_{(r+6)(r+6)} & \bar{K}^b_{(r+6)(r+7)} & \bar{K}^b_{(r+6)(r+8)} \\
\bar{K}^b_{(r+7)(r+6)} & \bar{K}^b_{(r+7)(r+7)} & \bar{K}^b_{(r+7)(r+8)} \\
\bar{K}^b_{(r+8)(r+6)} & \bar{K}^b_{(r+8)(r+7)} & \bar{K}^b_{(r+8)(r+8)}
\end{bmatrix}
\begin{Bmatrix}
\Delta_{r+3} \\
\Delta_{r+4} \\
\Delta_{r+5} \\
\Delta_{r+6} \\
\Delta_{r+7} \\
\Delta_{r+8}
\end{Bmatrix}
\tag{8.17}
$$

$$
\begin{Bmatrix}
\bar{F}^{cj}_1 \\
\bar{F}^{cj}_2 \\
\bar{M}^{cj}_3 \\
\bar{F}^{ck}_1 \\
\bar{F}^{ck}_2 \\
\bar{M}^{ck}_3
\end{Bmatrix}
=
\begin{bmatrix}
\bar{K}^c_{(r+3)(r+3)} & \bar{K}^c_{(r+3)(r+4)} & \bar{K}^c_{(r+3)(r+5)} \\
\bar{K}^c_{(r+4)(r+3)} & \bar{K}^c_{(r+4)(r+4)} & \bar{K}^c_{(r+4)(r+5)} \\
\bar{K}^c_{(r+5)(r+3)} & \bar{K}^c_{(r+5)(r+4)} & \bar{K}^c_{(r+5)(r+5)} \\
\bar{K}^c_{(r+9)(r+3)} & \bar{K}^c_{(r+9)(r+4)} & \bar{K}^c_{(r+9)(r+5)} \\
\bar{K}^c_{(r+10)(r+3)} & \bar{K}^c_{(r+10)(r+4)} & \bar{K}^c_{(r+10)(r+5)} \\
\bar{K}^c_{(r+11)(r+3)} & \bar{K}^c_{(r+11)(r+4)} & \bar{K}^c_{(r+11)(r+5)}
\end{bmatrix}
$$

$$
\begin{bmatrix}
\bar{K}^c_{(r+3)(r+9)} & \bar{K}^c_{(r+3)(r+10)} & \bar{K}^c_{(r+3)(r+11)} \\
\bar{K}^c_{(r+4)(r+9)} & \bar{K}^c_{(r+4)(r+10)} & \bar{K}^c_{(r+4)(r+11)} \\
\bar{K}^c_{(r+5)(r+9)} & \bar{K}^c_{(r+5)(r+10)} & \bar{K}^c_{(r+5)(r+11)} \\
\bar{K}^c_{(r+9)(r+9)} & \bar{K}^c_{(r+9)(r+10)} & \bar{K}^c_{(r+9)(r+11)} \\
\bar{K}^c_{(r+10)(r+9)} & \bar{K}^c_{(r+10)(r+10)} & \bar{K}^c_{(r+10)(r+11)} \\
\bar{K}^c_{(r+11)(r+9)} & \bar{K}^c_{(r+11)(r+10)} & \bar{K}^c_{(r+11)(r+11)}
\end{bmatrix}
\begin{Bmatrix}
\Delta_{r+3} \\
\Delta_{r+4} \\
\Delta_{r+5} \\
\Delta_{r+9} \\
\Delta_{r+10} \\
\Delta_{r+11}
\end{Bmatrix}
\tag{8.18}
$$

Notice that in the above relations we have replaced the global components of the nodal displacements of each element by the corresponding components of the displacements of the nodes of the structure to which the element is connected. This implies that the nodal displacements of the elements of the structure are compatible with the displacements of its nodes. Moreover, notice that in relations (8.16) to (8.18), the indices of the global stiffness coefficients for an element correspond to the indices of the components of displacements of the nodes to which this element is connected.

Referring to Fig. 8.1 from the equilibrium of node m, we have

$$
\sum \bar{F}_1 = 0 \qquad \hat{P}^E_{r+3} + R^{(m)}_1 = \bar{F}^{ak}_1 + \bar{F}^{bj}_1 + \bar{F}^{cj}_1
$$

$$
\sum \bar{F}_2 = 0 \qquad P^E_{r+4} + R^{(m)}_2 = \bar{F}^{ak}_2 + \bar{F}^{bj}_2 + \bar{F}^{cj}_2 \tag{8.19}
$$

$$
\sum \bar{M}_3 = 0 \qquad P^E_{r+5} + R^{(m)}_3 = \bar{M}^{ak}_3 + \bar{M}^{bj}_3 + \bar{M}^{cj}_3
$$

where P^E_{r+3}, P^E_{r+4}, and P^E_{r+5} are the global components of the known equivalent actions acting on node m. $R_1^{(m)}$, $R_2^{(m)}$, and $R_3^{(m)}$ are the unknown components of the reactions acting on mode m in case this node is restrained from moving.

The equilibrium equations (8.19) for node m can be converted into the $(r+3)$th, $(r+4)$th, and $(r+5)$th stiffness equations for the structure using stiffness equations (8.16), (8.17), and (8.18) for elements a, b, and c which are connected to node m. This is accomplished by replacing each one of the global components of nodal actions $\bar{F}_i^{ak}$; $\bar{F}_i^{bj}$, $\bar{F}_i^{cj}$ $(i = 1, 2)$, $\bar{M}_3^{ak}$, $\bar{M}_3^{bj}$, and $\bar{M}_3^{cj}$ of elements a, b, and c by its expression as a linear combination of the components of displacements of the nodes of the structure to which these elements are connected. For example, substituting in the first of equations (8.19) the values of $\bar{F}_1^{ak}$, $\bar{F}_1^{bj}$, and $\bar{F}_1^{cj}$ obtained from relations (8.16), (8.17), and (8.18), respectively, we obtain the following expression for the $(r+3)$th stiffness equation for the structure:

$$
\begin{aligned}
P_{r+3}^E + R_1^{(m)} = &[\bar{K}_{(r+3)r}^a \, \Delta_r + \bar{K}_{(r+3)(r+1)}^a \, \hat{\Delta}_{r+1} + \bar{K}_{(r+3)(r+2)}^a \, \Delta_{r+2} \\
&+ \bar{K}_{(r+3)(r+3)}^a \, \Delta_{r+3} + \bar{K}_{(r+3)(r+4)}^a \, \Delta_{(r+4)} + \bar{K}_{(r+3)(r+5)}^a \, \Delta_{(r+5)}] \\
&+ [\bar{K}_{(r+3)(r+3)}^b \, \Delta_{(r+3)} + \bar{K}_{(r+3)(r+4)}^b \, \Delta_{(r+4)} + \bar{K}_{(r+3)(r+5)}^b \, \Delta_{(r+5)} \\
&+ \bar{K}_{(r+3)(r+6)}^b \, \Delta_{(r+6)} + \bar{K}_{(r+3)(r+7)}^b \, \Delta_{(r+7)} + \bar{K}_{(r+3)(r+8)}^b \, \Delta_{(r+8)}] \\
&+ [\bar{K}_{(r+9)(r+3)}^c \, \Delta_{(r+3)} + \bar{K}_{(r+9)(r+4)}^c \, \Delta_{(r+4)} + \bar{K}_{(r+9)(r+5)}^c \, \Delta_{(r+5)} \\
&+ \bar{K}_{(r+9)(r+9)}^c \, \Delta_{(r+9)} + \bar{K}_{(r+9)(r+10)}^c \, \Delta_{(r+10)} + \bar{K}_{(r+9)(r+11)}^c \, \Delta_{(r+11)}]
\end{aligned}
$$

$$(8.20)$$

Equation (8.20) must be identical to the $(r+3)$th equation of the stiffness equations (8.2), which can be written as

$$
\begin{aligned}
P_{(r+3)}^E + R_1^{(m)} = &[S_{(r+3)1} \, \Delta_1 + S_{(r+3)2} \, \Delta_2 + \cdots + S_{(r+3)r} \, \Delta_r \\
&+ S_{(r+3)(r+1)} \, \Delta_{(r+1)} + S_{(r+3)(r+2)} \, \Delta_{r+2} + S_{(r+3)(r+3)} \, \Delta_{(r+3)} \\
&+ S_{(r+3)(r+4)} \, \Delta_{r+4} + S_{(r+3)(r+5)} \, \Delta_{r+5} + S_{(r+3)(r+6)} \, \Delta_{(r+6)} \\
&+ S_{(r+3)(r+7)} \, \Delta_{r+7} + S_{(r+3)(r+8)} \, \Delta_{r+8} + S_{(r+3)(r+9)} \, \Delta_{r+9} \\
&+ S_{(r+3)(r+10)} \, \Delta_{r+10} + S_{(r+3)(r+11)} \, \Delta_{(r+11)} + \cdots]
\end{aligned}
$$

$$(8.21)$$

Comparing Eqs. (8.20) and (8.21), it can be seen that

$$S_{(r+3)1} = 0$$

$$S_{(r+3)2} = 0$$

.

.

.

$$S_{(r+3)(r-1)} = 0$$

$$S_{(r+3)(r)} = \bar{K}^a_{(r+3)(r)}$$

$$S_{(r+3)(r+1)} = \bar{K}^a_{(r+3)(r+1)}$$

$$S_{(r+3)(r+2)} = \bar{K}^a_{(r+3)(r+2)}$$

$$S_{(r+3)(r+3)} = \bar{K}^a_{(r+3)(r+3)} + \bar{K}^b_{(r+3)(r+3)} + \bar{K}^c_{(r+3)(r+3)}$$

$$S_{(r+3)(r+4)} = \bar{K}^a_{(r+3)(r+4)} + \bar{K}^b_{(r+3)(r+4)} + \bar{K}^c_{(r+3)(r+4)}$$

$$S_{(r+3)(r+5)} = \bar{K}^a_{(r+3)(r+5)} + \bar{K}^b_{(r+3)(r+5)} - \bar{K}^c_{(r+3)(r+5)} \qquad (8.22)$$

$$S_{(r+3)(r+6)} = \bar{K}^b_{(r+3)(r+6)}$$

$$S_{(r+3)(r+7)} = \bar{K}^b_{(r+3)(r+7)}$$

$$S_{(r+3)(r+8)} = \bar{K}^b_{(r+3)(r+8)}$$

$$S_{(r+3)(r+9)} = \bar{K}^c_{(r+3)(r+9)}$$

$$S_{(r+3)(r+10)} = \bar{K}^c_{(r+3)(r+10)}$$

$$S_{(r+3)(r+11)} = \bar{K}^c_{(r+3)(r+11)}$$

$$S_{(r+3)j} = 0 \qquad [j = (r+12), (r+13), \ldots]$$

Thus, a stiffness coefficient S_{ij} for a structure is equal to the sum of the global stiffness coefficients for the elements of the structure having the same indices [see Eqs. (8.16) to (8.18)]. That is, for a structure having NE elements we have

$$S_{ij} = \sum_{n=1}^{NE} \bar{K}^n_{ij} \qquad (8.23)$$

In the above relation, the global stiffness coefficients for an element of the structure vanish if their indices do not correspond to the indices of the components of displacements of the nodes to which the element is connected [see Eqs. (8.16) to (8.18)].

On the basis of the foregoing, it is apparent that the stiffness matrix $[\hat{S}]$ for a structure may be established directly from the global stiffness matrices for its elements by adhering to following steps.

Step 1. The local stiffness matrices for the elements of the structure are established.

Step 2. The global stiffness matrices for the elements of the structure are computed from their local stiffness matrices using relation (6.74).

Step 3. The stiffness matrix for the structure is assembled from the global stiffness matrices for its elements. In order to accomplish this the indices of the global stiffness coefficients for each element of the structure are chosen so as to correspond to those of the components of displacements of the nodes of the structure to which the element is connected. The stiffness coefficients for the structure are computed on the basis of relation (8.23).

In the sequel we illustrate how to compute the stiffness matrix for a structure using the procedure described in this subsection by the following three examples.

Example 1. The stiffness matrix of a three-member truss is established.

Example 2. The stiffness matrix for a two-member beam is established.

Example 3. The stiffness matrices for two planar frames consisting of the same two elements is established. The one frame has an internal action release mechanism (hinge), while the other does not.

Example 1. Establish the stiffness matrix for the truss shown in Fig. a. The elements of the truss are made from the same isotropic linearly elastic material and have the same constant cross section.

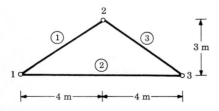

Figure a Geometry of the truss.

---

solution

STEP 1.   We compute the local stiffness matrix for each element of the truss.

That is, referring to relations (4.50) we have

$$[K^1] = \frac{EA}{5} \begin{bmatrix} 1 & -1 \\ -1 & 1 \end{bmatrix}$$

$$[K^2] = \frac{EA}{8} \begin{bmatrix} 1 & -1 \\ -1 & 1 \end{bmatrix} \tag{a}$$

$$[K^3] = \frac{EA}{5} \begin{bmatrix} 1 & -1 \\ -1 & 1 \end{bmatrix}$$

STEP 2.   From the local stiffness matrix for each element of the truss, we compute its global stiffness matrix. Referring to relation (6.74), we have

$$[\bar{K}^e] = [\Lambda^e]^T [K^e][\Lambda^e] \tag{b}$$

where the matrix $[\Lambda^e]$ is given by relation (6.42). That is,

$$[\Lambda^e] = \begin{bmatrix} \cos\phi_{11} & \sin\phi_{11} & 0 & 0 \\ 0 & 0 & \cos\phi_{11} & \sin\phi_{11} \end{bmatrix} \tag{c}$$

Substituting relations (a) and (c) into (b) we obtain

$$[\bar{K}^1] = \frac{1}{5}\begin{bmatrix} 4 & 0 \\ 3 & 0 \\ 0 & 4 \\ 0 & 3 \end{bmatrix} \frac{EA}{5}\begin{bmatrix} 1 & -1 \\ -1 & 1 \end{bmatrix}\frac{1}{5}\begin{bmatrix} 4 & 3 & 0 & 0 \\ 0 & 0 & 4 & 3 \end{bmatrix}$$

$$= \frac{EA}{125}\begin{bmatrix} 16 & 12 & -16 & -12 \\ 12 & 9 & -12 & -9 \\ -16 & -12 & 16 & 12 \\ -12 & -9 & 12 & 9 \end{bmatrix}$$

$$[\bar{K}^2] = \begin{bmatrix} 1 & 0 \\ 0 & 0 \\ 0 & 1 \\ 0 & 0 \end{bmatrix}\frac{EA}{8}\begin{bmatrix} 1 & -1 \\ -1 & 1 \end{bmatrix}\begin{bmatrix} 1 & 0 & 0 & 0 \\ 0 & 0 & 1 & 0 \end{bmatrix}$$

$$= \frac{EA}{8}\begin{bmatrix} 1 & 0 & -1 & 0 \\ 0 & 0 & 0 & 0 \\ -1 & 0 & 1 & 0 \\ 0 & 0 & 0 & 0 \end{bmatrix} \tag{d}$$

$$[\bar{K}^3] = \frac{1}{5}\begin{bmatrix} 4 & 0 \\ -3 & 0 \\ 0 & 4 \\ 0 & -3 \end{bmatrix}\frac{EA}{5}\begin{bmatrix} 1 & -1 \\ -1 & 1 \end{bmatrix}\frac{1}{5}\begin{bmatrix} 4 & -3 & 0 & 0 \\ 0 & 0 & 4 & -3 \end{bmatrix}$$

$$= \frac{EA}{125}\begin{bmatrix} 16 & -12 & -16 & 12 \\ -12 & 9 & 12 & -9 \\ -16 & 12 & 16 & -12 \\ 12 & -9 & -12 & 9 \end{bmatrix}$$

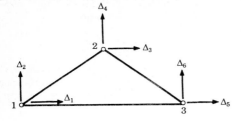

Figure b Numbering of the components of displacement of the nodes of the truss.

STEP 3.   We assemble the stiffness matrix for the structure from the global stiffness matrices for its elements. In order to accomplish this we choose the indices of the stiffness coefficients for each element of the truss so as to correspond to those of the components of displacements of the nodes of the truss to which the element is connected. Referring to Fig. b the global stiffness relations (6.69) for the elements of the truss may be expressed as

$$[\bar{A}^1] = [\bar{K}^1][\bar{D}^1] = \begin{bmatrix} \bar{K}_{11}^1 & \bar{K}_{12}^1 & \bar{K}_{13}^1 & \bar{K}_{14}^1 \\ \bar{K}_{21}^1 & \bar{K}_{22}^1 & \bar{K}_{23}^1 & \bar{K}_{24}^1 \\ \bar{K}_{31}^1 & \bar{K}_{32}^1 & \bar{K}_{33}^1 & \bar{K}_{34}^1 \\ \bar{K}_{41}^1 & \bar{K}_{42}^1 & \bar{K}_{43}^1 & \bar{K}_{44}^1 \end{bmatrix} \begin{Bmatrix} \Delta_1 \\ \Delta_2 \\ \Delta_3 \\ \Delta_4 \end{Bmatrix} \tag{e}$$

$$[\bar{A}^2] = [\bar{K}^2][\bar{D}^2] = \begin{bmatrix} \bar{K}_{11}^2 & \bar{K}_{12}^2 & \bar{K}_{15}^2 & \bar{K}_{16}^2 \\ \bar{K}_{21}^2 & \bar{K}_{22}^2 & \bar{K}_{25}^2 & \bar{K}_{26}^2 \\ \bar{K}_{51}^2 & \bar{K}_{52}^2 & \bar{K}_{55}^2 & \bar{K}_{56}^2 \\ \bar{K}_{61}^2 & \bar{K}_{62}^2 & \bar{K}_{65}^2 & \bar{K}_{66}^2 \end{bmatrix} \begin{Bmatrix} \Delta_1 \\ \Delta_2 \\ \Delta_5 \\ \Delta_6 \end{Bmatrix} \tag{f}$$

$$[\bar{A}^3] = [\bar{K}^3][\bar{D}^3] = \begin{bmatrix} \bar{K}_{33}^3 & \bar{K}_{34}^3 & \bar{K}_{35}^3 & \bar{K}_{36}^3 \\ \bar{K}_{43}^3 & \bar{K}_{44}^3 & \bar{K}_{45}^3 & \bar{K}_{46}^3 \\ \bar{K}_{53}^3 & \bar{K}_{54}^3 & \bar{K}_{55}^3 & \bar{K}_{56}^3 \\ \bar{K}_{63}^3 & \bar{K}_{64}^3 & \bar{K}_{65}^3 & \bar{K}_{66}^3 \end{bmatrix} \begin{Bmatrix} \Delta_3 \\ \Delta_4 \\ \Delta_5 \\ \Delta_6 \end{Bmatrix} \tag{g}$$

The stiffness matrix for the truss has the following form:

$$[\hat{S}] = \begin{bmatrix} S_{11} & S_{12} & S_{13} & S_{14} & S_{15} & S_{16} \\ S_{21} & S_{22} & S_{23} & S_{24} & S_{25} & S_{26} \\ S_{31} & S_{32} & S_{33} & S_{34} & S_{35} & S_{36} \\ S_{41} & S_{42} & S_{43} & S_{44} & S_{45} & S_{46} \\ S_{51} & S_{52} & S_{53} & S_{54} & S_{55} & S_{56} \\ S_{61} & S_{62} & S_{63} & S_{64} & S_{65} & S_{66} \end{bmatrix} \tag{h}$$

where the stiffness coefficient S_{pq} for the truss is equal to the sum of the stiffness coefficients $\bar{K}_{pq}$ for all the elements of the truss. Thus, referring to relations (d) to (g) and using relation (8.23) we have

$$S_{11} = \bar{K}_{11}^1 + \bar{K}_{11}^2 = \frac{EA}{125}\left(16 + \frac{125}{8}\right) = \frac{31.625EA}{125}$$

$$S_{12} = \bar{K}_{12}^1 + \bar{K}_{12}^2 = \frac{EA}{125}(12 + 0) = \frac{12EA}{125}$$

$$S_{13} = \bar{K}_{13}^1 = -\frac{16EA}{125}$$

$$S_{14} = \bar{K}_{14}^1 = -\frac{12EA}{125}$$

$$S_{15} = \bar{K}_{15}^2 = \frac{EA}{8}(-1) = -\frac{15.625EA}{125}$$

$$S_{16} = \bar{K}_{16} = 0$$

$$\vdots$$

$$S_{66} = \bar{K}_{66}^2 + \bar{K}_{66}^3 = \frac{9EA}{125}$$

(i)

Substituting relations (i) into (h) we obtain the stiffness matrix $[\hat{S}]$ for the truss. That is,

$$[\hat{S}] = \frac{EA}{125}\begin{bmatrix} 31.625 & 12 & -16 & -12 & -15.625 & 0 \\ 12 & 9 & -12 & -9 & 0 & 0 \\ -16 & -12 & 32 & 0 & -16 & 12 \\ -12 & -9 & 0 & 18 & 12 & -9 \\ -15.625 & 0 & -16 & 12 & 31.625 & -12 \\ 0 & 0 & 12 & -9 & -12 & 9 \end{bmatrix}$$

(j)

Notice that the matrix $[\hat{S}]$ is singular because its determinant vanishes. This becomes apparent by noting that the sum of rows 2 and 3 of this determinant is the negative of row 6. This was anticipated because in establishing the matrix $[\hat{S}]$ we did not take into account the supports of the structure and, thus, it can move as a rigid body. Consequently, if the external actions acting on the structure are known, the components of displacements of its nodes cannot be established from its stiffness equations (8.1).

Example 2. Assemble the stiffness matrix for the beam shown in Fig. a. The elements of the beam are made from the same isotropic linearly elastic material and have constant cross sections whose moments of inertia are indicated in the figure.

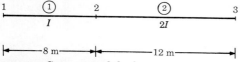

Figure a Geometry of the beam.

solution

STEP 1.   We compute the local stiffness matrix for each element of the beam. We do not include the axial components of the nodal forces in the matrices of nodal actions of the elements of the beam, and neither do we include the axial components of the nodal displacements in the matrices of nodal displacements of the elements of beam. Thus referring to relation (4.62), we have

$$[K^1] = EI \begin{bmatrix} 0.02344 & 0.09375 & -0.02344 & 0.09375 \\ 0.09375 & 0.50000 & -0.09375 & 0.25000 \\ -0.02344 & -0.09375 & 0.02344 & -0.09375 \\ 0.09375 & 0.25000 & -0.09375 & 0.50000 \end{bmatrix} \tag{a}$$

$$[K^2] = EI \begin{bmatrix} 0.01389 & 0.08333 & -0.01389 & 0.08333 \\ 0.08333 & 0.66667 & -0.08333 & 0.33333 \\ -0.01389 & -0.08333 & 0.01389 & -0.08333 \\ 0.08333 & 0.33333 & -0.08333 & 0.66667 \end{bmatrix} \tag{b}$$

STEP 2.   We compute the global stiffness matrix for each element of the beam. In this example it is equal to its local stiffness matrix.

STEP 3.   We assemble the stiffness matrix for the structure from the global stiffness matrices for its elements. In order to accomplish this we choose the indices of the stiffness coefficients for each element of the beam so as to correspond to those of the components of displacements of the nodes of the beam to which the element is connected. Referring to Fig. b the global stiffness relations (6.69) for the elements of the beam may be written as

$$\{\bar{A}^1\} = [\bar{K}^1]\{\bar{D}^1\} = \begin{bmatrix} \bar{K}^1_{11} & \bar{K}^1_{12} & \bar{K}^1_{13} & \bar{K}^1_{14} \\ \bar{K}^1_{21} & \bar{K}^1_{22} & \bar{K}^1_{23} & \bar{K}^1_{24} \\ \bar{K}^1_{31} & \bar{K}^1_{32} & \bar{K}^1_{33} & \bar{K}^1_{34} \\ \bar{K}^1_{41} & \bar{K}^1_{42} & \bar{K}^1_{43} & \bar{K}^1_{44} \end{bmatrix} \begin{Bmatrix} \Delta_1 \\ \Delta_2 \\ \Delta_3 \\ \Delta_4 \end{Bmatrix} \tag{c}$$

$$\{\bar{A}^2\} = [\bar{K}^2]\{\bar{D}^2\} = \begin{bmatrix} \bar{K}^2_{33} & \bar{K}^2_{34} & \bar{K}^2_{35} & \bar{K}^2_{36} \\ \bar{K}^2_{43} & \bar{K}^2_{44} & \bar{K}^2_{45} & \bar{K}^2_{46} \\ \bar{K}^2_{53} & \bar{K}^2_{54} & \bar{K}^2_{55} & \bar{K}^2_{56} \\ \bar{K}^2_{63} & \bar{K}^2_{64} & \bar{K}^2_{65} & \bar{K}^2_{66} \end{bmatrix} \begin{Bmatrix} \Delta_3 \\ \Delta_4 \\ \Delta_5 \\ \Delta_6 \end{Bmatrix} \tag{d}$$

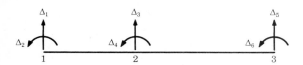

Figure b Numbering of the components of displacements of the nodes of the beam.

The stiffness matrix for the beam has the following form:

$$[\hat{S}] = \begin{bmatrix} S_{11} & S_{12} & S_{13} & S_{14} & S_{15} & S_{16} \\ S_{21} & S_{22} & S_{23} & S_{24} & S_{25} & S_{26} \\ S_{31} & S_{32} & S_{33} & S_{34} & S_{35} & S_{36} \\ S_{41} & S_{42} & S_{43} & S_{44} & S_{45} & S_{46} \\ S_{51} & S_{52} & S_{53} & S_{54} & S_{55} & S_{56} \\ S_{61} & S_{62} & S_{63} & S_{64} & S_{65} & S_{66} \end{bmatrix} \tag{e}$$

where the stiffness coefficient S_{ij} for the beam is obtained from the stiffness coefficients K_{ij} for its elements on the basis of relation (8.23). Thus, referring to relations (a) to (e) and using relation (8.23) the stiffness matrix for the beam of Fig. a is

$$[\hat{S}] = EI \begin{bmatrix} 0.02344 & 0.09375 & -0.02344 & 0.09375 & 0 & 0 \\ 0.09375 & 0.50000 & -0.09375 & 0.25000 & 0 & 0 \\ -0.02344 & -0.09375 & 0.03733 & -0.01042 & -0.01389 & 0.08333 \\ 0.09375 & 0.25000 & -0.01042 & 1.16667 & -0.08333 & 0.33333 \\ 0 & 0 & -0.01389 & -0.08333 & 0.01389 & -0.08333 \\ 0 & 0 & 0.08333 & 0.33333 & -0.08333 & 0.66667 \end{bmatrix} \tag{f}$$

Example 3. Assemble the stiffness matrix for the structures shown in Figs. a and b. The elements of these structures are made from the same isotropic linearly elastic material and have the same constant cross section with $I/A = 0.025 \text{ m}^2$. The elements of the structure of Fig. a are rigidly connected, while those of the structure of Fig. b are connected by a pin.

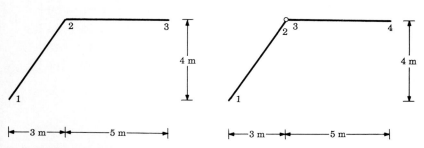

Figure a Geometry of the structure.   **Figure b** Geometry of the structure.

solution

STEP 1. We compute the local stiffness matrix for each element of the structure. That is, referring to relation (4.62) we have

$$[K^1] = [K^2] = \frac{EI}{125} \begin{bmatrix} 1000 & 0 & 0 & -1000 & 0 & 0 \\ 0 & 12 & 30 & 0 & -12 & 30 \\ 0 & 30 & 100 & 0 & -30 & 50 \\ -1000 & 0 & 0 & 1000 & 0 & 0 \\ 0 & -12 & -30 & 0 & 12 & -30 \\ 0 & 30 & 50 & 0 & -30 & 100 \end{bmatrix} \qquad (a)$$

STEP 2.   From the local stiffness matrix for each element of the structure, we compute its global stiffness matrix. Referring to relation (6.74), we have

$$[\bar{K}^e] = [\Lambda^e]^T [K^e][\Lambda^e] \qquad e = 1, 2 \qquad (b)$$

The transformation matrix $[\Lambda^1]$ of element 1 is given by relation (6.48). That is,

$$[\Lambda^1] = \begin{bmatrix} 0.6 & 0.8 & 0 & 0 & 0 & 0 \\ -0.8 & 0.6 & 0 & 0 & 0 & 0 \\ 0 & 0 & 1 & 0 & 0 & 0 \\ 0 & 0 & 0 & 0.6 & 0.8 & 0 \\ 0 & 0 & 0 & -0.8 & 0.6 & 0 \\ 0 & 0 & 0 & 0 & 0 & 1 \end{bmatrix} \qquad (c)$$

Substituting relations (c) and (a) into (b) we get

$$[\bar{K}^1] = \frac{EI}{125} \begin{bmatrix} 367.68 & 474.24 & -24 & -367.68 & -474.24 & -24 \\ 474.24 & 644.32 & 18 & -474.24 & -644.32 & 18 \\ -24.00 & 18.00 & 100 & 24.00 & -18.00 & 50 \\ -367.68 & -474.24 & 24 & 367.68 & 474.24 & 24 \\ -474.24 & -644.32 & -18 & 474.24 & 644.32 & -18 \\ -24.00 & 18.00 & 50 & 24.00 & -18.00 & 100 \end{bmatrix} \qquad (d)$$

Moreover,

$$[\bar{K}^2] = [K^2] \qquad (e)$$

Structure of Fig. a

STEP 3.   We assemble the stiffness matrix for the structure from the global stiffness matrices for its elements. To accomplish this we choose the indices of the stiffness coefficients for each element of the structure so as to correspond to those of the components of displacement of the nodes of the structure to which the element is connected. Referring to Fig. c the global

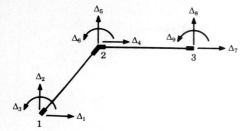

Figure c Numbering of the components of displacements of the nodes of the structure of Fig. a.

stiffness relations (6.69) for the elements of the structure may be written as

$$\{\bar{A}^1\} = [\bar{K}^1]\{\bar{D}^1\} = \begin{bmatrix} \bar{K}^1_{11} & \bar{K}^1_{12} & \bar{K}^1_{13} & \bar{K}^1_{14} & \bar{K}^1_{15} & \bar{K}^1_{16} \\ \bar{K}^1_{21} & \bar{K}^1_{22} & \bar{K}^1_{23} & \bar{K}^1_{24} & \bar{K}^1_{25} & \bar{K}^1_{26} \\ \bar{K}^1_{31} & \bar{K}^1_{32} & \bar{K}^1_{33} & \bar{K}^1_{34} & \bar{K}^1_{35} & \bar{K}^1_{36} \\ \bar{K}^1_{41} & \bar{K}^1_{42} & \bar{K}^1_{43} & \bar{K}^1_{44} & \bar{K}^1_{45} & \bar{K}^1_{46} \\ \bar{K}^1_{51} & \bar{K}^1_{52} & \bar{K}^1_{53} & \bar{K}^1_{54} & \bar{K}^1_{55} & \bar{K}^1_{56} \\ \bar{K}^1_{61} & \bar{K}^1_{62} & \bar{K}^1_{63} & \bar{K}^1_{64} & \bar{K}^1_{65} & \bar{K}^1_{66} \end{bmatrix} \begin{Bmatrix} \Delta_1 \\ \Delta_2 \\ \Delta_3 \\ \Delta_4 \\ \Delta_5 \\ \Delta_6 \end{Bmatrix} \qquad \text{(f)}$$

$$\{\bar{A}^2\} = [\bar{K}^2]\{\bar{D}^2\} = \begin{bmatrix} \bar{K}^2_{44} & \bar{K}^2_{45} & \bar{K}^2_{46} & \bar{K}^2_{47} & \bar{K}^2_{48} & \bar{K}^2_{49} \\ \bar{K}^2_{54} & \bar{K}^2_{55} & \bar{K}^2_{56} & \bar{K}^2_{57} & \bar{K}^2_{58} & \bar{K}^2_{59} \\ \bar{K}^2_{64} & \bar{K}^2_{65} & \bar{K}^2_{66} & \bar{K}^2_{67} & \bar{K}^2_{68} & \bar{K}^2_{69} \\ \bar{K}^2_{74} & \bar{K}^2_{75} & \bar{K}^2_{76} & \bar{K}^2_{77} & \bar{K}^2_{78} & \bar{K}^2_{79} \\ \bar{K}^2_{84} & \bar{K}^2_{85} & \bar{K}^2_{86} & \bar{K}^2_{87} & \bar{K}^2_{88} & \bar{K}^2_{89} \\ \bar{K}^2_{94} & \bar{K}^2_{95} & \bar{K}^2_{96} & \bar{K}^2_{97} & \bar{K}^2_{98} & \bar{K}^2_{99} \end{bmatrix} \begin{Bmatrix} \Delta_4 \\ \Delta_5 \\ \Delta_6 \\ \Delta_7 \\ \Delta_8 \\ \Delta_9 \end{Bmatrix} \qquad \text{(g)}$$

The stiffness matrix for the structure of Fig. a has the following form:

$$[\hat{S}] = \begin{bmatrix} S_{11} & S_{12} & S_{13} & S_{14} & S_{15} & S_{16} & S_{17} & S_{18} & S_{19} \\ S_{21} & S_{22} & S_{23} & S_{24} & S_{25} & S_{26} & S_{27} & S_{28} & S_{29} \\ S_{31} & S_{32} & S_{33} & S_{34} & S_{35} & S_{36} & S_{37} & S_{38} & S_{39} \\ S_{41} & S_{42} & S_{43} & S_{44} & S_{45} & S_{46} & S_{47} & S_{48} & S_{49} \\ S_{51} & S_{52} & S_{53} & S_{54} & S_{55} & S_{56} & S_{57} & S_{58} & S_{59} \\ S_{61} & S_{62} & S_{63} & S_{64} & S_{65} & S_{66} & S_{67} & S_{68} & S_{69} \\ S_{71} & S_{72} & S_{73} & S_{74} & S_{75} & S_{76} & S_{77} & S_{78} & S_{79} \\ S_{81} & S_{82} & S_{83} & S_{84} & S_{85} & S_{86} & S_{87} & S_{88} & S_{89} \\ S_{91} & S_{92} & S_{93} & S_{94} & S_{95} & S_{96} & S_{97} & S_{98} & S_{99} \end{bmatrix} \qquad \text{(h)}$$

where the stiffness coefficient S_{ij} for the structure is obtained from the stiffness coefficients K_{ij} for its elements on the basis of relation (8.23). Thus referring to relations (d) to (h) and using relation (8.23) the stiffness matrix for the structure of Fig. a is

$$[\hat{S}] = \frac{EI}{125} \begin{bmatrix} 367.68 & 474.24 & -24 & -367.68 & -474.24 & -24 & 0 & 0 & 0 \\ 474.24 & 644.32 & 18 & -474.24 & -644.32 & 18 & 0 & 0 & 0 \\ -24.00 & 18.00 & 100 & 24.00 & -18.00 & 50 & 0 & 0 & 0 \\ -367.68 & -474.24 & 24 & 1367.68 & 474.24 & 24 & -1000 & 0 & 0 \\ -474.24 & -644.32 & -18 & 474.24 & 656.32 & 12 & 0 & -12 & 30 \\ -24.00 & 18.00 & 50 & 24.00 & 12.00 & 200 & 0 & -30 & 50 \\ 0 & 0 & 0 & -1000.00 & 0 & 0 & 1000 & 0 & 0 \\ 0 & 0 & 0 & 0 & -12.00 & -30 & 0 & 12 & -30 \\ 0 & 0 & 0 & 0 & 30.00 & 50 & 0 & -30 & 100 \end{bmatrix}$$

(i)

Structure of Fig. b

In this structure we have two connected nodes adjacent to the internal hinge. These nodes can rotate and translate in any direction. However, their components of translation are equal. Thus, as shown in Fig. d, these two nodes together have four independent components of displacements.

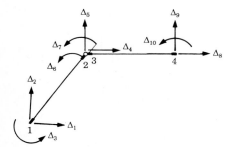

Figure d Numbering of the components of displacement of the nodes of the structure of Fig. b.

STEP 3.   We assemble the stiffness matrix for the structure from the global stiffness matrices for its elements. Referring to Fig. d, it is apparent that the matrix $[\bar{K}^1]$ can be written as in Eq. (f), while the matrix $[\bar{K}^2]$ can be written as

$$\{\bar{A}^2\} = [\bar{K}^2]\{\bar{D}^2\} = \begin{bmatrix} \bar{K}^2_{44} & \bar{K}^2_{45} & \bar{K}^2_{47} & \bar{K}^2_{48} & \bar{K}^2_{49} & \bar{K}^2_{4,10} \\ \bar{K}^2_{54} & \bar{K}^2_{55} & \bar{K}^2_{57} & \bar{K}^2_{58} & \bar{K}^2_{59} & \bar{K}^2_{5,10} \\ \bar{K}^2_{74} & \bar{K}^2_{75} & \bar{K}^2_{77} & \bar{K}^2_{78} & \bar{K}^2_{79} & \bar{K}^2_{7,10} \\ \bar{K}^2_{84} & \bar{K}^2_{85} & \bar{K}^2_{87} & \bar{K}^2_{88} & \bar{K}^2_{89} & \bar{K}^2_{8,10} \\ \bar{K}^2_{94} & \bar{K}^2_{95} & \bar{K}^2_{97} & \bar{K}^2_{98} & \bar{K}^2_{99} & \bar{K}^2_{9,10} \\ \bar{K}^2_{10,4} & \bar{K}^2_{10,5} & \bar{K}^2_{10,7} & \bar{K}^2_{10,8} & \bar{K}^2_{10,9} & \bar{K}^2_{10,10} \end{bmatrix} \begin{Bmatrix} \Delta_4 \\ \Delta_5 \\ \Delta_7 \\ \Delta_8 \\ \Delta_9 \\ \Delta_{10} \end{Bmatrix}$$

(j)

Thus referring to relations (f) and (j) and using relations (d) and (e), the stiffness matrix for the structure is

$$[\hat{S}] = \frac{EI}{125} = \begin{bmatrix}
367.68 & 474.24 & -24 & -367.68 & -474.24 & -24 & 0 & 0 & 0 & 0 \\
474.24 & 644.32 & 18 & -474.24 & -644.32 & 18 & 0 & 0 & 0 & 0 \\
-24.00 & 18.00 & 100 & 24.00 & -18.00 & 50 & 0 & 0 & 0 & 0 \\
-367.68 & -474.24 & 24 & 1367.68 & 474.24 & 24 & 0 & -1000 & 0 & 0 \\
-474.24 & -644.32 & -18 & 474.24 & 656.32 & -18 & 30 & 0 & -12 & 30 \\
-24.00 & 18.00 & 50 & 24.00 & -18 & 100 & 0 & 0 & 0 & 0 \\
0 & 0 & 0 & 0 & 30 & 0 & 100 & 0 & -30 & 50 \\
0 & 0 & 0 & -1000.00 & 0 & 0 & 0 & 1000 & 0 & 0 \\
0 & 0 & 0 & 0 & -12.00 & 0 & -30 & 0 & 12 & -30 \\
0 & 0 & 0 & 0 & 30.00 & 0 & 50 & 0 & -30 & 100
\end{bmatrix}$$

(k)

8.3.2 Procedure 2 for assembling the stiffness matrix for a structure

Consider node m of a planar structure, and as shown in Fig. 8.1 assume that three elements are connected to it. Referring to Fig. 8.1 the relations between the global components of the nodal displacements of the three elements are expressed as

$$\left\{ \begin{matrix} \{\bar{A}^{aj}\} \\ \{\bar{A}^{ak}\} \end{matrix} \right\} = \begin{bmatrix} [\bar{K}^a]_{(m-1)(m-1)} & [\bar{K}^a]_{(m-1)m} \\ [\bar{K}^a]_{m(m-1)} & [\bar{K}^a]_{mm} \end{bmatrix} \left\{ \begin{matrix} \{\Delta\}_{(m-1)} \\ \{\Delta\}_m \end{matrix} \right\} \qquad (8.24)$$

$$\left\{ \begin{matrix} \{\bar{A}^{bj}\} \\ \{\bar{A}^{bk}\} \end{matrix} \right\} = \begin{bmatrix} [\bar{K}^b]_{mm} & [\bar{K}^b]_{m(m+1)} \\ [\bar{K}^b]_{(m+1)m} & [\bar{K}^b]_{(m+1)(m+1)} \end{bmatrix} \left\{ \begin{matrix} \{\Delta\}_m \\ \{\Delta\}_{(m+1)} \end{matrix} \right\} \qquad (8.25)$$

$$\left\{ \begin{matrix} \{\bar{A}^{cj}\} \\ \{\bar{A}^{ck}\} \end{matrix} \right\} = \begin{bmatrix} [\bar{K}^c]_{mm} & [\bar{K}^c]_{m(m+2)} \\ [\bar{K}^c]_{(m+2)m} & [\bar{K}^c]_{(m+2)(m+2)} \end{bmatrix} \left\{ \begin{matrix} \{\Delta\}_m \\ \{\Delta\}_{(m+2)} \end{matrix} \right\} \qquad (8.26)$$

where $[\bar{A}^{eq}]$ $(e = a, b, c)$ $(q = j, k)$ is the matrix of the global components of the actions acting at the end q of element e; $\{\Delta\}_m$ is the matrix of the global components of displacement of node m. Notice that in relations (8.24) to (8.26) the indices of the four submatrices in which the global stiffness matrix for an element is partitioned represent the numbers of the nodes at the ends of the element. Referring to Fig. 8.1 from the equilibrium of node m we have

$$\{P^E\}_m + \{R\}_m = \{\bar{A}^{ak}\} + \{\bar{A}^{bj}\} + \{\bar{A}^{cj}\} \qquad (8.27)$$

where $\{P^E\}_m$ is the matrix of the global components of the equivalent actions acting on mode m, and $\{R\}_m$ is the matrix of the global components of the reactions acting on mode m if this node is restrained from moving. Moreover, using relations (8.24) to (8.26), relation (8.27) can be written as

$$\{P^E\}_m + \{R\}_m = [\bar{K}^a]_{m(m-1)}\{\Delta\}_{(m-1)} + [\bar{K}^a]_{mm}\{\Delta\}_m$$
$$+ [\bar{K}^b]_{mm}\{\Delta\}_m + [\bar{K}^b]_{m(m+1)}\{\Delta\}_{(m+1)}$$
$$+ [\bar{K}^c]_{mm}\{\Delta\}_m + [\bar{K}]_{m(m+2)}\{\Delta\}_{(m+2)} \qquad (8.28)$$

Furthermore, referring to relation (8.3) it is apparent that

$$\{P^E\}_m + \{R\}_m = [S]_{m1}\{\Delta\}_1 + [S]_{m2}\{\Delta\}_2 + \cdots + [S]_{mQ}\{\Delta\}_Q \qquad (8.29)$$

Comparing relations (8.28) and (8.29) it can be seen that

$$[S]_{m1} = [0]$$
$$[S]_{m2} = [0]$$
$$\vdots$$
$$[S]_{m(m-2)} = [0]$$
$$[S]_{m(m-1)} = [\bar{K}^a]_{m(m-1)}$$
$$[S]_{mm} = [\bar{K}^a]_{mm} + [\bar{K}^b]_{mm} + [\bar{K}^c]_{mm} \qquad (8.30)$$
$$[S]_{m(m+1)} = [\bar{K}^b]_{m(m+1)}$$
$$[S]_{m(m+2)} = [\bar{K}^c]_{m(m+2)}$$
$$[S]_{m(m+3)} = [0]$$
$$\vdots$$

Thus, the submatrix $[S]_{mn}$ of the stiffness matrix [see relation (8.3)] for a structure is equal to the sum of the submatrices of the global stiffness matrices for the elements of the structure having the same indices. That is, for a structure with NE elements we have

$$[S]_{mn} = \sum_{e=1}^{NE} [\bar{K}^e]_{mn} \qquad (8.31)$$

On the basis of the foregoing it is apparent that the stiffness matrix $[\hat{S}]$ for a structure may be established from the global stiffness matrices for its elements by adhering to the following steps.

Step 1. The local stiffness matrix for each element of the structure is established.

Step 2. The global stiffness matrix for each element of the structure is computed from its local stiffness matrix using relation (6.74).

Step 3. The stiffness matrix for the structure is established from the global stiffness matrices for its elements. In order to accomplish this the global stiffness matrix for each element of the structure is subdivided into four submatrices as follows:

$$[\bar{K}^e] = \begin{bmatrix} [\bar{K}^e]_{jj} & [\bar{K}^e]_{jk} \\ [\bar{K}^e]_{kj} & [\bar{K}^e]_{kk} \end{bmatrix} \tag{8.32}$$

where j and k are the numbers of the nodes at the end j and k of the element, respectively.

If a structure with n nodes does not have internal action release mechanisms, its stiffness matrix is subdivided into n submatrices of the same order [see relation (8.3)]. These submatrices are then established from the submatrices of the global stiffness matrices for the elements of the structure on the basis of relation (8.31). If a structure has an internal action release mechanism, it has two or more connected nodes (see Fig. 8.2a). In order to establish the stiffness matrix for such a structure a model is considered made from the actual structure by disconnecting its connected nodes (see Fig. 8.2b). The stiffness matrix for the model is established and is modified to obtain the stiffness matrix for the structure (see structure of Fig. b in Example 2 in this subsection).

In the sequel we illustrate how to compute the stiffness matrix for a structure using the procedure described in this subsection by the following two examples.

Example 1. The stiffness matrix for a three-member truss is established. The stiffness matrix for this truss has also been established in Example 1 of Sec. 8.3.1 using the procedure described in that subsection.

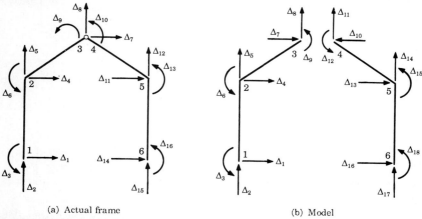

(a) Actual frame   (b) Model

Figure 8.2 Planar frame having an internal action release mechanism.

Example 2.   The stiffness matrices for two planar frames consisting of the same two elements is established. One frame has an internal action release mechanism (pin), while the other does not. The stiffness matrices of these frames have also been established in Example 3 of Sec. 8.3.1 using the procedure described in that subsection.

Example 1.   Establish the stiffness matrix for the truss shown in Fig. a. The elements of the truss are made from the same isotropic linearly elastic material and have the same constant cross section.

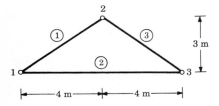

Figure a   Geometry of the truss.

---

solution

STEPS 1 AND 2.   These steps are identical to those of Example 1 of Sec. 8.3.1.

STEP 3.   We assemble the stiffness matrix for the structure from the global stiffness matrices for its elements. Referring to relation (8.32) we subdivide the global stiffness matrix for each element of the truss into four submatrices (see Fig. b). That is, using relations (d) of Example 1 of Sec. 8.3.1 we have

$$[\bar{K}^1] = \begin{bmatrix} [\bar{K}^1]_{11} & [\bar{K}^1]_{12} \\ [\bar{K}^1]_{21} & [\bar{K}^1]_{22} \end{bmatrix} = \frac{EI}{125} \begin{bmatrix} \begin{bmatrix} 16 & 12 \\ 12 & 9 \end{bmatrix} & \begin{bmatrix} -16 & -12 \\ -12 & -9 \end{bmatrix} \\ \begin{bmatrix} -16 & -12 \\ -12 & -9 \end{bmatrix} & \begin{bmatrix} 16 & 12 \\ 12 & 9 \end{bmatrix} \end{bmatrix}$$

$$[\bar{K}^2] = \begin{bmatrix} [\bar{K}^2]_{11} & [\bar{K}^2]_{13} \\ [\bar{K}^2]_{31} & [\bar{K}^2]_{33} \end{bmatrix} = \frac{EA}{8} \begin{bmatrix} \begin{bmatrix} 1 & 0 \\ 0 & 0 \end{bmatrix} & \begin{bmatrix} -1 & 0 \\ 0 & 0 \end{bmatrix} \\ \begin{bmatrix} -1 & 0 \\ 0 & 0 \end{bmatrix} & \begin{bmatrix} 1 & 0 \\ 0 & 0 \end{bmatrix} \end{bmatrix} \qquad \text{(a)}$$

$$[\bar{K}^3] = \begin{bmatrix} [\bar{K}^3]_{22} & [\bar{K}^3]_{23} \\ [\bar{K}^3]_{32} & [\bar{K}^3]_{33} \end{bmatrix} = \frac{EA}{125} \begin{bmatrix} \begin{bmatrix} 16 & -12 \\ -12 & 9 \end{bmatrix} & \begin{bmatrix} -16 & 12 \\ 12 & -9 \end{bmatrix} \\ \begin{bmatrix} -16 & 12 \\ 12 & -9 \end{bmatrix} & \begin{bmatrix} 16 & -12 \\ -12 & 9 \end{bmatrix} \end{bmatrix}$$

We establish the stiffness matrix for the truss directly from the global stiffness matrices for its elements. That is, referring to relation (8.3), we

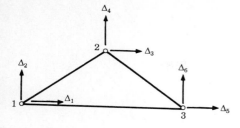

Figure b Numbering of the components of displacements of the nodes of the truss.

subdivide the stiffness matrix for the truss into the following submatrices:

$$[\hat{S}] = \begin{bmatrix} [S]_{11} & [S]_{12} & [S]_{13} \\ [S]_{21} & [S]_{22} & [S]_{23} \\ [S]_{31} & [S]_{32} & [S]_{33} \end{bmatrix} \tag{b}$$

Each submatrix of the stiffness matrix for the truss is equal to the sum of the corresponding submatrices of the stiffness matrices for the elements of the truss [see relation (8.31)]. That is,

$$[S]_{pq} = [\bar{K}^1]_{pq} + [\bar{K}^2]_{pq} + [\bar{K}^3]_{pq} \tag{c}$$

Referring to relations (a), from relation (c) we obtain

$$[S]_{11} = [\bar{K}^1]_{11} + [\bar{K}^2]_{11} = \frac{EA}{125}\begin{bmatrix} 16 & 12 \\ 12 & 9 \end{bmatrix} + \frac{EA}{125}\begin{bmatrix} \dfrac{125}{8} & 0 \\ 0 & 0 \end{bmatrix}$$

$$= \frac{EA}{125}\begin{bmatrix} 31.625 & 12 \\ 12 & 9 \end{bmatrix}$$

$$[S]_{12} = [\bar{K}^1]_{12} = \frac{EA}{125}\begin{bmatrix} -16 & -12 \\ -12 & -9 \end{bmatrix}$$

$$[S]_{13} = [\bar{K}^2]_{13} = \frac{EA}{125}\begin{bmatrix} -15.625 & 0 \\ 0 & 0 \end{bmatrix}$$

$$[S]_{21} = [\bar{K}^1]_{21} = \frac{EA}{125}\begin{bmatrix} -16 & -12 \\ -12 & -9 \end{bmatrix}$$

$$[S]_{22} = [\bar{K}^1]_{22} + [\hat{\bar{K}}^2]_{22} = \frac{EA}{125}\begin{bmatrix} 16 & 12 \\ 12 & 9 \end{bmatrix} + \frac{EA}{125}\begin{bmatrix} 16 & -12 \\ -12 & 9 \end{bmatrix}$$

$$= \frac{EA}{125}\begin{bmatrix} 32 & 0 \\ 0 & 18 \end{bmatrix}$$

$$[S]_{23} = [\bar{K}^3]_{23} = \frac{EA}{125}\begin{bmatrix} -16 & 12 \\ 12 & -9 \end{bmatrix}$$

$$[S]_{31} = [\bar{K}^2]_{31} = \frac{EA}{125} \begin{bmatrix} -15.625 & 0 \\ 0 & 0 \end{bmatrix}$$

$$[S]_{32} = [\bar{K}^3]_{32} = \frac{EA}{125} \begin{bmatrix} -16 & 12 \\ 12 & -9 \end{bmatrix}$$

$$[S]_{33} = [\bar{K}^2]_{33} + [\bar{K}^3]_{33} = \frac{EA}{125} \begin{bmatrix} 31.625 & -12 \\ -12 & 9 \end{bmatrix}$$

Substituting the above relations into (b) we obtain the stiffness matrix for the truss given by relation (j) of Example 1 of Sec. 8.3.1.

Example 2. Assemble the stiffness matrix for the structures shown in Figs. a and b. The elements of these structures are made from the same isotropic linearly elastic material and have the same constant cross section with $I/A = 0.025$ m^2. The elements of the structure of Fig. a are rigidly connected, while those of the structure of Fig. b are connected by a pin.

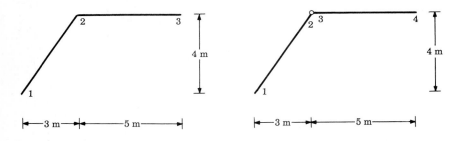

Figure a Geometry of the structure.　　　**Figure b** Geometry of the structure.

solution

STEPS 1 AND 2.   These steps are identical to those of Example 3 of Sec. 8.3.1.

Structure of Fig. a

STEP 3.   We assemble the stiffness matrix for the structure from the global stiffness matrices for its elements. Referring to relation (8.32), we subdivide the global stiffness matrix for each element of the structure into four submatrices. That is, using relations (d) and (e), of Example 3 of Sec. 8.3.1

and referring to Fig. a we have

$$[\bar{K}^1] = \begin{bmatrix} [\bar{K}^1]_{11} & [\bar{K}^1]_{12} \\ [\bar{K}^1]_{21} & [\bar{K}^1]_{22} \end{bmatrix}$$

$$= \frac{EI}{125} \begin{bmatrix} \begin{bmatrix} 367.68 & 474.24 & -24 \\ 474.24 & 644.32 & 18 \\ -24.00 & 18.00 & 100 \end{bmatrix} & \begin{bmatrix} -367.68 & -474.24 & -24 \\ -474.24 & -644.32 & 18 \\ 24.00 & -18.00 & 50 \end{bmatrix} \\ \begin{bmatrix} -367.68 & -474.24 & 24 \\ -474.24 & -644.32 & -18 \\ -24.00 & 18.00 & 50 \end{bmatrix} & \begin{bmatrix} 367.68 & 474.24 & 24 \\ 474.24 & 644.32 & -18 \\ 24.00 & -18.00 & 100 \end{bmatrix} \end{bmatrix}$$

(a)

$$[\bar{K}^2] = \begin{bmatrix} [\bar{K}^2]_{22} & [\bar{K}^2]_{23} \\ [\bar{K}^2]_{32} & [\bar{K}^2]_{33} \end{bmatrix}$$

$$= \frac{EI}{125} \begin{bmatrix} \begin{bmatrix} 1000 & 0 & 0 \\ 0 & 12 & 30 \\ 0 & 30 & 100 \end{bmatrix} & \begin{bmatrix} -1000 & 0 & 0 \\ 0 & -12 & 30 \\ 0 & -30 & 50 \end{bmatrix} \\ \begin{bmatrix} -1000 & 0 & 0 \\ 0 & -12 & -30 \\ 0 & 30 & 50 \end{bmatrix} & \begin{bmatrix} 1000 & 0 & 0 \\ 0 & 12 & -30 \\ 0 & -30 & 100 \end{bmatrix} \end{bmatrix}$$

The structure of Fig. a does not have an internal action release mechanism. Consequently, we establish its stiffness matrix directly from the global stiffness matrices for its elements. That is, referring to relation (8.3) we subdivide the stiffness matrix for the structure into the following submatrices:

$$[\hat{S}] = \begin{bmatrix} [S]_{11} & [S]_{12} & [S]_{13} \\ [S]_{21} & [S]_{22} & [S]_{23} \\ [S]_{31} & [S]_{32} & [S]_{33} \end{bmatrix}$$

(b)

Each submatrix of the stiffness matrix for the structure is equal to the sum of the corresponding submatrices of the stiffness matrices for the elements of the structure [see relation (8.31)]. That is,

$$[S]_{pq} = [\bar{K}^1]_{pq} + [\bar{K}^2]_{pq}$$

(c)

Referring to relations (a) from relation (c), we get

$$[S]_{11} = [\bar{K}^1]_{11}$$

$$[S]_{12} = [\bar{K}^1]_{12}$$

$$[S]_{13} = [0]$$

$$[S]_{21} = [\bar{K}^1]_{21}$$

$$[S]_{22} = [\bar{K}^1]_{22} + [\bar{K}^2]_{22} = \frac{EI}{125} \begin{bmatrix} 1367.68 & 474.24 & 24 \\ 474.24 & 656.32 & 12 \\ 24.00 & 12.00 & 200 \end{bmatrix}$$

(d)

$$[S]_{23} = [\bar{K}^2]_{23}$$
$$[S]_{31} = [0]$$
$$[S]_{32} = [\bar{K}^2]_{32}$$
$$[S]_{33} = [\bar{K}^2]_{33}$$

Substituting the above relations into (b) we obtain the stiffness matrix for the structure of Fig. a given by relation (i) of Example 3 of Sec. 8.3.1.

Structure of Fig. b

In this structure we have two connected nodes adjacent to the internal hinge. These nodes can rotate and translate in any direction. However, their components of translation are equal. Thus, these two nodes together have four independent components of displacements.

STEP 4.   We assemble the stiffness matrix for the structure from the global stiffness matrices for its elements. Referring to relation (8.32) we subdivide the global stiffness matrix for each element of the structure into four submatrices. That is, using relations (d) and (e) of Example 3 of Sec. 8.3.1 we have

$$[\bar{K}^1] = \begin{bmatrix} [\bar{K}^1]_{11} & [\bar{K}^1]_{12} \\ [\bar{K}^1]_{21} & [\bar{K}^1]_{22} \end{bmatrix}$$

$$= \frac{EI}{125} \begin{bmatrix} \begin{bmatrix} 367.68 & 474.24 & -24 \\ 474.24 & 644.32 & 18 \\ -24.00 & 18.00 & 100 \end{bmatrix} & \begin{bmatrix} -367.68 & -474.24 & -24 \\ -474.24 & -644.82 & 18 \\ 24.00 & -18.00 & 50 \end{bmatrix} \\ \begin{bmatrix} -367.68 & -474.24 & 24 \\ -474.24 & -644.32 & -18 \\ -24.00 & 18.00 & 50 \end{bmatrix} & \begin{bmatrix} 367.68 & 474.24 & 24 \\ 474.24 & 44.32 & -18 \\ 24.00 & -18.00 & 100 \end{bmatrix} \end{bmatrix}$$

(e)

$$[\bar{K}^2] = \begin{bmatrix} [\bar{K}^2]_{33} & [\bar{K}^2]_{34} \\ [\bar{K}^2]_{43} & [\bar{K}^2]_{44} \end{bmatrix}$$

$$= \frac{EI}{125} \begin{bmatrix} \begin{bmatrix} 1000 & 0 & 0 \\ 0 & 12 & 30 \\ 0 & 30 & 100 \end{bmatrix} & \begin{bmatrix} -1000 & 0 & 0 \\ 0 & -12 & 30 \\ 0 & -30 & 50 \end{bmatrix} \\ \begin{bmatrix} -1000 & 0 & 0 \\ 0 & -12 & -30 \\ 0 & 30 & 50 \end{bmatrix} & \begin{bmatrix} 1000 & 0 & 0 \\ 0 & 12 & -30 \\ 0 & -30 & 100 \end{bmatrix} \end{bmatrix}$$

STEP 5.   The structure of Fig. b has an internal action release mechanism. For this reason we form a model obtained from the structure by disconnecting nodes 2 and 3 (see Fig. c). We establish the stiffness matrix for this model from the global stiffness matrices for its elements. That is, referring to relation (8.3) we subdivide the stiffness matrix for the model of the structure into the following submatrices:

$$[\hat{S}] = \begin{bmatrix} [S]_{11} & [S]_{12} & [S]_{13} & [S]_{14} \\ [S]_{21} & [S]_{22} & [S]_{23} & [S]_{24} \\ [S]_{31} & [S]_{32} & [S]_{33} & [S]_{34} \\ [S]_{41} & [S]_{42} & [S]_{43} & [S]_{44} \end{bmatrix}$$

(f)

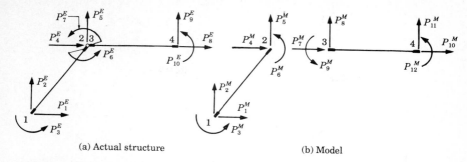

(a) Actual structure   (b) Model

Figure c Numbering of the equivalent actions acting on the nodes of the structure.

Each submatrix of the stiffness matrix for the model of the structure is equal to the sum of the corresponding submatrices of the stiffness matrices for the elements of the structure [see relation (8.31)]. That is,

$$[S]_{pq} = [\bar{K}^1]_{pq} + [\bar{K}^2]_{pq} \qquad \text{(g)}$$

Referring to relations (e) from relation (g), we obtain

$$[S]_{11} = [\bar{K}^1]_{11} \qquad [S]_{12} = [\bar{K}^1]_{12}$$

$$[S]_{13} = [0] \qquad [S]_{14} = [0]$$

$$[S]_{21} = [\bar{K}^1]_{21} \qquad [S]_{22} = [\bar{K}^1]_{22}$$

$$[S]_{23} = [0] \qquad [S]_{24} = [0]$$

$$[S]_{31} = [0] \qquad [S]_{32} = [\bar{K}^2]_{32} \qquad \text{(h)}$$

$$[S]_{33} = [\bar{K}^2]_{33} \qquad [S]_{34} = [\bar{K}^2]_{34}$$

$$[S]_{41} = [0] \qquad [S]_{42} = [0]$$

$$[S]_{43} = [\bar{K}^2]_{43} \qquad [S]_{44} = [\bar{K}^2]_{44}$$

Substituting the above relations into (f) we obtain the stiffness matrix for the model of the structure given by relation (i) (see p. 268).

We modify the stiffness matrix (i) in order to account for the presence of the connected nodes 2 and 3. That is, we translate column 7 into 4 and column 8 into 5 in order to take into account that $\Delta_1^{(2)} = \Delta_1^{(3)} = \Delta_4$ and $\Delta_2^{(2)} = \Delta_2^{(3)} = \Delta_5$ (see Fig. d of Example 3 of Sec. 8.3.1). Moreover, we translate row 7 into 4 and row 8 into 5 in order to take into account (see Fig. c) that

$$[\hat{S}] = \frac{EI}{125}$$

	1	2	3	4	5	6	7	8	9	10	11	12	
	367.68	474.24	-24	-367.68	-474.24	-24	0	0	0	0	0	0	1
	474.24	644.32	18	-474.24	644.32	18	0	0	0	0	0	0	2
	-24.00	18.00	100	24.00	-18.00	50	0	0	0	0	0	0	3
	-367.68	-474.24	24	367.68	474.24	24	0	0	0	0	0	0	4
	-474.24	-644.32	-18	474.24	644.32	-18	0	0	0	0	0	0	5
	-24.00	18.00	50	24.00	-18.00	100	0	0	0	0	0	0	6
	0	0	0	0	0	0	1000	0	0	-1000	0	0	7
	0	0	0	0	0	0	0	12	30	0	-12	30	8
	0	0	0	0	0	0	0	30	100	0	-30	50	9
	0	0	0	0	0	0	-1000	0	0	1000	0	0	10
	0	0	0	0	0	0	0	-12	-30	0	12	-30	11
	0	0	0	0	0	0	0	30	50	0	-30	100	12

(i)

$$P_4^E = P_4^M + P_7^M$$
$$P_5^E = P_5^M + P_8^M$$

This modification of the stiffness matrix (h) gives the stiffness matrix (k) of Example 3 of Sec. 8.3.1 for the structure of Fig. b.

8.4 Nature of the Stiffness Matrix for a Structure

The stiffness matrix for the truss given by relation (j) of Example 1 of Sec. 8.3.1 is densely populated. This is a result of the close interconnection of the nodes of the truss. The stiffness matrix for a structure with many nodes is sparsely populated and, moreover, it has a banded structure whose width depends on the numbering of the global components of displacements of the nodes of the structure. For example, a schematic representation of the stiffness matrix for the planar frame of Fig. 8.3a is given in Fig. 8.4. In this figure each square indicates a slot of the stiffness matrix for the structure. Moreover, a cross in a square indicates a nonvanishing stiffness

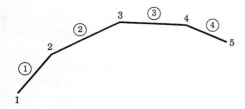

(a) Numbering of the nodes and elements

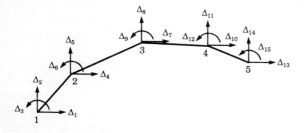

(b) Numbering of the global components of displacements

Figure 8.3 Planar frame.

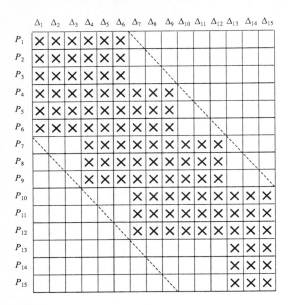

Figure 8.4 Schematic representation of the stiffness matrix for the frame of Fig. 8.3.

coefficient. It is apparent that the stiffness matrix for a "one-branch" structure as the one shown in Fig. 8.3*a* is a banded matrix as long as its nodes are numbered consecutively.

The semibandwidth b of a matrix is equal to the maximum number of nonvanishing coefficients on the one side of the diagonal in any row of the matrix plus 1 (the diagonal). The semibandwidth of the stiffness matrix for a structure can be expressed as

$$b = (d + 1)N \tag{8.33}$$

d is the maximum difference in the numbers assigned to the nodes at the ends of one element of the structure and N is the number of global components of displacement of each node of the structure. For a planar truss, N is equal to 2, whereas for a planar beam, a planar frame, or a space truss, N is equal to 3.

In general, the ratio of the semibandwidth to the size of the stiffness matrix for a structure depends on the number of degrees of freedom of the nodes of the structure and on the numbering of its nodes. This ratio is small for structures with many degrees of freedom, provided that proper attention is given in numbering of their nodes.

Methods are available, referred to as "band solvers" which facilitate

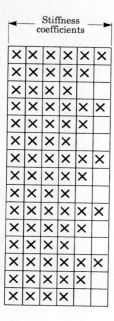

Figure 8.5 Storing of the coefficients of the stiffness matrix of the structure of Fig. 8.3a as a rectangular array.

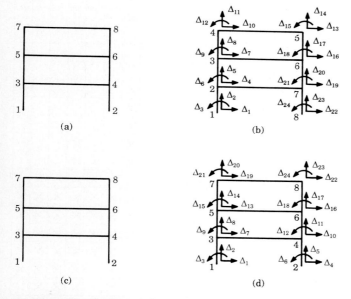

Figure 8.6 Numbering of the nodes and of the components of displacement of a frame.

the solution of systems of algebraic equations whose coefficient matrix is a banded matrix.[1] Moreover, in view of the symmetry of the stiffness matrix, only the stiffness coefficients located within its semibandwidth must be stored in the computer, thus rendering data handling more efficient. For instance, the stiffness coefficients of the matrix of Fig. 8.4 can be stored in the rectangular array shown in Fig. 8.5.

In the sequel, we illustrate by an example the dependence of the semibandwidth of the stiffness matrix for a structure on the numbering of its nodes.

Consider the planar frame of Fig. 8.6. Its stiffness matrix is a

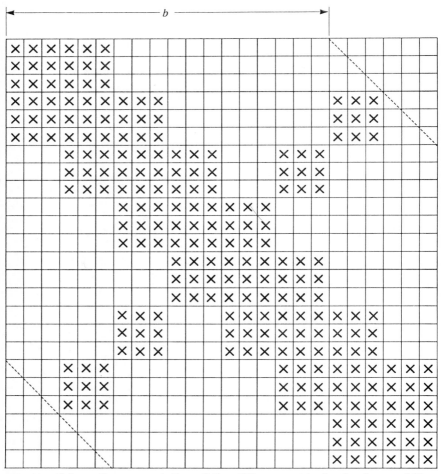

Figure 8.7 Schematic representation of the stiffness matrix for the frame of Fig. 8.6a.

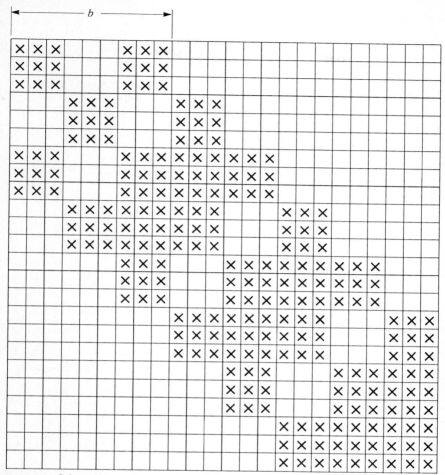

Figure 8.8 Schematic representation of the stiffness matrix for the frame of Fig. 8.6b.

24×24 matrix. For the numbering of the nodes of the structure shown in Fig. 8.6a, the maximum difference in the numbers of the nodes at the ends of one element is 5, while for the numbering of Fig. 8.6c it is only 2. Thus, for the numbering of Fig. 8.6a the semiband-width is 18 (see Fig. 8.7), while it is only 9 for the numbering of Fig. 8.6b (see Fig. 8.8).

8.5 Problems

1 to 6. Assemble the stiffness matrix of the structure shown in Fig. 8.P1. The elements of the structure are made of steel ($E = 210\,\text{kN/mm}^2$). Repeat with the structure of Figs. 8.P2 to 8.P6.

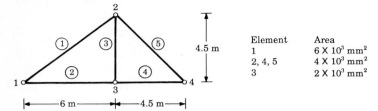

Element	Area
1	6×10^3 mm²
2, 4, 5	4×10^3 mm²
3	2×10^3 mm²

Figure 8.P1

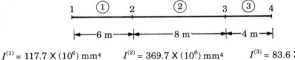

$I^{(1)} = 117.7 \times (10^6)$ mm⁴ $I^{(2)} = 369.7 \times (10^6)$ mm⁴ $I^{(3)} = 83.6 \times (10^6)$ mm⁴

$A_1 = 6.26 \times (10^3)$ mm² $A_2 = 13.2 \times (10^3)$ mm² $A_3 = 5.38 \times (10^3)$ mm²

Figure 8.P2

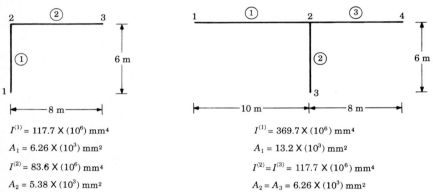

$I^{(1)} = 117.7 \times (10^6)$ mm⁴

$A_1 = 6.26 \times (10^3)$ mm²

$I^{(2)} = 83.6 \times (10^6)$ mm⁴

$A_2 = 5.38 \times (10^3)$ mm²

Figure 8.P3

$I^{(1)} = 369.7 \times (10^6)$ mm⁴

$A_1 = 13.2 \times (10^3)$ mm²

$I^{(2)} = I^{(3)} = 117.7 \times (10^6)$ mm⁴

$A_2 = A_3 = 6.26 \times (10^3)$ mm²

Figure 8.P4

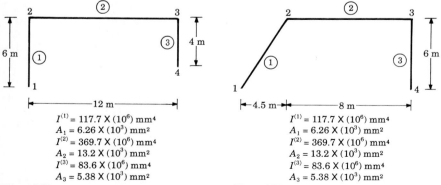

$I^{(1)} = 117.7 \times (10^6)$ mm⁴
$A_1 = 6.26 \times (10^3)$ mm²
$I^{(2)} = 369.7 \times (10^6)$ mm⁴
$A_2 = 13.2 \times (10^3)$ mm²
$I^{(3)} = 83.6 \times (10^6)$ mm⁴
$A_3 = 5.38 \times (10^3)$ mm²

Figure 8.P5

$I^{(1)} = 117.7 \times (10^6)$ mm⁴
$A_1 = 6.26 \times (10^3)$ mm²
$I^{(2)} = 369.7 \times (10^6)$ mm⁴
$A_2 = 13.2 \times (10^3)$ mm²
$I^{(3)} = 83.6 \times (10^6)$ mm⁴
$A_3 = 5.38 \times (10^3)$ mm²

Figure 8.P6

7 and 8.   Assemble the stiffness matrix of the space structure shown in Fig. 8.P7. The elements of the structure are steel tubes ($E = 210 \text{ kN/mm}^2$, $v = 0.33$) of the same constant cross section ($d_{\text{ext}} = 80 \text{ mm}$, $d_{\text{int}} = 60 \text{ mm}$). Repeat with the structure of Fig. 8.P8.

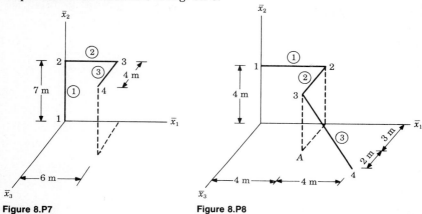

Figure 8.P7   **Figure 8.P8**

9.   Assemble the stiffness matrix for the frame shown in Fig. 8.P9. The elements of the frame are made of steel. The area of the cross section of the element which is pinned at both ends is $A_4 = 800 \text{ mm}^2$. The other elements of the frame have a constant cross section ($A = 13.2 \times 10^3 \text{ mm}^2$, $I = 369.7 \times 10^6 \text{ mm}^4$).

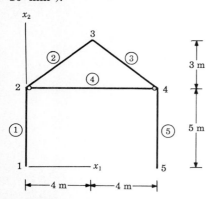

Figure 8.P9

Reference

1. C. Meyer, "Special Problems Related to Linear Equation Solvers," *Proc. ASCE J. Struct. Div.*, no. ST4, April 1975, pp. 869–890.

Chapter

9

Analysis of Framed Structures Without Skew Supports Using the Direct Stiffness Method

9.1 Solution of the Stiffness Equations for a Structure

The support conditions of a structure are referred to as its boundary conditions. They consist of specified values of certain components of displacement (translations or rotations) at some nodes (supports) of the structure. If we include the specified movements of the supports of a structure in the loading of the restrained structure (see Sec. 7.1), the supports of the structure subjected to the equivalent actions do not move. Notice, however, that we could consider the specified movements of the supports of the structure as boundary conditions of the structure subjected to the equivalent actions instead of including them in the loading of the restrained structure. This approach is employed in the slope deflection method[1] of classical structural analysis. Moreover, this approach is illustrated in Example 3 at the end of Sec. 9.3.

The roller supports of structures and the ball supports of space structures whose plane of rolling is not normal to one of their global axes are called *skew supports*. In this chapter we limit our attention to structures without skew supports. We consider structures with skew supports in Chap. 10.

The unknown quantities in the stiffness equations (8.1) for a structure are the components of its reactions and the components of displacement of its nodes. Notice that for every specified component of displacement there is a corresponding unknown component of reaction. Thus the number of unknown quantities in the stiffness equations for a structure does not change when its support conditions change. Moreover, the number of unknown quantities in the stiffness equations for a structure is equal to the number of stiffness equations. However, the stiffness matrix $[\hat{S}]$ for a structure is singular, and, consequently, the stiffness equations cannot be solved directly to yield the components of displacements of the nodes of the structure and its reactions. In order to be able to solve the stiffness equations for a structure its boundary conditions must be incorporated into them. This is accomplished by rearranging the rows and columns of the stiffness equations (8.1) as follows:

1. We move to the bottom of Eqs. (8.1) their rows corresponding to the reactions of the supports of the structure.
2. We move to the right-hand side of Eqs. (8.1) their columns corresponding to vanishing or, in general, to specified components of displacements of the supports of the structure.

An algorithm can be written for moving the rows and columns of the stiffness equations (8.1). The resulting stiffness equations for the structure can be partitioned as follows:

$$\left\{ \begin{matrix} \{P^{EF}\} \\ \{P^{ES}\} \end{matrix} \right\} + \left\{ \begin{matrix} \{0\} \\ \{R\} \end{matrix} \right\} = \left[\begin{matrix} [S^{FF}] & [S^{FS}] \\ [S^{SF}] & [S^{SS}] \end{matrix} \right] \left\{ \begin{matrix} \{\Delta^F\} \\ \{\Delta^S\} \end{matrix} \right\} \qquad (9.1)$$

The terms of matrix $\{P^{EF}\}$ are the known components of the equivalent actions which are not directly absorbed by the supports of the structure. The terms of matrix $\{P^{ES}\}$ are the known components of equivalent actions which are directly absorbed by the supports of the structure. The terms of matrix $\{R\}$ are the reactions of the structure. The terms of matrix $\{\Delta^F\}$ are the unknown components of displacements of the nodes of the structure subjected to the given loading; as discussed in Sec. 7.1 they are also equal to the components of displacements of the nodes of the structure subjected to the equivalent actions on its nodes. The terms of matrix $\{\Delta^S\}$ are the known (given) components of displacements of the supported nodes of the structure.

Relations (9.1) can be expanded to yield

$$\{P^{EF}\} = [S^{FF}]\{\Delta^F\} + [S^{FS}]\{\Delta^S\} \qquad (9.2)$$

$$\{P^{ES}\} + \{R\} = [S^{SF}]\{\Delta^F\} + [S^{SS}]\{\Delta^S\} \qquad (9.3)$$

If there are specified movements of the supports of a structure and they are not included as loading on the restrained structure, the matrix $[\Delta^S]$ is not a zero matrix. However, if the given movements of

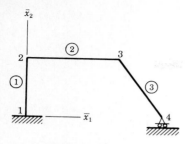

Figure 9.1 Planar frame.

the supports of a structure are included as loading on the restrained structure, the matrix $[\Delta^S]$ is a zero matrix. In this case, relations (9.2) and (9.3) reduce to

$$\{P^{EF}\} = [S^{FF}]\{\Delta^F\} \tag{9.4}$$

$$\{P^{ES}\} + \{R\} = [S^{SF}]\{\Delta^F\} \tag{9.5}$$

The matrix $[S^{FF}]$ is called the *basic stiffness matrix* for the structure. Its terms depend on the stiffness matrices for the elements of the structure, on how the elements are connected to form the structure, and on how the structure is supported. The matrix $[S^{FF}]$ is square and symmetric, and if the structure is not a mechanism, it is nonsingular. Thus relation (9.2) or (9.4) can be solved to yield the unknown components of displacements $\{\Delta^F\}$ of the nodes of the structure. These can be substituted in relation (9.3) or (9.5), respectively, to yield the reactions $\{R\}$ of the structure. If the basic stiffness matrix $[S^{FF}]$ for a structure is singular, *the structure is a mechanism* (see Sec. 1.11).

In what follows we write the stiffness equations for the structure of Fig. 9.1 and rearrange their rows and columns as discussed above. Thus, referring to Fig. 9.1 we obtain Eq. (9.6) on p. 280. Referring to Fig. 9.2b the boundary conditions for the frame of Fig. 9.1 are

$$\Delta_1 = \Delta_2 = \Delta_3 = \Delta_{11} = 0 \tag{9.7}$$

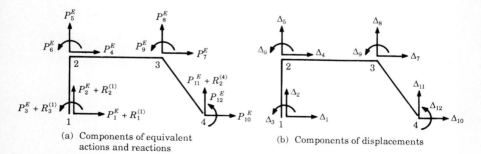

(a) Components of equivalent actions and reactions

(b) Components of displacements

Figure 9.2 Components of equivalent actions and reactions and components of displacements of the nodes of the frame of Fig. 9.1.

$$
\begin{Bmatrix}
R_1^{(1)} \\
R_2^{(1)} \\
R_3^{(1)} \\
0 \\
0 \\
0 \\
0 \\
0 \\
0 \\
0 \\
R_2^{(4)} \\
0
\end{Bmatrix}
+
\begin{Bmatrix}
P_1^E \\
P_2^E \\
P_3^E \\
P_4^E \\
P_5^E \\
P_6^E \\
P_7^E \\
P_8^E \\
P_9^E \\
P_{10}^E \\
P_{11}^E \\
P_{12}^E
\end{Bmatrix}
=
\begin{bmatrix}
S_{11} & S_{12} & S_{13} & S_{14} & S_{15} & S_{16} & S_{17} & S_{18} & S_{19} & S_{1,10} & S_{1,11} & S_{1,12} \\
S_{21} & S_{22} & S_{23} & S_{24} & S_{25} & S_{26} & S_{27} & S_{28} & S_{29} & S_{2,10} & S_{2,11} & S_{2,12} \\
S_{31} & S_{32} & S_{33} & S_{34} & S_{35} & S_{36} & S_{37} & S_{38} & S_{39} & S_{3,10} & S_{3,11} & S_{3,12} \\
S_{41} & S_{42} & S_{43} & S_{44} & S_{45} & S_{46} & S_{47} & S_{48} & S_{49} & S_{4,10} & S_{4,11} & S_{4,12} \\
S_{51} & S_{52} & S_{53} & S_{54} & S_{55} & S_{56} & S_{57} & S_{58} & S_{59} & S_{5,10} & S_{5,11} & S_{5,12} \\
S_{61} & S_{62} & S_{63} & S_{64} & S_{65} & S_{66} & S_{67} & S_{68} & S_{69} & S_{6,10} & S_{6,11} & S_{6,12} \\
S_{71} & S_{72} & S_{73} & S_{74} & S_{75} & S_{76} & S_{77} & S_{78} & S_{79} & S_{7,10} & S_{7,11} & S_{7,12} \\
S_{81} & S_{82} & S_{83} & S_{84} & S_{85} & S_{86} & S_{87} & S_{88} & S_{89} & S_{8,10} & S_{8,11} & S_{8,12} \\
S_{91} & S_{92} & S_{93} & S_{94} & S_{95} & S_{96} & S_{97} & S_{98} & S_{99} & S_{9,10} & S_{9,11} & S_{9,12} \\
S_{10,1} & S_{10,2} & S_{10,3} & S_{10,4} & S_{10,5} & S_{10,6} & S_{10,7} & S_{10,8} & S_{10,9} & S_{10,10} & S_{10,11} & S_{10,12} \\
S_{11,1} & S_{11,2} & S_{11,3} & S_{11,4} & S_{11,5} & S_{11,6} & S_{11,7} & S_{11,8} & S_{11,9} & S_{11,10} & S_{11,11} & S_{11,12} \\
S_{12,1} & S_{12,2} & S_{12,3} & S_{12,4} & S_{12,5} & S_{12,6} & S_{12,7} & S_{12,8} & S_{12,9} & S_{12,10} & S_{12,11} & S_{12,12}
\end{bmatrix}
\begin{Bmatrix}
\Delta_1 \\
\Delta_2 \\
\Delta_3 \\
\Delta_4 \\
\Delta_5 \\
\Delta_6 \\
\Delta_7 \\
\Delta_8 \\
\Delta_9 \\
\Delta_{10} \\
\Delta_{11} \\
\Delta_{12}
\end{Bmatrix}
\tag{9.6}
$$

In order to introduce the boundary conditions (9.7) of the frame of Fig. 9.1 into its stiffness equations (9.6), we rearrange their rows and columns and partition them as indicated by Eqs. (9.8) on p. 282.

9.2 Comments on the Solution of the Stiffness Equations for a Structure

The solution of the stiffness equation (9.2) or (9.4) may be obtained using one of the known matrix inversion techniques. Direct inversion of the stiffness matrix is not performed in practice because it requires considerable computation time. Generally, the stiffness equations (9.2) or (9.4) are solved using either the Gauss elimination method or the Cholesky method which is a variation of the Gauss elimination method suitable for symmetric matrices.[2]

The accuracy of the computed components of displacements of the nodes of the structure depends on the condition of the stiffness equations. If small errors in the stiffness coefficients or in the solution process do not affect the results significantly, we say that the stiffness equations are well conditioned. On the contrary, if small errors in the stiffness coefficients or in the solution process affect the results significantly, we say that the stiffness equations are ill conditioned. The stiffness equations may be ill conditioned if, among other things, the stiffness coefficients of adjacent elements vary widely, or if the structure has many degrees of freedom (a few thousand).

As discussed in Sec. 8.5, the stiffness matrix $[\hat{S}]$ of a structure with many nodes is symmetric, is sparsely populated, and has a banded structure whose width depends on the numbering of the global components of the displacements of the nodes of the structure. Consequently, the basic stiffness matrix $[S^{FF}]$ of a structure having many nodes which are free to translate and/or rotate is also sparsely populated and has a banded structure. For example, a schematic representation of the basic stiffness matrix $[S^{FF}]$ of the structure of Fig. 9.3 is shown in Fig. 9.4.

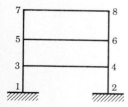

Figure 9.3 Planar frame.

$$
\begin{Bmatrix}
0 \\ 0 \\ 0 \\ 0 \\ 0 \\ 0 \\ 0 \\ 0 \\ \hline
R_1^{(1)} \\ R_2^{(1)} \\ R_3^{(1)} \\ R_2^{(4)}
\end{Bmatrix}
=
\left[
\begin{array}{cccccccc:cccc}
S_{44} & S_{45} & S_{46} & S_{47} & S_{48} & S_{49} & S_{4,10} & S_{4,12} & S_{41} & S_{42} & S_{43} & S_{4,11} \\
S_{54} & S_{55} & S_{56} & S_{57} & S_{58} & S_{59} & S_{5,10} & S_{5,12} & S_{51} & S_{52} & S_{53} & S_{5,11} \\
S_{64} & S_{65} & S_{66} & S_{67} & S_{68} & S_{69} & S_{6,10} & S_{6,12} & S_{61} & S_{62} & S_{63} & S_{6,11} \\
S_{74} & S_{75} & S_{76} & S_{77} & S_{78} & S_{79} & S_{7,10} & S_{7,12} & S_{71} & S_{72} & S_{73} & S_{7,11} \\
S_{84} & S_{85} & S_{86} & S_{87} & S_{88} & S_{89} & S_{8,10} & S_{8,12} & S_{81} & S_{82} & S_{83} & S_{8,11} \\
S_{94} & S_{95} & S_{96} & S_{97} & S_{98} & S_{99} & S_{9,10} & S_{9,12} & S_{91} & S_{92} & S_{93} & S_{9,11} \\
S_{10,4} & S_{10,5} & S_{10,6} & S_{10,7} & S_{10,8} & S_{10,9} & S_{10,10} & S_{10,12} & S_{10,1} & S_{10,2} & S_{10,3} & S_{10,11} \\
S_{12,4} & S_{12,5} & S_{12,6} & S_{12,7} & S_{12,8} & S_{12,9} & S_{12,10} & S_{12,12} & S_{12,1} & S_{12,2} & S_{12,3} & S_{12,11} \\
\hdashline
S_{14} & S_{15} & S_{16} & S_{17} & S_{18} & S_{19} & S_{1,10} & S_{1,12} & S_{11} & S_{12} & S_{13} & S_{1,11} \\
S_{24} & S_{25} & S_{26} & S_{27} & S_{28} & S_{29} & S_{2,10} & S_{2,12} & S_{21} & S_{22} & S_{23} & S_{2,11} \\
S_{34} & S_{35} & S_{36} & S_{37} & S_{38} & S_{39} & S_{3,10} & S_{3,12} & S_{31} & S_{32} & S_{33} & S_{3,11} \\
S_{11,4} & S_{11,5} & S_{11,6} & S_{11,7} & S_{11,8} & S_{11,9} & S_{11,10} & S_{11,12} & S_{11,1} & S_{11,2} & S_{11,3} & S_{11,11}
\end{array}
\right]
\begin{Bmatrix}
\Delta_4 \\ \Delta_5 \\ \Delta_6 \\ \Delta_7 \\ \Delta_8 \\ \Delta_9 \\ \Delta_{10} \\ \Delta_{12} \\ \hline
\Delta_1 = 0 \\ \Delta_2 = 0 \\ \Delta_3 = 0 \\ \Delta_{11} = 0
\end{Bmatrix}
+
\begin{Bmatrix}
P_4^E \\ P_5^E \\ P_6^E \\ P_7^E \\ P_8^E \\ P_9^E \\ P_{10}^E \\ P_{12}^E \\ \hline
P_1^E \\ P_2^E \\ P_3^E \\ P_{11}^E
\end{Bmatrix}
\tag{9.8}
$$

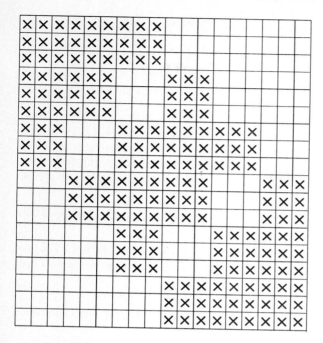

Figure 9.4 Stiffness matrix of the planar frame of Fig. 9.3.

A system of algebraic equations whose coefficient matrix is a banded matrix has the advantage that when it is solved with the aid of a computer, using elimination methods it can be subdivided into smaller groups of equations and its solution may be obtained by working with only two of these groups at a time. Thus, information from previously processed groups of equations that is no longer required may be overwritten, whereas information which is subsequently utilized may be stored in peripheral storage and called into the high-speed core only when it is required.

9.3 Analysis of Framed Structures

When we analyze a framed structure using the direct stiffness method, we adhere to the following steps.

Step 1. We compute the matrix of fixed-end actions $\{A^R\}$ of each element of the structure. The terms of this matrix represent the local components of nodal actions of the element subjected to the given loads with its ends fixed. Moreover, we transform the matrix $\{A^R\}$

from local to global. That is, referring to relation (6.65) we have

$$\{\bar{A}^R\} = [\Lambda]^T \{A^R\} \tag{9.9}$$

Step 2. We establish the matrix of equivalent action $\{\hat{P}^E\}$ to be applied on the nodes of the structure. To accomplish this we do the following.

1. We compute the matrix of the restraining actions $\{\widehat{RA}\}$ which must be applied to the nodes of the restrained structure in order to prevent them from rotating or translating. This is accomplished by considering the equilibrium of the nodes of the restrained structure (see Sec. 7.2).

2. We form the matrix of the given actions acting on the nodes of the structure $\{\hat{P}^G\}$.

3. We substitute the matrices $\{\hat{P}^G\}$ and $\{\widehat{RA}\}$ into relation (7.2) to obtain the matrix of the equivalent actions for the structure. That is,
$$\{\hat{P}^E\} = \{\hat{P}^G\} - \{\widehat{RA}\} \tag{9.10}$$

Step 3. We establish the local stiffness matrix $[K]$ for each element of the structure (see Sec. 4.4).

Step 4. We compute the hybrid $[\overset{\prime}{K}]$ and the global $[\bar{K}]$ stiffness matrices for each element of the structure using relations (6.72) and (6.74), respectively. That is,

$$[\overset{\prime}{K}] = [K][\Lambda] \tag{9.11}$$

$$[\bar{K}] = [\Lambda]^T [\overset{\prime}{K}] \tag{9.12}$$

Moreover, we write the stiffness equations for each element of the structure.

Step 5. We assemble the stiffness matrix for the structure from the global stiffness matrices of its elements (see Sec. 8.3). Moreover, we form the stiffness equations for the structure.

Step 6. We compute the components of displacements of the nodes of the structure and its reactions. To accomplish this we do the following.

1. We rearrange the rows and the columns of the stiffness equations for the structure in order to incorporate its boundary conditions in them (see Sec. 9.1).

2. We partition the modified stiffness matrix of the structure to obtain the matrices $[S^{FF}]$, $[S^{FS}]$, $[S^{SF}]$, and $[S^{SS}]$ [see relation (9.1)].

3. We use relations (9.2) and (9.3) to compute the components of displacement of the nodes and the reactions of the supports, respectively, of the structure subjected to the equivalent actions on its nodes and possibly to the given components of translation and rotation of the supports of the structure. They are equal to those of the structure subjected to the given loads.

Step 7. From the components of displacement of the nodes of the structure computed in step 6, we compute the global components of nodal displacements of each element of the structure. From these, we compute the local components of nodal actions of each element of the structure. That is, substituting relation (6.71) into (4.7) we obtain

$$\{A\} = [\hat{K}]\{\bar{D}\} + \{A^R\} \tag{9.13}$$

where $[\hat{K}]$ is the hybrid stiffness matrix of the element defined by relation (9.11). It transforms the global components of nodal displacements of the element to the local components of its nodal actions.

The direct stiffness method is a general method which can be employed in establishing the components of displacements of the nodes and the nodal actions of the elements of a structure subjected to given loads. The structure may be statically determinate or indeterminate. The direct stiffness method is the most satisfactory given for writing programs for analyzing groups of framed structures with the aid of an electronic computer. However, the direct stiffness method is not the most suitable for analyzing by hand calculation or by a desk calculator statically determinate or indeterminate structures having a large number of unknown components of displacements of their nodes.

In the sequel, we illustrate the direct stiffness method by the following three examples.

Example 1. The components of displacements of the nodes, the reactions, and the internal forces in the elements of a simply supported, statically determinate truss subjected to external forces are computed. The internal forces in the elements of statically determinate trusses can be easily established by considering the equilibrium of their nodes. Moreover, the components of translation of the nodes of trusses can be computed using the dummy load method.[3] This example illustrates that the direct stiffness method is not convenient for analyzing statically determinate structures by hand calculations. This is especially true if the number of the unknown displacements of the nodes of a statically determinate structure is large.

Example 2. The components of displacements of the nodes, the reactions, and the nodal actions of the elements of two beams subjected to external actions on their nodes are computed. The beams differ only in their support conditions. The one beam is statically determinate, while the other is statically indeterminate to the second degree. This example illustrates that the direct stiffness method is not convenient for analyzing statically determinate structures by hand calculations. Moreover, it illustrates that the direct stiffness method can be employed conveniently for analyzing statically indeterminate structures by hand calculations, particularly if the number of the unknown displacements of their nodes (degree of kinematic indeterminacy) does not exceed their degree of static indeterminacy. However, it should be emphasized that the direct stiffness method is the most suitable one available for writing programs for analyzing a group of framed structures with the aid of an electronic computer.

Example 3. The components of displacements of the nodes, the reactions, and the nodal actions of the elements of a statically indeterminate frame are computed. The frame is subjected to external forces acting along the length of its elements, to a change of temperature of its elements, and to settlement of a support. The settlement of the support of the frame is taken into account in the following two ways.

1. It is included in the loading of the restrained structure. In this case $[[\Delta^S] \equiv [0]]$, when analyzing the structure subjected to the equivalent actions on its nodes, relations (9.4) and (9.5) are used.

2. It is not included in the loading of the restrained structure, but it is included in the boundary conditions of the structure subjected to the equivalent actions on its nodes. That is, in this case $[[\Delta^S] \neq [0]]$, relations (9.2) and (9.3) are used. From this example, it appears to be preferable to consider the movement of the supports of a structure when analyzing the structure subjected to the equivalent actions on its nodes.

Example 1. Using the direct stiffness method, compute the components of displacements of the nodes, the reactions, and the internal forces in the elements of the truss subjected to the forces shown in Fig. a. The elements of the truss are made of the same isotropic linearly elastic material and have the same constant cross section $(AE = 20{,}000 \text{ kN})$.

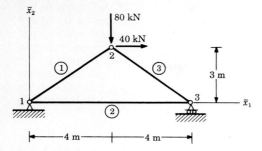

Figure a Geometry and loading of the truss.

---

solution

STEPS 1 TO 5.   The equivalent actions for the truss of Fig. a are the external forces shown in this figure. The stiffness matrix for the truss has been computed in Example 1 of Sec. 8.3.1. Referring to relations (j) of this example and to Figs. a and b the stiffness equations for the truss are

$$
\begin{Bmatrix} 0 \\ 0 \\ 40 \\ -80 \\ 0 \\ 0 \end{Bmatrix} + \begin{Bmatrix} R_1^{(1)} \\ R_2^{(1)} \\ 0 \\ 0 \\ 0 \\ R_2^{(3)} \end{Bmatrix} = \frac{EA}{125} \begin{bmatrix} 31.625 & 12 & -16 & -12 & -15.625 & 0 \\ 12 & 9 & -12 & -9 & 0 & 0 \\ -16 & -12 & 32 & 0 & -16 & 12 \\ -12 & -9 & 0 & 18 & 12 & -9 \\ -15.625 & 0 & -16 & 12 & 31.625 & -12 \\ 0 & 0 & 12 & -9 & -12 & 9 \end{bmatrix} \begin{Bmatrix} \Delta_1 \\ \Delta_2 \\ \Delta_3 \\ \Delta_4 \\ \Delta_5 \\ \Delta_6 \end{Bmatrix}
$$

(a)

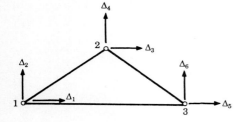

Figure b Numbering of the components of displacements of the nodes of the truss.

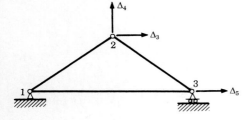

Figure c Degrees of freedom of the nodes of the truss.

STEP 6. We compute the components of displacements of the nodes of the truss and its reactions. In order to accomplish this we take into account the boundary conditions of the truss. That is, we rearrange the rows of the stiffness equations (a) in order to move to the bottom those corresponding to the reactions of the supports of the truss. Moreover, we rearrange the columns of the stiffness matrix $[\hat{S}]$ in order to move to the right the columns which are multiplied by the vanishing components of displacements of the supports of the truss (see Fig. c). Furthermore, we partition the resulting stiffness equations as indicated in relation (9.1). Thus,

$$\begin{Bmatrix} 40 \\ -80 \\ 0 \\ \hdashline 0 \\ 0 \\ 0 \end{Bmatrix} + \begin{Bmatrix} 0 \\ 0 \\ 0 \\ \hdashline R_1^{(1)} \\ R_2^{(1)} \\ R_2^{(3)} \end{Bmatrix} = \frac{EA}{125} \left[\begin{array}{ccc:ccc} 32 & 0 & -16 & -16 & -12 & 12 \\ 0 & 18 & 12 & -12 & -9 & -9 \\ -16 & 12 & 31.625 & -15.625 & 0 & -12 \\ \hdashline -16 & -12 & -15.625 & 31.625 & 12 & 0 \\ -12 & -9 & 0 & 12 & 9 & 0 \\ 12 & -9 & -12 & 0 & 0 & 9 \end{array} \right] \begin{Bmatrix} \Delta_3 \\ \Delta_4 \\ \Delta_5 \\ \Delta_1 \\ \Delta_2 \\ \Delta_6 \end{Bmatrix} \qquad \text{(b)}$$

Referring to relations (9.1) and (b) we have

$$\{P^{EF}\} = \begin{Bmatrix} 40 \\ -80 \\ 0 \end{Bmatrix} \qquad \text{(c)}$$

$$\{P^{ES}\} = \begin{Bmatrix} 0 \\ 0 \\ 0 \end{Bmatrix} \qquad \text{(d)}$$

$$\{R\} = \begin{Bmatrix} R_1^{(1)} \\ R_2^{(1)} \\ R_2^{(3)} \end{Bmatrix} \qquad \text{(e)}$$

$$\{\Delta^S\} = \begin{Bmatrix} 0 \\ 0 \\ 0 \end{Bmatrix} \qquad \text{(f)}$$

$$\{S^{FF}\} = \frac{EA}{125} \begin{bmatrix} 32 & 0 & -16 \\ 0 & 18 & 12 \\ -16 & 12 & 31.625 \end{bmatrix} \qquad \text{(g)}$$

$$[S^{SF}] = \frac{EA}{125} \begin{bmatrix} -16 & -12 & -15.625 \\ -12 & -9 & 0 \\ 12 & -9 & -12 \end{bmatrix} \qquad \text{(h)}$$

We compute the components of displacements of the nodes of the truss.

Substituting relations (c) and (g) in (9.4), we obtain

$$\{\Delta^F\} = \begin{Bmatrix} \Delta_3 \\ \Delta_4 \\ \Delta_5 \end{Bmatrix} = \frac{125}{EA} \begin{bmatrix} 32 & 0 & -16 \\ 0 & 18 & 12 \\ -16 & 12 & 31.625 \end{bmatrix}^{-1} \begin{Bmatrix} 40 \\ -80 \\ 0 \end{Bmatrix}$$

$$= \frac{1}{EA} \begin{Bmatrix} 449.58 \\ -946.67 \\ 586.67 \end{Bmatrix} = \begin{Bmatrix} -22.479 \\ -47.333 \\ 29.333 \end{Bmatrix} \text{mm} \tag{i}$$

We compute the reactions of the truss. Substituting relations (d), (h), and (i) into (9.5), we get

$$\{R\} = \begin{Bmatrix} R_1^{(1)} \\ R_2^{(1)} \\ R_2^{(3)} \end{Bmatrix} = \frac{1}{125} \begin{bmatrix} -16 & -12 & -15.625 \\ -12 & -9 & 0 \\ 12 & -9 & -12 \end{bmatrix} \begin{Bmatrix} 449.58 \\ -946.67 \\ 586.67 \end{Bmatrix} = \begin{Bmatrix} -40.0 \\ 25.0 \\ 55.0 \end{Bmatrix} \tag{j}$$

The reactions of the truss are shown in Fig. d. Their values can be checked by considering the equilibrium of the truss.

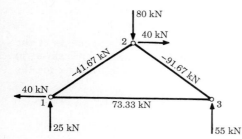

Figure d Results for the truss of Fig. a.

STEP 7. We compute the local components of internal forces in the elements of the truss. Referring to Fig. b and relation (i), the global components of the nodal displacements of the elements of the truss are

$$\{\bar{D}^1\} = \begin{Bmatrix} 0 \\ 0 \\ \Delta_3 \\ \Delta_4 \end{Bmatrix} = \frac{1}{EA} \begin{Bmatrix} 0 \\ 0 \\ 449.58 \\ -946.67 \end{Bmatrix}$$

$$\{\bar{D}^2\} = \begin{Bmatrix} 0 \\ 0 \\ \Delta_5 \\ 0 \end{Bmatrix} = \frac{1}{EA} \begin{Bmatrix} 0 \\ 0 \\ 586.67 \\ 0 \end{Bmatrix} \tag{k}$$

$$\{\bar{D}^3\} = \begin{Bmatrix} \Delta_3 \\ \Delta_4 \\ \Delta_5 \\ 0 \end{Bmatrix} = \frac{1}{EA} \begin{Bmatrix} 449.58 \\ -946.67 \\ 586.67 \\ 0 \end{Bmatrix}$$

We compute the hybrid stiffness matrices for the elements of the truss using relations (4.50), (6.42), and (9.11). Thus,

$$[\overset{\frown}{K^1}] = \frac{EA}{5}\begin{bmatrix} 1 & -1 \\ -1 & 1 \end{bmatrix}\begin{bmatrix} 0.8 & 0.6 & 0 & 0 \\ 0 & 0 & 0.8 & 0.6 \end{bmatrix}$$

$$= \frac{EA}{5}\begin{bmatrix} 0.8 & 0.6 & -0.8 & -0.6 \\ -0.8 & -0.6 & 0.8 & 0.6 \end{bmatrix}$$

$$[\overset{\frown}{K^2}] = \frac{EA}{8}\begin{bmatrix} 1 & -1 \\ -1 & 1 \end{bmatrix}\begin{bmatrix} 1 & 0 & 0 & 0 \\ 0 & 0 & 1 & 0 \end{bmatrix}$$

$$= \frac{EA}{8}\begin{bmatrix} 1 & 0 & -1 & 0 \\ -1 & 0 & 1 & 0 \end{bmatrix} \tag{1}$$

$$[\overset{\frown}{K^3}] = \frac{EA}{5}\begin{bmatrix} 1 & -1 \\ -1 & 1 \end{bmatrix}\begin{bmatrix} 0.8 & -0.6 & 0 & 0 \\ 0 & 0 & 0.8 & -0.6 \end{bmatrix}$$

$$= \frac{EA}{5}\begin{bmatrix} 0.8 & -0.6 & -0.8 & 0.6 \\ -0.8 & 0.6 & 0.8 & -0.6 \end{bmatrix}$$

Normally we compute the hybrid stiffness matrices for the elements of the truss in step 4 before computing their global stiffness matrices. However, in this example, steps 3 to 5 are omitted because the stiffness matrix for the truss is taken from Example 1 of Sec. 8.3.1.

We substitute relations (1) and (k) into (9.13). Thus,

$$\{A^1\} = \frac{EA}{5}\begin{bmatrix} 0.8 & 0.6 & -0.8 & -0.6 \\ -0.8 & -0.6 & 0.8 & 0.6 \end{bmatrix}\frac{1}{EA}\begin{Bmatrix} 0 \\ 0 \\ 449.58 \\ -946.67 \end{Bmatrix} = \begin{Bmatrix} 41.67 \\ -41.67 \end{Bmatrix}$$

$$[A^2] = \frac{EA}{8}\begin{bmatrix} 1 & 0 & -1 & 0 \\ -1 & 0 & 1 & 0 \end{bmatrix}\frac{1}{EA}\begin{Bmatrix} 0 \\ 0 \\ 586.67 \\ 0 \end{Bmatrix} = \begin{Bmatrix} -73.33 \\ 73.33 \end{Bmatrix} \tag{m}$$

$$\{A^3\} = \frac{EA}{5}\begin{bmatrix} 0.8 & -0.6 & -0.8 & 0.6 \\ -0.8 & 0.6 & 0.8 & -0.6 \end{bmatrix}\frac{1}{EA}\begin{Bmatrix} 449.58 \\ -946.67 \\ 586.67 \\ 0 \end{Bmatrix} = \begin{Bmatrix} 91.67 \\ -91.67 \end{Bmatrix}$$

The results are shown in Fig. d. They can be checked by considering the equilibrium of the nodes of the truss. It is apparent that the direct stiffness method is not suitable for analyzing trusses by hand calculations. It involves lengthy computations.

Example 2. Using the direct stiffness method, compute the components of displacements of the nodes, the reactions, and the nodal actions of the elements of the two beams subjected to the external actions shown in Figs. a and b. The elements of the beams are made from the same isotropic linearly elastic material and have a constant cross section whose moments of inertia are indicated in Figs. a and b.

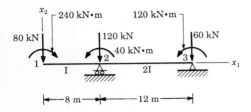

Figure a Geometry and loading of the beam.

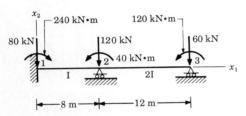

Figure b Geometry and loading of the beam.

solution

Beam of Fig. a

STEPS 1 TO 5. The equivalent actions for the beam are equal to the external actions shown in Fig. a. Moreover, the stiffness matrix for the beam of Fig. a has been computed in Example 2 of Sec. 8.3.1. Referring to relation (f) of this example and to Figs. a and c the stiffness equations for the beam of Fig. a are

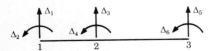

Figure c Numbering of the components of displacements of the nodes of the beams of Figs. a and b.

Figure d Degrees of freedom of the nodes of the beam of Fig. a.

$$\begin{Bmatrix} -80 \\ -240 \\ -120 \\ 40 \\ -60 \\ 120 \end{Bmatrix} + \begin{Bmatrix} 0 \\ 0 \\ R_2^{(2)} \\ 0 \\ R_2^{(3)} \\ 0 \end{Bmatrix}$$

$$= EI \begin{bmatrix} 0.02344 & 0.09375 & -0.02344 & 0.09375 & 0 & 0 \\ 0.09375 & 0.50000 & -0.09375 & 0.25000 & 0 & 0 \\ -0.02344 & -0.09375 & 0.03733 & -0.01042 & -0.01389 & 0.08333 \\ 0.09375 & 0.25000 & -0.01042 & 1.16667 & -0.08333 & 0.33333 \\ 0 & 0 & -0.01389 & -0.08333 & 0.01389 & -0.08333 \\ 0 & 0 & 0.08333 & 0.33333 & -0.08333 & 0.66667 \end{bmatrix} \begin{Bmatrix} \Delta_1 \\ \Delta_2 \\ \Delta_3 \\ \Delta_4 \\ \Delta_5 \\ \Delta_6 \end{Bmatrix} \quad \text{(a)}$$

STEP 6. We compute the components of displacements of the nodes of the beam and its reactions. In order to accomplish this we take into account the boundary conditions of the beam. That is, we rearrange the rows of the stiffness equations (a) in order to move the rows corresponding to the reactions of the supports of the beam to the bottom. Moreover, we rearrange the columns of the stiffness matrix $[\hat{S}]$ in order to move to the right its columns which are multiplied by the vanishing components of displacements of the supports of the beam. Furthermore, we partition the resulting stiffness equations for the beam as indicated in relation (9.1). Thus, referring to Figs. c and d we have

$$\begin{Bmatrix} -80 \\ -240 \\ 40 \\ 120 \\ \hdashline -120 \\ -60 \end{Bmatrix} + \begin{Bmatrix} 0 \\ 0 \\ 0 \\ 0 \\ \hdashline R_2^{(2)} \\ R_2^{(3)} \end{Bmatrix}$$

$$= EI \begin{bmatrix} 0.02344 & 0.09375 & 0.09375 & 0 & \vdots & -0.02344 & 0 \\ 0.09375 & 0.50000 & 0.25000 & 0 & \vdots & -0.09375 & 0 \\ 0.09375 & 0.25000 & 1.16667 & 0.33333 & \vdots & -0.01042 & -0.08333 \\ 0 & 0 & 0.33333 & 0.66667 & \vdots & 0.08333 & -0.08333 \\ \hdashline -0.02344 & -0.09375 & -0.01042 & 0.08333 & \vdots & 0.03733 & -0.01389 \\ 0 & 0 & -0.08333 & -0.08333 & \vdots & -0.01389 & 0.01389 \end{bmatrix} \begin{Bmatrix} \Delta_1 \\ \Delta_2 \\ \Delta_4 \\ \Delta_6 \\ \hdashline \Delta_3 \\ \Delta_5 \end{Bmatrix} \quad \text{(b)}$$

Referring to relations (9.1) and (b) we have

$$\{P^{EF}\} = \begin{Bmatrix} -80 \\ -240 \\ 40 \\ 120 \end{Bmatrix} \tag{c}$$

$$\{P^{ES}\} = \begin{Bmatrix} -120 \\ -60 \end{Bmatrix} \tag{d}$$

$$\{R\} = \begin{Bmatrix} R_2^{(2)} \\ R_2^{(3)} \end{Bmatrix} \tag{e}$$

$$\{\Delta^S\} = \begin{Bmatrix} 0 \\ 0 \end{Bmatrix} \tag{f}$$

$$[S^{FF}] = EI \begin{bmatrix} 0.02344 & 0.09375 & 0.09375 & 0 \\ 0.09375 & 0.50000 & 0.25000 & 0 \\ 0.09375 & 0.25000 & 1.16667 & 0.33333 \\ 0 & 0 & 0.33333 & 0.66667 \end{bmatrix} \tag{g}$$

and
$$[S^{SF}] = EI \begin{bmatrix} -0.02344 & -0.09375 & -0.01042 & 0.08333 \\ 0 & 0 & -0.08333 & -0.08333 \end{bmatrix} \tag{h}$$

STEP 7. We compute the components of displacements of the nodes of the beam. Substituting relations (c) and (g) into (9.4), we obtain

$$\{\Delta^F\} = \begin{Bmatrix} \Delta_1 \\ \Delta_2 \\ \Delta_4 \\ \Delta_6 \end{Bmatrix} = \frac{1}{EI} \begin{bmatrix} 0.02344 & 0.09375 & 0.09375 & 0 \\ 0.09375 & 0.50000 & 0.25000 & 0 \\ 0.09375 & 0.25000 & 1.16667 & 0.33333 \\ 0 & 0 & 0.33333 & 0.66667 \end{bmatrix}^{-1} \begin{Bmatrix} -80 \\ -240 \\ 40 \\ 120 \end{Bmatrix}$$

$$= \frac{1}{EI} \begin{Bmatrix} -12044.272 \\ 1398.5448 \\ 759.4952 \\ -199.7168 \end{Bmatrix} \tag{i}$$

We compute the reactions of the beam. Substituting relations (d), (h), and (i) we obtain

$$\begin{Bmatrix} -120 \\ -60 \end{Bmatrix} + \begin{bmatrix} R_2^{(2)} \\ R_2^{(3)} \end{bmatrix}$$

$$= \begin{bmatrix} -0.02344 & -0.09375 & -0.01042 & 0.08333 \\ 0 & 0 & -0.08333 & -0.08333 \end{bmatrix} \begin{Bmatrix} -12044.2720 \\ 1398.5448 \\ 759.4952 \\ -199.7168 \end{Bmatrix} = \begin{Bmatrix} 126.65 \\ -46.65 \end{Bmatrix}$$

Thus,
$$\begin{Bmatrix} R_2^{(2)} \\ R_2^{(3)} \end{Bmatrix} = \begin{Bmatrix} 246.65 \\ 13.35 \end{Bmatrix} \tag{j}$$

The reactions of the beam are shown in Fig. e. They can be checked by considering the equilibrium of the beam.

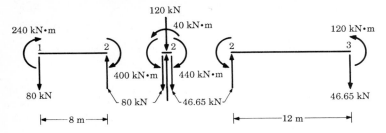

Figure e Free-body diagrams of the elements and node 2 of the beam of Fig. a.

STEP 8. We compute the nodal actions of the elements of the beam. Referring to Fig. b and relation (i), the global and local components of the nodal displacements of the elements of the beam are

$$\{D^1\} = \{\bar{D}^1\} = \begin{Bmatrix} \Delta_1 \\ \Delta_2 \\ 0 \\ \Delta_4 \end{Bmatrix} = \frac{1}{EI} \begin{Bmatrix} -12044.2720 \\ 1398.5448 \\ 0 \\ 759.4952 \end{Bmatrix}$$ (k)

$$\{D^2\} = \{\bar{D}^2\} = \begin{Bmatrix} 0 \\ \Delta_4 \\ 0 \\ \Delta_6 \end{Bmatrix} = \frac{1}{EI} \begin{Bmatrix} 0 \\ 759.4952 \\ 0 \\ -199.7168 \end{Bmatrix}$$

The relations between the nodal actions and the nodal displacements of the elements of the beam are

$$\{A^E\} = \begin{Bmatrix} F_2^{Ej} \\ M_3^{Ej} \\ F_2^{Ek} \\ M_3^{Ek} \end{Bmatrix} = [K]\{D\}$$ (l)

Substituting the local stiffness matrix $[K]$ of each element of the beam given by relations (a) and (b) of Example 2 of Sec. 8.3.1 and the matrix $\{D\}$ given by relations (k) into Eq. (l), we obtain

$$\{A^{E1}\} = \begin{bmatrix} 0.02344 & 0.09375 & -0.02344 & 0.09375 \\ 0.09375 & 0.50000 & -0.09375 & 0.25000 \\ -0.02344 & -0.09375 & 0.02344 & -0.09375 \\ 0.09375 & 0.25000 & -0.09375 & 0.50000 \end{bmatrix} \begin{Bmatrix} -12044.2720 \\ 1398.5448 \\ 0 \\ 759.4952 \end{Bmatrix}$$

$$= \begin{Bmatrix} -80.00 \\ -240.00 \\ 80.00 \\ -400.00 \end{Bmatrix}$$ (m)

$$\{A^{E2}\} = \begin{bmatrix} 0.01389 & 0.08333 & -0.01389 & 0.08333 \\ 0.08333 & 0.66667 & -0.08333 & 0.33333 \\ -0.01389 & -0.08333 & 0.01389 & -0.08333 \\ 0.08333 & 0.33333 & -0.08333 & 0.66667 \end{bmatrix} \begin{Bmatrix} 0 \\ 759.4952 \\ 0 \\ -199.7168 \end{Bmatrix}$$

$$= \begin{Bmatrix} 46.65 \\ 440.00 \\ -46.65 \\ 120.00 \end{Bmatrix}$$

The results are shown in Fig. e. They can be checked by considering the equilibrium of the nodes and elements of the beam.

Beam of Fig. b

STEPS 1 TO 5. The equivalent actions for the beam are equal to the external actions shown in Fig. b. Moreover, the stiffness matrix for the beams has been computed in Example 2 of Sec. 8.3.1. Referring to relation (f) of this example and to Figs. b and c the stiffness equations for the beam of Fig. b are

$$\begin{Bmatrix} -80 \\ -240 \\ -120 \\ 40 \\ -60 \\ 120 \end{Bmatrix} + \begin{Bmatrix} R_2^{(1)} \\ R_3^{(1)} \\ R_2^{(2)} \\ 0 \\ R_2^{(3)} \\ 0 \end{Bmatrix}$$

$$= EI \begin{bmatrix} 0.02344 & 0.09375 & -0.02344 & 0.09375 & 0 & 0 \\ 0.09375 & 0.50000 & -0.09375 & 0.25000 & 0 & 0 \\ -0.02344 & -0.09375 & 0.03733 & -0.01042 & -0.01389 & 0.08333 \\ 0.09375 & 0.25000 & -0.01042 & 1.16667 & -0.08333 & 0.33333 \\ 0 & 0 & -0.01389 & -0.08333 & 0.01389 & -0.08333 \\ 0 & 0 & 0.08333 & 0.33333 & -0.08333 & 0.66667 \end{bmatrix} \begin{Bmatrix} \Delta_1 \\ \Delta_2 \\ \Delta_3 \\ \Delta_4 \\ \Delta_5 \\ \Delta_6 \end{Bmatrix} \quad (n)$$

STEP 6. We compute the components of displacements of the nodes of the beam and its reactions. In order to accomplish this we take into account the boundary conditions of the beam. That is, we rearrange the rows of the stiffness equations (n) in order to move the rows corresponding to the reactions of the supports of the beam to the bottom. Moreover, we rearrange

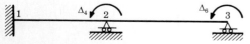

Figure f Degrees of freedom of the nodes of the beam of Fig. b.

the columns of the stiffness matrix $[\hat{S}]$ in order to move to its right the columns which are multiplied by the vanishing components of displacements of the supports of the beam. Furthermore, we partition the resulting stiffness equations for the beam as indicated in relation (9.1). Thus referring to Figs. c and f we have

$$
\begin{Bmatrix} 40 \\ 120 \\ \hline -80 \\ -240 \\ -120 \\ -60 \end{Bmatrix} + \begin{Bmatrix} 0 \\ 0 \\ \hline R_2^{(1)} \\ R_3^{(1)} \\ R_2^{(2)} \\ R_2^{(3)} \end{Bmatrix}
$$

$$
= EI \begin{bmatrix} 1.16667 & 0.33333 & 0.09375 & 0.25000 & -0.01042 & -0.08333 \\ 0.33333 & 0.66667 & 0 & 0 & 0.08333 & -0.08333 \\ 0.09375 & 0 & 0.02344 & 0.09375 & -0.02344 & 0 \\ 0.25000 & 0 & 0.09375 & 0.50000 & -0.09375 & 0 \\ -0.01042 & 0.08333 & -0.02344 & -0.09375 & 0.03733 & -0.01389 \\ -0.08333 & -0.08333 & 0 & 0 & -0.01389 & 0.01389 \end{bmatrix} \begin{Bmatrix} \Delta_4 \\ \Delta_6 \\ \Delta_1 \\ \Delta_2 \\ \Delta_3 \\ \Delta_5 \end{Bmatrix} \quad (o)
$$

Referring to relation (9.1) and (o) we have

$$
\{P^{EF}\} = \begin{Bmatrix} 40 \\ 120 \end{Bmatrix} \tag{p}
$$

$$
\{P^{ES}\} = \begin{Bmatrix} -80 \\ -240 \\ -120 \\ -60 \end{Bmatrix} \tag{q}
$$

$$
\{R\} = \begin{Bmatrix} R_2^{(1)} \\ R_3^{(1)} \\ R_2^{(2)} \\ R_2^{(3)} \end{Bmatrix} \tag{r}
$$

$$
(\Delta^S) = \begin{Bmatrix} 0 \\ 0 \\ 0 \\ 0 \end{Bmatrix} \tag{s}
$$

$$
[S^{FF}] = EI \begin{bmatrix} 1.16667 & 0.33333 \\ 0.33333 & 0.66667 \end{bmatrix} \tag{t}
$$

and

$$
[S^{SF}] = EI \begin{bmatrix} 0.09375 & 0 \\ 0.25000 & 0 \\ -0.01042 & 0.08333 \\ -0.08333 & -0.08333 \end{bmatrix} \tag{u}
$$

We compute the components of displacement of the nodes of the beam. Substituting relations (p) and (t) into (9.4), we have

$$\{\Delta^F\} = \begin{Bmatrix} \Delta_4 \\ \Delta_6 \end{Bmatrix} = \frac{1}{EI} \begin{bmatrix} 1.16667 & 0.33333 \\ 0.33333 & 0.66667 \end{bmatrix}^{-1} \begin{Bmatrix} 40 \\ 120 \end{Bmatrix} = \frac{1}{EI} \begin{Bmatrix} -20 \\ 190 \end{Bmatrix} \qquad \text{(v)}$$

We compute the reactions of the beam. Substituting relations (q), (u), and (v) into (9.5), we obtain

$$\begin{Bmatrix} -80 \\ -240 \\ -120 \\ -60 \end{Bmatrix} + \begin{Bmatrix} R_2^{(1)} \\ R_3^{(1)} \\ R_2^{(2)} \\ R_2^{(3)} \end{Bmatrix} = \begin{bmatrix} 0.09375 & 0 \\ 0.25000 & 0 \\ -0.01042 & 0.08333 \\ -0.08333 & -0.08333 \end{bmatrix} \begin{Bmatrix} -20.0 \\ 190.0 \end{Bmatrix} = \begin{Bmatrix} -1.875 \\ -5.000 \\ 16.040 \\ -14.166 \end{Bmatrix}$$

Thus,

$$\begin{Bmatrix} R_2^{(1)} \\ R_3^{(1)} \\ R_2^{(2)} \\ R_2^{(3)} \end{Bmatrix} = \begin{Bmatrix} 78.125 \\ 235.000 \\ 136.040 \\ 45.834 \end{Bmatrix} \qquad \text{(w)}$$

The reactions of the beam are shown in Fig. g.

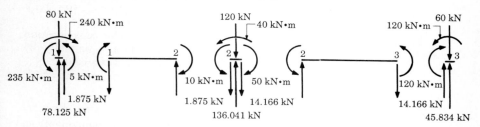

Figure g Free-body diagrams of the elements and the nodes of the beam of Fig. b.

STEP 7. We compute the nodal actions of the elements of the beam. Referring to Fig. c and relations (v), the local and global components of nodal displacements of the elements of the beam are

$$\{D^1\} = \{\bar{D}^1\} = \begin{Bmatrix} 0 \\ 0 \\ 0 \\ \Delta_4 \end{Bmatrix} = \frac{1}{EI} \begin{Bmatrix} 0 \\ 0 \\ 0 \\ -20.0 \end{Bmatrix}$$

$$\{D^2\} = \{\bar{D}^2\} = \begin{Bmatrix} 0 \\ \Delta_4 \\ 0 \\ \Delta_6 \end{Bmatrix} = \frac{1}{EI} \begin{Bmatrix} 0 \\ -20.0 \\ 0 \\ 190.0 \end{Bmatrix} \qquad \text{(x)}$$

Substituting the local stiffness matrix $[K]$ of an element of the beam given by relations (a) and (b) of Example 2 of Sec. 8.3.1, and the matrix $\{D^e\}$

$(e = 1, 2)$ given by relations (x) into Eq. (1), we obtain

$$\{A^{E1}\} = \begin{Bmatrix} F_2^{E1j} \\ M_3^{E1j} \\ F_2^{E1k} \\ M_3^{E1k} \end{Bmatrix}$$

$$= \begin{bmatrix} 0.02344 & 0.09375 & -0.02344 & 0.09375 \\ 0.09375 & 0.50000 & -0.09375 & 0.25000 \\ -0.02344 & -0.09375 & 0.02344 & -0.09375 \\ 0.09375 & 0.25000 & -0.09375 & 0.50000 \end{bmatrix} \begin{Bmatrix} 0 \\ 0 \\ 0 \\ -20.0 \end{Bmatrix} = \begin{Bmatrix} -1.875 \\ -5.000 \\ 1.875 \\ -10.000 \end{Bmatrix}$$

$$\text{(v)}$$

$$\{A^{E2}\} = \begin{Bmatrix} F_2^{E2j} \\ M_3^{E2j} \\ F_2^{E2k} \\ M_3^{E2k} \end{Bmatrix}$$

$$= \begin{bmatrix} 0.01389 & 0.08333 & -0.01389 & 0.08333 \\ 0.08333 & 0.66667 & -0.08333 & 0.33333 \\ -0.01389 & -0.08333 & 0.01389 & -0.08333 \\ 0.08333 & 0.33333 & -0.08333 & 0.66667 \end{bmatrix} \begin{Bmatrix} 0 \\ -20.0 \\ 0 \\ 190.0 \end{Bmatrix} = \begin{Bmatrix} 14.166 \\ 50.000 \\ -14.166 \\ 120.000 \end{Bmatrix}$$

The results are shown in Fig. g. Inasmuch as the beam under consideration is statically indeterminate the results can be checked only partially by considering the equilibrium of the nodes and the elements of the beam.

Example 3. Using the direct stiffness method, compute the components of displacement of the nodes of the frame of Fig. a subjected to the external actions shown in this figure, as well as to a change of temperature of its elements and to a settlement of 20 mm of support 1. Moreover, compute the reactions of the frame and the nodal actions of its elements. The temperature of the top and bottom fibers of the elements of the frame is $T_t = 25°C$ and

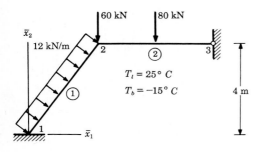

Figure a Geometry and loading of the frame.

$T_b = -15°C$, respectively. The temperature during construction was $T_0 = 5°C$; thus $\Delta T_c = 0$. The elements of the frame are made of the same isotropic linearly elastic material ($E = 210 \text{ kN/mm}^2$, $\alpha = 10^{-5}/°C$) and have the same constant cross section ($A = 16 \times 10^3 \text{ mm}^2$, $I = 400 \times 10^6 \text{ mm}^4$, $h = 420 \text{ mm}$).

---

solution 1 **The settlement of the support of the frame is included in the loading of the restrained structure**   We establish the nodal actions of the elements of the frame of Fig. a by superimposing the corresponding nodal actions of the elements of the restrained structure subjected to the given loading and the structure subjected to the equivalent actions on its nodes (see Fig. b).

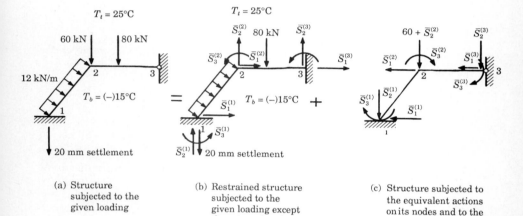

(a) Structure
subjected to the
given loading

(b) Restrained structure
subjected to the
given loading except
the external actions
acting on the nodes

(c) Structure subjected to
the equivalent actions
on its nodes and to the
settlement of its support 1

Figure b   Superposition of the restrained structure and the structure subjected to the equivalent actions.

STEP 1.
1. We establish the local matrix of fixed-end actions of each element of the structure by referring to the table on the inside of the back cover.

Computation of the fixed-end actions of element 1 due to the settlement of support 1

In order to accomplish this we decompose the 20-mm settlement of support 1 into a component parallel (16 mm) and a component normal (12 mm) to the axis of element 1. Thus the axial force required to elongate element 1 of the restrained structure by 16 mm is equal to

$$F_1^{R1j} = -\frac{0.016AE}{L} = -10,752 \text{ kN}$$

Moreover, referring to the table on the inside of the back cover, the shearing components of the fixed-end forces F_2^{R1j}, F_2^{R1k}, and the components of the fixed-end moments M_3^{R1j} and M_3^{R1k} due to the settlement of support 1 are equal to

$$F_2^{R1k} = -F_2^{R1j} = \frac{12EI(0.012)}{L^3}$$

$$= \frac{12(210)(10^6)(400)(10^{-6})(0.012)}{(5)^3} = 96.77 \text{ kN}$$

$$M_3^{R1k} = M_3^{R1j} = -\frac{6EI(0.012)}{L^2}$$

$$= -\frac{6(210)(10^6)(400)(10^{-6})(0.012)}{(5)^2} = -241.92 \text{ kN} \cdot \text{m}$$

Computation of the components of fixed-end moments of the elements of the structure due to the change of temperature

$$M^{R1k} = -M^{R1j} = M^{R2k} = -M^{R2j} = \frac{EI\alpha\delta T_2}{h}$$

$$= \frac{210.000(10^6) \times 400(10^{-6})(10^{-5})(40)}{0.420} = 80 \text{ kN} \cdot \text{m}$$

Computation of the fixed-end actions of the elements of the structure due to the external actions given in Fig. a

Referring to the table on the inside of the back cover, we have

$$M_3^{R1j} = -M_3^{R1k} = \frac{p_2 L^2}{12} = 25 \text{ kN} \cdot \text{m}$$

$$F_2^{R1j} = F_2^{R1k} = \frac{p_2 L}{2} = 30 \text{ kN}$$

$$M_3^{R2j} = -M_3^{R2k} = \frac{80(2.5)^3}{5^2} = 50 \text{ kN} \cdot \text{m}$$

$$F_2^{R2j} = F_2^{R2k} = 40 \text{ kN}$$

The results are shown in Fig. c. Referring to this figure the local matrices of fixed-end actions of the elements of the structure are

$$\{A^{R1}\} = \left\{ \begin{array}{c} -10752.00 \\ -66.77 \\ -296.92 \\ \hline 10752.00 \\ 126.77 \\ -186.92 \end{array} \right\} \qquad \{A^{R2}\} = \left\{ \begin{array}{c} 0 \\ 40 \\ -30 \\ \hline 0 \\ 40 \\ 30 \end{array} \right\} \qquad \text{(a)}$$

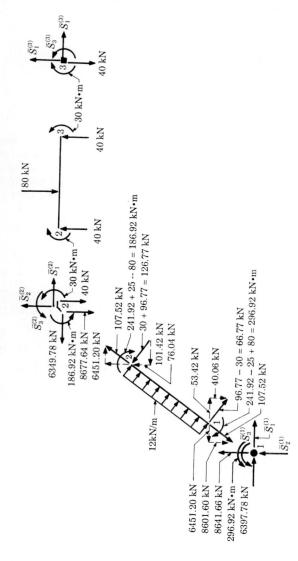

Figure c Free-body diagrams of the elements and nodes of the restrained structure.

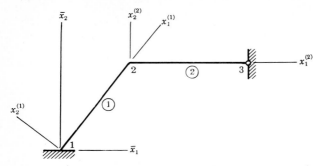

Figure d Global axes of the frame and local axes of its elements.

2. We transform the local matrix of fixed-end actions of each element of the structure to global. The local axes of the elements of the frame are shown in Fig. d. Referring to this figure we form the transformation matrix for each element of the frame [see relations (6.6) and (6.48)] and substitute it in relation (6.65) to obtain its global matrix of fixed-end actions. That is,

$$\{\bar{A}^{R1}\} = \left\{ \frac{\{\bar{A}^{R1j}\}}{\{\bar{A}^{R1k}\}} \right\}$$

$$= \begin{bmatrix} 0.6 & -0.8 & 0 & 0 & 0 & 0 \\ 0.8 & 0.6 & 0 & 0 & 0 & 0 \\ 0 & 0 & 1 & 0 & 0 & 0 \\ 0 & 0 & 0 & 0.6 & 0.8 & 0 \\ 0 & 0 & 0 & -0.8 & 0.6 & 0 \\ 0 & 0 & 0 & 0 & 0 & 1 \end{bmatrix} \left\{ \begin{array}{r} -10752.00 \\ -66.77 \\ -296.92 \\ 10752.00 \\ 126.77 \\ -186.92 \end{array} \right\} = \left\{ \begin{array}{r} -6397.79 \\ -8641.66 \\ -296.92 \\ \hline 6349.79 \\ 8677.66 \\ -186.92 \end{array} \right\} \quad \text{(b)}$$

$$\{\bar{A}^{R2}\} = \{\bar{A}^{R1}\}$$

STEP 2. We establish the matrix of equivalent actions for the frame. To accomplish this we do the following.

1. We establish the matrix of the restraining actions $\{\widehat{RA}\}$. That is, referring to Fig. b*b* from the equilibrium of the nodes of the restrained structure we have

$$\{\widehat{RA}\} = \left\{ \begin{array}{c} \{\bar{A}^{R1j}\} \\ \hline \{\bar{A}^{R1k}\} + \{\bar{A}^{R2j}\} \\ \hline \{\bar{A}^{R2k}\} \end{array} \right\} = \left\{ \begin{array}{r} -6397.79 \\ -8641.66 \\ -296.92 \\ \hline 6349.92 \\ 8717.66 \\ -216.92 \\ \hline 0 \\ 40 \\ 30 \end{array} \right\} \quad \text{(c)}$$

This result can be verified by referring to the free-body diagrams of the nodes of the restrained structure shown in Fig. c and considering their equilibrium.

2. We form the matrix of the given actions $\{\hat{P}^G\}$ acting on the nodes of the frame. Referring to Fig. a we have

$$\{\hat{P}^G\}^T = [0 \quad 0 \quad 0 \quad 0 \quad -60 \quad 0 \quad 0 \quad 0 \quad 0] \tag{d}$$

3. We establish the matrix of the equivalent actions $\{\hat{P}^E\}$. Substituting relations (c) and (d) into (7.2) we obtain

$$\{\hat{P}^E\} = \{\hat{P}^G\} - \{\widehat{RA}\} = \left\{ \begin{array}{c} 6397.79 \\ 8641.66 \\ 296.92 \\ \hline -6349.79 \\ -8777.66 \\ 216.92 \\ \hline 0 \\ -40 \\ -30 \end{array} \right\} \tag{e}$$

The structure subjected to the equivalent actions on its nodes is shown in Fig. e.

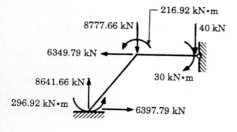

Figure e Structure subjected to the equivalent actions on its nodes.

STEPS 3 TO 5. The stiffness matrix for the frame has been computed in Example 3 of Sec. 8.3.1. Referring to relation (i) of this example and to Figs.

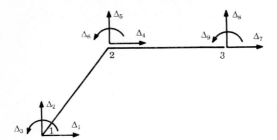

Figure f Numbering of the components of displacements of the nodes of the frame.

a and f the stiffness equations for the frame of Fig. a are

$$
\begin{Bmatrix}
6397.79 \\
8641.66 \\
296.92 \\
\hdashline
-6349.79 \\
-8777.69 \\
216.92 \\
\hdashline
0 \\
-40.00 \\
-30.00
\end{Bmatrix}
+
\begin{Bmatrix}
R_1^{(1)} \\
R_2^{(1)} \\
R_3^{(1)} \\
\hdashline
0 \\
0 \\
0 \\
\hdashline
R_1^{(3)} \\
R_2^{(3)} \\
0
\end{Bmatrix}
$$

$$
= \frac{EI}{125}
\begin{bmatrix}
367.68 & 474.24 & -24 & -367.68 & -474.24 & -24 & 0 & 0 & 0 \\
474.24 & 644.32 & 18 & -474.24 & -644.32 & 18 & 0 & 0 & 0 \\
-24.00 & 18.00 & 100 & 24.00 & -18.00 & 50 & 0 & 0 & 0 \\
-367.68 & -474.24 & 24 & 1367.68 & 474.24 & 24 & -1000 & 0 & 0 \\
-474.24 & -644.32 & -18 & 474.24 & 656.32 & 12 & 0 & -12 & 30 \\
-24.00 & 18.00 & 50 & 24.00 & 12.00 & 200 & 0 & -30 & 50 \\
0 & 0 & 0 & -1000.00 & 0 & 0 & 1000 & 0 & 0 \\
0 & 0 & 0 & 0 & -12.0 & -30 & 0 & 12 & -30 \\
0 & 0 & 0 & 0 & 30.0 & 50 & 0 & -30 & 100
\end{bmatrix}
\begin{Bmatrix}
\Delta_1 \\
\Delta_2 \\
\Delta_3 \\
\Delta_4 \\
\Delta_5 \\
\Delta_6 \\
\Delta_7 \\
\Delta_8 \\
\Delta_9
\end{Bmatrix}
\quad (f)
$$

STEP 6.   We compute the components of displacements of the nodes of the frame and its reactions. In order to accomplish this we take into account the boundary conditions of the frame. That is, we rearrange the rows of the

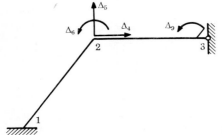

Figure g Degrees of freedom of the nodes of the frame.

stiffness equations (f) in order to move to the bottom the equations corresponding to the reactions of the supports of the frame. Moreover, we rearrange the columns of the stiffness matrix in order to move to its right the columns which are multiplied by the vanishing components of displacements of the supports of the frame. Furthermore, we partition the resulting stiffness equations as indicated in relation (9.1). Thus, referring to Fig. g we have

$$
\begin{Bmatrix} -6349.79 \\ -8777.69 \\ 216.92 \\ -30.00 \\ \hline 6397.79 \\ 8641.66 \\ 296.92 \\ 0 \\ -40.00 \end{Bmatrix} + \begin{Bmatrix} 0 \\ 0 \\ 0 \\ 0 \\ \hline R_1^{(1)} \\ R_2^{(1)} \\ R_3^{(1)} \\ R_1^{(3)} \\ R_2^{(3)} \end{Bmatrix}
$$

$$
= \frac{EI}{125} \left[\begin{array}{cccc|cccccc} 1367.68 & 474.24 & 24 & 0 & -367.68 & -474.24 & 24 & -1000 & 0 \\ 474.24 & 656.32 & 12 & 30 & -474.24 & -644.32 & -18 & 0 & -12 \\ 24.00 & 12.00 & 200 & 50 & -24.00 & 18.00 & 50 & 0 & -30 \\ 0 & 30.00 & 50 & 100 & 0 & 0 & 0 & 0 & -30 \\ \hline -367.68 & -474.24 & -24 & 0 & 367.68 & 474.24 & -24 & 0 & 0 \\ -474.24 & -644.32 & 18 & 0 & 474.24 & 644.32 & 18 & 0 & 0 \\ 24.00 & -18.00 & 50 & 0 & -24.00 & 18.00 & 100 & 0 & 0 \\ -1000.00 & 0 & 0 & 0 & 0 & 0 & 0 & 1000 & 0 \\ 0 & -12.00 & -30 & -30 & 0 & 0 & 0 & 0 & 12 \end{array} \right] \begin{Bmatrix} \Delta_4 \\ \Delta_5 \\ \Delta_6 \\ \Delta_9 \\ \Delta_1 \\ \Delta_2 \\ \Delta_3 \\ \Delta_7 \\ \Delta_8 \end{Bmatrix} \quad (g)
$$

Referring to relations (9.1) and (g) we have

$$
\{P^{EF}\} = \begin{Bmatrix} -6349.79 \\ -8777.66 \\ 216.92 \\ -30.00 \end{Bmatrix} \tag{h}
$$

$$
\{P^{ES}\} = \begin{Bmatrix} 6397.79 \\ 8641.66 \\ 296.92 \\ 0 \\ -40.00 \end{Bmatrix} \tag{i}
$$

$$
\{R\} = \begin{Bmatrix} R_1^{(1)} \\ R_2^{(1)} \\ R_3^{(1)} \\ R_1^{(3)} \\ R_2^{(3)} \end{Bmatrix} \tag{j}
$$

$$\{\Delta^S\} = \begin{Bmatrix} 0 \\ 0 \\ 0 \\ 0 \\ 0 \end{Bmatrix} \tag{k}$$

$$[S^{FF}] = \frac{EI}{125} \begin{bmatrix} 1367.68 & 474.24 & 24 & 0 \\ 474.24 & 656.32 & 12 & 30 \\ 24.00 & 12.00 & 200 & 50 \\ 0 & 30.00 & 50 & 100 \end{bmatrix} \tag{l}$$

$$[S^{SF}] = \frac{EI}{125} \begin{bmatrix} -367.68 & -474.24 & -24 & 0 \\ -474.24 & -644.32 & 18 & 0 \\ 24.00 & -18.00 & 50 & 0 \\ -1000.00 & 0 & 0 & 0 \\ 0 & -12.00 & -30 & -30 \end{bmatrix} \tag{m}$$

We compute the components of displacements of the nodes of the frame of Fig. e. They are identical to the corresponding components of displacements of the nodes of the frame of Fig. a. Substituting relations (h) and (l) into relation (9.4) we obtain

$$\{\Delta^F\} = \frac{125}{EI} \begin{bmatrix} 1367.68 & 474.24 & 24 & 0 \\ 474.24 & 656.32 & 12 & 30 \\ 24.00 & 12.00 & 200 & 50 \\ 0 & 30.00 & 50 & 100 \end{bmatrix}^{-1} \begin{Bmatrix} -6349.79 \\ -8777.66 \\ 216.92 \\ -30.00 \end{Bmatrix}$$

$$= \frac{125}{EI} \begin{Bmatrix} 0.04494 \\ -13.57399 \\ 1.08640 \\ 3.22900 \end{Bmatrix} \tag{n}$$

We compute the reactions of the frame of Fig. e. They are identical to the corresponding reactions of the frame of Fig. a. Substituting relations (m) and (n) into relation (9.5) we obtain

$$\begin{Bmatrix} R_1^{(1)} \\ R_2^{(1)} \\ R_3^{(1)} \\ R_1^{(3)} \\ R_2^{(3)} \end{Bmatrix} + \begin{Bmatrix} 6397.79 \\ 8641.66 \\ 296.92 \\ 0 \\ -40.00 \end{Bmatrix}$$

$$= \begin{Bmatrix} -367.68 & -474.24 & -24 & 0 \\ -474.24 & -644.32 & 18 & 0 \\ 24.00 & -18.00 & 50 & 0 \\ -1000.00 & 0 & 0 & 0 \\ 0 & -12.00 & -30 & -30 \end{Bmatrix} \begin{Bmatrix} 0.04494 \\ -13.57399 \\ 1.08640 \\ 3.22900 \end{Bmatrix} = \begin{Bmatrix} 6394.73 \\ 8744.23 \\ 299.73 \\ -44.94 \\ 33.43 \end{Bmatrix}$$

Thus the reactions of the frame of Figs. a and e are

$$\{R\} = \begin{Bmatrix} R_1^{(1)} \\ R_2^{(1)} \\ R_3^{(1)} \\ R_1^{(3)} \\ R_2^{(3)} \end{Bmatrix} = \begin{Bmatrix} -3.06 \\ 102.57 \\ 2.81 \\ -44.94 \\ 73.43 \end{Bmatrix} \qquad (o)$$

The reactions of the frame are shown in Fig. h.

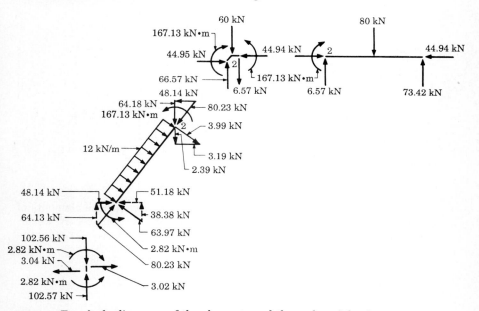

Figure h Free-body diagrams of the elements and the nodes of the frame.

STEP 7. We compute the nodal actions of the elements of the frame of Fig. a. Referring to Fig. f and relation (1), the global components of nodal displacements of the elements of the frame are

$$[\bar{D}^1] = \begin{Bmatrix} 0 \\ 0 \\ 0 \\ \Delta_4 \\ \Delta_5 \\ \Delta_6 \end{Bmatrix} = \frac{125}{EI} \begin{Bmatrix} 0 \\ 0 \\ 0 \\ \hline 0.04494 \\ -13.57399 \\ 1.08640 \end{Bmatrix} \qquad [\bar{D}^2] = \begin{Bmatrix} \Delta_4 \\ \Delta_5 \\ \Delta_6 \\ 0 \\ 0 \\ \Delta_5 \end{Bmatrix} = \frac{125}{EI} \begin{Bmatrix} 0.04494 \\ -13.57399 \\ 1.08640 \\ \hline 0 \\ 0 \\ 3.22900 \end{Bmatrix} \qquad (p)$$

Using relation (9.11) we compute the hybrid stiffness matrix for each element of the frame. The transformation matrix for the elements of the frame is given by relation (6.48). The local stiffness matrix for the elements of the frame is given by relation (a) of Example 3 of Sec. 8.3.1. Thus

$$[\overset{r}{K}{}^1] = \frac{125}{EI}\begin{bmatrix} 1000 & 0 & 0 & -1000 & 0 & 0 \\ 0 & 12 & 30 & 0 & -12 & 30 \\ 0 & 30 & 100 & 0 & -30 & 50 \\ -1000 & 0 & 0 & 1000 & 0 & 0 \\ 0 & -12 & -30 & 0 & 12 & -30 \\ 0 & 30 & 50 & 0 & -30 & 100 \end{bmatrix}\begin{bmatrix} 0.6 & 0.8 & 0 & 0 & 0 & 0 \\ -0.8 & 0.6 & 0 & 0 & 0 & 0 \\ 0 & 0 & 1 & 0 & 0 & 0 \\ 0 & 0 & 0 & 0.6 & 0.8 & 0 \\ 0 & 0 & 0 & -0.8 & 0.6 & 0 \\ 0 & 0 & 0 & 0 & 0 & 1 \end{bmatrix}$$

$$= \frac{125}{EI}\begin{bmatrix} 600.0 & 800.0 & 0 & -600.0 & -800.0 & 0 \\ -9.6 & 7.2 & 30 & 9.6 & -7.2 & 30 \\ -24.0 & 18.0 & 100 & 24.0 & -18.0 & 50 \\ -600.0 & -800.0 & 0 & 600.0 & 800.0 & 0 \\ 9.6 & -7.2 & -30 & -9.6 & 7.2 & -30 \\ -24.0 & 18.0 & 50 & 24.0 & -18.0 & 100 \end{bmatrix} \quad (q)$$

$$[\overset{r}{K}{}^2] = [K^2] = \frac{125}{EI}\begin{bmatrix} 1000 & 0 & 0 & -1000 & 0 & 0 \\ 0 & 12 & 30 & 0 & -12 & 30 \\ 0 & 30 & 100 & 0 & -30 & 50 \\ -1000 & 0 & 0 & 1000 & 0 & 0 \\ 0 & -12 & -30 & 0 & 12 & -30 \\ 0 & 30 & 50 & 0 & -30 & 100 \end{bmatrix}$$

It is convenient to compute the hybrid stiffness matrices for the elements of the frame in step 4 before computing their global stiffness matrices. However, in this example steps 3 to 5 are omitted because the stiffness matrix for the frame is taken from Example 3 of Sec. 8.3.1.

We substitute relations (p) and (q) into relation (9.13). Thus,

$$[A^1] = \begin{bmatrix} 600.0 & 800.0 & 0 & -600.0 & -800.0 & 0 \\ -9.6 & 7.2 & 30 & 9.6 & -7.2 & 30 \\ -24.0 & 18.0 & 100 & 24.0 & -18.0 & 50 \\ -600.0 & -800.0 & 0 & 600.0 & 800.0 & 0 \\ 9.6 & -7.2 & -30 & -9.6 & 7.2 & -30 \\ -24.0 & 18.0 & 50 & 24.0 & -18.0 & 100 \end{bmatrix}\begin{Bmatrix} 0 \\ 0 \\ 0 \\ \hline 0.04494 \\ -13.57399 \\ 1.08640 \end{Bmatrix} + \begin{Bmatrix} -10{,}752.00 \\ -66.77 \\ -296.92 \\ \hline 10{,}752.00 \\ 126.77 \\ -186.92 \end{Bmatrix}$$

$$= \begin{Bmatrix} 80.23 \\ 63.97 \\ 2.81 \\ \hline -80.23 \\ -3.99 \\ 167.13 \end{Bmatrix} \quad (r)$$

$$[A^2] = \begin{bmatrix} 1000 & 0 & 0 & -1000 & 0 & 0 \\ 0 & 12 & 30 & 0 & -12 & 30 \\ 0 & 30 & 100 & 0 & -30 & 50 \\ -1000 & 0 & 0 & 1000 & 0 & 0 \\ 0 & -12 & -30 & 0 & 12 & -30 \\ 0 & 30 & 50 & 0 & -30 & 100 \end{bmatrix}\begin{Bmatrix} 0.04494 \\ -13.57399 \\ 1.08640 \\ \hline 0 \\ 0 \\ 3.22900 \end{Bmatrix} + \begin{Bmatrix} 0 \\ 40 \\ -30 \\ \hline 0 \\ 40 \\ 30 \end{Bmatrix} = \begin{Bmatrix} 44.94 \\ 6.58 \\ -167.13 \\ \hline -44.94 \\ 73.42 \\ 0 \end{Bmatrix}$$

The results are shown in Fig. h. Referring to this figure we see that the results satisfy the equilibrium of the nodes and of the elements of the frame to within a very small error.

---

solution 2   The settlement of the support of the frame is not included in the loading of the restrained structure   We establish the nodal actions of the elements of the frame of Fig. a by superimposing the corresponding nodal actions of the elements of the restrained structure subjected to the given loading except the settlement of support 1 and the structure subjected to the equivalent actions on its nodes and the settlement of its support 1 (see Fig. i).

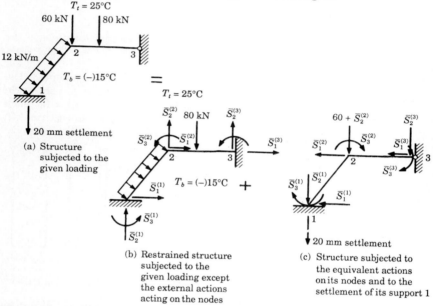

(a) Structure subjected to the given loading

(b) Restrained structure subjected to the given loading except the external actions acting on the nodes

(c) Structure subjected to the equivalent actions on its nodes and to the settlement of its support 1

Figure i Superposition of the restrained structure and the structure subjected to the equivalent actions and the settlement of support 1.

STEP 1.   We establish the fixed-end actions of the elements of the structure by referring to the table on the inside of the back cover. They are shown in Fig. j. Referring to Fig. j the local matrices of fixed-end actions of the elements of the structure are

$$\{A^{R1}\} = \begin{Bmatrix} 0 \\ 30 \\ -55 \\ \hline 0 \\ 30 \\ 55 \end{Bmatrix} \qquad \{A^{R2}\} = \begin{Bmatrix} 0 \\ 40 \\ -30 \\ \hline 0 \\ 40 \\ 30 \end{Bmatrix} \qquad \text{(t)}$$

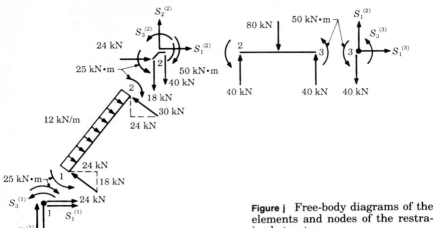

Figure j Free-body diagrams of the elements and nodes of the restrained structure.

We transform the local matrix of fixed-end actions of each element of the structure to global. The local axes of the elements of the frame are shown in Fig. d. Referring to this figure we form the transformation matrix for each element of the frame [see relations (6.6) and (6.48)], and we substitute it in relation (6.65) to obtain its global matrix of fixed-end actions. That is,

$$\{\bar{A}^{R1}\} = \left\{\frac{\{\bar{A}^{R1j}\}}{\{\bar{A}^{R1k}\}}\right\} = \begin{bmatrix} 0.6 & -0.8 & 0 & 0 & 0 & 0 \\ 0.8 & 0.6 & 0 & 0 & 0 & 0 \\ 0 & 0 & 1 & 0 & 0 & 0 \\ 0 & 0 & 0 & 0.6 & -0.8 & 0 \\ 0 & 0 & 0 & 0.8 & 0.6 & 0 \\ 0 & 0 & 0 & 0 & 0 & 1 \end{bmatrix} \begin{Bmatrix} 0 \\ 30 \\ -55 \\ \hline 0 \\ 30 \\ 55 \end{Bmatrix} = \begin{Bmatrix} -24 \\ 18 \\ -55 \\ \hline -24 \\ 18 \\ 55 \end{Bmatrix}$$

$$\{\bar{A}^{R2}\} = \{\bar{A}^{R1}\} \tag{u}$$

STEP 2.   We establish the matrix of equivalent actions. To accomplish this we do the following.

1. We establish the matrix of the restraining actions $\{\widehat{RA}\}$. That is, referring to Fig. i from the equilibrium of the nodes of the restrained structure we have

$$\{\widehat{RA}\} = \left\{\begin{array}{c} \{\bar{A}^{R1j}\} \\ \hline \{\bar{A}^{R1k}\} + \{\bar{A}^{R2j}\} \\ \hline \{\bar{A}^{R2k}\} \end{array}\right\} = \begin{Bmatrix} -24 \\ 18 \\ -55 \\ \hline -24 \\ 58 \\ 25 \\ \hline 0 \\ 40 \\ 30 \end{Bmatrix} \tag{v}$$

This result can be verified by referring to the free-body diagrams of the nodes of the restrained structure shown in Fig. j and considering their equilibrium.

2. We form the matrix of the given actions $\{\hat{P}^G\}$ acting on the nodes of the structure. Referring to Fig. a we have

$$\{\hat{P}^G\}^T = [0 \quad 0 \quad 0 \quad 0 \quad -60 \quad 0 \quad 0 \quad 0 \quad 0] \tag{w}$$

3. We establish the matrix of the equivalent actions $\{\hat{P}^E\}$. Substituting relations (v) and (w) into (7.2) we obtain

$$\{\hat{P}^E\} = \{\hat{P}^G\} - \widehat{\{RA\}} = \left\{ \begin{array}{c} 24 \\ -18 \\ 55 \\ \hline 24 \\ -118 \\ -25 \\ \hline 0 \\ -40 \\ -30 \end{array} \right\} \tag{x}$$

The structure subjected to the equivalent actions is shown in Fig. k.

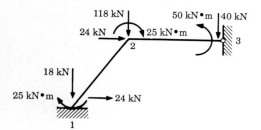

Figure k Structure subjected to the equivalent actions.

STEPS 3 TO 6.   The stiffness equations for the frame of Fig. k are obtained from Eqs. (f) by replacing the matrix of equivalent actions by relation (x). Thus the matrices $\{R\}$, $[S^{FF}]$, and $[S^{EF}]$ for the frame of Fig. k are given by relations (j), (l), and (m), respectively, while the matrices $\{P^{EF}\}$ and $\{P^{ES}\}$ are obtained from relation (x). That is,

$$\{P^{EF}\} = \left\{ \begin{array}{c} 24 \\ -118 \\ -25 \\ -30 \end{array} \right\} \qquad \{P^{ES}\} = \left\{ \begin{array}{c} 24 \\ -18 \\ 55 \\ 0 \\ -40 \end{array} \right\} \tag{y}$$

Moreover,

$$\{\Delta^S\} = \left\{ \begin{array}{c} 0 \\ -0.020 \\ 0 \\ 0 \end{array} \right\} \tag{z}$$

Furthermore, from the stiffness matrix of relation (g) we have

$$[S^{SS}] = \frac{EI}{125} \begin{bmatrix} 367.68 & 474.24 & -24 & 0 & 0 \\ 474.24 & 644.32 & 18 & 0 & 0 \\ -24.00 & 18.00 & 100 & 0 & 0 \\ 0 & 0 & 0 & 1000 & 0 \\ 0 & 0 & 0 & 0 & 12 \end{bmatrix} \qquad \text{(aa)}$$

Substituting relations (l), (m), (y), and (z) into relation (9.2) we obtain

$$\{\Delta^F\} = [S^{FF}]^{-1}[\{P^{EF}\} - [S^{FS}]\{\Delta^S\}] = \frac{125}{EI} \begin{bmatrix} 1367.68 & 474.24 & 24 & 0 \\ 474.24 & 656.32 & 12 & 30 \\ 24.00 & 12.00 & 200 & 50 \\ 0 & 30.00 & 50 & 100 \end{bmatrix}^{-1}$$

$$\times \left\{ \begin{array}{c} 24 \\ -118 \\ -25 \\ -30 \end{array} \right\} - \frac{EI}{125} \begin{bmatrix} -367.68 & -474.24 & 24 & -1000 & 0 \\ -474.24 & -644.32 & -18 & 0 & -12 \\ -24.00 & 18.00 & 50 & 0 & -30 \\ 0 & 0 & 0 & 0 & -30 \end{bmatrix} \left\{ \begin{array}{c} 0 \\ -0.020 \\ 0 \\ 0 \\ 0 \end{array} \right\}$$

$$= \frac{125}{EI} \left[\left\{ \begin{array}{c} 0.103735 \\ -0.244736 \\ -0.075565 \\ -0.188797 \end{array} \right\} - \left\{ \begin{array}{c} 0.058791 \\ 13.329256 \\ -1.161961 \\ -3.417797 \end{array} \right\} \right] = \frac{125}{EI} \left\{ \begin{array}{c} 0.04494 \\ -13.57399 \\ 1.08640 \\ 3.22900 \end{array} \right\} \qquad \text{(bb)}$$

Moreover, substituting relations (y), (aa), and (bb) into relation (9.3) we get

$$\left\{ \begin{array}{c} R_1^{(1)} \\ R_2^{(1)} \\ R_3^{(1)} \\ R_1^{(3)} \\ R_2^{(3)} \end{array} \right\} + \left\{ \begin{array}{c} 24 \\ -18 \\ 55 \\ 0 \\ -40 \end{array} \right\} = [S^{SF}]\{\Delta^F\} + [S^{SS}]\{\Delta^s\}$$

$$= \begin{bmatrix} -367.68 & -474.24 & -24 & 0 \\ -474.24 & -644.32 & 18 & 0 \\ 24 & -18 & 50 & 0 \\ -1000 & 0 & 0 & 0 \\ 0 & -12 & -30 & -30 \end{bmatrix} \left\{ \begin{array}{c} 0.04494 \\ -13.57399 \\ 1.08640 \\ 3.22900 \end{array} \right\}$$

$$+ \frac{EI}{125} \begin{bmatrix} 367.68 & 474.24 & -24 & 0 & 0 \\ 474.24 & 644.32 & 18 & 0 & 0 \\ -24 & 18 & 100 & 0 & 0 \\ 0 & 0 & 0 & 1000 & 0 \\ 0 & 0 & 0 & 0 & 12 \end{bmatrix} \left\{ \begin{array}{c} 0 \\ -0.02 \\ 0 \\ 0 \\ 0 \end{array} \right\}$$

$$
= \left\{\begin{array}{r} 6394.732 \\ 8744.236 \\ 299.730 \\ -44.940 \\ 33.426 \end{array}\right\} + \left\{\begin{array}{r} -6373.786 \\ -8659.661 \\ -241.920 \\ 0 \\ 0 \end{array}\right\} = \left\{\begin{array}{r} 20.946 \\ 84.575 \\ 57.810 \\ -44.940 \\ 33.426 \end{array}\right\}
$$

Thus
$$
\left\{\begin{array}{c} R_1^{(1)} \\ R_2^{(1)} \\ R_3^{(1)} \\ R_1^{(3)} \\ R_2^{(3)} \end{array}\right\} = \left\{\begin{array}{r} -3.05 \\ 102.57 \\ 2.81 \\ -44.94 \\ 73.43 \end{array}\right\}
\tag{cc}
$$

STEP 7. We compute the internal actions in the elements of the frame. Referring to Fig. f and relation (bb), the global components of the displacements of the ends of the elements of the frame are

$$
\{\bar{D}^1\} = \left\{\begin{array}{c} 0 \\ \Delta_2 \\ 0 \\ \hline \Delta_4 \\ \Delta_5 \\ \Delta_6 \end{array}\right\} = \frac{125}{EI}\left\{\begin{array}{c} 0 \\ -\dfrac{EI}{125}(0.02) \\ 0 \\ \hline 0.04494 \\ -13.57399 \\ 1.08640 \end{array}\right\} = \frac{125}{EI}\left\{\begin{array}{c} 0 \\ -13.44000 \\ 0 \\ \hline 0.04494 \\ -13.57399 \\ 1.08640 \end{array}\right\}
\tag{dd}
$$

$$
\{\bar{D}^2\} = \left\{\begin{array}{c} \Delta_4 \\ \Delta_5 \\ \Delta_6 \\ \hline 0 \\ 0 \\ \Delta_9 \end{array}\right\} = \frac{125}{EI}\left\{\begin{array}{c} 0.04494 \\ -13.57399 \\ 1.08640 \\ \hline 0 \\ 0 \\ 3.22900 \end{array}\right\}
$$

The hybrid stiffness matrices of the elements of the frame are given by relations (q). Substituting these relations into (9.13) we obtain

$$
\{A^1\} = \left[\begin{array}{rrrrrr} 600.0 & 800.0 & 0 & -600.0 & -800.0 & 0 \\ -9.6 & 7.2 & 30 & 9.6 & -7.2 & 30 \\ 24.0 & 18.0 & 100 & 24.0 & 18.0 & 50 \\ -600.0 & -800.0 & 0 & 600.0 & 800.0 & 0 \\ 9.6 & -7.2 & -30 & 9.6 & 7.2 & -30 \\ -24.0 & 18.0 & 50 & 24.0 & -18.0 & 100 \end{array}\right]\left\{\begin{array}{c} 0 \\ \dfrac{0.02EI}{125} \\ 0 \\ \hline 0.04494 \\ -13.57399 \\ 1.08640 \end{array}\right\}
$$

$$
+ \left\{ \begin{array}{c} 0 \\ 30 \\ -55 \\ \hline 0 \\ 30 \\ 55 \end{array} \right\} = \left\{ \begin{array}{c} 80.24 \\ 63.99 \\ 2.82 \\ \hline -80.24 \\ -3.99 \\ 167.15 \end{array} \right\}
$$

(ee)

$$
\{A^2\} = \begin{bmatrix} 1000 & 0 & 0 & -1000 & 0 & 0 \\ 0 & 12 & 30 & 0 & -12 & 30 \\ 0 & 30 & 100 & 0 & -30 & 50 \\ -1000 & 0 & 0 & 1000 & 0 & 0 \\ 0 & -12 & -30 & 0 & 12 & -30 \\ 0 & 30 & 50 & 0 & -30 & 100 \end{bmatrix} \left\{ \begin{array}{c} 0.04494 \\ -13.5740 \\ 1.0864 \\ \hline 0 \\ 0 \\ 3.2290 \end{array} \right\} + \left\{ \begin{array}{c} 0 \\ 40 \\ -30 \\ \hline 0 \\ 40 \\ 30 \end{array} \right\}
$$

$$
= \left\{ \begin{array}{c} 44.94 \\ 6.57 \\ -187.13 \\ \hline -44.94 \\ 73.43 \\ 0 \end{array} \right\}
$$

9.4   Problems

1 and 2.   Consider the truss subjected to the external forces and supported as shown in Fig. 9.P1. The elements of the truss are made from an isotropic

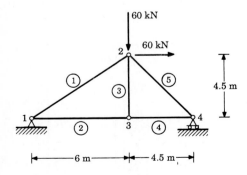

Figure 9.P1

linearly elastic material $(E = 210 \text{ kN/mm}^2)$ and their areas are given below.

Element	Area
1	$6 \times 10^3 \text{ mm}^2$
2, 4, 5	$4 \times 10^3 \text{ mm}^2$
3	$2 \times 10^3 \text{ mm}^2$

The stiffness matrix for the truss is assembled in Problem 1 at the end of Chap. 8. Compute

(a) The components of translation of the nodes of the truss
(b) The reactions of the supports of the truss
(c) The internal forces in the elements of the truss

Repeat with the same truss subjected to the external forces and supported as shown in Fig. 9.P2.

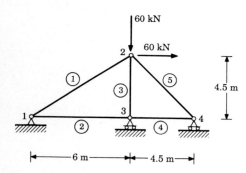

Figure 9.P2

3 to 6. Consider the beam subjected to the external actions and supported as shown in Fig. 9.P3. The elements of the beam are made from an isotropic linearly elastic material $(E = 210 \text{ kN/mm}^2)$ and the areas and moments of inertia of their cross sections are

$$I^{(1)} = 117.7 \times 10^6 \text{ mm}^4 \qquad I^{(2)} = 369.70 \times 10^6 \text{ mm}^4 \qquad I^{(3)} = 83.60 \times 10^6 \text{ mm}^4$$
$$A_1 = 6.26 \times 10^3 \text{ mm}^2 \qquad A^{(2)} = 13.2 \times 10^3 \text{ mm}^2 \qquad A_3 = 5.38 \times 10^3 \text{ mm}^2$$

The stiffness matrix for the beam is assembled in Problem 2 at the end of Chap. 8. Compute

(a) The components of displacements of the nodes of the beam
(b) The reactions of the supports of the beam
(c) The nodal actions of the elements of the beam

Repeat with the same beam subjected to the external actions and supported as shown in Figs. 9.P4 to 9.P6.

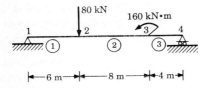

Figure 9.P3

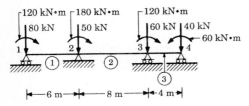

Figure 9.P4

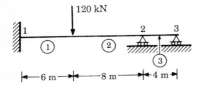

Figure 9.P5

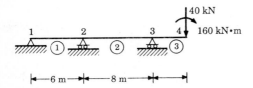

Figure 9.P6

7 and 8. Consider the frame subjected to the external actions and supported as shown in Fig. 9.P7. The elements of the frame are made from the same isotropic linearly elastic material ($E = 210$ kN/mm^2) and the areas and moments of inertia of their cross sections are

$$I^{(1)} = 117.7 \times 10^6 \text{ mm}^4 \qquad A_1 = 6.26 \times 10^3 \text{ mm}^3$$

$$I^{(2)} = 83.6 \times 10^6 \text{ mm}^4 \qquad A_2 = 5.38 \times 10^3 \text{ mm}^2$$

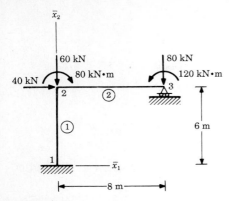

Figure 9.P7

The stiffness matrix for the frame is assembled in Problem 3 at the end of Chap. 8. Compute

    (a) The components of displacements of the nodes of the frame
    (b) The reactions of the supports of the frame
    (c) The nodal actions of the elements of the frame

Repeat with the same frame subjected to the external actions and supported as shown in Fig. 9.P8.

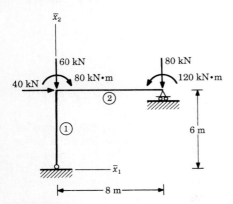

Figure 9.P8

9 and 10. Consider the frame subjected to the external actions and supported as shown in Fig. 9.P9. The elements of the frame are made from the same isotropic linearly elastic material ($E = 210 \text{ kN/mm}^2$) and the areas and moments of inertia of their cross sections are

$$I^{(1)} = 369.7 \times 10^6 \text{ mm}^4 \qquad A_1 = 13.2 \times 10^3 \text{ mm}^2$$

$$I^{(2)} = I^{(3)} = 117.7 \times 10^6 \text{ mm}^4 \qquad A_2 = A_3 = 6.26 \times 10^3 \text{ mm}^2$$

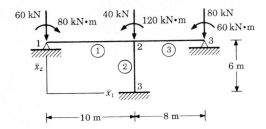

Figure 9.P9

The stiffness matrix for the frame is assembled in Problem 4 at the end of Chap. 8. Compute
(a) The components of displacements of the nodes of the frame
(b) The reaction of the supports of the frame
(c) The nodal actions of the elements of the frame

Repeat with the same frame subjected to the external actions and supported as shown in Fig. 9.P10.

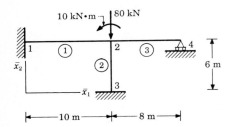

Figure 9.P10

11. Consider the frame subjected to the external actions and supported as shown in Fig. 9.P11. The elements of the frame are made from the same isotropic linearly elastic material $(E = 210 \text{ kN/mm}^2)$ and the areas and moments of inertia of their cross sections are

$$I^{(1)} = 117.7 \times 10^6 \text{ mm}^4 \qquad A_1 = 6.26 \times 10^3 \text{ mm}^2$$

$$I^{(2)} = 369.7 \times 10^6 \text{ mm}^4 \qquad A_2 = 13.2 \times 10^3 \text{ mm}^2$$

$$I^{(3)} = 83.6 \times 10^6 \text{ mm}^4 \qquad A_3 = 5.38 \times 10^3 \text{ mm}^2$$

The stiffness matrix for the frame is assembled in Problem 6 at the end of Chap. 8. Compute
(a) The components of displacements of the nodes of the frame
(b) The reactions of the supports of the frame
(c) The nodal actions of the elements of the frame

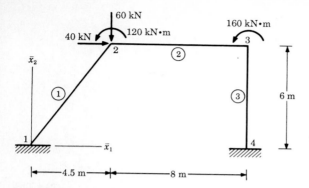

Figure 9.P11

12 and 13. Consider the frame subjected to the external actions and supported as shown in Fig. 9.P12. The elements of the frame are made from the same isotropic linearly elastic material ($E = 210 \text{ kN/mm}^2$) and the areas and moments of inertia of their cross sections are

$$I^{(1)} = 117.7 \times 10^6 \text{ mm}^4 \qquad A_1 = 6.26 \times 10^3 \text{ mm}^2$$

$$I^{(2)} = 369.7 \times 10^6 \text{ mm}^4 \qquad A_2 = 13.2 \times 10^3 \text{ mm}^2$$

$$I^{(3)} = 83.6 \times 10^6 \text{ mm}^4 \qquad A_3 = 5.38 \times 10^3 \text{ mm}^2$$

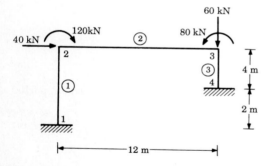

Figure 9.P12

The stiffness matrix for the frame is assembled in Problem 5 at the end of Chap. 8. Compute

    (a) The components of displacements of the nodes of the frame
    (b) The reactions of the supports of the frame
    (c) The nodal actions of the elements of the frame

Repeat with the same frame, subjected to the external actions and supported as shown in Fig. 9.P13.

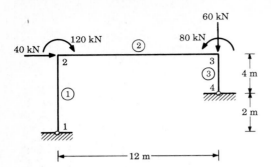

Figure 9.P13

14 and 15. Consider the space structure shown in the Fig. 9.P14. The elements of the structure are steel tubes ($E = 210 \text{ kN/mm}^2$, $v = 0.33$) of the same constant cross section ($d_{\text{ext}} = 80$ mm, $d_{\text{int}} = 60$ mm). Its stiffness matrix is assembled in Problem 7 at the end of Chap. 8. Compute

(a) The components of displacements of the nodes of the frame
(b) The reactions of the supports of the frame
(c) The nodal actions of the elements of the frame

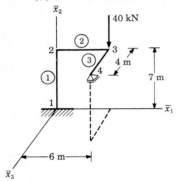

Figure 9.P14

Repeat with the same frame subjected to the external actions and supported as shown in Fig. 9.P15.

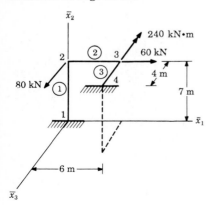

Figure 9.P15

16 and 17.   Consider the space structure shown in Fig. 9.P16. The elements of the structure are steel tubes ($E = 210\,\mathrm{kN/mm^2}$, $v = 0.33$) of the same constant cross section ($d_{\mathrm{ext}} = 80\,\mathrm{mm}$, $d_{\mathrm{int}} = 60\,\mathrm{mm}$). Its stiffness matrix is assembled in Problem 8 at the end of Chap. 8. Compute
(*a*) The components of displacements of the nodes of the frame
(*b*) The reactions of the supports of the frame
(*c*) The nodal actions of the elements of the frame

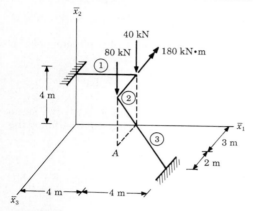

Figure 9.P16

Repeat with the same frame subjected to the external actions and supported as shown in Fig. 9.P17.

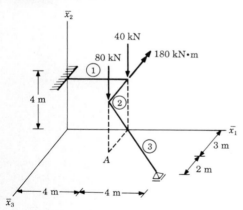

Figure 9.P17

18 and 19.   The beam shown in Fig. 9.P18 is subjected to the following loading:
(*a*) The external actions shown in the Fig. 9.P18
(*b*) Settlement of support 1 of 20 mm
(*c*) Temperature of upper fibers $T = 35°C$ and of lower fibers $T = -5°C$

The elements of the beam are made from an isotropic linearly elastic material ($E = 210 \text{ kN/mm}^2$, $\alpha = 10^{-5} \,^\circ\text{C}$) and the areas of their cross sections are

$$I^{(1)} = 117.7 \times 10^6 \text{ mm}^4 \qquad I^{(2)} = 369.70 \times 10^6 \text{ mm}^4 \qquad I^{(3)} = 83.60 \times 10^6 \text{ mm}^4$$

$$A_1 = 6.26 \times 10^3 \text{ mm}^2 \qquad A_2 = 13.2 \times 10^3 \text{ mm}^2 \qquad A_3 = 5.38 \times 10^3 \text{ mm}^2$$

The stiffness matrix for the beam is assembled in Problem 2 at the end of Chap. 8. Compute
     (a) The equivalent actions which must be applied on the nodes of the beam
     (b) The components of displacements of the nodes of the beam
     (c) The reactions of the supports of the beam
     (d) The nodal actions of the elements of the beam
Plot the shear and moment diagrams for the beam. Repeat with the same beam supported as shown in Fig. 9.P19 and subjected to the following loading:
     (a) The external forces shown in Fig. 9.P19
     (b) Settlement of support 2 of 20 mm
     (c) Temperature of upper fibers $T = 35^\circ\text{C}$ and of lower fibers $T = -5^\circ\text{C}$

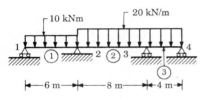

Figure 9.P18

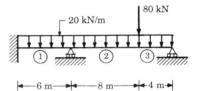

Figure 9.P19

20 and 21. For the beam subjected to the external actions shown in the Fig. 9.P20, compute
     (a) The components of displacements of its nodes
     (b) The reactions of its supports
     (c) The nodal actions of its elements

Repeat with the beam of Fig. 9.P21. The beams are made of steel ($E = 210 \text{ kN/mm}^2$) and have a constant cross section ($A = 6.26 \times 10^3 \text{ mm}^2$, $I = 117.7 \times 10^6 \text{ mm}^4$).

Figure 9.P20

Figure 9.P21

22 to 35. The frame shown in Fig. 9.P22 is subjected to the following loading:

    (*a*) The external actions shown in Fig. 9.P22.
    (*b*) Settlement of support of 20 mm.
    (*c*) Temperature of the upper or outside fibers $T_e = 35°C$ and of lower or inside fibers $T_i = -5°C$. The temperature during the construction of the frame was $T_0 = 15°C$.

Compute

    (*a*) The components of displacements of the nodes of the frame
    (*b*) The reactions of the supports of the frame
    (*c*) The nodal actions of the elements of the frame

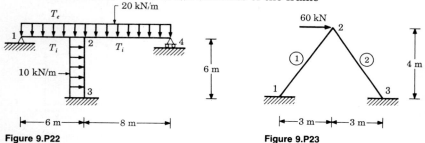

Figure 9.P22　　　　　　　　　　　　　　**Figure 9.P23**

Repeat with the frames of Figs. 9.P23 to 9.P35. The elements of the frames are made of steel $(E = 210\ kN/mm^2,\ \alpha = 10^{-5}°C)$ and unless otherwise indicated in the figures have the same constant cross section $(A = 13.2 \times 10^3\ mm^2,\ I = 369.7 \times 10^6\ mm^4,\ h = 425\ mm)$.

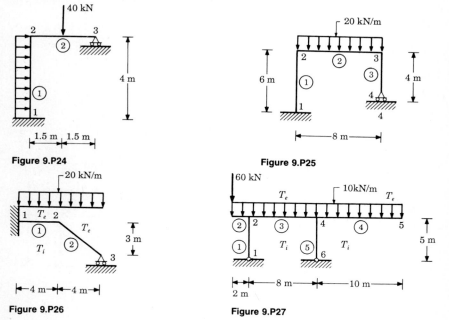

Figure 9.P24　　　　　　　　　　　**Figure 9.P25**

Figure 9.P26　　　　　　　　　　　**Figure 9.P27**

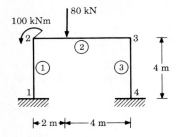

Figure 9.P28

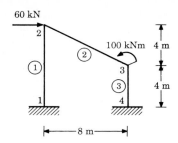

Figure 9.P29

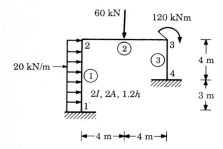

Figure 9.P30

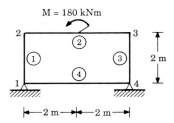

Figure 9.P31

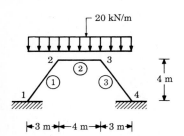

Figure 9.P32

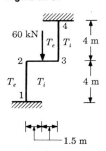

Figure 9.P33

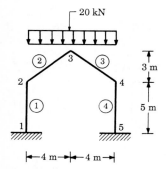

Figure 9.P34

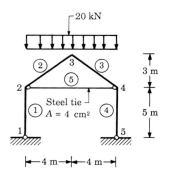

Figure 9.P35

36 to 40. The frame shown in Fig. 9.P36 is subjected to the following loading:

    (*a*) The external actions shown in Fig. 9.P36.
    (*b*) Settlement of support of 20 mm.
    (*c*) Temperature of the upper fibers $T_e = 35°C$ and of lower fibers $T_i = -5°C$. The temperature during the construction of the frame was $T_0 = 15°C$.

Compute:

    (*a*) The components of displacements of the nodes of the frame
    (*b*) The reactions of the supports of the frame
    (*c*) The nodal actions of the elements of the frame

Repeat with the frames of Figs. 9.P37 to 9.P40. The elements of the frames are steel tubes ($E = 210 \text{ kN/mm}^2$, $v = 0.33$, $\alpha = 10^{-5}°C$) and have a constant cross section ($d_{ext} = 80$ mm, $d_{int} = 60$ mm).

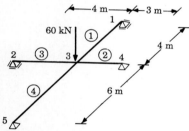

Figure 9.P36

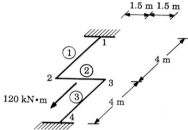

Figure 9.P37

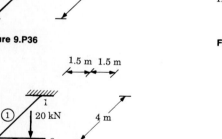

Figure 9.P38

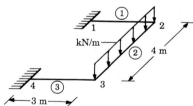

Figure 9.P39

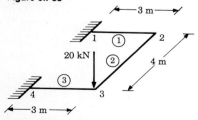

Figure 9.P40

References

1. A. E. Armenàkas, *Classical Structural Analysis: A Modern Approach,* McGraw-Hill, New York, 1988, Sec. 8.6.
2. See R. W. Southworth and S. L. De Leeuw, *Digital Computation and Numerical Methods,* McGraw-Hill, New York, 1965; and K.-J. Bathe and E. L. Wilson, *Numerical Methods in Finite Element Analysis,* Prentice-Hall, Englewood Cliffs, N.J., 1976.
3. A. E. Armenàkas, op. cit., Sec. 5.9.

10

Analysis of Structures Having Skew Supports or Other Constraints by the Direct Stiffness Method

10.1 Introduction

Consider the planar frame of Fig. 10.1a whose node 4 is on rollers. If we choose the $\bar{x}_1'$ and $\bar{x}_2'$ axes shown in Fig. 10.1a as the global axes for this frame, the direction normal to the plane of rolling of support 4 is parallel to global axis $\bar{x}_2'$. In this case the frame can be analyzed using the procedure described in Sec. 9.3. If, however, we choose the $\bar{x}_1$ and $\bar{x}_2$ axes shown in Fig. 10.1a as the global axes for the frame, the direction normal to the plane of rolling of support 4 is not parallel to a global axis of the frame, and in order to analyze it the procedure described in Sec. 9.3 must be modified.

We call the roller supports of structures and the ball supports of space structures, whose plane of rolling is not normal to one of the chosen global axes, *skew supports*. Notice that for structures with more than one roller or ball support it may not be possible to find a set of axes such that the planes of rolling of all their roller supports are normal to one of these axes. For example we cannot find such a set of axes for the structure of Fig. 10.1b. If we choose the $\bar{x}_1$ and $\bar{x}_2$ axes as the global axes, support 6 is a skew support, whereas if we choose the $\bar{x}_1'$ and $\bar{x}_2'$ axes as the global axes, support 4 is a skew support.

In this chapter we consider structures with one or more skew supports and present procedures for analyzing them. Moreover, in Sec. 10.3 we present a method for establishing the stiffness equations

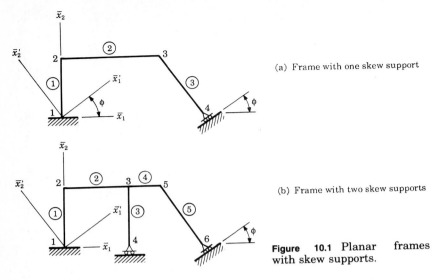

(a) Frame with one skew support

(b) Frame with two skew supports

Figure   10.1 Planar   frames with skew supports.

for framed structures subjected to other constraints in addition to their supports.

For a roller support of a planar or space structure the direction normal to the plane of rolling and the direction of rolling are specified. For a ball support of a space structure only the direction normal to the plane of rolling is specified. For each skew support of a structure, we choose an orthogonal system of axes $\bar{x}'_1$, $\bar{x}'_2$, and $\bar{x}'_3$ with the node connected to the skew support as the origin and the $\bar{x}'_2$ axis normal to the plane of rolling. Moreover, for a rolling support we choose the $\bar{x}'_1$ axis in the direction of rolling, while for a ball support of a space structure we choose the $\bar{x}'_1$ axis in any convenient direction. We call this system of axes the *skew axes* and the components of a force or a displacement in the direction of the skew axes the *skew components*. In order to specify the skew axes with respect to the global axes for planar structures we usually give the angle of inclination ϕ (see Fig. 10.1), while for space structures we usually give the coordinates of two points; one located on the $\bar{x}'_2$ axis and the other located somewhere in the $\bar{x}'_1\bar{x}'_2$ plane but not on the $\bar{x}'_2$ axis. From the coordinates of these points the direction cosines of the skew axes can be established. The transformation relations between the global $(\bar{x}_1, \bar{x}_2)$ and the skew $(\bar{x}'_1, \bar{x}'_2)$ coordinates of a planar structure can be written as

$$\{\bar{x}'_1\} = \begin{Bmatrix} \bar{x}'_1 \\ \bar{x}'_2 \end{Bmatrix} = [\Lambda'_P] \begin{Bmatrix} \bar{x}_1 \\ \bar{x}_2 \end{Bmatrix} \qquad (10.1a)$$

or

$$\{\bar{x}_1\} = \begin{Bmatrix} \bar{x}_1 \\ \bar{x}_2 \end{Bmatrix} = [\Lambda'_P]^T \begin{Bmatrix} \bar{x}'_1 \\ \bar{x}'_2 \end{Bmatrix} \qquad (10.1b)$$

where referring to Fig. 10.1 and relation (6.6) we have

$$[\Lambda_P'] = \begin{bmatrix} \cos\phi & \sin\phi \\ -\sin\phi & \cos\phi \end{bmatrix} \tag{10.2}$$

Referring to relations (B.26) and (B.27) the transformation relations between the global $\bar{x}_1$, $\bar{x}_2$, $\bar{x}_3$ and the skew $\bar{x}_1'$, $\bar{x}_2'$, $\bar{x}_3'$ coordinates of a space structure can be written as

$$\{\bar{x}'\} = \begin{Bmatrix} \bar{x}_1' \\ \bar{x}_2' \\ \bar{x}_3' \end{Bmatrix} = [\Lambda_S'] \begin{Bmatrix} \bar{x}_1 \\ \bar{x}_2 \\ \bar{x}_3 \end{Bmatrix} \tag{10.3a}$$

or

$$\{\bar{x}\} = \begin{Bmatrix} \bar{x}_1 \\ \bar{x}_2 \\ \bar{x}_3 \end{Bmatrix} = [\Lambda_S']^T \begin{Bmatrix} \bar{x}_1' \\ \bar{x}_2' \\ \bar{x}_3' \end{Bmatrix} \tag{10.3b}$$

where

$$[\Lambda_S'] = \begin{bmatrix} \lambda_{11}' & \lambda_{12}' & \lambda_{13}' \\ \lambda_{21}' & \lambda_{22}' & \lambda_{23}' \\ \lambda_{31}' & \lambda_{32}' & \lambda_{33}' \end{bmatrix} \tag{10.4}$$

and λ_{ij}' is the direction cosine of the axis $\bar{x}_i'$ with respect to the axis $\bar{x}_j$.

Consider the planar frame of Fig. 10.1a and assume that the x_1 and x_2 axes have been chosen as its global axes. In this case support 4 is a skew support. Referring to Fig. 10.2 the stiffness equations (8.1) for

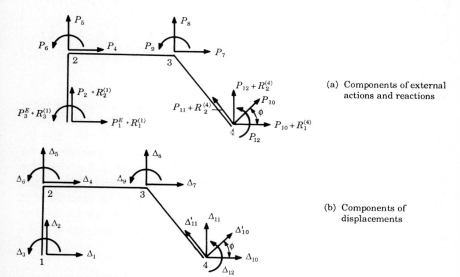

(a) Components of external actions and reactions

(b) Components of displacements

Figure 10.2 Components of external actions and reactions and components of displacements of the nodes of the frame.

this frame are

$$
\begin{Bmatrix}
P_1^E \\
P_2^E \\
P_3^E \\
P_4^E \\
P_5^E \\
P_6^E \\
P_7^E \\
P_8^E \\
P_9^E \\
P_{10}^E \\
P_{11}^E \\
P_{12}^E
\end{Bmatrix}
+
\begin{Bmatrix}
R^{(1)} \\
R^{(1)} \\
R^{(1)} \\
0 \\
0 \\
0 \\
0 \\
0 \\
0 \\
R_1^{(4)} \\
R_2^{(4)} \\
0
\end{Bmatrix}
$$

$$
=
\begin{bmatrix}
K_{11} & K_{12} & K_{13} & K_{14} & K_{15} & K_{16} & K_{17} & K_{18} & K_{19} & K_{1,10} & K_{1,11} & K_{1,12} \\
K_{21} & K_{22} & K_{23} & K_{24} & K_{25} & K_{26} & K_{27} & K_{28} & K_{29} & K_{2,10} & K_{2,11} & K_{2,12} \\
K_{31} & K_{32} & K_{33} & K_{34} & K_{35} & K_{36} & K_{37} & K_{38} & K_{39} & K_{3,10} & K_{3,11} & K_{3,12} \\
K_{41} & K_{42} & K_{43} & K_{44} & K_{45} & K_{46} & K_{47} & K_{48} & K_{49} & K_{4,10} & K_{4,11} & K_{4,12} \\
K_{51} & K_{52} & K_{53} & K_{54} & K_{55} & K_{56} & K_{57} & K_{58} & K_{59} & K_{5,10} & K_{5,11} & K_{5,12} \\
K_{61} & K_{62} & K_{63} & K_{64} & K_{65} & K_{66} & K_{67} & K_{68} & K_{69} & K_{6,10} & K_{6,11} & K_{6,12} \\
K_{71} & K_{72} & K_{73} & K_{74} & K_{75} & K_{76} & K_{77} & K_{78} & K_{79} & K_{7,10} & K_{7,11} & K_{7,12} \\
K_{81} & K_{82} & K_{83} & K_{84} & K_{85} & K_{86} & K_{87} & K_{88} & K_{89} & K_{8,10} & K_{8,11} & K_{8,12} \\
K_{91} & K_{92} & K_{93} & K_{94} & K_{95} & K_{96} & K_{97} & K_{98} & K_{99} & K_{9,10} & K_{9,11} & K_{9,12} \\
K_{10,1} & K_{10,2} & K_{10,3} & K_{10,4} & K_{10,5} & K_{10,6} & K_{10,7} & K_{10,8} & K_{10,9} & K_{10,10} & K_{10,11} & K_{10,12} \\
K_{11,1} & K_{11,2} & K_{11,3} & K_{11,4} & K_{11,5} & K_{11,6} & K_{11,7} & K_{11,8} & K_{11,9} & K_{11,10} & K_{11,11} & K_{11,12} \\
K_{12,1} & K_{12,2} & K_{12,3} & K_{12,4} & K_{12,5} & K_{12,6} & K_{12,7} & K_{12,8} & K_{12,9} & K_{12,10} & K_{12,11} & K_{12,12}
\end{bmatrix}
\begin{Bmatrix}
\Delta_1 \\
\Delta_2 \\
\Delta_3 \\
\Delta_4 \\
\Delta_5 \\
\Delta_6 \\
\Delta_7 \\
\Delta_8 \\
\Delta_9 \\
\Delta_{10} \\
\Delta_{11} \\
\Delta_{12}
\end{Bmatrix}
$$

$$(10.5)$$

Each row of Eqs. (10.5) represents an equation of equilibrium for a node of the frame in a global direction. For example the fifth row of Eqs. (10.5) represents the equations of equilibrium for the $\bar{x}_2$ component of the forces acting on node 2 ($\sum \bar{F}_2 = 0$). However, we know that at support 4 the component of the reacting force in the direction of rolling vanishes. In order to introduce this information into relation (10.5) we must convert rows 10 and 11 to equations of equilibrium in the direction of rolling and in the direction normal to it. Moreover, referring to Fig. 10.2b the boundary conditions for the frame are

$$
\Delta_1 = \Delta_2 = \Delta_3 = 0
$$
$$
\Delta_{11}' = 0
$$

$$(10.6)$$

In order to introduce the last boundary condition (10.6) into relation (10.4), the global components of displacement Δ_{10} and Δ_{11} in the matrix $\{\hat{\Delta}\}$ must be replaced with the components of displacement in the direction of rolling (Δ'_{10}) and in the direction normal to it (Δ'_{11}), respectively. This is accomplished by modifying columns 10 and 11 of the stiffness matrix for the structure. We write the modified stiffness equations for a structure with skew supports as follows:

$$\{\hat{P}'^E\} + \{\hat{R}'\} = [\hat{S}']\{\hat{\Delta}'\} \tag{10.7}$$

where $\{\hat{P}'^E\}$ = modified matrix of equivalent actions for structure
$\{\hat{\Delta}'\}$ = modified matrix of components of displacements of nodes of structure
$[\hat{S}']$ = modified stiffness matrix of structure

The modified stiffness equations (10.7) for a structure with skew supports can be solved to obtain the components of displacements of its nodes and its reactions following the procedure described in Sec. 9.3. Notice, however, that the solution gives the skew components of translation and the skew components of reacting forces of skew supports.

In certain structures there are internal action release mechanisms which allow the nodes adjacent to them to translate (roll) relative to each other in one direction. Moreover, in certain space structures there are internal action release mechanisms which allow the nodes adjacent to them to translate relative to each other in one plane. For example, the internal action release mechanism of the frame of Fig. 10.3 allows node 3 to translate relative to node 2 in the direction of element 1. In order to analyze this structure using the procedure

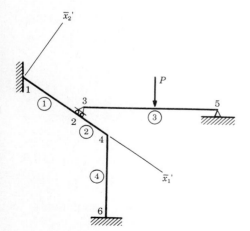

Figure 10.3 Frame with an internal action release mechanism.

described in Sec. 9.3 we must choose the axes $\bar{x}_1'$ and $\bar{x}_2'$ shown in Fig. 10.3 as the global axes, because with this choice the relative translation of nodes 2 and 3 is in a global direction. If none of the chosen global axes for the structure of Fig. 10.3 is in the direction of the relative translation of nodes 2 and 3 in order to analyze this structure we must use a procedure similar to the one described in the next section for structures with skew supports. Notice that for structures which have more than one internal action release mechanism of the type described previously it may not be possible to find a set of axes such that the relative motion of the nodes adjacent to each of these mechanisms is in the direction of one of these axes. In this case these structures must be analyzed using a procedure similar to the one described in the next section for structures with skew supports.

10.2 Modification of the Stiffness Equations for Structures Having Skew Supports

In this section we present two procedures for modifying the stiffness equations for structures having skew supports, that is, modifying their matrix of equivalent actions $\{\hat{P}^E\}$ and their stiffness matrix $[\hat{S}]$. Moreover, we apply these procedures to the analysis of a truss having a skew support.

Consider a framed structure having a node (say node n) connected to a skew support. Its modified matrices of equivalent actions $\{\hat{P}'^E\}$, of reactions $\{\hat{R}'\}$, and of displacements of its nodes $\{\hat{\Delta}'\}$ can be partitioned as follows

$$\{\hat{P}'^E\} = \left\{ \begin{array}{c} \{\bar{P}^{E(A)}\} \\ \{\bar{P}'^{E(n)}\} \\ \{\bar{P}^{E(B)}\} \end{array} \right\} \tag{10.8a}$$

$$\{\hat{R}'\} = \left\{ \begin{array}{c} \{\bar{R}^{(A)}\} \\ \{\bar{R}'^{(n)}\} \\ \{\bar{R}^{(B)}\} \end{array} \right\} \tag{10.8b}$$

$$\{\hat{\Delta}'\} = \left\{ \begin{array}{c} \{\bar{\Delta}^{(A)}\} \\ \{\bar{\Delta}'^{(n)}\} \\ \{\bar{\Delta}^{(B)}\} \end{array} \right\} \tag{10.8c}$$

where $\{\bar{P}'^{E(n)}\}$ = matrix whose terms are skew components of equivalent force acting on node n

$\{\bar{P}^{E(A)}\} =$ matrix whose terms are global components of equivalent actions acting on nodes 1 to $(n-1)$

$\{\bar{P}^{E(B)}\} =$ matrix whose terms are global components of equivalent moment acting on node n and of equivalent actions acting on node $(n+1)$ to last node of structure

$\{\bar{R}'^{(n)}\} =$ matrix whose terms are skew components of reacting force acting on node n

$\{\bar{R}^{(A)}\} =$ matrix whose terms are global components of reactions (forces and moments) acting on nodes 1 to $(n-1)$

$\{\bar{R}^{(B)}\} =$ matrix whose terms are global components of reacting moment acting on node n and of reacting actions (forces and moments) acting on nodes $(n+1)$ to last node of structure

$\{\bar{\Delta}'^{(n)}\} =$ matrix whose terms are skew components of translation of node n

$\{\bar{\Delta}^{(A)}\} =$ matrix whose terms are global components of displacements (translations and rotations) of nodes 1 to $(n-1)$

$\{\bar{\Delta}^{(B)}\} =$ matrix whose terms are global components of rotation of node n and of displacements (translations and rotations) of nodes $(n+1)$ to last node of structure

The matrices $\{\bar{P}'^{E(n)}\}$, $\{\bar{R}'^{(n)}\}$, and $\{\bar{\Delta}'^{(n)}\}$ are related to the matrices of global components of the force, the reacting force, and the translation of node n, respectively, by the following relations:

$$\{\bar{P}'^{E(n)}\} = [\Lambda']\{\bar{P}^{E(n)}\} \qquad (10.9a)$$

$$\{\bar{R}'^{(n)}\} = [\Lambda']\{\bar{R}^{(n)}\} \qquad (10.9b)$$

$$\{\bar{\Delta}'^{(n)}\} = [\Lambda']\{\bar{\Delta}^{(n)}\} \qquad (10.9c)$$

and

$$\{\bar{P}^{E(n)}\} = [\Lambda']^{T}\{\bar{P}'^{E(n)}\} \qquad (10.10a)$$

$$\{\bar{R}^{(n)}\} = [\Lambda']^{T}\{\bar{R}'^{(n)}\} \qquad (10.10b)$$

$$\{\bar{\Delta}^{(n)}\} = [\Lambda']^{T}\{\bar{\Delta}'^{(n)}\} \qquad (10.10c)$$

$[\Lambda']$ is either $[\Lambda'_P]$ [see Eq. (10.2)] or $[\Lambda'_S]$ [see Eq. (10.4)] depending on whether the structure under consideration is planar or space.

On the basis of the foregoing discussion the modified matrix of equivalent actions $\{\hat{P}'^E\}$ for the structure under consideration can be established as follows:

1. From the given loads we compute the global components of equivalent actions.

2. Using relation (10.9a) we transform the matrix $\{\bar{P}^{E(n)}\}$ to $\{\bar{P}'^{E(n)}\}$.

3. We assemble the matrix $\{\hat{P}'^E\}$ on the basis of relation (10.8a).

In order to establish the modified stiffness matrix for a structure we use one of the two procedures described below.

10.2.1   Procedure for establishing the modified stiffness matrix for a structure with skew supports

In this procedure we adhere to the following steps.

Step 1.   We establish the global stiffness matrix for each element of the structure which is not connected to a skew support and the modified global stiffness matrix for each element of the structure which is connected to a skew support.

Step 2.   We use the global or the modified global stiffness matrices of the elements of the structure established in step 1 to assemble the modified stiffness matrix of the structure by adhering to one of the procedures described in Sec. 8.3.

In what follows we establish the modified global stiffness matrix of an element connected to a skew support. Suppose end q ($q = j$ or k) of element e of a framed structure is connected to a skew support. The components in the direction of the $\bar{x}'_1$, $\bar{x}'_2$, and $\bar{x}'_3$ axes (skew components) of the action and the displacement at the end q ($q = j$ or k) of element e are related to their global components as follows:

$$\{\bar{A}'^{eq}\} = [\Lambda_{SKG}]\{\bar{A}^{eq}\} \qquad q = j \text{ or } k \qquad (10.11)$$

$$\{\bar{D}'^{eq}\} = [\Lambda_{SKG}]\{\bar{D}^{eq}\} \qquad q = j \text{ or } k \qquad (10.12)$$

The matrix $\{\bar{A}^{eq}\}$ contains the global components of nodal action at the end q ($q = j$ or k) of element e while the matrix $\{\bar{D}^{eq}\}$ contains the global components of displacement at the end q of element e. Moreover, for a planar truss

$$[\Lambda_{SKG}] = [\Lambda'_P] \qquad (10.13a)$$

where $[\Lambda'_P]$ is given by relation (10.2). For a planar beam or frame

$$[\Lambda_{SKG}] = \begin{bmatrix} [\Lambda'_P] & [0] \\ [0] & 1 \end{bmatrix} \qquad (10.13b)$$

For a space truss

$$[\Lambda_{\text{SKG}}] = [\Lambda'_S] \qquad (10.13c)$$

where $[\Lambda'_S]$ is given by relation (10.4). For a space beam or frame

$$[\Lambda_{\text{SKG}}] = \begin{bmatrix} [\Lambda'_S] & [0] & [0] & [0] \\ [0] & 1 & 0 & 0 \\ [0] & 0 & 1 & 0 \\ [0] & 0 & 0 & 1 \end{bmatrix} \qquad (10.13d)$$

Thus, the modified matrices of nodal actions $\{\bar{A}'^e\}^q$ and of nodal displacement $\{\bar{D}'^e\}^q$ of element e whose end q $(q = j$ or $k)$ is connected to a skew support are

$$\{\bar{A}'^e\}^j = \begin{bmatrix} \{\bar{A}'^{ej}\} \\ \{\bar{A}^{ek}\} \end{bmatrix} = [\Lambda_{\text{SKG}}]^j \begin{bmatrix} \{\bar{A}^{ej}\} \\ \{\bar{A}^{ek}\} \end{bmatrix} = [\Lambda_{\text{SKG}}]^j \{\bar{A}^e\} \quad (10.14a)$$

$$\{\bar{A}'^e\}^k = \begin{bmatrix} \{\bar{A}^{ej}\} \\ \{\bar{A}'^{ek}\} \end{bmatrix} = [\Lambda_{\text{SKG}}]^k \begin{bmatrix} \{\bar{A}^{ej}\} \\ \{\bar{A}^{ek}\} \end{bmatrix} = [\Lambda_{\text{SKG}}]^k \{\bar{A}^e\} \quad (10.14b)$$

and

$$\{\bar{D}'^e\}^j = \begin{bmatrix} \{\bar{D}'^{ej}\} \\ \{\bar{D}^{ek}\} \end{bmatrix} = [\Lambda_{\text{SKG}}]^j \begin{bmatrix} \{\bar{D}^{ej}\} \\ \{\bar{D}^{ek}\} \end{bmatrix} = [\Lambda_{\text{SKG}}]^j \{\bar{D}^e\} \quad (10.15a)$$

$$\{\bar{D}'^e\}^k = \begin{bmatrix} \{\bar{D}^{ej}\} \\ \{\bar{D}'^{ek}\} \end{bmatrix} = [\Lambda_{\text{SKG}}]^k \begin{bmatrix} \{\bar{D}^{ej}\} \\ \{\bar{D}^{ek}\} \end{bmatrix} = [\Lambda_{\text{SKG}}]^k \{\bar{D}^e\} \quad (10.15b)$$

where

$$[\Lambda_{\text{SKG}}]^j = \begin{bmatrix} [\Lambda_{\text{SKG}}] & [0] \\ [0] & [I] \end{bmatrix} \qquad (10.16a)$$

and

$$[\Lambda_{\text{SKG}}]^k = \begin{bmatrix} [I] & [0] \\ [0] & [\Lambda_{\text{SKG}}] \end{bmatrix} \qquad (10.16b)$$

Substituting relations (6.65) and (6.68) into (10.14) and (10.15), respectively, we obtain

$$\{\bar{A}'^e\}^q = [\Lambda^e_{\text{SKL}}]^q \{\bar{A}^e\} \qquad q = j \text{ or } k \qquad (10.17)$$

$$\{\bar{D}'^e\}^q = [\Lambda^e_{\text{SKL}}]^q \{\bar{D}^e\} \qquad q = j \text{ or } k \qquad (10.18)$$

where

$$[\Lambda^e_{\text{SKL}}]^q = [\Lambda_{\text{SKG}}]^q [\Lambda^e]^T \qquad (10.19)$$

$[\Lambda^e]$ is the transformation matrix for element e. For an element of a planar truss, a planar beam or frame, a space truss, and a space frame it is equal to $[\Lambda^e_{PT}]$, $[\Lambda^e_{PF}]$, $[\Lambda^e_{ST}]$, and $[\Lambda^e_{SF}]$ given by relations (6.42), (6.48), (6.51), and (6.57), respectively.

The work of the nodal actions of element e due to its deformation is equal to

$$W = \tfrac{1}{2}\{\{\bar{A}'^e\}^q\}^T \{\bar{D}'^e\}^q \qquad (10.20a)$$

or
$$W = \tfrac{1}{2}\{A^e\}^T \{D^e\} \qquad (10.20b)$$

Substituting relations (10.14) and (10.15) into (10.20a) we obtain

$$W = \tfrac{1}{2}\{[\Lambda^e_{\text{SKL}}]^q \{A^e\}\}^T [\Lambda^e_{\text{SKL}}]^q \{D^e\}$$

$$= \tfrac{1}{2}\{A^e\}^T [[\Lambda^e_{\text{SKL}}]^q]^T [\Lambda^e_{\text{SKL}}]^q \{D^e\} \qquad (10.21)$$

Comparing relations (10.20b) and (10.21) we get

$$[[\Lambda^e_{\text{SKL}}]^q]^T [\Lambda^e_{\text{SKL}}]^q = [I] \qquad (10.22)$$

or
$$[[\Lambda^e_{\text{SKL}}]^q]^{-1} = [[\Lambda^e_{\text{SKL}}]^q]^T \qquad (10.23)$$

Consequently,

$$\{A^e\} = [[\Lambda^e_{\text{SKL}}]^q]^T \{\bar{A}'^e\}^q \qquad q = j \text{ or } k \qquad (10.24)$$

$$\{D^e\} = [[\Lambda^e_{\text{SKL}}]^q]^T \{\bar{D}'^e\}^q \qquad q = j \text{ or } k \qquad (10.25)$$

Substituting relations (10.25) into (4.8) we obtain

$$\{A^{Ee}\} = [\overset{\Box}{K}'^e]^q \{\bar{D}'^e\}^q \qquad (10.26)$$

where $[\overset{\Box}{K}'^e]^q$ is the *modified hybrid stiffness matrix for element e* defined as

$$[\overset{\Box}{K}'^e]^q = [K^e][[\Lambda^e_{\text{SKL}}]^q]^T \qquad (10.27)$$

Substituting Eq. (10.24) into (10.26) we get

$$[[\Lambda^e_{\text{SKL}}]^q]^T \{\bar{A}'^e\}^q = [K^e][[\Lambda^e_{\text{SKL}}]^q]^T \{D'^e\}^q \qquad q = j \text{ or } k \quad (10.28)$$

Multiplying both sides of relation (10.28) by $[\Lambda^e_{\text{SKL}}]^q$ and using (10.22) we obtain

$$\{\bar{A}'^e\}^q = [\bar{K}'^e]^q \{\bar{D}'^e\}^q \qquad q = j \text{ or } k \qquad (10.29)$$

where $[\bar{K}'^e]^q$ is the modified global stiffness matrix for element e whose end q ($q = j$ or k) is connected to a skew support. It is obtained from the local stiffness matrix of the element using the following relation:

$$[\bar{K}'^e]^q = [\Lambda^e_{\text{SKL}}]^q [K^e][[\Lambda^e_{\text{SKL}}]^q]^T \qquad (10.30)$$

In what follows we apply the procedure described in this subsection to a truss with a skew support.

Example. Using the direct stiffness method, compute the components of displacements of the nodes of the truss shown in Fig. a. Moreover, compute the reactions of the truss and the internal forces in its elements. The elements of the truss are made of the same isotropic linearly elastic material and have the same constant cross section $(AE = 20{,}000 \text{ kN})$.

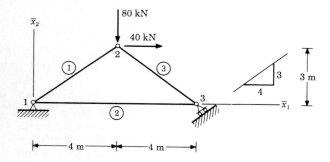

Figure a Geometry and loading of the truss.

---

solution

STEPS 1 AND 2. We establish the matrix of equivalent actions. Referring to Fig. a we see that there are no external forces acting on the skew support. Thus, the modified matrix of equivalent actions $\{\hat{P}'^E\}$ for the truss is equal to its matrix of equivalent actions. Hence, referring to Figs. a and b we have

$$[\hat{P}^E] = [\hat{P}'^E] = \begin{Bmatrix} P_1 \\ P_2 \\ P_3 \\ P_4 \\ P'_5 \\ P'_6 \end{Bmatrix} = \begin{Bmatrix} 0 \\ 0 \\ 40 \\ -80 \\ 0 \\ 0 \end{Bmatrix} \qquad (a)$$

STEP 3. We compute the local stiffness matrix for each element of the truss.

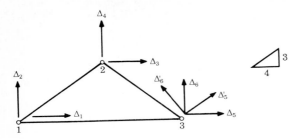

Figure b   Numbering of the components of displacements of the nodes of the truss of Fig. a.

That is, referring to relations (4.50) we have

$$[K^1] = \frac{EA}{5}\begin{bmatrix} 1 & -1 \\ -1 & 1 \end{bmatrix} \tag{b}$$

$$[K^2] = \frac{EA}{8}\begin{bmatrix} 1 & -1 \\ -1 & 1 \end{bmatrix} \tag{c}$$

$$[K^3] = \frac{EA}{5}\begin{bmatrix} 1 & -1 \\ -1 & 1 \end{bmatrix} \tag{d}$$

STEP 4.   From the local stiffness matrix of each element of the truss we compute its global or, in case the element is connected to a skew support, its modified global stiffness matrix. Element 1 is not connected to a skew support; thus, referring to Fig. a and to relations (6.42) and (6.74) and noting that for a truss $[\Lambda^e] = [\Lambda^e_{PT}]$ we have

$$[\bar{K}^1] = [\Lambda^1_{PT}]^T[K^1][\Lambda^1_{PT}] = \begin{bmatrix} 0.8 & 0 \\ 0.6 & 0 \\ 0 & 0.8 \\ 0 & 0.6 \end{bmatrix} \frac{EA}{5}\begin{bmatrix} 1 & -1 \\ -1 & 1 \end{bmatrix}\begin{bmatrix} 0.8 & 0.6 & 0 & 0 \\ 0 & 0 & 0.8 & 0.6 \end{bmatrix}$$

$$= \frac{EA}{125}\begin{bmatrix} 16 & 12 & -16 & -12 \\ 12 & 9 & -12 & -9 \\ -16 & -12 & 16 & 12 \\ -12 & -9 & 12 & 9 \end{bmatrix} \tag{e}$$

Elements 2 and 3 of the truss are connected to skew support 3 with their end k. Thus, we compute their modified global stiffness matrices. Referring to relation (10.2) and Fig. a the transformation matrix $[\Lambda'_p]$ for support 3 is equal to

$$[\Lambda'_p] = \begin{bmatrix} 0.8 & 0.6 \\ -0.6 & 0.8 \end{bmatrix} \tag{f}$$

Substituting relation (f) into (10.13a) and the resulting expression into

(10.16*b*) we obtain

$$[\Lambda_{\text{SKG}}]^k = \begin{bmatrix} 1 & 0 & 0 & 0 \\ 0 & 1 & 0 & 0 \\ 0 & 0 & 0.8 & 0.6 \\ 0 & 0 & -0.6 & 0.8 \end{bmatrix} \quad \text{(g)}$$

Substituting relation (g) into (10.19) and noting that for a truss $[\Lambda^e] = [\Lambda_{\text{PT}}^e]$ we get

$$[\Lambda_{\text{SKL}}^e]^k = [\Lambda_{\text{SKG}}]^k [\Lambda_{\text{PT}}^e]^T = \begin{bmatrix} 1 & 0 & 0 & 0 \\ 0 & 1 & 0 & 0 \\ 0 & 0 & 0.8 & 0.6 \\ 0 & 0 & -0.6 & 0.8 \end{bmatrix} [\Lambda_{\text{PT}}^e]^T \quad e = 2, 3 \quad \text{(h)}$$

where referring to relations (6.42) we have

$$[\Lambda_{\text{PT}}^2]^T = \begin{bmatrix} 1 & 0 \\ 0 & 0 \\ 0 & 1 \\ 0 & 0 \end{bmatrix}$$

$$[\Lambda_{\text{PT}}^3]^T = \begin{bmatrix} 0.8 & 0 \\ -0.6 & 0 \\ 0 & 0.8 \\ 0 & -0.6 \end{bmatrix} \quad \text{(i)}$$

Substituting relations (i) into (h) we get

$$[\Lambda_{\text{SKL}}^2]^k = \begin{bmatrix} 1 & 0 \\ 0 & 0 \\ 0 & 0.8 \\ 0 & -0.6 \end{bmatrix} \quad \text{(j)}$$

$$[\Lambda_{\text{SKL}}^3]^k = \begin{bmatrix} 0.8 & 0 \\ -0.6 & 0 \\ 0 & 0.28 \\ 0 & -0.96 \end{bmatrix} \quad \text{(k)}$$

Substituting relations (c) and (j) into (10.30) we obtain

$$[\bar{K}'^2]^k = \frac{EA}{8} \begin{bmatrix} 1 & 0 \\ 0 & 0 \\ 0 & 0.8 \\ 0 & -0.6 \end{bmatrix} \begin{bmatrix} 1 & -1 \\ -1 & 1 \end{bmatrix} \begin{bmatrix} 1 & 0 & 0 & 0 \\ 0 & 0 & 0.8 & -0.6 \end{bmatrix}$$

$$= \frac{EA}{8} \begin{bmatrix} 1 & 0 & -0.8 & 0.6 \\ 0 & 0 & 0 & 0 \\ -0.8 & 0 & 0.64 & -0.48 \\ 0.6 & 0 & -0.48 & 0.36 \end{bmatrix} \quad \text{(l)}$$

Substituting relations (d) and (k) into (10.30) we get

$$[\bar{K}'^3] = \frac{EA}{5} \begin{bmatrix} 0.8 & 0 \\ -0.6 & 0 \\ 0 & 0.28 \\ 0 & -0.96 \end{bmatrix} \begin{bmatrix} 1 & -1 \\ -1 & 1 \end{bmatrix} \begin{bmatrix} 0.8 & -0.6 & 0 & 0 \\ 0 & 0 & 0.28 & 0.96 \end{bmatrix}$$

$$= \frac{EA}{125} \begin{bmatrix} 16.0 & -12.0 & -5.60 & 9.20 \\ -12.0 & 9.0 & 4.20 & -14.40 \\ -5.6 & 4.2 & 1.96 & -6.72 \\ 19.2 & -14.4 & -6.72 & 23.04 \end{bmatrix} \tag{m}$$

STEP 5. We assemble the stiffness matrix for the structure from the global or the modified global stiffness matrices for its elements. In order to accomplish this we choose the indices of the stiffness coefficients for each element of the truss so as to correspond to those of the components of displacements of the nodes of the truss to which the element is connected. Referring to Fig. b the stiffness equations for the elements of the truss may be expressed as

$$[\bar{A}^1] = [\bar{K}^1][\bar{D}^1] = \begin{bmatrix} \bar{K}_{11}^1 & \bar{K}_{12}^1 & \bar{K}_{13}^1 & \bar{K}_{14}^1 \\ \bar{K}_{21}^1 & \bar{K}_{22}^1 & \bar{K}_{23}^1 & \bar{K}_{24}^1 \\ \bar{K}_{31}^1 & \bar{K}_{32}^1 & \bar{K}_{33}^1 & \bar{K}_{34}^1 \\ \bar{K}_{41}^1 & \bar{K}_{42}^1 & \bar{K}_{43}^1 & \bar{K}_{44}^1 \end{bmatrix} \begin{Bmatrix} \Delta_1 \\ \Delta_2 \\ \Delta_3 \\ \Delta_4 \end{Bmatrix} \tag{n}$$

$$[\bar{A}'^2] = [\bar{K}'^2][\bar{D}'^2] = \begin{bmatrix} \bar{K}_{11}'^2 & \bar{K}_{12}'^2 & \bar{K}_{15}'^2 & \bar{K}_{16}'^2 \\ \bar{K}_{21}'^2 & \bar{K}_{22}'^2 & \bar{K}_{25}'^2 & \bar{K}_{26}'^2 \\ \bar{K}_{51}'^2 & \bar{K}_{52}'^2 & \bar{K}_{55}'^2 & \bar{K}_{56}'^2 \\ \bar{K}_{61}'^2 & \bar{K}_{62}'^2 & \bar{K}_{65}'^2 & \bar{K}_{66}'^2 \end{bmatrix} \begin{Bmatrix} \Delta_1 \\ \Delta_2 \\ \Delta_5 \\ \Delta_6 \end{Bmatrix} \tag{o}$$

$$\{\bar{A}'^3\} = [\bar{K}'^3]\{\bar{D}'^3\} = \begin{bmatrix} \bar{K}_{33}'^3 & \bar{K}_{34}'^3 & \bar{K}_{35}'^3 & \bar{K}_{36}'^3 \\ \bar{K}_{43}'^3 & \bar{K}_{44}'^3 & \bar{K}_{45}'^3 & \bar{K}_{46}'^3 \\ \bar{K}_{53}'^3 & \bar{K}_{54}'^3 & \bar{K}_{55}'^3 & \bar{K}_{56}'^3 \\ \bar{K}_{63}'^3 & \bar{K}_{64}'^3 & \bar{K}_{65}'^3 & \bar{K}_{66}'^3 \end{bmatrix} \begin{Bmatrix} \Delta_3 \\ \Delta_4 \\ \Delta_5' \\ \Delta_6' \end{Bmatrix} \tag{p}$$

The modified stiffness matrix for the truss has the following form:

$$[\hat{S}'] = \begin{bmatrix} S_{11}' & S_{12}' & S_{13}' & S_{14}' & S_{15}' & S_{16}' \\ S_{21}' & S_{22}' & S_{23}' & S_{24}' & S_{25}' & S_{26}' \\ S_{31}' & S_{32}' & S_{33}' & S_{34}' & S_{35}' & S_{36}' \\ S_{41}' & S_{42}' & S_{43}' & S_{44}' & S_{45}' & S_{46}' \\ S_{51}' & S_{52}' & S_{53}' & S_{54}' & S_{55}' & S_{56}' \\ S_{61}' & S_{62}' & S_{63}' & S_{64}' & S_{65}' & S_{66}' \end{bmatrix} \tag{q}$$

where the stiffness coefficients S_{pq}' for the truss are equal to the sums of the stiffness coefficients $\bar{K}_{pq}$ or K_{pq} for all the elements of the truss. Thus, referring to relations (h) and (p) and using relation (8.22) we have

$$S'_{11} = \bar{K}^1_{11} + \bar{K}'^2_{11} = \frac{EA}{125}\left(16 + \frac{125}{8}\right) = \frac{31.625EA}{125}$$

$$S'_{12} = \bar{K}^1_{12} + \bar{K}'^2_{12} = \frac{EA}{125}(12 + 0) = \frac{12EA}{125}$$

$$S'_{13} = \bar{K}^1_{13} = -\frac{16EA}{125}$$

$$S'_{14} = \bar{K}^1_{14} = -\frac{12EA}{125}$$

$$S'_{15} = \bar{K}'^2_{15} = \frac{EA}{8}(-0.8) = -\frac{12.5EA}{125}$$

$$S'_{16} = \bar{K}'^2_{16} = \frac{EA}{8}(0.6) = \frac{9.375EA}{12.5}$$

$$S'_{21} = \bar{K}^1_{21} + \bar{K}'^2_{21} = \frac{12EA}{125}$$

$$S'_{22} = \bar{K}^1_{22} + \bar{K}'^2_{22} = \frac{9EA}{125}$$

$$S'_{23} = \bar{K}^1_{23} = -\frac{12EA}{125}$$

$$S'_{24} = \bar{K}^1_{24} = -\frac{9EA}{125}$$

$$S'_{25} = \bar{K}'^2_{25} = 0$$

$$S'_{26} = \bar{K}'^2_{26} = 0$$

$$S'_{31} = \bar{K}^1_{31} = -\frac{16EA}{125}$$

$$S'_{32} = \bar{K}^1_{32} = -\frac{12EA}{125}$$

$$S'_{33} = \bar{K}^1_{33} + \bar{K}'^3_{33} = \frac{EA}{125}(16 + 16) = \frac{32EA}{125}$$

$$S'_{34} = \bar{K}^1_{34} + \bar{K}'^3_{34} = \frac{EA}{125}(12 - 12) = 0$$

$$S'_{35} = \bar{K}'^3_{35} = -\frac{5.6EA}{125} \tag{r}$$

$$S'_{36} = \bar{K}'^3_{36} = \frac{19.2EA}{125}$$

$$S'_{41} = \bar{K}^1_{41} = -\frac{12EA}{125}$$

$$S'_{42} = \bar{K}^1_{42} = -\frac{9EA}{125}$$

$$S'_{43} = \bar{K}^1_{43} + \bar{K}'^3_{43} = \frac{EA}{234}(12 - 12) = 0$$

$$S'_{44} = \bar{K}^1_{44} + \bar{K}'^3_{44} = \frac{EA}{125}(9 + 9) = \frac{18EA}{125}$$

$$S'_{45} = \bar{K}'^3_{45} = \frac{4.2EA}{125}$$

$$S'_{46} = \bar{K}'^3_{46} = -\frac{14.4EA}{125}$$

$$S'_{51} = \bar{K}'^2_{51} = \frac{EA}{8}(-0.8) = -\frac{12.5EA}{125}$$

$$S'_{52} = \bar{K}'^2_{52} = 0$$

$$S'_{53} = \bar{K}'^2_{53} = -\frac{5.6EA}{125}$$

$$S'_{54} = \bar{K}'^3_{54} = \frac{4.2EA}{125}$$

$$S'_{55} = \bar{K}'^2_{55} + \bar{K}'^3_{55} = \frac{EA}{8}(0.64) + \frac{EA}{125}(1.96) = \frac{11.96EA}{125}$$

$$S'_{56} = \bar{K}'^2_{56} + \bar{K}'^3_{56} = \frac{EA}{8}(-0.48) + \frac{EA}{125}(-6.84) = -\frac{14.22EA}{125}$$

$$S'_{61} = \bar{K}'^2_{61} = \frac{EA}{8}(0.6) = \frac{9.375EA}{125}$$

$$S'_{62} = \bar{K}'^2_{62} = 0$$

$$S'_{63} = \bar{K}'^3_{63} = \frac{19.2EA}{125}$$

$$S'_{64} = \bar{K}'^3_{64} = \frac{14.4EA}{125}$$

$$S'_{65} = \bar{K}'^{2}_{65} + \bar{K}'^{3}_{65} = \frac{EA(-0.48)}{8} + \frac{EA(-6.72)}{125} = \frac{14.22EA}{125}$$

$$S'_{66} = \bar{K}'^{2}_{66} + \bar{K}'^{3}_{66} = \frac{EA(0.36)}{8} + \frac{EA(22.88)}{125} = \frac{28.665EA}{125}$$

Thus the modified stiffness equations for the truss of Fig. a are

$$\begin{Bmatrix} 0 \\ 0 \\ 40 \\ -80 \\ 0 \\ 0 \end{Bmatrix} + \begin{Bmatrix} R_1^{(1)} \\ R_2^{(1)} \\ 0 \\ 0 \\ 0 \\ R_2^{(3)} \end{Bmatrix} = \frac{EA}{125} \begin{bmatrix} 31.625 & 12 & -16 & -12 & -12.5 & 9.375 \\ 12 & 9 & -12 & -9 & 0 & 0 \\ -16 & -12 & 32 & 0 & -5.6 & 19.2 \\ -12 & -9 & 0 & 18 & 4.2 & -14.4 \\ -12.5 & 0 & -5.6 & 4.2 & 11.96 & -14.22 \\ 9.375 & 0 & 19.2 & -14.4 & -14.22 & 28.665 \end{bmatrix} \begin{Bmatrix} \Delta_1 \\ \Delta_2 \\ \Delta_3 \\ \Delta_4 \\ \Delta'_5 \\ \Delta'_6 \end{Bmatrix}$$

(s)

or

$$\{\hat{P}'^{E}\} + \{\hat{R}'\} = [\hat{S}']\{\hat{\Delta}'\}$$

(t)

STEP 6. We rearrange the rows of the stiffness matrix $[\hat{S}']$ in order to move to the bottom of Eq. (s) its rows corresponding to the reactions of the truss. Moreover, we rearrange the columns of the stiffness matrix $[\hat{S}']$ in order to move to its right its columns which are multiplied by the vanishing components of displacements at the supports of the truss. Furthermore, we partition the resulting stiffness equations for the truss as indicated by relation (9.1). Thus, referring to Figs. a and b we have

$$\begin{Bmatrix} 40 \\ -80 \\ 0 \\ 0 \\ 0 \\ 0 \end{Bmatrix} + \begin{Bmatrix} 0 \\ 0 \\ 0 \\ R_1^{(1)} \\ R_2^{(2)} \\ R_2^{(3)} \end{Bmatrix} = \frac{EA}{125} \begin{bmatrix} 32.0 & 0 & -5.6 & \vdots & -16.0 & -12 & 19.2 \\ 0 & 18.0 & 4.2 & \vdots & -12.0 & -9 & -14.4 \\ -5.6 & 4.2 & 11.96 & \vdots & -12.5 & 0 & -14.22 \\ \cdots & \cdots & \cdots & \vdots & \cdots & \cdots & \cdots \\ -16.0 & -12.0 & -12.5 & \vdots & 31.625 & 12 & 9.375 \\ -12.0 & -9.0 & 0 & \vdots & 12.0 & 9 & 0 \\ 19.2 & -14.4 & -14.22 & \vdots & 9.375 & 0 & 28.665 \end{bmatrix} \begin{Bmatrix} \Delta_3 \\ \Delta_4 \\ \Delta'_5 \\ \Delta_1 \\ \Delta_2 \\ \Delta'_6 \end{Bmatrix}$$

(u)

Referring to relations (9.1) and (u) we have

$$\{P^{EF}\} = \begin{Bmatrix} 40 \\ -80 \\ 0 \end{Bmatrix}$$

(v)

$$\{P^{ES}\} = \begin{Bmatrix} 0 \\ 0 \\ 0 \end{Bmatrix}$$

(w)

$$\{R\} = \begin{Bmatrix} R_1^{(1)} \\ R_2^{(2)} \\ R_2^{(3)} \end{Bmatrix} \tag{x}$$

$$\{\Delta^S\} = \begin{Bmatrix} 0 \\ 0 \\ 0 \end{Bmatrix} \tag{y}$$

$$[S^{FF}] = \frac{EA}{125} \begin{bmatrix} 32.0 & 0 & -5.60 \\ 0 & 18.0 & 4.20 \\ -5.6 & 4.2 & 11.96 \end{bmatrix} \tag{z}$$

and

$$[S^{SF}] = \frac{EA}{125} \begin{bmatrix} -16.0 & -12.0 & -12.50 \\ -12.0 & -9.0 & 0 \\ 19.2 & -14.4 & -14.22 \end{bmatrix} \tag{aa}$$

We compute the components of displacements of the nodes of the truss. Substituting relations (v) and (z) into (9.4), we obtain

$$\{\Delta^F\} = \begin{Bmatrix} \Delta_3 \\ \Delta_4 \\ \Delta_5' \end{Bmatrix} = \frac{125}{EA} \begin{bmatrix} 32.0 & 0 & -5.60 \\ 0 & 18.0 & 4.20 \\ -5.6 & 4.2 & 11.96 \end{bmatrix}^{-1} \begin{Bmatrix} 40 \\ -80 \\ 0 \end{Bmatrix}$$

$$= \frac{1}{EA} \begin{Bmatrix} 212.40 \\ -630.42 \\ 320.83 \end{Bmatrix} = \begin{Bmatrix} 10.62 \\ -31.52 \\ 16.04 \end{Bmatrix} \text{mm} \tag{bb}$$

where Δ_5' is the component of translation of node 3 in the direction of rolling.

We compute the reactions of the truss. Substituting relations (aa) and (bb) into (9.5), we get

$$\{R\} = \begin{Bmatrix} R_1^{(1)} \\ R_2^{(1)} \\ R_2'^{(3)} \end{Bmatrix} = \frac{1}{125} \begin{bmatrix} -16.0 & -12.0 & 12.5 \\ -12.0 & -9.0 & 0 \\ 19.2 & -14.4 & -14.22 \end{bmatrix} \begin{Bmatrix} 212.40 \\ -630.42 \\ 320.83 \end{Bmatrix} = \begin{Bmatrix} -1.25 \\ 25.00 \\ 68.75 \end{Bmatrix} \text{kN}$$

$$\tag{cc}$$

The reactions of the truss are shown in Fig. c. Their values can be checked by considering the equilibrium of the truss.

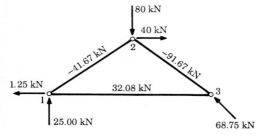

Figure c Results for the truss of Fig. a.

STEP 7. We compute the local components of internal actions in the elements of the truss. Referring to Fig. b and relation (bb), the global components of the nodal displacements of the elements of the truss are

$$\{\bar{D}^1\} = \begin{Bmatrix} 0 \\ 0 \\ \Delta_3 \\ \Delta_4 \end{Bmatrix} = \frac{1}{EA} \begin{Bmatrix} 0 \\ 0 \\ 212.40 \\ -630.42 \end{Bmatrix}$$

$$\{\bar{D}^2\} = \begin{Bmatrix} 0 \\ 0 \\ \dfrac{4\Delta_5'}{5} \\ \dfrac{3\Delta_5'}{5} \end{Bmatrix} = \frac{1}{EA} \begin{Bmatrix} 0 \\ 0 \\ 256.66 \\ 192.50 \end{Bmatrix} \qquad \{\bar{D}^3\} = \begin{Bmatrix} \Delta_3 \\ \bar{\Delta}_4 \\ \dfrac{4\Delta_5'}{5} \\ \dfrac{3\Delta_5'}{5} \end{Bmatrix} = \frac{1}{EA} \begin{Bmatrix} 212.40 \\ -630.42 \\ 256.66 \\ 192.50 \end{Bmatrix} \qquad (dd)$$

We compute the hybrid stiffness matrices of the elements of the truss. To accomplish this we substitute into Eq. (9.13) the transformation matrix and the local stiffness matrix for each element. The first is given by relation (6.42), while the second is given by relations (b) to (d). Thus,

$$[\bar{K}^1] = \frac{EA}{5} \begin{bmatrix} 1 & -1 \\ -1 & 1 \end{bmatrix} \begin{bmatrix} 0.8 & 0.6 & 0 & 0 \\ 0 & 0 & 0.8 & 0.6 \end{bmatrix}$$

$$= \frac{EA}{5} \begin{bmatrix} 0.8 & 0.6 & -0.8 & -0.6 \\ -0.8 & -0.6 & 0.8 & 0.6 \end{bmatrix}$$

$$[\bar{K}^2] = \frac{EA}{8} \begin{bmatrix} 1 & -1 \\ -1 & 1 \end{bmatrix} \begin{bmatrix} 1 & 0 & 0 & 0 \\ 0 & 0 & 1 & 0 \end{bmatrix}$$

$$= \frac{EA}{8} \begin{bmatrix} 1 & 0 & -1 & 0 \\ -1 & 0 & 1 & 0 \end{bmatrix} \qquad (ee)$$

$$[\bar{K}^3] = \frac{EA}{5} \begin{bmatrix} 1 & -1 \\ -1 & 1 \end{bmatrix} \begin{bmatrix} 0.8 & -0.6 & 0 & 0 \\ 0 & 0 & 0.8 & -0.6 \end{bmatrix}$$

$$= \frac{EA}{5} \begin{bmatrix} 0.8 & -0.6 & -0.8 & 0.6 \\ -0.8 & 0.6 & 0.8 & -0.6 \end{bmatrix}$$

We substitute relations (dd) and (ee) into Eq. (9.13). Thus

$$\{A^1\} = \frac{EA}{5} \begin{bmatrix} 0.8 & 0.6 & -0.8 & -0.6 \\ -0.8 & -0.6 & 0.8 & 0.6 \end{bmatrix} \frac{1}{EA} \begin{Bmatrix} 0 \\ 0 \\ 212.40 \\ -630.42 \end{Bmatrix} = \begin{Bmatrix} 41.67 \\ -41.67 \end{Bmatrix}$$

$$\{A^2\} = \frac{EA}{8} \begin{bmatrix} 1 & 0 & -1 & 0 \\ -1 & 0 & 1 & 0 \end{bmatrix} \frac{1}{EA} \begin{Bmatrix} 0 \\ 0 \\ 256.66 \\ 192.50 \end{Bmatrix} = \begin{Bmatrix} -32.08 \\ 32.08 \end{Bmatrix} \qquad \text{(ff)}$$

$$\{A^3\} = \frac{EA}{5} \begin{bmatrix} 0.8 & -0.6 & -0.8 & 0.6 \\ -0.8 & 0.6 & 0.8 & -0.6 \end{bmatrix} \frac{1}{EA} \begin{Bmatrix} 212.40 \\ -630.42 \\ 256.66 \\ 192.50 \end{Bmatrix} = \begin{Bmatrix} 91.67 \\ -91.67 \end{Bmatrix}$$

The results are shown in Fig. c. They can be checked by considering the equilibrium of the nodes of the truss. Comparing the internal forces in the elements of the truss with the plane of rolling of support 3 normal to the $\bar{x}_2$ axis (see Fig. d of Example 1 of Sec. 9.3) with those of the elements of the truss with the plane of rolling of support 3 inclined to the global axes (see Fig. c), we can see that

1. The magnitudes of the internal forces in elements 1 and 3 do not change when an angle of inclination of support 3 is changed. The forces in these elements can be established by considering the equilibrium of node 2 and consequently are not affected by the boundary conditions at support 3.

2. The magnitude of the internal force in element 2 can change considerably by changing the angle of inclination of support 3.

10.2.2 Procedure for establishing the modified stiffness matrix of a structure with skew supports

In this procedure we adhere to the following steps.

Step 1. We assemble the global stiffness matrix of the structure from the global stiffness matrices of its elements.

Step 2. We modify the global stiffness matrix of the structure as indicated below.

Consider a structure whose node n is connected to a skew support. The skew components of the equivalent force of the reacting force and of the translation at node n are related to their global components by relations (10.10). The modified matrices of equivalent actions and reactions $\{\hat{P}'^E\} + \{\hat{R}'\}$ and of displacements $\{\hat{\Delta}'\}$ of the structure under consideration are related to its global matrices $\{\{\hat{P}^E\} + \{\hat{R}\}\}$ and $\{\hat{\Delta}\}$, respectively, by the following relations:

$$\{\hat{P}'^E\} + \{\hat{R}'\} = \left\{ \begin{array}{c} \{\bar{P}^{E(A)}\} \\ \{\bar{P}^{E'(n)}\} \\ \{\bar{P}^{E(B)}\} \end{array} \right\} + \left\{ \begin{array}{c} \{\bar{R}^{(A)}\} \\ \{\bar{R}'^{(n)}\} \\ \{\bar{R}^{(B)}\} \end{array} \right\}$$

$$= \begin{bmatrix} [I_1] & [0] & [0] \\ [0] & [\Lambda'] & [0] \\ [0] & [0] & [I_2] \end{bmatrix} \left\{ \left\{ \begin{array}{c} \{\bar{P}^{E(A)}\} \\ \{\bar{P}^{E(n)}\} \\ \{\bar{P}^{E(B)}\} \end{array} \right\} + \left\{ \begin{array}{c} \{\bar{R}^{(A)}\} \\ \{\bar{R}^{(n)}\} \\ \{\bar{R}^{(B)}\} \end{array} \right\} \right\}$$

$$\{\hat{\Delta}'\} = \left\{ \begin{array}{c} \{\bar{\Delta}^{(A)}\} \\ \{\bar{\Delta}'^{(n)}\} \\ \{\bar{\Delta}^{(B)}\} \end{array} \right\} = \begin{bmatrix} [I_1] & [0] & [0] \\ [0] & [\Lambda'] & [0] \\ [0] & [0] & [I_2] \end{bmatrix} \left\{ \begin{array}{c} \{\bar{\Delta}^{(A)}\} \\ \{\bar{\Delta}^{(n)}\} \\ \{\bar{\Delta}^{(B)}\} \end{array} \right\}$$

$$(10.31)$$

where the terms of the matrices $\{\bar{P}^{E(n)}\}$, $\{\bar{R}^{(n)}\}$, and $\{\bar{\Delta}^{(n)}\}$ represent the global components of the equivalent action, of the reaction, and of the displacement, respectively, of node n of the structure. Relations (10.31) can be rewritten as

$$\{\hat{P}'^E\} + \{\hat{R}'\} = [V]\{\{\hat{P}^E\} + \{\hat{R}\}\} \tag{10.32}$$

$$\{\hat{\Delta}'\} = [V]\{\hat{\Delta}\} \tag{10.33}$$

where

$$[V] = \begin{bmatrix} [I_1] & [0] & [0] \\ [0] & [\Lambda'] & [0] \\ [0] & [0] & [I_2] \end{bmatrix} \tag{10.34}$$

The work of the external actions acting on a structure is equal to

$$W = \tfrac{1}{2}\{\{\hat{P}'^E\} + \{\hat{R}'\}\}^T\{\hat{\Delta}'\} \tag{10.35a}$$

or

$$W = \tfrac{1}{2}\{\{\hat{P}^E\} + \{\hat{R}\}\}^T\{\hat{\Delta}\} \tag{10.35b}$$

Substituting relations (10.32) and (10.33) into (10.35a) we obtain

$$W = \tfrac{1}{2}\{[V]\{\{\hat{P}^E\} + \{\hat{R}\}\}\}^T[V]\{\hat{\Delta}\}$$

$$= \tfrac{1}{2}\{\{\hat{P}^E\} + \{\hat{R}\}\}^T[V]^T[V]\{\hat{\Delta}\} \tag{10.36}$$

Comparing relations (10.35b) and (10.36), we get

$$[V]^T[V] = [I] \tag{10.37}$$

or

$$[V]^{-1} = [V]^T \tag{10.38}$$

Consequently

$$\{\hat{P}^E\} + \{\hat{R}\} = [V]^T\{\{\hat{P}'^E\} + \{\hat{R}'\}\} \tag{10.39}$$

$$[\hat{\Delta}] = [V]^T[\hat{\Delta}'] \tag{10.40}$$

Substituting relation (8.1) into (10.32) and using relation (10.40), we have

$$\{\hat{P}'^E\} + \{\hat{R}'\} = [V][\hat{S}][\hat{\Delta}] = [V][\hat{S}][V]^T[\hat{\Delta}'] \qquad (10.41)$$

Comparing relations (10.7) and (10.41) we get

$$[\hat{S}'] = [V][\hat{S}][V]^T \qquad (10.42)$$

This relation may be rewritten as

$$[\hat{S}'] = \begin{bmatrix} [I_1] & [0] & [0] \\ [0] & [\Lambda'] & [0] \\ [0] & [0] & [I_2] \end{bmatrix} \begin{bmatrix} [S_1] & [S_2] & [S_3] \\ [S_4] & [S_5] & [S_6] \\ [S_7] & [S_8] & [S_9] \end{bmatrix} \begin{bmatrix} [I_1] & [0] & [0] \\ [0] & [\Lambda']^T & [0] \\ [0] & [0] & [I_2] \end{bmatrix}$$

$$= \begin{bmatrix} [S_1] & [S_2][\Lambda']^T & [S_3] \\ [\Lambda'][S_4] & [\Lambda'][S_5][\Lambda']^T & [\Lambda'][S_6] \\ [S_7] & [S_8][\Lambda']^T & [S_9] \end{bmatrix} \qquad (10.43)$$

where for the frame of Fig. 10.1 $[I_1]$ is a 9×9 unit matrix; $[I_2]$ is 1; $[S_1]$ is a 9×9 matrix; $[S_2]$ is a 9×2 matrix; $[S_3]$ is a 9×1 matrix; $[S_4]$ is a 2×9 matrix; $[S_5]$ is a 2×2 matrix; $[S_6]$ is a 2×1 matrix; $[S_7]$ is a 1×9 matrix; $[S_8]$ is a 1×2 matrix, and $[S_9]$ is a 1×1 matrix.

An algorithm can be written for assembling the matrix $[V]$ of a structure; this matrix can then be employed in relation (10.42) to obtain the modified stiffness matrix $[\hat{S}']$ of the structure from its stiffness matrix $[\hat{S}]$. However, only a small number (two for a planar structure, three for a space structure) of rows and columns of the matrix $[\hat{S}]$ are affected by the multiplication indicated in relation (10.43).

Moreover, for certain structures the matrices $[V]$ and $[\hat{S}]$ can be very large and the multiplication indicated by relation (10.42) can consume considerable computer time. In this case the stiffness matrix of a planar structure with a skew support can be modified by referring to relation (10.43) and adhering to the following steps.

Step 1. The $n \times 2$ matrix $[S_c]$ is formed containing the two columns of the matrix $[\hat{S}]$ corresponding to the global components of translation of the node which is connected to the skew support. That is, for the frame of Fig. 10.1a, the matrix $[S_c]$ contains columns 10 and 11 of the matrix $[\hat{S}]$.

Step 2. The matrix $[S_c]$ is postmultiplied by the matrix $[\Lambda']^T$, and the resulting matrix is denoted by $[S_c^1]$. The two columns of the matrix $[\hat{S}]$ corresponding to the global components of translation of the node which is connected to the skew support are replaced by the columns of the matrix $[S_c^1]$, and the resulting matrix is denoted by $[\hat{S}^1]$.

Step 3. The $2 \times n$ matrix $[S_r]$ is formed containing the two rows of the matrix $[\hat{S}^1]$ corresponding to the global components of the external force and the reaction acting on the node which is connected to the skew support. That is, for the frame of Fig. 10.1a the matrix $[S_r]$ contains rows 10 and 11 of the matrix $[\hat{S}^1]$.

Step 4. The matrix $[S_r]$ is premultiplied by the matrix $[\Lambda']$, and the resulting matrix is denoted by $[S_r^1]$. The two rows of the matrix $[\hat{S}^1]$ corresponding to the global components of the external forces and the reactions acting on the node which is connected to the skew support are replaced by the rows of the matrix $[S_r^1]$. The resulting matrix is the modified stiffness matrix $[\hat{S}']$ of the structure under consideration.

In what follows we apply the procedure described in this subsection to the same truss with a skew support to which we have applied the procedure described in the previous subsection.

Example. Using the method described in this section establish the modified stiffness matrix for the truss shown in Fig. a. The elements of the truss are made of the same isotropic linearly elastic material and have the same constant cross section ($AE = 20,000$ kN).

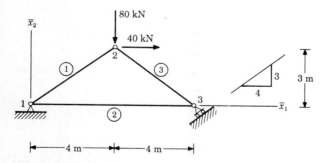

Figure a Geometry and loading of the truss.

---

solution

STEPS 1 TO 6. The stiffness matrix for the truss has been computed in Example 1 of Sec. 8.3.1. Referring to relation (j) of this example the stiffness equations for the truss are

$$
\begin{Bmatrix} 0 \\ 0 \\ 40 \\ -80 \\ 0 \\ 0 \end{Bmatrix} + \begin{Bmatrix} R_1^{(1)} \\ R_2^{(1)} \\ 0 \\ 0 \\ R_1^{(3)} \\ R_2^{(3)} \end{Bmatrix} = \frac{EA}{125} \begin{bmatrix} 31.625 & 12 & -16 & -12 & -15.625 & 0 \\ 12 & 9 & -12 & -9 & 0 & 0 \\ -16 & -12 & 32 & 0 & -16 & 12 \\ -12 & -9 & 0 & 18 & 12 & -9 \\ -15.625 & 0 & -16 & 12 & 31.625 & -12 \\ 0 & 0 & 12 & -9 & -12 & 9 \end{bmatrix} \begin{Bmatrix} \Delta_1 \\ \Delta_2 \\ \Delta_3 \\ \Delta_4 \\ \Delta_5 \\ \Delta_6 \end{Bmatrix} = [\hat{S}]\{\hat{\Delta}\}
$$

$$(a)$$

We modify the last two columns and the last two rows of the stiffness matrix $[\hat{S}]$ in order to account for the roller at support 3 whose direction of rolling is not normal to one of the global axes of the truss. Referring to Eqs. (a) the last two columns of the stiffness matrix $[\hat{S}]$ are

$$
[S_c] = \begin{bmatrix} -15.625 & 0 \\ 0 & 0 \\ -16 & 12 \\ 12 & -8 \\ 31.625 & -12 \\ -12 & 9 \end{bmatrix}
$$

$$(b)$$

Postmultiplying the matrix $[S_c]$ with the matrix $[\Lambda']^T$, we get

$$
[S_c^1] = [S_c][\Lambda']^T = \begin{bmatrix} -12.5 & 9.375 \\ 0 & 0 \\ -5.6 & 19.2 \\ 4.2 & -14.4 \\ 18.1 & -18.575 \\ -4.2 & 14.4 \end{bmatrix}
$$

$$(c)$$

We replace the last two columns of the matrix $[\hat{S}]$ [relation (a)] by the two columns of the matrix $[S_c^1]$. Thus

$$
[\hat{S}^1] = \frac{EA}{125} \begin{bmatrix} 31.625 & 12 & -16 & -12 & -12.5 & 9.375 \\ 12 & 9 & -12 & -9 & 0 & 0 \\ -16 & -12 & 32 & 0 & -5.6 & 19.2 \\ -12 & -9 & 0 & 18 & 4.2 & -14.4 \\ -15.625 & 0 & -16 & 12 & 18.1 & -28.575 \\ 0 & 0 & 12 & -9 & -4.2 & 14.4 \end{bmatrix}
$$

$$(d)$$

We denote the last two rows of the matrix $[\hat{S}^1]$ by $[S_r]$. Premultiplying the matrix $[S_r]$ with the matrix $[\Lambda']$ we obtain

$$[S_r^1] = [\Lambda'][S_r] = \begin{bmatrix} 0.8 & 0.6 \\ -0.6 & 0.8 \end{bmatrix} \begin{bmatrix} -15.625 & 0 & -16 & 12 & 18.1 & -28.575 \\ 0 & 0 & 12 & -9 & -4.2 & 14.4 \end{bmatrix}$$

$$= \begin{bmatrix} -12.500 & 0 & -5.6 & 4.2 & 11.96 & -14.220 \\ 9.375 & 0 & 19.2 & -14.4 & -14.22 & 28.665 \end{bmatrix} \qquad \text{(e)}$$

We replace the last two rows of the matrix $[\hat{S}^1]$ given by relation (d) by the rows of the matrix $[\hat{S}_r^1]$ to obtain the modified stiffness matrix for the truss. That is,

$$[\hat{S}'] = \frac{EA}{125} \begin{bmatrix} 31.625 & 12.00 & -16.00 & -12.00 & -12.50 & 9.375 \\ 12.00 & 9.00 & -12.00 & -9.00 & 0 & 0 \\ -16.00 & -12.00 & 32.00 & 0 & -5.60 & 19.20 \\ -12.00 & -9.00 & 0 & 18.00 & 4.20 & -14.40 \\ -12.50 & 0 & -5.60 & 4.20 & 11.96 & -14.22 \\ 9.375 & 0 & 19.20 & -14.40 & -24.22 & 28.665 \end{bmatrix} \qquad \text{(f)}$$

10.3 Analysis of Structures Having General Constraints

In Sec. 9.1, we presented a method for solving the stiffness equations for a framed structure. However, our presentation was limited only to structures whose constraints are their supports, that is, structures whose constraints consist of a number of specified components of displacements of some of their nodes (supports). In this section, we present a method for establishing the stiffness equations for a framed structure when it is subjected to more general constraints. Such constraints could involve specified components of displacement of some nodes of the structure (single-point constraints) as well as related components of displacements of some nodes of the structure (multipoint constraints). Multipoint constraints are encountered for example when the effect of the axial deformation of the elements of a frame is disregarded, when some elements of a framed structure are considered rigid, or when a structure has a skew support (see Sec. 10.2). Consider a structure whose nodes have n possible components of displacements and r constraints specified by the following constraint equations:

$$C_{11}\Delta_1 + C_{12}\Delta_2 + \cdots + C_{1n}\Delta_n = G_1$$
$$C_{21}\Delta_1 + C_{22}\Delta_2 + \cdots + C_{2n}\Delta_n = G_2$$
$$\vdots \qquad \vdots \qquad \qquad \vdots \qquad \vdots$$
$$C_{r1}\Delta_1 + C_{r2}\Delta_2 + \cdots + C_{rn}\Delta_n = G_r \qquad \text{(10.44)}$$

or $$[C][\hat{\Delta}] = [G]$$

In Eq. (10.44) C_{ij} and G_i are known coefficients. In order to simplify our presentation we limit our attention only to cases wherein G_i $(i = 1, 2, \ldots, r)$ are equal to zero. Thus, the constraint equations are of the following form

$$[C]\{\hat{\Delta}\} = 0 \tag{10.45}$$

We will apply Eq. (10.45) to the structure subjected to equivalent actions on its nodes. Thus the assumption that G_i $(i = 1, 2, \ldots, r)$ are equal to zero implies that the components of displacement of the supports of the structure, subjected to the equivalent actions on its nodes, must vanish. That is, due to this assumption, we cannot take into account the effect of the given movement of the supports of a structure when analyzing the structure subjected to the equivalent actions at its nodes. This effect must be considered when analyzing the restrained structure (see Example 3 of Sec. 9.3).

Referring to Fig. 10.4b, the constraints of the structure of Fig.

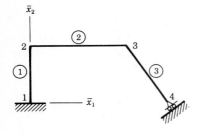

(b) Components of displacements

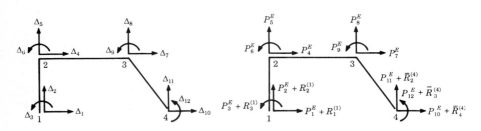

(c) Components of external actions and reactions

Figure 10.4 Planar frame.

10.4a are

$$\left.\begin{matrix} \Delta_1 = 0 \\ \Delta_2 = 0 \\ \Delta_3 = 0 \end{matrix}\right\} \quad \begin{matrix}\text{zero displacement of} \\ \text{fixed support 1}\end{matrix}$$

$$\left.\Delta_{11}\cos\theta - \Delta_{10}\sin\theta = 0 \right\} \quad \begin{matrix}\text{zero translation in} \\ \text{direction normal to} \\ \text{direction of rolling of} \\ \text{support 4}\end{matrix} \qquad (10.46)$$

Relation (10.46) can be written as

$$\begin{bmatrix} 1 & 0 & 0 & 0 & 0 & 0 & 0 & 0 & 0 & 0 & 0 & 0 \\ 0 & 1 & 0 & 0 & 0 & 0 & 0 & 0 & 0 & 0 & 0 & 0 \\ 0 & 0 & 1 & 0 & 0 & 0 & 0 & 0 & 0 & 0 & 0 & 0 \\ 0 & 0 & 0 & 0 & 0 & 0 & 0 & 0 & 0 & -\sin\theta & \cos\theta & 0 \end{bmatrix} \begin{Bmatrix} \Delta_1 \\ \Delta_2 \\ \Delta_3 \\ \Delta_4 \\ \Delta_5 \\ \Delta_6 \\ \Delta_7 \\ \Delta_8 \\ \Delta_9 \\ \Delta_{10} \\ \Delta_{11} \\ \Delta_{12} \end{Bmatrix} = 0 \quad (10.47)$$

If the effect of the axial deformation of the elements of the structure of Fig. 10.4a is disregarded, referring to Fig. 10.4b, its constraint equations are

$$\left.\begin{matrix} \Delta_1 = 0 \\ \Delta_2 = 0 \\ \Delta_3 = 0 \end{matrix}\right\} \quad \begin{matrix}\text{zero displacement of fixed} \\ \text{support 1}\end{matrix}$$

$$\left.\Delta_{11}\cos\theta - \Delta_{10}\sin\theta = 0 \right\} \quad \begin{matrix}\text{zero translation in} \\ \text{direction normal to} \\ \text{direction of rolling of support 4}\end{matrix} \qquad (10.48)$$

$$\left.\begin{matrix} \Delta_5 = 0 \\ \Delta_6 = \Delta_7 \\ \Delta_7\sin\theta - \Delta_8\cos\theta = 0 \end{matrix}\right\} \quad \begin{matrix}\text{negligible change of} \\ \text{length of elements} \\ \text{of frame}\end{matrix}$$

Relations (10.48) can be written as

$$
\begin{bmatrix}
1 & 0 & 0 & 0 & 0 & 0 & 0 & 0 & 0 & 0 & 0 & 0 \\
0 & 1 & 0 & 0 & 0 & 0 & 0 & 0 & 0 & 0 & 0 & 0 \\
0 & 0 & 1 & 0 & 0 & 0 & 0 & 0 & 0 & 0 & 0 & 0 \\
0 & 0 & 0 & 1 & 0 & 0 & -1 & 0 & 0 & 0 & 0 & 0 \\
0 & 0 & 0 & 0 & 1 & 0 & 0 & 0 & 0 & 0 & 0 & 0 \\
0 & 0 & 0 & 0 & 0 & 0 & \sin\theta & -\cos\theta & 0 & 0 & 0 & 0 \\
0 & 0 & 0 & 0 & 0 & 0 & 0 & 0 & 0 & -\sin\theta & \cos\theta & 0
\end{bmatrix}
\begin{Bmatrix}
\Delta_1 \\ \Delta_2 \\ \Delta_3 \\ \Delta_4 \\ \Delta_5 \\ \Delta_6 \\ \Delta_7 \\ \Delta_8 \\ \Delta_9 \\ \Delta_{10} \\ \Delta_{11} \\ \Delta_{12}
\end{Bmatrix} = 0
$$

(10.49)

The r constraint equations of a structure can be used to eliminate r components of displacements, including those which vanish, from its stiffness equations (8.1). We store the components of displacements of the nodes of the structure which are to be eliminated in a matrix which we denote as $\{\Delta_e\}$. Moreover, we store the remaining components of displacements of the nodes of the structure in a matrix which we denote as $\{\Delta_c\}$. For instance, if the effect of axial deformation of the elements of the frame of Fig. 10.4a is disregarded, referring to relations (10.48) we include the following components of displacements in the matrix $\{ \lambda_e \}$:

1. Δ_1, Δ_2, Δ_3, Δ_5 because they vanish

2. Δ_4, Δ_8 because they can be computed from Δ_7

3. One of the global components of translation Δ_{10} or Δ_{11} at support 4 because the other can be computed from the fourth constraint equation (10.48)

That is,

$$
\{\Delta_e\} = \begin{Bmatrix} \Delta_1 \\ \Delta_2 \\ \Delta_3 \\ \Delta_4 \\ \Delta_5 \\ \Delta_8 \\ \Delta_{10} \end{Bmatrix}
\qquad
\{\Delta_c\} = \begin{Bmatrix} \Delta_6 \\ \Delta_7 \\ \Delta_9 \\ \Delta_{11} \\ \Delta_{12} \end{Bmatrix}
\qquad (10.50)
$$

We store the components of external actions and reactions corresponding to the components of displacement $\{\Delta_e\}$ and $\{\Delta_c\}$ in the matrices $\{P_e\}$ and $\{P_c\}$, respectively. Moreover, we store the components of the reactions of the supports of the structure corresponding to the components of displacements $\{\Delta_e\}$ and $\{\Delta_c\}$ in the matrices $\{R_e\}$ and $\{R_c\}$, respectively. That is, referring to Fig. 10.4c for the frame of Fig. 10.4a we have

$$\{P_e\} = \begin{Bmatrix} P_1^E \\ P_2^E \\ P_3^E \\ P_4^E \\ P_5^E \\ P_8^E \\ P_{10}^E \end{Bmatrix} \quad \{R_e\} = \begin{Bmatrix} R_1^{(1)} \\ R_2^{(1)} \\ R_3^{(1)} \\ 0 \\ 0 \\ 0 \\ R_1^{(4)} \end{Bmatrix} \quad [P_c] = \begin{Bmatrix} P_6^E \\ P_7^E \\ P_9^E \\ P_{11}^E \\ P_{12}^E \end{Bmatrix} \quad \{R_c\} = \begin{Bmatrix} 0 \\ 0 \\ 0 \\ R_2^{(4)} \\ 0 \end{Bmatrix} \qquad (10.51)$$

Notice that the matrix $\{R_c\}$ includes the global component of the reaction acting on the skew support of the structure which was chosen as the independent. Thus, if a structure does not have skew supports, the matrix $\{R_c\}$ is a zero matrix.

In order to eliminate the components of displacement $\{\Delta_e\}$ from the stiffness equation (8.1), we do the following:

1. We partition Eq. (10.45) as follows:

$$[[C_e] \quad [C_c]]\begin{Bmatrix} \{\Delta_e\} \\ \{\Delta_c\} \end{Bmatrix} = 0 \qquad (10.52)$$

Notice that $[C_e]$ is an $r \times r$ nonsingular matrix, while $[C_c]$ is an $r \times (n-r)$ matrix. Thus from relation (10.52) we obtain

$$\{\Delta_e\} = [C_{ec}]\{\Delta_c\} \qquad (10.53)$$

where
$$[C_{ec}] = -[C_e]^{-1}[C_c] \qquad (10.54)$$

2. We partition the stiffness equation (8.1) as follows:

$$\begin{Bmatrix} \{P_e\} \\ \{P_c\} \end{Bmatrix} + \begin{Bmatrix} \{R_e\} \\ \{R_c\} \end{Bmatrix} = \begin{bmatrix} [S_{ee}] & [S_{ec}] \\ [S_{ce}] & [S_{cc}] \end{bmatrix} \begin{Bmatrix} \{\Delta_e\} \\ \{\Delta_c\} \end{Bmatrix} \qquad (10.55)$$

or

$$\{P_e\} + \{R_e\} = [S_{ee}][\Delta_e] + [S_{ec}][\Delta_c]$$
$$\{P_c\} + \{R_c\} = [S_{ce}][\Delta_e] + [S_{cc}][\Delta_c] \qquad (10.56)$$

3. We substitute relation (10.53) into relations (10.56) to obtain

$$\{P_e\} + \{R_e\} = [[S_{ee}][C_{ec}] + [S_{ec}]][\Delta_c] \qquad (10.57)$$

$$\{P_c\} + \{R_c\} = [[S_{ce}][C_{ec}] + [S_{cc}]][\Delta_c] \qquad (10.58)$$

Notice that if a structure does not have skew supports $[\{R_c\} = 0]$, the components of displacement $[\Delta_c]$ can be established from relation (10.58). However, the stiffness matrix $[[S_{ce}][C_{ec}] + [S_{ce}]]$ is not symmetric, and consequently its inversion by a computer is time-consuming. In order to establish stiffness equations with a symmetric stiffness matrix which do not involve the reactions of the structure, we multiply relation (10.57) by $[C_{ec}]^T$ and add the resulting expression to (10.58). That is,

$$\{P_c\} + [C_{ec}]^T\{P_e\} + \{R_c\} + [C_{ec}]^T\{R_e\} = [S_c]\{\Delta_c\} \qquad (10.59)$$

where $[S_c]$ is the stiffness matrix of the constrained structure and is equal to

$$[S_c] = [S_{ce}][C_{ec}] + [S_{ce}] + [C_{ec}]^T[[S_{ee}][C_{ec}] + [S_{ec}]] \qquad (10.60)$$

We have stipulated that the specified movements of the supports of a structure are to be taken into account when analyzing the restrained structure. Thus the supports of the structure subjected to the equivalent actions on its nodes do not move and consequently the work of its reactions vanishes. That is,

$$\{\hat{R}\}^T\{\hat{\Delta}\} = \{\hat{\Delta}\}^T\{\hat{R}\} = 0 \qquad (10.61)$$

or $$\left\{ \begin{matrix} \{\Delta_e\} \\ \{\Delta_c\} \end{matrix} \right\}^T \left\{ \begin{matrix} \{R_e\} \\ \{R_c\} \end{matrix} \right\} = [\{\Delta_e\}^T, \{\Delta_c\}^T] \left\{ \begin{matrix} \{R_e\} \\ \{R_c\} \end{matrix} \right\} = 0 \qquad (10.62)$$

Hence $$\{\Delta_e\}^T\{R_e\} + \{\Delta_c\}^T\{R_c\} = 0 \qquad (10.63)$$

Substituting relation (10.53) into (10.63) we get

$$\{\Delta_c\}^T[[C_{ec}]^T\{R_e\} + \{R_c\}] \qquad (10.64)$$

Consequently $$\{R_c\} + [C_{ec}]^T\{R_e\} = 0 \qquad (10.65)$$

Thus relation (10.59) reduces to

$$\{\tilde{P}_c\} = [S_c]\{\Delta_c\} \qquad (10.66)$$

where $$\{\tilde{P}_c\} = \{P_c\} + [C_{ec}]^T\{P_e\} \qquad (10.67)$$

$\{\tilde{P}_c\}$ is referred to as the *matrix of effective actions*. The matrix

$[C_{ec}]^T\{P_e\}$ represents the actions which must be added to $\{P_c\}$ in order to account for the effect on the components of displacements $\{\Delta_c\}$ of the actions $\{P_e\}$.

Equation (10.66) can be solved to obtain the components of displacements $\{\Delta_c\}$. The components of displacements $\{\Delta_e\}$ may then be obtained from relation (10.53). Moreover, the nodal actions of the elements of the structure may be established from the components of displacements of its nodes (see examples of Sec. 9.3).

When the effect of axial deformation of the elements of a structure is disregarded, the axial components of nodal displacements of the elements of the structure obtained from the components of displacement $\{\Delta_c\}$ and $\{\Delta_e\}$ of the nodes of the structure are equal to zero. Consequently, in this case, the axial component of nodal forces of the elements of the structure cannot be obtained from the components of displacements $\{\Delta_c\}$ and $\{\Delta_e\}$. Neither can the reactions of a structure be established using relation (10.57). The axial components of nodal forces of the elements of the structure and its reactions are obtained by considering the equilibrium of its nodes (see Example 2 at the end of this section.)

As discussed in Sec. 9.1, the basic stiffness matrix of a structure $[S^{FF}]$ may be obtained from its stiffness matrix $[\hat{S}]$ by taking into account the conditions at its supports. The procedure described above can also be used to eliminate from the basic stiffness Eqs. (9.4) $[\{P^{EF}\} = [S^{FF}]\{\Delta^F\}]$ the additional constraints imposed by the assumption that the effect of the axial components of displacement of the elements of a structure is negligible. In this case, in relation (10.45) we include only the constraints imposed by the aforementioned assumption. The constraints imposed by the supports of the structure have been taken into account in obtaining the basic stiffness equations for the structure.

In the sequel, we illustrate the method presented in this section by the following two examples:

Example 1. The stiffness matrix $[S_c]$ of a simple truss is established from its stiffness matrix $[\hat{S}]$. The truss has a skew support and, thus, the global components of translation at this skew support are related.

Example 2. The stiffness matrix $[S_c]$ of a frame is established from its stiffness matrix $[\hat{S}]$. The effect of the axial deformation of the frame is disregarded.

Example 1. Using the procedure described in this section, establish the components of translation of the nodes of the truss shown in Fig. a. The elements of the truss are made from the same isotropic linearly elastic material and have the same constant cross section ($AE = 20,000$ kN).

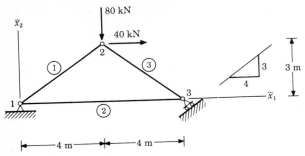

Figure a Geometry and loading of the truss.

solution

STEPS 1 TO 5. The stiffness matrix of the truss has been computed in Example 1 of Sec. 8.3.1 as

$$[\hat{S}] = \frac{EA}{125}\begin{bmatrix} 31.625 & 12 & -16 & -12 & -15.625 & 0 \\ 12 & 9 & -12 & -9 & 0 & 0 \\ -16 & -12 & 32 & 0 & -16 & 12 \\ -12 & -9 & 0 & 18 & 12 & -9 \\ -15.625 & 0 & -16 & 12 & 31.625 & -12 \\ 0 & 0 & 12 & -9 & -12 & 9 \end{bmatrix} \quad \text{(a)}$$

STEP 6. We incorporate the constraining conditions of the truss into its stiffness equations. To accomplish this we form the constraint equations for the truss. That is, referring to Fig. b we have

$$\begin{bmatrix} 1 & 0 & 0 & \vdots & 0 & 0 & 0 \\ 0 & 1 & 0 & \vdots & 0 & 0 & 0 \\ 0 & 0 & 0.8 & \vdots & 0 & 0 & -0.6 \end{bmatrix}\begin{Bmatrix} \Delta_1 \\ \Delta_2 \\ \Delta_6 \\ \text{----} \\ \Delta_3 \\ \Delta_4 \\ \Delta_5 \end{Bmatrix} = 0 \quad \text{(b)}$$

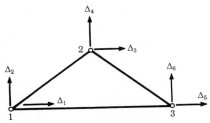

Figure b Numbering of the components of displacements of the nodes of the truss.

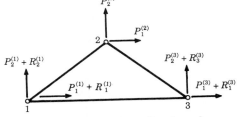

Figure c Components of external actions and reactions of the truss.

Referring to relation (10.52) from the above relation we have

$$[C_e] = \begin{bmatrix} 1 & 0 & 0 \\ 0 & 1 & 0 \\ 0 & 0 & 0.8 \end{bmatrix}$$ (c)

$$[C_c] = \begin{bmatrix} 0 & 0 & 0 \\ 0 & 0 & 0 \\ 0 & 0 & -0.6 \end{bmatrix}$$ (d)

$$\{\Delta_e\} = \begin{Bmatrix} \Delta_1 \\ \Delta_2 \\ \Delta_6 \end{Bmatrix}$$ (e)

$$\{\Delta_c\} = \begin{Bmatrix} \Delta_3 \\ \Delta_4 \\ \Delta_5 \end{Bmatrix}$$ (f)

Referring to Figs. a and c and relations (b) and (a) the stiffness equations for the truss can be written as

$$\begin{Bmatrix} 0 \\ 0 \\ 0 \\ \hline 40 \\ -80 \\ 0 \end{Bmatrix} = \begin{Bmatrix} R_1^{(1)} \\ R_2^{(1)} \\ R_1^{(3)} \\ \hline 0 \\ 0 \\ R_2^{(3)} \end{Bmatrix} = 160 \begin{bmatrix} 31.625 & 12 & 0 & -16 & -12 & -15.625 \\ 12 & 9 & 0 & -12 & -9 & 0 \\ 0 & 0 & 9 & 12 & -9 & -12 \\ \hline -16 & -12 & 12 & 32 & 0 & -16 \\ -12 & -9 & -9 & 0 & 18 & 12 \\ -15.625 & 0 & -12 & -16 & 12 & 31.625 \end{bmatrix} \begin{Bmatrix} \Delta_1 \\ \Delta_2 \\ \Delta_6 \\ \hline \Delta_3 \\ \Delta_4 \\ \Delta_5 \end{Bmatrix}$$ (g)

Hence, referring to relations (g) and (10.55) we have

$$\{P_e\} = \begin{Bmatrix} 0 \\ 0 \\ 0 \end{Bmatrix}$$ (h)

$$\{R_e\} = \begin{Bmatrix} R_1^{(1)} \\ R_2^{(1)} \\ R_1^{(3)} \end{Bmatrix}$$ (i)

$$\{P_c\} = \begin{Bmatrix} 40 \\ -80 \\ 0 \end{Bmatrix}$$ (j)

$$\{R_c\} = \begin{Bmatrix} 0 \\ 0 \\ R_2^{(3)} \end{Bmatrix}$$ (k)

$$[S_{ee}] = 160 \begin{bmatrix} 31.625 & 12 & 0 \\ 12 & 9 & 0 \\ 0 & 0 & 9 \end{bmatrix} \tag{l}$$

$$[S_{ec}] = 160 \begin{bmatrix} -16 & -12 & -15.625 \\ -12 & -9 & 0 \\ 12 & -9 & -12 \end{bmatrix} \tag{m}$$

$$[S_{ce}] = 160 \begin{bmatrix} -16 & -12 & 12 \\ -12 & -9 & -9 \\ -15.625 & 0 & -12 \end{bmatrix} \tag{n}$$

$$[S_{cc}] = 160 \begin{bmatrix} 32 & 0 & -12 \\ 0 & 18 & 12 \\ -16 & 12 & 31.625 \end{bmatrix} \tag{o}$$

Substituting relations (c) and (d) into (10.54) we get

$$[C_{ec}] = -[C_e]^{-1}[C_c] = \begin{bmatrix} 0 & 0 & 0 \\ 0 & 0 & 0 \\ 0 & 0 & 0.75 \end{bmatrix} \tag{p}$$

Substituting relations (h), (j), and (p) into (10.67) we obtain

$$\{\tilde{P}_c\} = \{P_c\} + [C_{ec}]^T\{P_e\} = \begin{Bmatrix} 40 \\ -80 \\ 0 \end{Bmatrix} + \begin{Bmatrix} 0 \\ 0 \\ 0 \end{Bmatrix} = \begin{Bmatrix} 40 \\ -80 \\ 0 \end{Bmatrix} \tag{q}$$

Moreover, using relations (i), (k), and (p) we get

$$\{R_c\} + [C_{ec}]^T\{R_e\} = \begin{Bmatrix} 0 \\ 0 \\ R_2^{(3)} \end{Bmatrix} + \begin{bmatrix} 0 & 0 & 0 \\ 0 & 0 & 0 \\ 0 & 0 & 0.75 \end{bmatrix} \begin{Bmatrix} R_1^{(1)} \\ R_2^{(1)} \\ R_1^{(3)} \end{Bmatrix}$$

$$= \begin{Bmatrix} 0 \\ 0 \\ R_2^{(3)} + 0.75R_1^{(3)} \end{Bmatrix} = \begin{Bmatrix} 0 \\ 0 \\ 0 \end{Bmatrix} \tag{r}$$

The last matrix in relation (r) was obtained by noting that the reactions at support 4 are normal to the direction of rolling. Thus from geometric considerations we have $R_1^{(3)} = -\frac{4}{3}R_2^{(3)}$

Substituting relation (l) to (p) into relation (10.60) we get

$$[S_c] = 160 \begin{bmatrix} 32 & 0 & -7 \\ 0 & 18 & 5.25 \\ -7 & 5.25 & 18.6875 \end{bmatrix} \tag{s}$$

STEP 7. We compute the components of displacements of the nodes of the truss. Substituting relations (f), (j), and (s) into (10.59) we obtain

$$\{\Delta_c\} = \begin{Bmatrix} \Delta_3 \\ \Delta_4 \\ \Delta_5 \end{Bmatrix} = \frac{1}{160} \begin{bmatrix} 32 & 0 & -7 \\ 0 & 18 & 5.25 \\ -7 & 5.25 & 18.687 \end{bmatrix}^{-1} \begin{Bmatrix} 40 \\ -80 \\ 0 \end{Bmatrix} = \begin{Bmatrix} 0.01062 \\ -0.03152 \\ 0.01283 \end{Bmatrix} \tag{t}$$

Substituting relations (p) and (t) into (10.53), we have

$$\{\Delta_e\} = \begin{Bmatrix} \Delta_1 \\ \Delta_2 \\ \Delta_6 \end{Bmatrix} = \begin{bmatrix} 0 & 0 & 0 \\ 0 & 0 & 0 \\ 0 & 0 & 0.75 \end{bmatrix} \begin{Bmatrix} 0.01062 \\ -0.03152 \\ 0.01283 \end{Bmatrix} = \begin{Bmatrix} 0 \\ 0 \\ 0.0096225 \end{Bmatrix} \text{ m} \qquad (u)$$

The components of displacement of point 3 in the direction of rolling is equal to

$$\Delta_5' = \tfrac{4}{5}\Delta_5 + \tfrac{3}{5}\Delta_6 = 0.01604 \text{ m} \qquad (v)$$

STEP 8.   The local components of internal forces in the elements of the truss are computed in the example of Sec. 10.2.

Example 2. Using the procedure described in this subsection, compute the components of displacement of the nodes and the nodal actions of the elements of the planar frame loaded as shown in Fig. a. The elements of the frame are made from the same isotropic linearly elastic material ($E = 210\,\text{kN/mm}^2$) and have the same constant cross section with moment of inertia to area ratio $I/A = 0.02\,\text{m}^2$. Disregard the effect of axial deformation of the elements of the frame.

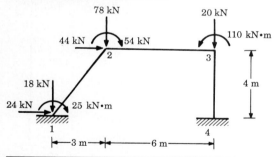

Figure a Geometry and loading of the frame.

solution

STEPS 1 AND 2.   We form the matrix of equivalent actions for the frame. Referring to Fig. a we have

$$\{P^E\} = \begin{Bmatrix} 24 \\ -18 \\ -25 \\ 44 \\ -78 \\ -5 \\ 0 \\ -20 \\ 110 \\ 0 \\ 0 \\ 0 \end{Bmatrix} \qquad (a)$$

STEP 3. We compute the local stiffness matrix of each element of the structure. That is, referring to relation (4.62) we have

$$[K^1] = EI \begin{bmatrix} 10 & 0 & 0 & -10 & 0 & 0 \\ 0 & 0.096 & 0.24 & 0 & -0.096 & 0.24 \\ 0 & 0.240 & 0.80 & 0 & -0.240 & 0.40 \\ -10 & 0 & 0 & 10 & 0 & 0 \\ 0 & -0.096 & -0.24 & 0 & 0.096 & -0.24 \\ 0 & 0.240 & 0.40 & 0 & -0.240 & 0.80 \end{bmatrix}$$

$$[K^2] = EI \begin{bmatrix} 8.3333 & 0 & 0 & -8.3333 & 0 & 0 \\ 0 & 0.0556 & 0.1667 & 0 & -0.0566 & 0.1667 \\ 0 & 0.1667 & 0.6667 & 0 & -0.1667 & 0.3333 \\ -8.3333 & 0 & 0 & 8.3333 & 0 & 0 \\ 0 & -0.0556 & -0.1667 & 0 & 0.0556 & -0.1667 \\ 0 & 0.1667 & 0.3333 & 0 & -0.1667 & 0.6667 \end{bmatrix}$$

$$[K^3] = EI \begin{bmatrix} 12.50 & 0 & 0 & -12.50 & 0 & 0 \\ 0 & 0.1875 & 0.375 & 0 & -0.1875 & 0.375 \\ 0 & 0.3750 & 1.000 & 0 & -0.3750 & 0.500 \\ -12.50 & 0 & 0 & 12.50 & 0 & 0 \\ 0 & -0.1875 & -0.375 & 0 & 0.1875 & -0.375 \\ 0 & 0.3750 & 0.500 & 0 & -0.3750 & 1.000 \end{bmatrix}$$

(b)

STEP 4. From the local stiffness matrix of each element of the structure, we compute its global stiffness matrix using relation (6.74). That is,

$$[\bar{K}] = [\Lambda]^T [K][\Lambda] \tag{c}$$

Referring to relation (6.48) the transformation matrices for the elements of the structure are

$$[\Lambda^1] = \begin{bmatrix} 0.6 & 0.8 & 0 & 0 & 0 & 0 \\ -0.8 & 0.6 & 0 & 0 & 0 & 0 \\ 0 & 0 & 1 & 0 & 0 & 0 \\ 0 & 0 & 0 & 0.6 & 0.8 & 0 \\ 0 & 0 & 0 & -0.8 & 0.6 & 0 \\ 0 & 0 & 0 & 0 & 0 & 1 \end{bmatrix}$$

(d)

$[\Lambda^2]$ = unit matrix

$$[\Lambda^3] = \begin{bmatrix} 0 & -1 & 0 & 0 & 0 & 0 \\ 1 & 0 & 0 & 0 & 0 & 0 \\ 0 & 0 & 1 & 0 & 0 & 0 \\ 0 & 0 & 0 & 0 & -1 & 0 \\ 0 & 0 & 0 & 1 & 0 & 0 \\ 0 & 0 & 0 & 0 & 0 & 1 \end{bmatrix}$$

Substituting relations (b) and (d) into (c), we obtain

$$[\bar{K}^1] = EI \begin{bmatrix} 3.6614 & 4.7539 & -0.192 & -3.6614 & -4.7539 & -0.192 \\ 4.7539 & 6.4346 & 0.144 & -4.7539 & -6.4346 & 0.144 \\ -0.1920 & 0.1440 & 0.800 & 0.1920 & -0.1440 & 0.400 \\ -3.6614 & -4.7539 & 0.192 & 3.6614 & 4.7539 & 0.192 \\ -4.7539 & -6.4346 & -0.144 & 4.7539 & 6.4346 & -0.144 \\ -0.1920 & 0.1440 & 0.400 & 0.1920 & -0.1440 & 0.800 \end{bmatrix}$$

$$[\bar{K}^2] = [K^2] \tag{e}$$

$$[\bar{K}^3] = EI \begin{bmatrix} 0.1875 & 0 & 0.375 & -0.1875 & 0 & 0.375 \\ 0 & 12.50 & 0 & 0 & -12.50 & 0 \\ 0.3750 & 0 & 1.00 & -0.3750 & 0 & 0.500 \\ -0.1875 & 0 & -0.375 & 0.1875 & 0 & -0.375 \\ 0 & -12.50 & 0 & 0 & 12.50 & 0 \\ 0.3750 & 0 & 0.500 & -0.3750 & 0 & 1.00 \end{bmatrix}$$

STEP 5. Using the global stiffness matrices of the elements of the structure, we assemble its stiffness matrix by adhering to one of the procedures described in Sec. 8.3. Thus,

$$[\hat{S}] = EI$$

$$\begin{bmatrix}
3.6614 & 4.7539 & -0.192 & -3.6614 & -4.7539 & -0.1920 & 0 & 0 & 0 & 0 & 0 & 0 \\
4.7539 & 6.4346 & 0.144 & -4.7539 & -6.4346 & 0.1440 & 0 & 0 & 0 & 0 & 0 & 0 \\
-0.1920 & 0.1440 & 0.800 & 0.1920 & -0.1440 & 0.4000 & 0 & 0 & 0 & 0 & 0 & 0 \\
-3.6614 & -4.7539 & 0.192 & 11.9947 & 4.7539 & 0.1920 & -8.3333 & 0 & 0 & 0 & 0 & 0 \\
-4.7539 & -6.4346 & -0.144 & 4.7539 & 6.4902 & 0.0227 & 0 & -0.0556 & 0.16667 & 0 & 0 & 0 \\
-0.1920 & 0.1440 & 0.400 & 0.1920 & 0.0227 & 1.4667 & 0 & -0.1667 & 0.3333 & 0 & 0 & 0 \\
0 & 0 & 0 & -8.3333 & 0 & 0 & 8.5208 & 0 & 0.3750 & -0.1875 & 0 & 0.375 \\
0 & 0 & 0 & 0 & -0.0556 & -0.1667 & 0 & 12.5556 & -0.1667 & 0 & -12.50 & 0 \\
0 & 0 & 0 & 0 & 0.1667 & 0.3333 & 0.3750 & -0.1667 & 1.6667 & -0.3750 & 0 & 0.500 \\
0 & 0 & 0 & 0 & 0 & 0 & -0.1875 & 0 & -0.3750 & 0.1875 & 0 & -0.375 \\
0 & 0 & 0 & 0 & 0 & 0 & 0 & -12.5000 & 0 & 0 & 12.50 & 0 \\
0 & 0 & 0 & 0 & 0 & 0 & 0.3750 & 0 & 0.5000 & -0.3750 & 0 & 1.00
\end{bmatrix} \quad (f)$$

STEP 6. We incorporate the constraining conditions of the frame into its stiffness equations. To accomplish this we form the constraint equations for the frame.

That is, referring to Fig. b, we have

$$\left. \begin{aligned} \Delta_1 &= 0 \\ \Delta_2 &= 0 \\ \Delta_3 &= 0 \\ \Delta_{10} &= 0 \\ \Delta_{11} &= 0 \\ \Delta_{12} &= 0 \end{aligned} \right\} \quad \begin{aligned} &\text{zero displacement of} \\ &\text{fixed supports} \end{aligned}$$

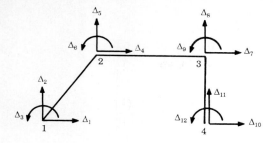

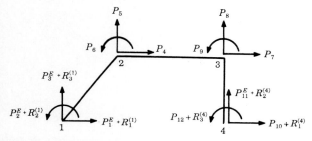

Figure c Components of external actions and reactions of the frame.

$$\left.\begin{array}{r} \Delta_4 \cos\theta - \Delta_5 \sin\theta = 0 \\ \Delta_4 = \Delta_7 \\ \Delta_8 = 0 \end{array}\right\} \quad \begin{array}{l} \text{negligible change of length} \\ \text{of elements of frame} \end{array} \qquad (g)$$

Relations (g) can be rewritten as

$$\begin{bmatrix} 1 & 0 & 0 & 0 & 0 & 0 & 0 & 0 & 0 & \vdots & 0 & 0 & 0 \\ 0 & 1 & 0 & 0 & 0 & 0 & 0 & 0 & 0 & \vdots & 0 & 0 & 0 \\ 0 & 0 & 1 & 0 & 0 & 0 & 0 & 0 & 0 & \vdots & 0 & 0 & 0 \\ 0 & 0 & 0 & 1 & 0 & 0 & 0 & 0 & 0 & \vdots & 0 & -1 & 0 \\ 0 & 0 & 0 & 0.6 & 0.8 & 0 & 0 & 0 & 0 & \vdots & 0 & 0 & 0 \\ 0 & 0 & 0 & 0 & 0 & 1 & 0 & 0 & 0 & \vdots & 0 & 0 & 0 \\ 0 & 0 & 0 & 0 & 0 & 0 & 1 & 0 & 0 & \vdots & 0 & 0 & 0 \\ 0 & 0 & 0 & 0 & 0 & 0 & 0 & 1 & 0 & \vdots & 0 & 0 & 0 \\ 0 & 0 & 0 & 0 & 0 & 0 & 0 & 0 & 1 & \vdots & 0 & 0 & 0 \end{bmatrix} \begin{Bmatrix} \Delta_1 \\ \Delta_2 \\ \Delta_3 \\ \Delta_4 \\ \Delta_5 \\ \Delta_8 \\ \Delta_{10} \\ \Delta_{11} \\ \Delta_{12} \\ \text{-----} \\ \Delta_6 \\ \Delta_7 \\ \Delta_9 \end{Bmatrix} = 0 \qquad (h)$$

Thus referring to relation (10.52) we have

$$[C_e] = \begin{bmatrix} 1 & 0 & 0 & 0 & 0 & 0 & 0 & 0 & 0 \\ 0 & 1 & 0 & 0 & 0 & 0 & 0 & 0 & 0 \\ 0 & 0 & 1 & 0 & 0 & 0 & 0 & 0 & 0 \\ 0 & 0 & 0 & 1 & 0 & 0 & 0 & 0 & 0 \\ 0 & 0 & 0 & 0.6 & 0.8 & 0 & 0 & 0 & 0 \\ 0 & 0 & 0 & 0 & 0 & 1 & 0 & 0 & 0 \\ 0 & 0 & 0 & 0 & 0 & 0 & 1 & 0 & 0 \\ 0 & 0 & 0 & 0 & 0 & 0 & 0 & 1 & 0 \\ 0 & 0 & 0 & 0 & 0 & 0 & 0 & 0 & 1 \end{bmatrix} \qquad \text{(i)}$$

$$[C_c] = \begin{bmatrix} 0 & 0 & 0 \\ 0 & 0 & 0 \\ 0 & 0 & 0 \\ 0 & -1 & 0 \\ 0 & 0 & 0 \\ 0 & 0 & 0 \\ 0 & 0 & 0 \\ 0 & 0 & 0 \\ 0 & 0 & 0 \end{bmatrix} \qquad \text{(j)}$$

$$\{\Delta_e\} = \begin{Bmatrix} \Delta_1 \\ \Delta_2 \\ \Delta_3 \\ \Delta_4 \\ \Delta_5 \\ \Delta_8 \\ \Delta_{10} \\ \Delta_{11} \\ \Delta_{12} \end{Bmatrix} \qquad \text{(k)}$$

$$\{\Delta_c\} = \begin{Bmatrix} \Delta_6 \\ \Delta_7 \\ \Delta_9 \end{Bmatrix} \qquad \text{(l)}$$

Substituting relations (i) and (j) into (10.54) we have

$$[C_{ec}] = -[C_e]^{-1}[C_c] = \begin{bmatrix} 0 & 0 & 0 \\ 0 & 0 & 0 \\ 0 & 0 & 0 \\ 0 & 1 & 0 \\ 0 & -0.75 & 0 \\ 0 & 0 & 0 \\ 0 & 0 & 0 \\ 0 & 0 & 0 \\ 0 & 0 & 0 \end{bmatrix} \quad \text{(m)}$$

$$\begin{Bmatrix} 24 \\ -18 \\ -25 \\ 44 \\ 78 \\ -20 \\ 0 \\ 0 \\ 0 \\ \hdashline -5 \\ 0 \\ 110 \end{Bmatrix} + \begin{Bmatrix} R_1^{(1)} \\ R_2^{(1)} \\ R_3^{(1)} \\ 0 \\ 0 \\ 0 \\ R_1^{(4)} \\ R_2^{(4)} \\ R_3^{(4)} \\ \hdashline 0 \\ 0 \\ 0 \end{Bmatrix} = EI \begin{bmatrix} 3.6614 & 4.7539 & -0.192 & -3.6614 & -4.7539 \\ 4.7539 & 6.4346 & 0.144 & -4.7539 & -6.4346 \\ -0.1920 & 0.1440 & 0.800 & 0.1920 & -0.1440 \\ -3.6614 & -4.7539 & 0.192 & 11.9947 & 4.7539 \\ -4.7539 & -6.4346 & -0.144 & 4.7539 & 6.4902 \\ 0 & 0 & 0 & 0 & -0.0556 \\ 0 & 0 & 0 & 0 & 0 \\ 0 & 0 & 0 & 0 & 0 \\ 0 & 0 & 0 & 0 & 0 \\ \hdashline -0.1920 & 0.1440 & 0.400 & 0.1920 & 0.0227 \\ 0 & 0 & 0 & -8.3333 & 0 \\ 0 & 0 & 0 & 0 & 0.1667 \end{bmatrix}$$

$$\begin{bmatrix} 0 & 0 & 0 & 0 & \vdots & -0.1920 & 0 & 0 \\ 0 & 0 & 0 & 0 & \vdots & 0.1440 & 0 & 0 \\ 0 & 0 & 0 & 0 & \vdots & 0.4000 & 0 & 0 \\ 0 & 0 & 0 & 0 & \vdots & 0.1920 & -8.3333 & 0 \\ -0.0556 & 0 & 0 & 0 & \vdots & 0.0227 & 0 & 0.1667 \\ 12.5556 & 0 & -12.50 & 0 & \vdots & -0.1667 & 0 & -0.1667 \\ 0 & 0.1875 & 0 & -0.375 & \vdots & 0 & -0.1875 & -0.3750 \\ -12.5000 & 0 & 12.50 & 0 & \vdots & 0 & 0 & 0 \\ 0 & -0.3750 & 0 & 1.000 & \vdots & 0 & 0.3750 & 0.5000 \\ \hdashline -0.1667 & 0 & 0 & 0 & \vdots & 1.4667 & 0 & 0.3333 \\ 0 & -0.1875 & 0 & 0.375 & \vdots & 0 & 8.5208 & 0.3750 \\ -0.1667 & -0.3750 & 0 & 0.500 & \vdots & 0.3333 & 0.3750 & 1.6667 \end{bmatrix} \begin{Bmatrix} \Delta_1 \\ \Delta_2 \\ \Delta_3 \\ \Delta_4 \\ \Delta_5 \\ \Delta_8 \\ \Delta_{10} \\ \Delta_{11} \\ \Delta_{12} \\ \hdashline \Delta_6 \\ \Delta_7 \\ \Delta_9 \end{Bmatrix} \quad \text{(n)}$$

Referring to Figs. a and c and relations (a) and (h) we write the stiffness equations (n) for the structure. From these equations and (10.55) we have

$$\{P_e\} = \begin{Bmatrix} 24 \\ -18 \\ -25 \\ 44 \\ -78 \\ -20 \\ 0 \\ 0 \\ 0 \end{Bmatrix} \tag{o}$$

$$\{R_e\} = \begin{Bmatrix} R_1^{(1)} \\ R_2^{(1)} \\ R_3^{(1)} \\ 0 \\ 0 \\ 0 \\ R_1^{(4)} \\ R_2^{(4)} \\ R_3^{(4)} \end{Bmatrix} \tag{p}$$

$$\{P_c\} = \begin{Bmatrix} -5 \\ 0 \\ 110 \end{Bmatrix} \tag{q}$$

$$\{R_c\} = \begin{Bmatrix} 0 \\ 0 \\ 0 \end{Bmatrix} \tag{r}$$

and

$$[S_{ee}] = EI \begin{bmatrix} 3.6614 & 4.7539 & -0.192 & -3.6614 & -4.7539 & 0 & 0 & 0 & 0 \\ 4.7539 & 6.4346 & 0.144 & -4.7539 & -6.4346 & 0 & 0 & 0 & 0 \\ -0.1920 & 0.1440 & 0.800 & 0.1920 & -0.1440 & 0 & 0 & 0 & 0 \\ -3.6614 & -4.7539 & 0.192 & 11.9947 & 4.7539 & 0 & 0 & 0 & 0 \\ -4.7539 & -6.4346 & -0.144 & 4.7539 & 6.4902 & -0.0556 & 0 & 0 & 0 \\ 0 & 0 & 0 & 0 & -0.0556 & 12.5556 & 0 & -12.5 & 0 \\ 0 & 0 & 0 & 0 & 0 & 0 & 0.1875 & 0 & -0.375 \\ 0 & 0 & 0 & 0 & 0 & -12.5000 & 0 & 12.5 & 0 \\ 0 & 0 & 0 & 0 & 0 & 0 & -0.3750 & 0 & 1.000 \end{bmatrix} \tag{s}$$

$$[S_{ec}] = EI \begin{bmatrix} -0.1920 & 0 & 0 \\ 0.1440 & 0 & 0 \\ 0.4000 & 0 & 0 \\ 0.1920 & -8.3333 & 0 \\ 0.0227 & 0 & 0.1667 \\ -0.1667 & 0 & -0.1667 \\ 0 & -0.1875 & -0.3750 \\ 0 & 0 & 0 \\ 0 & 0.3750 & 0.5000 \end{bmatrix} \tag{t}$$

$$[S_{ce}] = EI \begin{bmatrix} -0.192 & 0.144 & 0.4 & 0.1920 & 0.0227 & -0.1667 & 0 & 0 & 0 \\ 0 & 0 & 0 & -8.3333 & 0 & 0 & -0.1875 & 0 & 0.375 \\ 0 & 0 & 0 & 0 & 0.1667 & -0.1667 & -0.3750 & 0 & 0.500 \end{bmatrix} \tag{u}$$

$$[S_{cc}] = EI \begin{bmatrix} 1.4667 & 0 & 0.3333 \\ 0 & 8.5208 & 0.3750 \\ 0.3333 & 0.3750 & 1.6667 \end{bmatrix} \tag{v}$$

Substituting relations (m), (o), and (q), into (10.67) we obtain

$$\{\tilde{P}_c\} = \begin{Bmatrix} -5 \\ 0 \\ 110 \end{Bmatrix} + \begin{bmatrix} 0 & 0 & 0 & 0 & 0 & 0 & 0 & 0 & 0 \\ 0 & 0 & 0 & 1 & -0.75 & 0 & 0 & 0 & 0 \\ 0 & 0 & 0 & 0 & 0 & 0 & 0 & 0 & 0 \end{bmatrix} \begin{Bmatrix} 0 \\ 0 \\ 0 \\ 44 \\ -78 \\ -20 \\ 0 \\ 0 \\ 0 \end{Bmatrix} = \begin{Bmatrix} -5.0 \\ 102.5 \\ 110.0 \end{Bmatrix} \tag{w}$$

Moreover, using relations (m), (p), and (r) as expected we get

$$\{R_c\} + [C_{ec}]^T \{R_e\} = \begin{Bmatrix} 0 \\ 0 \\ 0 \end{Bmatrix} + \begin{bmatrix} 0 & 0 & 0 & 0 & 0 & 0 & 0 & 0 & 0 \\ 0 & 0 & 0 & 1 & -0.75 & 0 & 0 & 0 & 0 \\ 0 & 0 & 0 & 0 & 0 & 0 & 0 & 0 & 0 \end{bmatrix} \begin{Bmatrix} R_1^{(1)} \\ R_2^{(1)} \\ R_3^{(1)} \\ 0 \\ 0 \\ 0 \\ R_1^{(4)} \\ R_2^{(4)} \\ R_3^{(4)} \end{Bmatrix} = 0 \tag{x}$$

Substituting relations (m) and (s) to (v) into (10.60) we obtain

$$
[S_c] = EI \begin{bmatrix} 0 & 0.175 & 0 \\ 0 & -8.333 & 0 \\ 0 & -0.125 & 0 \end{bmatrix} + EI \begin{bmatrix} 1.4667 & 0 & 0.3333 \\ 0 & 8.5208 & 0.3750 \\ 0.3333 & 0.3750 & 1.6667 \end{bmatrix}
$$

$$
+ EI \begin{bmatrix} 0 & 0 & 0 \\ 0.1750 & 0.1814 & -0.1250 \\ 0 & 0 & 0 \end{bmatrix} = EI \begin{bmatrix} 1.4667 & 0.1750 & 0.3333 \\ 0.1750 & 0.3689 & 0.2500 \\ 0.3333 & 0.2500 & 1.6667 \end{bmatrix}
$$

$$(y)$$

STEP 7.   We compute the components of displacement of the nodes of the frame. Substituting relations (1), (w), and (y) into (10.59) we get

$$
\{\Delta_c\} = \begin{Bmatrix} \Delta_6 \\ \Delta_7 \\ \Delta_8 \end{Bmatrix} = [S_c]^{-1}\{\tilde{P}_c\} = \frac{1}{EI} \begin{bmatrix} 1.4667 & 0.1750 & 0.3333 \\ 0.1750 & 0.3695 & 0.2500 \\ 0.3333 & 0.2500 & 0.6667 \end{bmatrix}^{-1} \begin{Bmatrix} -5.0 \\ 102.5 \\ 110.0 \end{Bmatrix}
$$

$$
= \frac{1}{EI} \begin{Bmatrix} -43.93 \\ 276.08 \\ 33.38 \end{Bmatrix}
$$

$$(z)$$

Moreover, substituting relations (m) and (z) into (10.53) we have

$$
\{\Delta_e\} = \begin{Bmatrix} \Delta_1 \\ \Delta_2 \\ \Delta_3 \\ \Delta_4 \\ \Delta_5 \\ \Delta_8 \\ \Delta_{10} \\ \Delta_{11} \\ \Delta_{12} \end{Bmatrix} = \begin{bmatrix} 0 & 0 & 0 \\ 0 & 0 & 0 \\ 0 & 0 & 0 \\ 0 & 1 & 0 \\ 0 & -0.75 & 0 \\ 0 & 0 & 0 \\ 0 & 0 & 0 \\ 0 & 0 & 0 \\ 0 & 0 & 0 \end{bmatrix} \frac{1}{EI} \begin{Bmatrix} -43.93 \\ 276.08 \\ 33.38 \end{Bmatrix} = \frac{1}{EI} \begin{Bmatrix} 0 \\ 0 \\ 0 \\ 206.08 \\ -207.06 \\ 0 \\ 0 \\ 0 \\ 0 \end{Bmatrix} \quad (aa)
$$

STEP 8.   We compute the local components of nodal actions of the elements of the frame. Referring to Fig. b and relations (z) and (aa) the global components of the nodal displacements of the elements of the frame are

$$
\{\bar{D}^1\} = \frac{1}{EI}
\begin{Bmatrix}
0 \\
0 \\
0 \\
276.08 \\
-207.06 \\
-43.93
\end{Bmatrix}
\qquad
\{\bar{D}^2\} = \frac{1}{EI}
\begin{Bmatrix}
276.08 \\
-207.06 \\
-43.93 \\
276.08 \\
0 \\
33.38
\end{Bmatrix}
\qquad
\{\bar{D}^3\} = \frac{1}{EI}
\begin{Bmatrix}
276.08 \\
0 \\
33.38 \\
0 \\
0 \\
0
\end{Bmatrix}
$$

$$(bb)$$

Using relation (6.67) we transform the matrices $[\bar{D}^e]$ $(e = 1, 2, 3)$ to the matrices $\{D^e\}$. Thus

$$
\{D^1\} =
\begin{bmatrix}
0.6 & 0.8 & 0 & 0 & 0 & 0 \\
-0.8 & 0.6 & 0 & 0 & 0 & 0 \\
0 & 0 & 1 & 0 & 0 & 0 \\
0 & 0 & 0 & 0.6 & 0.8 & 0 \\
0 & 0 & 0 & -0.8 & 0.6 & 0 \\
0 & 0 & 0 & 0 & 0 & 1
\end{bmatrix}
\frac{1}{EI}
\begin{Bmatrix}
0 \\
0 \\
0 \\
276.08 \\
-207.06 \\
-43.93
\end{Bmatrix}
= \frac{1}{EI}
\begin{Bmatrix}
0 \\
0 \\
0 \\
0 \\
-345.11 \\
-43.93
\end{Bmatrix}
$$

$$\{D^2\} = [\bar{D}^2]$$

$$(cc)$$

$$
\{D^3\} =
\begin{bmatrix}
0 & -1 & 0 & 0 & 0 & 0 \\
1 & 0 & 0 & 0 & 0 & 0 \\
0 & 0 & 1 & 0 & 0 & 0 \\
0 & 0 & 0 & -1 & 0 & 0 \\
0 & 0 & 0 & 1 & 0 & 0 \\
0 & 0 & 0 & 0 & 0 & 1
\end{bmatrix}
\frac{1}{EI}
\begin{Bmatrix}
276.08 \\
0 \\
33.38 \\
0 \\
0 \\
0
\end{Bmatrix}
= \frac{1}{EI}
\begin{Bmatrix}
0 \\
276.08 \\
33.38 \\
0 \\
0 \\
0
\end{Bmatrix}
$$

The axial components of nodal displacements of the elements of the frame were assumed a priori equal to zero. However, the axial components of nodal forces of the elements of the frame are not equal to zero. Thus, these components cannot be obtained from the components of nodal displacements (cc). Consequently, we can only use the components of nodal displacements (cc) to obtain the shearing components of the forces and the components of the moments acting at the ends of the elements of the frame. Thus

$$
\{A^e\} =
\begin{Bmatrix}
F_2^{ej} \\
M_3^{ej} \\
F_2^{ek} \\
M_3^{ek}
\end{Bmatrix}
= [\hat{K}^e]
\begin{Bmatrix}
u_2^{ej} \\
\theta_3^{ej} \\
u_2^{ek} \\
\theta_3^{ek}
\end{Bmatrix}
$$

$$(dd)$$

where $[\hat{K}^e]$ is obtained from the local stiffness matrix of element e of the frame by omitting its rows and columns corresponding to the axial components of the nodal forces and the axial components of the nodal translations of element e. Thus, using relations (cc) and (b), relation (dd) gives

$$\{A^1\} = EI \begin{bmatrix} 0.096 & 0.24 & -0.096 & 0.24 \\ 0.240 & 0.80 & -0.240 & 0.40 \\ -0.096 & -0.24 & 0.096 & -0.24 \\ 0.240 & 0.40 & -0.240 & 0.80 \end{bmatrix} \frac{1}{EI} \begin{Bmatrix} 0 \\ 0 \\ -345.11 \\ -43.93 \end{Bmatrix} = \begin{Bmatrix} 22.59 \\ 65.25 \\ \hdashline -22.59 \\ 47.68 \end{Bmatrix}$$

$$\{A^2\} = EI \begin{bmatrix} 0.0556 & 0.1667 & -0.0556 & 0.1667 \\ 0.1667 & 0.6667 & -0.1667 & 0.3333 \\ -0.0556 & -0.1667 & 0.556 & -0.1667 \\ 0.1667 & 0.3333 & -0.1667 & 0.6667 \end{bmatrix} \frac{1}{EI} \begin{Bmatrix} -207.06 \\ -43.93 \\ 0 \\ 33.38 \end{Bmatrix} = \begin{Bmatrix} -13.27 \\ -52.68 \\ 13.27 \\ -26.91 \end{Bmatrix}$$

$$\{A^3\} = EI \begin{bmatrix} 0.1875 & 0.375 & -0.1875 & 0.375 \\ 0.3750 & 1.00 & -0.375 & 0.500 \\ -0.1875 & -0.375 & 0.1875 & -0.375 \\ 0.3750 & 0.500 & -0.375 & 1.000 \end{bmatrix} \frac{1}{EI} \begin{Bmatrix} 276.08 \\ 33.38 \\ 0 \\ 0 \end{Bmatrix} = \begin{Bmatrix} 64.28 \\ 136.91 \\ \hdashline -64.28 \\ 120.22 \end{Bmatrix}$$

$$\text{(ee)}$$

In order to compute the axial components of nodal forces in the elements of the frame we must consider the equilibrium of its nodes. Thus, using the results (ee) and referring to the free-body diagrams of the elements and the nodes of the frame shown in Fig. d, we obtain by considering the equilibrium of node 2

$$\sum \bar{F}_1 = 0 \qquad 44 - 0.6F_1^{1k} - 18.07 + F_1^{2k} = 0$$

$$\sum \bar{F}_2 = 0 \qquad 78 + 0.8F_1^{1k} - 13.55 - 13.27 = 0$$

or $\qquad F_1^{1k} = -63.91 \text{ kN} \qquad F_1^{2k} = -64.31 \text{ kN}$ $\qquad$ (ff)

Moreover, by considering the equilibrium of node 3 we get

$$\sum F_h = 0 \qquad F_1^{2k} = -64.28 \text{ kN check}$$

$$\text{(gg)}$$

$$\sum F_v = 0 \qquad F_1^{3k} = -33.27 \text{ kN}$$

If the effect of axial deformation was taken into account, the matrices of nodal actions of the elements of the frame are

$$\{A^1\} = \begin{Bmatrix} 63.64 \\ 22.93 \\ 66.28 \\ \hdashline -63.64 \\ -22.93 \\ 48.39 \end{Bmatrix} \qquad \{A^2\} = \begin{Bmatrix} 63.83 \\ -13.33 \\ -53.39 \\ \hdashline -63.83 \\ 13.33 \\ -26.59 \end{Bmatrix} \qquad \{A^3\} = \begin{Bmatrix} 33.33 \\ 63.83 \\ 136.59 \\ \hdashline -33.33 \\ -63.83 \\ 118.75 \end{Bmatrix} \qquad \text{(hh)}$$

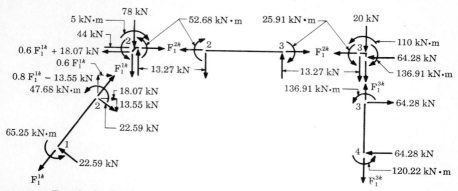

Figure d Free-body diagrams of the elements and nodes 2 and 3 of the frame of Fig. a.

Comparing results (ee) to (gg) with (hh) we see that when the effect of axial deformation of the elements of the frame is not taken into account, the error in the values of the nodal actions of its elements is less than 2 percent.

Notice that the reactions of the frame cannot be obtained by substituting the computed values of the components of displacements of the nodes of the frame into relation (10.57). To compute the reactions of the frame the equilibrium of nodes 1 and 4 must be considered.

10.4 Problems

1 and 2. Consider the truss described in Problem 1 at the end of Chap. 8 subjected to the external forces and supported as shown in Fig. 10.P1. Its stiffness matrix is assembled in Problem 1 at the end of Chap. 8. Using the two procedures described in Sec. 10.2 compute

(a) The components of translation of the nodes of the truss
(b) The reactions of the supports of the truss
(c) The internal forces in the elements of the truss

Repeat with the same truss subjected to the external forces and supported as shown in Fig. 10.P2.

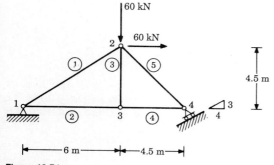

Figure 10.P1

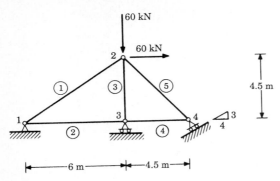

Figure 10.P2

3 and 4. Consider the beam described in Problem 2 at the end of Chap. 8 subjected to the external actions and supported as shown in Fig. 10.P3. Its stiffness matrix is assembled in Problem 2 at the end of Chap. 8. Using the procedure described in Sec. 10.2.1 compute

(a) The components of displacements of the nodes of the beam
(b) The reactions of the supports of the beam
(c) The nodal actions of the elements of the beam

Repeat with the same beam subjected to the external actions and supported as shown in Fig. 10.P4.

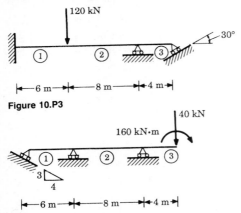

Figure 10.P3

Figure 10.P4

5. Consider the frame described in Problem 3 at the end of Chap. 8 subjected to the external actions and supported as shown in Fig. 10.P5. Its stiffness matrix is assembled in Problem 3 at the end of Chap. 8. Using the procedure described in Sec. 10.2.1 compute

(a) The components of displacements of the nodes of the frame
(b) The reactions of the supports of the frame
(c) The nodal actions of the elements of the frame

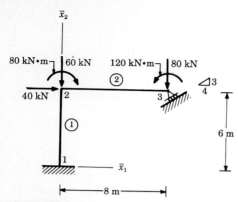

Figure 10.P5

6. Consider the frame described in Problem 4 at the end of Chap. 8 subjected to the external actions and supported as shown in Fig. 10.P6. Its stiffness matrix is assembled in Problem 4 at the end of Chap. 8. Using both procedures described in Sec. 10.2 compute

    (a) The components of displacements of the nodes of the frame
    (b) The reactions of the supports of the frame
    (c) The nodal actions of the elements of the frame

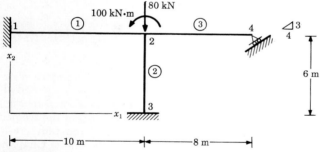

Figure 10.P6

7 to 8. The frame shown in Fig. 10.P7 is subjected to the following loading:

    (a) The external actions shown in Fig. 10.P7.
    (b) Settlement of support 1 of 20 mm.
    (c) Temperature of the upper or outside fibers $T_e = 35°C$ and of lower or inside fibers $T_i = -5°C$. The temperature during the construction of the frame was $T_0 = 15°C$.

Using the procedures described in Sec. 10.2 compute

    (a) The components of displacements of the nodes of the frame
    (b) The reactions of the supports of the frame
    (c) The nodal actions of the elements of the frame

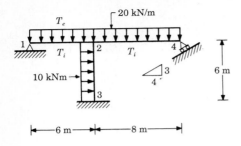

Figure 10.P7

Repeat with the frame of Fig. 10.P8. The elements of the frames are made of steel $(E = 210\ \text{kN/mm}^2)$ and have the same constant cross section $(A = 13.2 \times 10^3\ \text{mm}^2,\ I = 369.7 \times 10^6\ \text{mm}^4,\ h = 425\ \text{mm})$.

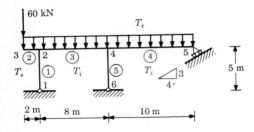

Figure 10.P8

9 and 10. Using the procedure described in Sec. 10.3 compute the displacements of the nodes of the frame described in Problem 3 at the end of Chap. 8 subjected to the external actions and supported as shown in Fig. 10.P5. Disregard the effect of axial deformation of the elements of the frame. The elements of the frame are made of steel and have a constant cross section. Repeat with the frame described in Problem 4 at the end of Chap. 8 subjected to the external actions shown in Fig. 10.P6.

11 to 12. The frame shown in Fig. 10.P11 is subjected to the following loading:
(a) The external actions shown in Fig. 10.P11.
(b) Settlement of support 1 of 20 mm.
(c) Temperature of the upper fibers $T_e = 35°\text{C}$ and of lower fibers $T_i = -5°\text{C}$. The temperature during the construction of the frame was $T_0 = 15°\text{C}$.

Using the procedure described in Sec. 10.3 compute
(a) The components of displacements of the nodes of the frame
(b) The reactions of the supports of the frame
(c) The nodal actions of the elements of the frame

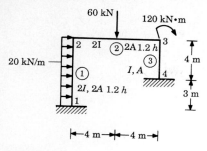

Figure 10.P11

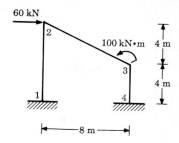

Figure 10.P12

Disregard the effect of axial deformation of the elements of the frame. Repeat with the frame of Fig. 10.P12. The elements of the frames are made of steel ($E = 210\ \text{kN/mm}^2$, $\alpha = 10^{-5}/°C$) and have a constant cross section ($A = 13.2 \times 10^3\ \text{mm}^2$, $I = 369.7 \times 10^6\ \text{mm}^4$, $h = 425\ \text{mm}$).

11

Procedure for Programming the Analysis of Framed Structures Using the Direct Stiffness Method

11.1 Introduction

A computer program for analyzing framed structures has the following parts:

1. Data input.

2. Formulation of the stiffness matrices for each element of the structure. This includes calculation of their transformation matrices.

3. Assemblage of the submatrices $[S^{FF}]$, $[S^{FS}]$, $[S^{SF}]$, and $[S^{SS}]$ of the stiffness matrix of the structure [see Eqs. (9.1)]. We use one of the procedures described in Sec. 10.2 (preferably that of Sec. 10.2.1) to obtain the modified submatrices of the stiffness matrix of structures having skew supports.

4. Calculation and assemblage of the matrices of equivalent actions.

5. Solution of the stiffness equations—computation of the components of displacements of the nodes of the structure and of its reactions.

6. Computation of the nodal actions of each element of the structure.

11.2 Data Input

The following data is supplied to the computer.

1. Node data
   a. The global coordinates of each node.

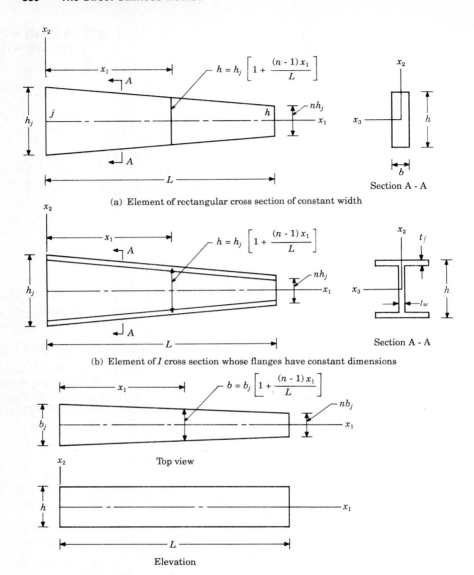

(a) Element of rectangular cross section of constant width

(b) Element of I cross section whose flanges have constant dimensions

(c) Element of rectangular cross section of constant depth and variable width

Figure 11.1 Tapered elements.

    b. The constraint conditions for each node. For example, in Tables 11.1, 11.4, and 11.6 we use a zero in the Δ_i $(i = 1, 2, 3)$ column to indicate that the node is free to translate in the x_i direction. Moreover, we use 1 in the Θ_i $(i = 1, 2, 3)$ column to indicate that the node is not free to rotate about the x_i axis. We use a 2 when a component of displacement does represent a possible motion for a node of the structure (i.e., for a planar structure in the $x_1 x_2$ plane we use a 2 for Δ_3, Θ_1, and Θ_2). Furthermore we use a 3 in the Δ_i ($i = 1, 2, 3$) column for node n and $(n + 1)$ to indicate that nodes $(n - 1)$, n, and $(n + 1)$ are adjacent to a hinge and that their components of translation Δ_i are equal. The nodes adjacent to a hinge are numbered consecutively.

    c. The nodes connected to a skew support. In Tables 11.1 and 11.6 we indicate these nodes by placing a number from 1 to N in the column entitled "skew coordinate system."

    d. The coordinates of a point on the $\bar{x}_1'$ skew axis for each node of a structure connected to a skew support (see Tables 11.2 and 11.7). The coordinates of a second point located in the $\bar{x}_1' \bar{x}_2'$ plane for each node of a space structure connected to a skew support. The origin of the skew axes is the node connected to the skew support. The $\bar{x}_2'$ axis is normal to the plane of rolling. The $\bar{x}_1'$ axis is in the direction of rolling if the support is a roller or in any convenient direction if the support is a ball support. In Tables 11.2 and 11.7 the coordinates of three points are given. The first point is the node connected to the skew support, the second is a point on the $\bar{x}_1'$ axis, and the third is a point on the $\bar{x}_1' \bar{x}_2'$ plane.

2. Element data

    a. The type of each element. We distinguish the following types of elements:

        (1) Type P1. Axial deformation element of constant cross section of a planar structure

        (2) Type P2. General planar element of constant cross section

        (3) Type P3. General planar tapered element (see Fig. 11.1)

        (4) Type P4. General planar element of complex geometry

        (5) Type S1. Axial deformation element of constant cross section of a space structure

        (6) Type S2. General space element of constant cross section

        (7) Type S2. General space tapered element

        (8) Type S4. General space element of complex geometry

    b. The quantities which characterize the geometry of the cross sections of each element.

        (1) For an element of constant cross section of a planar structure in the $\bar{x}_1 \bar{x}_2$ plane, these quantities are

            (*a*) The area A of the cross section

            (*b*) The moment of inertia I_3 of the cross section about the x_3 axis

            (*c*) The depth $h^{(2)}$ of the cross section (see Fig. 11.2)

        (2) For an element of constant cross section of a space

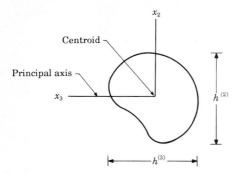

Figure 11.2 Cross section of an element.

structure these qualities are
(a) The area A of the cross section
(b) The moments of inertia I_2 and I_3 of the cross section about the x_2 and x_3 axes, respectively
(c) The depths $h^{(2)}$ and $h^{(3)}$ of the cross section (see Fig. 11.2)
(3) For the tapered elements of Fig. 11.1 used in a space structure these quantities are
(a) The area of the cross sections at the end j of the element
(b) The moments of inertia $I_2^{(j)}$ and $I_3^{(j)}$ of the cross section at the end j of the element
(c) The dimensions $h_j^{(2)}$ and $h_j^{(3)}$ of the cross section at the end j of the element
(d) The ratio n of the area of the cross section of the element at its end k to the area at the cross section at its end j, that is,

$$n = \frac{A_k}{A_j}$$

(e) The thickness of the web for the element of Fig. 11.1b only
c. The material properties for each element. For elements made of isotropic linearly elastic materials these properties are
(1) The modulus of elasticity E
(2) The coefficient of linear expansion α
(3) The shear modulus G for elements of space beams and frames only
d. For each element of a space structure the coordinates of a point on the $x_1 x_2$ or $x_1 x_3$ plane. This point should not be located on the x_1 axis of the element (see Sec. 6.3). We denote this point as node i

e. The connectivity of the elements and nodes of the structure, that is, the number assigned to the nodes to which the ends of each element are connected

3. **Load data.** The loading applied to a framed structure consists of one or more of the following disturbances.

a. Components of external concentrated actions applied to the nodes of the structure. These actions are specified by

(1) The number of the node on which they act

(2) The type of each of their nonvanishing global components. In Tables 11.5, 11.10, and 11.17 we indicate the global components of an external concentrated action $\bar{P}_1$ by 1, $\bar{P}_2$ by 2, $\bar{P}_3$ by 3, $\bar{\mathcal{M}}_1$ by 4, $\bar{\mathcal{M}}_2$ by 5, and $\bar{\mathcal{M}}_3$ by 6.

(3) The magnitude of each of their nonvanishing components. The magnitude of a component of an action acting in the negative $\bar{x}_i$ $(i = 1, 2, 3, \dots)$ direction is considered negative

b. The components of concentrated external actions acting at some points along the length of the elements of the structure. For each element these actions are specified by

(1) The number of the element

(2) The x_1 coordinate of the point on which each of them acts

(3) The type of each of their nonvanishing local components. We indicate the local components of an external concentrated action P_1 by 1, P_2 by 2, P_3 by 3, $\mathcal{M}_1$ by 4, $\mathcal{M}_2$ by 5, and $\mathcal{M}_3$ by 6 (see Table 11.11)

(4) The magnitude of each of their nonvanishing components. The magnitude of a component of an external action acting in the negative x_i $(i = 1, 2, 3, \dots)$ direction is considered negative

c. Components of distributed external actions acting on a portion of the length of each element of the structure. We assume that these actions can be either piecewise constant or piecewise linear functions of the axial coordinate of the element on which they act. Thus for each element these actions are specified (see Table 11.11) by

(1) The number of the element

(2) The x_1 coordinates of the points where each continuous distribution of external actions begins and ends

(3) The type of the nonvanishing local components of each continuous distribution of external actions. We indicate the local components of an external distribution of external actions p_1 by 1, p_2 by 2, p_3 by 3, m_1 by 4, m_2 by 5, and m_3 by 6.

(4) The magnitude of each nonvanishing component of each continuous distribution of external actions at the points

where it begins and ends. The magnitude of a component acting in the negative x_i $(i = 1, 2, 3, \ldots)$ direction is considered negative.

d. Changes of temperature. We assume that the changes of temperature are either constant throughout the length of an element or linear functions of the axial coordinate of the element. Thus, for each element they are specified by (see Table 11.12)

(1) The number of element

(2) The changes ΔT_c [see relation (2.13)] at the ends of the element

(3) The changes δT_2 [see relation (2.11)] at the ends of the element

(4) The changes δT_3 [see relation (2.12)] at the ends of the element

e. Specified movement of the supports of the structure. This loading is specified by (see Table 11.13)

(1) The number of each node connected to a support whose movement in one or more directions is specified.

(2) The type of each nonvanishing global component of displacement of each node connected to such support. We indicate the global components of displacement $\Delta_1^{(n)}$ by 1, $\Delta_2^{(n)}$ by 2, $\Delta_3^{(n)}$ by 3, $\theta_2^{(n)}$ by 4, $\theta_2^{(n)}$ by 5, $\theta_3^{(n)}$ by 6.

(3) The magnitude of each specified nonvanishing global component of displacement of each node connected to such support. The magnitude of a component of displacement in the negative x_i $(i = 1, 2, 3, \ldots)$ direction is considered negative.

f. Initial strain of the elements of the structure (see Sec. 1.3). We assume that this loading does not vary along the length of the elements of the structure. Thus, for each element it is specified by

(1) The number of the element

(2) The axial component of initial strain e_{11}^{Ic} (see Sec. 2.3)

(3) The components of initial curvature k_2^I and k_3^I (see Sec. 2.3)

Notice that in a computer program the effect of the initial strain of the elements of a structure can be taken into account by converting them to equivalent temperature changes. This becomes apparent by referring to relations (2.37) and (2.40) or (2.45) and (2.53).

Tables 11.1 to 11.5, 11.6 to 11.13, and 11.14 to 11.17 present the node, element, and load data which must be supplied to the computer for the truss of Fig. 11.3, the planar frame of Fig. 11.4, and the space structure of Fig. 11.5, respectively.

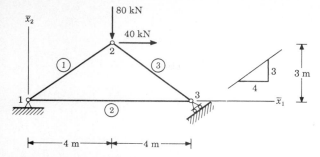

Figure 11.3 Three-bar truss with inclined support.

TABLE 11.1   Node Coordinates and Constraints for the Truss of Fig. 11.3

Node no. (n)	$\bar{x}_1$	$\bar{x}_2$	$\bar{x}_3$	Skew coord. system	$\Delta_1^{(n)}$	$\Delta_2^{(n)}$	$\Delta_3^{(n)}$	$\Theta_1^{(n)}$	$\Theta_2^{(n)}$	$\Theta_3^{(n)}$
					Nodal constraints					
1	0	0	0		1	1	2	2	2	2
2	4	3	0		0	0	2	2	2	2
3	8	0	0	1	0	1	2	2	2	2

TABLE 11.2   Coordinate System for Skew Support

Coord. system	$\bar{x}_1'$	$\bar{x}_2'$	$\bar{x}_3'$
1	8	0	0
	12	3	0
	8	10	0

TABLE 11.3   Element Connectivity and Cross-sectional Properties

Element no.	Element type	Material ID	End j	End k	Node i	A, m^2	I_2 m^4	I_3 m^4	K m^4
1	P1	1	1	2		10^{-2}			
2	P1	1	1	3		10^{-2}			
3	P1	1	2	3		10^{-2}			

TABLE 11.4   Material Data

Material ID	E, kN/mm^2	G, kN/mm^2	α, /°C
1	200		

TABLE 11.5 Concentrated Actions Acting on the Nodes of the Truss

Node	Component type	Magnitude, kN
2	1	40
2	2	−80
3	1	12

TABLE 11.6 Node Coordinates and Constraints for the Planar Frame of Fig. 11.4

Node no. (n)	$\bar{x}_1$	$\bar{x}_2$	$\bar{x}_3$	Skew coord. system	Nodal constraints $\Delta_1^{(n)}$	$\Delta_2^{(n)}$	$\Delta_3^{(n)}$	$\Theta_1^{(n)}$	$\Theta_2^{(n)}$	$\Theta_3^{(n)}$
1	0	0	0		1	1	2	2	2	1
2	0	5	0		0	0	2	2	2	0
3	8	7	0		0	0	2	2	2	0
4	8	7	0		3	3	2	2	2	0
5	16	5	0		0	0	2	2	2	0
6	16	0	0	1	0	1	2	2	2	0
7	19	5	0		0	0	2	2	2	0

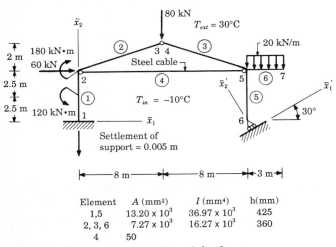

Element	A (mm²)	I (mm⁴)	h(mm)
1,5	13.20 x 10³	36.97 x 10³	425
2, 3, 6	7.27 x 10³	16.27 x 10³	360
4	50		

Figure 11.4 Geometry and loading of the frame.

TABLE 11.7 Coordinate System for Skew Support

Coord. system	$\bar{x}_1'$	$\bar{x}_2'$	$\bar{x}_3'$
1	16	0	0
	17	0.5774	0
	16	10	0

TABLE 11.8   **Element Connectivity and Cross-sectional Properties**

Element no.	Element type	Material ID	Node numbers End j	End k	Node i	$A,$ m^2	$I_2,$ m^4	$I_3,$ $K,$ m^4 m^4
1	P_2	1	1	2		13.2×10^{-3}	36.97×10^{-6}	
2	P_2	1	2	3		7.27×10^{-3}	16.27×10^{-6}	
3	P_2	1	4	5		7.27×10^{-3}	16.27×10^{-6}	
4	P_1	1	2	5		5.0×10^{-5}		
5	P_2	1	5	6		13.2×10^{-3}	39.97×10^{-6}	
6	P_2	1	5	7		7.27×10^{-3}	16.27×10^{-6}	

TABLE 11.9   **Material Data**

Material ID	$E,$ kN/mm^2	$G,$ kN/mm^2	$\alpha,$ $/°C$
1	210		10^{-5}

TABLE 11.10   **Concentrated Actions Acting on the Nodes of the Frame**

Node no.	Component type	Magnitude
2	1	60
2	6	-180
3	2	-80

TABLE 11.11   **Actions Acting along the Length of the Elements of the Frame**

Element no.	Component type	Location x_1 (A)	Magnitude at A	Location x_1 (B)	Magnitude at B
1	6	2.5	120		
6	2	0.0	-20	3.0	-20

TABLE 11.12   **Temperature Changes**

Element no.	$(\Delta T_c)_j,$ $°C$	$(\Delta T_c)_k,$ $°C$	$(\delta T_2)_j,$ $°C$	$(\delta T_2)_k,$ $°C$	$h^{(2)},$ m	$(\delta T_3)_j,$ $°C$	$(\delta T_3)_k,$ $°C$	$h^{(3)},$ m
1	-10	-10	40	40	0.425			
2	-10	-10	40	40	0.360			
3	-10	-10	40	40	0.360			
4	-30	-30						
5	-10	-10	40	40	0.425			
6	10	10	0	0	0.360			

TABLE 11.13   **Movement of Support**

Node no.	Component type	Magnitude, m
1	2	-0.005

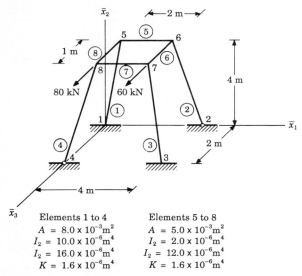

Figure 11.5 Space structure.

TABLE 11.14 Node Coordinates and Constraints for the Space Frame of Fig. 11.5

Node no.	$\bar{x}_1$	$\bar{x}_2$	$\bar{x}_3$	Skew coord. system	Nodal constraints $\Delta_1^{(n)}$	$\Delta_2^{(n)}$	$\Delta_3^{(n)}$	$\Theta_1^{(n)}$	$\Theta_2^{(n)}$	$\Theta_3^{(n)}$
1	0	0	0		1	1	1	1	1	1
2	4	0	0		1	1	1	0	0	0
3	4	0	2		1	1	1	1	1	1
4	0	0	2		1	1	1	0	1	1
5	1	4	0.5		0	0	0	0	0	0
6	3	4	0.5		0	0	0	0	0	0
7	3	4	1.5		0	0	0	0	0	0
8	1	4	1.5		0	0	0	0	0	0

TABLE 11.15 Element Connectivity and Cross-sectional Properties

Elem. no.	Elem. type	Mat. ID	Node numbers j	k	l	$A,$ m^2	$I_2,$ m^4	$I_3,$ m^4	$K,$ m^4
1	S2	1	1	5	2	8.0×10^{-3}	10.0×10^{-6}	16.0×10^{-6}	1.6×10^{-6}
2	S2	1	2	6	1	8.0×10^{-3}	10.0×10^{-6}	16.0×10^{-6}	1.6×10^{-6}
3	S2	1	3	7	4	8.0×10^{-3}	10.0×10^{-6}	16.0×10^{-6}	1.6×10^{-6}
4	S2	1	4	8	3	8.0×10^{-3}	10.0×10^{-6}	16.0×10^{-6}	1.6×10^{-6}
5	S2	1	5	6	8	5.0×10^{-3}	2.0×10^{-6}	12.0×10^{-6}	1.6×10^{-6}
6	S2	1	6	7	5	5.0×10^{-3}	2.0×10^{-6}	12.0×10^{-6}	1.6×10^{-6}
7	S2	1	7	8	6	5.0×10^{-3}	2.0×10^{-6}	12.0×10^{-6}	1.6×10^{-6}
8	S2	1	8	5	7	5.0×10^{-3}	2.0×10^{-6}	12.0×10^{-6}	1.6×10^{-6}

TABLE 11.16   Material Data

Material ID	E, kN/mm^2	G, kN/mm^2	α, /°C
1	210	80.8	10^{-5}

TABLE 11.17   Concentrated Actions Acting on the Nodes of the Space Structure

Node no.	Component type	Magnitude
7	3	60
8	3	80

11.3   Calculation of the Stiffness Matrices for the Elements of a Structure

11.3.1   Formation of the transformation matrices for the elements of a structure

From the node and element data the program forms the following transformation matrices for each element of a structure and for each of its nodes connected to a skew support.

1. The transformation matrix $[\Lambda_P]$ or $[\Lambda_S]$ for each element of a structure depending on whether the structure is planar or space, respectively. This matrix transforms the local coordinates of a vector to global [see relation (6.1) or (6.4)]. The matrix $[\Lambda_P]$ or $[\Lambda_S]$ for each element of a structure is established from the direction cosines of its local axes with respect to the global axes of the structure [see relations (6.6) and (6.3), respectively]. The direction cosines of the local axes of an element of a planar structure are established from the global coordinates of the nodes to which each element is connected (see Sec. 6.2). The direction cosines of the local axes of each element of a space structure, with respect to its global axes, are established from the global coordinates of the nodes to which each element is connected and from the global coordinates of a point in the $x_1 x_2$ or $x_1 x_3$ plane of the element which is not located on its x_1 axis (see Sec. 6.3). We denote this node as node i.

2. The transformation matrix $[\Lambda_P']$ or $[\Lambda_S']$ for each node of a structure connected to a skew support depending on whether the structure is planar or space, respectively. This matrix transforms the global components of a vector to components along the skew axes [see

relation (10.1a) or (10.3a)]. The matrix $[\Lambda'_P]$ or $[\Lambda'_S]$ for a node connected to a skew support is established from the direction cosines of the skew axes for this node with respect to the global axes of the structure [see relations (10.2) and (10.4), respectively]. The direction cosines of the skew $\bar{x}'_1$ and $\bar{x}'_2$ axes of a node of a planar structure connected to a skew support are established from the global coordinates of a point located on the $\bar{x}'_1$ axis. The direction cosines of the skew $\bar{x}'_1$, $\bar{x}'_2$, and $\bar{x}'_3$ axes of a node of a space structure connected to a skew support are established from the global coordinates of two points; one located on the $\bar{x}'_1$ axis and the other on the $\bar{x}'_2\bar{x}'_1$ plane. The origin of these axes is the node connected to the skew support. The $\bar{x}'_2$ axis is normal to the plane of rolling. The $\bar{x}'_1$ axis is in the direction of rolling if the support is a roller or in any convenient direction if the support is a ball support.

3. The transformation matrix $[\Lambda]$ for each element of a structure where

$$[\Lambda] = \begin{cases} [\Lambda_{PT}] & \text{for each element of a planar truss [see relation (6.42)]} \\ [\Lambda_{PF}] & \text{for each element of a planar beam or frame [see relation (6.48)]} \\ [\Lambda_{ST}] & \text{for each element of a space truss [see relation (6.51)]} \\ [\Lambda_{SF}] & \text{for each element of a space beam or frame [see relations (6.54) and (6.57)]} \end{cases}$$

The matrix $[\Lambda]$ transforms the global components of nodal actions or displacements of an element to local and vice versa. That is, referring to relations (6.64), (6.65), (6.67), and (6.68) we have

$$[A] = [\Lambda][\bar{A}]$$

$$[D] = [\Lambda][\bar{D}]$$

$$[\bar{A}] = [\Lambda]^T[A] \tag{11.1}$$

$$[\bar{D}] = [\Lambda]^T[D]$$

4. The transformation matrix $[\Lambda_{\text{SKL}}]^q$ ($q = j$ or k) for each element connected to a skew support with its end q ($q = j$ or k). This matrix transforms the local matrix of nodal actions or displacements of an element connected to a skew support with its end q ($q = j$ or k) to a matrix whose terms are the global components of nodal actions or displacements at the end p ($p \neq q$; q, $p = j$ or k) of the element and the skew components of nodal actions or displacements at the end q ($q = j$ or k) of the element. The matrix $[\Lambda_{\text{SKL}}]^j$ is obtained using relations (10.13), (10.16a), and (10.19), while the matrix $[\Lambda_{\text{SKL}}]^k$ is obtained using relations (10.13), (10.16b), and (10.19).

11.3.2   Calculation of the stiffness matrices for the elements of a structure

The program computes the following stiffness matrices for the elements of a structure.

1. *The local stiffness matrix [K] for each element of a structure.* For elements of constant cross section this matrix is computed using the formulas given by relations (4.50), (4.62), or (4.69). These formulas are stored in the computer. The local stiffness matrix of the tapered elements of Fig. 11.1 is computed from their flexibility matrix using relation (5.93). The flexibility matrix of tapered elements is established using the formula given in Table 19.1. These formulas are stored in the computer. The local stiffness matrix of elements of more complex geometry are computed by a subroutine using one of the following approximate methods.

    a. A method involving a model of the element consisting of a number of segments of constant cross section (see Sec. 12.3). The model can be regarded as a structure made from elements (its segments) of constant cross section, and thus the program can be employed to compute its stiffness matrix. The stiffness matrix for the model can be contracted to yield an approximation of the local stiffness matrix for the element (see Sec. 12.3).

    b. The finite-element method as described in Sec. 19.2.

2. *The hybrid stiffness matrix $[\overset{\square}{K}]$ for each element of the structure not connected to a skew support.* Referring to relation (9.11) we have

$$[\overset{\square}{K}] = [K][\Lambda] \tag{11.2}$$

3. The modified hybrid stiffness matrix $[\overset{\square}{K}']$ for each element connected to a skew support with its end q ($q = j$ or k). Referring to relation (10.27) we have

$$[\overset{\square}{K}']^q = [K][[\Lambda_{\text{SKL}}]^q]^T \tag{11.3}$$

4. The global stiffness matrix $[\bar{K}]$ for each element not connected to a skew support. Referring to relations (6.74) we have

$$[\bar{K}] = [\Lambda]^T[\overset{\square}{K}] \tag{11.4}$$

5. The modified global stiffness matrix $[\bar{K}']^q$ for each element connected to a skew support with its end q ($q = j$ or k). Referring to relations (10.27) and (10.30) we get

$$[\bar{K}']^q = [\Lambda_{\text{SKL}}]^q[\overset{\square}{K}']^q \tag{11.5}$$

11.4 Assemblage of the Submatrices of the Stiffness Matrix of the Structure

In this section we present a procedure for programming the assemblage of the submatrices $[S^{FF}]$, $[S^{SF}]$, $[S^{FS}]$, and $[S^{SS}]$ [see relation (9.1)] of the stiffness matrix for a framed structure from the global stiffness matrices of those of its elements which are not connected to skew supports and the modified global stiffness matrices of those of its elements which are connected to skew supports.

Starting with node 1 and proceeding to the other nodes of the structure successively, the program forms the *identification of constraining conditions (ICC) array* for the structure using the input node data described in Sec. 11.2. In this array the unconstrained components of displacements are numbered with a sequence of positive numbers, while the constrained components of displacements are numbered with another sequence of negative numbers. For example, the program forms the ICC arrays shown in Tables 11.18, 11.19, and 11.20 for the structures of Figs. 11.3, 11.4, and 11.5,

TABLE 11.18 Identification of Constraining Conditions Array for the Planar Truss of Fig. 11.3

Node no. (n)	$\Delta_1^{(n)}$	$\Delta_2^{(n)}$	$\Delta_3^{(n)}$	$\Delta_4^{(n)} = \theta_1^{(n)}$	$\Delta_5^{(n)} = \theta_2^{(n)}$	$\Delta_6^{(n)} = \theta_3^{(n)}$
1	-1	-2	0	0	0	0
2	1	2	0	0	0	0
3	3	-3	0	0	0	0

TABLE 11.19 Identification of Constraining Conditions Array for the Planar Frame of Fig. 11.4

Node no. (n)	$\Delta_1^{(n)}$	$\Delta_2^{(n)}$	$\Delta_3^{(n)}$	$\Delta_4^{(n)} = \theta_1^{(n)}$	$\Delta_5^{(n)} = \theta_2^{(n)}$	$\Delta_6^{(n)} = \theta_3^{(n)}$
1	-1	-2	0	0	0	-3
2	1	2	0	0	0	3
3	4	5	0	0	0	6
4	4	5	0	0	0	7
5	8	9	0	0	0	10
6	11	-4	0	0	0	12
7	13	14	0	0	0	15

TABLE 11.20 Identification of Constraining Conditions Array for the Space Frame of Fig. 11.5

Node no. (n)	$\Delta_1^{(n)}$	$\Delta_2^{(n)}$	$\Delta_3^{(n)}$	$\Delta_4^{(n)} = \theta_1^{(n)}$	$\Delta_5^{(n)} = \theta_2^{(n)}$	$\Delta_6^{(n)} = \theta_3^{(n)}$
1	-1	-2	-3	-4	-5	-6
2	-7	-8	-9	1	2	3
3	-10	-11	-12	-13	-14	-15
4	-16	-17	-18	4	-19	-20
5	5	6	7	8	9	10
6	11	12	13	14	15	16
7	17	18	19	20	21	22
8	-23	24	15	26	27	28

respectively. Notice that the program assigns one number (4) to the $\bar{x}_1$ components of translation and another number (5) to the $\bar{x}_2$ components of translation of the two connected nodes 3 and 4 of the frame of Fig. 11.4.

The program forms the *location array KK(I)* for each element of the structure. This array relates the global components of nodal displacements of the element to the constrained and unconstrained components of displacements of the nodes of the structure to which the element is connected. For example, the global stiffness matrix for an element of a truss has the following form:

$$[\bar{K}] = \begin{bmatrix} \bar{K}_{11} & \bar{K}_{12} & \bar{K}_{13} & \bar{K}_{14} \\ \bar{K}_{21} & \bar{K}_{22} & \bar{K}_{23} & \bar{K}_{24} \\ \bar{K}_{31} & \bar{K}_{32} & \bar{K}_{33} & \bar{K}_{34} \\ \bar{K}_{41} & \bar{K}_{42} & \bar{K}_{43} & \bar{K}_{44} \end{bmatrix} \qquad (11.6)$$

The location arrays for an element of a truss have four rows $(I = 1, 2, 3, 4)$, one for each global component of nodal displacements. Referring to the ICC array of Table 11.18 the location arrays for the elements of the truss of Fig. 11.3 are

Element 1 *(connected to* *nodes 1 and 2)*	Element 2 *(connected to* *nodes 1 and 3)*	Element 3 *(connected to* *nodes 2 and 3)*	
$KK(1) = -1$	$KK(1) = -1$	$KK(1) = 1$	(11.7)
$KK(2) = -2$	$KK(2) = -2$	$KK(2) = 2$	
$KK(3) = 1$	$KK(3) = 3$	$KK(3) = 3$	
$KK(4) = 2$	$KK(4) = -3$	$KK(4) = -3$	

The truss of Fig. 11.3 has three nodes and consequently six independent components of displacement. Thus its stiffness matrix is a 6×6 matrix. Of the six components of displacement three are constrained and three are free. Thus $[S^{FF}]$, $[S^{SF}]$, and $[S^{SS}]$ are 3×3 submatrices.

Referring to the location array for element 1 [see relation (11.7)] we can write its global stiffness matrix as

$$[\bar{K}^1] = \begin{bmatrix} \bar{K}_{-1,-1} & \bar{K}_{-1,-2} & \bar{K}_{-1,2} & \bar{K}_{-1,2} \\ \bar{K}_{-2,-1} & \bar{K}_{-2,-2} & \bar{K}_{-2,1} & \bar{K}_{-2,2} \\ \bar{K}_{1,-1} & \bar{K}_{1,-2} & \bar{K}_{1,1} & \bar{K}_{1,2} \\ \bar{K}_{2,-1} & \bar{K}_{2,-2} & \bar{K}_{2,1} & \bar{K}_{2,2} \end{bmatrix} \qquad (11.8)$$

The stiffness coefficients with positive subscripts in relation (11.8) go to the slots of the submatrix $[S^{FF}]$ indicated by their subscripts.

The stiffness coefficients with negative subscripts go to the slots of the submatrix $[S^{SS}]$ indicated by their subscripts, while the stiffness coefficients with the first subscript negative and second subscript positive go to the submatrix $[S^{SF}]$. The coefficients with the first subscript positive and the second subscript negative go to the submatrix $[S^{FS}]$. However, we do not form this submatrix because it is the transpose of $[S^{SF}]$. Referring to relations (11.8) we see that the stiffness coefficients $\bar{K}_{33}^{(1)}$, $\bar{K}_{43}^{(1)}$, $\bar{K}_{34}^{(1)}$, and $\bar{K}_{44}^{(1)}$ for element 1 are placed in the slots of submatrix $[S^{FF}]$. For example, $\bar{K}_{33}^{(1)}$ is placed in slot $(1, 1)$, that is, the first row and first column of the matrix $[S^{FF}]$, because the subscripts of the term of the third row and third column of the matrix of relation (11.8) are $(1, 1)$; $\bar{K}_{43}^{(1)}$ is placed in slot $(2, 1)$ of the matrix $[S^{FF}]$ because the subscripts of the term of the fourth row and third column of the matrix of relation (11.8) are $(2, 1)$. Thus

$$[S^{FF}] = \begin{bmatrix} \bar{K}_{33}^{(1)} + \bar{K}_{11}^{\prime(3)} & \bar{K}_{34}^{(1)} + \bar{K}_{12}^{\prime(3)} & \bar{K}_{13}^{\prime(3)} \\ \bar{K}_{43}^{(1)} + \bar{K}_{21}^{\prime(3)} & \bar{K}_{44}^{(1)} + \bar{K}_{22}^{\prime(3)} & \bar{K}_{23}^{\prime(3)} \\ \bar{K}_{31}^{\prime(3)} & \bar{K}_{32}^{\prime(3)} & \bar{K}_{33}^{\prime(3)} + \bar{K}_{33}^{(2)} \end{bmatrix} \qquad (11.9)$$

Moreover, the stiffness coefficients $\bar{K}_{13}^{(1)}$, $\bar{K}_{23}^{(1)}$, $\bar{K}_{14}^{(1)}$, and $\bar{K}_{24}^{(1)}$ for element 1 are placed in the slots of the submatrix $[S^{SF}]$. For example, $\bar{K}_{13}^{(1)}$ is placed in slot $(1, 1)$ of the matrix $[S^{SF}]$ because the subscripts of the term of the first row and fourth column of the matrix of relation (11.8) are $(-1, 1)$. Thus

$$[S^{SF}] = \begin{bmatrix} \bar{K}_{13}^{(1)} & \bar{K}_{14}^{(1)} & \bar{K}_{13}^{\prime(2)} \\ \bar{K}_{23}^{(1)} & \bar{K}_{24}^{(1)} & \bar{K}_{23}^{\prime(2)} \\ \bar{K}_{41}^{\prime(3)} & \bar{K}_{42}^{\prime(3)} & \bar{K}_{43}^{\prime(3)} + \bar{K}_{43}^{\prime(2)} \end{bmatrix} \qquad (11.10)$$

Furthermore the stiffness coefficients $\bar{K}_{11}^{(1)}$, $\bar{K}_{12}^{(1)}$, $\bar{K}_{21}^{(1)}$, and $\bar{K}_{22}^{(1)}$ of element 1 are placed in the slots of the submatrix $[S^{SS}]$. For example, as shown in relation (11.11) $\bar{K}_{21}^{(1)}$ is placed in slot $(2, 1)$ of the matrix $[S^{SS}]$ because the subscripts of the term of the second row and first column of the matrix of relation (11.8) are $(-2, -1)$. Thus

$$[S^{SS}] = \begin{bmatrix} \bar{K}_{11}^{(1)} + \bar{K}_{11}^{(2)} & \bar{K}_{12}^{(1)} + \bar{K}_{12}^{(2)} & \bar{K}_{14}^{(2)} \\ \bar{K}_{21}^{(1)} + \bar{K}_{21}^{(2)} & \bar{K}_{22}^{(1)} + \bar{K}_{22}^{(2)} & \bar{K}_{24}^{(2)} \\ \bar{K}_{41}^{(2)} & \bar{K}_{42}^{(2)} & \bar{K}_{44}^{(3)} + \bar{K}_{44}^{(2)} \end{bmatrix} \qquad (11.11)$$

The global stiffness matrix for an element of a planar frame has the following form:

$$[\bar{K}] = \begin{bmatrix} \bar{K}_{11} & \bar{K}_{12} & \bar{K}_{13} & \bar{K}_{14} & \bar{K}_{15} & \bar{K}_{16} \\ \bar{K}_{21} & \bar{K}_{22} & \bar{K}_{23} & \bar{K}_{24} & \bar{K}_{25} & \bar{K}_{26} \\ \bar{K}_{31} & \bar{K}_{32} & \bar{K}_{33} & \bar{K}_{34} & \bar{K}_{35} & \bar{K}_{36} \\ \bar{K}_{41} & \bar{K}_{42} & \bar{K}_{43} & \bar{K}_{44} & \bar{K}_{45} & \bar{K}_{46} \\ \bar{K}_{51} & \bar{K}_{52} & \bar{K}_{53} & \bar{K}_{54} & \bar{K}_{55} & \bar{K}_{56} \\ \bar{K}_{61} & \bar{K}_{62} & \bar{K}_{63} & \bar{K}_{64} & \bar{K}_{65} & \bar{K}_{66} \end{bmatrix} \tag{11.12}$$

The location array for an element of a planar frame has six rows $(I = 1, 2, 3, 4, 5, 6)$, one for each global component of nodal displacements. Referring to the ICC array of Table 11.19 the location arrays for the elements of the planar frame of Fig. 11.4 are

Element 1 (connected to nodes 1 and 2)	Element 2 (connected to nodes 2 and 3)	Element 3 (connected to nodes 4 and 5)
$KK(1) = -1$	$KK(1) = 1$	$KK(1) = 4$
$KK(2) = -2$	$KK(2) = 2$	$KK(2) = 5$
$KK(3) = -3$	$KK(3) = 3$	$KK(3) = 7$
$KK(4) = 1$	$KK(4) = 4$	$KK(4) = 8$
$KK(5) = 2$	$KK(5) = 5$	$KK(5) = 9$
$KK(6) = 3$	$KK(6) = 6$	$KK(6) = 10$

$$\tag{11.13}$$

Element 4 (connected to nodes 2 and 5)	Element 5 (connected to nodes 5 and 6)	Element 6 (connected to nodes 5 and 7)
$KK(1) = 1$	$KK(1) = 8$	$KK(1) = 8$
$KK(2) = 2$	$KK(2) = 9$	$KK(2) = 9$
$KK(3) = 3$	$KK(3) = 10$	$KK(3) = 10$
$KK(4) = 8$	$KK(4) = 11$	$KK(4) = 13$
$KK(5) = 9$	$KK(5) = -4$	$KK(5) = 14$
$KK(6) = 10$	$KK(6) = 12$	$KK(6) = 15$

The frame of Fig. 11.4 has seven nodes, two of which are connected by a hinge. The components of translation of the connected nodes are not independent. Consequently, the nodes of the frame under consideration have nineteen $(21 - 2 = 19)$ independent components of displacements. The stiffness matrix of the frame is a 19×19 matrix. From the 19 independent components of displacements of the nodes of the frame 4 are constrained (three at support 1 and one at support 6). Hence, $[S^{FF}]$ is a 15×15 submatrix; $[S^{SF}]$ is a 4×15 submatrix; and $[S^{SS}]$ is a 4×4 submatrix.

Element 5 of the frame is connected to a skew support. Referring to the location arrays for elements 1 and 5 given in relations (11.13) we can write the global stiffness matrix of element 1 and the modified global stiffness matrix [see Sec. (10.4)] of element 5 as follows:

$$[\bar{K}^1] = \begin{bmatrix} \bar{K}_{-1,-1} & \bar{K}_{-1,-2} & \bar{K}_{-1,-3} & \bar{K}_{-1,1} & \bar{K}_{-1,2} & \bar{K}_{-1,3} \\ \bar{K}_{-2,-1} & \bar{K}_{-2,-2} & \bar{K}_{-2,-3} & \bar{K}_{-2,1} & \bar{K}_{-2,2} & \bar{K}_{-2,3} \\ \bar{K}_{-3,-1} & \bar{K}_{-3,-2} & \bar{K}_{-3,-3} & \bar{K}_{-3,1} & \bar{K}_{-3,2} & \bar{K}_{-3,3} \\ \bar{K}_{1,-1} & \bar{K}_{1,-2} & \bar{K}_{1,-3} & \bar{K}_{1,1} & \bar{K}_{1,2} & \bar{K}_{1,3} \\ \bar{K}_{2,-1} & \bar{K}_{2,-2} & \bar{K}_{2,-3} & \bar{K}_{2,1} & \bar{K}_{2,2} & \bar{K}_{2,3} \\ \bar{K}_{3,-1} & \bar{K}_{3,-2} & \bar{K}_{3,-3} & \bar{K}_{3,1} & \bar{K}_{3,2} & \bar{K}_{3,3} \end{bmatrix} \quad (11.14a)$$

$$[\bar{K}'^5] = \begin{bmatrix} \bar{K}'_{8,8} & \bar{K}'_{8,9} & \bar{K}'_{8,10} & \bar{K}'_{8,11} & \bar{K}'_{8,-4} & \bar{K}'_{8,12} \\ \bar{K}'_{9,8} & \bar{K}'_{9,9} & \bar{K}'_{9,10} & \bar{K}'_{9,11} & \bar{K}'_{9,-4} & \bar{K}'_{9,12} \\ \bar{K}'_{10,8} & \bar{K}'_{10,9} & \bar{K}'_{10,10} & \bar{K}'_{10,11} & \bar{K}'_{10,-4} & \bar{K}'_{10,12} \\ \bar{K}'_{11,8} & \bar{K}'_{11,9} & \bar{K}'_{11,10} & \bar{K}'_{11,11} & \bar{K}'_{11,-4} & \bar{K}'_{11,12} \\ \bar{K}'_{-4,8} & \bar{K}'_{-4,9} & \bar{K}'_{-4,10} & \bar{K}'_{-4,11} & \bar{K}'_{-4,-4} & \bar{K}'_{-4,12} \\ \bar{K}'_{12,8} & \bar{K}'_{12,9} & \bar{K}'_{12,10} & \bar{K}'_{12,11} & \bar{K}'_{12,-4} & \bar{K}'_{12,12} \end{bmatrix} \quad (11.14b)$$

The coefficients with positive subscripts in relations (11.14) go to the slots of the submatrix $[S^{FF}]$ indicated by their subscripts. The coefficients with negative subscripts go to the slots of the submatrix $[S^{SS}]$ indicated by their subscripts, while the coefficients with the first subscript negative and second subscript positive go to the slots of the submatrix $[S^{SF}]$ indicated by their subscripts. The coefficients with the first subscript positive and the second subscript negative go to the submatrix $[S^{FS}]$. However, we do not form this submatrix because it is the transpose of $[S^{SF}]$. Referring to the location arrays (11.13) the coefficients of the first and fifth rows of the submatrix $[S^{FF}]$ are equal to

$$S^{FF}_{11} = K^{(1)}_{44} + K^{(2)}_{11} + K^{(4)}_{11} \qquad S^{FF}_{51} = K^{(2)}_{51}$$

$$S^{FF}_{12} = K^{(1)}_{45} + K^{(2)}_{12} + K^{(4)}_{12} \qquad S^{FF}_{52} = K^{(2)}_{52}$$

$$S_{13}^{FF} = K_{46}^{(1)} + K_{13}^{(2)} + K_{13}^{(4)} \qquad S_{53}^{FF} = K_{53}^{(2)}$$

$$S_{14}^{FF} = K_{14}^{(2)} \qquad S_{54}^{FF} = K_{54}^{(2)} + K_{21}^{(3)}$$

$$S_{15}^{FF} = K_{15}^{(2)} \qquad S_{55}^{FF} = K_{55}^{(2)} + K_{22}^{(3)}$$

$$S_{16}^{FF} = K_{16}^{(2)} \qquad S_{56}^{FF} = K_{56}^{(2)}$$

$$S_{17}^{FF} = 0 \qquad S_{57}^{FF} = K_{23}^{(3)}$$

$$S_{18}^{FF} = K_{14}^{(4)} \qquad S_{58}^{FF} = K_{24}^{(3)}$$

$$S_{19}^{FF} = K_{15}^{(4)} \qquad S_{59}^{FF} = K_{25}^{(3)} \qquad (11.15)$$

$$S_{1,10}^{FF} = K_{16}^{(4)} \qquad S_{5,10}^{FF} = K_{26}^{(3)}$$

$$S_{1,11}^{FF} = 0 \qquad S_{5,11}^{FF} = 0$$

$$S_{1,12}^{FF} = 0 \qquad S_{5,12}^{FF} = 0$$

$$S_{1,13}^{FF} = 0 \qquad S_{5,13}^{FF} = 0$$

$$S_{1,14}^{FF} = 0 \qquad S_{5,14}^{FF} = 0$$

$$S_{1,15}^{FF} = 0 \qquad S_{5,15}^{FF} = 0$$

Moreover, referring to the location arrays (11.13) and to relations (11.14) the submatrices $[S^{SF}]$ and $[S^{SS}]$ of the stiffness matrix for the frame are

$$[S^{SF}] =$$

$$\begin{bmatrix} \bar{K}_{14}^{(1)} & \bar{K}_{15}^{(1)} & \bar{K}_{16}^{(1)} & 0 & 0 & 0 & 0 & 0 & 0 & 0 & 0 & 0 & 0 & 0 & 0 \\ \bar{K}_{24}^{(1)} & \bar{K}_{25}^{(1)} & \bar{K}_{26}^{(1)} & 0 & 0 & 0 & 0 & 0 & 0 & 0 & 0 & 0 & 0 & 0 & 0 \\ \bar{K}_{34}^{(1)} & \bar{K}_{35}^{(1)} & \bar{K}_{36}^{(1)} & 0 & 0 & 0 & 0 & 0 & 0 & 0 & 0 & 0 & 0 & 0 & 0 \\ 0 & 0 & 0 & 0 & 0 & 0 & 0 & \bar{K}_{15}^{\prime(5)} & \bar{K}_{25}^{\prime(5)} & \bar{K}_{35}^{\prime(5)} & \bar{K}_{45}^{\prime(5)} & \bar{K}_{65}^{\prime(5)} & 0 & 0 & 0 \end{bmatrix}$$

$$(11.16)$$

$$[S^{SS}] = \begin{bmatrix} \bar{K}_{11}^{(1)} & \bar{K}_{12}^{(1)} & \bar{K}_{13}^{(1)} & 0 \\ \bar{K}_{21}^{(1)} & \bar{K}_{22}^{(1)} & \bar{K}_{23}^{(1)} & 0 \\ \bar{K}_{31}^{(1)} & \bar{K}_{32}^{(1)} & \bar{K}_{33}^{(1)} & 0 \\ 0 & 0 & 0 & \bar{K}_{55}^{\prime(5)} \end{bmatrix}$$

11.5 Assemblage of the Submatrices of the Matrix of Equivalent Actions for a Structure

In this section we present a procedure for programming the assemblage of the submatrices $\{P^{EF}\}$ and $\{P^{ES}\}$ [see relations (9.1)] of the matrix of equivalent actions $\{\hat{P}^E\}$ for a structure from the global components of the given concentrated actions acting on its nodes and the global components of the global matrix of fixed-end actions $\{\bar{A}^R\}$ of its elements.

11.5.1 Concentrated actions acting on the nodes of a structure

The program places the ith $(i = 1, 2, \ldots, 6)$ global component of the given external actions acting on node n of a structure $(P_1^{(n)} = 1,$ $P_2^{(n)} = 2,$ $P_3^{(n)} = 3,$ $P_4^{(n)} = \mathcal{M}_1^{(n)} = 4,$ $P_5^{(n)} = \mathcal{M}_2^{(n)} = 5,$ $P_6^{(n)} = \mathcal{M}_3^{(n)} = 6)$ in the row of the submatrix $\{P^{EF}\}$ or $\{P^{ES}\}$ on the basis of the number, say k, in the ICC array for the structure, which corresponds to the ith component of displacement of node $n(\Delta_i^{(n)})$ $(i = 1, 2, \ldots, 6)$ (see Tables 11.18–11.20). If k is a positive number, the program places the ith component $P_i^{(n)}$ $(i = 1, 2, \ldots, 6)$ of external action acting on node n on the kth row of the matrix $\{P^{EF}\}$. If k is a negative number, the program places it in the kth row of the matrix $\{P^{ES}\}$. For example, referring to Table 11.5 we see that there is no force acting on node 1 of the planar truss of Fig. 11.3. Moreover, referring to Table 11.18 we see that the numbers in the ICC array for this truss corresponding to node 1 and components of displacement $\Delta_1^{(n)}$ and $\Delta_2^{(n)}$ are -1 and -2, respectively. Consequently, the program places zeros in rows 1 and 2 of submatrix $\{P^{ES}\}$ for the truss. Similarly referring to Table 11.5 we see that the force acting on node 2 has two components $[P_1^{(2)} = 40 \text{ kN and } P_2^{(2)} = -80 \text{ kN}]$. Moreover, referring to Table 11.18 we see that the numbers in the ICC array for the truss corresponding to node 2 and components of displacement $\Delta_1^{(n)}$ and $\Delta_2^{(n)}$ are 1 and 2, respectively. Consequently, the program places 40 in row 1 and -80 in row 2 of the matrix $\{P^{EF}\}$ for the truss. Similarly, referring to Table 11.5 we see that the force acting on node 3 has one component $(P_1^{(3)} = 12 \text{ kN})$. Moreover, we note that node 3 is connected to a skew support. Consequently, the program transforms the global components of the force acting on it to skew using relation $(10.9a)$. That is,

$$\{\bar{P}'^{E(3)}\} = [\Lambda_P']^T \{\bar{P}^{E(3)}\} = \begin{bmatrix} 0.8 & 0.6 \\ -0.6 & 0.8 \end{bmatrix} \begin{Bmatrix} 12 \\ 0 \end{Bmatrix} = \begin{Bmatrix} 9.6 \\ -7.2 \end{Bmatrix} \quad (11.17)$$

Furthermore, referring to Table 11.18 we see that the numbers in the ICC array for the truss corresponding to node 3 and components of displacement $\Delta_1^{(3)}$ and $\Delta_2^{(3)}$ are 3 and -3, respectively. Consequently, the program places 9.6 in row 3 of matrix $\{P'^{EF}\}$ and -7.6 in row 3 of matrix $\{P'^{ES}\}$. Thus the submatrices $\{P^{EF}\}$ and $\{P^{ES}\}$ of the matrix of equivalent actions $\{P^E\}$ for the truss of Fig. 11.3 are

$$\{P'^{EF}\} = \begin{Bmatrix} 40 \\ -80 \\ 9.6 \end{Bmatrix} \qquad \{P'^{ES}\} = \begin{Bmatrix} 0 \\ 0 \\ -7.2 \end{Bmatrix} \quad (11.18)$$

Referring to Tables 11.6 and 11.19 the submatrices $\{P^{EF}\}$ and $\{P^{ES}\}$ for the planar frame of Fig. 11.4 corresponding only to the given external actions acting on its nodes are

$$\{P^{EF}\} = \begin{Bmatrix} 60 \\ 0 \\ -180 \\ 0 \\ -80 \\ 0 \\ 0 \\ 0 \\ 0 \\ 0 \\ 0 \\ 0 \\ 0 \\ 0 \\ 0 \end{Bmatrix} \quad \{P^{ES}\} = \begin{Bmatrix} 0 \\ 0 \\ 0 \\ 0 \end{Bmatrix} \tag{11.19}$$

11.5.2 Loads acting along the length of the elements of a structure

The program computes the local components of nodal actions of each element of a structure subjected to the given loads with its ends fixed (fixed-end actions). For elements of constant cross section subjected to concentrated actions or to uniform or linearly varying external forces and to uniform changes of temperature, this is accomplished using the formulas given in the table on the inside of the back cover. These formulas are included in the program. The fixed-end actions of elements of more complex loading and/or variable cross sections can be computed by a subroutine using one of the following approximate methods:

1. The finite-element method as described in Sec. 19.2
2. The method described in Sec. 12.4

The program transforms the local matrix of fixed-end actions of each element which is not connected to a skew support to global using relation (6.65). That is,

$$\{\bar{A}^R\} = [\Lambda]^T \{A^R\} \tag{11.20}$$

Moreover, the program transforms the local matrix of fixed-end actions of each element which is connected to a skew support to skew using relation (10.17).

The program places the negative of the global components of the fixed-end actions of each element in the submatrices $\{P^{EF}\}$ and $\{P^{ES}\}$ of the matrix of equivalent actions $\{\hat{P}^{E}\}$ for the structure, on the basis of the location array for the element. A positive (or negative) number, say k, in the mth row of the location array for an element indicates that the negative of the global component of nodal action in the mth row of the matrix $\{\bar{A}^{R}\}$ for this element must be placed in the kth row of the matrix $\{P^{EF}\}$ [or $\{P^{ES}\}$]. For example, referring to the table on the inside of the back cover the local matrices of fixed-end actions of the elements of the planar frame of Fig. 11.4 are

$$\{A^{R1}\} = \begin{Bmatrix} 0 \\ 36 \\ 30 \\ 0 \\ -36 \\ 30 \end{Bmatrix}$$

$$\{A^{R2}\} = \{A^{R3}\} = \{A^{R4}\} = \{A^{R5}\} = \{0\} \qquad (11.21)$$

$$\{A^{R6}\} = \begin{Bmatrix} 0 \\ 30 \\ 15 \\ 0 \\ 30 \\ -15 \end{Bmatrix}$$

The transformation matrices for elements 1 and 6 of the frame are obtained using relation (6.48). That is,

$$[\Lambda^1] = \begin{bmatrix} 0 & 1 & 0 & 0 & 0 & 0 \\ -1 & 0 & 0 & 0 & 0 & 0 \\ 0 & 0 & 1 & 0 & 0 & 0 \\ 0 & 0 & 0 & 0 & 1 & 0 \\ 0 & 0 & 0 & -1 & 0 & 0 \\ 0 & 0 & 0 & 0 & 0 & 1 \end{bmatrix}$$

$$\qquad (11.22)$$

$$[\Lambda^6] = \text{unit matrix}$$

Substituting relations (11.21) and (11.22) into (11.20) we obtain

$$[\bar{A}^{R1}] = \left\{ \begin{array}{r} -36 \\ 0 \\ 30 \\ 36 \\ 0 \\ 30 \end{array} \right\}$$

$$\{\bar{A}^{R6}\} = \{A^{R6}\}$$

(11.23)

Notice that element 5 is connected to a skew support. Consequently, if $\{A^{R5}\}$ was not a zero matrix it would have been necessary to transform it to skew using relation (10.14). Referring to the location arrays (11.13) for the elements of the frame, the submatrices $\{P^{EF}\}$ and $\{P^{ES}\}$ for the planar frame of Fig. 11.4 corresponding to the given actions acting along the length of its elements are

$$\{P^{EF}\} = \left\{ \begin{array}{r} -36 \\ 0 \\ -30 \\ 0 \\ 0 \\ 0 \\ 0 \\ 0 \\ -30 \\ -15 \\ 0 \\ 0 \\ 0 \\ -30 \\ 15 \end{array} \right\} \qquad \{P^{ES}\} = \left\{ \begin{array}{r} 36 \\ 0 \\ -30 \\ 0 \end{array} \right\}$$

(11.24)

11.6   Solution of the Stiffness Equations for a Structure and Computation of the Nodal Actions of its Elements

The program substitutes the submatrices $\{P^{EF}\}$ and $\{P^{ES}\}$ of the matrix of equivalent actions of the structure (see Sec. 11.5), the

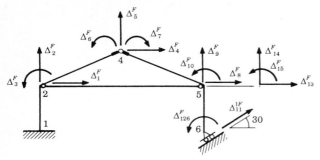

Figure 11.6 Numbering of the unconstrained components of displacements of the frame of Fig. 11.4.

submatrix $\{\Delta^S\}$ (see Sec. 9.1), and the submatrices $[S^{FF}]$, $[S^{SF}]$, $[S^{FS}]$, and $[S^{SS}]$ of the stiffness matrix for the structure (see Sec. 11.4) in relations (9.2) and (9.3) to obtain the unconstrained components of displacements $\{\Delta^F\}$ of the nodes of the structure and its reactions $\{R\}$. From the components of displacements of the nodes of a structure the program establishes the global components of nodal displacements of its elements by referring to their location array. A row $KK(J) = X$ of the location array for element e of a structure indicates that the Xth row of the matrix $\{\Delta^F\}$ of this structure must be placed in the Jth row of the matrix of nodal displacements of element e. If X is a negative number, a zero must be placed in the Jth row of the matrix of nodal displacements of element e. For example, referring to the location array (11.13) and Fig. 11.6 for the planar frame of Fig. 11.4 the components of nodal displacements of its elements are

$$\{\bar{D}^1\} = \begin{Bmatrix} 0 \\ 0 \\ 0 \\ \Delta_1^F \\ \Delta_2^F \\ \Delta_3^F \end{Bmatrix} \qquad \{\bar{D}^2\} = \begin{Bmatrix} \Delta_1^F \\ \Delta_2^F \\ \Delta_3^F \\ \Delta_4^F \\ \Delta_5^F \\ \Delta_6^F \end{Bmatrix} \qquad \{\bar{D}^3\} = \begin{Bmatrix} \Delta_4^F \\ \Delta_5^F \\ \Delta_7^F \\ \Delta_8^F \\ \Delta_9^F \\ \Delta_{10}^F \end{Bmatrix}$$

$$\tag{11.25}$$

$$\{\bar{D}^4\} = \begin{Bmatrix} \Delta_1^F \\ \Delta_2^F \\ \Delta_3^F \\ \Delta_8^F \\ \Delta_9^F \\ \Delta_{10}^F \end{Bmatrix} \qquad \{\bar{D}'^5\} = \begin{Bmatrix} \Delta_8^F \\ \Delta_9^F \\ \Delta_{10}^F \\ 0 \\ \Delta_{11}^F \\ \Delta_{12}^F \end{Bmatrix} \qquad \{\bar{D}^6\} = \begin{Bmatrix} \Delta_8^F \\ \Delta_9^F \\ \Delta_{10}^F \\ \Delta_{13}^F \\ \Delta_{14}^F \\ \Delta_{15}^F \end{Bmatrix}$$

The program substitutes the global matrix of nodal displacements, the local matrix of fixed-end actions $\{A^R\}$, and the hybrid stiffness matrix for each element which is not connected to a skew support into relation (9.13) to obtain its local matrix of nodal actions. Moreover, the program substitutes the skew matrix of nodal displacements and the modified hybrid stiffness matrix [see relation (10.27)] for each element which is connected to a skew support into relation (10.26) to obtain the local matrix of nodal actions $\{A^E\}$. The program then adds the matrices $\{A^E\}$ and $\{A^R\}$ to obtain the local matrix of nodal actions $\{A\}$ [see relation (4.7)].

12

A Method for Obtaining the Approximate Response of Elements of Complex Geometry and/or Loading

12.1 Introduction

In Sec. 4.4 we discuss briefly various methods for establishing the response of an element of a framed structure. In this section we present one of these methods which is suitable for obtaining the approximate response of elements of constant cross section subjected to a complex load distribution or of elements of variable cross section subjected to any loading. In this method an element is approximated by a model consisting of a finite number of subelements of constant cross section which has the following attributes.

1. It consists of n subelements of constant cross section (see Fig. 12.1).

2. Each subelement represents a segment of the element of the same length. The depth and width of a subelement are equal to the depth and width, respectively, of the midpoint of the segment of the element which is represented in the model by this subelement.

3. Each subelement is subjected to the given concentrated actions and to a load distribution which is an approximation of the actual load distribution acting on the corresponding segment of the actual element (see Fig. 12.1). Usually the load distribution acting on a segment of an element is approximated by a uniform or by a

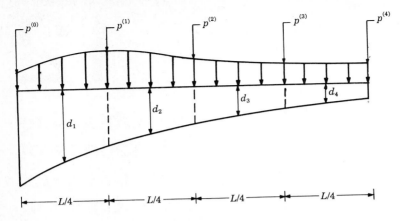

(a) Element of variable cross section

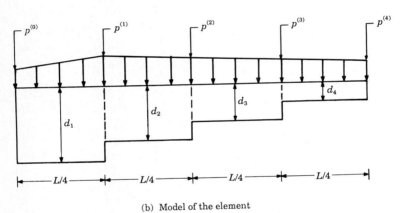

(b) Model of the element

Figure 12.1 Element of variable cross section subjected to a complex distribution of external forces and its model.

linearly varying load distribution acting on the corresponding subelement of the model.

The model is analyzed using the direct stiffness method presented in Chap. 9.

In Sec. 12.2 we present a method for condensing the stiffness matrix for a structure. In Sec. 12.3 we use this method to establish approximations for the stiffness matrix of elements of variable cross section, by condensing the stiffness matrix of their model. In Sec. 12.4 we establish approximations for the fixed-end actions of elements of constant cross section subjected to a complex load distribution or of elements of variable cross section. We accomplish this by finding the fixed-end reactions of their model.

12.2 Condensation of the Stiffness Equations for a Structure

We consider a set of stiffness equations relating n components of external actions acting at some nodes of a structure to the corresponding n components of displacements; that is,

$$
\begin{array}{cccc}
\{P^E\} & + & \{R\} & = & [S] & \{\Delta\} \\
(n \times 1) & & (n \times 1) & & (n \times n) & (n \times n)
\end{array}
\tag{12.1}
$$

The matrices $\{P\} + \{R\}$ and $\{\Delta\}$ are conjugate. That is, the term in the ith row of matrix $\{\Delta\}$ is the component of displacement over which the component of action represented by the term in the ith row of the matrix $\{P^E\}$ performs work. In general, the matrix $\{P^E\}$ could include any number of equivalent actions acting on the nodes of a structure. For instance, $\{P^E\}$ could include all possible components of equivalent actions acting on the nodes of the structure, or it could include only the components of equivalent actions which are not directly absorbed by the supports of the structure. We assume that $p < n$ of the components of displacement $\{\Delta\}$ are unconstrained, while the remaining $n - p$ are constrained. Frequently, it is necessary to eliminate $r < p$ *unconstrained* components of displacements $\{\Delta_e\}$ of the nodes of the structure from Eq. (12.1) and obtain a smaller number $n - r$ of equations involving only the remaining components of displacements $\{\Delta_c\}$. That is,

$$
\begin{array}{cccc}
\{P_c\} & + & \{R_c\} & = & [S_c] & \{\Delta_c\} \\
(n - r) \times 1 & & (n - r) \times 1 & & (n - r) \times (n - r) & (n - r) \times 1
\end{array}
\tag{12.2}
$$

where the matrices $\{P_c\}$ and $\{R_c\}$ are conjugate to the matrix $\{\Delta_c\}$. We say that Eqs. (12.1) *have been condensed to* Eqs. (12.2). In order to establish the matrices $[S_c]$ and $[P_c]$ of the condensed Eq. (12.2) we rearrange the rows and columns and partition Eqs. (12.1) as follows:

$$
\left\{ \begin{array}{c} \{P_e^E\} \\ \{P_c^E\} \end{array} \right\} + \left\{ \begin{array}{c} \{0\} \\ \{R_c\} \end{array} \right\} = \left[\begin{array}{cc} [S_{ee}] & [S_{ec}] \\ [S_{ce}] & [S_{cc}] \end{array} \right] \left\{ \begin{array}{c} \{\Delta_e\} \\ \{\Delta_c\} \end{array} \right\}
\tag{12.3}
$$

The matrix $\{P_e^E\}$ is conjugate to $\{\Delta_e\}$. Relation (12.3) can be expanded to give

$$\{P_e^E\} = [S_{ee}]\{\Delta_e\} + [S_{ec}]\{\Delta_c\} \tag{12.4}$$

$$\{P_c^E\} + \{R_c\} = [S_{ce}]\{\Delta_e\} + [S_{cc}]\{\Delta_c\} \tag{12.5}$$

If the structure is not a mechanism (see Sec. 1.11), the matrix $[S_{ee}]$ is nonsingular, and consequently from relation (12.4) we get

$$\{\Delta_e\} = [S_{ee}]^{-1}[\{P_e^E\} - [S_{ec}]\{\Delta_c\}] \tag{12.6}$$

Substituting relation (12.6) into (12.5), we obtain

$$\{P_c^E\} + \{R_c\} - [S_{ce}][S_{ee}]^{-1}\{P_e^E\} = [[S_{cc}] - [S_{ce}][S_{ee}]^{-1}[S_{ec}]]\{\Delta_c\} \tag{12.7}$$

Comparing Eqs. (12.2) and (12.7), we get

$$[S_c] = [S_{cc}] - [S_{ce}][S_{ee}]^{-1}[S_{ec}] \tag{12.8}$$

$$\{P_c\} = \{P_c^E\} - \{\overset{\square}{P}_c\} \tag{12.9}$$

where
$$\{\overset{\square}{P}_c\} = [S_{ce}][S_{ee}]^{-1}\{P_e^E\} \tag{12.10}$$

The matrix $\{\overset{\square}{P}_c\}$ represents the actions which must be added to the equivalent actions $\{P_c^E\}$ in order to take into account the effect of the equivalent actions $\{P_e^E\}$. Notice that when the stiffness equations (12.1) for a structure are given and the matrix $[S_{ee}]$ is nonsingular, the condensed stiffness matrix $[S_c]$ can be established from relation (12.8). If Eqs. (12.1) are the stiffness equations for the structure, the matrix $[S_{ee}]$ can be either singular or nonsingular. If Eqs. (12.1) are the basic stiffness equations for the structure, the stiffness matrix $[S] = [S^{FF}]$ is nonsingular and the condensed stiffness matrix $[S_c]$ is nonsingular. Consequently, the components of displacements $\{\Delta_c\}$ can be established from relation (12.2).

Example. We consider an element of constant cross section (A = constant, EI = constant) of a planar structure, having a hinge at a point along its length (see Fig. a), and we establish its stiffness matrix by adhering to the following procedure.

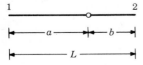

Figure a Geometry of the element.

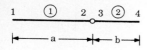

Figure b Numbering of the elements of the structure.

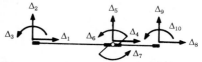

Figure c Numbering of the components of displacement of the nodes of the structure of Fig. b.

STEP 1. We consider the element as a structure made of two general planar elements (see Fig. b), and we assemble its stiffness matrix from the stiffness matrices of its elements.

STEP 2. We condense the stiffness matrix of the structure of Fig. b by eliminating the components of displacements of nodes 2 and 3. Referring to Fig. c, the condensed stiffness equations of the structure of Fig. b have the following form:

$$\{P_c\} = \begin{Bmatrix} P_1 \\ P_2 \\ P_3 \\ P_8 \\ P_9 \\ P_{10} \end{Bmatrix} = [S_c] \begin{Bmatrix} \Delta_1 \\ \Delta_2 \\ \Delta_3 \\ \Delta_8 \\ \Delta_9 \\ \Delta_{10} \end{Bmatrix} \tag{a}$$

It is apparent that the condensed stiffness matrix $[S_c]$ is the stiffness matrix of the element of Fig. a.

---

solution

STEP 1   We establish the stiffness matrix for the structure of Fig. b. For this purpose we compute the stiffness matrix for each of its elements; that is, referring to relation (4.62) we have

$$[\bar{K}^1] = [K^1] = EI \begin{bmatrix} \dfrac{A}{Ia} & 0 & 0 & -\dfrac{A}{Ia} & 0 & 0 \\[2ex] 0 & \dfrac{12}{a^3} & \dfrac{6}{a^2} & 0 & -\dfrac{12}{a^3} & \dfrac{6}{a^2} \\[2ex] 0 & \dfrac{6}{a^2} & \dfrac{4}{a} & 0 & -\dfrac{6}{a^2} & \dfrac{2}{a} \\[2ex] -\dfrac{A}{Ia} & 0 & 0 & \dfrac{A}{Ia} & 0 & 0 \\[2ex] 0 & -\dfrac{12}{a^3} & -\dfrac{6}{a^2} & 0 & \dfrac{12}{a^3} & -\dfrac{6}{a^2} \\[2ex] 0 & \dfrac{6}{a^2} & \dfrac{2}{a} & 0 & -\dfrac{6}{a^2} & \dfrac{4}{a} \end{bmatrix} \tag{b}$$

$$[\bar{K}^2] = [K^2] = EI \begin{bmatrix} \dfrac{A}{Ib} & 0 & 0 & -\dfrac{A}{Ib} & 0 & 0 \\[2mm] 0 & \dfrac{12}{b^3} & \dfrac{6}{b^2} & 0 & -\dfrac{12}{b^3} & \dfrac{6}{b^2} \\[2mm] 0 & \dfrac{6}{b^2} & 0 & 0 & -\dfrac{6}{b^2} & \dfrac{2}{b} \\[2mm] -\dfrac{A}{Ib} & 0 & \dfrac{A}{Ib} & 0 & 0 \\[2mm] 0 & -\dfrac{12}{b^3} & -\dfrac{6}{b^2} & 0 & \dfrac{12}{b^3} & -\dfrac{6}{b^2} \\[2mm] 0 & \dfrac{6}{b^2} & \dfrac{2}{b} & 0 & -\dfrac{6}{b^2} & \dfrac{4}{b} \end{bmatrix} \quad\text{(c)}$$

We assemble the stiffness matrix for the structure of Fig. b. Referring to Fig. c and using relations (b) and (c) we have

$$[\hat{S}] = EI \begin{bmatrix} \dfrac{A}{Ia} & 0 & 0 & -\dfrac{A}{Ia} & 0 & 0 & 0 & 0 & 0 & 0 \\[2mm] 0 & \dfrac{12}{a^3} & \dfrac{6}{a^2} & 0 & -\dfrac{12}{a^3} & \dfrac{6}{a^2} & 0 & 0 & 0 & 0 \\[2mm] 0 & \dfrac{6}{a^2} & \dfrac{4}{a} & 0 & -\dfrac{6}{a} & \dfrac{2}{a} & 0 & 0 & 0 & 0 \\[2mm] -\dfrac{A}{Ia} & 0 & 0 & \dfrac{A}{I}\left(\dfrac{1}{a}+\dfrac{1}{b}\right) & 0 & 0 & 0 & -\dfrac{A}{Ib} & 0 & 0 \\[2mm] 0 & -\dfrac{12}{a^3} & -\dfrac{6}{a^2} & 0 & 12\left(\dfrac{1}{a^3}+\dfrac{1}{b^3}\right) & -\dfrac{6}{a^2} & \dfrac{6}{b^2} & 0 & -\dfrac{12}{b^3} & \dfrac{6}{b^2} \\[2mm] 0 & \dfrac{6}{a^2} & \dfrac{2}{a} & 0 & -\dfrac{6}{a^2} & \dfrac{4}{a} & 0 & 0 & 0 & 0 \\[2mm] 0 & 0 & 0 & 0 & \dfrac{6}{b^2} & 0 & \dfrac{4}{b} & 0 & -\dfrac{6}{b^2} & \dfrac{2}{b} \\[2mm] 0 & 0 & 0 & -\dfrac{A}{Ib} & 0 & 0 & 0 & \dfrac{A}{Ib} & 0 & 0 \\[2mm] 0 & 0 & 0 & 0 & -\dfrac{12}{b^3} & 0 & -\dfrac{6}{b^2} & 0 & \dfrac{12}{b^3} & -\dfrac{6}{b^2} \\[2mm] 0 & 0 & 0 & 0 & \dfrac{6}{b^2} & 0 & \dfrac{2}{b} & 0 & -\dfrac{6}{b^2} & \dfrac{4}{b} \end{bmatrix}$$

$$\text{(d)}$$

STEP 2. We rearrange the rows and columns of the stiffness matrix $[\hat{S}]$ so that in the stiffness equations $[\{\hat{P}^E\} + \{\hat{R}\} = [\hat{S}]\{\hat{\Delta}\}]$ for the structure of Fig. b, the components of equivalent actions and the components of displacements associated with nodes 2 and 3 of the structure of Fig. b are transferred to the top of the matrices $\{\hat{P}^E\}$ and $\{\hat{\Delta}\}$, respectively. That is, rows 4 to 7 of the matrix $[\hat{S}]$ are transferred to its top, while columns 4 to 7 are transferred to its right part. Moreover, referring to relation (12.3) we partition the resulting stiffness matrix $[\hat{S}_m]$. Thus

$$[\hat{S}_m] = EI \begin{bmatrix}
\frac{A}{I}\left(\frac{1}{a}+\frac{1}{b}\right) & 0 & 0 & 0 & -\frac{A}{Ia} & 0 & 0 & -\frac{A}{Ib} & 0 & 0 \\
0 & 12\left(\frac{1}{a^3}+\frac{1}{b^3}\right) & -\frac{6}{a^2} & \frac{6}{b^2} & 0 & -\frac{12}{a^3} & \frac{6}{a^2} & 0 & -\frac{12}{b^3} & \frac{6}{b^2} \\
0 & -\frac{6}{a^2} & \frac{4}{a} & 0 & 0 & \frac{6}{a^2} & \frac{2}{a} & 0 & 0 & 0 \\
0 & \frac{6}{b^2} & 0 & \frac{4}{b} & 0 & 0 & 0 & 0 & -\frac{6}{b^2} & \frac{2}{b} \\
-\frac{A}{Ia} & 0 & 0 & 0 & \frac{A}{Ia} & 0 & 0 & 0 & 0 & 0 \\
0 & -\frac{12}{a^3} & \frac{6}{a^2} & 0 & 0 & \frac{12}{a^3} & \frac{6}{a^2} & 0 & 0 & 0 \\
0 & -\frac{6}{a^2} & \frac{2}{a} & 0 & 0 & \frac{6}{a^2} & \frac{4}{a} & 0 & 0 & 0 \\
-\frac{A}{Ib} & 0 & 0 & 0 & 0 & 0 & 0 & \frac{A}{Ib} & 0 & 0 \\
0 & -\frac{12}{b^3} & 0 & -\frac{6}{b^2} & 0 & 0 & 0 & 0 & \frac{12}{b^3} & -\frac{6}{b} \\
0 & \frac{6}{b^2} & 0 & \frac{2}{b} & 0 & 0 & 0 & 0 & -\frac{6}{b^2} & \frac{4}{b}
\end{bmatrix} \tag{e}$$

Referring to relation (12.3) from the above matrix, we obtain

$$[S_{ee}] = EI \begin{bmatrix}
\frac{A}{I}\left(\frac{1}{a}+\frac{1}{b}\right) & 0 & 0 & 0 \\
0 & 12\left(\frac{1}{a^3}+\frac{1}{b^3}\right) & -\frac{6}{a^2} & \frac{6}{b^2} \\
0 & -\frac{6}{a^2} & \frac{4}{a} & 0 \\
0 & \frac{6}{a^2} & 0 & \frac{4}{b}
\end{bmatrix} \tag{f}$$

$$[S_{ec}] = EI \begin{bmatrix}
-\frac{A}{Ia} & 0 & 0 & -\frac{A}{Ib} & 0 & 0 \\
0 & -\frac{12}{a^3} & -\frac{6}{a^2} & 0 & -\frac{12}{b^3} & \frac{6}{b^2} \\
0 & \frac{6}{a^2} & \frac{2}{a} & 0 & 0 & 0 \\
0 & 0 & 0 & 0 & -\frac{6}{b^2} & \frac{2}{b}
\end{bmatrix} \tag{g}$$

$$[S_{ce}] = EI \begin{bmatrix} -\dfrac{A}{Ia} & 0 & 0 & 0 \\[2ex] 0 & -\dfrac{12}{a^3} & \dfrac{6}{a^2} & 0 \\[2ex] 0 & -\dfrac{6}{a^2} & \dfrac{2}{a} & 0 \\[2ex] -\dfrac{A}{Ib} & 0 & 0 & 0 \\[2ex] 0 & -\dfrac{12}{b^3} & 0 & -\dfrac{6}{b^2} \\[2ex] 0 & \dfrac{6}{b^2} & 0 & \dfrac{2}{b} \end{bmatrix} \qquad \text{(h)}$$

$$[S_{cc}] = EI \begin{bmatrix} \dfrac{A}{Ia} & 0 & 0 & 0 & 0 & 0 \\[2ex] 0 & \dfrac{12}{a^3} & \dfrac{6}{a^2} & 0 & 0 & 0 \\[2ex] 0 & \dfrac{6}{a^2} & \dfrac{4}{a} & 0 & 0 & 0 \\[2ex] 0 & 0 & 0 & \dfrac{A}{Ib} & 0 & 0 \\[2ex] 0 & 0 & 0 & 0 & \dfrac{12}{b^3} & -\dfrac{6}{b} \\[2ex] 0 & 0 & 0 & 0 & -\dfrac{6}{b^2} & \dfrac{4}{b} \end{bmatrix} \qquad \text{(i)}$$

and

$$[S_{ee}]^{-1} = EI \begin{bmatrix} \dfrac{I}{A\left(\dfrac{1}{a}+\dfrac{1}{b}\right)} & 0 & 0 & 0 \\[4ex] 0 & \dfrac{1}{3\left(\dfrac{1}{a^3}+\dfrac{1}{b^3}\right)} & \dfrac{1}{2a\left(\dfrac{1}{a^3}+\dfrac{1}{b^3}\right)} & -\dfrac{1}{2b\left(\dfrac{1}{a^3}+\dfrac{1}{b^3}\right)} \\[4ex] 0 & \dfrac{1}{2a\left(\dfrac{1}{a^3}+\dfrac{1}{b^3}\right)} & \dfrac{a}{4}+\dfrac{3}{4a^2\left(\dfrac{1}{a^3}+\dfrac{1}{b^3}\right)} & -\dfrac{1}{4ab\left(\dfrac{1}{a^3}+\dfrac{1}{b^3}\right)} \\[4ex] 0 & -\dfrac{1}{2b\left(\dfrac{1}{a^3}+\dfrac{1}{b^3}\right)} & -\dfrac{1}{4ab\left(\dfrac{1}{a^3}+\dfrac{1}{b^3}\right)} & \dfrac{b}{4}+\dfrac{3}{4b^2\left(\dfrac{1}{a^3}+\dfrac{1}{b^3}\right)} \end{bmatrix}$$

$$\text{(j)}$$

We compute the condensed stiffness matrix $[S_c]$ by substituting relations (g) to (j) into (12.8). Thus

$$[S_c] = [S_{cc}] - [S_{ce}][S_{ee}]^{-1}[S_{ec}]$$

$$= \frac{3EI}{a^3 + b^2} \begin{bmatrix} m & 0 & 0 & -m & 0 & 0 \\ 0 & 1 & a & 0 & -1 & b \\ 0 & a & a^2 & 0 & -a & ab \\ -m & 0 & 0 & m & 0 & 0 \\ 0 & -1 & -a & 0 & 1 & -b \\ 0 & b & ab & 0 & -b & b^2 \end{bmatrix} \qquad (k)$$

where

$$m = \frac{A(a^3 + b^3)}{3IL} \qquad (l)$$

The matrix $[S_c]$ is the stiffness matrix $[K]$ of the element of Fig. a. Notice that as expected $[S_c]$ is a singular matrix.

12.3   An Approximate Method for Establishing the Stiffness Matrix for an Element of Variable Cross Section

In this section, we present a method for establishing the local stiffness matrix of an element of variable cross section by adhering to the following steps.

Step 1. The element is approximated by a model consisting of n subelements of constant cross section. Each subelement represents a segment of the element of the same length. The depth and width of a subelement are equal to the depth and width, respectively, of the midpoint of the segment of the element which is represented in the model by this subelement (see the example at the end of this section). The stiffness matrix of each subelement is formed.

Step 2. The stiffness matrix of the model with $n + 1$ nodes is assembled from the stiffness matrices of its subelements.

Step 3. The stiffness matrix of the model is condensed by eliminating from its stiffness equations the components of displacements associated with its $n - 1$ intermediate nodes. The resulting stiffness matrix is an approximation to the local stiffness matrix of the element.

It is apparent, that as the element is divided into a larger number of segments, the local stiffness matrix obtained on the basis of this method approaches its actual local stiffness matrix.

Example. Establish an approximate expression for the local stiffness matrix of the general planar element of constant width b shown in Fig. a by approximating the element by a model consisting of three subelements of equal length. The depth of each subelement is constant (see Fig. b). Compare the results with those obtained on the basis of the exact analysis.

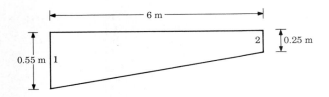

Figure a Geometry of the element.

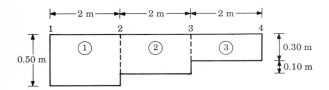

Figure b Model for the element of Fig. a.

solution
STEP 1. Referring to relation (4.62), we form the stiffness matrix of each subelement. That is,

$$[\bar{K}^1] = [K^1] = \frac{Eb}{24}\begin{bmatrix} 6.0 & 0 & 0 & -6.0 & 0 & 0 \\ 0 & 0.375 & 0.375 & 0 & -0.375 & 0.375 \\ 0 & 0.375 & 0.500 & 0 & -0.375 & 0.250 \\ -6.0 & 0 & 0 & 6.0 & 0 & 0 \\ 0 & -0.375 & -0.375 & 0 & 0.375 & -0.375 \\ 0 & 0.375 & 0.250 & 0 & -0.375 & 0.500 \end{bmatrix} \quad \text{(a)}$$

$$[\bar{K}^2] = [K^2] = \frac{Eb}{24}\begin{bmatrix} 4.8 & 0 & 0 & -4.8 & 0 & 0 \\ 0 & 0.192 & 0.192 & 0 & -0.192 & 0.192 \\ 0 & 0.192 & 0.256 & 0 & -0.192 & 0.128 \\ -4.8 & 0 & 0 & 4.8 & 0 & 0 \\ 0 & -0.192 & -0.192 & 0 & 0.192 & -0.192 \\ 0 & 0.192 & 0.128 & 0 & -0.192 & 0.256 \end{bmatrix} \quad \text{(b)}$$

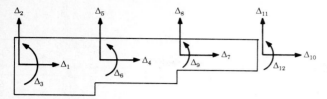

Figure c Components of displacements of the nodes of the model of Fig. b.

$$[\bar{K}^3] = [K^3] = \frac{Eb}{24} \begin{bmatrix} 3.6 & 0 & 0 & -3.6 & 0 & 0 \\ 0 & 0.081 & 0.081 & 0 & -0.081 & 0.081 \\ 0 & 0.081 & 0.108 & 0 & -0.081 & 0.054 \\ -3.6 & 0 & 0 & 3.6 & 0 & 0 \\ 0 & -0.081 & -0.081 & 0 & 0.081 & -0.081 \\ 0 & 0.081 & 0.054 & 0 & -0.081 & 0.108 \end{bmatrix} \quad \text{(c)}$$

STEP 2.   We assemble the stiffness matrix of the four-node model of Fig. b. Referring to Fig. c, using relations (a) to (c) and adhering to one of the procedures described in Sec. 9.3, we obtain the stiffness matrix for the model of Fig. b [see Eq. (d) on p. 416]. The stiffness equations for the model of Fig. b can be written as shown by Eqs. (e) on p. 416. Thus, referring to relations (12.3) and Eqs. (e), we have

$$[S_{ee}] = \frac{Eb}{24} \begin{bmatrix} 10.8 & 0 & 0 & -4.8 & 0 & 0 \\ 0 & 0.567 & -0.183 & 0 & -0.192 & 0.192 \\ 0 & -0.183 & 0.756 & 0 & -0.192 & 0.128 \\ -4.8 & 0 & 0 & 8.4 & 0 & 0 \\ 0 & -0.192 & -0.192 & 0 & 0.273 & -0.111 \\ 0 & 0.192 & 0.128 & 0 & -0.111 & 0.364 \end{bmatrix} \quad \text{(f)}$$

$$[S_{ec}] = \frac{Eb}{24} \begin{bmatrix} -6.0 & 0 & 0 & 0 & 0 & 0 \\ 0 & -0.375 & -0.375 & 0 & 0 & 0 \\ 0 & 0.375 & 0.250 & 0 & 0 & 0 \\ 0 & 0 & 0 & -3.6 & 0 & 0 \\ 0 & 0 & 0 & 0 & -0.081 & 0.081 \\ 0 & 0 & 0 & 0 & -0.081 & 0.054 \end{bmatrix} \quad \text{(g)}$$

$$[S_{ce}] = \frac{Eb}{24} \begin{bmatrix} -6.0 & 0 & 0 & 0 & 0 & 0 \\ 0 & -0.375 & 0.375 & 0 & 0 & 0 \\ 0 & -0.375 & 0.250 & 0 & 0 & 0 \\ 0 & 0 & 0 & -3.6 & 0 & 0 \\ 0 & 0 & 0 & 0 & -0.081 & -0.081 \\ 0 & 0 & 0 & 0 & 0.081 & 0.054 \end{bmatrix} \quad \text{(h)}$$

(d)

$$[\hat{S}]_{model} = \frac{Eb}{24}$$

6.0	0	0	-6.0	0	0	0	0	0	0	0	0
0	0.375	0.375	0	-0.375	0.375	0	0	0	0	0	0
0	0.375	0.500	0	-0.375	0.250	0	0	0	0	0	0
-6.0	0	0	10.8	0	0	-4.8	0	0	0	0	0
0	-0.375	-0.375	0	0.567	-0.183	0	-0.192	-0.192	0	-0.081	-0.081
0	0.375	0.250	0	-0.183	0.756	0	0.192	0.128	0	-0.081	0.054
0	0	0	-4.8	0	0	8.4	0	0	-3.6	0	0
0	0	0	0	-0.192	0.192	0	0.273	-0.111	0	-0.081	-0.081
0	0	0	0	-0.192	0.128	0	-0.111	0.364	0	0.081	0.054
0	0	0	0	0	0	-3.6	0	0	3.6	0	0
0	0	0	0	-0.081	-0.081	0	-0.081	0.081	0	0.081	-0.081
0	0	0	0	-0.081	0.054	0	-0.081	0.054	0	-0.081	0.108

(e)

$$\begin{Bmatrix} P \end{Bmatrix} = \frac{Eb}{24} \,[\,\ldots\,]\, \begin{Bmatrix} \Delta \end{Bmatrix}$$

	Δ_4	Δ_5	Δ_6	Δ_7	Δ_8	Δ_9	Δ_1	Δ_2	Δ_3	Δ_{10}	Δ_{11}	Δ_{12}
P_4	10.8	0	0	-4.8	0	0	-6.0	0	0	0	0	0
P_5	0	0.567	-0.183	0	-0.192	-0.192	0	-0.375	-0.375	0	-0.081	-0.081
P_6	0	-0.183	0.756	0	0.192	0.128	0	0.375	0.250	0	-0.081	0.054
P_7	-4.8	0	0	8.4	0	0	0	0	0	-3.6	0	0
P_8	0	-0.192	0.192	0	0.273	-0.111	0	0	0	0	-0.081	-0.081
P_9	0	-0.192	0.128	0	-0.111	0.364	0	0	0	0	0.081	0.054
P_1	-6.0	0	0	0	0	0	6.0	0	0	0	0	0
P_2	0	-0.375	0.375	0	0	0	0	0.375	0.375	0	0	0
P_3	0	-0.375	0.250	0	0	0	0	0.375	0.500	0	0	0
P_{10}	0	0	0	-3.6	0	0	0	0	0	3.6	0	0
P_{11}	0	-0.081	-0.081	0	-0.081	0.081	0	0	0	0	0.081	-0.081
P_{12}	0	-0.081	0.054	0	-0.081	0.054	0	0	0	0	-0.081	0.108

$$[S_{cc}] = \frac{Eb}{24}\begin{bmatrix} 6.0 & 0 & 0 & 0 & 0 & 0 \\ 0 & 0.375 & 0.375 & 0 & 0 & 0 \\ 0 & 0.375 & 0.500 & 0 & 0 & 0 \\ 0 & 0 & 0 & 3.6 & 0 & 0 \\ 0 & 0 & 0 & 0 & 0.081 & -0.081 \\ 0 & 0 & 0 & 0 & -0.081 & 0.108 \end{bmatrix} \quad \text{(i)}$$

Moreover,

$$[S_{ee}]^{-1} = \frac{24}{Eb}\begin{bmatrix} 0.1241 & 0 & 0 & 0.0709 & 0 & 0 \\ 0 & 4.8253 & 2.6434 & 0 & 4.3831 & -2.1387 \\ 0 & 2.6434 & 3.0790 & 0 & 3.4443 & -1.4273 \\ 0.0709 & 0 & 0 & 0.1596 & 0 & 0 \\ 0 & 4.3831 & 3.4443 & 0 & 8.8302 & -0.8309 \\ 0 & -2.1387 & -1.4273 & 0 & -0.8309 & 4.1245 \end{bmatrix}$$

$$\text{(j)}$$

Substituting relations (f) and (j) into (12.8), we obtain an approximation to the stiffness matrix of the element. That is,

$$[K]_{\text{appr}} = [S_c] = [S_{cc}] - [S_{ce}][S_{ee}]^{-1}[S_{ec}]$$

$$= \frac{Eb}{24}\begin{bmatrix} 1.5314 & 0 & 0 & -1.5314 & 0 & 0 \\ 0 & 0.0069 & 0.0273 & 0 & -0.0069 & 0.0141 \\ 0 & 0.0273 & 0.1246 & 0 & -0.0273 & 0.0393 \\ -1.5314 & 0 & 0 & 1.5314 & 0 & 0 \\ 0 & -0.0069 & -0.0273 & 0 & 0.0069 & -0.0141 \\ 0 & 0.0141 & 0.0393 & 0 & -0.0141 & 0.0453 \end{bmatrix}$$

$$\text{(k)}$$

The stiffness matrix of the element of Fig. a obtained on the basis of exact analysis is

$$[S_c]_{\text{exact}} = \frac{Eb}{24}\begin{bmatrix} 1.5219 & 0 & 0 & -1.5219 & 0 & 0 \\ 0 & 0.0069 & 0.0272 & 0 & -0.0069 & 0.0141 \\ 0 & 0.0272 & 0.1249 & 0 & -0.0272 & 0.0389 \\ -1.5219 & 0 & 0 & 1.5219 & 0 & 0 \\ 0 & -0.0069 & -0.0272 & 0 & 0.0069 & -0.0141 \\ 0 & 0.0141 & 0.0389 & 0 & -0.0141 & 0.0457 \end{bmatrix}$$

$$\text{(l)}$$

12.4   An Approximate Method for Establishing the Fixed-End Actions of an Element

In this section, we present an approximate method which can be employed to compute the fixed-end actions of an element of constant cross section subjected to a complex distribution of external loads or of an element of variable cross section subjected to any given load.

Step 1. The element is approximated by a model consisting of n subelements of constant cross section. Each subelement represents a segment of the element of the same length. The depth and width of a subelement are equal to the depth and width, respectively, of the midpoint of the segment of the element which is represented in the model by this subelement (see Fig. b of the example at the end of this section). Each subelement is subjected to the given concentrated actions and to a load distribution which is an approximation of the actual load distribution acting on the corresponding segment of the element (see Fig. 12.1). Usually, the load distribution acting on a segment of an element is approximated by a uniform or by a linearly varying load distribution acting on the corresponding subelement of the model. The model is fixed at both ends.

Step 2. The fixed-end actions of each subelement of the model subjected to the load distribution described in step 1 are obtained directly by referring to the table on the inside of the back cover. They are employed to establish the equivalent actions to be placed on the nodes of the model.

Step 3. The stiffness matrix of the model is assembled from the stiffness matrices of its subelements. The latter are obtained using the formulas given in Sec. 4.4.1.

Step 4. The stiffness equations for the model are written as follows:

$$\left\{ \begin{array}{c} \{P^E_{in}\} \\ \{P^E_{en}\} \end{array} \right\} + \left\{ \begin{array}{c} \{0\} \\ \{R\} \end{array} \right\} = \left[\begin{array}{cc} [S_{ii}] & [S_{ie}] \\ [S_{ei}] & [S_{ee}] \end{array} \right] \left\{ \begin{array}{c} \{\Delta_{in}\} \\ \{0\} \end{array} \right\} \tag{12.11}$$

where $[\Delta_{in}]$ = matrix of components of displacements of intermediate nodes of model

$\{P^E_{in}\}$ = matrix of components of equivalent actions acting at intermediate nodes of model

$\{P^E_{en}\}$ = matrix of components of equivalent actions acting on end nodes of model

$\{R\}$ = matrix of reactions of the model

Relation (12.11) can be rewritten as

$$\{P^E_{in}\} = [S_{ii}]\{\Delta_{in}\}$$
$$\{P^E_{en}\} + \{R\} = [S_{ei}]\{\Delta_{in}\} \tag{12.12}$$

Eliminating $\{\Delta_{in}\}$ from the above relations we get

$$\{R\} = [S_{ei}][S_{ii}]^{-1}\{P^E_{in}\} - \{P^E_{en}\} \qquad (12.13)$$

It is apparent that as the number of segments into which an element is divided increases, the reactions of the model approach those of the element subjected to the given loads with its ends fixed.

Example. Establish an approximate expression for the fixed-end actions of the element of constant width b shown in Fig. a by approximating the element by a model consisting of three subelements of equal length. The depth of each subelement is constant (see Fig. b).

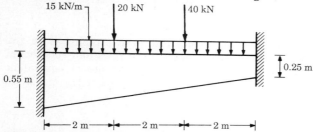

Figure a   Geometry and loading of the element.

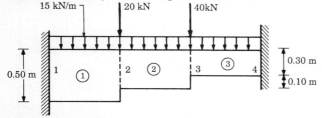

Figure b   Model for the element of Fig. a.

solution
STEP 1.   The element is approximated by the model of Fig. b.

STEP 2.   Referring to the table on the inside of the back cover we compute the fixed-end actions of the subelements of the model of Fig. b and we employ them to establish the equivalent actions to be placed on the nodes of the model. They are shown in Fig. c. Referring to this figure and omitting the

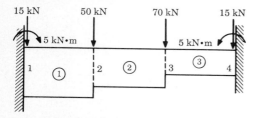

Figure c   Model of the element of Fig. a subjected to the equivalent actions.

axial forces, the matrix of equivalent actions for the model is

$$\{\hat{P}^E\} = \begin{Bmatrix} P_2^{(1)} \\ M_3^{(1)} \\ P_2^{(2)} \\ M_3^{(2)} \\ P_2^{(3)} \\ M_3^{(3)} \\ P_2^{(4)} \\ M_3^{(4)} \end{Bmatrix} = \begin{Bmatrix} -15 \\ -5 \\ -50 \\ 0 \\ -70 \\ 0 \\ -15 \\ 5 \end{Bmatrix} \tag{a}$$

STEP 3. The stiffness matrix for the model has been established in the example of Sec. 12.3 [see Eq. (d)]. Thus omitting the rows and columns of the stiffness matrix corresponding to the axial components of equivalent forces and axial components of displacements of the nodes of the model we have

$$[\hat{S}] = \frac{Eb}{24} \begin{bmatrix} 0.375 & 0.375 & -0.375 & 0.375 & 0 & 0 & 0 & 0 \\ 0.375 & 0.500 & -0.375 & 0.250 & 0 & 0 & 0 & 0 \\ -0.375 & -0.375 & 0.567 & -0.183 & -0.192 & 0.192 & 0 & 0 \\ 0.375 & 0.250 & -0.183 & 0.756 & -0.192 & 0.128 & 0 & 0 \\ 0 & 0 & -0.192 & -0.192 & 0.273 & -0.111 & -0.081 & 0.081 \\ 0 & 0 & 0.192 & 0.128 & -0.111 & 0.364 & -0.081 & 0.054 \\ 0 & 0 & 0 & 0 & -0.081 & -0.081 & 0.081 & -0.081 \\ 0 & 0 & 0 & 0 & 0.081 & 0.054 & -0.081 & 0.108 \end{bmatrix}$$

$$\tag{b}$$

STEP 4. Referring to relations (a) and (b) the stiffness equations for the model can be written as

$$\begin{Bmatrix} -50 \\ 0 \\ -70 \\ 0 \\ \hdashline -15 \\ -5 \\ 15 \\ 5 \end{Bmatrix} + \begin{Bmatrix} 0 \\ 0 \\ 0 \\ 0 \\ \hdashline R_2^{(1)} \\ R_3^{(1)} \\ R_2^{(4)} \\ R_3^{(4)} \end{Bmatrix}$$

$$= \frac{Eb}{24} \begin{bmatrix} 0.567 & -0.183 & -0.192 & 0.192 & -0.375 & -0.375 & 0 & 0 \\ -0.183 & 0.756 & -0.192 & 0.128 & 0.375 & 0.250 & 0 & 0 \\ -0.192 & -0.192 & 0.273 & -0.111 & 0 & 0 & -0.081 & 0.081 \\ 0.192 & 0.128 & -0.111 & 0.364 & 0 & 0 & -0.081 & 0.054 \\ -0.375 & 0.375 & 0 & 0 & 0.375 & 0.375 & 0 & 0 \\ -0.375 & 0.250 & 0 & 0 & 0.375 & 0.500 & 0 & 0 \\ 0 & 0 & -0.081 & -0.081 & 0 & 0 & 0.081 & -0.081 \\ 0 & 0 & 0.081 & 0.054 & 0 & 0 & -0.081 & 0.108 \end{bmatrix} \begin{Bmatrix} \Delta_2^{(2)} \\ \Delta_3^{(2)} \\ \Delta_2^{(3)} \\ \Delta_3^{(3)} \\ 0 \\ 0 \\ 0 \\ 0 \end{Bmatrix}$$

(c)

Thus, referring to relations (12.11) we have

$$\{P_{\text{in}}\} = \begin{Bmatrix} -50 \\ 0 \\ -70 \\ 0 \end{Bmatrix} \qquad \{P_{\text{en}}\} = \begin{Bmatrix} -15 \\ -5 \\ 15 \\ 5 \end{Bmatrix} \qquad \{R\} = \begin{Bmatrix} R_2^{(1)} \\ R_3^{(1)} \\ R_2^{(4)} \\ R_3^{(4)} \end{Bmatrix}$$

(d)

$$[S_{ei}] = \frac{Eb}{24} \begin{bmatrix} -0.375 & 0.375 & 0 & 0 \\ -0.375 & 0.250 & 0 & 0 \\ 0 & 0 & -0.081 & -0.081 \\ 0 & 0 & 0.081 & 0.054 \end{bmatrix}$$

(e)

$$[S_{ii}] = \frac{Eb}{24} \begin{bmatrix} 0.567 & -0.183 & -0.192 & 0.192 \\ -0.183 & 0.756 & -0.192 & 0.128 \\ -0.192 & -0.192 & 0.273 & -0.111 \\ 0.192 & 0.128 & -0.111 & 0.364 \end{bmatrix}$$

(f)

and

$$[S_{ii}]^{-1} = \frac{24}{Eb} \begin{bmatrix} 4.8253 & 2.6434 & 4.3831 & -2.1387 \\ 2.6434 & 3.0790 & 3.4443 & -1.4273 \\ 4.3831 & 3.4443 & 8.8302 & -0.8309 \\ -2.1387 & -1.4273 & -0.8309 & 4.1245 \end{bmatrix}$$

(g)

Substituting relations (d), (e), and (g) into (12.13), we obtain

$$\{R\} = [S_{ei}][S_{ii}]^{-1}\{P_{\text{in}}\} - \{P_{\text{en}}^E\}$$

$$= \begin{bmatrix} -0.8182 & 0.1633 & -0.3520 & 0.2667 \\ -1.1486 & -0.2215 & -0.7825 & 0.4451 \\ -0.1817 & -0.1633 & -0.6479 & -0.2667 \\ 0.2395 & 0.20191 & 0.6703 & 0.1554 \end{bmatrix} \begin{bmatrix} -50 \\ 0 \\ -70 \\ 0 \end{bmatrix} - \begin{Bmatrix} -15 \\ -5 \\ -15 \\ 5 \end{Bmatrix}$$

$$= \begin{bmatrix} 65.55 \\ 112.20 \\ 54.43 \\ -58.89 \end{bmatrix} - \begin{Bmatrix} -15 \\ -5 \\ -15 \\ 5 \end{Bmatrix} = \begin{Bmatrix} 80.55 \\ 117.20 \\ 39.43 \\ -63.39 \end{Bmatrix}$$

(h)

Thus, an approximate expression for the matrix of fixed-end actions of the element subjected to the loads shown in Fig. a is

$$\{A^R\} = \begin{Bmatrix} 0 \\ 80.55 \\ 117.20 \\ 0 \\ 69.43 \\ -63.39 \end{Bmatrix}$$

12.5   Problems

1.   Using the procedure described in Sec. 12.3 compute the stiffness matrix of the element shown in Fig. 12.P1. The cross section of the element has a constant width of 0.20 m. The element is made from an isotropic linearly elastic material. Subdivide the element into three subelements of equal length.

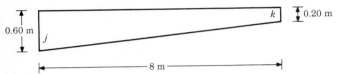

Figure 12.P1

2.   Using the procedure described in Sec. 12.3 compute the stiffness matrix of the element shown in Fig. 12.P2. The element is made from an isotropic linearly elastic material. The cross section of the element has a constant width of 0.30 m. Subdivide the portion of the element which has a variable cross section into two subelements of equal length.

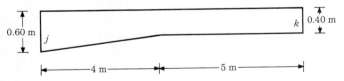

Figure 12.P2

3.   Using the procedure described in Sec. 12.3 compute the stiffness matrix of the element shown in Fig. 12.P3. The area of the cross section of each flange is equal to $A_f = 4 \times 10^3 \text{ mm}^2$. The element is made from an isotropic linearly elastic material. Neglect the effect of the web on the moment of inertia of the cross sections of the element. Subdivide the element into three subelements of equal length.

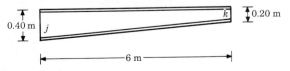

Figure 12.P3

4.   Using the procedure described in Sec. 12.3 compute the stiffness matrix of the element shown in Fig. 12.P4. The area of the cross section of each flange of the element is equal to $A_f = 4 \times 10^3 \text{ mm}^2$. The element is made from an isotropic linearly elastic material. Neglect the effect of the moment of inertia of the cross sections of the element. Subdivide the portion of the element which has a variable cross section into two subelements of equal length.

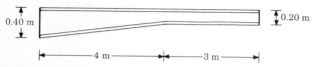

0.40 m   0.20 m

|← 4 m →|← 3 m →|

Figure 12.P4

5.   Using the procedure described in Sec. 12.4 compute the reactions of the fixed-end element loaded as shown in Fig. 12.P5. The element is made from an isotropic linearly elastic material and its cross section has a constant width of 0.20 m. Subdivide the element into three subelements of equal length.

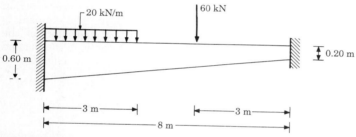

20 kN/m   60 kN

0.60 m   0.20 m

|← 3 m →|← 3 m →|

|← 8 m →|

Figure 12.P5

6.   Using the procedure described in Sec. 12.4 compute the reactions of the fixed-end element loaded as shown in Fig. 12.P6. The element is made from an isotropic linearly elastic material and its cross section has a constant width of 0.30 m. Subdivide the portion of the element which has a variable cross section into two subelements of equal length.

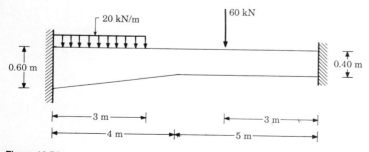

20 kN/m   60 kN

0.60 m   0.40 m

|← 3 m →|← 3 m →|

|← 4 m →|← 5 m →|

Figure 12.P6

7. Using the procedure described in Sec. 12.4 compute the reactions of the fixed-end element loaded as shown in Fig. 12.P7. The area of the cross section of each flange is equal to $A_f = 4 \times 10^3 \, mm^2$. The element is made from an isotropic linearly elastic material. Neglect the effect of the web on the moment of inertia of the cross sections of the element. Subdivide the element into three subelements of equal length.

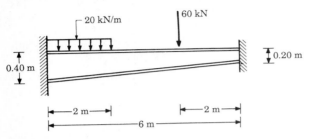

Figure 12.P7

8. Using the procedure described in Sec. 12.4 compute the reactions of the fixed-end element loaded as shown in Fig. 12.P8. The area of the cross section of each flange of the element is equal to $A_f = 4 \times 10^3 \, mm^2$. The element is made from an isotropic linearly elastic material. Neglect the effect of the moment of inertia of the cross sections of the element. Subdivide the portion of the element which has a variable cross section into two subelements of equal length.

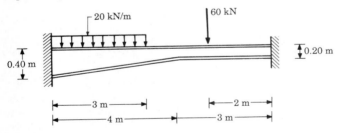

Figure 12.P8

13

The Method of Substructures

13.1 Introduction

Many real structures have a large number of degrees of freedom, and, consequently, their stiffness matrix is large. The analysis of a large structure may be expedited by subdividing it into smaller parts referred to as *substructures*. Each substructure may then be analyzed by a different team of analysts, and the results may be combined to yield the nodal actions of the elements and the components of displacements of the nodes of the actual structure. Moreover, the size of the stiffness matrix of a structure may exceed the maximum which can be inverted by a computer. One way of avoiding a large stiffness matrix is to partition the structure into substructures. This approach is referred to as the *method of substructures*.

Many techniques are available for analyzing structures by the method of substructures. These techniques may be classified as *single-level* or *multilevel*. In the single-level techniques the structure is subdivided into a number of substructures which are used as elements of the total structure. In the multilevel techniques the substructures are also subdivided into smaller parts which are analyzed first, and their properties are used to establish the properties of the substructures. In the subsequent section we present a single-level technique for analyzing structures by the method of substructures, using the direct stiffness method.

13.2 Assemblage of the Condensed Basic Stiffness Matrix of a Structure from that of Its Substructures

Consider the structure shown in Fig. 13.1 subjected to equivalent actions on its nodes and suppose that it is partitioned into three

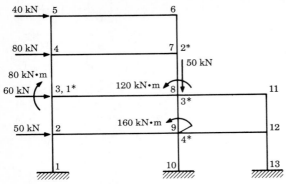

Figure 13.1 Geometry and loading of the frame.

substructures by cutting it slightly above nodes 3 and 7 and slightly to the left of nodes 8 and 9. The nodes of each substructure adjacent to the cuts are termed its *interface nodes,* while the remaining nodes of each substructure are termed its *internal nodes.* Notice that the structure is partitioned in such a way that each interface node connects only two substructures. We visualize a structure as consisting of the substructures into which it is partitioned connected to the

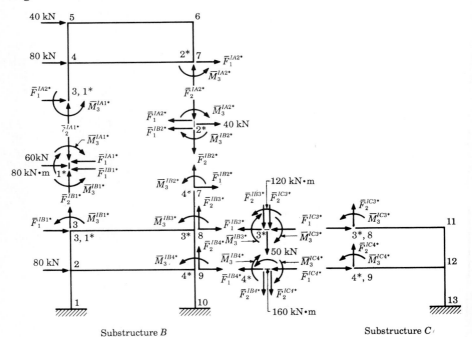

Substructure *B*   Substructure *C*

Figure 13.2 Free-body diagrams of the substructures and the interface nodes of the frame of Fig. 13.1.

interface nodes. The free-body diagrams of the substructures and the interface nodes are shown in Fig. 13.2. In this figure the interface nodes are numbered consecutively as 1*, 2*, 3*, and 4*. Each substructure is subjected to

1. Known equivalent external actions acting on its internal nodes. We store the global components of equivalent actions acting on the internal nodes of substructure i which are not directly absorbed by the supports of the structure in a matrix which we denote by $\{P_e^{EFi}\}$.

2. Unknown interface internal actions. These actions are exerted by each interface node on the ends of the elements of the substructures which are connected to this interface node. We denote the global component in the direction of the $\bar{x}_j$ ($j = 1, 2, 3$) axis of the interface internal force exerted by node n^* on substructure i by $\bar{F}_j^{Iin*}$. Moreover, we denote the global component in the direction of the $\bar{x}_j$ ($j = 1, 2, 3$) axis of the interface internal moment exerted by node n^* of substructure i by $\bar{M}_j^{Iin*}$. We store the global components of all the interface internal actions acting on substructure i in a matrix which we denote by $\{\bar{A}^{Ii}\}$.

Referring to Fig. 13.2 the matrices $\{P_e^{EFi}\}$ and $\{\bar{A}^{Ii}\}$ of the three substructures of the framed structure of Fig. 13.1 are

$$\{P_e^{EFA}\} = \begin{Bmatrix} P_1^{(4)} \\ P_2^{(4)} \\ \mathcal{M}_3^{(4)} \\ P_1^{(5)} \\ P_2^{(5)} \\ \mathcal{M}_3^{(5)} \\ P_1^{(6)} \\ P_2^{(6)} \\ \mathcal{M}_3^{(6)} \end{Bmatrix} = \begin{Bmatrix} 80 \\ 0 \\ 0 \\ 40 \\ 0 \\ 0 \\ 0 \\ 0 \\ 0 \end{Bmatrix} \qquad \{P_e^{EFB}\} = \begin{Bmatrix} P_1^{(2)} \\ P_2^{(2)} \\ \mathcal{M}_3^{(2)} \end{Bmatrix} = \begin{Bmatrix} 50 \\ 0 \\ 0 \end{Bmatrix}$$

$$\{P_e^{EFC}\} = \begin{Bmatrix} P_1^{(11)} \\ P_2^{(11)} \\ \mathcal{M}_3^{(11)} \\ P_1^{(12)} \\ P_2^{(12)} \\ \mathcal{M}_3^{(12)} \end{Bmatrix} = \begin{Bmatrix} 0 \\ 0 \\ 0 \\ 0 \\ 0 \\ 0 \end{Bmatrix}$$

$$(13.1)$$

and

$$\{\bar{A}^{IA}\} = \begin{Bmatrix} \bar{F}_1^{IA1*} \\ \bar{F}_2^{IA1*} \\ \bar{M}_3^{IA1*} \\ \bar{F}_1^{IA2*} \\ \bar{F}_2^{IA2*} \\ \bar{M}_3^{IA2*} \end{Bmatrix} = \begin{Bmatrix} \{\bar{A}^{IA}\}_{1*} \\ \{\bar{A}^{IA}\}_{2*} \end{Bmatrix} \qquad \{\bar{A}^{IB}\} = \begin{Bmatrix} \bar{F}_1^{IB1*} \\ \bar{F}_2^{IB1*} \\ \bar{M}_3^{IB1*} \\ \bar{F}_1^{IB2*} \\ \bar{F}_2^{IB2*} \\ \bar{M}_3^{IB2*} \\ \bar{F}_1^{IB3*} \\ \bar{F}_2^{IB3*} \\ \bar{M}_3^{IB3*} \\ \bar{F}_1^{IB4*} \\ \bar{F}_2^{IB4*} \\ \bar{M}_3^{IB4*} \end{Bmatrix} = \begin{Bmatrix} \{\bar{A}^{IB}\}_{1*} \\ \{\bar{A}^{IB}\}_{2*} \\ \{\bar{A}^{IB}\}_{3*} \\ \{\bar{A}^{IB}\}_{4*} \end{Bmatrix}$$

$$(13.2)$$

$$\{\bar{A}^{IC}\} = \begin{Bmatrix} \bar{F}_1^{IC3*} \\ \bar{F}_2^{IC3*} \\ \bar{M}_3^{IC3*} \\ \bar{F}_1^{IC4*} \\ \bar{F}_2^{IC4*} \\ \bar{M}_3^{IC4*} \end{Bmatrix} = \begin{Bmatrix} \{\bar{A}^{IC}\}_{3*} \\ \{\bar{A}^{IC}\}_{4*} \end{Bmatrix}$$

The basic stiffness equations for each substructure can be written as

$$\begin{Bmatrix} \{P_e^{EFi}\} \\ \hline \{\bar{A}^{Ii}\} \end{Bmatrix} = [K^i]\{\Delta^{Fi}\} = \begin{bmatrix} [K_{ee}^i][K_{ec}^i] \\ [K_{ce}^i][K_{cc}^i] \end{bmatrix} \begin{Bmatrix} \{\Delta_e^{Fi}\} \\ \{\Delta_c^{Fi}\} \end{Bmatrix} \qquad (i = A, B, C) \quad (13.3)$$

where $[K^i]$ is either the stiffness matrix or the basic stiffness matrix of substructure i. For instance, $[K^A]$ is the stiffness matrix of substructure A, while $[K^B]$ and $[K^C]$ are the basic stiffness matrices of substructures B and C, respectively. That is, the support conditions of a structure are incorporated in the stiffness equation of its substructures. $\{\Delta_e^{Fi}\}$ and $\{\Delta_c^{Fi}\}$ are the matrices of the components of displacements of the internal and the interface nodes of substructure i, respectively. We use the superscript F to indicate that only components of displacement which are not inhibited by the supports of a structure (free) are included in these matrices. The matrices

$\{\Delta_c^{Fi}\}$ $(i = A, B, C)$ of the substructures of Fig. 13.2 are

$$
\{\Delta_e^{FA}\} = \begin{Bmatrix} \Delta_1^{(4)} \\ \Delta_2^{(4)} \\ \Delta_3^{(4)} \\ \Delta_1^{(5)} \\ \Delta_2^{(5)} \\ \Delta_3^{(5)} \\ \Delta_1^{(6)} \\ \Delta_2^{(6)} \\ \Delta_3^{(6)} \end{Bmatrix} \qquad
\{\Delta_e^{FB}\} = \begin{Bmatrix} \Delta_1^{(2)} \\ \Delta_2^{(2)} \\ \Delta_3^{(2)} \end{Bmatrix} \qquad
\{\Delta_e^{FC}\} = \begin{Bmatrix} \Delta_1^{(11)} \\ \Delta_2^{(11)} \\ \Delta_3^{(11)} \\ \Delta_1^{(12)} \\ \Delta_2^{(12)} \\ \Delta_3^{(12)} \end{Bmatrix} \qquad (13.4)
$$

The matrices $\{\Delta_c^{Fi}\}$ of the substructures of Fig. 13.2 are

$$
\{\Delta_c^{FA}\} = \begin{Bmatrix} \Delta_1^{(3)} \\ \Delta_2^{(3)} \\ \Delta_3^{(3)} \\ \Delta_1^{(7)} \\ \Delta_2^{(7)} \\ \Delta_3^{(7)} \end{Bmatrix} = \begin{Bmatrix} \{\Delta_c\}_{1*} \\ \{\Delta_c\}_{2*} \end{Bmatrix} \qquad
\{\Delta_c^{FB}\} = \begin{Bmatrix} \Delta_1^{(3)} \\ \Delta_2^{(3)} \\ \Delta_3^{(3)} \\ \hdashline \Delta_1^{(7)} \\ \Delta_2^{(7)} \\ \Delta_3^{(7)} \\ \hdashline \Delta_1^{(8)} \\ \Delta_2^{(8)} \\ \Delta_3^{(8)} \\ \hdashline \Delta_1^{(9)} \\ \Delta_2^{(9)} \\ \Delta_3^{(9)} \end{Bmatrix} = \begin{Bmatrix} \{\Delta_c\}_{1*} \\ \{\Delta_c\}_{2*} \\ \{\Delta_c\}_{3*} \\ \{\Delta_c\}_{4*} \end{Bmatrix}
$$

$$ (13.5) $$

$$
\{\Delta_c^{FC}\} = \begin{Bmatrix} \Delta_1^{(8)} \\ \Delta_2^{(8)} \\ \Delta_3^{(8)} \\ \Delta_1^{(9)} \\ \Delta_2^{(9)} \\ \Delta_3^{(9)} \end{Bmatrix} = \begin{Bmatrix} \{\Delta_c\}_{3*} \\ \{\Delta_c\}_{4*} \end{Bmatrix}
$$

where the matrix $\{\Delta_c\}_{n*}$ represents the components of displacement of the interface node $n*$ of the structure. The matrix $[K_{ee}^i]$ for any substructure of a structure which does not form a mechanism defined by Eq. (13.3) is nonsingular.

13.3   Analysis of Framed Structures Using the Method of Substructures

When we analyze a structure by the direct stiffness method without partitioning it into substructures, we visualize it as consisting of elements connected to nodes, and we establish its response from that of its elements. In order to accomplish this we regard the response of a structure as the sum of its response when subjected to the given loads acting along the length of its elements with its nodes fixed and its response when subjected to equivalent actions on its nodes. The components of displacements of the nodes of a structure subjected to equivalent actions on its nodes are equal to the corresponding components of displacements of the nodes of the structure subjected to the given loads.

In what follows we describe the steps to which we adhere when we analyze a structure subjected to given loads by the direct stiffness method.

Step 1. We establish the response of each one of its elements. We regard the response of an element as the sum of its response when subjected to the given loads with its ends fixed and its response when subjected only to the displacements of its nodes. In the first case we describe the response of an element by its matrix of fixed-end actions. In the second case we describe the response of an element by its stiffness equations (see Sec. 4.3).

Step 2. We use the matrix of fixed-end actions of its elements to assemble directly its matrix of equivalent actions (see Chap. 7).

Step 3. We use the global stiffness matrix for its elements to assemble directly its stiffness matrix (see Chap. 8) and we form its stiffness equations.

Step 4. We solve its stiffness equations to obtain the global components of displacements of its nodes and its reactions (see Chap. 9).

Step 5. We use the global components of displacements of its nodes to compute the local components of nodal actions of its elements.

When we analyze a framed structure subjected to equivalent actions on its nodes by the method of substructures, we visualize it as consisting of substructures connected to interface nodes and we establish its response from that of its substructures. In order to accomplish this we regard the response of a structure as the sum of its response when subjected to the known equivalent actions on its internal nodes with its interface nodes fixed and its response when subjected to the interface equivalent actions on its interface nodes.

The components of displacements of the interface nodes of a structure subjected to the interface equivalent actions are equal to the corresponding components of displacements of the interface nodes of the structure subjected to the known equivalent actions on its nodes.

In what follows we describe the steps to which we adhere when analyzing a framed structure by the method of substructures.

Step 1. We establish the response of each one of its substructures. We regard the response of a substructure as the sum of its response when subjected only to the known equivalent actions on its internal nodes with its interface nodes fixed and its response when subjected only to the components of displacements of its interface nodes. In the first case the response of a substructure is described by its interface fixed-end actions. In the second case the response of a substructure is described by its condensed stiffness equations.

Step 2. We use the interface fixed-end actions of each substructure to assemble directly the matrix of interface equivalent actions of the structure.

Step 3. We use the condensed stiffness matrix for each substructure to assemble directly the condensed stiffness matrix for the structure. In the condensed stiffness equations for a structure the components of the interface equivalent actions established in step 2 are expressed as a linear combination of the global components of displacements of the structure's interface nodes.

Step 4. We solve the condensed stiffness equations for the structure to obtain the global components of displacements of the structure's interface nodes. Moreover, we compute the global components of displacements for the internal nodes of all the substructures of the structure from the components of displacements of the structure's interface nodes, computed in step 4.

Step 5. We use the global components of displacements of the nodes of the structure to compute the local components of nodal actions of the structure's elements.

13.3.1 Computation of the condensed stiffness equations and the matrix of interface fixed-end actions for a substructure

We condense the stiffness equations (13.3) for substructure i by eliminating the components of displacements of its internal nodes. Thus, referring to relations (12.2) and (12.9) the condensed stiffness equations of substructure i can be written as

$$\{\overset{\square}{P}{}_c^{Ii}\} + \{\bar{A}^{Ii}\} = [K_c^i]\{\Delta_c^{Fi}\} \tag{13.6}$$

where the matrix $[K_c^i]$ is the condensed stiffness matrix for substructure i. It is given by relation (12.8) as

$$[K_c^i] = [K_{cc}^i] - [K_{ce}^i][K_{ee}^i]^{-1}[K_{ec}^i] \qquad (i = A, B, C) \tag{13.7}$$

The matrices in the above relation are defined by relation (13.3). Moreover, the matrix $\{\overset{\square}{P}{}_c^{Ii}\}$ is given by relation (12.10) as

$$\{\overset{\square}{P}{}_c^{Ii}\} = -[K_{ce}^i][K_{ee}^i]^{-1}\{P_e^{EFi}\} \qquad (i = A, B, C) \tag{13.8}$$

Inasmuch as the matrix $\{P_e^{EFi}\}$ of each substructure is known, the matrix $\{\overset{\square}{P}{}_c^{Ii}\}$ of each substructure can be established using relation (13.8). The terms of the matrix $\{\overset{\square}{P}{}_c^{Ii}\}$ represent the global components of the actions which when applied together with the interface internal actions $\{\bar{A}^{Ii}\}$ to the interface nodes of substructure i, the components of displacements of its interface nodes are equal to $\{\Delta^{Fi}\}$. That is, they are equal to the components of displacement of the interface nodes of substructure i when it is subjected to the known equivalent actions $\{P_e^{EFi}\}$ on its internal nodes and the unknown interface internal actions $\{\bar{A}^{Ii}\}$ on its interface nodes. Assuming that the interface nodes of substructure i are $q^*, p^*, \ldots, n^*, \ldots, r^*$, relation (13.6) may be rewritten as

$$\begin{Bmatrix} \{\overset{\square}{P}{}_c^{Ii}\}_{p^*} \\ \{\overset{\square}{P}{}_c^{Ii}\}_{q^*} \\ \vdots \\ \{\overset{\square}{P}{}_c^{Ii}\}_{n^*} \\ \vdots \\ \{\overset{\square}{P}{}_c^{Ii}\}_{r^*} \end{Bmatrix} + \begin{Bmatrix} \{\bar{A}^{Ii}\}_{p^*} \\ \{\bar{A}^{Ii}\}_{q^*} \\ \vdots \\ \{\bar{A}^{Ii}\}_{n^*} \\ \vdots \\ \{\bar{A}^{Ii}\}_{r^*} \end{Bmatrix}$$

$$= \begin{bmatrix} [K^i]_{p^*p^*} & [K^i]_{p^*q^*} & \cdots & [K^i]_{p^*n^*} & \cdots & [K^i]_{p^*r^*} \\ [K^i]_{q^*p^*} & [K^i]_{q^*q^*} & \cdots & [K^i]_{q^*n^*} & \cdots & [K^i]_{q^*r^*} \\ \vdots & \vdots & & \vdots & & \vdots \\ [K^i]_{n^*p^*} & [K^i]_{n^*q^*} & \cdots & [K^i]_{n^*n^*} & \cdots & [K^i]_{n^*r^*} \\ \vdots & \vdots & & \vdots & & \vdots \\ [K^i]_{r^*p^*} & [K^i]_{r^*q^*} & \cdots & [K^i]_{r^*n^*} & \cdots & [K^i]_{r^*r^*} \end{bmatrix} \begin{Bmatrix} \{\Delta_c\}_{p^*} \\ \{\Delta_c\}_{q^*} \\ \vdots \\ \{\Delta_c\}_{n^*} \\ \vdots \\ \{\Delta_c\}_{r^*} \end{Bmatrix}$$

$$\tag{13.9}$$

where the matrix $\{\bar{A}^{Ii}\}_{n*}$ represents the interface internal actions acting at the interface node $n*$ of substructure i.

Notice that if the interface nodes of substructure i were fixed ($\{\Delta_c\}_{m*} = \{0\}$, $m* = p*, q*, \ldots, n*, \ldots, r*$), the matrix of interface internal actions $\{\bar{A}^{Ii}\}$ becomes identical to the matrix of interface fixed-end actions $\{\bar{A}^{IRi}\}$. Thus relation (13.6) reduces to

$$-\{\overset{\Box}{P}{}_c^{Ii}\} = \{\bar{A}^{IRi}\} \tag{13.10}$$

or

$$-\begin{Bmatrix} \{\overset{\Box}{P}{}_c^{Ii}\}_{p*} \\ \{\overset{\Box}{P}{}_c^{Ii}\}_{q*} \\ \vdots \\ \{\overset{\Box}{P}{}_c^{Ii}\}_{n*} \\ \vdots \\ \{\overset{\Box}{P}{}_c^{Ii}\}_{r*} \end{Bmatrix} = \begin{Bmatrix} \{\bar{A}^{IRi}\}_{p*} \\ \{\bar{A}^{IRi}\}_{q*} \\ \vdots \\ \{\bar{A}^{IRi}\}_{n*} \\ \vdots \\ \{\bar{A}^{IRi}\}_{r*} \end{Bmatrix} \tag{13.11}$$

Relation (13.10) may be employed to establish the matrix of interface fixed-end actions $\{\bar{A}^{IRi}\}$ of substructure i. The matrix $\{\overset{\Box}{P}{}_c^{Ii}\}$ in relation (13.10) may be obtained from relation (13.8).

13.3.2 Computation of the interface equivalent actions of a structure

In Fig. 13.3a the structure of Fig. 13.1 is shown with only its interface nodes numbered. The structure of Fig. 13.3b is subjected to known equivalent actions on its internal nodes with its interface nodes fixed by applying to them restraining actions. It is apparent that the internal actions in the elements of the structure of Fig. 13.3a may be established by superimposing the corresponding actions in the elements of the restrained structure of Fig. 13.3b and of the structure subjected to the interface equivalent actions on its interface nodes (see Fig. 13.3c). Moreover, the components of displacements of the interface nodes of the structure of Fig. 13.3a are equal to the corresponding components of displacements of the nodes of the structure of Fig. 13.3c. However, the components of displacements of the internal nodes of the structure of Fig. 13.3a are not in general equal to the corresponding components of displacements of the nodes of the structure of Fig. 13.3c. This is due to the fact that in general the components of displacement of the internal nodes of the restrained structure of Fig. 13.3b do not vanish.

The restraining actions which must be applied to the interface nodes of the restrained structure may be obtained by first establishing the interface fixed-end actions $\{\bar{A}^{IRi}\}$ of each substructure of the

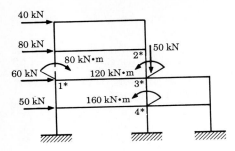

(a) Structure subjected to known
equivalent actions on its nodes

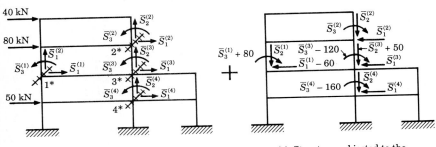

(b) Restrained structure subjected
to the known equivalent actions
on its internal nodes

(c) Structure subjected to the
interface equivalent actions
on its interface nodes

Figure 13.3 Superposition of the restrained structure and the structure subjected
to the interface equivalent actions.

restrained structure using relation (13.10). Assuming that node n^* is
one of the interface nodes connecting substructures B and C from the
equilibrium of node n^* of the restraint structure, the interface
restraining actions $\{\bar{S}^I\}_{n^*}$ which must be applied to it are equal to

$$\{\bar{S}^I\}_{n^*} = \sum_{i=B,C} \{\bar{A}^{IRi}\}_{n^*} = -\sum_{i=B,C} \{\bar{P}^{Ii}_c\}_{n^*} \qquad (13.12)$$

Thus referring to Fig. 13.3c the matrix of the interface equivalent
actions $\{P^{IE}\}_{n^*}$ which are applied to the interface node n^* of the
structure is equal to

$$\{P^{IE}\}_{n^*} = \{P^E\}_{n^*} - \{\bar{S}^I\}_{n^*} = \sum_{i=B,C} \{\bar{P}^{Ii}_c\}_{n^*} + \{P^E\}_{n^*} \qquad (13.13)$$

where the matrix $\{P^E\}_{n^*}$ represents the known equivalent actions
acting on interface node n^* of the actual structure.

13.3.3 Assemblage of the condensed stiffness matrix of a structure from the condensed stiffness matrices of its substructures

In the sequel we establish the condensed stiffness equations for the structure subjected to the interface equivalent actions on its interface nodes, from the condensed stiffness equations (13.6) of its substructures. The condensed stiffness equations for a structure can be written in the following form:

$$
\begin{Bmatrix} \{P^{IE}\}_{1*} \\ \{P^{IE}\}_{2*} \\ \vdots \\ \{P^{IE}\}_{n*} \\ \vdots \\ \{P^{IE}\}_{\bar{N}*} \end{Bmatrix}
=
\begin{bmatrix}
[S]_{1*1*} & [S]_{1*2*} & \cdots & [S]_{1*n*} & \cdots & [S]_{1*\bar{N}*} \\
[S]_{2*1*} & [S]_{2*2*} & \cdots & [S]_{2*n*} & \cdots & [S]_{2*\bar{N}*} \\
\vdots & \vdots & & \vdots & & \vdots \\
[S]_{n*1*} & [S]_{n*2*} & \cdots & [S]_{n*n*} & \cdots & [S]_{n*\bar{N}*} \\
\vdots & \vdots & & \vdots & & \vdots \\
[S]_{\bar{N}*1*} & [S]_{\bar{N}*2*} & \cdots & [S]_{N*n*} & \cdots & [S]_{\bar{N}*\bar{N}*}
\end{bmatrix}
\begin{Bmatrix} [\Delta_c]_{1*} \\ [\Delta_c]_{2*} \\ \vdots \\ [\Delta_c]_{n*} \\ \vdots \\ [\Delta_c]_{\bar{N}*} \end{Bmatrix}
$$

$$\text{(13.14)}$$

or
$$\{P^{IE}\} = [S_c][\Delta_c] \tag{13.14a}$$

where $\bar{N}*$ is the total number of interface nodes of the structure.

Considering the equilibrium of the interface node $n*$ of the structure subjected to the interface equivalent actions on its interface nodes and assuming that substructures B and C are connected to node $n*$ we obtain

$$\{P^{IE}\}_{n*} = \sum_{i=B,C} \{\bar{A}^{IEi}\}_{n*} \tag{13.15}$$

Referring to relation (13.9) for the structure subjected to the interface equivalent actions on its nodes $[\{P_e^{EFi}\} = \{0\}$, thus $\{\bar{P}_c^{Ii}\} = \{0\}]$ we have

$$
\{\bar{A}^{Ii}\}_{n*} = [[K^i]_{n*p*}\ [K^i]_{n*q*}, \ldots\ [K^i]_{n*n*}, \ldots\ [K^i]_{n*r*}]
\begin{Bmatrix} \{\Delta_c\}_{p*} \\ \{\Delta_c\}_{q*} \\ \vdots \\ \{\Delta_c\}_{n*} \\ \vdots \\ \{\Delta_c\}_{r*} \end{Bmatrix}
$$

$$\text{(13.16)}$$

For example, referring to Fig. 13.3 for substructures B and C, relation (13.16) reduces to

$$\{\bar{A}^{IB}\}_{n*} = [[K^B]_{n*1*} \ [K^B]_{n*2*} \ [K^B]_{n*3*} \ [K^B]_{n*4*} \ [K^B]_{n*5*}] \begin{Bmatrix} \{\Delta_c\}_{1*} \\ \{\Delta_c\}_{2*} \\ \{\Delta_c\}_{3*} \\ \{\Delta_c\}_{4*} \\ \{\Delta_c\}_{5*} \end{Bmatrix}$$

(13.17)

$$\{\bar{A}^{IC}\}_{n*} = [[K^c]_{n*3*} \ [K^c]_{n*4*} \ [K^c]_{n*5*}] \begin{Bmatrix} \{\Delta_c\}_{3*} \\ \{\Delta_c\}_{4*} \\ \{\Delta_c\}_{5*} \end{Bmatrix}$$

Substituting relations (13.17) into (13.15) we obtain

$$\{P_{int}^E\}_{n*} = [K^B]_{n*1*}\{\Delta_c\}_{1*} + [K^B]_{n*2*}\{\Delta_c\}_{2*} + [[K^B]_{n*3*} + [K^C]_{n*3*}]\{\Delta_c\}_{3*}$$
$$+ [[K^B]_{n*4*} + [K^C]_{n*4*}]\{\Delta_c\}_{4*}$$

(13.18)

Comparing relations (13.18) and (13.14) we get

$$[S]_{n*1*} = [K^B]_{n*1*}$$

$$[S]_{n*2*} = [K^B]_{n*2*}$$

$$[S]_{n*3*} = [K^B]_{n*3*} + [K^C]_{n*3*}$$

$$[S]_{n*4*} = [K^B]_{n*4*} + [K^C]_{n*4*}$$

(13.19)

That is, in general when node $n*$ is an interface node of substructures B and C, the submatrices of the stiffness equations (13.14) are equal to

$$[\bar{S}]_{n*m*} = \sum_{i=B,C} [K^i]_{n*m*} \qquad (n*, m* = 1*, 2*, \ldots, \bar{N}*) \quad (13.20)$$

where $\bar{N}$ is the number of the interface nodes of the structure. In relation (13.20) the stiffness submatrix $[K^i]_{n*m*}$ of substructure i vanishes if its indices $n*$ and $m*$ do not correspond to numbers of the interface nodes to which this substructure is connected.

13.3.4 Computation of the components of displacements of the nodes of a structure

The stiffness matrix on the right-hand side of relation (13.14) is nonsingular because the structure of Fig. 13.3c is not a mechanism, and, moreover, its support conditions have been taken into account in establishing this relation. Thus, relation (13.14) may be solved to yield the matrices $\{\Delta_c\}_{n*}$ $(n* = 1*, 2*, \ldots, \bar{N}*)$.

In order to compute the components of displacements for the internal nodes of a structure we write the basic stiffness equations (13.3) for each of its substructures as follows

$$\{P_e^{EFi}\} = [K_{ee}^i]\{\Delta_e^{Fi}\} + [K_{ec}^i]\{\Delta_c^{Fi}\} \tag{13.21a}$$

$$\{\bar{A}^{Ii}\} = [K_{ce}^i]\{\Delta_e^{Fi}\} + [K_{cc}^i]\{\Delta_c^{Fi}\} \qquad (i = A, B, C) \quad (13.21b)$$

For a structure which is not a mechanism the stiffness matrix $[K_{ee}^i]$ is nonsingular. Thus relation (13.21a) gives

$$\{\Delta_e^{Fi}\} = [K_{ee}^i]^{-1}\{\{P_e^{EFi}\} - [K_{ec}^i]\{\Delta_c^{Fi}\}\} \qquad (i = A, B, C) \quad (13.22)$$

Hence, the components of displacements of all the nodes of the structure subjected to equivalent actions at its nodes can be established. From those, the components of nodal actions of the elements of the structure may be computed using the procedure described in Sec. 9.3.

13.3.5 Application of the method of substructures

In this subsection we illustrate the method of substructures by applying it to two examples. The first is a fixed-end beam that is subdivided into two substructures. Obviously, it is easier to analyze this structure without subdividing it into substructures. Therefore, its analysis by the method of substructures is presented solely for the purpose of illustrating this method. The second example is a planar frame that is subdivided into two substructures.

Example 1. Compute the components of displacement of the nodes of the fixed-end beam subjected to the external actions shown in Fig. a using the method of substructures. Moreover, compute the nodal actions of the elements of the beam. The beam is made of steel $(E = 210 \text{ kN/mm}^2)$ and has a constant cross section $(I = 360 \times 10^{-6} \text{ m}^4, A = 18 \times 10^{-3} \text{ m}^2)$.

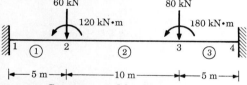

Figure a Geometry and loading of the beam.

---

solution

STEP 1.   We compute the matrix of fixed-end actions for each element of the beam.

$$[A^R] = 0$$

STEP 2.   We compute the equivalent actions. The beam of Fig. a is subjected only to external actions on its nodes. Thus the equivalent actions are the given actions.

STEP 3.   We establish the local, the hydrid, and the global stiffness matrices for each element of the beam. That is,

$$[\bar{K}^3] = [\bar{K}^1] = [\overset{\square}{\bar{K}}{}^3] = [\overset{\square}{\bar{K}}{}^1] = [K^3] = [K^1]$$

$$= EI \begin{bmatrix} 0.096 & 0.240 & -0.096 & 0.240 \\ 0.240 & 0.800 & -0.240 & 0.400 \\ -0.096 & -0.240 & 0.096 & -0.240 \\ 0.240 & 0.400 & -0.240 & 0.800 \end{bmatrix} \quad \text{(a)}$$

$$[\bar{K}^2] = [\overset{\square}{\bar{K}}{}^2] = [K^2] = EI \begin{bmatrix} 0.012 & 0.060 & -0.012 & 0.060 \\ 0.060 & 0.400 & -0.060 & 0.200 \\ -0.012 & -0.060 & 0.012 & -0.060 \\ 0.060 & 0.200 & -0.060 & 0.400 \end{bmatrix} \quad \text{(b)}$$

STEP 4.   We partition the beam into the two substructures shown in Fig. b and assemble the stiffness matrix for each substructure using the global stiffness matrices of its elements. That is,

$$[K^A] = EI \begin{bmatrix} 0.096 & 0.240 & -0.096 & 0.240 & 0 & 0 \\ 0.240 & 0.800 & -0.240 & 0.400 & 0 & 0 \\ -0.096 & -0.240 & 0.108 & -0.180 & -0.012 & 0.060 \\ 0.240 & 0.400 & -0.180 & 1.200 & -0.060 & 0.200 \\ 0 & 0 & -0.012 & -0.060 & 0.012 & -0.060 \\ 0 & 0 & 0.060 & 0.200 & -0.060 & 0.400 \end{bmatrix}$$

$$[K^B] = [\bar{K}^3] \quad \text{(c)}$$

Moreover, we write the basic stiffness equations for each substructure and

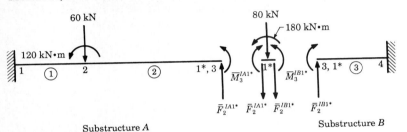

Substructure A   Substructure B

Figure b Beam partitioned into substructures.

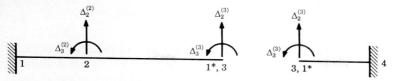

Figure c Components of displacements of nodes of the substructure.

partition them as indicated by relation (13.3). Thus, referring to Fig. c we have the following.

Substructure A

$$\left\{\begin{array}{c} -60 \\ 120 \\ \hline \bar{F}_2^{IA1*} \\ \bar{M}_3^{IA1*} \end{array}\right\} = \left[\begin{array}{cc|cc} 0.108 & -0.180 & -0.012 & 0.060 \\ -0.180 & 1.200 & -0.060 & 0.200 \\ \hline -0.012 & -0.060 & 0.012 & -0.060 \\ 0.060 & 0.200 & -0.060 & 0.400 \end{array}\right] \left\{\begin{array}{c} \Delta_2^{(2)} \\ \Delta_3^{(2)} \\ \Delta_2^{(3)} \\ \Delta_3^{(3)} \end{array}\right\} \tag{d}$$

Thus,

$$\{P_e^{EFA}\} = \left\{\begin{array}{c} -60 \\ 120 \end{array}\right\} \tag{e}$$

$$[K_{ee}^A] = EI\left[\begin{array}{cc} 0.108 & -0.180 \\ -0.180 & 1.200 \end{array}\right] \tag{f}$$

$$[K_{ec}^A] = EI\left[\begin{array}{cc} -0.012 & 0.060 \\ -0.060 & 0.200 \end{array}\right] \tag{g}$$

$$[K_{ce}^A] = EI\left[\begin{array}{cc} -0.012 & -0.060 \\ 0.060 & 0.200 \end{array}\right] \tag{h}$$

$$[K_{cc}^A] = EI\left[\begin{array}{cc} 0.012 & -0.060 \\ -0.060 & 0.400 \end{array}\right] \tag{i}$$

Substructure B

$$\left\{\begin{array}{c} \bar{F}_2^{IB1*} \\ \bar{M}_3^{IB1*} \end{array}\right\} = EI\left[\begin{array}{cc} 0.096 & 0.240 \\ 0.240 & 0.800 \end{array}\right] \left\{\begin{array}{c} \Delta_2^{(3)} \\ \Delta_3^{(3)} \end{array}\right\} \tag{j}$$

Thus,

$$\{P_e^{EFB}\} = [0] \tag{k}$$

$$[K_{ec}^B] = [K_{ce}^B] = [K_{ee}^B] = [0] \tag{l}$$

$$[K_{cc}^B] = EI\left[\begin{array}{cc} 0.096 & 0.240 \\ 0.240 & 0.800 \end{array}\right] \tag{m}$$

STEP 5. We compute the condensed stiffness matrix of each substructure. Substituting relations (f) to (l) into (13.7) we get

$$[K_c^A] = [K_{cc}^A] - [K_{ce}^A][K_{ee}^A]^{-1}[K_{ec}^A] = EI\left[\begin{array}{cc} 0.012 & -0.060 \\ -0.060 & 0.400 \end{array}\right]$$

$$- EI\left[\begin{array}{cc} -0.012 & -0.060 \\ 0.060 & 0.200 \end{array}\right]\left[\begin{array}{cc} 0.108 & -0.180 \\ -0.180 & 1.200 \end{array}\right]^{-1}\left[\begin{array}{cc} -0.012 & 0.060 \\ -0.060 & 0.200 \end{array}\right]$$

$$= EI\left[\begin{array}{cc} 0.003556 & -0.02667 \\ -0.02667 & 0.26667 \end{array}\right] \tag{n}$$

$$[K_c^B] = EI \begin{bmatrix} 0.096 & 0.240 \\ 0.240 & 0.800 \end{bmatrix} \tag{o}$$

STEP 6.   We establish the interface equivalent actions to be applied to each interface node of the structure. Substituting relations (e), (h), and (j) into (13.8) we get

$$\{\overset{\square}{P_c^{IA}}\} = \begin{bmatrix} -0.012 & -0.060 \\ 0.060 & 0.200 \end{bmatrix} \begin{bmatrix} 0.108 & -0.180 \\ -0.180 & 1.200 \end{bmatrix}^{-1} \begin{Bmatrix} -60 \\ 120 \end{Bmatrix}$$

$$= -\begin{Bmatrix} 4.8889 \\ -26.6665 \end{Bmatrix} \tag{p}$$

$$\{\overset{\square}{P_c^{IB}}\} = [0] \tag{q}$$

Substituting relations (e) and (p) into relation (13.13) and referring to Fig. a we get

$$\{P^{IE}\} = \{P^{IE}\}_{1*} = -\begin{Bmatrix} 4.8889 \\ -26.6665 \end{Bmatrix} + \begin{Bmatrix} -80 \\ 180 \end{Bmatrix} = \begin{Bmatrix} -84.8889 \\ 206.6665 \end{Bmatrix} \tag{r}$$

STEP 7.   We compute the condensed stiffness matrix for the structure using relation (13.20). That is,

$$[S_c] = [S]_{1*1*} = [K^A] + [K^B] = EI\left[\begin{bmatrix} 0.003556 & -0.02667 \\ -0.02667 & 0.26667 \end{bmatrix} + \begin{bmatrix} 0.096 & 0.240 \\ 0.240 & 0.800 \end{bmatrix}\right]$$

$$= EI\begin{bmatrix} 0.099556 & 0.21333 \\ 0.21333 & 1.06667 \end{bmatrix} \tag{s}$$

STEP 8.   We compute the components of displacement of the interface node of the structure. Referring to Fig. c and to relations (r) and (s) the condensed stiffness equations (13.14a) for the structure are

$$\{P^{IE}\} = \begin{Bmatrix} -84.8889 \\ 206.6665 \end{Bmatrix} = [S_c]\{\Delta_c\} = EI\begin{bmatrix} 0.099556 & 0.21333 \\ 0.213333 & 1.06667 \end{bmatrix} \begin{Bmatrix} \Delta_2^{(3)} \\ \Delta_3^{(3)} \end{Bmatrix} \tag{t}$$

Thus,

$$\{\Delta_c\} = -\frac{1}{EI}\begin{bmatrix} 0.099556 & 0.21333 \\ 0.213333 & 1.06667 \end{bmatrix}^{-1} \begin{Bmatrix} -84.8889 \\ 206.6665 \end{Bmatrix} = \frac{1}{EI}\begin{Bmatrix} -2218.662 \\ 637.4733 \end{Bmatrix}$$

$$= \{\Delta_c^{FA}\} = \{\Delta_c^{FB}\} \tag{u}$$

STEP 9.   We compute the components of displacements of the interior nodes of each substructure. Substituting relations (e) to (g) and (u) into (13.22), we get

$$\{\Delta_e^{FA}\} = [K_{ee}^A]^{-1}\{\{P_e^A\} - [K_{ec}^A]\{\Delta_c^{FA}\}\}$$

$$= \frac{1}{EI}\begin{bmatrix} 0.108 & -0.180 \\ -0.180 & 1.200 \end{bmatrix}\begin{Bmatrix} -60 \\ 120 \end{Bmatrix} - \begin{bmatrix} -0.012 & 0.060 \\ -0.060 & 0.200 \end{bmatrix}\begin{Bmatrix} -2218.662 \\ 637.473 \end{Bmatrix}$$

$$= \frac{1}{EI}\begin{Bmatrix} -1802.021 \\ -387.481 \end{Bmatrix} \tag{v}$$

Thus

$$\{\Delta\} = \begin{Bmatrix} \Delta_2^{(2)} \\ \Delta_3^{(2)} \\ \Delta_2^{(3)} \\ \Delta_3^{(3)} \end{Bmatrix} = \frac{1}{EI} \begin{Bmatrix} -1802.021 \\ -387.481 \\ -2218.662 \\ 637.473 \end{Bmatrix} \tag{w}$$

STEP 10.   We compute the components of nodal actions of the elements of the beam. Referring to Fig. c and to relation (w) the components of nodal displacements of the elements of the beam are

$$\{D^1\} = \{\bar{D}^1\} = \begin{Bmatrix} 0 \\ 0 \\ \Delta_2^{(2)} \\ \Delta_3^{(2)} \end{Bmatrix} = \frac{1}{EI} \begin{Bmatrix} 0 \\ 0 \\ -1802.021 \\ -387.481 \end{Bmatrix} \tag{x}$$

$$\{D^2\} = \{\bar{D}^2\} = \begin{Bmatrix} \bar{\Delta}_2^{(2)} \\ \bar{\Delta}_3^{(2)} \\ \bar{\Delta}_2^{(3)} \\ \bar{\Delta}_3^{(3)} \end{Bmatrix} = \frac{1}{EI} \begin{Bmatrix} -1802.021 \\ -387.481 \\ -2218.662 \\ 637.473 \end{Bmatrix}$$

$$\{D^3\} = \{\bar{D}^3\} = \begin{Bmatrix} \bar{\Delta}_2^{(3)} \\ \bar{\Delta}_3^{(3)} \\ 0 \\ 0 \end{Bmatrix} = \frac{1}{EI} \begin{Bmatrix} -2218.662 \\ 637.473 \\ 0 \\ 0 \end{Bmatrix}$$

The relations between the local components of nodal actions and nodal displacements of an element of the beam are

$$\{A^e\} = \begin{Bmatrix} F_2^{ej} \\ M_3^{ej} \\ F_2^{ek} \\ M_3^{ej} \end{Bmatrix} = \{K^e\}\{D^e\} \tag{y}$$

Substituting the stiffness matrix of each element of the beam given by relations (a) or (b), and the matrix $\{D^e\}$ given by relations (x) into Eq. (y) we obtain the local matrices of nodal actions of the elements of the beam of Fig. a. Thus,

$$\{A^1\} = \begin{bmatrix} 0.096 & 0.240 & -0.096 & 0.240 \\ 0.240 & 0.800 & -0.240 & 0.400 \\ -0.096 & -0.240 & 0.096 & -0.240 \\ 0.240 & 0.400 & -0.240 & 0.800 \end{bmatrix} \begin{Bmatrix} 0 \\ 0 \\ -1802.021 \\ -387.481 \end{Bmatrix} = \begin{Bmatrix} 80.00 \\ 277.50 \\ -80.00 \\ 122.50 \end{Bmatrix}$$

$$\{A^2\} = \begin{bmatrix} 0.012 & 0.060 & -0.012 & 0.060 \\ 0.060 & 0.400 & -0.060 & 0.200 \\ -0.012 & -0.060 & 0.012 & -0.060 \\ 0.060 & 0.200 & -0.060 & 0.400 \end{bmatrix} \begin{Bmatrix} -1802.021 \\ -387.481 \\ -2218.662 \\ 637.473 \end{Bmatrix} = \begin{Bmatrix} 20.00 \\ -2.50 \\ -20.00 \\ 202.50 \end{Bmatrix}$$

$$\{A^3\} = \begin{bmatrix} 0.096 & 0.240 & -0.096 & 0.240 \\ 0.240 & 0.800 & -0.240 & 0.400 \\ -0.096 & -0.240 & 0.096 & -0.240 \\ 0.240 & 0.400 & -0.240 & 0.800 \end{bmatrix} \begin{Bmatrix} -2218.662 \\ 637.173 \\ 0 \\ 0 \end{Bmatrix} = \begin{Bmatrix} -60.00 \\ -22.50 \\ 60.00 \\ -277.50 \end{Bmatrix}$$

$$\tag{z}$$

The results are shown in Fig. d.

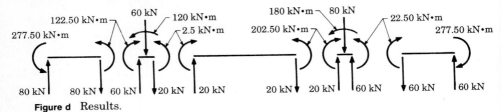

Figure d Results.

Example 2. Compute the components of displacement of the nodes of the frame shown in Fig. a using the method of substructures. Moreover, compute the nodal actions of the elements of the frame. The frame is made of steel and has a constant cross section $(I = 396.7 \times 10^{-6} \, \text{m}^4, \, A = 13.2 \times 10^{-3} \, \text{m}^2)$. Subdivide the frame into two substructures by cutting it just above node 5.

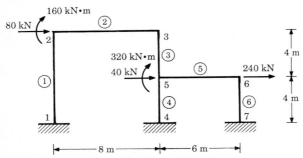

Figure a Geometry and loading of the frame.

---

solution

STEP 1. We compute the matrix of fixed-end actions for each element of the frame.

$$\{A^R\} = 0$$

STEP 2. We compute the equivalent actions. The frame of Fig. a is subjected only to external actions on its nodes. Thus the equivalent actions are the given actions.

STEP 3. We establish the local, the hybrid, and the global stiffness matrices for each element of the frame. Referring to relation (4.62) we assemble the local stiffness matrix of each element of the frame.

$$[K^1] = [K^2] = EI \begin{bmatrix} 4.463 & 0 & 0 & -4.463 & 0 & 0 \\ 0 & 0.02344 & 0.09375 & 0 & -0.02344 & 0.09375 \\ 0 & 0.09375 & 0.50000 & 0 & -0.09375 & 0.25000 \\ -4.463 & 0 & 0 & 4.463 & 0 & 0 \\ 0 & -0.02344 & -0.09375 & 0 & 0.02344 & -0.09375 \\ 0 & 0.09375 & 0.25000 & 0 & -0.09375 & 0.50000 \end{bmatrix}$$

$$[K^3] = [K^4] = [K^6] = EI \begin{bmatrix} 8.926 & 0 & 0 & -8.926 & 0 & 0 \\ 0 & 0.1875 & 0.375 & 0 & -0.1875 & 0.375 \\ 0 & 0.3750 & 1.000 & 0 & -0.3750 & 0.500 \\ -8.926 & 0 & 0 & 8.926 & 0 & 0 \\ 0 & -0.1875 & -0.375 & 0 & 0.1875 & -0.375 \\ 0 & 0.3750 & 0.500 & 0 & -0.3750 & 1.000 \end{bmatrix} \quad \text{(a)}$$

$$[K^5] = EI \begin{bmatrix} 5.951 & 0 & 0 & -5.951 & 0 & 0 \\ 0 & 0.0556 & 0.1667 & 0 & -0.0556 & 0.1667 \\ 0 & 0.1667 & 0.6667 & 0 & -0.1667 & 0.3333 \\ -5.951 & 0 & 0 & 5.951 & 0 & 0 \\ 0 & -0.0556 & -0.1667 & 0 & 0.0556 & -0.1667 \\ 0 & 0.1667 & 0.3333 & 0 & -0.1667 & 0.6667 \end{bmatrix}$$

The transformation matrix $[\Lambda^e]$ of an element of the frame is given by relation (6.48). That is,

$$[\Lambda^1] = [\Lambda^4] \begin{bmatrix} 0 & 1 & 0 & 0 & 0 & 0 \\ -1 & 0 & 0 & 0 & 0 & 0 \\ 0 & 0 & 1 & 0 & 0 & 0 \\ 0 & 0 & 0 & 0 & 1 & 0 \\ 0 & 0 & 0 & -1 & 0 & 0 \\ 0 & 0 & 0 & 0 & 0 & 1 \end{bmatrix}$$

$$[\Lambda^2] = [\Lambda^5] = \text{unit matrix} \quad \text{(b)}$$

$$[\Lambda^3] = [\Lambda^6] \begin{bmatrix} 0 & -1 & 0 & 0 & 0 & 0 \\ 1 & 0 & 0 & 0 & 0 & 0 \\ 0 & 0 & 1 & 0 & 0 & 0 \\ 0 & 0 & 0 & 0 & -1 & 0 \\ 0 & 0 & 0 & 1 & 0 & 0 \\ 0 & 0 & 0 & 0 & 0 & 1 \end{bmatrix}$$

We compute the hybrid stiffness matrices of the elements of the frame. Substituting the matrices (a) and (b) in relation (6.72) we get

$$[\bar{K}^1] = [K^1][\Lambda^1]$$

$$= EI \begin{bmatrix} 0 & 4.463 & 0 & 0 & -4.463 & 0 \\ -0.02344 & 0 & 0.09375 & 0.02344 & 0 & 0.09375 \\ -0.09375 & 0 & 0.50000 & 0.09375 & 0 & 0.25000 \\ 0 & -4.463 & 0 & 0 & 4.463 & 0 \\ 0.02344 & 0 & -0.09375 & -0.02344 & 0 & -0.09375 \\ -0.09375 & 0 & 0.25000 & 0.09375 & 0 & 0.50000 \end{bmatrix}$$

$$[\bar{K}^2] = [K^2][\Lambda^2] = [K^2] \quad \text{(c)}$$

$[\overset{\square}{K^3}] = [K^3][\Lambda^3]$

$$= EI \begin{bmatrix} 0 & -8.926 & 0 & 0 & 8.926 & 0 \\ 0.1875 & 0 & 0.375 & -0.1875 & 0 & 0.375 \\ 0.3750 & 0 & 1.000 & -0.3750 & 0 & 0.500 \\ 0 & 8.926 & 0 & 0 & -8.926 & 0 \\ -0.1875 & 0 & -0.375 & 0.1875 & 0 & -0.375 \\ 0.3750 & 0 & 0.500 & -0.3750 & 0 & 1.000 \end{bmatrix}$$

$[\overset{\square}{K^4}] = [K^4][\Lambda^4]$

$$= EI \begin{bmatrix} 0 & 8.926 & 0 & 0 & -8.926 & 0 \\ -0.1875 & 0 & 0.375 & 0.1875 & 0 & 0.375 \\ -0.3750 & 0 & 1.00 & 0.3750 & 0 & 0.500 \\ 0 & -8.926 & 0 & 0 & 8.926 & 0 \\ 0.1875 & 0 & -0.375 & -0.1875 & 0 & -0.375 \\ -0.3750 & 0 & 0.500 & 0.3750 & 0 & 1.000 \end{bmatrix}$$

$[\overset{\square}{K^5}] = [K^5][\Lambda^5] = [K^5]$

$[\overset{\square}{K^6}] = \overset{\square}{K^4}]$

We compute the global stiffness matrices of the elements of the frame. Substituting relations (a) and (b) into (6.74) we get

$$[\check{K}^1] = EI \begin{bmatrix} 0.02344 & 0 & -0.09375 & -0.02344 & 0 & -0.09375 \\ 0 & 4.463 & 0 & 0 & -4.463 & 0 \\ -0.09375 & 0 & 0.50000 & 0.09375 & 0 & 0.25000 \\ -0.02344 & 0 & 0.09375 & 0.02344 & 0 & 0.09375 \\ 0 & -4.463 & 0 & 0 & 4.463 & 0 \\ -0.09375 & 0 & 0.25000 & 0.09375 & 0 & 0.50000 \end{bmatrix}$$

$$[\check{K}^2] = EI \begin{bmatrix} 4.463 & 0 & 0 & -4.463 & 0 & 0 \\ 0 & 0.02344 & 0.09375 & 0 & -0.02344 & 0.09375 \\ 0 & 0.09375 & 0.50000 & 0 & -0.09375 & 0.25000 \\ -4.463 & 0 & 0 & 4.463 & 0 & 0 \\ 0 & -0.02344 & -0.09375 & 0 & 0.02344 & -0.09375 \\ 0 & 0.09375 & 0.25000 & 0 & -0.09375 & 0.50000 \end{bmatrix}$$

(d)

$$[K^3] = [\bar{K}^6] = EI \begin{bmatrix} 0.1875 & 0 & 0.375 & -0.1875 & 0 & 0.375 \\ 0 & 8.926 & 0 & 0 & -8.926 & 0 \\ 0.3750 & 0 & 1.000 & -0.3750 & 0 & 0.500 \\ -0.1875 & 0 & -0.375 & 0.1875 & 0 & -0.375 \\ 0 & -8.926 & 0 & 0 & 8.926 & 0 \\ 0.3750 & 0 & 0.500 & -0.3750 & 0 & 1.000 \end{bmatrix}$$

$$[K^4] = EI \begin{bmatrix} 0.1875 & 0 & -0.375 & -0.1875 & 0 & -0.375 \\ 0 & 8.926 & 0 & 0 & -8.926 & 0 \\ -0.3750 & 0 & 1.000 & 0.3750 & 0 & 0.500 \\ -0.1875 & 0 & 0.375 & 0.1875 & 0 & 0.375 \\ 0 & -8.926 & 0 & 0 & 8.926 & 0 \\ -0.3750 & 0 & 0.500 & 0.3750 & 0 & 1.000 \end{bmatrix}$$

$$[\bar{K}^5] = EI \begin{bmatrix} 5.951 & 0 & 0 & -5.951 & 0 & 0 \\ 0 & 0.0556 & 0.1667 & 0 & -0.0556 & 0.1667 \\ 0 & 0.1667 & 0.6667 & 0 & -0.1667 & 0.3333 \\ -5.951 & 0 & 0 & 5.951 & 0 & 0 \\ 0 & -0.0556 & -0.1667 & 0 & 0.0556 & -0.1667 \\ 0 & 0.1667 & 0.3333 & 0 & -0.1667 & 0.6667 \end{bmatrix}$$

STEP 4.   We partition the frame into the two substructures shown in Fig. b and compute the stiffness matrix of each substructure using the global stiffness matrices for its elements. They are given by relations (e) and (f) on pp. 446 and 447.

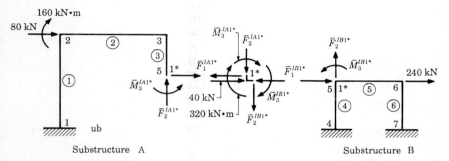

Figure b   Frame partitioned into substructures.

$$[K^A] =$$

$$
\begin{bmatrix}
0.02344 & 0 & -0.09375 & -0.02344 & 0 & -0.09375 & 0 & 0 & 0 & 0 & 0 & 0 \\
0 & 4.463 & 0 & 0 & -4.46300 & 0 & 0 & 0 & 0 & 0 & 0 & 0 \\
-0.09375 & 0 & 0.50000 & 0.09375 & 0 & 0.25000 & 0 & 0 & 0 & 0 & 0 & 0 \\
-0.02344 & 0 & 0.09375 & 4.48644 & 0 & 0.09375 & -4.4630 & 0 & 0 & 0 & 0 & 0 \\
0 & -4.463 & 0 & 0 & 4.48644 & 0.09375 & 0 & -0.02344 & 0.09375 & 0 & 0 & 0 \\
-0.09375 & 0 & 0.25000 & 0.09375 & 0.09375 & 1.00000 & 0 & -0.09375 & 0.25000 & 0 & 0 & 0 \\
0 & 0 & 0 & -4.46300 & 0 & 0 & 4.6505 & 0 & 0.37500 & -0.1875 & 0 & 0.375 \\
0 & 0 & 0 & 0 & -0.02344 & -0.09375 & 0 & 8.94944 & -0.09375 & 0 & -8.926 & 0 \\
0 & 0 & 0 & 0 & 0.09375 & 0.25000 & 0.3750 & -0.09375 & 1.50000 & -0.3780 & 0 & 0.500 \\
0 & 0 & 0 & 0 & 0 & 0 & -0.1875 & 0 & -0.37500 & 0.1875 & 0 & -0.375 \\
0 & 0 & 0 & 0 & 0 & 0 & 0 & -8.92600 & 0 & 0 & 8.926 & 0 \\
0 & 0 & 0 & 0 & 0 & 0 & 0.3750 & 0 & 0.50000 & -0.3750 & 0 & 1.000 \\
\end{bmatrix}
\qquad \text{(e)}
$$

$$[K^B] = \begin{bmatrix}
0.1875 & 0 & -0.375 & -0.1875 & 0 & -0.3750 & 0 & 0 & 0 & 0 & 0 & 0 \\
0 & 8.926 & 0 & 0 & -8.9260 & 0 & 0 & 0 & 0 & 0 & 0 & 0 \\
-0.3750 & 0 & 1.000 & 0.3750 & 0 & 0.5000 & 0 & 0 & 0 & 0 & 0 & 0 \\
-0.1875 & 0 & 0.375 & 6.1385 & 0 & 0.3750 & -5.951 & 0 & 0 & 0 & 0 & 0 \\
0 & -8.926 & 0 & 0 & 8.9816 & 0.1667 & 0 & -0.0556 & 0.1667 & 0 & 0 & 0 \\
-0.3750 & 0 & 0.500 & 0.3750 & 0.1667 & 1.6667 & 0 & -0.1667 & 0.3333 & 0 & 0 & 0 \\
0 & 0 & 0 & -5.9510 & 0 & 0 & 6.1385 & 0 & 0.3750 & -0.1875 & 0 & 0.375 \\
0 & 0 & 0 & 0 & -0.0556 & -0.1667 & 0 & 8.9816 & -0.1667 & 0 & -8.926 & 0 \\
0 & 0 & 0 & 0 & 0.1667 & 0.3333 & 0.3750 & -0.1667 & 1.6667 & -0.3750 & 0 & 0.500 \\
0 & 0 & 0 & 0 & 0 & 0 & -0.1875 & 0 & -0.3750 & 0.1875 & 0 & -0.375 \\
0 & 0 & 0 & 0 & 0 & 0 & 0 & -8.926 & 0 & 0 & 8.926 & 0 \\
0 & 0 & 0 & 0 & 0 & 0 & 0.3750 & 0 & 0.5000 & -0.3750 & 0 & 1.000
\end{bmatrix}$$

(f)

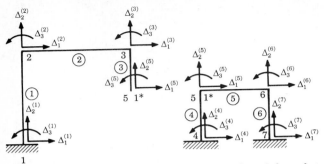

Figure c Components of displacements of nodes of the substructures.

Moreover, we write the basic stiffness equations for each substructure and partition them as indicated by relations (13.3). Thus, referring to Fig. c we have the following.

Substructure A

The basic stiffness equations for substructure A are given on p. 449. Referring to these equations we have

$$\{P_e^{EFA}\} = \begin{Bmatrix} 80 \\ 0 \\ -160 \\ 0 \\ 0 \\ 0 \end{Bmatrix} \tag{h}$$

$[K_{ee}^A]$

$$= EI \begin{bmatrix} 4.48644 & 0 & 0.09375 & -4.4630 & 0 & 0 \\ 0 & 4.48644 & 0.09375 & 0 & -0.02344 & 0.09375 \\ 0.09375 & 0.09375 & 1.0000 & 0 & -0.09375 & 0.25000 \\ -4.46300 & 0 & 0 & 4.6505 & 0 & 0.37500 \\ 0 & -0.02344 & -0.09375 & 0 & 8.94944 & -0.09375 \\ 0 & 0.09375 & 0.25000 & 0.3750 & -0.09375 & 1.50000 \end{bmatrix} \tag{i}$$

$$[K_{ce}^A] = EI \begin{bmatrix} 0 & 0 & 0 & -0.1875 & 0 & -0.375 \\ 0 & 0 & 0 & 0 & -8.926 & 0 \\ 0 & 0 & 0 & 0.3750 & 0 & 0.500 \end{bmatrix} \tag{j}$$

$$[K_{ec}^A] = EI \begin{bmatrix} 0 & 0 & 0 \\ 0 & 0 & 0 \\ 0 & 0 & 0 \\ -0.1875 & 0 & 0.3750 \\ 0 & -8.926 & 0 \\ -0.3750 & 0 & 0.5000 \end{bmatrix} \tag{k}$$

$$
\begin{Bmatrix}
80 \\
0 \\
160 \\
0 \\
0 \\
0 \\
\hdashline
\bar{F}_1^{IA1*} \\
\bar{F}_2^{IA1*} \\
\bar{M}_3^{IA1*}
\end{Bmatrix}
= EI
\left[
\begin{array}{cccccc:ccc}
4.48644 & 0 & 0.09375 & -4.4630 & 0 & 0 & 0 & 0 & 0 \\
0 & 4.48644 & 0.09375 & 0 & -0.02344 & 0.09375 & 0 & 0 & 0 \\
0.09375 & 0.09375 & 1.00000 & 0 & -0.09375 & 0.25000 & 0 & 0 & 0 \\
-4.46300 & 0 & 0 & 4.6505 & 0 & 0.37500 & -0.1875 & 0 & 0.375 \\
0 & -0.02344 & -0.09375 & 0 & 8.94944 & -0.9375 & 0 & -8.926 & 0 \\
0 & 0.09375 & 0.25000 & 0.3750 & -0.09375 & 1.5000 & -0.3750 & 0 & 0.500 \\
\hdashline
0 & 0 & 0 & -0.1875 & 0 & -0.3750 & 0.1875 & 0 & -0.375 \\
0 & 0 & 0 & 0 & -8.92600 & 0 & 0 & 8.926 & 0 \\
0 & 0 & 0 & 0.3750 & 0 & 0.5000 & -0.3750 & 0 & 1.000
\end{array}
\right]
\begin{Bmatrix}
\Delta_1^{(2)} \\
\Delta_2^{(2)} \\
\Delta_3^{(2)} \\
\Delta_1^{(3)} \\
\Delta_2^{(3)} \\
\Delta_3^{(3)} \\
\hdashline
\Delta_1^{(5)} \\
\Delta_2^{(5)} \\
\Delta_3^{(5)}
\end{Bmatrix}
$$

(g)

$$[K_{cc}^A] = EI \begin{bmatrix} 0.1875 & 0 & -0.375 \\ 0 & 8.926 & 0 \\ -0.3750 & 0 & 1.000 \end{bmatrix} \tag{1}$$

Substructure B

$$\begin{Bmatrix} 240 \\ 0 \\ 0 \\ \hdashline \bar{F}_1^{IB1*} \\ \bar{F}_2^{IB1*} \\ \bar{M}_3^{IB1*} \end{Bmatrix}$$

$$= EI \left[\begin{array}{ccc:ccc} 6.1385 & 0 & 0.3750 & -5.9510 & 0 & 0 \\ 0 & 8.9816 & -0.1667 & 0 & -0.0556 & -0.1667 \\ 0.3750 & -0.1667 & 1.6667 & 0 & 0.1667 & 0.3333 \\ \hdashline -5.9510 & 0 & 0 & 6.1385 & 0 & 0.3750 \\ 0 & -0.0556 & 0.16667 & 0 & 8.9816 & 0.1667 \\ 0 & -0.1667 & 0.3333 & 0.3750 & 0.1667 & 1.6667 \end{array} \right] \begin{Bmatrix} \Delta_1^{(6)} \\ \Delta_2^{(6)} \\ \Delta_3^{(6)} \\ \Delta_1^{(5)} \\ \Delta_2^{(5)} \\ \Delta_3^{(5)} \end{Bmatrix}$$

$$\tag{m}$$

Hence,
$$\{P_e^{EFB}\} = \begin{Bmatrix} 240 \\ 0 \\ 0 \end{Bmatrix} \tag{n}$$

$$[K_{ee}^B] = EI \begin{bmatrix} 6.1385 & 0 & 0.3750 \\ 0 & 8.9816 & -0.1667 \\ 0.3750 & -0.1667 & 1.6667 \end{bmatrix} \tag{o}$$

$$[K_{ce}^B] = EI \begin{bmatrix} -5.951 & 0 & 0 \\ 0 & -0.0556 & 0.1667 \\ 0 & -0.1667 & 0.3333 \end{bmatrix} \tag{p}$$

$$[K_{ec}^B] = EI \begin{bmatrix} -5.951 & 0 & 0 \\ 0 & -0.0556 & -0.1667 \\ 0 & 0.1667 & 0.3333 \end{bmatrix} \tag{q}$$

$$[K_{cc}^B] = EI \begin{bmatrix} 6.1385 & 0 & 0.3750 \\ 0 & 8.9816 & 0.1667 \\ 0.3750 & 0.1667 & 1.6667 \end{bmatrix} \tag{r}$$

STEP 5. We compute the condensed stiffness matrix of each substructure. Substituting relations (i) to (l) and (o) to (r) into (13.7) we get

$$[K_c^A] = [K_{cc}^A] - [K_{ce}^A][K_{ee}^A]^{-1}[K_{ec}^A] = EI \begin{bmatrix} 0.010664 & 0.0030704 & -0.036079 \\ 0.0030704 & 0.005389 & -0.031988 \\ -0.036079 & -0.031988 & 0.275692 \end{bmatrix}$$

(s)

$$[K_c^B] = [K_{cc}^B] - [K_{ce}^B][K_{ee}^B]^{-1}[K_{ec}^B] = EI \begin{bmatrix} 0.28872 & -0.036709 & 0.30183 \\ -0.036709 & 8.96453 & 0.13233 \\ 0.30183 & 0.13233 & 1.59715 \end{bmatrix}$$

(t)

STEP 6. We compute the interface equivalent actions of the structure. Substituting relations (h) to (j) and (n) to (q) into (13.8) we obtain

$$\{\overset{\Box}{P}{}_c^{IA}\} = -[K_{ce}^A][K_{ee}^A]^{-1}\{P_e^A\} = -\begin{Bmatrix} -80.5404 \\ 32.1022 \\ 198.9867 \end{Bmatrix}$$

(u)

$$\{\overset{\Box}{P}{}_c^{IB}\} = -\begin{Bmatrix} -235.9180 \\ -1.4804 \\ -2.9508 \end{Bmatrix}$$

(v)

Substituting the above results into relation (13.13) and referring to Fig. b we get

$$\{P^{IE}\} = \{P^{IE}\}_{1*} = \{P_c^{IA}\} + \{P_c^{IB}\} + \{P^E\}_{1*}$$

$$= -\begin{Bmatrix} -80.5404 \\ 32.1022 \\ 198.0867 \end{Bmatrix} - \begin{Bmatrix} -235.9180 \\ -1.4804 \\ -2.9508 \end{Bmatrix} + \begin{Bmatrix} 40 \\ 0 \\ -320 \end{Bmatrix} = \begin{Bmatrix} 356.458 \\ -30.622 \\ -516.036 \end{Bmatrix}$$

(w)

STEP 7. We compute the condensed stiffness matrix for the structure using relation (13.20). That is,

$$[S_c] = [S]_{1*1*} = [K_c^A] + [K_c^B] = EI \begin{bmatrix} 0.29938 & -0.03364 & 0.26576 \\ -0.03364 & 8.96991 & 0.10034 \\ 0.26576 & 0.10035 & 1.87284 \end{bmatrix}$$

(x)

STEP 8. We compute the components of displacement of the interface node of the structure. Referring to Fig. c and using relations (t) and (w) we write and solve the condensed stiffness equations (13.14) for the frame. That is,

$$\{\Delta_c\} = \{\Delta_c^{FA}\} = \{\Delta_c^{FB}\} = \begin{Bmatrix} \Delta_1^{(5)} \\ \Delta_2^{(5)} \\ \Delta_3^{(5)} \end{Bmatrix} = [S_c]^{-1} \begin{Bmatrix} 356.458 \\ -30.622 \\ -516.036 \end{Bmatrix} = \frac{1}{EI} \begin{bmatrix} 1643.619 \\ 8.446 \\ -509.222 \end{bmatrix}$$

(y)

STEP 9. We compute the components of displacements of the interior nodes of each substructure. Substituting relations (h), (i), (k), and (y) and relations (n), (o), (q), and (y) into (13.23) we get

$$
\{\Delta_e^{FA}\} = [K_{ee}^A]^{-1}\{\{P_e^A\} - [K_{ec}^A]\{\Delta_c^A\}\} =
\begin{Bmatrix}
\Delta_1^{(2)} \\
\Delta_2^{(2)} \\
\Delta_3^{(2)} \\
\Delta_1^{(3)} \\
\Delta_2^{(3)} \\
\Delta_3^{(3)}
\end{Bmatrix}
= \frac{1}{EI}
\begin{Bmatrix}
3206.909 \\
11.984 \\
-424.431 \\
3196.912 \\
2.4556 \\
-148.440
\end{Bmatrix}
\quad (z)
$$

$$
\{\Delta_e^{FB}\} = [K_{ee}^B]^{-1}\{\{P_e^B\} - [K_{ec}^B]\{\Delta_c^B\}\} =
\begin{Bmatrix}
\Delta_1^{(6)} \\
\Delta_2^{(6)} \\
\Delta_3^{(6)}
\end{Bmatrix}
= \frac{1}{EI}
\begin{Bmatrix}
1649.098 \\
-14.438 \\
-271.496
\end{Bmatrix}
\quad (aa)
$$

STEP 10. We compute the components of nodal actions of the elements of the frame. Referring to Fig. c and relations (y), (z), and (aa) the global components of nodal displacements of the elements of the frame are

$$
\{\bar{D}^1\} = \frac{1}{EI}
\begin{Bmatrix}
0 \\
0 \\
0 \\
3206.909 \\
11.984 \\
-424.431
\end{Bmatrix}
\qquad
\{\bar{D}^2\} = \frac{1}{EI}
\begin{Bmatrix}
3206.909 \\
11.984 \\
-424.431 \\
3196.912 \\
2.455 \\
-148.440
\end{Bmatrix}
$$

$$
\{\bar{D}^3\} = \frac{1}{EI}
\begin{Bmatrix}
3196.912 \\
2.455 \\
-148.440 \\
1643.619 \\
8.446 \\
-509.222
\end{Bmatrix}
\qquad
\{\bar{D}^4\} = \frac{1}{EI}
\begin{Bmatrix}
0 \\
0 \\
0 \\
1643.619 \\
8.446 \\
-509.222
\end{Bmatrix}
\quad (bb)
$$

$$
\{\bar{D}^5\} = \frac{1}{EI}
\begin{Bmatrix}
1643.619 \\
8.446 \\
-509.222 \\
1649.098 \\
-14.438 \\
-271.496
\end{Bmatrix}
\qquad
\{\bar{D}^6\} = \frac{1}{EI}
\begin{Bmatrix}
1649.098 \\
-14.438 \\
-271.496 \\
0 \\
0 \\
0
\end{Bmatrix}
$$

The relation between the local components of nodal actions and the global components of nodal displacements is

$$\{A^e\} = [\bar{K^e}]\{\bar{D}^e\} \qquad e = 1, 2, \ldots, 5 \tag{cc}$$

Substituting relations (c) and (bb) into relation (cc) we get

$$\{A^1\} = \begin{Bmatrix} -53.48 \\ 35.38 \\ 194.54 \\ 53.48 \\ -35.38 \\ 88.43 \end{Bmatrix} \qquad \{A^2\} = \begin{Bmatrix} 44.62 \\ -53.48 \\ -248.43 \\ -44.62 \\ 53.48 \\ -179.44 \end{Bmatrix}$$

$$\{A^3\} = \begin{Bmatrix} 53.48 \\ 44.62 \\ 179.43 \\ -53.48 \\ -44.62 \\ -0.96 \end{Bmatrix} \qquad \{A^4\} = \begin{Bmatrix} -75.39 \\ 117.22 \\ 361.75 \\ 75.39 \\ -117.22 \\ 107.14 \end{Bmatrix}$$

$$= \quad \{A^5\} = \begin{Bmatrix} -32.61 \\ -128.87 \\ -426.17 \\ 32.61 \\ 128.87 \\ -34.92 \end{Bmatrix} \qquad \{A^6\} = \begin{Bmatrix} 128.87 \\ 207.40 \\ 346.92 \\ -128.87 \\ -207.40 \\ 482.66 \end{Bmatrix}$$

13.4 Problems

1. Using the method of substructures analyze the structure loaded as shown in Fig. 13.P1. Subdivide the structure into two substructures by

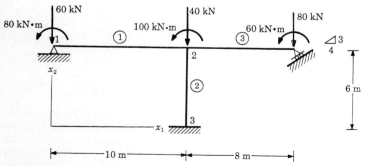

Figure 13.P1

cutting it just below node 2. The elements of the structure have the same
constant cross section and are made from the same isotropic linearly elastic
material.

2. Using the method of substructures analyze the structure loaded as
shown in Fig. 13.P2. Subdivide the structure into two substructures by
cutting it just to the right of node 4. The elements of the structure have the
same constant cross section and are made from the same isotropic linearly
elastic material.

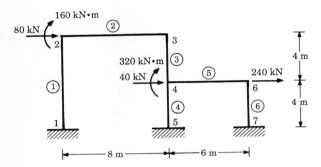

Figure 13.P2

3. Using the method of substructures analyze the structure loaded as
shown in Fig. 13.P3. Subdivide the structure into two substructures by
cutting it just below node 3. The elements of the structure have the same
constant cross section and are made from the same isotropic linearly elastic
material.

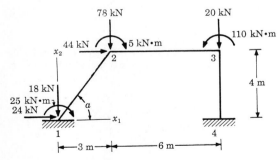

Figure 13.P3

4. Using the method of substructures analyze the structure loaded as
shown in Fig. 13.P4. Subdivide the structure into three substructures by
cutting it just above nodes 2, 3, and 6. The elements of the structure have the
same constant cross section and are made from the same isotropic linearly
elastic material.

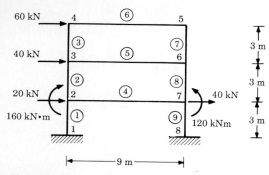

Figure 13.P4

5. Using the method of substructures analyze the structure loaded as shown in Fig. 13.P2. Subdivide the structure into three substructures by cutting it just to the right of nodes 2 and 4. The elements of the structure have the same constant cross section and are made from the same isotropic linearly elastic material.

14

Procedure for Analyzing a Redesigned Structure

14.1 Introduction†

In designing a structure, the designer might need to change the dimensions of the cross sections of a number of its elements until the internal actions in all the elements of the structure approach their allowed values. When the dimensions of the cross sections of an element of a structure change, the structure's stiffness is altered and, consequently, if the structure is statically indeterminate, the internal actions in its elements are also altered. Thus, it is important to be able to establish readily the change in the internal actions of the elements of a structure when the stiffness of a number of its elements is altered. Moreover, when there is doubt about the condition of some of the supports of the structure, it may be necessary to analyze the structure for two or more support conditions.

In this chapter, we present one of the procedures available for analyzing a redesigned structure, using the results of the analysis of the original structure. When a structure has a large number of elements and nodes and is redesigned by modifying the geometry of a relatively small number of its elements, the procedure described in this section may save computational cost.‡ Moreover, in this chapter, we present a procedure for reanalyzing a structure after some of its

---

† See the articles listed in the References at the end of this chapter for techniques for analyzing redesigned structures.

‡ The analysis of a redesigned structure using the results of the analysis of the original structure may not cost less in computer time than its direct analysis, that is, without using the results of the analysis of the original structure (see Ref. 4).

fixed supports have been released, using the results of the analysis of the structure with the original supports.

14.2   Analysis of a Redesigned Structure

In this section we limit our attention to structures whose elements have constant cross section and are made from the same isotropic linearly elastic material. Moreover, we consider structures subjected only to external actions along the length of their elements and on their nodes. The equivalent actions applied on the nodes of a structure subjected only to external actions along the length of its elements and on its nodes are independent of the geometry of its elements or the material from which they are made. This becomes apparent by referring to the table on the inside of the back cover and noting that the fixed-end actions of an element subjected only to external actions along its length are independent of EA and EI. Only when an element is subjected to a change of temperature or to relative movement of its ends does its fixed-end actions depend on EA or EI.

Consider a structure whose elements have a known geometry. We refer to this structure as the *original structure*. Its basic stiffness equations (9.4) can be written as

$$\{\Delta_0^F\} = [F^0]\{P^{EF}\} \tag{14.1}$$

where $[F^0] = [S_0^{FF}]^{-1}$

$\{\Delta_0^F\}$ = matrix of components of displacements of nodes of original structure which are not inhibited by its supports

$\{P^{EF}\}$ = matrix of components of equivalent actions, applied to nodes of original and redesigned structures, which are not directly absorbed by its supports

$\{S_0^{FF}\}$ = basic stiffness matrix for original structure

$\{F^0\}$ = flexibility matrix for original structure

Suppose that the structure under consideration is redesigned. That is, the dimensions of the cross sections of a number of its elements are changed and, consequently, their stiffness is altered. We assume that the effect of the change of the weight of the modified elements of the redesigned structure on the equivalent actions acting on its nodes is negligible. Therefore, the matrix of the equivalent actions acting on the nodes of the redesigned structure is identical to that of the original structure. The basic stiffness equations (9.2) for the redesigned structure are written as

$$\{\Delta^F\} = [[S_0^{FF}] + [\delta S_0]]^{-1}\{P^{EF}\} \tag{14.2}$$

where $[\delta S_0]$ represents the change of the matrix $[S_0^{FF}]$ of the original structure due to the alteration of the stiffness of its elements. Relation (14.2) may be rewritten as

$$\{\Delta^F\} = [[I] + [S_0^{FF}]^{-1}[\delta S_0^{FF}]]^{-1}[S_0^{FF}]^{-1}\{P^{EF}\} \qquad (14.3)$$

Substituting relations (14.1) into (14.3) we get

$$\{\Delta^F\} = [[I] + [F^0][\delta S_0^{FF}]]^{-1}\{\Delta_0^F\} \qquad (14.4)$$

We are interested in establishing the matrix of the node displacements $\{\Delta^F\}$ of the redesigned structure from the matrix of node displacements $\{\Delta_0^F\}$ of the original structure. In order to simplify the required matrix algebra, the rows and columns of relation (14.1) are arranged so that the components of the equivalent actions acting on the nodes to which the modified elements are connected and the components of displacements of these nodes are placed in the upper rows of the matrices $\{P_0^{EF}\}$ and $\{\Delta_0^F\}$, respectively. That is,

$$\left\{\frac{\{\Delta^F\}_a}{\{\Delta^F\}_u}\right\} = \left[\begin{bmatrix} [S_0^{FF}]_{aa} & [S_0^{FF}]_{au} \\ [S_0^{FF}]_{ua} & [S_0^{FF}]_{uu} \end{bmatrix} + \begin{bmatrix} [\delta S_0]_{aa} & [0] \\ [0] & [0] \end{bmatrix}\right]^{-1} \left\{\frac{\{P^{EF}\}_a}{\{P^{EF}\}_u}\right\} \qquad (14.5)$$

where $[\delta S_0]_{aa}$ = condensed change of basic stiffness matrix $[S_0^{FF}]$ due to alteration of stiffness of some elements of original structure

$\{\Delta^F\}_a$ = matrix of uninhibited components of displacements of nodes of redesigned structure to which one or more altered elements are connected

$\{\Delta^F\}_u$ = matrix of uninhibited components of displacements of nodes of redesigned structure to which no altered element is connected

$\{P^{EF}\}_a$ = matrix of components of equivalent actions, not directly absorbed by supports of structure, which act on nodes of original and redesigned structures to which one or more altered elements are connected

$\{P^{EF}\}_u$ = matrix of components of equivalent actions, not directly absorbed by supports of structure, which act on nodes of original and redesigned structures to which no altered element is connected

Moreover, we can write

$$[I] + [F^0][\delta S^0] = \begin{bmatrix} [I] & [0] \\ [0] & [I] \end{bmatrix} + \begin{bmatrix} [F_{aa}^0] & [F_{au}^0] \\ [F_{ua}^0] & [F_{uu}^0] \end{bmatrix}\begin{bmatrix} [\delta S_0]_{aa} & [0] \\ [0] & [0] \end{bmatrix}$$

$$= \begin{bmatrix} [I] + [F_{aa}^0][\delta S_0]_{aa} & [0] \\ [F_{ua}^0][\delta S_0]_{aa} & [I] \end{bmatrix} \qquad (14.6)$$

It can be shown that from relation (14.6) we obtain

$$[[I] + [F^0][\delta S_0]_{aa}]^{-1} = \begin{bmatrix} [[I] + [F^0_{aa}][\delta S_0]_{aa}]^{-1} & [0] \\ -[F^0_{ua}][\delta S_0]_{aa}[[I] + [F^0_{aa}][\delta S_0]_{aa}]^{-1} & [I] \end{bmatrix}$$

$$(14.7)$$

Substituting relation (14.7) into (14.4) we get

$$\{\Delta^F\} = \begin{bmatrix} \dfrac{[N]\{\Delta^F_0\}_a}{\{\Delta^F_0\}_u - [F^0_{ua}][\delta S_0]_{aa}[N]\{\Delta^F_0\}_a} \end{bmatrix} \qquad (14.8)$$

where $$[N] = [[I] + [F^0_{aa}][\delta S_0]_{aa}]^{-1} \qquad (14.9)$$

On the basis of the foregoing presentation, we can establish directly the matrix of node displacements of a redesigned structure from the basic stiffness matrix and the matrix of node displacements of the original structure by adhering to the following procedure.

Part I. We analyze the original structure following the procedure described in Sec. 9.3. That is, we establish its stiffness matrix and compute the components of displacements of its nodes and the nodal actions of its elements.

Part II. We analyze the redesigned structure by adhering to the following steps.

Step 1. We arrange the rows and columns of the basic stiffness equations (14.1) for the original structure so that the components of equivalent actions acting on the nodes of the structure to which its altered elements are connected and the components of displacements of these nodes are placed in the upper rows of the matrices $\{P^{EF}_0\}$ and $\{\Delta^F_0\}$, respectively.

Step 2. We establish and partition the flexibility matrix of the structure $[F^0] = [S^{FF}_0]^{-1}$ as indicated in relation (14.6) and obtain the matrices $[F^0_{ua}]$ and $[F^0_{aa}]$.

Step 3. We compute the matrix $[\delta S_0]_{aa}$.

Step 4. Using relation (14.8) we compute the matrix of the components of displacements $\{\Delta^F\}$ of the nodes of the redesigned structure, which are not inhibited by its supports.

Step 5. Using the matrix $\{\Delta^F\}$ obtained in step 4 we establish the matrices of nodal actions of the elements of the redesigned structure as discussed in Sec. 9.3.

In the sequel, we illustrate the application of the procedures presented in this section by redesigning (modifying the stiffness of one of its elements) and reanalyzing the frame analyzed in Example 3 of Sec. 9.3. This frame consists of two elements and has four unknown components of displacements. Thus, the stiffness matrix of the redesigned frame could be readily obtained directly without employing the procedure presented in this section. The efficiency of this procedure increases as the number of unknown displacements of the nodes of the structure to which it is applied increases.

Example. Consider the frame of Fig. a subjected to the external actions shown in the figure. Originally the elements of the frame have the same constant cross section $(A_0 = 16 \times 10^3 \text{ mm}^2, I_0 = 400 \times 10^6 \text{ mm}^4)$ and are made from the same isotropic linearly elastic material $(E = 210 \text{ kN/mm}^2)$. The structure is redesigned by changing only the dimensions of the cross section of element 1 so that

$$A^{(1)} = 15 \times 10^3 \qquad I^{(1)} = 300 \times 10^6 \text{ mm}^4$$

Compute the internal actions in the elements of the redesigned structure.

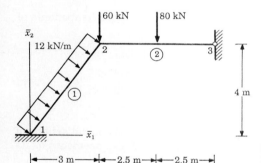

Figure a Geometry and loading of the frame.

---

solution

Part I. Analysis of the Original Structure

We analyze the original structure following the procedure described in Sec. 9.3, referring to the table on the inside of the back cover. The matrices of fixed-end actions of the elements of the frame are

$$[A^{R1}] = \begin{Bmatrix} 0 \\ 30 \\ 25 \\ \hline 0 \\ 30 \\ -25 \end{Bmatrix} \qquad [A^{R2}] = \begin{Bmatrix} 0 \\ 40 \\ 50 \\ \hline 0 \\ 40 \\ -50 \end{Bmatrix} \qquad \text{(a)}$$

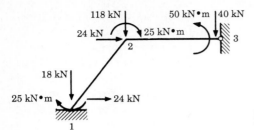

118 kN

24 kN   25 kN•m   50 kN•m   40 kN

2   3

18 kN

25 kN•m   24 kN

1

Figure b Structure subjected to the equivalent actions.

The structure subjected to the equivalent actions is shown in Fig. b. Moreover its basic stiffness matrix is given by relation (1) of Example 3 of Sec. 9.3. Thus, referring to Fig. a its basic stiffness equations are

$$\{P^{EF}\} = \begin{Bmatrix} 24 \\ -118 \\ -25 \\ \hline 50 \end{Bmatrix} = [S_0^{FF}]\{\Delta_0^F\} = \frac{EI_0}{125} \begin{bmatrix} 1367.68 & 474.24 & 24 & 0 \\ 474.24 & 656.32 & 12 & 30 \\ 24.00 & 12.00 & 200 & 50 \\ \hline 0 & 30.00 & 50 & 100 \end{bmatrix} \begin{Bmatrix} \Delta_1^{(2)} \\ \Delta_2^{(2)} \\ \Delta_3^{(2)} \\ \hline \Delta_3^{(3)} \end{Bmatrix}$$

(b)

We solve the stiffness equations (b) to obtain the matrix of the components of displacements of the nodes of the structure. That is,

$$\{\Delta_0^F\} = \begin{Bmatrix} \{\Delta_0^F\}_a \\ \{\Delta_0^F\}_u \end{Bmatrix} = [S^{FF}]^{-1}\{P^{EF}\} = \frac{125}{EI_0} \begin{Bmatrix} 0.126944735 \\ -0.299893435 \\ -0.308264778 \\ \hline 0.744100409 \end{Bmatrix}$$

(c)

Following the procedure described in Sec. 9.3 we compute the nodal actions of the elements of the structure, from the components of displacements of its nodes. That is,

$$\{A^1\} = \begin{Bmatrix} 163.75 \\ 24.13 \\ 18.03 \\ \hline -163.75 \\ 35.70 \\ 47.81 \end{Bmatrix} \qquad \{A^2\} = \begin{Bmatrix} 126.95 \\ 49.48 \\ 47.38 \\ \hline -126.95 \\ 30.52 \\ 0 \end{Bmatrix}$$

(d)

The resulting distribution of internal actions in the elements of the frame could be considered satisfactory. However, for illustrative purposes, the frame is redesigned in order to decrease the moment of $M_3^{1j} = 18.03$ kN · m transferred to the ground. This is accomplished by reducing the stiffness of element 1.

Part II. Analysis of the Redesigned Structure

STEP 1.   The rows and columns of the basic stiffness equations for the frame do not have to be rearranged. The components of external actions acting on node 2 into which the altered element 1 frames are at the top of matrices $\{P^{EF}\}$ and $\{\Delta_0^F\}$.

STEP 2.   We establish and partition the flexibility matrix $[F^0]$ as indicated in relation (14.6). Thus referring to relation (14.6) we have

$$[F^0] = [S_0^{FF}]^{-1}$$

$$= \frac{125}{EI_0} \left[\begin{array}{ccc:c} 0.00098394 & -0.00072154 & -0.00014731 & 0.00029012 \\ -0.00072154 & 0.00207410 & 0.00013451 & -0.00068948 \\ -0.00014731 & 0.00013451 & 0.00573684 & -0.00290870 \\ \hdashline 0.00029012 & -0.00068948 & -0.00290870 & 0.01166100 \end{array} \right] \quad \text{(e)}$$

$$[F_{aa}^0] = \frac{125}{EI_0} \left[\begin{array}{ccc} 0.00098394 & -0.00072154 & -0.00014731 \\ -0.00072154 & 0.00207410 & 0.00013451 \\ -0.00014731 & 0.00013451 & 0.00573680 \end{array} \right] \quad \text{(f)}$$

and $$[F_{ua}^0] = \frac{125}{EI_0} [0.00029012 \quad -0.00068948 \quad -0.00290870] \quad \text{(g)}$$

STEP 3.   We compute the matrix $[\delta S_0]_{aa}$. Referring to relation (4.62), the stiffness matrix for element 1 of the redesigned structure is equal to

$$[K^1] = \frac{EI}{125} \left[\begin{array}{cccccc} 1250 & 0 & 0 & -1250 & 0 & 0 \\ 0 & 12 & 30 & 0 & -12 & 30 \\ 0 & 30 & 100 & 0 & -30 & 50 \\ -1250 & 0 & 0 & 1250 & 0 & 0 \\ 0 & -12 & -30 & 0 & 12 & -30 \\ 0 & 30 & 50 & 0 & -30 & 100 \end{array} \right] \quad \text{(h)}$$

where $I = 300 \times 10^{-6}\,\text{m}^4$. Moreover, the stiffness matrix of element 1 of the original structure is given by relation (a) of Example 3 of Sec. 8.3.1 as

$$[K_0^1] = \frac{EI_0}{125} \left[\begin{array}{cccccc} 1000 & 0 & 0 & -1000 & 0 & 0 \\ 0 & 12 & 30 & 0 & -12 & 30 \\ 0 & 30 & 100 & 0 & -30 & 50 \\ -1000 & 0 & 0 & 1000 & 0 & 0 \\ 0 & -12 & -30 & 0 & 12 & -30 \\ 0 & 30 & 50 & 0 & -30 & 100 \end{array} \right] \quad \text{(i)}$$

where $I_0 = 400 \times 10^{-6} \, \text{m}^4$. From relations (h) and (i) we obtain

$$[\delta K^1] = \frac{E(I - I_0)}{125} \begin{bmatrix} a & 0 & 0 & -a & 0 & 0 \\ 0 & 12 & 30 & 0 & -12 & 30 \\ 0 & 30 & 100 & 0 & -30 & 50 \\ -a & 0 & 0 & a & 0 & 0 \\ 0 & -12 & 30 & 0 & 12 & -30 \\ 0 & 30 & 50 & 0 & -30 & 100 \end{bmatrix} \qquad (j)$$

$$[\delta K^2] = [0]$$

where

$$a = \frac{1250I - 1000I_0}{I - I_0} = 250$$

The transformation matrix $[\Lambda^1]$ of element 1 is given by relation (6.48) as

$$[\Lambda^1] = \begin{bmatrix} 0.6 & 0.8 & 0 & 0 & 0 & 0 \\ -0.8 & 0.6 & 0 & 0 & 0 & 0 \\ 0 & 0 & 1 & 0 & 0 & 0 \\ 0 & 0 & 0 & 0.6 & 0.8 & 0 \\ 0 & 0 & 0 & -0.8 & 0.6 & 0 \\ 0 & 0 & 0 & 0 & 0 & 1 \end{bmatrix} \qquad (k)$$

Referring to relation (6.74) and using relations (j) and (k), we compute the global change of the stiffness matrix of element 1. That is,

$$[\delta \bar{K}^1] = [\Lambda^1]^T [\delta K^1][\Lambda^1]$$

$$= \frac{E(I - I_0)}{125} \begin{bmatrix} 97.68 & 114.24 & -24 & -97.68 & -114.24 & -24 \\ 114.24 & 164.32 & 18 & -114.24 & -164.32 & 18 \\ -24.00 & 18.00 & 100 & 24.00 & -18.00 & 50 \\ -97.68 & -114.24 & 24 & 97.68 & 114.24 & 24 \\ -114.24 & -164.32 & -18 & 114.24 & 164.32 & -18 \\ -24.00 & 18.00 & 50 & 24.00 & -18.00 & 100 \end{bmatrix} \qquad (l)$$

$$[\delta \bar{K}^2] = [0] \qquad (m)$$

We assemble the change of the stiffness matrix $[\delta S_0]$ of the structure from the global change of the stiffness matrices of its elements. Thus, using relations (l) and (m) we obtain

$$[\delta S_0]$$

$$= \frac{E(I - I_0)}{125} \begin{bmatrix} 97.68 & 114.24 & -24 & -97.68 & -114.24 & -24 & 0 & 0 & 0 \\ 114.24 & 164.32 & 18 & -114.24 & -164.32 & 18 & 0 & 0 & 0 \\ -24.00 & 18.00 & 100 & 24.00 & -18.00 & 50 & 0 & 0 & 0 \\ -97.68 & -114.24 & -24 & 97.68 & 114.24 & 24 & 0 & 0 & 0 \\ -114.24 & -164.32 & 18 & 114.24 & 164.32 & -18 & 0 & 0 & 0 \\ -24.00 & 18.00 & 50 & 24.00 & -18.00 & 100 & 0 & 0 & 0 \\ 0 & 0 & 0 & 0 & 0 & 0 & 0 & 0 & 0 \\ 0 & 0 & 0 & 0 & 0 & 0 & 0 & 0 & 0 \\ 0 & 0 & 0 & 0 & 0 & 0 & 0 & 0 & 0 \end{bmatrix}$$

$$\text{(n)}$$

We rearrange the rows and columns of the change of the stiffness matrix $[\delta S_0]$ of the structure in order to separate the rows and columns which correspond to components of displacements which are inhibited by the supports of the structure. Thus, we obtain

$$[\delta S_0]_m = \frac{E(I - I_0)}{125} \begin{bmatrix} 97.68 & 114.24 & 24 & 0 & -97.68 & -114.24 & 24 & 0 & 0 \\ 114.24 & 164.32 & -18 & 0 & -114.24 & -164.32 & -18 & 0 & 0 \\ 24.00 & -18.00 & 100 & 0 & -24.00 & 18.00 & 50 & 0 & 0 \\ 0 & 0 & 0 & 0 & 0 & 0 & 0 & 0 & 0 \\ -97.68 & -114.24 & -24 & 0 & 97.68 & 114.24 & -24 & 0 & 0 \\ -114.24 & -164.32 & 18 & 0 & 114.24 & 164.24 & 18 & 0 & 0 \\ 24.00 & -18.00 & 50 & 0 & -24.00 & 18.00 & 100 & 0 & 0 \\ 0 & 0 & 0 & 0 & 0 & 0 & 0 & 0 & 0 \\ 0 & 0 & 0 & 0 & 0 & 0 & 0 & 0 & 0 \end{bmatrix}$$

$$\text{(o)}$$

Hence,

$$[\delta S_0^{FF}] = \frac{E(I - I_0)}{125} \begin{bmatrix} 97.68 & 114.24 & 24 & 0 \\ 114.24 & 164.32 & -18 & 0 \\ 24.00 & -18.00 & 100 & 0 \\ \hline 0 & 0 & 0 & 0 \end{bmatrix} \qquad \text{(p)}$$

and

$$[\delta S_0]_{aa} = \frac{E(I - I_0)}{125} \begin{bmatrix} 97.68 & 114.24 & 24 \\ 114.24 & 164.32 & -18 \\ 24.00 & -18.00 & 100 \end{bmatrix} \qquad \text{(q)}$$

STEP 4. We compute the matrix $\{\Delta^F\}$. Using relations (f) and (q), we obtain

$$
[I] + [F^0_{aa}][\delta S_0]_{aa} = \begin{bmatrix} 0.997463227 & 0.000876639 & -0.00546782 \\ -0.042422492 & 0.936009681 & 0.010299805 \\ -0.034665095 & 0.024497102 & 0.858069155 \end{bmatrix} \quad \text{(r)}
$$

Substituting relation (r) into (14.10), we get

$$
[N] = [[I] + [F_{aa}][\delta S_0]_{aa}]^{-1} = \begin{bmatrix} 1.00271868 & -0.00110669 & 0.00640285 \\ 0.04501431 & 1.06865107 & -0.01254068 \\ 0.03922367 & -0.03055374 & 1.16602393 \end{bmatrix} \quad \text{(s)}
$$

Using relations (c), (g), (q), and (s), we obtain

$$
[[N]\{\Delta^F_0\}_a]^T = \frac{125}{EI_0}[0.12564797 \quad -0.31090126 \quad -0.34530200] \quad \text{(t)}
$$

Using relations (q) and (g) we get

$$
[F^0_{ua}][\delta S^{FF}_0]_{aa} = [0.030059014 \quad -0.0069488605 \quad -0.067874125] \quad \text{(u)}
$$

From relations (t) and (u) we obtain

$$
[F_{ua}][\delta S_0]_{aa}[N]\{\Delta^F_0\}_a = \frac{125}{EI_0}[-0.02182063] \quad \text{(v)}
$$

Substituting relations (t), (u), and $\{\Delta^F_0\}_a$ from (c) into (14.9), we have

$$
[\Delta^F] = \frac{125}{EI_0}\begin{Bmatrix} 0.125648 \\ -0.310901 \\ -0.345302 \\ \hline 0.765921 \end{Bmatrix} \quad \text{(w)}
$$

STEP 5. We compute the components of nodal actions of the elements of the frame. Referring to relation (q), the global components of displacements of the elements of the frame are

$$
\{\bar D^1\} = \frac{125}{EI_0}\begin{Bmatrix} 0 \\ 0 \\ 0 \\ 0.125648 \\ -0.310901 \\ -0.345302 \end{Bmatrix} \qquad \{\bar D^2\} = \frac{125}{EI_0}\begin{Bmatrix} 0.125648 \\ -0.310901 \\ -0.345302 \\ 0 \\ 0 \\ 0.765921 \end{Bmatrix} \quad \text{(x)}
$$

Using relation (6.67), we transform the matrices $\{\bar D^e\}$ $(e = 1, 2)$ to the matrices $\{D^e\}$. Thus,

$$\{D^1\} = [\Lambda_{PF}^1]\{\bar{D}^1\} = \frac{125}{EI_0} \begin{bmatrix} 0.6 & 0.8 & 0 & 0 & 0 & 0 \\ -0.8 & 0.6 & 0 & 0 & 0 & 0 \\ 0 & 0 & 1 & 0 & 0 & 0 \\ 0 & 0 & 0 & 0.6 & 0.8 & 0 \\ 0 & 0 & 0 & -0.8 & 0.6 & 0 \\ 0 & 0 & 0 & 0 & 0 & 1 \end{bmatrix} \begin{Bmatrix} 0 \\ 0 \\ 0 \\ 0.125648 \\ -0.310901 \\ -0.345302 \end{Bmatrix} = \begin{Bmatrix} 0 \\ 0 \\ 0 \\ -0.17333 \\ -0.28706 \\ -0.34530 \end{Bmatrix}$$

$$\{D^2\} = \{\bar{D}^2\} \tag{y}$$

The relations between the components of nodal actions and of nodal displacements of the elements of the frame are

$$\{A^{Ee}\} = \begin{Bmatrix} F_1^{ej} \\ F_2^{ej} \\ M_3^{ej} \\ F_1^{ek} \\ F_2^{ek} \\ M_3^{ej} \end{Bmatrix} = [K^e][D^e] \tag{z}$$

Substituting the local stiffness matrix of element e of the frame given by relation (h) of this example and relation (a) of Example 3 of Sec. 8.4 and the matrix $[D^e]$ given by relations (x) into Eq. (z), we obtain the matrices of nodal actions of the elements of the structure subjected to the equivalent actions on its nodes. Thus,

$$\{A^{E1}\} = \begin{bmatrix} -1250 & 0 & 0 & -1250 & 0 & 0 \\ 0 & 12 & 30 & 0 & -12 & 30 \\ 0 & 30 & 100 & 0 & -30 & 50 \\ -1250 & 0 & 0 & 1250 & 0 & 0 \\ 0 & -12 & -30 & 0 & 12 & -30 \\ 0 & 30 & 50 & 0 & -30 & 100 \end{bmatrix} \begin{Bmatrix} 0 \\ 0 \\ 0 \\ -0.17333 \\ -0.28706 \\ -0.34530 \end{Bmatrix} \left(\frac{3}{4}\right) = \begin{Bmatrix} 162.50 \\ -5.19 \\ -6.49 \\ -162.50 \\ 5.19 \\ -19.44 \end{Bmatrix}$$

$$\tag{aa}$$

$$\{A^{E2}\} = \begin{bmatrix} 1000 & 0 & 0 & -1000 & 0 & 0 \\ 0 & 12 & 30 & 0 & -12 & 30 \\ 0 & 30 & 100 & 0 & -30 & 50 \\ -1000 & 0 & 0 & 1000 & 0 & 0 \\ 0 & -12 & -30 & 0 & 12 & -30 \\ 0 & 30 & 50 & 0 & -30 & 100 \end{bmatrix} \begin{Bmatrix} 0.12565 \\ -0.31090 \\ -0.34530 \\ 0 \\ 0 \\ 0.76592 \end{Bmatrix} = \begin{Bmatrix} 125.65 \\ 8.89 \\ -5.56 \\ -125.65 \\ -8.89 \\ 50.00 \end{Bmatrix}$$

Substituting relations (a) and (aa) into (4.7) we obtain the matrices of nodal actions of the elements of the structure of Fig. a. Thus,

$$\{A^1\} = \{A^{E1}\} + \{A^{R1}\} = \begin{Bmatrix} 162.50 \\ -5.19 \\ -6.49 \\ -162.50 \\ 5.19 \\ -19.44 \end{Bmatrix} + \begin{Bmatrix} 0 \\ 30 \\ 25 \\ 0 \\ 30 \\ -25 \end{Bmatrix} = \begin{Bmatrix} 162.50 \\ 24.81 \\ 18.51 \\ -162.50 \\ 35.19 \\ -44.44 \end{Bmatrix}$$

$$\{A^2\} = \{A^{E2}\} + \{A^{R2}\} = \begin{Bmatrix} 125.65 \\ 8.89 \\ -5.56 \\ -125.65 \\ -8.89 \\ 50.00 \end{Bmatrix} + \begin{Bmatrix} 0 \\ 40 \\ 50 \\ 0 \\ 40 \\ -50 \end{Bmatrix} = \begin{Bmatrix} 125.65 \\ 48.89 \\ 44.44 \\ -125.65 \\ 31.11 \\ 0 \end{Bmatrix}$$

14.3 Modification of the Support Conditions of a Framed Structure

Consider a framed structure subjected to a general loading, that is, subjected to

1. External actions acting along the length of its elements and on its nodes
2. A change of temperature of its elements
3. A settlement of its supports

We refer to this structure as the *original structure*. Its basic stiffness equations can be written as

$$\{P_0^{EF}\} = [S_0^{FF}]\{\Delta_0^F\} \tag{14.11}$$

We assume that the supports of the structure are modified by introducing internal action release mechanisms at the nodes connected to them. Thus the number of components of displacements of the nodes of the modified structure which are not inhibited by its supports is larger than that of the original structure. We write the stiffness equations of the modified structure as

$$\{P^{EF}\} = \begin{Bmatrix} \{P_0^{EF}\} \\ \hline \{P_0^{ER}\} \end{Bmatrix} = \begin{bmatrix} [S_0^{FF}] & \vdots & [S^{MR}] \\ \hline [S^{RM}] & \vdots & [S^{RR}] \end{bmatrix} \begin{bmatrix} \{\Delta^{FM}\} \\ \{\Delta^{FR}\} \end{bmatrix} \tag{14.12}$$

In matrix $\{\Delta^{FR}\}$ we store the free components of displacements of the nodes of the modified structure which are inhibited by the supports of the original structure. We call this matrix the matrix of the released components of displacements of the nodes of the modified structure. In matrix $\{\Delta^{FM}\}$ we store the free components of displacements of the nodes of the modified structure which are not inhibited by the supports of the original structure. In matrix $\{P_0^{ER}\}$ we store the components of equivalent actions which correspond to the components of displacements $\{\Delta^{FR}\}$.

Relation (14.12) may be expanded to yield

$$\{P_0^{EF}\} = [S_0^{FF}]\{\Delta^{FM}\} + [S^{MR}]\{\Delta^{FR}\} \tag{14.13}$$

$$\{P_0^{ER}\} = [S^{RM}]\{\Delta^{FM}\} + [S^{RR}]\{\Delta^{FR}\} \tag{14.14}$$

From relation (14.13) we get

$$\{\Delta^{FM}\} = [S_0^{FF}]^{-1}[\{P_0^{EF}\} - [S^{MR}]\{\Delta^{FR}\}] \tag{14.15}$$

Substituting relation (14.15) into (14.14), we obtain

$$\{\Delta^{FR}\} = [[S^{RR}] - [S^{RM}][S_0^{FF}]^{-1}[S^{MR}]]^{-1}$$
$$\times [\{P_0^{ER}\} - [S^{RM}][S_0^{FF}]^{-1}[P_0^{EF}]] \tag{14.16}$$

On the basis of the foregoing presentation we can establish directly the matrix of the components of displacements of the nodes of a structure after the conditions of some of its supports have been modified from the stiffness equations and the matrix of the components of displacements of the nodes of the original structure, by adhering to the following procedure.

Part I. We analyze the original structure following the procedure described in Sec. 9.3. That is, we establish its stiffness matrix, we form its stiffness equations, and we compute the components of displacements of its nodes and the nodal actions of its elements.

Part II. We analyze the modified structure by adhering to the following procedure.

Step 1. We form the stiffness equations of the modified structure and rearrange its rows and columns so that the components of equivalent actions and the components of displacements of the nodes of the original structure are placed at the top rows of the matrices $\{P^{EF}\}$ and $\{\Delta^F\}$, respectively, of the modified structure. We partition the stiffness equations of the modified structure as indicated in relation (14.12).

Step 2. We compute the matrix of node displacement $\{\Delta^{FR}\}$ using relation (14.16). Notice, that the inverse of the matrix $[S_0^{FF}]$ is known from the analysis of the original structure. Therefore, in relation (14.16), we only have to invert the $n \times n$ matrix $[[S^{RR}] - [S^{RM}][S_0^{FF}]^{-1}[S^{MR}]]$, where n is the number of the released components of displacements of the supports of the structure. Thus, when only a few components of displacements of the supports of a structure are released, n is a small number.

Step 3. We compute the matrix $\{\Delta^{FM}\}$ of the modified structure by substituting the matrix $\{\Delta^{FR}\}$ obtained in step 2 into relation (14.15).

Step 4. We use the components of displacements of the nodes of the modified structure to establish the matrices of nodal actions of its elements following the procedure described in Sec. 9.3.

In the sequel, we illustrate the application of the procedures presented in this section by reanalyzing the frame analyzed in Example 3 of Sec. 9.3 after its support conditions are modified.

Example.   Consider the frame of Fig. a subjected to the external actions shown in the figure, to a settlement of its support 1 of 0.02 m, and to a change of temperature $\delta T_2 = 40°C$. The elements of the frame have the same constant cross section $(A = 16 \times 10^3 \text{ mm}^2, I = 400 \times 10^6 \text{ mm}^4)$ and are made

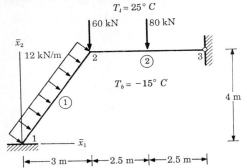

Figure a Geometry and loading of the frame.

from the same isotropic linearly elastic material $(E = 210 \text{ kN/mm}^2)$. Using the procedure described in this section, reanalyze the structure when its support 1 is changed from fixed to hinged.

solution

Part I. Analysis of the Original Structure

The original structure under consideration and its loading are identical to those of Example 3 of Sec. 9.3. Thus its stiffness equations are given by Eqs. (f) of that example.

Part II. Analysis of the Structure with the Modified Support Conditions

STEP 1. We form the basic stiffness equations of the modified structure. Referring to Eqs. (f) of Example 3 of Sec. 9.3 the basic stiffness equations of the structure of Fig. a with pinned support 1 are

$$
\begin{Bmatrix} P_1^{(2)} \\ P_2^{(2)} \\ M_3^{(2)} \\ M_3^{(3)} \\ M_3^{(1)} \end{Bmatrix} = \begin{Bmatrix} -6349.79 \\ -8777.66 \\ 216.92 \\ -30.00 \\ 296.92 \end{Bmatrix} = \frac{EI}{125} \begin{bmatrix} 1367.68 & 474.24 & 24 & 0 & \vdots & 24 \\ 474.24 & 656.32 & 12 & 30 & \vdots & -18 \\ 24.00 & 12.00 & 200 & 50 & \vdots & 50 \\ 0 & 30.00 & 50 & 100 & \vdots & 0 \\ \cdots & \cdots & \cdots & \cdots & \vdots & \cdots \\ 24.00 & -18.00 & 50 & 0 & \vdots & 100 \end{bmatrix} \begin{Bmatrix} \Delta_1^{(2)} \\ \Delta_2^{(2)} \\ \Delta_3^{(2)} \\ \Delta_3^{(3)} \\ \Delta_3^{(1)} \end{Bmatrix}
$$

(a)

In these equations the settlement of the support of the structure has been applied to the restrained structure (see Example 3 of Sec. 9.3).

STEP 2. We compute the components of displacements of the nodes of the structure. Referring to Eqs. (14.12) and (a) we have

$$
\{P_0^{EF}\} = \begin{Bmatrix} -6349.79 \\ -8777.66 \\ 216.92 \\ 30.00 \end{Bmatrix} \qquad \{P_0^{ER}\} = 296.92
$$

(b)

$$
[S_0^{FF}]^{-1} = \frac{125}{EI} \begin{bmatrix} 0.00098394 & -0.00072154 & -0.00014731 & 0.00029012 \\ -0.00072154 & 0.00207407 & 0.00013451 & -0.00068948 \\ -0.00014731 & 0.00013451 & 0.00573680 & -0.00290870 \\ 0.00029012 & 0.00068948 & -0.00290870 & 0.01166120 \end{bmatrix}
$$

(c)

$$
[S^{MR}] = \frac{EI_0}{125} \begin{Bmatrix} 24 \\ -18 \\ 50 \\ 0 \end{Bmatrix}
$$

(d)

$$
[S^{RM}] = \frac{EI_0}{125} [24 \quad -18 \quad 50 \quad 0]
$$

(e)

$$
[S^{RR}] = \frac{EI(100)}{125}
$$

(f)

and

$$
[S^{RM}][S_0^{FF}]^{-1} = [0.0292237 \quad -0.0479247 \quad 0.2808834 \quad 0.1260615]
$$

Thus

$$
[S^{RM}][S_0^{FF}]^{-1}\{P_0^{EF}\} = 299.730
$$

$$
[S^{RM}][S_0^{FF}]^{-1}[S^{MR}] = \frac{EI(15.6085)}{125}
$$

and

$$\{P_0^{ER}\}[S^{RM}][S_0^{FF}]^{-1}\{P_0^{FF}\} = 296.92 - 299.730 = -2.81 \qquad \text{(g)}$$

$$[S^{RR}] - [S^{RM}][S_0^{FF}]^{-1}[S^{MR}] = \frac{EI_0}{125}[100 - 15.6085] = \frac{EI_0}{125}(84.3915) \qquad \text{(h)}$$

Substituting relations (g) and (h) into (14.16), we get

$$\{\Delta^{FR}\} = \frac{125}{EI_0}[-0.03330] \qquad \text{(i)}$$

Substituting relations (b) to (d) and (i) into (14.15), we obtain

$$[\Delta^{00}] = \begin{bmatrix} 0.00098394 & -0.00072154 & -0.00014731 & 0.00029012 \\ -0.00072154 & 0.00207407 & 0.00013451 & -0.00068948 \\ -0.00014731 & 0.00013451 & 0.00573680 & -0.00290870 \\ 0.00029012 & 0.00068948 & -0.00290870 & 0.01166120 \end{bmatrix}$$

$$\times \left[\begin{Bmatrix} -6349.7 \\ -8777.6 \\ 216.92 \\ 50.00 \end{Bmatrix} - \begin{Bmatrix} 24 \\ -18 \\ 50 \\ 0 \end{Bmatrix}[-0.03330] \right] \frac{125}{EI_0}$$

$$= \frac{125}{EI_0} \begin{Bmatrix} 0.04594 \\ -13.57529 \\ 1.09578 \\ 3.22483 \end{Bmatrix} \qquad \text{(j)}$$

Thus, the matrix of the components of displacements of the nodes of the frame is

$$\{\Delta^F\} = \begin{Bmatrix} \Delta_3^{(1)} \\ \Delta_1^{(2)} \\ \Delta_2^{(2)} \\ \Delta_3^{(2)} \\ \hdashline \Delta_3^{(3)} \end{Bmatrix} = \frac{125}{EI} \begin{Bmatrix} -0.03330 \\ -0.04594 \\ -13.57529 \\ 1.07578 \\ \hdashline 3.22483 \end{Bmatrix} \qquad \text{(k)}$$

STEP 3.   We compute the components of nodal actions of the elements of the modified structure. Referring to relation (k) we have

$$\{\bar{D}^1\} = \begin{Bmatrix} 0 \\ -0.03330 \\ 0 \\ \hdashline 0.04594 \\ -13.57529 \\ 1.09578 \end{Bmatrix} \qquad \text{(l)}$$

$$\{\bar{D}^2\} = \left\{ \begin{array}{c} 0.04594 \\ -13.57529 \\ 1.09578 \\ \hline 0 \\ 0 \\ 3.22483 \end{array} \right\}$$

The hybrid stiffness matrices of the elements of the structure are given by relations (q) of Example 3 of Sec. 9.3. Moreover, the matrix of fixed-end actions of the elements of the structure are given by relations (a) of Example 3 of Sec. 9.3. Substituting these matrices and Eq. (1) into relations (9.13) we obtain

$$[A^1] = \begin{bmatrix} 600.0 & 800.0 & 0 & -600.0 & -800.0 & 0 \\ -9.6 & 7.2 & 30 & 9.6 & -7.2 & 30 \\ -24.0 & 18.0 & 100 & 24.0 & -18.0 & 50 \\ -600.0 & -800.0 & 0 & 600.0 & 800.0 & 0 \\ 9.6 & -7.2 & -30 & -9.6 & 7.2 & -30 \\ -24.0 & 18.0 & 50 & 24.0 & -18.0 & 100 \end{bmatrix} \left\{ \begin{array}{c} 0 \\ 0 \\ -0.03330 \\ \hline 0.04594 \\ -13.57529 \\ 1.09578 \end{array} \right\}$$

$$+ \left\{ \begin{array}{c} -10752.00 \\ -66.77 \\ -296.92 \\ \hline 10752.00 \\ 126.77 \\ -186.92 \end{array} \right\} = \left\{ \begin{array}{c} 80.69 \\ 63.36 \\ 0.29 \\ \hline 80.84 \\ -3.34 \\ -116.52 \end{array} \right\}$$

$$[A^2] = \begin{bmatrix} 1000 & 0 & 0 & -1000 & 0 & 0 \\ 0 & 12 & 30 & 0 & -12 & 30 \\ 0 & 30 & 100 & 0 & -30 & 50 \\ -1000 & 0 & 0 & 1000 & 0 & 0 \\ 0 & -12 & -30 & 0 & 12 & -30 \\ 0 & 30 & 50 & 0 & -30 & 100 \end{bmatrix} \left\{ \begin{array}{c} 0.04594 \\ -13.57529 \\ 1.09578 \\ \hline 0 \\ 0 \\ 3.22483 \end{array} \right\}$$

$$+ \left\{ \begin{array}{c} 0 \\ 40 \\ -30 \\ \hline 0 \\ 40 \\ -30 \end{array} \right\} = \left\{ \begin{array}{c} 45.94 \\ 6.70 \\ -166.52 \\ \hline 45.83 \\ 73.30 \\ 0 \end{array} \right\}$$

14.4 Problems

1. The beam loaded as shown in Fig. 9.P4 has been redesigned. The cross section of elements 1 and 3 of the redesigned beam are identical to those of the original beam. The cross section of element 2 was changed to $I^{(2)} = 117.7 \times 10^6$ mm^4, $A_2 = 6.26 \times 10^3$ mm^2. Reanalyze the redesigned beam using the procedure described in Sec. 14.2.

2. The planar truss loaded as shown in Fig. 9.P2 has been redesigned. The cross section of elements 2 to 5 of the redesigned truss are identical to those of the original truss. The cross section of element 1 of the redesigned truss was changed to $A_1 = 5 \times 10^3$ mm^2. Reanalyze the redesigned truss using the procedure described in Sec. 14.2.

3 to 5. The planar frame loaded as shown in Fig. 9.P8 has been redesigned. The cross section of elements 1 and 3 of the redesigned frame did not change. The cross section of element 2 of the redesigned frame was changed to $I^{(2)} = 117.7 \times 10^6$ mm^4, $A_2 = 6.26 \times 10^3$ mm^2. Reanalyze the redesigned frame using the procedure described in Sec. 14.2. Repeat with the frames loaded as shown in Figs. 9.P11 and 9.P12.

6. The space frame loaded as shown in Fig. 9.P14 has been redesigned. The cross section of elements 1 and 3 of the redesigned frame did not change. The cross section of element 2 was changed to $d_{ext} = 70$ mm, $d_{int} = 50$ mm. Reanalyze the redesigned frame using the procedure described in Sec. 14.2.

7 to 10. Consider the frame loaded as shown in Fig. 9.P7 and assume that its fixed support 1 is altered to hinge. Reanalyze the modified frame using the procedure described in Sec. 14.3. Repeat with the frames loaded as shown in Figs. 9.P10, 9.P12, and 9.P16.

References

1. R. L. Sack, W. C. Carpenter, and G. L. Hatch, "Modification of Elements in the Displacement Method," *Journal of AIAA,* vol. 5 no. 9, Sept. 1967, pp. 1708–1710.
2. R. J. Melosh and R. Luik, "Multiple Configuration Analysis of Structures," *Journal of Structural Engineering,* Nov. 1968, pp. 2581–2596.
3. J. Sobieszczanski, "Structural Modification by the Perturbation Method," *Journal of Structural Engineering,* Dec. 1968, pp. 2799–2816.
4. D. Kavlie and G. H. Powell, "Efficient Reanalysis of Modified Structures," *Journal of Structural Engineering,* Jan. 1971, pp. 377–392.
5. J. H. Argyris and J. R. Roy, "General Treatment of Structural Modifications," *Journal of Structural Engineering,* Feb. 1972, pp. 465–492.
6. U. Kirsch and M. F. Rubenstein, "Reanalysis for Limited Structural Design Modifications," *Journal of Engineering Mechanics,* Feb. 1972, pp. 61–70.
7. A. K. Noor and H. E. Lowder, "Approximate Techniques of Structural Reanalysis," *Journal of Computers and Structures,* no. 4, 1974, pp. 801–812.

Photographs and a Brief Description of Certain Features of the Design of Military Aircraft

From a structural point of view aircraft are distinguished from other structures by the fact that the factors of safety used in their design are low. In order to accomplish this materials with a high strength-to-weight ratio and precise design procedures are used. One of the significant improvements in aeronautical design in recent times is the application of composite materials to aircraft structures. These materials are composed of fibers, whiskers, or particles embedded in a matrix and could have properties and behavior not achievable by any of their components. Thus a composite can be prepared which satisfies a particular design requirement from components which do not.

The most widely used matrices are polymers, mainly epoxies. However, there is increasing interest in metal and ceramic (including glass) matrices. The cheapest and most extensively used fibers are glass (E glass or S glass). However, boron, graphite, and aramid fibers or whiskers have been used primarily in aircraft and space vehicles. The first production application of a composite on an aircraft component was on the U.S. Navy's F-14 (see Fig. 14.1) manufactured by the Grumman Aerospace Corporation. Its horizontal stabilizer is made of boron-epoxy composite laminates on a titanium substructure.

Boron-epoxy composites are used for the fin, rudder, and stabilator skins, and graphite-epoxy composites are used for

Figure 14.1 The F-14 superiority fighter. (*Courtesy Grumman Corporation, Bethpage, N.Y.*)

the speed brake of the U.S. Air Force F-15 plane manufactured by McDonnell–Douglas. Composites comprise 1 percent of the structural weight of this aircraft.

Graphite-epoxy composites are used for the wing skins in the horizontal and vertical tail boxes, wing and tail control surfaces, speed brake, leading edge extension, and miscellaneous doors of the U.S. Navy/Marine Corps. F/A-18 plane manufactured by McDonnell–Douglas. Composites make up 10 percent of the structural weight of this aircraft. Graphite-epoxy composites are also used in the wing box skins and substructure, forward fuelage horizontal stabilizer, elevators, rudder, overwing fairing, allerons, and flaps of the Navy/Marine Corps. AV-88 plane manufactured by McDonnell–Douglas. Composites comprise about 25 percent of the structural weight of this aircraft.

The F-14 air superiority fighter, shown in Fig. 14.1, is designed and built by the Grumman Aerospace Corporation for the Navy. It made its first flight in 1970. Today it is the mainstay of the Navy's carrier force. Its capabilities extend from long-range search and missile kill to close-in fighter effectiveness. The F-14 achieves these diverse functions through the use of an advanced airframe design with a variable sweep wing and the long-range Phoenix missile system which together with the AWG-9 computer system can track 24 targets and simultaneously fire at 6 of them. It is powered by two Pratt & Whitney TF-30 turbofan engines which provide more than 20,000 lb of thrust each and enable the F-14 to fly faster than Mach 2.

The F-14 weighs over 50,000 lb at take off and can sustain 7.5 times the pull of gravity. It is designed to withstand the severe catapult and landing loads encountered for operations on an aircraft carrier. A maximum catapult load of 160,000 lb accelerates the aircraft to 130 mi/h in $2\frac{1}{2}$ s. During landing, the landing gear must withstand sink speeds as high as 26 ft/s.

The variable geometry wing which maximizes the performance of the F-14 in various flight conditions is not new. It dates back to the German P1101 which was followed by the Bell X-5, the Grumman XF-10F, and the F-111B. This technology was optimized on the F-14. The wing position, which is controlled by a computer, varies from full aft (68° sweep) for reduction of the drag at supersonic speeds, to full forward (20° sweep) for maximum lift during takeoff and landing. Attaching the movable wings to the fuselage presented a difficult engineering problem. This problem was solved by designing a 22-ft-long structure made of titanium to which the wings are attached. This structure is called the wing box and constitutes the center portion of the fuselage. Titanium has a higher strength-to-weight ratio than steel and excellent fatigue properties. It is, however, difficult to work with. The wings pivot on ball races in a pair of hinges at each end of the wing box.

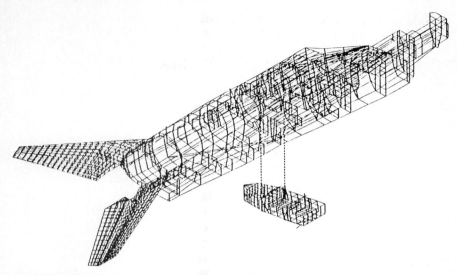

Figure 14.2 Finite-element model for the F-14A superiority fighter. (*Courtesy Grumman Corporation, Bethpage, N.Y.*)

An intermediate-complexity finite-element model used to analyze the F-14 structure is shown in Fig. 14.2. Only one-half of the airplane structure is modeled, and symmetric and antisymmetric analyses are performed and combined to obtain the stresses in the complete aircraft. This is standard procedure and allows the use of a smaller model resulting in savings

Figure 14.3 The F-14 superiority fighter during construction. (*Courtesy Grumman Corporation, Bethpage, N.Y.*)

in the cost of the analysis. The major components of the airplane, i.e., fuselage, wings, and tail, consist of thin metal (or composite) sheet wrapped around a framework of spars and ribs (see Fig. 14.3). Bending effects are neglected in the thin skin, and it is modeled with membrane elements that can resist only in plane tension compression or shear. For the framework supporting the skin, axial deformation elements and general space elements which resist axial and shearing forces and bending and torsional moments, as well as membrane elements, are used. The overall structural model, shown in Fig. 14.2 is used to assess the primary distribution of internal forces in the elements of the framework. In addition, detailed local models are used to study areas of concern such as the wing pivots.

The Modern Flexibility Method

15

The Equations of Equilibrium and the Equations of Compatibility for Structures Subjected to Equivalent Actions on Their Nodes

15.1 The Equations of Equilibrium for Structures Subjected to Equivalent Actions on Their Nodes

In this section, we consider structures subjected to equivalent actions on their nodes, and we establish the equations of equilibrium for their nodes. For statically determinate structures we can solve these equations to obtain the basic nodal actions of their elements and their reactions.

Consider the statically determinate truss shown in Fig. 15.1a. Referring to Fig. 15.1b and considering the equilibrium of the nodes of the truss, we have

Node 1

$$\sum \bar{F}_1 = 0 = R_1^{(1)} + F_1^{1k} + P_1^{(1)} + \frac{4F_1^{2k}}{5} \qquad (15.1a)$$

$$\sum \bar{F}_2 = 0 = R_2^{(1)} + P_2^{(1)} + \frac{3F_1^{2k}}{5} \qquad (15.1b)$$

Node 2

$$\sum \bar{F}_1 = 0 = R_1^{(2)} + F_1^{3k} \qquad (15.1c)$$

$$\sum \bar{F}_2 = 0 = R_2^{(2)} + P_2^{(2)} \qquad (15.1d)$$

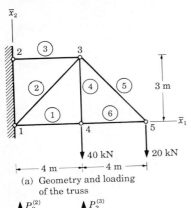

(a) Geometry and loading
of the truss

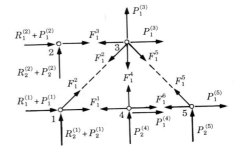

(b) Free-body diagrams of the
nodes of the truss

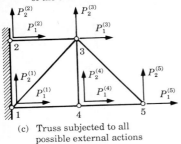

(c) Truss subjected to all
possible external actions

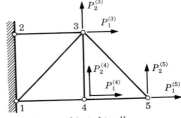

(d) Truss subjected to all
possible external actions
not directly absorbed by
its supports

Figure 15.1 Analysis of a statically determinate truss.

Node 3

$$\sum \bar{F}_1 = 0 = P_1^{(3)} - \frac{4F_1^{2k}}{5} - F_1^{3k} + \frac{4}{5}F_1^{5k} \qquad (15.1e)$$

$$\sum \bar{F}_2 = 0 = P_2^{(3)} - \frac{3F_1^{2k}}{5} - F_1^{4k} - \frac{3}{5}F_1^{5k} \qquad (15.1f)$$

Node 4

$$\sum \bar{F}_1 = 0 = P_1^{(4)} - F_1^{1k} + F_1^{6k} \qquad (15.1g)$$

$$\sum \bar{F}_2 = 0 = P_2^{(4)} + F_1^{4k} \qquad (15.1h)$$

Node 5

$$\sum \bar{F}_1 = 0 = P_1^{(5)} - \frac{4F_1^{5k}}{5} - F_1^{6k} \qquad (15.1i)$$

$$\sum \bar{F}_2 = 0 = P_2^{(5)} + \frac{3F_1^{5k}}{5} \qquad (15.1j)$$

These equations can be written in matrix form as

$$
\begin{array}{l}
\text{Node 1} \begin{cases} \sum \bar{F}_1 = 0 \\ \sum \bar{F}_2 = 0 \end{cases} \\[4pt]
\text{Node 2} \begin{cases} \sum \bar{F}_1 = 0 \\ \sum \bar{F}_2 = 0 \end{cases} \\[4pt]
\text{Node 3} \begin{cases} \sum \bar{F}_1 = 0 \\ \sum \bar{F}_2 = 0 \end{cases} \\[4pt]
\text{Node 4} \begin{cases} \sum \bar{F}_1 = 0 \\ \sum \bar{F}_2 = 0 \end{cases} \\[4pt]
\text{Node 5} \begin{cases} \sum \bar{F}_1 = 0 \\ \sum \bar{F}_2 = 0 \end{cases}
\end{array}
\quad
\left\{ \begin{array}{c}
P_1^{(1)} \\ P_2^{(1)} \\ \hline P_1^{(2)} \\ P_2^{(2)} \\ \hline P_1^{(3)} \\ P_2^{(3)} \\ \hline P_1^{(4)} \\ P_2^{(4)} \\ \hline P_1^{(5)} \\ P_2^{(5)}
\end{array} \right\}
$$

$$
\begin{bmatrix}
-1 & -0.8 & 0 & 0 & 0 & 0 & 1 & 0 & 0 & 0 \\
0 & -0.6 & 0 & 0 & 0 & 0 & 0 & 1 & 0 & 0 \\
0 & 0 & -1 & 0 & 0 & 0 & 0 & 0 & 1 & 0 \\
0 & 0 & 0 & 0 & 0 & 0 & 0 & 0 & 0 & 1 \\
0 & 0.8 & 1 & 0 & -0.8 & 0 & 0 & 0 & 0 & 0 \\
0 & 0.6 & 0 & 1 & 0.6 & 0 & 0 & 0 & 0 & 0 \\
1 & 0 & 0 & 0 & 0 & -1 & 0 & 0 & 0 & 0 \\
0 & 0 & 0 & -1 & 0 & 0 & 0 & 0 & 0 & 0 \\
0 & 0 & 0 & 0 & 0.8 & 1 & 0 & 0 & 0 & 0 \\
0 & 0 & 0 & 0 & -0.6 & 0 & 0 & 0 & 0 & 0
\end{bmatrix}
\left\{ \begin{array}{c}
F_1^{1k} \\ F_1^{2k} \\ F_1^{3k} \\ F_1^{4k} \\ F_1^{5k} \\ F_1^{6k} \\ \hline -R_1^{(1)} \\ -R_2^{(1)} \\ -R_1^{(2)} \\ -R_2^{(2)}
\end{array} \right\} = 0
$$

$$(15.2)$$

In general the equations of equilibrium for all the nodes of a structure can be written as

$$
\{\hat{P}^E\} - [\hat{B}] \left\{ \begin{array}{c} \{\hat{a}^E\} \\ -\{R\} \end{array} \right\} = 0 \tag{15.3}
$$

Notice that matrix $\{\hat{P}^E\}$ includes the global components of all possible external actions acting on the nodes of the truss, including those directly absorbed by its supports (see Fig. 15.1c). For a truss, the matrix of equivalent actions $\{\hat{P}^E\}$ is equal to the matrix of the given actions $\{\hat{P}^G\}$ acting on its nodes. The matrix $\{\hat{a}^E\}$ includes the basic nodal actions of all the elements of the truss. The matrix $\{R\}$ includes the independent components of the reactions of the truss. The matrix $[\hat{B}]$ is square and nonsingular. Its terms depend on the geometry of the truss. Consequently, Eqs. (15.3) can be solved to yield the matrices $\{\hat{a}^E\}$ and $\{R\}$. Thus, referring to Fig. 15.1a we

have

$$\begin{Bmatrix} \{\hat{a}^E\} \\ -\{R\} \end{Bmatrix} = [\hat{B}]^{-1}\{\hat{P}^E\}$$

$$= \begin{bmatrix} 0 & 0 & 0 & 0 & 0 & 0 & 1 & 0 & 1 & 1.333 \\ 0 & 0 & 0 & 0 & 0 & 1.667 & 0 & 1.667 & 0 & 1.667 \\ 0 & 0 & 0 & 0 & 1 & -1.333 & 0 & -1.333 & 0 & -2.667 \\ 0 & 0 & 0 & 0 & 0 & 0 & 0 & -1.000 & 0 & 0 \\ 0 & 0 & 0 & 0 & 0 & 0 & 0 & 0 & 0 & -1.667 \\ 0 & 0 & 0 & 0 & 0 & 0 & 0 & 0 & 1 & 1.333 \\ 1 & 0 & 0 & 0 & 0 & 1.333 & 1 & 1.333 & 1 & 2.667 \\ 0 & 1 & 0 & 0 & 1 & 1.000 & 0 & 1.000 & 0 & 1.000 \\ 0 & 0 & 1 & 0 & 0 & -1.333 & 0 & -1.333 & 0 & -2.667 \\ 0 & 0 & 0 & 1 & 0 & 0 & 0 & 0 & 0 & 0 \end{bmatrix} \begin{Bmatrix} 0 \\ 0 \\ \hline 0 \\ 0 \\ 0 \\ 0 \\ \hline 0 \\ -40 \\ \hline 0 \\ -20 \end{Bmatrix}$$

$$= \begin{Bmatrix} -26.66 \\ -100.00 \\ 106.64 \\ 40.00 \\ 33.34 \\ -26.66 \\ \hline -106.64 \\ -60.00 \\ 106.64 \\ 0.00 \end{Bmatrix} \text{ kN} \quad (15.4)$$

The six basic nodal actions of the six elements of the truss can also be obtained from the six equations of equilibrium [(15.1e) to (15.1j)]. These equations can be written in matrix form as

$$\begin{matrix} \text{Node 3} \begin{cases} \sum \bar{F}_1 = 0 \\ \sum \bar{F}_2 = 0 \end{cases} \\ \text{Node 4} \begin{cases} \sum \bar{F}_1 = 0 \\ \sum \bar{F}_2 = 0 \end{cases} \\ \text{Node 5} \begin{cases} \sum \bar{F}_1 = 0 \\ \sum \bar{F}_2 = 0 \end{cases} \end{matrix} \begin{Bmatrix} P_1^{(3)} \\ P_2^{(3)} \\ \hline P_1^{(4)} \\ P_2^{(4)} \\ \hline P_1^{(5)} \\ P_2^{(5)} \end{Bmatrix} - \begin{bmatrix} 0 & \frac{4}{5} & 1 & 0 & -\frac{4}{5} & 0 \\ 0 & \frac{3}{5} & 0 & 1 & \frac{3}{5} & 0 \\ 1 & 0 & 0 & 0 & 0 & -1 \\ 0 & 0 & 0 & -1 & 0 & 0 \\ 0 & 0 & 0 & 0 & \frac{4}{5} & 1 \\ 0 & 0 & 0 & 0 & -\frac{3}{5} & 0 \end{bmatrix} \begin{Bmatrix} F_1^{1k} \\ F_1^{2k} \\ F_1^{3k} \\ F_1^{4k} \\ F_1^{5k} \\ F_1^{6k} \end{Bmatrix} = 0$$

$$(15.5)$$

Thus in general the equations of equilibrium for the nodes of a structure can also be written as

$$\{P^{EF}\} - [B]\{\hat{a}^E\} = 0 \tag{15.6}$$

The matrix $\{P^{EF}\}$ includes the components of external actions acting on the nodes of the structure which are not directly absorbed by its supports (see Fig. 15.1d). $[B]$ is a square, nonsingular matrix. It can be obtained from the matrix $[\hat{B}]$, defined by Eqs. (15.3), by canceling its rows and columns which correspond to rows and columns of relation (15.3) involving reactions of the structure. For example, the matrix $[B]$ for the truss of Fig. 15.1a is obtained by canceling rows 1 to 4 and columns 7 to 10 of the matrix $[\hat{B}]$ in Eqs. (15.2). For statically determinate structures, Eqs. (15.6) can be solved to yield the matrix $\{\hat{a}^E\}$. Thus, referring to Fig. 15.1a, from relations (15.5) we have

$$\{\hat{a}^E\} = [B]^{-1}\{P^{EF}\}$$

$$= \begin{bmatrix} 0 & 0 & 1 & 0 & 1 & 1.333 \\ 0 & 1.667 & 0 & 1.667 & 0 & 1.667 \\ 1.0 & -1.333 & 0 & -1.333 & 0 & -2.667 \\ 0 & 0 & 0 & -1.000 & 0 & 0 \\ 0 & 0 & 0 & 0 & 0 & -1.667 \\ 0 & 0 & 0 & 0 & 1 & 1.333 \end{bmatrix} \begin{Bmatrix} 0 \\ 0 \\ 0 \\ -40 \\ 0 \\ -20 \end{Bmatrix} = \begin{Bmatrix} -26.66 \\ -100.00 \\ 106.64 \\ 40.00 \\ 33.34 \\ -26.66 \end{Bmatrix} \text{ kN}$$

$$\tag{15.7}$$

For any structure (statically determinate or indeterminate), Eqs. (15.3) can be established by considering the equilibrium of its nodes. That is, each row of Eqs. (15.3) represents an equilibrium equation for a node. It expresses either the requirement that the sum of the components along the $\bar{x}_1$, $\bar{x}_2$, or $\bar{x}_3$ axis of the forces acting on a node is equal to zero, or that the sum of the components along the $\bar{x}_1$, $\bar{x}_2$, or $\bar{x}_3$ axis of the moments acting on a node vanishes.

For any structure which does not have skew supports the matrix $[B]$ can be obtained from its matrix $[\hat{B}]$ by eliminating its rows and columns that correspond to rows and columns of Eqs. (15.3) which involve the reactions of the structure. If a structure has skew supports, in order to obtain the matrix $[B]$ from $[\hat{B}]$ Eqs. (15.3) must be modified prior to eliminating the appropriate rows and columns of the matrix $[\hat{B}]$ (see Sec. 15.3).

In general, matrices $[\hat{B}]$ and $[B]$ of statically determinate structures, which do not form a mechanism, are square nonsingular matrices. Thus, for such structures, Eqs. (15.3) and (15.6) can be solved to yield

$$\left\{ \begin{matrix} \{\hat{a}^E\} \\ -\{R\} \end{matrix} \right\} = [\hat{B}]^{-1}\{\hat{P}^E\} = [\hat{b}]\{\hat{P}^E\} \tag{15.8}$$

and

$$\{\hat{a}^E\} = [B]^{-1}\{P^{EF}\} = [b]\{P^{EF}\} \tag{15.9}$$

where

$$[\hat{b}] = [\hat{B}]^{-1} \tag{15.10}$$

and

$$[b] = [B]^{-1} \tag{15.11}$$

The matrix $[\hat{b}]$ is referred to as the action transformation matrix. It is apparent that the elements of the nth column of the matrix $[\hat{b}]$ represent the basic nodal actions of all the elements of the structure when it is subjected to a unit value of the external action located in the nth row of the matrix $\{\hat{P}^E\}$.

In a computer program for analyzing statically determinate structures the matrix $[\hat{B}]$ is used to calculate the basic nodal actions of their elements and their reactions [see Eqs. (15.8)]. Moreover the matrix $[B]$ is used to calculate the components of displacements of their nodes (see Chap. 16).

In general, matrices $[\hat{B}]$ and $[B]$ for statically indeterminate structures are not square matrices. Consequently, Eqs. (15.3) or (15.6) cannot be solved to yield the internal actions in the elements of statically indeterminate structures. That is, as anticipated, the internal actions of statically indeterminate structures cannot be established by considering only the equilibrium for their nodes. However, the computation of the matrix $[\hat{B}]$ of statically indeterminate structures constitutes the first step of their analysis using the modern flexibility method. As becomes apparent in Chap. 17, it is not necessary to compute the matrix $[B]$ for statically indeterminate structures when analyzing them by the modern flexibility method.

15.2 Systematic Procedure for Obtaining the Matrices $[\hat{B}]$ and $[B]$ for Framed Structures

In this section, we describe a systematic procedure for obtaining the matrix $[\hat{B}]$ for statically determinate or indeterminate, framed structures subjected to equivalent actions on their nodes, as well as the matrix $[B]$ for statically determinate, framed structures. This procedure is suitable for programming on an electronic computer. In this procedure, we adhere to the following steps.

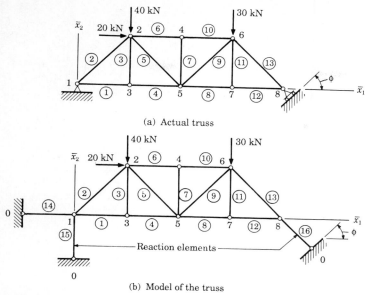

(a) Actual truss

(b) Model of the truss

Figure 15.2 Planar truss and its model.

Step 1. We consider a model for the structure obtained from the actual structure by replacing its supports with elements referred to as *reaction elements*. Each reaction element is connected with its end j to the supporting body (ground) and with its end k to a supported node of the structure. A roller support of a planar structure is replaced by an axial deformation reaction element whose axis is normal to the plane of rolling (see Figs. 15.2 and 15.3). A hinged support of a planar structure is replaced by two, mutually perpendicular, axial deformation reaction elements. The direction and sense of the x_1 axis of each of these reaction elements is the same as that of a global axis of the structure (see Figs. 15.2 and 15.3). A fixed support of a planar structure is replaced by a general planar reaction element whose local axes are parallel to the global axes of the structure (see Fig. 15.3). A ball support of a space structure is replaced by an axial deformation reaction element whose axis is normal to the plane of rolling. A ball-and-socket support of a space structure which can translate only in a specified direction is replaced by two, mutually perpendicular, axial deformation reaction elements located in the plane normal to the plane of rolling which contains the direction of rolling. A nontranslating ball-and-socket support of a space structure is replaced by three, mutually perpendicular axial deformation elements. The local axes of each of these elements are parallel to the global axis of the structure (see Fig. 15.4b). A fixed support of a space structure is replaced by a general space element whose local axes are

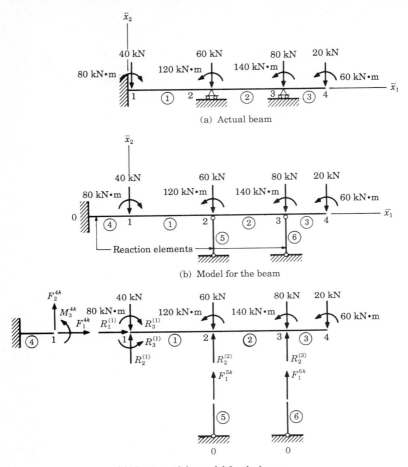

Figure 15.3 Planar beam and its model.

parallel to the global axes of the structure (see Fig. 15.4*b*). The model is subjected to the equivalent actions acting on the nodes of the structure.

Referring to Fig. 15.3*c*, it is apparent that each component of the basic nodal action of a reaction element of the model is equal and opposite to the corresponding component of a reaction of the actual structure.

Step 2. We form the transformation matrix $[\Lambda]$ for each element of the model which transforms its local matrix of nodal actions $\{A^E\}$ to global $\{\bar{A}^E\}$. Depending on the type of structure $[\Lambda]$ is one of the

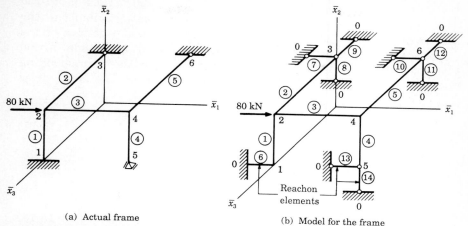

(a) Actual frame

(b) Model for the frame

Figure 15.4 Space frame and its model.

matrices $[\Lambda_{PT}]$, $[\Lambda_{PF}]$, $[\Lambda_{ST}]$, $[\Lambda_{SF}]$, or $[\Lambda_{GG}]$ given by relations (6.42), (6.48), (6.51), (6.57), or (6.63), respectively. Moreover, we form the transformation matrix $[T]$ for each element of the model which transforms its matrix of basic nodal actions $\{a^E\}$ to its local matrix of nodal actions $\{A^E\}$. That is,

$$\{A^E\} = [T]\{a^E\} \tag{15.12}$$

Depending on the type of structure $[T]$ is given by relations (5.10) to (5.14). Furthermore, we compute the matrix $[H]$ for each element which transforms matrix $\{a^E\}$ to $\{\bar{A}^E\}$. That is,

$$\{\bar{A}^E\} = \left\{\frac{\{\bar{A}^{Ej}\}}{\{\bar{A}^{Ek}\}}\right\} = \left[\frac{[H^j]}{[H^k]}\right]\{a^E\} = [H]\{a^E\} \tag{15.13}$$

Referring to relations (6.65) and (15.12), we have

$$\{\bar{A}^E\} = [\Lambda]^T[A^E] = [\Lambda]^T[T]\{a^E\} = [H]\{a^E\} \tag{15.14}$$

Thus

$$[H] = [\Lambda]^T[T] \tag{15.15}$$

Step 3. We write the equations of equilibrium for the unsupported nodes of the model of the structure described in step 1. That is,

$$\{P^{EFM}\} = [B^M]\{\hat{a}^{EM}\} \tag{15.16}$$

The matrix $\{P^{EFM}\}$ includes the global components of all possible external (equivalent) actions acting at the unsupported nodes of the model. Referring to Figs. 15.2 to 15.4, it is apparent that the matrix $\{P^{EFM}\}$ of the model is identical to the matrix $\{\hat{P}^E\}$ of the structure.

The matrix $\{\hat{a}^{EM}\}$ is the matrix of the basic nodal actions of all the elements of the model. It includes the basic nodal actions of all the elements of the actual structure and the basic nodal actions of the reaction elements. As mentioned previously, each basic nodal action of a reaction element is equal and opposite to the corresponding component of a reaction of the actual structure. Therefore, it is apparent that

$$\{\hat{a}^{EM}\} = \left\{ \begin{array}{c} \{\hat{a}^E\} \\ -\{R\} \end{array} \right\} \tag{15.17}$$

where for a structure with NE elements

$$\{\hat{a}^E\} = \left\{ \begin{array}{c} \{a^{E1}\} \\ \{a^{E2}\} \\ \vdots \\ \{a^{ENE}\} \end{array} \right\} \tag{15.18}$$

$\{a^{Ee}\}$ is the matrix of basic nodal actions of element e.

On the basis of the foregoing discussion the matrix $[B^M]$ of the model is identical to the matrix $[\hat{B}]$ of the actual structure. The matrix $[B^M]$ of the model can be constructed by placing the matrices $[H^j]$ and $[H^k]$ of each of its elements in the appropriate locations. This is accomplished by noting that the product of the matrices $[H^j]$ and $\{a^E\}$ for an element represents the global components of the force and the moment acting at the end j of this element. Moreover the product of the matrices $[H^k]$ and $\{a^E\}$ of an element represents the global components of the force and moment acting at the end k of this element.

Step 4. For statically determinate structures we establish the matrix $[B]$ from the matrix $[\hat{B}] = [B^M]$. The matrix $[B]$ for a structure which does not have a skew support can be obtained by eliminating the rows and columns of its matrix $[\hat{B}]$ corresponding to the rows and columns of Eqs. (15.3) which involve the reactions of the structure. If the structure has a skew support, in order to obtain its matrix $[B]$ from its matrix $[\hat{B}]$, the latter must be modified prior to eliminating its rows and columns (see Sec. 15.3).

We illustrate by the following three examples the procedure described in this section for establishing the matrix $[\hat{B}]$ for a structure as well as the matrix $[B]$ for a statically determinate structure.

Example 1. The matrices $[\hat{B}]$, $[B]$, and $\{\hat{a}^E\}$ of a statically determinate, simply supported planar truss are established.

Example 2. The matrices $[\hat{B}]$, $[B]$, and $\{\hat{a}^E\}$ of a statically determinate, simply supported planar beam with an overhang are established.

Example 3. The matrix $[\hat{B}]$ for a statically indeterminate space frame is established.

Example 1. Consider the planar statically determinate truss shown in Fig. a. Following the procedure described in this section compute matrices $[\hat{B}]$, $[B]$, and $\{\hat{a}^E\}$ for this truss, as well as the basic nodal actions of its elements.

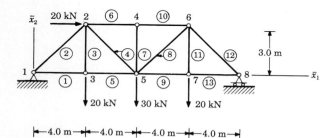

Figure a Geometry and loading of the truss.

---

solution
STEP 1. We consider the model of the truss shown in Fig. b.

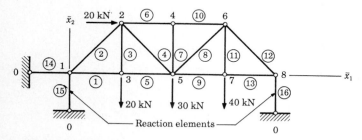

Figure b Model of the truss of Fig. a.

STEP 2. We compute the matrix $[H]$ for each element of the model shown in Fig. b. Referring to relation (5.10), for an element of a planar truss we have

$$\{T\} = \begin{Bmatrix} -1 \\ 1 \end{Bmatrix} \qquad\qquad \text{(a)}$$

Substituting relations (6.42) and (a) into (15.15) we get

$$\{H\} = [\Lambda_{PT}]^T \{T\} = \begin{Bmatrix} -\cos \phi_{11} \\ -\sin \phi_{11} \\ \hline \cos \phi_{11} \\ \sin \phi_{11} \end{Bmatrix} \qquad (b)$$

Noting that the number of the node adjacent to the end j of an element is smaller than the number of the node adjacent to its end k, we have

For elements 1, 5, 6, 9, 10, 13, and 14 ($\phi_{11} = 0$)

$$\{H\} = \begin{Bmatrix} -1 \\ 0 \\ \hline 1 \\ 0 \end{Bmatrix} \qquad (c)$$

For elements 2 and 8 ($\cos \phi_{11} = 0.8$, $\sin \phi_{11} = 0.6$)

$$\{H\} = \begin{Bmatrix} -0.8 \\ -0.6 \\ \hline 0.8 \\ 0.6 \end{Bmatrix} \qquad (d)$$

For elements 3, 7, and 11 ($\phi_{11} = -90°$)

$$\{H\} = \begin{Bmatrix} 0 \\ 1 \\ \hline 0 \\ -1 \end{Bmatrix} \qquad (e)$$

For elements 4 and 12 ($\cos \phi_{11} = 0.8$, $\sin \phi_{11} = -0.6$)

$$\{H\} = \begin{Bmatrix} -0.8 \\ 0.6 \\ \hline 0.8 \\ -0.6 \end{Bmatrix} \qquad (f)$$

For elements 15 and 16 ($\phi_{11} = 90°$)

$$\{H\} = \left\{ \begin{array}{c} 0 \\ -1 \\ \hline 0 \\ 1 \end{array} \right\} \tag{g}$$

STEP 3.   We form the matrix $[B^M]$ for the model of Fig. b using the matrices $[H]$ for its elements. That is, we write, consecutively, the equations of equilibrium for all the unsupported nodes of the model. For the model of Fig. b there are 16 equations of equilibrium (two for each unsupported node). As can be seen from Eqs. (h) (see p. 494), in order to form the matrix $[B^M]$ in a systematic manner, we assign one row to each equation of equilibrium, and one column to the internal force of each element of the model. Equations (h) can be rewritten as

$$\{P^{EFM}\} - [B^M]\{\hat{a}^{EM}\} \tag{i}$$

where the matrix $\{P^{EFM}\}$ includes the global components of all possible external actions acting on the unsupported nodes of the model. It is apparent that the matrix $\{P^{EFM}\}$ of the model of Fig. b is identical to the matrix $\{\hat{P}^E\}$ of the actual truss of Fig. a. The matrix $\{\hat{a}^{EM}\}$ includes the basic nodal actions of all the elements of the model, and as discussed previously, it is related to the matrix of basic nodal actions $\{\hat{a}^E\}$ of all the elements of the actual structure by relation (15.17). The matrix $[B^M]$ for the model is identical to the matrix $[\hat{B}]$ for the actual structure.

From Fig. b it is apparent that elements 1 and 2 of the model are connected to node 1 with their end j, while elements 14 and 15 are connected to node 1 with their end k. As discussed previously, the product of the matrices $[H^j]$ and $\{a^E\}$ for an element represents the global components of the action acting at the end j of this element. Thus, as shown in Eqs. (h), the upper half of matrix (c) is placed in rows 1 and 2 and column 1 of matrix $[B^M]$ of the model, the upper half of matrix (d) is placed in rows 1 and 2 and column 2, the lower half of matrix (c) is placed in rows 1 and 2 and column 14, and the lower half of matrix (g) is placed in rows 1 and 2 and column 15 of matrix $[B^M]$. Since other elements of the truss are not connected to node 1, zeros are placed in the remaining slots of rows 1 and 2 of the matrix $[B^M]$.

From Fig. b it is apparent that element 2 of the model is connected to node 2 with its end k, while elements 3, 4, and 6 are connected to node 2 with their end j. Thus, as shown in Eqs. (h), the bottom half of matrix (d) is placed into rows 3 and 4 and column 2 of the matrix $[B^M]$, the upper half of matrix (e) is placed into rows 3 and 4 and column 3, the upper half of matrix (f) is placed into rows 3 and 4 and column 4, and the upper half of matrix (c) is placed into rows 3 and 4 and column 6. Since other elements of the model are not connected to node 2, zeros are placed in the remaining slots of rows 3 and 4 of the matrix $[B^M]$. We proceed in a similar fashion to fill the remaining rows of the matrix $[B^M]$ in Eqs. (h).

STEP 4.   We establish the matrix $[B]$ for the truss from its matrix $[\hat{B}] = [B^M]$ by removing its columns and rows corresponding to the columns and rows of Eqs. (h) involving the basic nodal actions of the reaction

Element

Node	Equation		1	2	3	4	5	6	7	8	9	10	11	12	13	14	15	16	
1 $\big\{$ $\sum \bar{F}_1 = 0$	1	0	-1	-0.8	0	0	0	0	0	0	0	0	0	0	0	0	0	0	F_1^{1k}
$\quad \sum \bar{F}_2 = 0$	2	0	0	-0.6	0	0	0	0	0	0	0	0	0	0	0	0	0	0	F_1^{2k}
2 $\big\{$ $\sum \bar{F}_1 = 0$	3	20	0	0.8	0	-0.8	0	-1	0	0	0	0	0	0	0	0	0	0	F_1^{3k}
$\quad \sum \bar{F}_2 = 0$	4	0	0	0.6	1	0.6	0	0	0	0	0	0	0	0	0	0	0	0	F_1^{4k}
3 $\big\{$ $\sum \bar{F}_1 = 0$	5	0	1	0	0	0	-1	0	0	0	0	0	0	0	0	0	0	0	F_1^{5k}
$\quad \sum \bar{F}_2 = 0$	6	-20	0	0	-1	0	0	0	0	0	-1	0	0	0	0	0	0	0	F_1^{6k}
4 $\big\{$ $\sum \bar{F}_1 = 0$	7	0	0	0	0	0	0	1	1	0	0	0	0	0	0	0	0	0	F_1^{7k}
$\quad \sum \bar{F}_2 = 0$	8	0	0	0	0	0	1	0	0	-0.8	0	0	0	0	0	0	0	0	F_1^{8k}
5 $\big\{$ $\sum \bar{F}_1 = 0$	9	0	0	0	0	0.8	0	0	-1	-0.6	-1	0	0	0	0	0	0	0	F_1^{9k}
$\quad \sum \bar{F}_2 = 0$	10	-30	0	0	0	-0.6	0	0	0	0.8	0	1	0	0	0	0	0	0	F_1^{10k}
6 $\big\{$ $\sum \bar{F}_1 = 0$	11	0	0	0	0	0	0	0	0	0.6	0	0	-1	-0.8	0	0	0	0	F_1^{11k}
$\quad \sum \bar{F}_2 = 0$	12	0	0	0	0	0	0	0	0	0	0	0	0	0.6	0	0	0	0	F_1^{12k}
7 $\big\{$ $\sum \bar{F}_1 = 0$	13	0	0	0	0	0	0	0	0	0	0	1	0	0	-1	0	0	0	F_1^{13k}
$\quad \sum \bar{F}_2 = 0$	14	-20	0	0	0	0	0	0	0	0	0	0	-1	0	0	1	0	0	F_1^{14k}
8 $\big\{$ $\sum \bar{F}_1 = 0$	15	0	0	0	0	0	0	0	0	0	0	0	0	0.8	0	0	0.8	0	F_2^{15k}
$\quad \sum \bar{F}_2 = 0$	16	0	0	0	0	0	0	0	0	0	0	0	0	-0.6	0	0	0	-0.6	F_2^{16k}

$$= 0 \qquad \text{(h)}$$

elements of the model (columns 14 to 16, and rows 1, 2, and 16). Thus,

$$
[B] =
\begin{bmatrix}
0 & 0.8 & 0 & -0.8 & 0 & -1 & 0 & 0 & 0 & 0 & 0 & 0 & 0 \\
0 & 0.6 & 1 & 0.6 & 0 & 0 & 0 & 0 & 0 & 0 & 0 & 0 & 0 \\
1 & 0 & 0 & 0 & -1 & 0 & 0 & 0 & 0 & 0 & 0 & 0 & 0 \\
0 & 0 & -1 & 0 & 0 & 0 & 0 & 0 & 0 & 0 & 0 & 0 & 0 \\
0 & 0 & 0 & 0 & 0 & 1 & 0 & 0 & 0 & -1 & 0 & 0 & 0 \\
0 & 0 & 0 & 0 & 0 & 0 & 1 & 0 & 0 & 0 & 0 & 0 & 0 \\
0 & 0 & 0 & 0.8 & 1 & 0 & 0 & -0.8 & -1 & 0 & 0 & 0 & 0 \\
0 & 0 & 0 & -0.6 & 0 & 0 & -1 & -0.6 & 0 & 0 & 0 & 0 & 0 \\
0 & 0 & 0 & 0 & 0 & 0 & 0 & 0.8 & 0 & 1 & 0 & -0.8 & 0 \\
0 & 0 & 0 & 0 & 0 & 0 & 0 & 0.6 & 0 & 0 & 1 & 0.6 & 0 \\
0 & 0 & 0 & 0 & 0 & 0 & 0 & 0 & 1 & 0 & 0 & 0 & -1 \\
0 & 0 & 0 & 0 & 0 & 0 & 0 & 0 & 0 & 0 & -1 & 0 & 0 \\
0 & 0 & 0 & 0 & 0 & 0 & 0 & 0 & 0 & 0 & 0 & 0.8 & 1
\end{bmatrix}
$$

$$(j)$$

STEP 5. We compute the basic nodal actions of the elements of the truss. In order to accomplish this we form its matrix $\{P^{EF}\}$ wherein we include all possible external (equivalent) actions not absorbed directly by the supports of the truss. That is, referring to Figs. a and c we have

$$
\{P^{EF}\} =
\begin{Bmatrix}
P_1^{(2)} \\
P_2^{(2)} \\
\hline
P_1^{(3)} \\
P_2^{(3)} \\
\hline
P_1^{(4)} \\
P_2^{(4)} \\
\hline
P_1^{(5)} \\
P_2^{(5)} \\
\hline
P_1^{(6)} \\
P_2^{(6)} \\
\hline
P_1^{(7)} \\
P_2^{(7)} \\
\hline
P_1^{(8)}
\end{Bmatrix}
=
\begin{Bmatrix}
20 \\
0 \\
\hline
0 \\
-20 \\
\hline
0 \\
0 \\
\hline
0 \\
-30 \\
\hline
0 \\
0 \\
\hline
0 \\
-20 \\
\hline
0
\end{Bmatrix}
$$

$$(k)$$

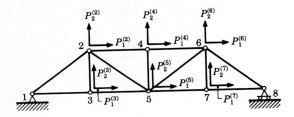

Figure c Possible external actions not directly absorbed by the supports of the truss.

Substituting matrices (j) and (k) into relation (15.9), we get

$$\{\hat{a}^E\} = \begin{Bmatrix} F_1^{1k} \\ F_1^{2k} \\ F_1^{3k} \\ F_1^{4k} \\ F_1^{5k} \\ F_1^{6k} \\ F_1^{7k} \\ F_1^{8k} \\ F_1^{9k} \\ F_1^{10k} \\ F_1^{11k} \\ F_1^{12k} \\ F_1^{13k} \end{Bmatrix} = [B]^{-1}\{P^{EF}\} = \begin{Bmatrix} 61.67 \\ -52.08 \\ 20 \\ 18.75 \\ 61.67 \\ -76.67 \\ 0 \\ -31.25 \\ 51.67 \\ -76.67 \\ 20 \\ -64.59 \\ 51.67 \end{Bmatrix} \tag{1}$$

Example 2. Consider the beam of constant cross section shown in Fig. a. Following the procedure described in this section, derive the matrices $[\hat{B}]$ and $[B]$ for the beam as well as the matrix of basic nodal actions of its elements when $L = 4m$.

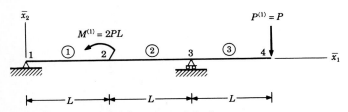

Figure a Geometry and loading of the beam.

solution

STEP 1.   We consider the model for the beam shown in Fig. b. In order to have all the external actions acting on nodes, we choose points 1 to 4 as nodes. Notice that we could have chosen only points 1, 3, and 4 as nodes. In this case, we would have to analyze the restrained structure and the structure subjected to the equivalent actions acting on its nodes (see Sec. 7.1).

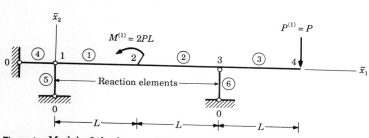

Figure b   Model of the beam of Fig. a.

STEP 2.   We compute the matrix $[H]$ for each element of the model of Fig. b. Referring to relation (5.11) for a general planar element we have

$$
[T] = \begin{bmatrix}
-1 & 0 & 0 \\
0 & -1 & 0 \\
0 & -L & -1 \\
\hdashline
1 & 0 & 0 \\
0 & 1 & 0 \\
0 & 0 & 1
\end{bmatrix}
\tag{a}
$$

Substituting relations (6.48) and (a) into (15.15), the matrix $[H]$ for elements 1, 2, and 3 ($\phi_{11} = 0$) of the model of Fig. b is equal to

$$
[H^1] = [H^2] = [H^3] = [\Lambda_{PF}]^T [T] = \begin{bmatrix}
-1 & 0 & 0 \\
0 & -1 & 0 \\
0 & -L & -1 \\
\hdashline
1 & 0 & 0 \\
0 & 1 & 0 \\
0 & 0 & 1
\end{bmatrix}
\tag{b}
$$

Moreover, referring to Eq. (5.12) for an axial deformation element of a planar beam or frame we have

$$[T] = \left\{ \begin{array}{c} -1 \\ 0 \\ 0 \\ \hline 1 \\ 0 \\ 0 \end{array} \right\} \tag{c}$$

Substituting relations (6.48) and (c) into (15.15) we obtain

$$[H] = \left\{ \begin{array}{c} -\cos\phi_{11} \\ -\sin\phi_{11} \\ 0 \\ \hline \cos\phi_{11} \\ \sin\phi_{11} \\ 0 \end{array} \right\} \tag{d}$$

Thus, the matrix $[H]$ for element 4 of the model of Fig. b ($\phi_{11} = 0$) is equal to

$$\{H^4\} = \left\{ \begin{array}{c} -1 \\ 0 \\ 0 \\ \hline 1 \\ 0 \\ 0 \end{array} \right\} \tag{e}$$

while the matrix $[H]$ for elements 5 and 6 of the model of Fig. b ($\phi_{11} = 90°$) is equal to

$$\{H^5\} = \{H^6\} = \left\{ \begin{array}{c} 0 \\ -1 \\ 0 \\ \hline 0 \\ 1 \\ 0 \end{array} \right\} \tag{f}$$

STEP 3.   We form the matrix $[B^M]$ of the model of Fig. b using the matrices $[H]$ of its elements. That is, we write consecutively the equations of equilibrium for all the unsupported nodes of the model of Fig. b. Referring to Fig. b, it is apparent that there are twelve equations of equilibrium for the nodes of the model. That is,

Node		Equation		
1	$\sum \bar{F}_1 = 0$	1	0	
	$\sum \bar{F}_2 = 0$	2	0	
	$\sum \bar{M}_3 = 0$	3	0	
2	$\sum \bar{F}_1 = 0$	4	0	
	$\sum \bar{F}_2 = 0$	5	0	
	$\sum \bar{M}_3 = 0$	6	2PL	
3	$\sum \bar{F}_1 = 0$	7	0	
	$\sum \bar{F}_2 = 0$	8	0	
	$\sum \bar{M}_3 = 0$	9	0	
4	$\sum \bar{F}_1 = 0$	10	0	
	$\sum \bar{F}_2 = 0$	11	$-P$	
	$\sum \bar{M}_3 = 0$	12	0	

Element

$$-\begin{bmatrix}
-1 & 0 & 0 & 0 & 0 & 0 & 0 & 0 & 0 & -1 & 0 & 0 \\
0 & -1 & 0 & 0 & 0 & 0 & 0 & 0 & 0 & 0 & -1 & 0 \\
0 & -L & 1 & 0 & 0 & 0 & 0 & 0 & 0 & 0 & 0 & 0 \\
1 & 0 & 0 & -1 & 0 & 0 & 0 & 0 & 0 & 0 & 0 & 0 \\
0 & 1 & 0 & 0 & -1 & 0 & 0 & 0 & 0 & 0 & 0 & 0 \\
0 & 0 & 1 & 0 & -L & -1 & 0 & 0 & 0 & 0 & 0 & 0 \\
0 & 0 & 0 & 1 & 0 & 0 & -1 & 0 & 0 & 0 & 0 & 0 \\
0 & 0 & 0 & 0 & 1 & 0 & 0 & -1 & 0 & 0 & 0 & -1 \\
0 & 0 & 0 & 0 & 0 & 1 & 0 & -L & -1 & 0 & 0 & 0 \\
0 & 0 & 0 & 0 & 0 & 0 & 1 & 0 & 0 & 0 & 0 & 0 \\
0 & 0 & 0 & 0 & 0 & 0 & 0 & 1 & 0 & 0 & 0 & 0 \\
0 & 0 & 0 & 0 & 0 & 0 & 0 & 0 & 1 & 0 & 0 & 0
\end{bmatrix}
\begin{Bmatrix}
F_1^{1k} \\ F_2^{1k} \\ M_3^{1k} \\ F_1^{2k} \\ F_2^{2k} \\ M_3^{2k} \\ F_1^{3k} \\ F_2^{3k} \\ M_3^{3k} \\ F_1^{4k} \\ F_1^{5k} \\ F_1^{6k}
\end{Bmatrix} = 0$$

(with column groups: 1 (cols 1–3), 2 (cols 4–6), 3 (cols 7–9), 4, 5, 6)

(g)

or
$$\{P^{EFM}\} - [B^M]\{a^{EM}\} = 0 \qquad \text{(h)}$$

As evident from relations (g), we assign three rows for each node (one row to each equation of equilibrium), three columns for each of the elements 1 to 3, and one column for each of the elements 4 to 6 of the model of Fig. b. In Eqs. (h) the matrix $\{P^{EFM}\}$ includes the global components of all possible external actions acting at the unsupported nodes of the model. It is apparent, that the matrix $\{P^{EFM}\}$ of the model of Fig. b is identical to the matrix $\{\hat{P}^E\}$ of the actual beam of Fig. a. The matrix $\{\hat{a}^{EM}\}$ includes the basic nodal actions of all the elements of the model, and as discussed previously, it is related to the matrix $\{\hat{a}^E\}$ of all the elements of the actual structure by relation (15.17). The matrix $[B^M]$ of the model is identical to the matrix $[\hat{B}]$ of the actual structure.

From Fig. b, it can be seen that element 1 is connected to node 1 with its end j, while elements 4 and 5 are connected to node 1 with their end k. As was discussed previously, the product of the matrices $[H^q]$ ($q = j$ or k), and $\{a^E\}$ of an element represents the global components of the actions acting at the end q of this element. Consequently, in Eqs. (g) matrix $[H^{1j}]$ is placed in rows 1 to 3 and columns 1 to 3 of matrix $[B^M]$ of the model, matrix $[H^{4k}]$ is placed in rows 1 to 3 and column 10, and matrix $[H^{5k}]$ is placed in rows 1 to 3 and column 11. Since other elements are not connected to node 1, zeros are placed in the remaining slots of rows 1 to 3 of the matrix $[B^M]$ of the model.

From Fig. b, it can be seen that element 1 of the model is connected to node 2 with its end k, while element 2 is connected to node 2 with its end j. Thus, as shown in Eqs. (g), matrix $[H^{1k}]$ is placed in rows 4 to 6 and columns 1 to 3, while matrix $[H^{2j}]$ is placed in rows 4 to 6 and columns 4 to 6 of the matrix $[B^M]$. Since other elements are not connected to node 2, zeros are placed in the remaining slots of rows 4 to 6 of the matrix $[B^M]$.

From Fig. b, it can be seen that elements 2 and 6 are connected to node 3 with their end k, while element 3 is connected to node 3 with its end j. Thus, as shown in Eqs. (g), matrix $[H^{2k}]$ is placed in rows 7 to 9 and columns 4 to 6, matrix $[H^{6k}]$ is placed in rows 7 to 9 and column 12 and matrix $[H^{3j}]$ is placed in rows 7 to 9 and columns 7 to 9 of the matrix $[B^M]$. Since other elements are not connected to node 3, zeros are placed in the remaining slots of rows 7 to 9 of the matrix $[B^M]$.

From Fig. b it can be seen that element 3 is connected to node 4 with its end k. Thus, matrix $[H^{3k}]$ is placed in rows 10 to 12 and columns 7 to 9 of matrix $[B^M]$. Since other elements are not connected to node 4, zeros are placed in the remaining slots of rows 10 to 12 of the matrix $[B^M]$.

Notice that for beams, Eqs. (g) can be further simplified by omitting the axial components of the forces from the matrices $\{P^{EFM}\}$ and $\{\hat{a}^{EM}\}$, and annulling rows and columns 1, 4, 7, and 10 of the matrix $[B^M]$.

STEP 4.   We establish the matrix $[B]$ of the beam from its matrix $[\hat{B}] = [B^M]$ by removing its rows and columns corresponding to rows and columns of Eqs. (g) involving the basic nodal actions of the reaction elements of the model (columns 10 to 12 and rows 1, 2, and 8). Thus,

$$[B] = \begin{bmatrix} 0 & -L & -1 & 0 & 0 & 0 & 0 & 0 & 0 \\ 1 & 0 & 0 & -1 & 0 & 0 & 0 & 0 & 0 \\ 0 & 1 & 0 & 0 & -1 & 0 & 0 & 0 & 0 \\ 0 & 0 & 1 & 0 & -L & -1 & 0 & 0 & 0 \\ 0 & 0 & 0 & 1 & 0 & 0 & -1 & 0 & 0 \\ 0 & 0 & 0 & 0 & 0 & 1 & 0 & -L & -1 \\ 0 & 0 & 0 & 0 & 0 & 0 & 1 & 0 & 0 \\ 0 & 0 & 0 & 0 & 0 & 0 & 0 & 1 & 0 \\ 0 & 0 & 0 & 0 & 0 & 0 & 0 & 0 & 1 \end{bmatrix} \qquad \text{(i)}$$

STEP 5. We compute the basic nodal actions of the elements of the beam. In order to accomplish this we form its matrix $\{P^{EF}\}$ wherein we include all possible external (equivalent) actions not absorbed directly by the supports of the beam. That is, referring to Fig. a for $L = 4$ m we have

$$
\{P^{EF}\} = \left\{ \begin{array}{c} \mathcal{M}_3^{(1)} \\ \bar{P}_1^{(2)} \\ \bar{P}_2^{(2)} \\ \bar{\mathcal{M}}_3^{(2)} \\ \bar{P}_1^{(3)} \\ \mathcal{M}_3^{(3)} \\ \bar{P}_1^{(4)} \\ P_2^{(4)} \\ \mathcal{M}_3^{(4)} \end{array} \right\} = \left\{ \begin{array}{c} 0 \\ 0 \\ 0 \\ 8 \\ 0 \\ 0 \\ 0 \\ -1 \\ 0 \end{array} \right\} P \tag{j}
$$

Substituting relations (i) and (j) into relation (15.9), we obtain

$$
\{\hat{a}^E\} = \left\{ \begin{array}{c} F_1^{1k} \\ F_2^{1k} \\ M_3^{1k} \\ F_1^{2k} \\ F_2^{2k} \\ M_3^{2k} \\ F_1^{3k} \\ F_2^{3k} \\ M_3^{3k} \end{array} \right\} = [B]^{-1}\{P^{EF}\} = \left\{ \begin{array}{c} 0 \\ -0.5 \\ 2.0 \\ \hline 0 \\ -0.5 \\ -4.0 \\ \hline 0 \\ -1 \\ 0 \end{array} \right\} P \tag{k}
$$

Example 3. Consider the space frame shown in Fig. a. Following the procedure described in this section, derive its matrix $[B]$.

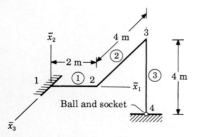

Figure a Geometry and loading of the frame.

solution

STEP 1.   We consider the model of the frame shown in Fig. b. Elements 1 to 4 of this model are general space elements, while elements 6 to 8 are axial deformation elements.

STEP 2.   We compute the matrix $[H]$ for each element of the model. We choose the global axes of the model and the local axes of its elements as shown in Fig. c ($x_1^{(e)}$ is the local x_1 axis of element e). Referring to Fig. c the

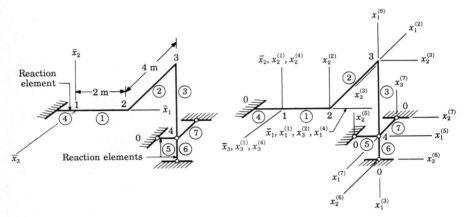

Figure b Model of the frame of Fig. a.

Figure c Global axes of the model of Fig. b and local axes of its elements.

direction cosines of the elements of the frame are

Elements 1 and 4

$$\lambda_{11} = 1 \qquad \lambda_{12} = 0 \qquad \lambda_{13} = 0$$

$$\lambda_{21} = 0 \qquad \lambda_{22} = 1 \qquad \lambda_{23} = 0 \qquad \text{(a)}$$

$$\lambda_{31} = 0 \qquad \lambda_{32} = 0 \qquad \lambda_{33} = 1$$

Element 2

$$\lambda_{11} = 0 \qquad \lambda_{12} = 0 \qquad \lambda_{13} = -1$$

$$\lambda_{21} = 0 \qquad \lambda_{22} = 1 \qquad \lambda_{23} = 0 \qquad \text{(b)}$$

$$\lambda_{31} = 1 \qquad \lambda_{32} = 0 \qquad \lambda_{33} = 0$$

Element 3

$$\lambda_{11} = 0 \qquad \lambda_{12} = -1 \qquad \lambda_{13} = 0$$

$$\lambda_{21} = 0 \qquad \lambda_{22} = 1 \qquad \lambda_{23} = 0 \qquad \text{(c)}$$

$$\lambda_{31} = 0 \qquad \lambda_{32} = 0 \qquad \lambda_{33} = 1$$

Element 5

$$\lambda_{11} = 0 \qquad \lambda_{12} = 0 \qquad \lambda_{13} = 0$$

$$\lambda_{21} = 0 \qquad \lambda_{22} = 1 \qquad \lambda_{23} = 0 \qquad\qquad (d)$$

$$\lambda_{31} = 0 \qquad \lambda_{32} = 0 \qquad \lambda_{33} = 1$$

Element 6

$$\lambda_{11} = 0 \qquad \lambda_{12} = 1 \qquad \lambda_{13} = 0$$

$$\lambda_{21} = -1 \qquad \lambda_{22} = 0 \qquad \lambda_{23} = 0 \qquad\qquad (e)$$

$$\lambda_{31} = 0 \qquad \lambda_{32} = 0 \qquad \lambda_{33} = 1$$

Element 7

$$\lambda_{11} = 0 \qquad \lambda_{12} = 0 \qquad \lambda_{13} = 1$$

$$\lambda_{21} = 1 \qquad \lambda_{22} = 0 \qquad \lambda_{23} = 0 \qquad\qquad (f)$$

$$\lambda_{31} = 0 \qquad \lambda_{32} = 1 \qquad \lambda_{33} = 0$$

Referring to relation (5.13) for a general space element, we have

$$[T] = \left[\begin{array}{cccccc}
-1 & 0 & 0 & 0 & 0 & 0 \\
0 & -1 & 0 & 0 & 0 & 0 \\
0 & 0 & -1 & 0 & 0 & 0 \\
0 & 0 & 0 & -1 & 0 & 0 \\
0 & 0 & L & 0 & -1 & 0 \\
0 & -L & 0 & 0 & 0 & -1 \\
\hline
1 & 0 & 0 & 0 & 0 & 0 \\
0 & 1 & 0 & 0 & 0 & 0 \\
0 & 0 & 1 & 0 & 0 & 0 \\
0 & 0 & 0 & 1 & 0 & 0 \\
0 & 0 & 0 & 0 & 1 & 0 \\
0 & 0 & 0 & 0 & 0 & 1
\end{array}\right] \qquad\qquad (g)$$

Moreover, referring to relation (5.14) for an axial deformation element of a space beam or frame we have

$$[T] = \left\{ \begin{array}{c} -1 \\ 0 \\ 0 \\ 0 \\ 0 \\ 0 \\ \hline 1 \\ 0 \\ 0 \\ 0 \\ 0 \\ 0 \end{array} \right\} \tag{h}$$

Substituting relations (g) and (6.57) with (a), (b), or (c) into relation (15.15), we obtain

$$[H^1] = [H^4] = \left[\begin{array}{cccccc} -1 & 0 & 0 & 0 & 0 & 0 \\ 0 & -1 & 0 & 0 & 0 & 0 \\ 0 & 0 & -1 & 0 & 0 & 0 \\ 0 & 0 & 0 & -1 & 0 & 0 \\ 0 & 0 & 2 & 0 & -1 & 0 \\ 0 & -2 & 0 & 0 & 0 & -1 \\ \hline 1 & 0 & 0 & 0 & 0 & 0 \\ 0 & 1 & 0 & 0 & 0 & 0 \\ 0 & 0 & 1 & 0 & 0 & 0 \\ 0 & 0 & 0 & 1 & 0 & 0 \\ 0 & 0 & 0 & 0 & 1 & 0 \\ 0 & 0 & 0 & 0 & 0 & 1 \end{array} \right] \tag{i}$$

$$[H^2] = \left[\begin{array}{cccccc} 0 & 0 & -1 & 0 & 0 & 0 \\ 0 & -1 & 0 & 0 & 0 & 0 \\ 1 & 0 & 0 & 0 & 0 & 0 \\ 0 & -4 & 0 & 0 & 0 & -1 \\ 0 & 0 & 4 & 0 & -1 & 0 \\ 0 & 0 & 0 & 1 & 0 & 0 \\ \hline 0 & 0 & 1 & 0 & 0 & 0 \\ 0 & 1 & 0 & 0 & 0 & 0 \\ -1 & 0 & 0 & 0 & 0 & 0 \\ 0 & 0 & 0 & 0 & 0 & 1 \\ 0 & 0 & 0 & 0 & 1 & 0 \\ 0 & 0 & 0 & -1 & 0 & 0 \end{array} \right] \tag{j}$$

$$[H^3] = \begin{bmatrix} 0 & -1 & 0 & 0 & 0 & 0 \\ 1 & 0 & 0 & 0 & 0 & 0 \\ 0 & 0 & -1 & 0 & 0 & 0 \\ 0 & 0 & 4 & 0 & -1 & 0 \\ 0 & 0 & 0 & 1 & 0 & 0 \\ 0 & -4 & 0 & 0 & 0 & -1 \\ \hdashline 0 & -1 & 0 & 0 & 0 & 0 \\ 1 & 0 & 0 & 0 & 0 & 0 \\ 0 & 0 & -1 & 0 & 0 & 0 \\ 0 & 0 & 0 & 0 & -1 & 0 \\ 0 & 0 & 0 & 1 & 0 & 0 \\ 0 & 0 & 0 & 0 & 0 & -1 \end{bmatrix} \tag{k}$$

Moreover, substituting relations (h) and (6.57) with (d), (e), or (f) into relation (15.15), we get

$$\{H^5\} = \left\{ \begin{array}{c} -1 \\ 0 \\ 0 \\ 0 \\ 0 \\ 0 \\ \hdashline 1 \\ 0 \\ 0 \\ 0 \\ 0 \\ 0 \end{array} \right\} \tag{l}$$

$$\{H^6\} = \left\{ \begin{array}{c} 0 \\ -1 \\ 0 \\ 0 \\ 0 \\ 0 \\ \hdashline 0 \\ 1 \\ 0 \\ 0 \\ 0 \\ 0 \end{array} \right\} \tag{m}$$

$$\{H^7\} = \left\{ \begin{array}{c} 0 \\ 0 \\ -1 \\ 0 \\ 0 \\ 0 \\ \hdashline 0 \\ 0 \\ 1 \\ 0 \\ 0 \\ 0 \end{array} \right\} \tag{n}$$

STEP 3.   We form the matrix $[B^M]$ of the model of Fig. b using the matrices $[H^-]$ for its elements. That is, we write consecutively the equations of equilibrium of all the unsupported nodes of the model. For the model of Fig. b, there are 24 equations of equilibrium (six for each node). We assign one row for each equation of equilibrium, six columns for each of elements 1 to 4, and one column for each of elements 5 to 7 of the model. Thus, the equations of equilibrium for the unsupported nodes of the model of Fig. b are given in Eqs. (o) on p. 507. These equations can be rewritten as

$$\{P^{EFM}\} = [B^M]\{\hat{a}^{EM}\} \tag{p}$$

In Eqs. (p), matrix $\{P^{EFM}\}$ includes the global components of all possible external actions not directly absorbed by the supports of the model of Fig. b. Referring to Figs. a and b, it is apparent that matrix $\{P^{EFM}\}$ of the model is identical to the matrix $\{\hat{P}^E\}$ of the actual frame. The matrix $\{\hat{a}^{EM}\}$ includes the basic nodal actions of all the elements of the model of Fig. b.

From Fig. b, it can be seen that element 1 is connected to node 1 with its end j, while element 4 is connected to node 1 with its end k. Since the product of the matrices $[H^q]$ ($q = j$ or k) and $\{a^E\}$ for an element represents the global components of the actions acting at its end q, as can be seen in Eqs. (o), the matrix $[H^{1j}]$ is placed in rows 1 to 6 and columns 1 to 6, while matrix $[H^{4k}]$ is placed in rows 1 to 6 and columns 19 to 24 of the matrix $[B^M]$. Inasmuch as other elements are not connected to node 1, zeros are placed in the remaining slots of rows 1 to 6 of the matrix $[B^M]$. Similarly, the other slots of the matrix $[B^M]$ are filled, as shown in Eqs. (o).

STEP 4.   The matrix $[\hat{B}]$ for a statically indeterminate structure must be computed when the structure is analyzed by the modern flexibility method. Moreover, the matrix $[B]$ can be used to compute the basic stiffness matrix of a structure. However, for large structures this approach requires more computer time and storage than the direct stiffness method. For this reason we do not usually compute the matrix $[B]$ when analyzing a statically indeterminate structure.

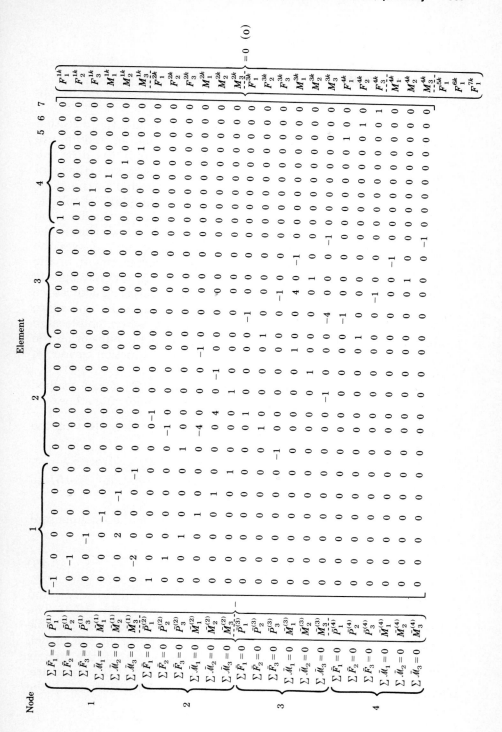

15.3 Computations of the Matrices $[\hat{B}]$ and $[B]$ for Structures Having Skew Supports

In this section we consider planar structures having roller supports or space structures having roller or ball supports whose plane of rolling is not normal to a global axis. We refer to these supports as *skew supports*.

For a roller support of a planar or space structure the direction normal to the plane of rolling and the direction of rolling are specified. For a ball support of a space structure only the direction normal to the plane of rolling is specified. We choose an orthogonal system of axes $\bar{x}'_1$, $\bar{x}'_2$, and $\bar{x}'_3$ with the node connected to the skew support as the origin and the $\bar{x}'_2$ axis normal to the plane of rolling. Moreover, for a rolling support we choose the $\bar{x}'_1$ axis in the direction of rolling, while for a ball support of a space structure we choose the $\bar{x}'_1$ axis in any convenient direction. We call this system of axes the *skew axes* and the components of a force or a displacement in the direction of the skew axes the *skew components*. In order to specify the skew axes with respect to the global axes for planar structures we usually give the angle of inclination ϕ_{11} of the x_1 axis. For space structures we usually give the coordinates of two points; one located on the $\bar{x}'_2$ axis and the other located on the $\bar{x}'_1 \bar{x}'_2$ plane but not on the $\bar{x}'_2$ axis. From the coordinates of these points the direction cosines of the skew axes can be established.

In order to establish the matrices $[\hat{B}]$ and $[B]$ for a framed structure having skew supports we adhere to the following steps.

Step 1. We consider the model of the structure described in the previous section.

Step 2. We form the transformation matrices $[\Lambda]$ and $[T]$ for each element of the model and substitute them in relation (15.15) to obtain its matrix $[H]$.

Step 3. We write the equations of equilibrium for the unsupported nodes of the model. That is,

$$\{P^{EFM}\} = [B^M]\{\hat{a}^{EM}\} \tag{15.19}$$

We construct the matrix $[B^M]$ for the model using matrices $[H^j]$ and $[H^k]$ for its elements (see the examples at the end of Sec. 15.2). The matrix $\{P^{EFM}\}$ includes the global components of all possible equivalent actions acting on the unsupported nodes of the model. Referring to Figs. 15.3 and 15.4, it is apparent that the matrix $\{P^{EFM}\}$ of the model is identical to the matrix $[\hat{P}^E]$ of the structure. The matrix $\{\hat{a}^{EM}\}$ is the matrix of the basic nodal actions of all the elements of the model. It includes the basic nodal actions of all the elements of the actual structure, in addition to the basic nodal actions of the reaction elements. As mentioned previously, each basic nodal action of a reaction element is equal and opposite to the corresponding component of a reaction of the actual structure. Therefore, Eqs. (15.19) are identical to the equations of equilibrium of all the nodes of the actual structure. That is,

$$\{\hat{P}^E\} = [\hat{B}]\begin{Bmatrix} \{a^E\} \\ -\{R\} \end{Bmatrix} \tag{15.20}$$

For a statically determinate structure Eqs. (15.20) can be solved to obtain its basic nodal actions and its reactions.

Step 4. We modify Eqs. (15.19) for the model of the structure by converting the global equations of equilibrium for the forces acting on its nodes which are connected to skew supports to equations of equilibrium in the directions of the skew axes. To accomplish this we do the following.

1. We form the transformation matrix $[\Lambda']$ for each skew support. This matrix converts the global components of a force acting on the skew support to components in the direction of the skew axes. For planar structures $[\Lambda']$ is equal to $[\Lambda_P]$ given by relation (6.6), while for space structures $[\Lambda']$ is equal to $[\Lambda_S]$ given by relation (6.3).

2. We form the matrix $[B_r]$ for each skew support. This matrix consists of the rows of the matrix $[\hat{B}]$ which correspond to the global equations of equilibrium for the forces acting on a node connected to a skew support. For planar structures $[B_r]$ has two rows, while for space structures it has three.

3. We compute the matrix $[B_r^1] = [\Lambda'][B_r]$ for each skew support. We replace the rows of the matrix $[\hat{B}]$ corresponding to the equations of equilibrium for the forces acting on the nodes which are connected to the skew supports by the rows of the matrices $[B_r^1]$. The resulting matrix is the matrix $[\hat{B}_m]$.

The modified equations of equilibrium for the structure can be written as

$$\{\hat{P}_m^E\} = [\hat{B}_m]\begin{Bmatrix} \{\hat{a}^E\} \\ -\{R_m\} \end{Bmatrix} \tag{15.21}$$

Step 5. We establish the matrix $[B]$ for the structure from its matrix $[\hat{B}_m]$ by canceling the rows and columns of Eqs. (15.21) which involve the reactions of the structure.

In what follows we establish the matrices $[\hat{B}]$ and $[B]$ of a statically determinate planar frame having a skew support.

Example. Consider the planar frame shown in Fig. a. Following the procedure described in this section, form the matrices $[\hat{B}]$ and $[B]$ of this frame and compute the basic nodal actions of its elements.

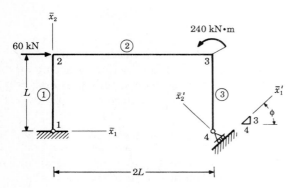

Figure a Geometry of the frame.

---

solution

STEP 1.   We consider the model of the frame shown in Fig. b.

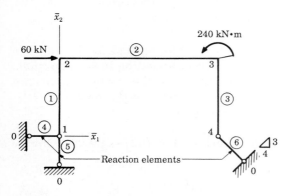

Figure b Model of the frame.

STEP 2. We compute the matrix $[H]$ for each element of the model of Fig. b. Elements 1 to 4 of this model are general planar elements. Thus referring to relations (5.11) we have

$$[T] = \left[\begin{array}{ccc} -1 & 0 & 0 \\ 0 & -1 & 0 \\ 0 & -L & -1 \\ \hline 1 & 0 & 0 \\ 0 & 1 & 0 \\ 0 & 0 & 1 \end{array}\right] \tag{a}$$

Substituting relations (6.48) and (a) into (15.15) we have

$$[H] = [\Lambda_{PT}]^T[T] = \left[\begin{array}{ccc} -\cos \phi_{11} & \sin \phi_{11} & 0 \\ -\sin \phi_{11} & -\cos \phi_{11} & 0 \\ 0 & -L & -1 \\ \cos \phi_{11} & -\sin \phi_{11} & 0 \\ \sin \phi_{11} & \cos \phi_{11} & 0 \\ 0 & 0 & 1 \end{array}\right] \tag{b}$$

Noting that $(\phi_{11}^{(1)} = 90, L_1 = L)$, $(\phi_{11}^{(2)} = 0, L_2 = 2L)$, $(\phi_{11}^{(3)} = -90, L_3 = L)$, $(\phi_{11}^{(4)} = 0, L_4 = \text{unspecified})$ from relation (b), we obtain

$$[H^1] = \left[\begin{array}{ccc} 0 & 1 & 0 \\ -1 & 0 & 0 \\ 0 & -L & -1 \\ \hline 0 & -1 & 0 \\ 1 & 0 & 0 \\ 0 & 0 & 1 \end{array}\right] \tag{c}$$

$$[H^2] = \left[\begin{array}{ccc} -1 & 0 & 0 \\ 0 & -1 & 0 \\ 0 & -2L & 0 \\ \hline 1 & 0 & 0 \\ 0 & 1 & 0 \\ 0 & 0 & 1 \end{array}\right] \tag{d}$$

$$[H^3] = \left[\begin{array}{ccc} 0 & -1 & 0 \\ 1 & 0 & 0 \\ 0 & -L & -1 \\ \hline 0 & 1 & 0 \\ -1 & 0 & 0 \\ 0 & 0 & 1 \end{array}\right] \tag{e}$$

Moreover, referring to relation (d) of Example 2 of Sec. 15.2 ($\phi_{11}^{(4)} = 0$, $\phi_{11}^{(5)} = 90$, $\cos\phi_{11}^{(6)} = -\frac{3}{5}$, $\sin\phi_{11}^{(6)} = \frac{4}{5}$) we obtain

$$\{H^4\} = \begin{Bmatrix} -1 \\ 0 \\ 0 \\ \hline 1 \\ 0 \\ 0 \end{Bmatrix} \tag{f}$$

$$\{H^5\} = \begin{Bmatrix} 0 \\ -1 \\ 0 \\ \hline 0 \\ 1 \\ 0 \end{Bmatrix} \tag{g}$$

$$\{H^6\} = \begin{Bmatrix} 0.6 \\ -0.8 \\ 0 \\ \hline -0.6 \\ 0.8 \\ 0 \end{Bmatrix} \tag{h}$$

STEP 3.   We form the matrix $[B^M]$ of the model of Fig. b using the matrices $[H]$ of its elements. That is, we write, consecutively, the equations of equilibrium of all the unsupported nodes of the model. For the model of Fig. b, there are twelve equations (three for each node) [see Eqs. (i) on p. 513]. As can be seen from Eqs. (i), we assign one row for each equation of equilibrium, three columns for each of the elements 1 to 4 and one column for element 5 of the model. Equations (i) can be rewritten as

$$\{P^{EFM}\} = [B^M]\{\hat{a}^{EM}\} \tag{j}$$

In relation (j), the matrix $\{P^{EFM}\}$ includes the global components of all possible internal actions not directly absorbed by the supports of the model. It is apparent, that the matrix $\{P^{EFM}\}$ of the model of Fig. b is identical to the matrix $[\hat{P}^E]$ of the actual frame of Fig. a. The matrix $[\hat{a}^M]$ includes the basic nodal actions of all the elements of the model of Fig. b.

From Fig. b it can be seen that element 1 is connected to node 1 with its end j, while elements 4 and 5 are connected to node 1 with their ends k. Since the product of the matrices $[H^q]$ ($q = j$ or k) and $\{a^E\}$ of an element represents the global components of the actions acting at end q of this element, as shown in relation (j), matrix $[H^{1j}]$ is placed in rows 1 to 3 and columns 1 and 3, while matrices $\{H^{4k}\}$ and $\{H^{5k}\}$ are placed in rows 1 to 3 and columns 10 and 11, respectively, of the matrix $[B^M]$ of the model.

Node equations, columns grouped by Element 1, Element 2, Element 3 and reaction columns 4, 5, 6:

$$
\underbrace{\begin{Bmatrix}
\sum \bar F_1 = 0 \\ \sum \bar F_2 = 0 \\ \sum \bar M_3 = 0 \\
\sum \bar F_1 = 0 \\ \sum \bar F_2 = 0 \\ \sum \bar M_3 = 0 \\
\sum \bar F_1 = 0 \\ \sum \bar F_2 = 0 \\ \sum \bar M_3 = 0 \\
\sum \bar F_1 = 0 \\ \sum \bar F_2 = 0 \\ \sum \bar M_3 = 0
\end{Bmatrix}}_{\text{Node } 1,\,2,\,3,\,4}
\;\;
\begin{Bmatrix}
0 \\ 0 \\ 0 \\ 60 \\ 0 \\ 0 \\ 0 \\ 0 \\ 240 \\ 0 \\ 0 \\ 0
\end{Bmatrix}
-
\begin{bmatrix}
0 & 1 & 0 & 0 & 0 & 0 & 0 & 0 & 0 & 1 & 0 & 0 \\
-1 & 0 & 0 & 0 & 0 & 0 & 0 & 0 & 0 & 0 & 1 & 0 \\
0 & -L & -1 & 0 & 0 & 0 & 0 & 0 & 0 & 0 & 0 & 0 \\
0 & -1 & 0 & -1 & 0 & 0 & 0 & 0 & 0 & 0 & 0 & 0 \\
1 & 0 & 0 & 0 & -1 & 0 & 0 & 0 & 0 & 0 & 0 & 0 \\
0 & 0 & 1 & 0 & -2L & -1 & 0 & 0 & 0 & 0 & 0 & 0 \\
0 & 0 & 0 & 1 & 0 & 0 & 0 & -1 & 0 & 0 & 0 & 0 \\
0 & 0 & 0 & 0 & 1 & 0 & -1 & 0 & 0 & 0 & 0 & 0 \\
0 & 0 & 0 & 0 & 0 & 1 & 0 & -L & -1 & 0 & 0 & 0 \\
0 & 0 & 0 & 0 & 0 & 0 & 0 & 1 & 0 & 0 & 0 & -0.6 \\
0 & 0 & 0 & 0 & 0 & 0 & 1 & 0 & 0 & 0 & 0 & 0.8 \\
0 & 0 & 0 & 0 & 0 & 0 & 0 & 0 & 1 & 0 & 0 & 0
\end{bmatrix}
\begin{Bmatrix}
F_1^{1k} \\ F_2^{1k} \\ M_3^{1k} \\
F_1^{2k} \\ F_2^{2k} \\ M_3^{2k} \\
F_1^{3k} \\ F_2^{3k} \\ M_3^{3k} \\
F_1^{4k} \\ F_1^{5k} \\ F_1^{6k}
\end{Bmatrix}
= 0
$$

where the matrix columns are grouped as Element 1 ($F_1^{1k},F_2^{1k},M_3^{1k}$), Element 2 ($F_1^{2k},F_2^{2k},M_3^{2k}$), Element 3 ($F_1^{3k},F_2^{3k},M_3^{3k}$), and columns 4, 5, 6 ($F_1^{4k},F_1^{5k},F_1^{6k}$).

(i)

Inasmuch as other elements are not connected to node 1, zeros are placed in the remaining slots of rows 1 to 3 of the matrix $[B^M]$.

From Fig. b, it can be seen that element 1 is connected to node 2 with its end k, while element 2 is connected to node 2 with its end j. Thus, as shown in Eqs. (i), the matrix $[H^{1k}]$ is placed in rows 4 to 6 and columns 1 to 3 while the matrix $[H^{2j}]$ is placed in rows 4 to 6 and columns 4 to 6 of the matrix $[B^M]$. Inasmuch as other elements are not connected to node 2, zeros are placed in the remaining slots of rows 4 to 6 of matrix $[B^M]$.

From Fig. b it can be seen that element 2 is connected to node 3 with its end k, while element 3 is connected to node 3 with its end j. Thus, as shown in Eqs. (i), the matrix $[H^{2k}]$ is placed in rows 7 to 9 and columns 4 to 6 while the matrix $[H^{3j}]$ is placed in rows 7 to 9 and columns 7 to 9 of the matrix $[B^M]$.

From Fig. b it can be seen that elements 3 and 6 are connected to node 4 with their end k. Thus, as shown in Eqs. (i), the matrix $[H^{3k}]$ is placed in rows 10 to 12 and columns 7 to 9, while the matrix $[H^{6k}]$ is placed in rows 10 to 12 and column 12 of the matrix $[B^M]$. As expected, the matrix $[\hat{B}]$ of the statically determinate structure under consideration is a square matrix.

STEP 4. We establish the matrix $[B]$ of the structure from its matrix $[\hat{B}] = [B^M]$. In order to accomplish this we first modify the matrix $[\hat{B}]$ by transforming rows 10 and 11 of Eqs. (i) to equations of equilibrium of the components of the forces acting on node 4 in the direction of rolling and normal to that direction. Denoting by $P_1'^{(4)}$ and $P_2'^{(4)}$ the components of the external equivalent actions on node 4 in the direction of rolling and normal to this direction, respectively, and referring to Fig. a, we have

$$P_1'^{(4)} = P_1^{(4)} \cos \phi_{11} + P_2^{(4)} \sin \phi_{11}$$
$$P_2'^{(4)} = -P_1^{(4)} \sin \phi_{11} + P_2^{(4)} \cos \phi_{11}$$
(k)

or
$$\left\{ \begin{matrix} P_1'^{(4)} \\ P_2'^{(4)} \end{matrix} \right\} = [\Lambda'] \left\{ \begin{matrix} P_1^{(4)} \\ P_2^{(4)} \end{matrix} \right\} = [\Lambda'][B_r] \left\{ \begin{matrix} \{\hat{a}\} \\ -\{R\} \end{matrix} \right\} = [B_r^1] \left\{ \begin{matrix} \{\hat{a}\} \\ -\{R\} \end{matrix} \right\}$$
(l)

where
$$[\Lambda'] = \begin{bmatrix} \cos \phi_{11} & \sin \phi_{11} \\ -\sin \phi_{11} & \cos \phi_{11} \end{bmatrix} = \begin{bmatrix} 0.8 & 0.6 \\ -0.6 & 0.8 \end{bmatrix}$$
(m)

The matrix $[B_r]$ consists of the two rows of the matrix $[B^M]$ corresponding to the equation of equilibrium for the forces acting on node 4. Hence from Eqs. (i) we get

$$[B_r] = \begin{bmatrix} 0 & 0 & 0 & 0 & 0 & 0 & 0 & 1 & 0 & 0 & 0 & -0.6 \\ 0 & 0 & 0 & 0 & 0 & 0 & -1 & 0 & 0 & 0 & 0 & 0.8 \end{bmatrix}$$
(n)

Thus

$$[B_r^1] = [\Lambda'][B_r]$$

$$= \begin{bmatrix} 0 & 0 & 0 & 0 & 0 & 0 & -0.6 & 0.8 & 0 & 0 & 0 & 0 \\ 0 & 0 & 0 & 0 & 0 & 0 & -0.8 & -0.6 & 0 & 0 & 0 & 1 \end{bmatrix}$$
(o)

The rows of the matrix $[B^M]$ corresponding to the global equations of equilibrium of the forces acting on node 4 (rows 10 and 11) are replaced by the rows of the matrix $[B_r^1]$ to obtain Eqs. (p) given on p. 515. In these equations $R^{(4)}$ is the total reaction at support 4 of the frame. Equations (p) can be rewritten as

The matrix equation (p):

$$\{P\} \;-\; [A]\,\{X\} \;=\; 0$$

where the unknown (variable) vector $\{X\}$ and load vector $\{P\}$ are

Variable	Element 1			Element 2			Element 3			4	5	6	$\{P\}$
F_1^{1k}	0	1	0	0	0	0	0	0	0	1	0	0	0
F_2^{1k}	0	0	-1	0	0	0	0	0	0	0	1	0	0
M_3^{1k}	-1	$-L$	0	0	0	0	0	0	0	0	0	0	0
F_1^{2k}	0	-1	0	0	0	-1	0	0	0	0	0	0	60
F_2^{2k}	0	0	1	0	-1	0	0	0	0	0	0	0	0
M_3^{2k}	1	0	0	-1	$-2L$	0	0	0	0	0	0	0	0
F_1^{3k}	0	0	0	0	0	1	0	-1	0	0	0	0	0
F_2^{3k}	0	0	0	0	1	0	0	0	1	0	0	0	0
M_3^{3k}	0	0	0	1	0	0	-1	$-L$	0	0	0	0	240
$-R_1^{(1)}$	0	0	0	0	0	0	0	0.8	-0.6	0	0	0	0
$-R_2^{(1)}$	0	0	0	0	0	0	0	-0.6	-0.8	0	0	1	0
$-R^{(4)}$	0	0	0	0	0	0	1	0	0	0	0	0	0

(p)

$$[\hat{P}_m^E] = [\hat{B}_m] \left\{ \begin{matrix} \{\hat{a}_m^E\} \\ -\{R\} \end{matrix} \right\} \tag{q}$$

STEP 5.   We establish the matrix $[B]$ of the frame from its matrix $[\hat{B}_m]$ by canceling all the rows and columns of Eqs. (p) which involve reactions of the frame. Thus,

$$[B] = \begin{bmatrix} 0 & -L & -1 & 0 & 0 & 0 & 0 & 0 & 0 \\ 0 & -1 & 0 & -1 & 0 & 0 & 0 & 0 & 0 \\ 1 & 0 & 0 & 0 & -1 & 0 & 0 & 0 & 0 \\ 0 & 0 & 1 & 0 & -2L & -1 & 0 & 0 & 0 \\ 0 & 0 & 0 & 1 & 0 & 0 & 0 & -1 & 0 \\ 0 & 0 & 0 & 0 & 1 & 0 & 1 & 0 & 0 \\ 0 & 0 & 0 & 0 & 0 & 1 & 0 & -L & -1 \\ 0 & 0 & 0 & 0 & 0 & 0 & -0.6 & 0.8 & 0 \\ 0 & 0 & 0 & 0 & 0 & 0 & 0 & 0 & 1 \end{bmatrix} \tag{r}$$

STEP 6.   We include in the matrix $\{P^{EF}\}$ all possible external actions not absorbed directly by the supports of the frame. That is,

$$\{P^{EF}\} = \left\{ \begin{matrix} \mathcal{M}_3^{(1)} \\ P_1^{(2)} \\ P_2^{(2)} \\ \mathcal{M}_3^{(2)} \\ P_1^{(3)} \\ P_2^{(3)} \\ \mathcal{M}_3^{(3)} \\ P_1'^{(4)} \\ \mathcal{M}_3^{(4)} \end{matrix} \right\} = \left\{ \begin{matrix} 0 \\ 60 \\ 0 \\ 0 \\ 0 \\ 0 \\ 240 \\ 0 \\ 0 \end{matrix} \right\} \tag{s}$$

where $P_1'^{(4)}$ is the component of the known external force in the direction of the skew axis $\tilde{x}_1'$. Substituting relation (r) and (s) into (15.20) for $L = 3$ we get

$$\{\hat{a}^E\} = \left\{ \begin{matrix} F_1^{1k} \\ F_2^{1k} \\ F_3^{1k} \\ \hdashline F_1^{2k} \\ F_2^{2k} \\ M_3^{2k} \\ \hdashline F_1^{3k} \\ F_2^{3k} \\ F_3^{3k} \end{matrix} \right\} = [B]^{-1}\{P^{EF}\} = \left\{ \begin{matrix} -30 \\ -82.5 \\ -165 \\ \hdashline 22.5 \\ -30 \\ 285 \\ \hdashline 30 \\ 22.5 \\ 0 \end{matrix} \right\} \tag{t}$$

15.4   Tree Structures

Tree structures result when a statically indeterminate structure is cut in order to reduce it to statically determinate parts. The matrix $[b]$ for tree structures can be obtained directly without having to compute the matrix $[B]$ and invert it.

In this section, we establish a formula for the matrix $[b]$ of the single-branch, space tree structure shown in Fig. 15.5. We denote by $\{\bar{P}\}_i$ the 6×1 matrix of the global components of equivalent actions acting at node i of this structure. The matrix of the equivalent actions $\{P^{EF}\}$ acting on all the nodes of the structure, not directly absorbed by its support is given as

$$\{P^{EF}\} = \begin{Bmatrix} \{\bar{P}\}_1 \\ \{\bar{P}\}_2 \\ \{\bar{P}\}_3 \\ \{\bar{P}\}_4 \end{Bmatrix} \tag{15.22}$$

Referring to Fig. 15.5, from geometric considerations it is apparent that the global components of nodal actions at the end k of element e are given as

$$\{\bar{A}^{ek}\} = \begin{Bmatrix} \bar{F}^{ek}_1 \\ \bar{F}^{ek}_2 \\ \bar{F}^{ek}_3 \\ \bar{M}^{ek}_1 \\ \bar{M}^{ek}_2 \\ \bar{M}^{ek}_3 \end{Bmatrix} = [U_e]\{\{\bar{P}\}_e + \{\bar{A}^{(e-1)k}\}\} \tag{15.23}$$

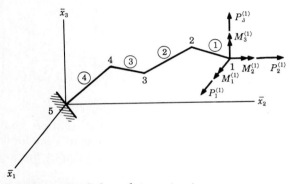

Figure 15.5 Single-branch tree structure.

where

$$[U_e] = \begin{bmatrix} -1 & 0 & 0 & 0 & 0 & 0 \\ 0 & -1 & 0 & 0 & 0 & 0 \\ 0 & 0 & -1 & 0 & 0 & 0 \\ 0 & \bar{x}_3^{ejk} & -\bar{x}_2^{ejk} & -1 & 0 & 0 \\ -\bar{x}_3^{ejk} & 0 & \bar{x}_2^{ejk} & 0 & -1 & 0 \\ \bar{x}_2^{ejk} & -\bar{x}_1^{ejk} & 0 & 0 & 0 & -1 \end{bmatrix} \qquad (15.24)$$

Denoting by $\bar{x}_p^{ej}$ and $\bar{x}_p^{ek}$ ($p = 1, 2, 3$) the global coordinates of the ends j and k, respectively, of element e the quantities $\bar{x}_p^{ejk}$ are defined as

$$\bar{x}_p^{ejk} = \bar{x}_p^{ej} - \bar{x}_p^{ek} \qquad (15.25)$$

Referring to relation (6.50), the matrices of the basic nodal actions of the elements of the tree structure are given as

$$\{a^{E1}\} = [A^{1k}] = [\hat{\Lambda}_{SF}^1][\bar{A}^{1k}] = [\hat{\Lambda}_{SF}^1][U_1]\{\bar{P}\}_1$$

$$\{a^{E2}\} = [A^{2k}] = [\hat{\Lambda}_{SF}^2][U_2][\{\bar{P}\}_2 + [U_1]\{\bar{P}\}_1]$$

$$\{a^{E3}\} = [A^{3k}] = [\hat{\Lambda}_{SF}^3][U_3][\{\bar{P}\}_3 + [U_2]\{\bar{P}\}_2 + [U_2][U_1]\{\bar{P}\}_1] \quad (15.26)$$

$$\{a^{E4}\} = [\hat{\Lambda}_{SF}^4][U_4][\{\bar{P}\}_4 + [U_3]\{\bar{P}\}_3 + [U_3][U_2]\{\bar{P}\}_2$$
$$+ [U_3][U_2][U_1]\{\bar{P}\}_1]$$

or
$$\{\hat{a}^E\} = \begin{Bmatrix} \{a^{E1}\} \\ \{a^{E2}\} \\ \{a^{E3}\} \\ \{a^{E4}\} \end{Bmatrix} = [b] \begin{Bmatrix} \{\bar{P}\}_1 \\ \{\bar{P}\}_2 \\ \{\bar{P}\}_3 \\ \{\bar{P}\}_4 \end{Bmatrix} = [b]\{P^{EF}\} \qquad (15.27)$$

where

$$[b] = \begin{bmatrix} [\hat{\Lambda}_{SF}^1][U_1] & 0 & 0 & 0 \\ [\hat{\Lambda}_{SF}^2][U_2][U_1] & [\hat{\Lambda}_{SF}^2][U_2] & 0 & 0 \\ [\hat{\Lambda}_{SF}^3][U_3][U_2][U_1] & [\hat{\Lambda}_{SF}^3][U_3][U_2] & [\hat{\Lambda}_{SF}^3][U_3] & 0 \\ [\hat{\Lambda}_{SF}^4][U_4][U_3][U_2][U_1] & [\hat{\Lambda}_{SF}^4][U_4][U_3][U_2] & [\hat{\Lambda}_{SF}^4][U_4][U_3] & [\hat{\Lambda}_{SF}^4][U_4] \end{bmatrix}$$
$$(15.28)$$

The matrices $[\hat{\Lambda}_{SP}^e]$ ($e = 1, 2, 3, 4$) are defined by relation (6.54). Referring to relations (15.24) to (15.28) it can be seen that for tree structures, the matrix $[b]$ can be obtained directly without having to compute the matrix $[B]$ and finding its inverse.

15.5   Equation of Compatibility for a Structure Subjected to Equivalent Actions on Its Nodes

In this section, we establish the equations of compatibility for a framed structure subjected to known equivalent actions on its nodes and to specified components of displacements of its supports. These equations can be expressed in the following form:

$$\left\{ \begin{matrix} \{\hat{d}\} \\ \{\Delta^S\} \end{matrix} \right\} = [\hat{C}] \left\{ \begin{matrix} \{\Delta^F\} \\ \{\Delta^S\} \end{matrix} \right\} = [\hat{C}]\{\hat{\Delta}\} = \left[\begin{matrix} [C][C^S] \\ [O][I] \end{matrix} \right] \left\{ \begin{matrix} \{\Delta^F\} \\ \{\Delta^S\} \end{matrix} \right\} \quad (15.29)$$

where

$$\{\hat{d}\} = \left\{ \begin{matrix} \{d^1\} \\ \{d^2\} \\ \vdots \\ \{d^{NE}\} \end{matrix} \right\} \quad (15.29a)$$

and $\{d^e\}$ = matrix of basic deformation parameters of element e
  $(e = 1, 2, \ldots, NE)$

  $\{\Delta^F\}$ = matrix of components of displacements of nodes of structure which are not inhibited by its supports

  $\{\Delta^S\}$ = matrix of components of displacements of nodes of structure which are inhibited by its supports

  NE = number of elements of structure

When a structure is subjected only to known equivalent actions on its nodes, $\{\Delta^S\}$ is a zero matrix and relation (15.29) reduces to

$$\{\hat{d}\} = [\hat{C}]\{\Delta^F\} \quad (15.30)$$

From relation (15.30) it can be deduced that the nth column of the matrix $[C]$ represents the basic deformation parameters of all the elements of the structure when subjected to a unit value of the component of displacement, located in the nth row of the matrix $\{\Delta^F\}$, while the other components of displacements of the nodes of the structure are kept equal to zero.

In the flexibility method presented in Chap. 17, we use the equations of equilibrium (15.16) for the model of a structure obtained from the actual structure by replacing its supports with reaction elements (see Sec. 15.2). Since for this model $\{\Delta^S\} = 0$ its equations of compatibility (15.30) are written as

$$\{d^M\} = [C^M]\{\Delta^{FM}\} \quad (15.31)$$

The matrix $\{\Delta^{FM}\}$ includes the global components of displacements of the nodes of the model which are not inhibited by its supports; that is, it includes all the global components of displacements of the nodes of the actual structure including those inhibited by its supports. Thus

$$\{\Delta^{FM}\} = \{\hat{\Delta}\} = \left\{ \begin{matrix} \{\Delta^F\} \\ \{\Delta^S\} \end{matrix} \right\} \tag{15.32}$$

The matrix $\{\hat{d}^M\}$ includes the basic deformation parameters of all the elements of the actual structure and of the reaction elements. That is,

$$\{\hat{d}^M\} = \left\{ \begin{matrix} \{\hat{d}\} \\ \{\Delta^S\} \end{matrix} \right\} \tag{15.33}$$

Thus, referring to Eqs. (15.29), (15.31), (15.32), and (15.33), it is apparent that

$$[\hat{C}^M] = [\hat{C}] \tag{15.34}$$

The matrix $[C]$ of a structure which does not have skew supports can be obtained from its matrix $[C^M]$ by eliminating its rows which correspond to the rows of Eq. (15.31) which involve the basic deformation parameters of the reaction elements and its columns which correspond to the columns of Eqs. (15.31) which are multiplied by the known components of displacements of the supports of the structure. If the structure has a roller or a ball support whose plane of rolling is not normal to a global axis (skew support), in order to obtain its matrix $[C]$ from the matrix $[C^M]$, the latter must be modified prior to eliminating its previously described rows and columns (see Example 2 at the end of this section).

The matrix $\{\hat{d}\}$ or $\{d^M\}$ is of the same rank as the matrix $\{\hat{a}^E\}$ or $\{\hat{a}^{EM}\}$, respectively. Moreover, the matrix $\{\Delta^F\}$ or $\{\Delta^{FM}\}$ is of the same rank as the matrix $\{P^{EF}\}$ or $\{\hat{P}^{EFM}\}$, respectively. Furthermore, for statically determinate structures, the matrix $\{\hat{a}^E\}$ or $\{\hat{a}^{EM}\}$ is of the same rank as the matrix $\{P^{EF}\}$ or $\{P^{EFM}\}$, respectively. Thus, it is apparent that for statically determinate structures, $[C]$ and $[\hat{C}] = [C^M]$ are square matrices. Consequently, for statically determinate structures relations (15.29) and (15.30) can be solved to yield

$$\{\hat{\Delta}\} = [\hat{c}] \left\{ \begin{matrix} \{\hat{d}\} \\ \{\Delta^S\} \end{matrix} \right\} \tag{15.35}$$

and

$$[\Delta^F] = [c]\{\hat{d}\} \tag{15.36}$$

where

$$[\hat{c}] = [\hat{C}]^{-1} \tag{15.37a}$$

and

$$[\hat{c}] = [C]^{-1} \tag{15.37b}$$

By the examples which we solve in this section we show that the matrices $[\hat{C}]$ and $[\hat{B}]$ and the matrices $[C]$ and $[B]$ of a structure (statically determinate or indeterminate) are related by the following relations:

$$[\hat{C}] = [\hat{B}]^T \tag{15.38a}$$

and
$$[C] = [B]^T \tag{15.38b}$$

Moreover, we show that the matrices $[c]$ and $[b]$ of a statically determinate structure are related by the following relations:

$$[\hat{c}] = [\hat{b}]^T \tag{15.39a}$$

and
$$[c] = [b]^T \tag{15.39b}$$

where the matrices $[\hat{b}]$ and $[b]$ are defined by relations (15.10) and (15.11), respectively.

Relations (15.38) are called the *contragradient law*. They indicate that for any structure the conditions of equilibrium (15.3) or (15.6) and the conditions of compatibility (15.29) or (15.30) are not independent. In Sec. (20.3) we prove the contragradient law using the principle of virtual work.

In the sequel, we describe a systematic procedure for generating the matrices $[\hat{C}]$ and $[C]$ of framed structures suitable for programming on an electronic computer. In this procedure, we adhere to the following steps.

Step 1. We consider a model of the structure obtained from the actual structure by replacing its supports with reaction elements as described in Sec. 15.2.

Step 2. We form the transformation matrix $[\Lambda]$ for each element of the model which transforms its global matrix of nodal displacements $\{\bar{D}\}$ to its local $\{D\}$. That is, referring to relation (6.67)

$$\{D\} = [\Lambda]\{\bar{D}\} \tag{15.40}$$

Moreover, we form the transformation matrix $[T]^T$ for each element of the model which transforms its local matrix of nodal displacements $\{D\}$ to its matrix of basic deformation parameters $\{d\}$. That is, referring to relation (5.39) we have

$$\{d\} = [T]^T\{D\} \tag{15.41}$$

Furthermore, we compute the matrix $[H]^T$ for each element of the model which transforms its matrix $\{\bar{D}\}$ to $\{d\}$. Substituting relation (15.40) into (15.41) we get

$$\{d\} = [H]^T\{\bar{D}\} \tag{15.42}$$

where $[H]$ is defined by relation (15.15). Thus we have shown that if the relation between the global matrix of nodal actions and the matrix of basic nodal actions of an element of a framed structure subjected to equivalent actions on its nodes is written as

$$\{\bar{A}^E\} = [H]\{a^E\} \tag{15.43}$$

the relation between its matrix of basic deformation parameters and its global matrix of nodal displacements is given by relation (15.42)

Step 3. We assemble the matrix $[C^M]$ of the model by considering the compatibility of the components of displacements of the ends of its elements with the components of displacement of its unsupported nodes, that is, by requiring that the global components of displacement of an end of an element of the model are equal to the components of displacement of the node of the model to which the element is connected. For instance, if the end j of an element is connected to node 3, and its end k is connected to node 4, the upper half of its matrix $\{\bar{D}\}$ represents the components of displacement of node 3, while the lower half of its matrix $\{\bar{D}\}$ represents the components of displacement of node 4. Consequently, it is apparent that if the end j of a general planar element of the model of a planar frame is connected to node 3, while its end k is connected to node 4, the first three columns of its matrix $[H]^T$ will occupy a 3×3 block of the matrix $[C^M]$ consisting of its three rows corresponding to the basic deformation parameters of the element and its three columns which are multiplied by the three components of displacement of node 3. Moreover, the last three columns of matrix $[H^T]$ will occupy a 3×3 block of matrix $[C^M]$ consisting of its three rows corresponding to the basic deformation parameters of the element and its three columns multiplied by the three components of displacement of node 4.

Step 4. We establish the matrix $[C]$ of the structure from the matrix $[C^M]$ of its model. The matrix $[C]$ of a structure which does not have skew supports can be obtained by eliminating the rows of its matrix $[C^M]$ corresponding to the rows of Eqs. (15.31) involving the basic deformation parameters of the reaction elements and the columns of the matrix $[C^M]$ corresponding to the columns of Eqs. (15.31) which involve the global components of displacements inhibited by the supports of the actual structure. If the structure has a roller or a ball support whose plane of rolling is not normal to a global axis (skew support), in order to obtain its matrix $[C]$ from the matrix $[C^M]$ of its model, the latter must be modified prior to eliminating its rows and columns (see Example 2 at the end of this section). The modified matrix $[C^M]$ is denoted by $[C^M_m]$ and is defined by the following equilibrium equations for the nodes of the model

$$\{\hat{d}^M\} = [C_m^M]\{\Delta_m^{FM}\} \qquad (15.44)$$

The matrix $\{\Delta_m^{FM}\}$ is obtained from the matrix $\{\Delta^{FM}\}$ by transforming the global components of translations of the nodes of the structure which are connected to inclined roller or ball supports (skew supports) to components in the direction of rolling and normal to that direction.

The matrix $[C]$ or $[B]$ for a structure can be used to compute its basic stiffness matrix (see Sec. 16.1). We illustrate the procedure described previously by the following two examples.

Example 1. The matrices $[\hat{C}] = [C^M]$ and $[C]$ for the same statically determinate, simply supported planar truss considered in Example 1 of Sec. 15.2 are assembled. Comparing the results of this example with those of Example 1 of Sec. 15.2, it can be seen that for this truss, the following relations are valid:

$$[C^M] = [B^M]^T$$
$$[C] = [B]^T$$

Example 2. The matrices $[C^M]$ and $[C]$ for the same statically determinate frame considered in the example of Sec. 15.3 are assembled. The one support of the frame is a roller whose plane of rolling is not normal to a global axis. Comparing the results of this example with those of the example of Sec. 15.3, it can be seen that for this frame the following relations are valid:

$$[C^M] = [B^M]^T$$
$$[C] = [B]^T$$

Example 1. Consider the planar truss shown in Fig. a. Following the procedure described in this section, construct its matrices $[\hat{C}] = [C^M]$ and $[C]$.

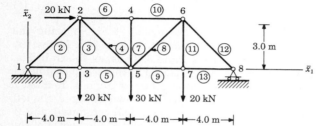

Figure a Geometry of the truss.

solution

STEP 1.   We consider the model of the truss shown in Fig. b.

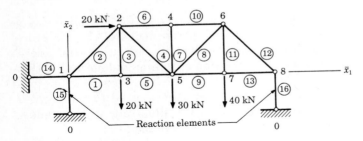

Figure b   Model of the truss of Fig. a.

STEP 2.   We compute matrix $[H]^T$ for each element of the truss. For a planar truss, matrix $[H]^T$ is given by relation (b) of Example 1 of Sec. 15.2. That is,

$$[H]^T = [-\cos\phi_{11} \quad -\sin\phi_{11} \quad \cos\phi_{11} \quad \sin\phi_{11}] \tag{a}$$

For elements 1, 5, 6, 8, 9, 10, 13, and 14 ($\cos\phi_{11} = 1$, $\sin\phi_{11} = 0$), we have

$$[H]^T = [-1 \quad 0 \quad 1 \quad 0] \tag{b}$$

For elements 3, 7, and 11 ($\cos\phi_{11} = 0$, $\sin\phi_{11} = -1$), we have

$$[H]^T = [0 \quad 1 \quad 0 \quad -1] \tag{c}$$

For elements 2 and 8 ($\cos\phi_{11} = 0.8$, $\sin\phi_{11} = 0.6$), we have

$$[H]^T = [-0.8 \quad -0.6 \quad 0.8 \quad 0.6] \tag{d}$$

For elements 4 and 12 ($\cos\phi_{11} = 0.8$, $\sin\phi_{11} = 0.6$), we get

$$[H]^T = [-0.8 \quad 0.6 \quad 0.8 \quad -0.6] \tag{e}$$

For elements 15 and 16, ($\cos\phi_{11} = 0$, $\sin\phi_{11} = 1$), we obtain

$$[H]^T = [0 \quad -1 \quad 0 \quad 1] \tag{f}$$

STEP 3.   We construct the matrix $[C^M]$ of the model of Fig. b using the matrices $[H]^T$ obtained in step 2. We note that $[C^M]$ is a 16×16 matrix. Member 1 of the model is connected to nodes 1 and 3. Thus, as shown in Fig. c, the left half of matrix $[H^1]^T$ [relation (b)] is placed in row 1 and columns 1 and 2, while its right half is placed in row 1 and columns 5 and 6. Member 2 of the model is connected to nodes 1 and 2. Thus, as shown in Fig. c, the left half of matrix $[H^2]^T$ [relation (d)] is placed in row 2 and columns 1 and 2, while its right half is placed in row 2 and columns 3 and 4. Element 3 of the model is connected to nodes 2 and 3. Thus, as shown in Fig. c, the left half of

| | $\Delta_1^{(1)}$ | $\Delta_2^{(1)}$ | $\Delta_1^{(2)}$ | $\Delta_2^{(2)}$ | $\Delta_1^{(3)}$ | $\Delta_2^{(3)}$ | $\Delta_1^{(4)}$ | $\Delta_2^{(4)}$ | $\Delta_1^{(5)}$ | $\Delta_2^{(5)}$ | $\Delta_1^{(6)}$ | $\Delta_2^{(6)}$ | $\Delta_1^{(7)}$ | $\Delta_2^{(7)}$ | $\Delta_1^{(8)}$ | $\Delta_2^{(8)}$ |
	1	2	3	4	5	6	7	8	9	10	11	12	13	14	15	16
$d_1'^{(1)}$	-1	0			1	0										
$d_1'^{(2)}$	-0.8	-0.6	0.8	0.6												
$d_1'^{(3)}$			0	1	0	-1										
$d_1'^{(4)}$			-0.8	0.6					0.8	0.6						
$d_1'^{(5)}$					0		-1	0			1	0				
$d_1'^{(6)}$				-1	0				1	0			-1			
$d_1'^{(7)}$									0	1	0	-1				
$d_1'^{(8)}$											-0.8	0.6	0.8	0.6		
$d_1'^{(9)}$							-1	0					1	0		
$d_1'^{(10)}$							-1	0			0	0				
$d_1'^{(11)}$									0	1	0	1				
$d_1'^{(12)}$											-0.8	0.6			0.8	0.6
$d_1'^{(13)}$													-1	0	1	0
$d_1'^{(14)}$	1	0														
$d_1'^{(15)}$	0	1														
$d_1'^{(16)}$															0	1

Figure c Construction of the matrix $[C^M]$ for the model for the truss of Fig. a.

matrix $[H^3]^T$ [relation (c)] is placed in row 3 and columns 3 and 4, while its right half is placed in row 3 and columns 5 and 6. Element 4 of the model is connected to nodes 2 and 5. Thus, as shown in Fig. c, the left half of matrix $[H^4]^T$ [relation (e)] is placed in row 4 and columns 3 and 4, while its right half is placed in row 4 and columns 9 and 10. Element 14 of the model is connected to a support and to node 1. Thus, as shown in Fig. c, the right half of matrix $[H^{14}]^T$ is placed in row 14 and columns 1 and 2. The remaining slots of the matrix of Fig. c are filled in in an analogous fashion.

Comparing the matrix $[B^M]$ in relation (h) of Example 1 of Sec. 15.2 and the matrix $[C^M]$ of Fig. c, it is apparent that

$$[C^M] = [B^M]^T \qquad (g)$$

STEP 4. We establish the matrix $[C]$ of the truss from the matrix $[C^M]$ of the model by omitting its rows corresponding to the rows of Eqs. (15.31) which involve the basic deformation parameters of the reaction elements and its columns corresponding to the columns of Eqs. (15.31) which involve the global components of displacements inhibited by the supports of the actual structure. That is, rows 14, 15, and 16 and columns 1, 2, and 16. Referring to Fig. c, the matrix $[C]$ is equal to

$$[C] = \begin{bmatrix}
0 & 0 & 1 & 0 & 0 & 0 & 0 & 0 & 0 & 0 & 0 & 0 & 0 \\
0.8 & 0.6 & 0 & 0 & 0 & 0 & 0 & 0 & 0 & 0 & 0 & 0 & 0 \\
0 & -1 & 0 & 1 & 0 & 0 & 0 & 0 & 0 & 0 & 0 & 0 & 0 \\
-0.8 & 0.6 & 0 & 0 & 0 & 0 & 0.8 & -0.6 & 0 & 0 & 0 & 0 & 0 \\
0 & 0 & -1 & 0 & 0 & 0 & 1 & 0 & 0 & 0 & 0 & 0 & 0 \\
-1 & 0 & 0 & 0 & 1 & 0 & 0 & 0 & 0 & 0 & 0 & 0 & 0 \\
0 & 0 & 0 & 0 & 0 & -1 & 0 & 1 & 0 & 0 & 0 & 0 & 0 \\
0 & 0 & 0 & 0 & -1 & 0 & 0 & 0 & 1 & 0 & 0 & 0 & 0 \\
0 & 0 & 0 & 0 & 0 & 0 & -1 & 0 & 0 & 1 & 0 & 0 \\
0 & 0 & 0 & 0 & 0 & 0 & -0.8 & -0.6 & 0.8 & 0.6 & 0 & 0 & 0 \\
0 & 0 & 0 & 0 & 0 & 0 & 0 & 0 & 0 & -1 & 0 & 1 & 0 \\
0 & 0 & 0 & 0 & 0 & 0 & 0 & 0 & -0.8 & 0.6 & 0 & 0 & 0.8 \\
0 & 0 & 0 & 0 & 0 & 0 & 0 & 0 & 0 & 0 & -1 & 0 & 1
\end{bmatrix}$$

(h)

Comparing the matrix $[B]$ given by relation (j) of Example 1 of Sec. 15.2 and the matrix $[C]$ above, it is apparent that

$$[C] = [B]^T$$

Example 2. Consider the planar frame shown in Fig. a. Following the procedure described in this section, assemble its matrices $[\hat{C}] = [C^M]$ and $[C]$.

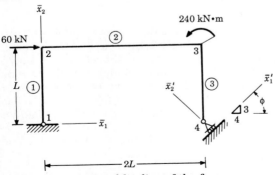

Figure a Geometry and loading of the frame.

solution

STEP 1.   We consider the model of the frame shown in Fig. b.

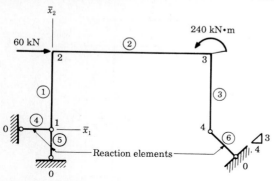

Figure b   Model of the frame of Fig. a.

STEP 2.   We compute the matrices $[H]^T$ for each element of the model of Fig. b. The matrices $[H]^T$ for the model of Fig. b are given by relations (c) to (i) of the example of Sec. 15.3. That is,

$$[H^1]^T = \begin{bmatrix} 0 & -1 & 0 & \vdots & 0 & 1 & 0 \\ 1 & 0 & -L & \vdots & -1 & 0 & 0 \\ 0 & 0 & -1 & \vdots & 0 & 0 & 1 \end{bmatrix} \tag{a}$$

$$[H^2]^T = \begin{bmatrix} -1 & 0 & 0 & \vdots & 1 & 0 & 0 \\ 0 & -1 & -2L & \vdots & 0 & 1 & 0 \\ 0 & 0 & -1 & \vdots & 0 & 0 & 1 \end{bmatrix} \tag{b}$$

$$[H^3]^T = \begin{bmatrix} 0 & 1 & 0 & \vdots & 0 & -1 & 0 \\ -1 & 0 & -L & \vdots & 1 & 0 & 0 \\ 0 & 0 & -1 & \vdots & 0 & 0 & 1 \end{bmatrix} \tag{c}$$

$$[H^4]^T = \begin{bmatrix} -1 & 0 & 0 & \vdots & 1 & 0 & 0 \end{bmatrix} \tag{d}$$

$$[H^5]^T = \begin{bmatrix} 0 & -1 & 0 & \vdots & 0 & -1 & 0 \end{bmatrix} \tag{e}$$

$$[H^6]^T = \begin{bmatrix} 0.6 & -0.8 & \vdots & 0 & -0.6 & 0.8 & 0 \end{bmatrix} \tag{f}$$

STEP 3.   We construct the matrix $[C^M]$ of the model of Fig. b using the matrices $[H]^T$ of its elements obtained in step 2. We note that $[C^M]$ is a 13×12 matrix. Element 1 of the model is connected to nodes 1 and 2. Thus, as shown in Fig. c, the left half of matrix $[H^1]^T$ [relation (a)] is placed in rows 1 to 3 and columns 1 to 3, while its right half is placed in rows 1 to 3 and columns 4 to 6. Element 2 of the model is connected to nodes 2 and 3. Thus, as shown in Fig. c, the left half of matrix $[H^2]^T$ [relation (b)] is placed in rows 4 to 6 and columns 4 to 6, while its right half is placed in rows 4 to 6 and columns 7 to 9. Element 3 of the model is connected to nodes 3 and 4.

| | $\Delta_1^{(1)}$ | $\Delta_2^{(1)}$ | $\Delta_3^{(1)}$ | $\Delta_1^{(2)}$ | $\Delta_2^{(2)}$ | $\Delta_3^{(2)}$ | $\Delta_1^{(3)}$ | $\Delta_2^{(3)}$ | $\Delta_3^{(3)}$ | $\Delta_1^{(4)}$ | $\Delta_2^{(4)}$ | $\Delta_3^{(4)}$ |
	1	2	3	4	5	6	7	8	9	10	11	12
$d_1^{(1)}$	−0	−1	0	0	1	0						
$d_2^{(1)}$	1	0	−L	−1	0	0						
$d_3^{(1)}$	0	0	−1	0	0	1						
$d_1^{(2)}$				−1	0	0	1	0	0			
$d_2^{(2)}$				0	−1	−2L	0	1	0			
$d_3^{(2)}$					0	−1	0	0	1			
$d_1^{(3)}$							0	1	0	0	−1	0
$d_2^{(3)}$							−1	0	−L	1	0	0
$d_3^{(3)}$							0	0	−1	0	0	1
$d_1^{(4)}$	1	0	0									
$d_1^{(5)}$	0	1	0									
$d_1^{(6)}$										−0.6	0.8	0

Figure c Construction of the matrix $[C^M]$ for the model for the frame of Fig. a.

Thus, referring to Fig. c, the left half of matrix $[H^3]^T$ [relation (c)] is placed in rows 7 to 9 and columns 7 to 9, while its right half is placed in rows 7 to 9 and columns 10 to 12. Element 4 of the model is connected to a support and node 1. Thus, referring to Fig. c, the right half of matrix $[H^4]^T$ [relation (d)] is placed in row 10 and columns 1 to 3. Element 5 of the model is connected to a support and node 1. Thus, referring to Fig. c, the right half of matrix $[H^5]^T$ [relation (e)] is placed in row 1 and columns 1 to 3. Element 6 of the model is connected to a support and node 4. Thus referring to Fig. c the right half of matrix $[H^6]^T$ [relation (f)] is placed in row 12 and columns 10 and 12.

Comparing the matrix $[B^M]$ in Eqs. (i) of the example of Sec. 15.3 and the matrix $[C^M]$ of Fig. c, it is apparent that for the model of Fig. b we have

$$[C^M] = [B^M]^T \tag{g}$$

STEP 4. We establish the matrix $[C]$ of the frame from its matrix $[\hat{C}] = [C^M]$. In order to accomplish this we first modify the matrix $[C^M]$. That is, we establish the matrix $[C_m^M]$, defined by Eqs. (15.44), as follows.

1. We form the transformation matrix $[\Lambda']$ which transforms the global components of translation of node 4 to skew components. Denoting by $\Delta_1'^{(4)}$ and $\Delta_2'^{(4)}$ the components of displacement of node 4 in the direction of rolling and normal to this direction, respectively and referring to Fig. a,

we have

$$\Delta_1^{(4)} = \Delta_1'^{(4)} \cos \phi_{11} - \Delta_2'^{(4)} \sin \phi_{11}$$

$$\Delta_2^{(4)} = \Delta_1'^{(4)} \sin \phi_{11} + \Delta_2'^{(4)} \cos \phi_{11} \tag{h}$$

or

$$\left\{ \begin{matrix} \Delta_1^{(4)} \\ \Delta_2^{(4)} \end{matrix} \right\} = [\Lambda']^T \left\{ \begin{matrix} \Delta_1'^{(4)} \\ \Delta_2'^{(4)} \end{matrix} \right\} \tag{i}$$

where

$$[\Lambda'] = \begin{bmatrix} \cos \phi_{11} & \sin \phi_{11} \\ -\sin \phi_{11} & \cos \phi_{11} \end{bmatrix} = \begin{bmatrix} 0.8 & 0.6 \\ -0.6 & 0.8 \end{bmatrix} \tag{j}$$

2. We form the matrix $[C_c]$. This matrix consists of columns 10 and 11 of the matrix of Fig. c. That is,

$$[C_c] = \left\{ \begin{matrix} 0 & 0 \\ 0 & 0 \\ 0 & 0 \\ 0 & 0 \\ 0 & 0 \\ 0 & 0 \\ 0 & -1 \\ 1 & 0 \\ 0 & 0 \\ 0 & 0 \\ 0 & 0 \\ -0.6 & 0.8 \end{matrix} \right\} \tag{k}$$

3. We compute the matrix $[C_c^1]$ defined as

$$[C_c^1] = [C_c][\Lambda']^T = \begin{bmatrix} 0 & 0 \\ 0 & 0 \\ 0 & 0 \\ 0 & 0 \\ 0 & 0 \\ 0 & 0 \\ -0.6 & -0.8 \\ 0.8 & -0.6 \\ 0 & 0 \\ 0 & 0 \\ 0 & 0 \\ 0 & 1 \end{bmatrix} \tag{l}$$

4. The matrix $[C_m^M]$ is obtained from the matrix $[C^M]$ by replacing columns 10 and 11 of the latter with the columns of the matrix $[C_c^1]$. Thus, referring to Fig. c. we have

$$[C_m^M] = \begin{bmatrix} 0 & -1 & 0 & 0 & 1 & 0 & 0 & 0 & 0 & 0 & 0 & 0 \\ 1 & 0 & -L & -1 & 0 & 0 & 0 & 0 & 0 & 0 & 0 & 0 \\ 0 & 0 & -1 & 0 & 0 & 1 & 0 & 0 & 0 & 0 & 0 & 0 \\ 0 & 0 & 0 & -1 & 0 & 0 & 1 & 0 & 0 & 0 & 0 & 0 \\ 0 & 0 & 0 & 0 & -1 & -2L & 0 & 1 & 0 & 0 & 0 & 0 \\ 0 & 0 & 0 & 0 & 0 & -1 & 0 & 0 & 1 & 0 & 0 & 0 \\ 0 & 0 & 0 & 0 & 0 & 0 & 0 & 1 & 0 & -0.6 & -0.8 & 0 \\ 0 & 0 & 0 & 0 & 0 & 0 & -1 & 0 & -L & 0.8 & -0.6 & 0 \\ 0 & 0 & 0 & 0 & 0 & 0 & 0 & 0 & -1 & 0 & 0 & 1 \\ 1 & 0 & 0 & 0 & 0 & 0 & 0 & 0 & 0 & 0 & 0 & 0 \\ 0 & 1 & 0 & 0 & 0 & 0 & 0 & 0 & 0 & 0 & 0 & 0 \\ 0 & 0 & 0 & 0 & 0 & 0 & 0 & 0 & 0 & 0 & 1 & 0 \end{bmatrix} \tag{m}$$

5. The matrix $[C]$ of the frame is obtained from the matrix $[C_m^M]$ of its model by canceling its rows corresponding to the rows of Eqs. (15.44) involving the basic deformation parameters of the reaction elements and its columns corresponding to the columns of Eqs. (15.44) which are multiplied by the components of displacements inhibited by the supports of the structure; that is, columns 1, 2, 3, and 11 and rows 10 to 12 of the matrix $[C_m^M]$. Thus, referring to relation (m) we have

$$[C] = \begin{bmatrix} 0 & 1 & 0 & 0 & 0 & 0 & 0 & 0 \\ -1 & 0 & 0 & 0 & 0 & 0 & 0 & 0 \\ 0 & 0 & 1 & 0 & 0 & 0 & 0 & 0 \\ -1 & 0 & 0 & 1 & 0 & 0 & 0 & 0 \\ 0 & -1 & -2L & 0 & 1 & 0 & 0 & 0 \\ 0 & 0 & -1 & 0 & 0 & 1 & 0 & 0 \\ 0 & 0 & 0 & 0 & 1 & 0 & -0.6 & 0 \\ 0 & 0 & 0 & -1 & 0 & -L & 0.8 & 0 \\ 0 & 0 & 0 & 0 & 0 & -1 & 0 & 1 \end{bmatrix} \tag{n}$$

Comparing the matrix $[C^M]$ given in Fig. c with the matrix $[B^M]$ of Eqs. (j) of the example of Sec. 15.3, and the matrix $[C]$ given by relation (n) with the matrix $[B]$ given by relation (r) of the example of Sec. 15.3, it is apparent that

$$[C^M] = [B^M]^T \tag{o}$$

$$[C] = [B]^T \tag{p}$$

15.6 Problems

1 to 6. Use the procedure described in Sec. 15.2 to form the matrix $[\hat{B}] = [\hat{B}^M]$ of the structure of Fig. 15.P1. Repeat with the structures of Figs 15.P2 to 15.P6.

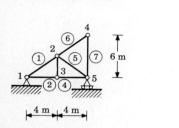

Figure 15.P1

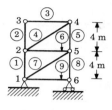

Figure 15.P2

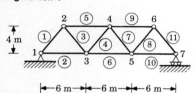

Figure 15.P3

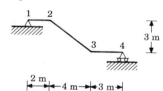

Figure 15.P4

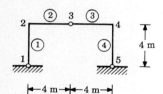

Figure 15.P5

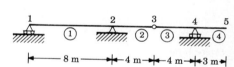

Figure 15.P6

7 and 8. Using the procedure described in Sec. 15.3, form the matrices $[\hat{B}] = [B^M]$ and $[B]$ for the structure of Fig. 15.P7. Repeat with the structure of Fig. 15.P8.

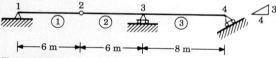

Figure 15.P7

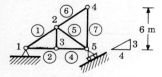

Figure 15.P8

9 to 18. Using the procedure described in Sec. 15.2, form the matrices $[\hat{B}] = [B^M]$ for the structure of Fig. 15.P9. Repeat with the structures of Figs. 15.P10 to 15.P18.

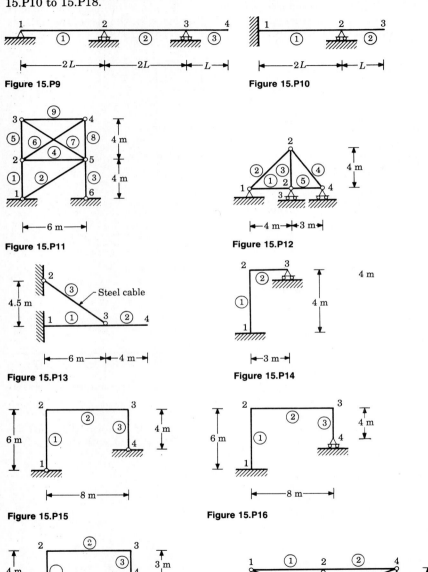

Figure 15.P9

Figure 15.P10

Figure 15.P11

Figure 15.P12

Figure 15.P13

Figure 15.P14

Figure 15.P15

Figure 15.P16

Figure 15.P17

Figure 15.P18

19 to 26. Compute the matrix of the basic nodal actions $\{a^E\}$ of all the elements of the structure loaded as shown in Fig. 15.P19. Repeat with the structures of Figs. 15.P20 to 15.P26.

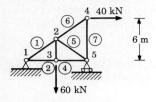

Figure 15.P19

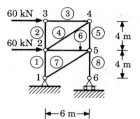

Figure 15.P20

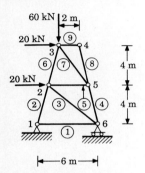

Figure 15.P21

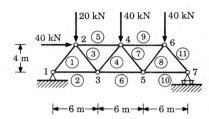

Figure 15.P22

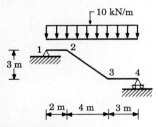

Figure 15.P23

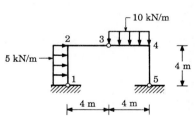

Figure 15.P24

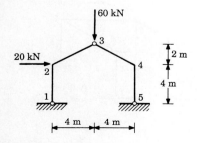

Figure 15.P25

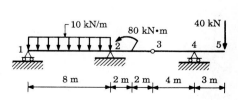

Figure 15.P26

27 to 30.   Using the procedure described in Sec. 15.5, form the matrices $[\hat{C}]$ and $[C]$ for the structure of Fig. 15.P1. Repeat with the structures of Figs. 15.P3, 15.P5, and 15.P7.

31 to 34.   Using the procedure described in Sec. 15.4, form the matrix $[\hat{C}]$ for the structure of Fig. 15.P11. Repeat with the structures of Figs. 15.P13, 15.P15, and 15.P17.

35 to 38.   Using the procedure described in Sec. 15.2, form the matrix $[\hat{B}]$ of the structure of Fig. 15.P35. Repeat with the structures of Figs. 15.P36 to 15.P38.

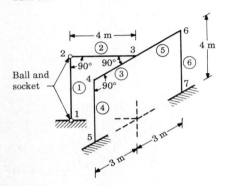

Figure 15.P35

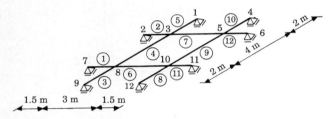

Figure 15.P36

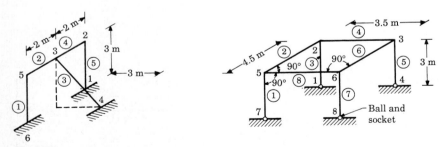

Figure 15.P37   **Figure 15.P38**

Computation of the Components of Displacements of the Nodes of Statically Determinate Framed Structures

16.1 Introduction

Consider a structure subjected only to equivalent actions on its nodes. For this structure the following fundamental relations are valid.

1. The equations of equilibrium of the nodes of a structure. That is,

$$\{P^{EF}\} = [B]\{\hat{a}^{E}\} \qquad (16.1)$$

For statically determinate structures which do not form a mechanism, $[B]$ is a square, nonsingular matrix, and thus relation (16.1) can be solved to give the basic nodal actions of all their elements. That is,

$$\{\hat{a}^{E}\} = [b]\{P^{EF}\} \qquad (16.2)$$

where
$$[b] = [B]^{-1} \qquad (16.3)$$

2. The equations of compatibility of the basic deformation parameters of all the elements of a structure and the components of displacements of its nodes. Since the structure is subjected only to equivalent actions on its nodes $[\{\Delta^{S}\} = 0]$ Eqs. (15.30) are valid; that is,

$$\{\hat{d}\} = [C]\{\Delta^{F}\} \qquad (16.4)$$

For statically determinate structures which do not form a mechanism, $[C]$ is a square, nonsingular matrix, and thus relation (16.4) can be written as

$$[\Delta^{F}] = [c]\{\hat{d}\} \qquad (16.5)$$

where
$$[c] = [C]^{-1} \tag{16.6}$$

3. The relations between the basic nodal actions and the basic deformation parameters of all the elements of a structure subjected to the equivalent actions on its nodes. For a structure with NE elements made from isotropic linearly elastic materials we have

$$\{d^1\} = [f^1]\{a^{E1}\}$$
$$\{d^2\} = [f^2]\{a^{E2}\} \tag{16.7}$$
$$\vdots$$
$$\{d^{NE}\} = [f^{NE}]\{a^{ENE}\}$$

These relations may be rewritten as

$$\{\hat{d}\} = [\hat{f}]\{\hat{a}^E\} \tag{16.8}$$

where $\{\hat{d}\}$ and $\{\hat{a}^E\}$ are defined by relations (15.30) and (15.18), respectively, and

$$[\hat{f}] = \begin{bmatrix} [f^1] & & & \\ & [f^2] & & \\ & & \ddots & \\ & & & [f^{NE}] \end{bmatrix} \tag{16.9}$$

Relations (16.7) may be rewritten in stiffness form. That is,

$$\{a^{E1}\} = [k^1]\{d^1\}$$
$$\{a^{E2}\} = [k^2]\{d^2\} \tag{16.10}$$
$$\vdots$$
$$\{a^{ENE}\} = [k^{NE}]\{d^{NE}\}$$

These relations may be rewritten as

$$\{\hat{a}^E\} = [\hat{k}]\{\hat{d}\} \tag{16.11}$$

where
$$[\hat{k}] = \begin{bmatrix} [k^1] & & & \\ & [k^2] & & \\ & & \ddots & \\ & & & [k^{NE}] \end{bmatrix} \tag{16.12}$$

Comparing relations (16.8) and (16.11) we obtain

$$[\hat{k}] = [\hat{f}]^{-1} \tag{16.13}$$

Moreover, in the examples of Sec. 15.5, we have shown that

$$[C] = [B]^T \tag{16.14}$$

Substituting relation (16.6) into (16.5) and using relations (16.2), (16.3), (16.8), and (16.14) for a statically determinate structure subjected only to equivalent actions on its nodes $[\{\Delta^S\} = 0]$, we get

$$\{\Delta^F\} = [C]^{-1}\{\hat{d}\} = [[B]^T]^{-1}\{\hat{d}\} = [[B]^{-1}]^T\{\hat{d}\} = [b]^T\{\hat{d}\}$$
$$= [b]^T[\hat{f}]\{a^E\} = [b]^T[\hat{f}]\{b\}\{P^{EF}\}$$

or
$$\{\Delta^F\} = [F]\{P^{EF}\} \tag{16.15}$$

where
$$[F] = [b]^T[\hat{f}][b] \tag{16.16}$$

Equations (16.15) are called the *flexibility equations* and $[F]$ is called the *flexibility matrix of the statically determinate structure.*

Substituting relation (16.12) into (16.1) and using (16.4) and (16.14) for any structure (statically determinate or indeterminate) subjected only to equivalent actions on its nodes $[\{\Delta^S\} = 0]$ we get

$$\{P^{EF}\} = [B]\{\hat{a}^E\} = [B][\hat{k}]\{\hat{d}\} = [B][\hat{k}][C]\{\Delta^F\} = [B][\hat{k}][B]^T\{\Delta^F\} \tag{16.17}$$

or
$$\{P^{EF}\} = [S^{FF}]\{\Delta^F\} \tag{16.18}$$

where
$$[S^{FF}] = [B][\hat{k}][B]^T = [C]^T[\hat{k}][C] \tag{16.19}$$

Referring to relation (9.4) it is apparent that $[S^{FF}]$ is the basic stiffness matrix of the structure.

Thus the basic stiffness matrix of a structure (statically determinate or indeterminate) can be obtained from its matrix $[B]$ or $[C]$ and its matrix $[\hat{k}]$. However, for large structures this approach, although conceptually straightforward, requires considerably more computer time and storage than the direct stiffness method.

Comparing relations (16.15) and (16.18) we have

$$[F] = [S^{FF}]^{-1} \tag{16.20}$$

16.2 Computation of the Components of Displacements of the Nodes of Statically Determinate Structures

In this section we present two methods for computing the components of displacements of the nodes of statically determinate framed structures—the flexibility method and the stiffness method. The flexibility method entails the following steps.

1. The matrix of equivalent actions $\{P^{EF}\}$ is established (see Sec. 7.2).
2. The matrix $[B]$ for the structure is established following the procedure described in Sec. 15.3 or 15.2 depending on whether the structure has or does not have skew supports, respectively.

3. The matrix $[b]$ of the structure is obtained using relation (16.3).

4. The flexibility matrix $[\hat{f}]$ of all the elements of the structure is assembled [see relation (16.9)].

5. The matrices $[b]$ and $[\hat{f}]$ are substituted into relation (16.16) to obtain the flexibility matrix $[F]$ of the structure.

6. The matrix of global components of displacements of the nodes of the structure $\{\Delta^F\}$ is computed using relation (16.15). As discussed in Sec. 7.2, the components of displacements of the nodes of the structure subjected to equivalent actions on its nodes are identical to those of the structure subjected to the given loads.

The stiffness method entails the following steps.

1. The matrix of equivalent actions $\{P^{EF}\}$ is established (see Sec. 7.2).

2. The matrix $[B]$ for the structure is established following the procedure presented in Sec. 15.3 or 15.2 depending on whether the structure has or does not have skew supports.

3. The stiffness matrix $[\hat{k}]$ of all the elements of the structure is assembled [see relation (16.12)].

4. The matrices $[B]$ and $[\hat{k}]$ are substituted into relation (16.19) to obtain the basic stiffness matrix $[S^{FF}]$ of the structure.

5. The matrix $[\Delta^F]$ for the structure is computed using relation (16.18). In what follows, we present two examples of how to compute the components of displacements of the nodes of statically determinate structures.

Example 1. Compute the components of displacements of the nodes of the truss of Fig. a using the flexibility method. All the elements of the truss have the same constant cross section and are made from the same isotropic linearly elastic material.

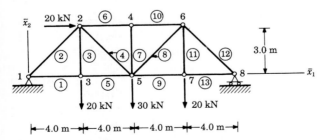

Figure a Geometry and loading of the truss.

solution

STEP 1.   Since the truss is subjected only to external forces on its nodes, its matrix of equivalent actions $\{P^{EF}\}$ is obtained by referring to Fig. a. Thus

$$\{P^{EF}\} = \begin{Bmatrix} 20 \\ 0 \\ 0 \\ -20 \\ 0 \\ 0 \\ 0 \\ -30 \\ 0 \\ 0 \\ 0 \\ -20 \\ 0 \end{Bmatrix} \qquad \text{(a)}$$

STEPS 2 AND 3.   The matrices $[B]$ and $[b]$ of the truss are established in Example 1 of Sec. 15.2. Referring to relations (j) of this example, we have matrix $[b]$, given on p. 540.

STEP 4.   We assemble the flexibility matrix of all the elements of the truss. Referring to relation (5.58) the flexibility matrix of an element of a truss is

$$[f] = \frac{L}{EA} \qquad \text{(c)}$$

Thus, referring to Fig. a, the flexibility matrix $[\hat{f}]$ for all the elements of the truss is

$$[\hat{f}] = \frac{1}{EA} \begin{bmatrix} 4 & 0 & 0 & 0 & 0 & 0 & 0 & 0 & 0 & 0 & 0 & 0 & 0 \\ 0 & 5 & 0 & 0 & 0 & 0 & 0 & 0 & 0 & 0 & 0 & 0 & 0 \\ 0 & 0 & 3 & 0 & 0 & 0 & 0 & 0 & 0 & 0 & 0 & 0 & 0 \\ 0 & 0 & 0 & 5 & 0 & 0 & 0 & 0 & 0 & 0 & 0 & 0 & 0 \\ 0 & 0 & 0 & 0 & 4 & 0 & 0 & 0 & 0 & 0 & 0 & 0 & 0 \\ 0 & 0 & 0 & 0 & 0 & 4 & 0 & 0 & 0 & 0 & 0 & 0 & 0 \\ 0 & 0 & 0 & 0 & 0 & 0 & 3 & 0 & 0 & 0 & 0 & 0 & 0 \\ 0 & 0 & 0 & 0 & 0 & 0 & 0 & 5 & 0 & 0 & 0 & 0 & 0 \\ 0 & 0 & 0 & 0 & 0 & 0 & 0 & 0 & 4 & 0 & 0 & 0 & 0 \\ 0 & 0 & 0 & 0 & 0 & 0 & 0 & 0 & 0 & 4 & 0 & 0 & 0 \\ 0 & 0 & 0 & 0 & 0 & 0 & 0 & 0 & 0 & 0 & 3 & 0 & 0 \\ 0 & 0 & 0 & 0 & 0 & 0 & 0 & 0 & 0 & 0 & 0 & 5 & 0 \\ 0 & 0 & 0 & 0 & 0 & 0 & 0 & 0 & 0 & 0 & 0 & 0 & 4 \end{bmatrix} \qquad \text{(d)}$$

STEP 5.   We compute the flexibility matrix of the truss by substituting relations (d) and (b) into (16.16). It is given on p. 541.

(b)

$$[b] = [B]^{-1} =$$

0.750	−1	1	−1	0.750	−0.667	1	−0.667	0.750	−0.333	1	−0.333	1
0.313	1.250	0	1.250	0.313	0.833	0	0.833	0.313	0.417	0	0.417	0
0	0	0	0	0	0	0	0	0	0	0	0	0
−0.313	0.417	0	0.417	−0.313	−0.833	0	−0.833	−0.313	−0.417	0	−0.417	0
0.750	−1	0	−1	0.750	−0.667	1	−0.667	0.750	−0.333	1	−0.333	0
−0.500	0.667	0	0.667	0.500	1.333	0	1.333	0.500	0.667	0	0.667	1
0	0	0	0	0	0	1	0	0	0	0	0	0
0.313	−0.417	0	−0.417	0.313	−0.833	0	−0.833	0.313	0.417	0	0.417	0
0.250	−0.333	0	−0.333	0.250	−0.667	0	−0.667	0.250	−1	1	−1	0
−0.500	0.667	0	0.667	−0.500	1.333	0	1.333	0.500	0.667	0	0.667	1
0	0	0	0	0	0	0	0	0	−1	0	−1	0
−0.313	0.417	0	0.417	−0.313	0.833	0	0.833	−0.313	1.250	0	1.250	0
0.250	−0.333	0	−0.333	0.250	−0.667	0	−0.667	0.250	−1	0	−1	1

$$[F] = [b]^T[\hat{f}][b]$$

$$= \frac{1}{EA}
\begin{bmatrix}
8.9594 & -9.3356 & 3 & -9.3356 & 6.9594 & -10.6680 & 6 & -10.6680 & 4.9594 & -6.6644 & 7 & -6.6644 & 8 \\
-9.3356 & -22.8671 & -4 & 22.8671 & -6.6676 & 21.1688 & -8 & 21.1688 & -3.9996 & 12.3607 & -9.3320 & 12.3607 & -10.6640 \\
3 & -4 & 4 & -4 & 3 & -2.6680 & 4 & -2.6680 & 3 & -1.3320 & 4 & -1.3320 & 4 \\
-9.3356 & 22.8671 & -4 & 25.8671 & -6.6676 & 21.1688 & -8 & 21.1688 & -3.9996 & 12.3607 & -9.3320 & 12.3607 & -10.6640 \\
6.9594 & -6.6676 & 3 & -6.6676 & 8.9594 & -5.3360 & 6 & -5.3360 & 6.9594 & -3.9964 & 7 & -3.9964 & 8 \\
-10.6680 & 21.1688 & -2.6680 & 21.1688 & -5.3360 & 38.2111 & -5.3360 & 35.2111 & -0.0040 & 21.1688 & -8.0040 & 21.1688 & -10.6720 \\
6 & -8 & 4 & -8 & 6 & -5.3360 & 8 & -5.3360 & 6 & -2.6640 & 8 & -2.6640 & 8 \\
-10.6680 & 21.1688 & -2.6680 & 21.1688 & -5.3360 & 35.2111 & -5.3360 & 35.2111 & -0.0040 & 21.1688 & -8.0040 & 21.1688 & -10.6720 \\
4.9594 & -3.9996 & 3 & -3.9996 & 6.9594 & -0.004 & 6 & -0.0040 & 8.9594 & -1.3284 & 7 & -1.3284 & 8 \\
-6.6644 & 12.3607 & -1.3320 & 12.3607 & -3.9964 & 21.1688 & -2.6640 & 21.1688 & -1.3284 & 22.8671 & -6.6640 & 22.8671 & -10.6640 \\
7 & -9.3320 & 4 & -9.3320 & 7 & -8.004 & 8 & -8.0040 & 7 & -6.6640 & 12 & -6.6640 & 12 \\
-6.6644 & 12.3607 & -1.3320 & 12.3607 & -3.9964 & 21.1688 & -2.6640 & 21.1688 & -1.3284 & 22.8671 & -6.6640 & 25.8671 & -10.6640 \\
8 & -10.6640 & 4 & -10.672 & 8 & -10.672 & 8 & -10.6720 & 8 & -10.6640 & 12 & -10.6640 & 16
\end{bmatrix}$$

(e)

STEP 6. We compute the matrix of the components of displacements of the nodes of the truss. Substituting relations (a) and (e) into (16.15) we obtain

$$\{\Delta^F\} = [F]\{P^{EF}\} = \frac{1}{EA}
\begin{Bmatrix}
819.228 \\
-1526.332 \\
246.68 \\
-1586.332 \\
512.548 \\
-2116.445 \\
493.360 \\
-2116.445 \\
205.868 \\
-1472.908 \\
700.040 \\
-1532.908 \\
906.720
\end{Bmatrix} \tag{f}$$

Example 2. Compute the deflection of point 2 of the simply supported beam loaded as shown in Fig. a. Use both the flexibility and the stiffness methods. The beam is made from an isotropic linearly elastic material and has a constant cross section.

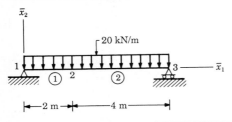

Figure a Geometry and loading of the beam.

solution In addition to the two endpoints of the beam we choose as a node the point whose deflection is desired. Thus as shown in Fig. a we consider the beam as consisting of two elements connected to three nodes.

STEP 1. We compute the equivalent actions which must be applied on the nodes of the beam. As discussed in Sec. 7.2, the components of displacements of the nodes of a structure, subjected to a general loading, are equal to the corresponding components of displacements of the nodes of the structure subjected to the equivalent actions on its nodes.

As shown in Fig. b, the restrained structure is formed by restraining the nodes of the beam from translating and rotating by applying restraining actions to them. The nodal actions of the elements of the restrained structure are established by referring to the table on the inside of the back cover and

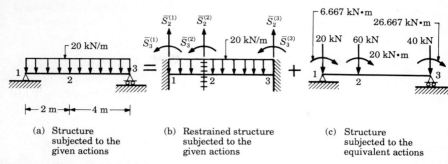

(a) Structure
subjected to the
given actions

(b) Restrained structure
subjected to the
given actions

(c) Structure
subjected to the
equivalent actions

Figure b Superposition of the restrained structure and the structure subjected to the equivalent actions.

are shown in Fig. c. Referring to this figure, we have

$$\{\hat{a}^R\} = \begin{Bmatrix} 20 \\ -6.667 \\ 40 \\ -26.667 \end{Bmatrix} \qquad (a)$$

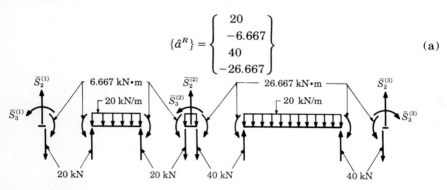

Figure c Free-body diagram of the elements and node 2 of the restrained structure.

Moreover, referring to Fig. c from the equilibrium of the nodes of the restrained structure, the restraining actions are obtained as

$$\bar{S}_2^{(1)} = 20 \text{ kN} \qquad \bar{S}_2^{(2)} = 60 \text{ kN} \qquad \bar{S}_2^{(3)} = 40 \text{ kN}$$
$$\bar{S}_3^{(1)} = 6.667 \text{ kN} \cdot \text{m} \qquad \bar{S}_3^{(2)} = 20 \text{ kN} \cdot \text{m} \qquad \bar{S}_3^{(3)} = -26.667 \text{ kN} \cdot \text{m} \qquad (b)$$

Using the above results the equivalent actions acting on the nodes of the beam are shown in Fig. bc. Referring to this figure, the matrix of equivalent actions for the beam is

$$\{P^{EF}\} = \begin{Bmatrix} P_3^{(1)} \\ P_2^{(2)} \\ P_3^{(2)} \\ P_3^{(3)} \end{Bmatrix} = \begin{Bmatrix} -6.667 \\ -60 \\ -20 \\ 26.667 \end{Bmatrix} \qquad (c)$$

Inasmuch as the beam is subjected only to transverse forces, we have not included the axial components of external forces in its matrix of equivalent actions. For this reason in the next step we will not include the axial components of the nodal forces in the matrix of the basic nodal actions of the elements of the beam.

Part I. Computation of the Deflection of Point 2 of the Beam Using the Flexibility Method

STEPS 2 AND 3.   We compute the matrix $[b]$ for the beam. The matrix $[b]$ of this simple beam may be established directly. Referring to Fig. d from the

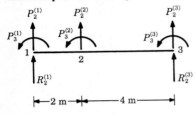

Figure d   Free-body diagram of the beam subjected to all possible external actions.

equilibrium of the beam we have

$$\sum \bar{M}_3^{(1)} = 0 \qquad R_2^{(3)} = -\frac{1}{6}[6P_3^{(1)} + 2P_2^{(2)} + P_3^{(2)} + 6P_2^{(3)} + P_3^{(3)}]$$

$$\sum \bar{F}_2 = 0 \qquad R_2^{(1)} = \frac{1}{6}[P_3^{(1)} - 4P_2^{(2)} + P_3^{(2)} + P_3^{(3)} - 6P_2^{(1)}]$$

Moreover, from the equilibrium of element 1 and node 1 we have

$$F_2^{1k} = -F_2^{1j} = -R_2^{(1)} - P_2^{(1)} = -\frac{1}{6}[P_3^{(1)} - 4P_2^{(2)} + P_3^{(2)} + P_3^{(3)}]$$

Furthermore, from the equilibrium of element 1 we get

$$\sum \bar{M}_3^{(1)} = 0$$

$$M_3^{1k} = -2F_2^{1k} - M_3^{1j} = -2F_2^{1k} - P_3^{(1)} = \frac{1}{6}[2P_3^{(1)} - 8P_2^{(2)} + 2P_3^{(2)} + 2P_3^{(3)}] - P_3^{(1)}$$

Finally, from the equilibrium of node 3 we have

$$\sum \bar{F}_2 = 0 \qquad F_2^{2k} = P_2^{(3)} + R_2^{(3)} = \frac{1}{6}[-P_3^{(1)} - 2P_2^{(2)} - P_3^{(2)} - P_3^{(3)}]$$

$$\sum \bar{M}_3 = 0 \qquad M_3^{2k} = P_3^{(3)}$$

The above relations may be rewritten as

$$\{\hat{a}^E\} = \begin{Bmatrix} F_2^{1k} \\ M_3^{1k} \\ F_2^{2k} \\ M_3^{2k} \end{Bmatrix} = \frac{1}{6} \begin{bmatrix} -1 & 4 & -1 & -1 \\ -4 & -8 & 2 & 2 \\ -1 & -2 & -1 & -1 \\ 0 & 0 & 0 & 6 \end{bmatrix} \begin{Bmatrix} P_3^{(1)} \\ P_2^{(2)} \\ P_3^{(2)} \\ P_3^{(3)} \end{Bmatrix} \tag{d}$$

Thus,
$$[b] = \frac{1}{6} \begin{bmatrix} -1 & 4 & -1 & -1 \\ -4 & -8 & 2 & 2 \\ -1 & -2 & -1 & -1 \\ 0 & 0 & 0 & 1 \end{bmatrix} \tag{e}$$

Substituting relation (c) into (d) we get

$$\{\hat{a}^E\} = \begin{Bmatrix} -40 \\ 86.667 \\ 20 \\ 26.667 \end{Bmatrix} \tag{f}$$

Notice that $[\hat{a}^E]$ is the matrix of basic nodal actions of all the elements of the beam when subjected to the equivalent actions shown in Fig. bc. Using results (a) and (f), the matrix of the basic nodal actions of all the elements of the beam of Fig. a is equal to

$$\{\hat{a}\} = \{\hat{a}^R\} + \{\hat{a}^E\} = \begin{Bmatrix} 20 \\ -6.667 \\ 40 \\ -26.667 \end{Bmatrix} + \begin{Bmatrix} -40 \\ 86.667 \\ 20 \\ 26.667 \end{Bmatrix} = \begin{Bmatrix} -20 \\ 80 \\ 60 \\ 0 \end{Bmatrix} \tag{g}$$

STEP 4. We assemble the flexibility matrix $[\hat{f}]$ for all the elements of the beam. That is, referring to relations (5.67) and (16.9) we have

$$[\hat{f}] = \frac{1}{3EI} \begin{bmatrix} 8 & 6 & 0 & 0 \\ 6 & 6 & 0 & 0 \\ 0 & 0 & 64 & 24 \\ 0 & 0 & 24 & 12 \end{bmatrix} \tag{h}$$

STEP 5. We compute the flexibility matrix for the beam. Substituting relations (e) and (h) into (16.16), we get

$$[F] = [b]^T[\hat{f}][b]$$

$$= \frac{1}{108EI} \begin{bmatrix} -1 & -4 & -1 & 0 \\ 4 & -8 & -2 & 0 \\ -1 & 2 & -1 & 0 \\ -1 & 2 & -1 & 6 \end{bmatrix} \begin{bmatrix} 8 & 6 & 0 & 0 \\ 6 & 6 & 0 & 0 \\ 0 & 0 & 64 & 24 \\ 0 & 0 & 24 & 12 \end{bmatrix} \begin{bmatrix} -1 & 4 & -1 & -1 \\ -4 & -8 & 2 & 2 \\ -1 & -2 & -1 & -1 \\ 0 & 0 & 0 & 6 \end{bmatrix}$$

$$= \frac{1}{108EI} \begin{bmatrix} 216 & 240 & 36 & -108 \\ 240 & 384 & 96 & -192 \\ 36 & 96 & 72 & -72 \\ -108 & -192 & -72 & 216 \end{bmatrix} \tag{i}$$

STEP 6. We compute the matrix of node displacements of the beam.

Substituting relations (c) and (i) into (16.15), we get

$$\{\Delta^F\} = \begin{Bmatrix} \bar{\Delta}_3^{(1)} \\ \bar{\Delta}_2^{(2)} \\ \bar{\Delta}_3^{(2)} \\ \bar{\Delta}_3^{(3)} \end{Bmatrix} = [F]\{P^{EF}\} = \frac{1}{108EI} \begin{bmatrix} 216 & 240 & 36 & -108 \\ 240 & 384 & 96 & -192 \\ 36 & 96 & 72 & -72 \\ -108 & -192 & -72 & 216 \end{bmatrix} \begin{Bmatrix} -6.667 \\ -60 \\ -20 \\ 26.667 \end{Bmatrix}$$

$$= \frac{1}{EI} \begin{Bmatrix} -180.00 \\ -293.34 \\ -86.67 \\ 180.00 \end{Bmatrix} \qquad (j)$$

Thus the deflection of point 2 of the beam are equal to

$$u_2^{(2)} = \bar{\Delta}_2^{(2)} = -\frac{293.34}{EI} \quad \text{or} \quad \frac{293.34}{EI} \downarrow \qquad (k)$$

Part II. Computation of the Deflection of Point 2 of the Beam Using the Stiffness Method

STEP 2. We compute the matrix $[B]$ of the beam. This may be accomplished by adhering to the procedure described in Sec. (15.2). However, we can establish the matrix $[B]$ of the simply supported beam of Fig. a by referring to Fig. e and considering the equilibrium of the nodes of the beam. Thus,

$$\begin{Bmatrix} \bar{P}_3^{(1)} \\ \bar{P}_2^{(2)} \\ \bar{P}_3^{(2)} \\ \bar{P}_3^{(3)} \end{Bmatrix} = \begin{bmatrix} -2 & -1 & 0 & 0 \\ 1 & 0 & -1 & 0 \\ 0 & 1 & -4 & -1 \\ 0 & 0 & 0 & 1 \end{bmatrix} \begin{Bmatrix} F_2^{1k} \\ M_3^{1k} \\ F_2^{2k} \\ M_3^{2k} \end{Bmatrix} \qquad (1)$$

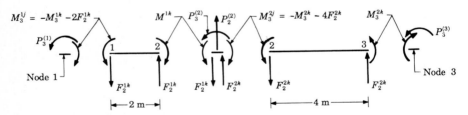

$M_3^{1j} = -M_3^{1k} - 2F_2^{1k}$ M^{1k} $P_3^{(2)}$ $P_2^{(2)}$ $M_3^{2j} = -M_3^{2k} - 4F_2^{2k}$ M_3^{2k}

$P_3^{(1)}$ $P_3^{(3)}$

Node 1   Node 3

F_2^{1k} F_2^{1k} F_2^{1k} F_2^{2k} F_2^{2k} F_2^{2k}

|←— 2 m —→| |←———— 4 m ————→|

Figure e Free-body diagram of the elements and node 2 of the beam.

Hence,

$$[B] = \begin{bmatrix} -2 & -1 & 0 & 0 \\ 1 & 0 & -1 & 0 \\ 0 & 1 & -4 & -1 \\ 0 & 0 & 0 & 1 \end{bmatrix} \qquad (m)$$

STEP 3. We assemble the stiffness matrix $[\hat{k}]$ of all the elements of the beam. That is, referring to relations (5.50), (5.67), and (16.12), we have

$$[\hat{k}] = \frac{EI}{16} \begin{bmatrix} 24 & -24 & 0 & 0 \\ -24 & 32 & 0 & 0 \\ 0 & 0 & 3 & -6 \\ 0 & 0 & -6 & 16 \end{bmatrix} \tag{n}$$

STEP 4. We compute the basic stiffness matrix $[S^{FF}]$ of the beam. Substituting relations (m) and (n) into (16.19), we get

$$[S^{FF}] = [B][\hat{k}][B]^T$$

$$= \frac{EI}{16} \begin{bmatrix} -2 & -1 & 0 & 0 \\ 1 & 0 & -1 & 0 \\ 0 & 1 & -4 & -1 \\ 0 & 0 & 0 & 1 \end{bmatrix} \begin{bmatrix} 24 & -24 & 0 & 0 \\ -24 & 32 & 0 & 0 \\ 0 & 0 & 3 & -6 \\ 0 & 0 & -6 & 16 \end{bmatrix} \begin{bmatrix} -2 & 1 & 0 & 0 \\ -1 & 0 & 1 & 0 \\ 0 & -1 & -4 & 0 \\ 0 & 0 & -1 & 1 \end{bmatrix}$$

or

$$[S^{FF}] = \frac{EI}{16} \begin{bmatrix} 32 & -24 & 16 & 0 \\ -24 & 27 & -18 & 6 \\ 16 & -18 & 48 & 8 \\ 0 & 6 & 8 & 16 \end{bmatrix} \tag{o}$$

STEP 5. We compute the matrix of the components of displacements of the nodes of the beam. Substituting relation (o) into (16.18) we obtain

$$\{\Delta^F\} = \begin{Bmatrix} \bar{\Delta}_3^{(1)} \\ \bar{\Delta}_2^{(2)} \\ \bar{\Delta}_3^{(2)} \\ \bar{\Delta}_3^{(3)} \end{Bmatrix} = [S^{FF}]^{-1}\{P^{EF}\}$$

$$= \frac{16}{EI} \begin{bmatrix} 0.1250 & 0.1389 & 0.0208 & -0.0625 \\ 0.1389 & 0.2222 & 0.0556 & -0.01111 \\ 0.0208 & 0.0556 & 0.0417 & -0.0417 \\ -0.0625 & -0.1111 & -0.0417 & 0.1250 \end{bmatrix} \begin{Bmatrix} -6.667 \\ -60 \\ -20 \\ 26.667 \end{Bmatrix}$$

$$= \frac{1}{EI} \begin{Bmatrix} -180.00 \\ -293.33 \\ -86.73 \\ 180.00 \end{Bmatrix} \tag{p}$$

16.3 Problems

1. Using both the flexibility and the stiffness methods compute the deflection of points 2 and 3 of the cantilever beam shown in Fig. 16.P1. The

beam is made of steel ($E = 210\,\mathrm{kN/mm^2}$) and has a constant cross section ($I = 57 \times 10^6\,\mathrm{mm^4}$).

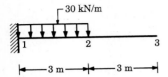

Figure 16.P1

2. Using both the flexibility and the stiffness methods compute the deflection of points 2 and 4 of the beam shown in Fig. 16.P2. The beam is

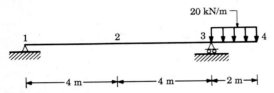

Figure 16.P2

made of steel ($E = 210\,\mathrm{kN/mm^2}$) and has a constant cross section ($I = 57 \times 10^6\,\mathrm{mm^4}$).

3 to 5. Using the flexibility method compute the components of displacements of the nodes of the truss of Fig. 16.P3. The elements of the truss are made of the same material ($E = 210\,\mathrm{kN/mm^2}$). Repeat using the trusses of Figs. 16.P4 and 16.P5.

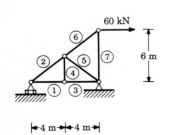

Member	Cross-sectional area
1, 2, 5, 6	$3 \times 10^3\,\mathrm{mm^2}$
3, 4	$2 \times 10^3\,\mathrm{mm^2}$
7	$6 \times 10^3\,\mathrm{mm^2}$

Figure 16.P3

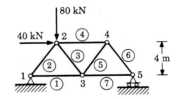

Member	Cross-sectional area
1, 3, 4	$8 \times 10^3\,\mathrm{mm^2}$
2, 6	$6 \times 10^3\,\mathrm{mm^2}$
5, 7	$10 \times 10^3\,\mathrm{mm^2}$

Figure 16.P4

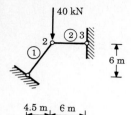

Member	Cross-sectional area
1	3 x 10³ mm²
2	2 x 10³ mm²

Figure 16.P5

6 to 12. Using the flexibility method compute the components of displacements of the nodes of the frame loaded as shown in Fig. 16.P6. The elements of the frame are made of the same material ($E = 210 \text{ kN/mm}^2$) and have the same constant cross section ($I = 56.46 \times 10^6 \text{ mm}^4$, $A = 16.3 \times 10^3 \text{ mm}^2$). Repeat using the frames subjected to the loads shown in Figs. 16.P7 to 16.P12.

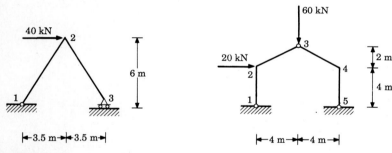

Figure 16.P6   **Figure 16.P7**

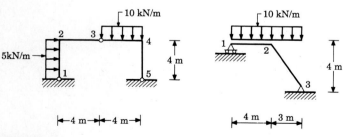

Figure 16.P8   **Figure 16.P9**

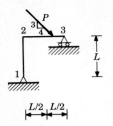

Figure 16.P10

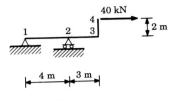

Figure 16.P11

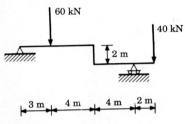

Figure 16.P12

Analysis of Statically Indeterminate Structures Using the Modern Flexibility Method

17.1 Introduction

When analyzing a statically indeterminate structure using the classical flexibility or force method, we first reduce the structure to a statically determinate one which is not a mechanism, referred to as the *primary structure*. This is accomplished by introducing action release mechanisms in the structure which eliminate enough of its reactions or internal actions to render it statically determinate. The eliminated reactions or internal actions are referred to as the *redundants* of the structure. The choice of the redundants of a structure is not unique. For example, the frame of Fig. 17.1a is statically indeterminate to the third degree. One way that it can be reduced to a statically determinate structure is by cutting it just above node 4 and, thus, releasing the three components of internal action $X_1 = F_1^{3k}$, $X_2 = F_2^{3k}$, and $X_3 = M_3^{3k}$ (see Fig. 17.1b). Another way of reducing this frame to a statically determinate structure is by disconnecting one of its elements from one of the nodes to which it is connected (see Fig. 17.1c). A third way of reducing this frame to a statically determinate structure is by introducing hinges at one end of three of its elements (see Fig. 17.1d). It can be shown that the choice of the redundants affects the condition of the matrix which must be inverted in analyzing a structure by the flexibility method. Thus, it is desirable to choose redundants which do not result in a poorly conditioned matrix, that is, a matrix whose inverse is sensitive to roundoff errors of its terms.

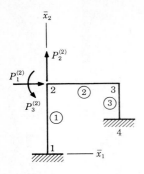

(a) Actual structure subjected
to the given external actions

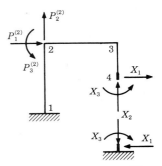

(b) Primary structure subjected
to the given external actions
and to the redundants

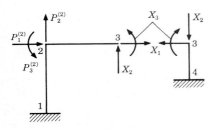

(c) Primary structure subjected
to the given external actions
and to the redundants

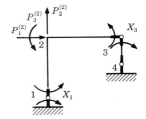

(d) Primary structure subjected
to the given external actions
and to the redundants

Figure 17.1   Primary structures of a statically indeterminate frame.

Consider a statically indeterminate to the Nth degree structure subjected to given loads and choose N nodal actions of its elements X_i $(i = 1, 2, \ldots, N)$ as its redundants. For instance, in Fig. 17.1b we have chosen as redundants the basic nodal actions of element 3, while in Fig. 17.1c we have chosen as redundants the basic nodal actions of element 2. We form the primary structure corresponding to the chosen redundants by introducing in the actual structure action release mechanisms which eliminate the redundants. We subject the primary structure to the given loads acting on the actual structure. We denote by Δ_i^{PL} $(i = 1, 2, \ldots, N)$ the component of the relative displacement corresponding to the redundant X_i of the two nodes of the primary structure adjacent to the action release mechanism introduced for eliminating the redundant X_i. For example, for the

primary structure of Fig. 17.1c, Δ_i^{PL} ($i = 1, 2$) is the relative translation in the direction $\bar{x}_i$ ($i = 1, 2$), while Δ_3^{PL} is the relative rotation
about the $\bar{x}_3$ axis of the two nodes of the primary structure numbered
3. Moreover, we subject the primary structure only to the chosen
redundants, and we denote by Δ_i^X ($i = 1, 2, \ldots, N$) the component of
the relative displacement corresponding to Δ_i^{PL} ($i = 1, 2, \ldots, N$).
Since the response of the structures which we are considering is
linear, the components of displacements Δ_i^X ($i = 1, 2, \ldots, N$) can be
expressed as a linear combination of the redundants X_i ($i = 1, 2, \ldots, N$). That is,

$$\begin{Bmatrix} \Delta_1^X \\ \Delta_2^X \\ \vdots \\ \Delta_N^X \end{Bmatrix} = \begin{bmatrix} F_{11} & F_{12} & \cdots & F_{1N} \\ F_{21} & F_{22} & \cdots & F_{2N} \\ \vdots & \vdots & & \vdots \\ F_{N1} & F_{N2} & \cdots & F_{NN} \end{bmatrix} \begin{Bmatrix} X_1 \\ X_2 \\ \vdots \\ X_N \end{Bmatrix} \tag{17.1}$$

or
$$\{\Delta^X\} = [F][X]$$

where $[F]$ is called the flexibility matrix of the structure corresponding to the chosen redundants X_i ($i = 1, 2, \ldots, N$).

Inasmuch as the internal actions of the elements and the components of displacement of the points of the primary structure
subjected to the given loads and to the redundants are equal to the
corresponding quantities of the actual structure subjected to the
given loads, the following relations are valid.

$$\begin{Bmatrix} \Delta_1^{PL} \\ \Delta_2^{PL} \\ \vdots \\ \Delta_N^{PL} \end{Bmatrix} + \begin{Bmatrix} \Delta_1^X \\ \Delta_2^X \\ \vdots \\ \Delta_N^X \end{Bmatrix} = 0$$

Using relation (17.1) the above relation can be rewritten as

$$\{\Delta^{PL}\} + \{\Delta^X\} = \{\Delta^{PL}\} + [F]\{X\} = 0 \tag{17.2}$$

Equations (17.2) are called the *compatibility equations for the
statically indeterminate structure* under consideration and are
employed in the classical flexibility method to compute its redundants $\{X\}$.

In this chapter, we present the modern flexibility method for
analyzing statically indeterminate framed structures. A brief description of this method is presented in Chap. 3.

In the direct stiffness method, the components of displacements of the nodes of a structure (statically determinate or indeterminate) are first established, and they are used to compute the reactions of the structure and the nodal actions of its elements. That is, in the direct stiffness method, the analysis of a structure is formulated in terms of the components of displacements of its nodes. Moreover, the relations between the components of the external actions acting on the nodes of a structure and the components of displacements of its nodes [see Eqs. (8.1)] are obtained by considering the equilibrium of its nodes. In the flexibility method, the reactions and the basic nodal actions of the elements of a structure subjected to the equivalent actions on its nodes are computed without first finding the displacements of its nodes. That is, in the flexibility method, the analysis of a structure is formulated in terms of some of its reactions and/or basic nodal actions of its elements which are chosen as the redundants. Moreover, the equations from which the redundants of a structure are established represent conditions imposed by the requirement that the displacements of the primary structure at the points where action release mechanisms have been introduced must be continuous when the primary structure is subjected to the given loads acting on the actual structure and to the redundants.

17.2   Analysis of Statically Indeterminate Structures Using the Modern Flexibility Method

Consider a statically indeterminate structure to the Nth degree, subjected to a general loading (external actions, change of temperature, initial strain of some of its elements, movement of supports). As discussed in Sec. 7.1, the components of nodal actions of the elements of a structure subjected to given loads may be obtained by superimposing the corresponding components of internal actions of the structure subjected to the following two loading cases.

1. The structure subjected to the given loads acting along the length of its elements and to the unknown restraining actions which must be applied to its nodes in order to restrain them from moving (translating and rotating). We call the structure subjected to these loads the *restrained structure*.

2. The structure subjected to the equivalent actions on its nodes.

In order to analyze a structure subjected to equivalent actions on its nodes, using the modern flexibility method, we consider a model made from the structure by replacing its supports with reaction elements (see Sec. 15.2). As discussed in Sec. 15.2, the matrix $\{\hat{a}^{EM}\}$

of this model includes the basic nodal actions of the reaction elements of the model in addition to the basic nodal actions of all the elements of the structure. Each basic nodal action of a reaction element is equal and opposite to the corresponding component of a reaction of the structure. As redundants of the model of a structure, we choose basic nodal actions of its elements which, if eliminated by introducing internal action release mechanisms, the resulting primary structure is statically determined and is not a mechanism.

When analyzing a structure by hand calculations using the classical flexibility method, the selection of the redundants is performed by the analyst.[1] However, when writing a computer program for analyzing a group of structures using the modern flexibility method, a procedure should be programmed for the selection of the redundants of a structure by computer.[2]

Consider the equations of equilibrium of the model and let us modify them by interchanging the columns of the matrix $[B^M]$ and the rows of the matrix $\{\hat{a}^{EM}\}$ so that the chosen redundants are in the lower part of the matrix $\{\hat{a}^{EM}\}$. We denote the matrix of the redundant basic nodal actions of the model by $\{a_x^{EM}\}$ and the matrix of the remaining basic nodal actions of the element of the model by $\{a_0^{EM}\}$. With this notation the equations of equilibrium for the unsupported nodes of the model can be written as

$$\{P^{EFM}\} = \{[B_0^M] \; \vdots \; [B_x^M]\} \begin{Bmatrix} \{a_0^{EM}\} \\ \{a_x^{EM}\} \end{Bmatrix} \tag{17.3}$$

or $\qquad\qquad \{P^{EFM}\} = [B_0^M]\{a_0^{EM}\} + [B_x^M]\{a_x^{EM}\}$

where $[B_0^M]$ is a square matrix and is not singular when the choice of the redundants is such that the primary structure is not a mechanism. Thus, relation (17.3) can be solved for $\{a_0^{EM}\}$ to give

$$\{a_0^{EM}\} = [F_0]\{P^{EFM}\} + [F_x]\{a_x^{EM}\} \tag{17.4}$$

where $\qquad\qquad\qquad [F_0] = [B_0^M]^{-1} \tag{17.5a}$

$$[F_x] = -[B_0^M]^{-1}[B_x^M] \tag{17.5b}$$

Consider an auxiliary structure obtained from the model of a structure by removing the constraints which cause the chosen redundants. This auxiliary structure is the primary structure for the model. For example, the model for the frame of Fig. 17.2a is shown in Fig. 17.2b. If we choose as redundants the basic nodal actions M_3^{1k} and M_3^{2k}, the primary structure for the model of Fig. 17.2b is shown in Fig. 17.2c.

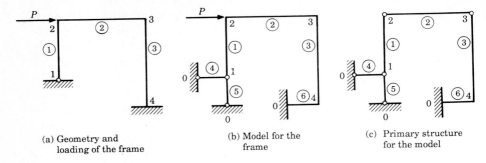

(a) Geometry and
    loading of the frame

(b) Model for the
    frame

(c) Primary structure
    for the model

Figure 17.2 Model for a frame and its primary structure.

From relations (17.4) it can be seen that the matrix $[F_0][P^{EFM}]$ represents the basic nodal actions of the elements of the primary structure for the model, when subjected to the known equivalent actions acting on the nodes of the structure. Moreover, the matrix $[F_x]\{a_X^{EM}\}$ represents the basic nodal actions of the elements of the primary structure for the model when subjected to the redundants. Furthermore, each column of the matrix $[F_X]$ represents the matrix of the basic nodal actions of the elements of the primary structure for the model when subjected to a unit value of one redundant. That is, in relation (17.4) the internal actions in the elements of the model are obtained by superimposing the internal actions in the elements of the primary structure subjected to the following loading cases.

1. The known equivalent actions applied on the nodes of the structure (see Fig. 17.3b).

2. The chosen redundants (see Fig. 17.3c). The internal actions due

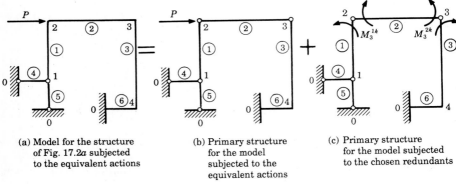

(a) Model for the structure
    of Fig. 17.2a subjected
    to the equivalent actions

(b) Primary structure
    for the model
    subjected to the
    equivalent actions

(c) Primary structure
    for the model subjected
    to the chosen redundants

Figure 17.3 Superposition of the primary structure for the model subjected (a) to the equivalent actions and (b) to the redundants.

to this loading may be established by superimposing the products of each redundant and the internal actions in the primary structure subjected to a unit value of this redundant (see Fig. 17.4).

Relation (17.4) may be combined with $\{a_x^{EM}\} = [I]\{a_x^{EM}\}$ to yield

$$\left\{ \begin{array}{c} \{a_0^{EM}\} \\ \{a_x^{EM}\} \end{array} \right\} = \left[\begin{array}{cc} [F_0] & [F_x] \\ [0] & [I] \end{array} \right] \left\{ \begin{array}{c} \{P^{EFM}\} \\ \{a_x^M\} \end{array} \right\} \qquad (17.6)$$

Let us consider the equations of compatibility $[\{\hat{d}^M\} = [C^M]\{\Delta^{FM}\}]$ for the model of the structure, and let us modify them by interchanging the rows of the matrices $[C^M]$ and $\{\hat{d}^M\}$ so that the basic deformation parameters of the elements of the model which correspond to the chosen redundant basic nodal actions are in the lower part of the matrix $\{\hat{d}^M\}$. This part of the matrix $\{\hat{d}^M\}$ is denoted by $\{d_x^M\}$, while the upper part of the matrix $\{\hat{d}^M\}$ is denoted by $\{d_0^M\}$. Thus, the equations of compatibility of the model of the structure can

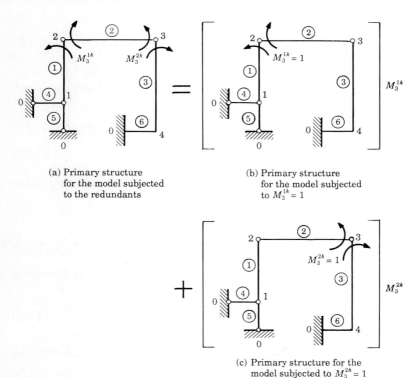

(a) Primary structure
for the model subjected
to the redundants

(b) Primary structure
for the model subjected
to $M_3^{1k} = 1$

(c) Primary structure for the
model subjected to $M_3^{2k} = 1$

Figure 17.4 Analysis of the primary structure for the model subjected to the chosen redundants.

be written as

$$\{d^M\} = \left\{ \frac{\{d_0^M\}}{\{d_x^M\}} \right\} = \left[\frac{[C_0^M]}{[C_x^M]} \right] \{\Delta^{FM}\} \tag{17.7}$$

or

$$[d_0^M] = [C_0^M]\{\Delta^{FM}\} \tag{17.8}$$

$$[d_x^M] = [C_x^M]\{\Delta^{FM}\} \tag{17.9}$$

Notice that when the choice of the redundants is such that the primary structure is not a mechanism the matrix $[C_0^M]$ is a square, nonsingular matrix. In the examples of Sec. 15.5 we have shown that

$$[C^M] = [B^M]^T \tag{17.10}$$

Thus

$$\left[\frac{[C_0^M]}{[C_x^M]} \right] = [[B_0^M] \ \vdots \ [B_x^M]]^T = \left[\frac{[B_0^M]^T}{[B_x^M]^T} \right] \tag{17.11}$$

Hence,

$$[C_0^M] = [B_0^M]^T \tag{17.12}$$

$$[C_x^M] = [B_x^M]^T \tag{17.13}$$

Substituting relation (17.12) into (17.8), noting that the inverse of the transpose of a matrix is equal to the transpose of its inverse, and using relation (17.12) we get

$$\{\Delta^{FM}\} = [C_0^M]^{-1}\{d_0^M\} = [[B_0^M]^T]^{-1}\{d_0^M\} = [[B_0^M]^{-1}]^T\{d_0^M\}$$
$$= [F_0]^T\{d_0^M\} \tag{17.14}$$

Moreover, substituting relation (17.14) into (17.9) and using relations (17.5) and (17.13), we obtain

$$\{d_x^M\} = [C_x^M]\{\Delta^{FM}\} = [B_x^M]^T[F_0]^T\{d_0^M\}$$
$$= [[F_0][B_x^M]]^T\{d_0^M\} = [[B_0^M]^{-1}[B_x^M]]^T\{d_0^M\} = -[F_x]^T\{d_0^M\} \tag{17.15}$$

Relations (17.14) and (17.15) may be combined to yield the following relation:

$$\left\{ \frac{\{\Delta^{FM}\}}{\{0\}} \right\} = \left[\begin{matrix} [F_0]^T & [0] \\ [F_x]^T & [I] \end{matrix} \right] \left\{ \frac{\{d_0^M\}}{\{d_x^M\}} \right\} \tag{17.16}$$

The relations between the basic deformation parameters and the basic nodal actions (16.8) of the model can be written as

$$\{\hat{d}^M\} = \left\{ \frac{\{d_0^M\}}{\{d_x^M\}} \right\} = [f^M]\{\hat{a}^{EM}\} = \left[\begin{matrix} [f_{00}^M][f_{0x}^M] \\ [f_{x0}^M][f_{xx}^M] \end{matrix} \right] \left\{ \frac{\{a_0^{EM}\}}{\{a_x^{EM}\}} \right\} \tag{17.17}$$

Using the symmetry of the flexibility matrix it can be shown that

$$[f_{0x}^M] = [f_{x0}^M]^T \qquad (17.18)$$

Substituting relation (17.6) into (17.17) and the resulting relation into (17.16) we get

$$\left\{ \begin{array}{c} \{\Delta^{FM}\} \\ \{0\} \end{array} \right\} = \begin{bmatrix} [F_{11}] & [F_{12}] \\ [F_{21}] & [F_{22}] \end{bmatrix} \left\{ \begin{array}{c} \{P^{EFM}\} \\ \{a_x^{EM}\} \end{array} \right\} \qquad (17.19)$$

where

$$\begin{bmatrix} [F_{11}] & [F_{12}] \\ [F_{21}] & [F_{22}] \end{bmatrix} = \begin{bmatrix} [F_0]^T & [0] \\ [F_x]^T & [I] \end{bmatrix} \begin{bmatrix} [f_{00}^M] & [f_{0x}^M] \\ [f_{x0}^M] & [f_{xx}^M] \end{bmatrix} \begin{bmatrix} [F_0] & [F_x] \\ [0] & [I] \end{bmatrix}$$

Carrying out the multiplication of the matrices indicated in the above relation and using relation (17.18), we have

$$[F_{11}] = [F_0]^T [f_{00}^M][F_0]$$

$$[F_{12}] = [F_{21}]^T = [F_0]^T [f_{00}^M][F_x] + [F_0]^T [f_{0x}^M] \qquad (17.20)$$

$$[F_{22}] = [F_x]^T [f_{00}^M][F_x] + [F_x]^T [f_{0x}^M] + [f_{x0}^M][F_x] + [f_{xx}^M]$$

Relation (17.19) can be written as

$$\{\Delta^{FM}\} = [F_{11}]\{P^{EFM}\} + [F_{12}]\{a_x^{EM}\} \qquad (17.21)$$

$$[0] = [F_{21}]\{P^{EFM}\} + [F_{22}]\{a_x^{EM}\} \qquad (17.22)$$

Equations (17.22) are the compatibility equations for the structure. They are equivalent to Eqs. (17.2). The first term on the right-hand side of Eq. (17.22) is equal to $\{\Delta^{PL}\}$, while the second term is equal to $\{\Delta^X\}$. $[F_{22}]$ is the flexibility matrix of the model corresponding to the chosen redundants $\{a_x^{EM}\}$.

Relation (17.22) can be solved for the redundants $\{a_x^{EM}\}$ to give

$$\{a_x^{EM}\} = -[F_{22}]^{-1}[F_{21}]\{P^{EFM}\} \qquad (17.23)$$

Substituting relation (17.23) into (17.4) we obtain

$$\{a_0^{EM}\} = [F_0]\{P^{EFM}\} + [F_x]\{a_x^{EM}\} = [[F_0] - [F_x][F_{22}]^{-1}[F_{21}]]\{P^{EFM}\} \qquad (17.24)$$

Substituting relation (17.23) into (17.21) we get

$$\{\Delta^{FM}\} = [[F_{11}] - [F_{12}][F_{22}]^{-1}[F_{21}]]\{P^{EFM}\} \qquad (17.25)$$

In order to simplify the matrix algebra involved in the analysis of a structure we rewrite relations (17.4) and (17.14) as

$$\{a_0^{EM}\} = [F_0^*]\{P^{EFM^*}\} + [F_x]\{a_x^{EM}\} \qquad (17.26)$$

$$\{\Delta^{FM^*}\} = [F_0^*]^T[d_0^M] \qquad (17.27)$$

Moreover, we substitute matrix $[F_0^*]$ for $[F_0]$, $\{P^{EFM^*}\}$ for $\{P^{EFM}\}$ and $\{\Delta^{FM^*}\}$ for $\{\Delta^{FM}\}$ in relations (17.16) to (17.25). The matrix $\{P^{EFM^*}\}$ includes only the nonvanishing components of the equivalent actions acting on the nodes of the model. The matrix $\{\Delta^{FM^*}\}$ is conjugate to $\{P^{EFM^*}\}$. The matrix $[F_0^*]$ is obtained from the matrix $[F_0]$ by canceling its columns corresponding to zero components of equivalent actions.

On the basis of the foregoing, a systematic procedure suitable for programming the analysis of statically indeterminate framed structures on an electronic computer, using the modern flexibility method, involves the following steps.

Step 1. We establish the matrix of basic nodal actions $\{\hat{a}^R\}$ of each element of the structure subjected to the given loads with its ends fixed. Moreover, we compute the matrix of equivalent actions $\{\hat{P}^E\}$ to be placed on the nodes of the structure (see Chap. 7).

Step 2. We form a model of the structure by replacing its supports with reaction elements (see Sec. 15.2). The model is subjected to the equivalent actions acting on the nodes of the actual structure $\{P^{EFM}\} = \{\hat{P}^E\}$. Moreover, we form the matrix $[B^M]$ of the model following the procedure described in Sec. 15.2, and we write the equations of equilibrium for the unsupported nodes of the model.

Step 3. We choose as redundants of the model, basic nodal actions of its elements. Moreover, we modify the equations of equilibrium for the model by interchanging the columns of the matrix $[B^M]$ and the rows of the matrix $\{\hat{a}^{EM}\}$ so that the chosen redundants are in the lower part of the matrix $\{\hat{a}^{EM}\}$. Furthermore, we establish the matrices $[B_0^M]$, $[B_x^M]$, $\{a_0^{EM}\}$, and $\{a_x^{EM}\}$ by referring to relation (17.3).

Step 4. We establish flexibility matrices $[f]$ for each element of the model and use them to form the flexibility matrix $[\hat{f}^M]$ of all the elements of the model using relation (16.9). Moreover, we modify the flexibility matrix $[\hat{f}^M]$ by rearranging its rows and columns so that in relation $\{\hat{d}^M\} = [\hat{f}^M]\{\hat{a}^{EM}\}$ the chosen redundant basic nodal actions and the corresponding basic deformation parameters are at the lower part of the matrices $\{\hat{a}^{EM}\}$ and $\{\hat{d}^M\}$, respectively. Furthermore, we partition the flexibility matrix $[\hat{f}^M]$ as indicated in relation (17.17) and establish the matrices $[f_{00}^M]$, $[f_{0x}^M]$, $[f_{x0}^M]$, and $[f_{xx}^M]$.

Step 5. We compute the matrices $[F_0]$ and $[F_x]$ using relations (17.5). Moreover, if required, we establish the matrices $[F_0^*]$ and $\{P^{EFM^*}\}$. Furthermore, we compute the matrices $[F_{11}]$, $[F_{12}]$, $[F_{21}]$, and $[F_{22}]$ using relations (17.20).

Step 6. We establish the basic nodal actions of the elements and the reactions of the structure subjected to the equivalent actions on its nodes. In order to accomplish this we do the following.

1. We compute the matrix of the redundant basic nodal actions $\{a_x^{EM}\}$ of the elements of the model using relation (17.23).

2. We compute the matrix of the remaining basic nodal actions of the elements of the model $\{a_0^{EM}\}$ using relations (17.24).

3. We establish the matrix of nodal actions of all the elements of the model $\{\hat{a}^{EM}\}$. This matrix is given as

$$\{\hat{a}^{EM}\} = \left\{ \begin{array}{c} \{a_0^{EM}\} \\ \{a_x^{EM}\} \end{array} \right\} \tag{17.28}$$

4. We arrange the terms of the matrix $\{a^{EM}\}$ so that the basic nodal actions of the reaction elements are at the lower part of the matrix $\{\hat{a}^{EM}\}$. Thus

$$\{\hat{a}^{EM}\} = \left\{ \begin{array}{c} \{\hat{a}^E\} \\ -\{R\} \end{array} \right\} \tag{17.29}$$

where $\{\hat{a}^E\}$ is the matrix of the basic nodal actions of the elements of the structure subjected to the equivalent actions on its nodes, and $\{R\}$ is the matrix of the reactions of the structure, subjected to the equivalent actions on its nodes. It is equal to the matrix of the reactions of the structure subjected to the given loads.

Step 7. We establish the matrix $\{\hat{a}\}$ of the basic nodal actions of each element of the structure subjected to the given loads. That is,

$$\{\hat{a}\} = \{\hat{a}^R\} + \{\hat{a}^E\} \tag{17.30}$$

Step 8. We compute the matrix $\{\Delta^{FM}\} = \{\hat{\Delta}\}$ of the components of displacements of the nodes of the structure using relations (17.25).

In the sequel, we illustrate the flexibility method by the following three examples.

Example 1. An internally statically indeterminate to the first degree truss is analyzed. The truss is subjected only to a horizontal force at one of its nodes. The analysis of the truss is simplified by employing the matrices $\{P^{EFM^*}\}$ and $[F_0^*]$. This truss was analyzed in Chap. 3

using the direct stiffness and the modern flexibility methods, in order to demonstrate the salient features of these methods without delving into details. We repeat the analysis of the truss employing the systematic procedure described in this chapter.

Example 2. A statically indeterminate to the second degree beam is analyzed. The beam is subjected to external forces acting along the length of its elements and to settlement of a support.

Example 3. A statically indeterminate to the third degree frame is analyzed. The frame is subjected to external actions along the length of its elements and on its nodes.

Example 1. Compute the internal forces in the elements of the truss, loaded as shown in Fig. a, as well as the horizontal component of translation of node 3. The elements of the truss have the same constant cross section and are made from the same isotropic linearly elastic material.

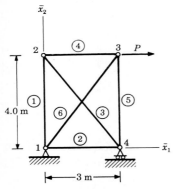

Figure a Geometry and loading of the truss.

---

solution
STEP 1.   Since the external actions are applied to the nodes of the truss, we have

$$\{\hat{a}^R\} = 0$$

Referring to Fig. a and to relation (7.2) we obtain

$$\{\hat{P}^E\} = \{\hat{P}^G\} = \begin{Bmatrix} 0 \\ 0 \\ 0 \\ 0 \\ P \\ 0 \\ 0 \\ 0 \end{Bmatrix} \tag{a}$$

STEP 2. We make a model of the truss by replacing its supports with reaction elements (see Fig. b). We form the matrix $[B^M]$ of the model of Fig. b following the procedure described in Sec. 15.2.

1. We form the matrix $[H^e]$ for each element of the model of Fig. b. Thus, referring to relation (b) of Example 1 of Sec. 15.2 ($\phi_{11}^1 = 90°$, $L_1 = 4$ m, $\phi_{11}^2 = 0$, $L_2 = 3$ m, $\cos \phi_{11}^3 = 0.6$, $\sin \phi_{11}^3 = -0.8$, $L_3 = 5$ m, $\phi_{11}^4 = 0$, $L_4 = 3$ m, $\phi_{11}^5 = -90$, $L_5 = 4$ m, $\cos \phi_{11}^6 = 0.6$, $\sin \phi_{11}^6 = 0.8$, $L_6 = 5$ m, $\phi_{11}^7 = 0$, $\phi_{11}^8 = 90°$, $\phi_{11}^9 = 90°$) we

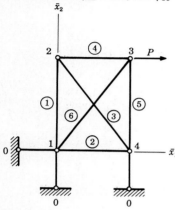

Figure b Model of the truss of Fig. a.

have

$$\{H^1\} = \{H^8\} = \{H^9\} = \begin{Bmatrix} 0 \\ -1 \\ \hline 0 \\ 1 \end{Bmatrix} \qquad \{H^2\} = \{H^4\} = \{H^7\} = \begin{Bmatrix} -1 \\ 0 \\ \hline 1 \\ 0 \end{Bmatrix}$$

$$\{H^3\} = \begin{Bmatrix} -0.6 \\ 0.8 \\ \hline 0.6 \\ -0.8 \end{Bmatrix} \qquad \{H^5\} = \begin{Bmatrix} 0 \\ 1 \\ \hline 0 \\ -1 \end{Bmatrix} \qquad \{H^6\} = \begin{Bmatrix} -0.6 \\ -0.8 \\ \hline 0.6 \\ 0.8 \end{Bmatrix} \qquad \text{(b)}$$

2. We form the matrix $[B^M]$ of the model of Fig. b using the matrices $\{H^e\}$ of its elements. Thus,

Element

Node			1	2	3	4	5	6	7	8	9	
1	$\sum \bar{F}_1 = 0$	$\begin{pmatrix} 0 \end{pmatrix}$	0	−1	0	0	0	−0.6	1	0	0	F_1^{1k}
	$\sum \bar{F}_2 = 0$	0	−1	0	0	0	0	−0.8	0	1	0	F_1^{2k}
2	$\sum \bar{F}_1 = 0$	0	0	0	−0.6	−1	0	0	0	0	0	F_1^{3k}
	$\sum F_2 = 0$	0	1	0	0.8	0	0	0	0	0	0	F_1^{4k}
3	$\sum \bar{F}_1 = 0$	P	0	0	0	1	0	0.6	0	0	0	F_1^{5k}
	$\sum F_2 = 0$	0	0	0	0	0	1	0.8	0	0	0	F_1^{6k}
4	$\sum F_1 = 0$	0	0	1	0.6	0	0	0	0	0	0	F_1^{7k}
	$\sum F_2 = 0$	0	0	0	−0.8	0	−1	0	0	0	1	F_1^{8k}

$= 0$

F_1^{9k}

(c)

or
$$\{P^{EFM}\} - [B^M]\{\hat{a}^{EM}\} = 0 \tag{d}$$

STEP 3.   The model of Fig. b is statically indeterminate to the first degree. We choose the internal force in element 6 as the redundant of the model. Moreover, we modify the equations of equilibrium (c) of the model by interchanging the columns of the matrix $[B^M]$ and the rows of the matrix $\{\hat{a}^{EM}\}$ so that the chosen redundant is in the lower part of the matrix $\{\hat{a}^{EM}\}$. Thus, referring to relation (c) we have

$$[B^M] = \begin{bmatrix} 0 & -1 & 0 & 0 & 0 & 1 & 0 & 0 & -0.6 \\ -1 & 0 & 0 & 0 & 0 & 0 & 1 & 0 & -0.8 \\ 0 & 0 & -0.6 & -1 & 0 & 0 & 0 & 0 & 0 \\ 1 & 0 & 0.8 & 0 & 0 & 0 & 0 & 0 & 0 \\ 0 & 0 & 0 & 1 & 0 & 0 & 0 & 0 & 0.6 \\ 0 & 0 & 0 & 0 & 1 & 0 & 0 & 0 & 0.8 \\ 0 & 1 & 0.6 & 0 & 0 & 0 & 0 & 0 & 0 \\ 0 & 0 & -0.8 & 0 & -1 & 0 & 0 & 1 & 0 \end{bmatrix} \tag{e}$$

and
$$\{\hat{a}^{EM}\} = \begin{Bmatrix} F_1^{1k} \\ F_1^{2k} \\ F_1^{3k} \\ F_1^{4k} \\ F_1^{5k} \\ F_1^{7k} \\ F_1^{8k} \\ F_1^{9k} \\ \hline F_1^{6k} \end{Bmatrix} \tag{f}$$

Moreover, referring to relations (e) and (17.3) we get

$$[B_o^M] = \begin{bmatrix} 0 & -1 & 0 & 0 & 0 & 1 & 0 & 0 \\ -1 & 0 & 0 & 0 & 0 & 0 & 1 & 0 \\ 0 & 0 & -0.6 & -1 & 0 & 0 & 0 & 0 \\ 1 & 0 & 0.8 & 0 & 0 & 0 & 0 & 0 \\ 0 & 0 & 0 & 1 & 0 & 0 & 0 & 0 \\ 0 & 0 & 0 & 0 & 1 & 0 & 0 & 0 \\ 0 & 1 & 0.6 & 0 & 0 & 0 & 0 & 0 \\ 0 & 0 & -0.8 & 0 & -1 & 0 & 0 & 1 \end{bmatrix} \tag{g}$$

$$[B_x^M] = \begin{bmatrix} -0.6 \\ -0.8 \\ 0 \\ 0 \\ 0.6 \\ 0.8 \\ 0 \\ 0 \end{bmatrix} \qquad (h)$$

Furthermore, referring to relations (f) and (17.3) we obtain

$$\{a_0^{EM}\} = \begin{Bmatrix} F_1^{1k} \\ F_1^{2k} \\ F_1^{3k} \\ F_1^{4k} \\ F_1^{5k} \\ F_1^{7k} \\ F_1^{8k} \\ F_1^{9k} \end{Bmatrix} \qquad (i)$$

$$\{a_x^{EM}\} = F_1^{6k} \qquad (j)$$

STEP 4. We form the flexibility matrix $[\hat{f}^M]$ of all the elements of the model of Fig. b using relation (16.9). We take into account that the flexibility of the reaction elements is equal to zero. Moreover, we rearrange the rows and columns of the matrix $[\hat{f}^M]$ so that in relation $\{\hat{d}^M\} = [\hat{f}^M]\{\hat{a}^{EM}\}$ the chosen redundant basic nodal action and the corresponding basic deformation parameter are at the lower part of the matrices $\{\hat{a}^{EM}\}$ and $\{\hat{d}^M\}$, respectively. Furthermore, we partition the matrix $[\hat{f}^M]$ as indicated in relation (17.17). Thus,

			Element						
1	2	3	4	5	7	8	9	6	

$$[\hat{f}^M] = \frac{1}{EA} \begin{bmatrix} 4 & 0 & 0 & 0 & 0 & 0 & 0 & 0 & 0 \\ 0 & 3 & 0 & 0 & 0 & 0 & 0 & 0 & 0 \\ 0 & 0 & 5 & 0 & 0 & 0 & 0 & 0 & 0 \\ 0 & 0 & 0 & 3 & 0 & 0 & 0 & 0 & 0 \\ 0 & 0 & 0 & 0 & 4 & 0 & 0 & 0 & 0 \\ 0 & 0 & 0 & 0 & 0 & 0 & 0 & 0 & 0 \\ 0 & 0 & 0 & 0 & 0 & 0 & 0 & 0 & 0 \\ 0 & 0 & 0 & 0 & 0 & 0 & 0 & 0 & 0 \\ 0 & 0 & 0 & 0 & 0 & 0 & 0 & 0 & 5 \end{bmatrix} \begin{matrix} 1 \\ 2 \\ 3 \\ 4 \\ 5 \\ 7 \\ 8 \\ 9 \\ 6 \end{matrix} \qquad (k)$$

Hence,

$$[f^M_{00}] = \frac{1}{EA} \begin{bmatrix} 4 & 0 & 0 & 0 & 0 & 0 & 0 & 0 \\ 0 & 3 & 0 & 0 & 0 & 0 & 0 & 0 \\ 0 & 0 & 5 & 0 & 0 & 0 & 0 & 0 \\ 0 & 0 & 0 & 3 & 0 & 0 & 0 & 0 \\ 0 & 0 & 0 & 0 & 4 & 0 & 0 & 0 \\ 0 & 0 & 0 & 0 & 0 & 0 & 0 & 0 \\ 0 & 0 & 0 & 0 & 0 & 0 & 0 & 0 \end{bmatrix} \tag{1}$$

$$[f^M_{0x}] = \begin{bmatrix} 0 \\ 0 \\ 0 \\ 0 \\ 0 \\ 0 \\ 0 \\ 0 \end{bmatrix} \tag{m}$$

$$[f^M_{x0}] = [0 \quad 0 \quad 0 \quad 0 \quad 0 \quad 0 \quad 0 \quad 0] \tag{n}$$

$$[f^M_{xx}] = \frac{5}{EA} \tag{o}$$

STEP 5.   We compute the matrices $[F_0]$ and $[F_x]$ using relations (17.5). That is,

$$[F_0] = [B^M_0]^{-1} = \begin{bmatrix} 0 & 0 & 1.333 & 1 & 1.333 & 0 & 0 & 0 \\ 0 & 0 & 1 & 0 & 1 & 0 & 1 & 0 \\ 0 & 0 & -1.667 & 0 & -1.667 & 0 & 0 & 0 \\ 0 & 0 & 0 & 0 & 1 & 0 & 0 & 0 \\ 0 & 0 & 0 & 0 & 0 & 1 & 0 & 0 \\ 1 & 0 & 1 & 0 & 1 & 0 & 1 & 0 \\ 0 & 1 & 1.333 & 1 & 1.333 & 0 & 0 & 0 \\ 0 & 0 & -1.333 & 0 & -1.333 & 1 & 0 & 1 \end{bmatrix} \tag{p}$$

$$[F_x] = -[B^M_0]^{-1}[B^M_x] = \begin{Bmatrix} -0.7998 \\ -0.6 \\ 1.0002 \\ -0.6 \\ -0.8 \\ 0 \\ 0.0002 \\ -0.0002 \end{Bmatrix} \tag{q}$$

Moreover, we establish the matrices $\{P^{EFM^*}\}$ and $[F_0^*]$ in order to simplify the required matrix algebra. That is, referring to Fig. a and to relations (p) and (17.4) we have

$$\{P^{EFM^*}\} = P \tag{r}$$

$$[F_0^*] = \begin{Bmatrix} 1.333 \\ 1 \\ -1.667 \\ 1 \\ 0 \\ 1 \\ 1.333 \\ -1.333 \end{Bmatrix} \tag{s}$$

Furthermore, we compute the matrices $[F_{11}^*]$, $[F_{12}^*]$, $[F_{21}^*]$, and $[F_{22}^*]$ using relations (17.20). That is,

$$[F_{11}^*] = [F_0^*]^T[f_{00}^M][F_0^*] = \frac{27.002}{EA} \tag{t}$$

$$[F_{12}^*] = [F_{21}^*]^T = [F_0^*]^T[f_{00}^M][F_x] + [F_0^*]^T[f_{0x}^M] = -\frac{16.2012}{EA} \tag{u}$$

$$[F_{22}^*] = [F_x]^T[f_{00}^M][F_x] + [F_x]^T[f_{0x}^M] + [f_{x0}^M][F_x] + [f_{xx}^M] = \frac{17.2807}{EA} \tag{v}$$

STEP 6.   We compute the matrix of the basic nodal actions of the elements of the truss and the matrix of its reactions. Substituting relations (r), (u), and (v) into (17.23) we obtain

$$\{a_x^{EM}\} = -[F_{22}]^{-1}[F_{21}^*]\{P^{EFM^*}\} = 0.938P \tag{w}$$

Moreover, substituting relations (q) to (s) and (w) into (17.24) we get

$$\{a_0^{EM}\} = [F_0^*]\{P^{EFM^*}\} + [F_x]\{a_x^{EM}\} = \begin{Bmatrix} 0.5832 \\ 0.4375 \\ -0.7293 \\ 0.4375 \\ -0.750 \\ 1 \\ 1.3332 \\ -1.3332 \end{Bmatrix} P \tag{x}$$

From results (w) and (x) we have

$$\{\hat{a}^{EM}\} = \left\{ {\begin{array}{c} \{a_0^{EM}\} \\ \{a_x^{EM}\} \end{array}} \right\} = \left\{ \begin{array}{c} F_1^{1k} \\ F_1^{2k} \\ F_1^{3k} \\ F_1^{4k} \\ F_1^{5k} \\ F_1^{7k} \\ F_1^{8k} \\ F_1^{9k} \\ F_1^{6k} \end{array} \right\} = \left\{ \begin{array}{r} 0.583 \\ 0.438 \\ -0.729 \\ 0.438 \\ -0.750 \\ 1.000 \\ 1.333 \\ -1.333 \\ 0.938 \end{array} \right\} P \tag{y}$$

or

$$\{\hat{a}^{E}\} = \left\{ \begin{array}{c} F_1^{1k} \\ F_1^{2k} \\ F_1^{3k} \\ F_1^{4k} \\ F_1^{5k} \\ F_1^{6K} \end{array} \right\} = \left\{ \begin{array}{r} 0.583 \\ 0.438 \\ -0.729 \\ 0.438 \\ -0.750 \\ 0.938 \end{array} \right\} P \tag{z}$$

and

$$\{R\} = \left\{ \begin{array}{r} -1.000 \\ -1.333 \\ 1.333 \end{array} \right\} P \tag{aa}$$

STEP 7.

$$\{\hat{a}\} = \{\hat{a}^{E}\}$$

STEP 8. We compute the horizontal component of translation of node 3. Substituting relations (r), (t), (u), and (v) into (17.25), we have

$$[\Delta^{FM^*}] = \Delta_1^{(3)} = [[F_{11}] - [F_{12}][F_{22}]^{-1}[F_{21}]][P^{EFM^*}] = \frac{11.81P}{EA} \tag{bb}$$

Example 2. Compute the basic nodal actions of the beam subjected to the external actions shown in Fig. a and to a settlement of support 2 of 20 mm. The elements of the beam have the same constant cross section and are made of the same isotropic linearly elastic material ($EI = 40,000 \text{ kN} \cdot \text{m}^2$). Plot the moment and shear diagrams for the beam.

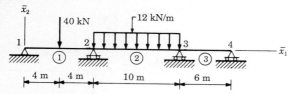

Figure a Geometry and loading of the beam.

---

solution   Since the beam is not subjected to axial components of force, we do not include these components in its matrix $\{\hat{P}^E\}$. Moreover, we do not include the axial components of nodal actions in the matrices $\{a\}$ and $\{\bar{A}\}$ or the terms representing axial deformation in the flexibility matrices $[f]$ of the elements of the beam. Moreover, we consider only the equilibrium of the transverse components of the forces and the equilibrium of the moments about the $\bar{x}_3$ axis when we write the equations of equilibrium of the nodes of the beam.

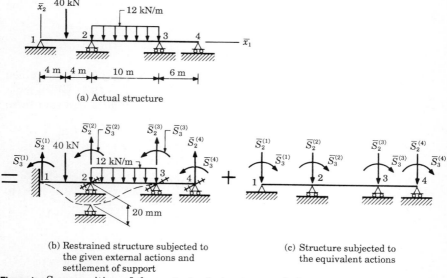

(a) Actual structure

(b) Restrained structure subjected to the given external actions and settlement of support

(c) Structure subjected to the equivalent actions

Figure b Superposition of the restrained structure and the structure subjected to the equivalent actions.

STEP 1.   We analyze the restrained structure (see Fig. bb) and establish the matrix of the basic nodal actions of all its elements $\{\hat{a}^R\}$ as well as the equivalent actions to be applied to the nodes of the beam. The nodal actions of the elements of the restrained structure are established by referring to the table on the inside of the back cover and are given in the free-body diagrams of the elements and nodes of the restrained structure shown in Fig. c. Referring to Fig. c, the matrix of the basic nodal actions of all the elements of

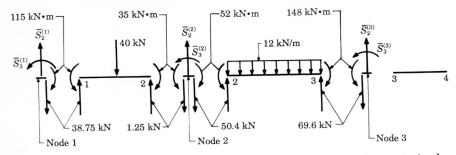

Figure c Free-body diagrams of the elements and nodes of the restrained structure.

the restrained beam is

$$\{\hat{a}^R\} = \begin{Bmatrix} F_2^{1k} \\ M_3^{1k} \\ \hline F_2^{2k} \\ M_3^{2k} \\ \hline F_2^{3k} \\ M_3^{3k} \end{Bmatrix} = \begin{Bmatrix} 1.25 \\ 35.00 \\ \hline 69.6 \\ -148.0 \\ \hline 0 \\ 0 \end{Bmatrix} \qquad (a)$$

Referring to Fig. c, we consider the equilibrium of the nodes of the restrained structure and obtain the matrix of the restraining actions $\{\widehat{RA}\}$ which must be placed on the nodes of the restrained structure in order to restrain them from translating and rotating. That is,

$$\{\widehat{RA}\} = \begin{Bmatrix} \bar{S}_2^{(1)} \\ \bar{S}_3^{(1)} \\ \hline \bar{S}_2^{(2)} \\ \bar{S}_3^{(2)} \\ \hline \bar{S}_2^{(3)} \\ \bar{S}_3^{(3)} \end{Bmatrix} = \begin{Bmatrix} 38.75 \\ 115.00 \\ \hline 175.40 \\ 87.00 \\ \hline 69.60 \\ -148.00 \end{Bmatrix} \qquad (b)$$

Referring to Fig. a we see that $\{P^G\} = 0$. Thus relation (7.2) reduces to

$$\{\hat{P}^E\} = -\{\widehat{RA}\} \qquad (c)$$

The structure subjected to the equivalent actions on its nodes is shown in Fig. d.

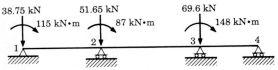

Figure d Structure subjected to the equivalent action on its nodes.

STEP 2.   We make a model of the beam of Fig. d by replacing its supports with reaction elements (see Fig. e). We form the matrix $[B^M]$ of the model of

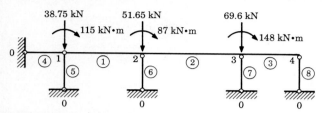

Figure e   Model of the beam of Fig. d.

Fig. e following the procedure described in Sec. 15.2. That is,

1. We establish the matrix $[H]$ for each element of the model of the beam shown in Fig. e. Thus, referring to relation (b) of Example 2 of Sec. 15.2 for elements 1, 2, and 3 $[\phi_{11}^1 = \phi_{11}^2 = \phi_{11}^3 = 0, L_1 = 8 \text{ m}, L_2 = 10 \text{ m}, L_3 = 6 \text{ m}]$ we have

$$[H^1] = \begin{bmatrix} -1 & 0 & 0 \\ 0 & -1 & 0 \\ 0 & -8 & -1 \\ \hline 1 & 0 & 0 \\ 0 & 1 & 0 \\ 0 & 0 & 1 \end{bmatrix} \quad [H^2] = \begin{bmatrix} -1 & 0 & 0 \\ 0 & -1 & 0 \\ 0 & -10 & -1 \\ \hline 1 & 0 & 0 \\ 0 & 1 & 0 \\ 0 & 0 & 1 \end{bmatrix} \quad [H^3] = \begin{bmatrix} -1 & 0 & 0 \\ 0 & -1 & 0 \\ 0 & -6 & -1 \\ \hline 1 & 0 & 0 \\ 0 & 1 & 0 \\ 0 & 0 & 1 \end{bmatrix}$$

$$(d)$$

Moreover, referring to relation (b) of Example 1 of Sec. 15.2 for elements 4 to 8 $[\phi_{11}^4 = \phi_{11}^5 = \phi_{11}^6 = \phi_{11}^7 = \phi_{11}^8 = 90°]$, we get

$$[H^4] = \begin{Bmatrix} -1 \\ 0 \\ 0 \\ \hline 1 \\ 0 \\ 0 \end{Bmatrix} \quad [H^5] = [H^6] = [H^7] = [H^8] = \begin{Bmatrix} 0 \\ -1 \\ 0 \\ \hline 0 \\ 1 \\ 0 \end{Bmatrix} \quad (e)$$

2. We form the matrix $[B^M]$ of the model of Fig. e using the matrices $[H^e]$ of its elements. The equations of equilibrium for the nodes of the model are given on p. 572. These equations can be written as

$$\{P^{EFM}\} - [B^M]\{\hat{a}^{EM}\} = 0 \qquad (g)$$

STEP 3.   The model of Fig. e is statically indeterminate to the second degree. We choose the internal moments at the end k of elements 1 and 2 as the

Node	Equation	value	F_1^{1k}	F_2^{1k}	M_3^{1k}	F_1^{2k}	F_2^{2k}	M_3^{2k}	F_1^{3k}	F_2^{3k}	M_3^{3k}	F_1^{4k}	F_1^{5k}	F_1^{6k}	F_1^{7k}	F_1^{8k}
1	$\sum \bar{F}_1 = 0$	0	-1	0	0	0	0	0	0	0	0	1	0	0	0	0
1	$\sum \bar{F}_2 = 0$	-38.75	0	-1	0	0	0	0	0	0	0	0	1	0	0	0
1	$\sum \bar{M}_3 = 0$	-115.00	0	-8	-1	0	0	0	0	0	0	0	0	0	0	0
2	$\sum \bar{F}_1 = 0$	0	1	0	0	-1	0	0	0	0	0	0	0	0	0	0
2	$\sum \bar{F}_2 = 0$	-51.65	0	1	0	0	-1	0	0	0	0	0	0	1	0	0
2	$\sum \bar{M}_3 = 0$	-87.00	0	0	1	0	-10	-1	0	0	0	0	0	0	0	0
3	$\sum \bar{F}_1 = 0$	0	0	0	0	1	0	0	-1	0	0	0	0	0	1	0
3	$\sum \bar{F}_2 = 0$	-69.60	0	0	0	0	1	0	0	-1	0	0	0	0	0	0
3	$\sum \bar{M}_3 = 0$	148.00	0	0	0	0	0	1	0	-6	-1	0	0	0	0	0
4	$\sum \bar{F}_1 = 0$	0	0	0	0	0	0	0	1	0	0	0	0	0	0	0
4	$\sum \bar{F}_2 = 0$	0	0	0	0	0	0	0	0	1	0	0	0	0	0	1
4	$\sum \bar{M}_3 = 0$	0	0	0	0	0	0	0	0	0	1	0	0	0	0	0

Element columns: Element 1 = columns $F_1^{1k}, F_2^{1k}, M_3^{1k}$; Element 2 = columns $F_1^{2k}, F_2^{2k}, M_3^{2k}$; Element 3 = columns $F_1^{3k}, F_2^{3k}, M_3^{3k}$; columns 4–8 = $F_1^{4k}, F_1^{5k}, F_1^{6k}, F_1^{7k}, F_1^{8k}$.

(f)

redundants of the model. Moreover, we modify the equations of equilibrium (f) of the model as follows.

1. We omit all the rows of Eqs. (f) which represent equations of equilibrium for the axial components of the forces acting on the nodes of the beam, that is, rows 1, 4, 7, and 10. Moreover, we omit all columns of Eqs. (f) which are multiplied by the axial component of nodal force in elements 1 to 4, that is, columns 1, 4, 7, and 10.

2. We interchange the columns of the matrix $[B^M]$ and the rows of the matrix $\{\hat{a}^{EM}\}$ so that the chosen redundants are in the lower part of the matrix $\{\hat{a}^{EM}\}$. Thus,

$$[B^M] = \begin{bmatrix} -1 & 0 & 0 & 0 & 1 & 0 & 0 & 0 & \vdots & 0 & 0 \\ -8 & 0 & 0 & 0 & 0 & 0 & 0 & 0 & \vdots & -1 & 0 \\ 1 & -1 & 0 & 0 & 0 & 1 & 0 & 0 & \vdots & 0 & 0 \\ 0 & -10 & 0 & 0 & 0 & 0 & 0 & 0 & \vdots & 1 & -1 \\ 0 & 1 & -1 & 0 & 0 & 0 & 1 & 0 & \vdots & 0 & 0 \\ 0 & 0 & -6 & -1 & 0 & 0 & 0 & 0 & \vdots & 0 & 1 \\ 0 & 0 & 1 & 0 & 0 & 0 & 0 & 1 & \vdots & 0 & 0 \\ 0 & 0 & 0 & 1 & 0 & 0 & 0 & 0 & \vdots & 0 & 0 \end{bmatrix} \qquad \text{(h)}$$

and

$$\{\hat{a}^{EM}\} = \begin{Bmatrix} F_2^{1k} \\ F_2^{2k} \\ F_2^{3k} \\ M_3^{3k} \\ F_1^{5k} \\ F_1^{6k} \\ F_1^{7k} \\ F_1^{8k} \\ \hdashline M_3^{1k} \\ M_3^{2k} \end{Bmatrix} \qquad \text{(i)}$$

$$\{P^{EFM}\} = \begin{Bmatrix} -38.75 \\ -115.00 \\ -51.65 \\ -87.00 \\ -69.60 \\ 148.00 \\ 0 \\ 0 \end{Bmatrix} \qquad \text{(j)}$$

Moreover, referring to relations (h) and (17.3) we obtain

$$[B_0^M] = \begin{bmatrix} -1 & 0 & 0 & 0 & 1 & 0 & 0 & 0 \\ -8 & 0 & 0 & 0 & 0 & 0 & 0 & 0 \\ 1 & -1 & 0 & 0 & 0 & 1 & 0 & 0 \\ 0 & -10 & 0 & 0 & 0 & 0 & 0 & 0 \\ 0 & 1 & -1 & 0 & 0 & 0 & 1 & 0 \\ 0 & 0 & -6 & -1 & 0 & 0 & 0 & 0 \\ 0 & 0 & 1 & 0 & 0 & 0 & 0 & 1 \\ 0 & 0 & 0 & 1 & 0 & 0 & 0 & 0 \end{bmatrix} \tag{k}$$

and

$$[B_x^M] = \begin{bmatrix} 0 & 0 \\ -1 & 0 \\ 0 & 0 \\ 1 & -1 \\ 0 & 0 \\ 0 & 1 \\ 0 & 0 \\ 0 & 0 \end{bmatrix} \tag{l}$$

Furthermore, referring to relations (i) and (17.3) we get

$$\{a_0^{EM}\} = \begin{Bmatrix} F_2^{1k} \\ F_2^{2k} \\ F_2^{3k} \\ M_3^{3k} \\ F_1^{5k} \\ F_1^{6k} \\ F_1^{7k} \\ F_1^{8k} \end{Bmatrix} \tag{m}$$

$$\{a_x^{EM}\} = \begin{Bmatrix} M_3^{1k} \\ M_3^{2k} \end{Bmatrix} \tag{n}$$

STEP 4. We form the matrix $[\hat{f}^M]$ of all the elements of the model of Fig. e using relation (16.9). Notice that since we have omitted the axial components of the basic nodal actions and of the basic deformation parameters of elements 1 to 3, their flexibility matrix is given by relation (5.47). Moreover, the rows and columns of relation $[\hat{d}^M] = [\hat{f}][\hat{a}^{EM}]$ corresponding to element 4 are omitted. Taking into account that the flexibility of elements 5 to 8 is equal to zero, we have

$$[\hat{f}^M] = \frac{1}{EI} \begin{bmatrix} 170.667 & 32 & 0 & 0 & 0 & 0 & 0 & 0 & 0 & 0 \\ 32 & 8 & 0 & 0 & 0 & 0 & 0 & 0 & 0 & 0 \\ 0 & 0 & 333.333 & 50 & 0 & 0 & 0 & 0 & 0 & 0 \\ 0 & 0 & 50 & 10 & 0 & 0 & 0 & 0 & 0 & 0 \\ 0 & 0 & 0 & 0 & 72 & 18 & 0 & 0 & 0 & 0 \\ 0 & 0 & 0 & 0 & 18 & 6 & 0 & 0 & 0 & 0 \\ 0 & 0 & 0 & 0 & 0 & 0 & 0 & 0 & 0 & 0 \\ 0 & 0 & 0 & 0 & 0 & 0 & 0 & 0 & 0 & 0 \\ 0 & 0 & 0 & 0 & 0 & 0 & 0 & 0 & 0 & 0 \\ 0 & 0 & 0 & 0 & 0 & 0 & 0 & 0 & 0 & 0 \\ 0 & 0 & 0 & 0 & 0 & 0 & 0 & 0 & 0 & 0 \\ 0 & 0 & 0 & 0 & 0 & 0 & 0 & 0 & 0 & 0 \end{bmatrix} \qquad (o)$$

We rearrange the rows and columns of the matrix $[\hat{f}^M]$ so that in relation $[\hat{d}^M] = [\hat{f}^M][\hat{a}^{EM}]$ the chosen redundant basic nodal actions and the corresponding basic deformation parameters are at the lower part of the matrices $[\hat{a}^{EM}]$ and $[\hat{d}^M]$, respectively. Moreover, we partition the matrix $[\hat{f}^M]$ as indicated in relation (17.17). That is,

$$[\hat{f}^M] = \frac{1}{EI} \left[\begin{array}{cccccccc:cc} 170.667 & 0 & 0 & 0 & 0 & 0 & 0 & 0 & 32 & 0 \\ 0 & 333.333 & 0 & 0 & 0 & 0 & 0 & 0 & 0 & 50 \\ 0 & 0 & 72 & 18 & 0 & 0 & 0 & 0 & 0 & 0 \\ 0 & 0 & 18 & 6 & 0 & 0 & 0 & 0 & 0 & 0 \\ 0 & 0 & 0 & 0 & 0 & 0 & 0 & 0 & 0 & 0 \\ 0 & 0 & 0 & 0 & 0 & 0 & 0 & 0 & 0 & 0 \\ 0 & 0 & 0 & 0 & 0 & 0 & 0 & 0 & 0 & 0 \\ \hdashline 32 & 0 & 0 & 0 & 0 & 0 & 0 & 0 & 8 & 0 \\ 0 & 50 & 0 & 0 & 0 & 0 & 0 & 0 & 0 & 10 \end{array} \right] \qquad (p)$$

Hence,

$$[\hat{f}^M_{oo}] = \frac{1}{EI} \begin{bmatrix} 170.667 & 0 & 0 & 0 & 0 & 0 & 0 & 0 \\ 0 & 333.333 & 0 & 0 & 0 & 0 & 0 & 0 \\ 0 & 0 & 72 & 18 & 0 & 0 & 0 & 0 \\ 0 & 0 & 18 & 6 & 0 & 0 & 0 & 0 \\ 0 & 0 & 0 & 0 & 0 & 0 & 0 & 0 \\ 0 & 0 & 0 & 0 & 0 & 0 & 0 & 0 \\ 0 & 0 & 0 & 0 & 0 & 0 & 0 & 0 \\ 0 & 0 & 0 & 0 & 0 & 0 & 0 & 0 \end{bmatrix} \qquad (q)$$

$$[f^M_{0x}] = \frac{1}{EI} \begin{bmatrix} 32 & 0 \\ 0 & 50 \\ 0 & 0 \\ 0 & 0 \\ 0 & 0 \\ 0 & 0 \\ 0 & 0 \\ 0 & 0 \end{bmatrix} \tag{r}$$

$$[f^M_{x0}] = \frac{1}{EI} \begin{bmatrix} 32 & 0 & 0 & 0 & 0 & 0 & 0 & 0 \\ 0 & 50 & 0 & 0 & 0 & 0 & 0 & 0 \end{bmatrix} \tag{s}$$

$$[f^M_{xx}] = \frac{1}{EI} \begin{bmatrix} 8 & 0 \\ 0 & 10 \end{bmatrix} \tag{t}$$

STEP 5. We compute the matrices $[F_0]$ and $[F_x]$ using relation (17.5). That is,

$$[F_0] = [B^M_0]^{-1} = \begin{bmatrix} 0 & -0.125 & 0 & 0 & 0 & 0 & 0 & 0 \\ 0 & 0 & 0 & -0.1 & 0 & 0 & 0 & 0 \\ 0 & 0 & 0 & 0 & 0 & -0.1667 & 0 & -0.1667 \\ 0 & 0 & 0 & 0 & 0 & 0 & 0 & 1 \\ 1 & -0.125 & 0 & 0 & 0 & 0 & 0 & 0 \\ 0 & 0.125 & 1 & -0.1 & 0 & 0 & 0 & 0 \\ 0 & 0 & 0 & 0.1 & 1 & -0.1667 & 0 & -0.1667 \\ 0 & 0 & 0 & 0 & 0 & 0.1667 & 1 & 0.1667 \end{bmatrix} \tag{u}$$

$$[F_x] = -[B^M_0]^{-1}[B^M_x] = \begin{bmatrix} -0.125 & 0 \\ 0.1 & -0.1 \\ 0 & 0.1667 \\ 0 & 0 \\ -0.125 & 0 \\ 0.225 & -0.1 \\ -0.1 & 0.2667 \\ 0 & -0.1667 \end{bmatrix} \tag{v}$$

Furthermore, we compute the matrices $[F_{11}]$, $[F_{12}]$, $[F_{21}]$, and $[F_{22}]$ employing relations (17.20). Thus, using relations (q) to (v) we get

$$[F_{11}] = [F_0]^T[f_{00}^M][F_0]$$

$$= \begin{bmatrix} 0 & 0 & 0 & 0 & 0 & 0 & 0 & 0 \\ 0 & 2.6666 & 0 & 0 & 0 & 0 & 0 & 0 \\ 0 & 0 & 0 & 0 & 0 & 0 & 0 & 0 \\ 0 & 0 & 0 & 3.3333 & 0 & 0 & 0 & 0 \\ 0 & 0 & 0 & 0 & 0 & 0 & 0 & 0 \\ 0 & 0 & 0 & 0 & 0 & 2.0008 & 0 & -0.9998 \\ 0 & 0 & 0 & 0 & 0 & 0 & 0 & 0 \\ 0 & 0 & 0 & 0 & 0 & -0.9998 & 0 & 1.9996 \end{bmatrix} \quad \textbf{(w)}$$

$$[F_{12}] = [F_{21}]^T = [F_0]^T[f_{00}^M][F_x] + [F_0]^T[f_{0x}^M] = \begin{bmatrix} 0 & 0 \\ -1.3334 & 0 \\ 0 & 0 \\ -3.3333 & -1.6667 \\ 0 & 0 \\ 0 & -2.008 \\ 0 & 0 \\ 0 & 0.9998 \end{bmatrix} \quad \textbf{(x)}$$

$$[F_{22}] = [F_x]^T[f_{00}^M][F_x] + [F_x]^T[f_{0x}^M] + [f_{x0}][F_x] + [f_{xx}^M] = \begin{bmatrix} 6.0000 & 1.6667 \\ 1.6667 & 5.3341 \end{bmatrix} \quad \textbf{(y)}$$

STEP 6. We compute the matrix of the basic nodal actions of the elements of the beam of Fig. d and the matrix of its reactions. Substituting relations (j), (x), and (y) into (17.23) we obtain

$$\{\hat{a}_x^{EM}\} = -[F_{22}]^{-1}[F_{21}]\{P^{EFM}\} = \begin{Bmatrix} -89.58 \\ 56.51 \end{Bmatrix} \quad \textbf{(z)}$$

Moreover, substituting relations (j), (u), and (v) into (17.24) we get

$$\{a_0^{EM}\} = [F_0]\{P^{EFM}\} + [F_x]\{a_x^{EM}\} = \begin{Bmatrix} 25.57 \\ -5.91 \\ -15.25 \\ 0 \\ -13.18 \\ -83.13 \\ -78.94 \\ -15.25 \end{Bmatrix} \quad \textbf{(aa)}$$

Thus,

$$\{\hat{a}^{EM}\} = \left\{ \frac{\{a_0^{EM}\}}{\{a_x^{EM}\}} \right\} = \left\{ \begin{array}{c} F_2^{1k} \\ F_2^{2k} \\ F_3^{3k} \\ M_3^{3k} \\ F_1^{5k} \\ F_1^{6k} \\ F_1^{7k} \\ F_1^{8k} \\ \hline M_3^{1k} \\ M_3^{2k} \end{array} \right\} = \left\{ \begin{array}{c} 25.57 \\ -5.91 \\ -15.25 \\ 0 \\ -13.18 \\ -83.13 \\ -78.94 \\ 15.25 \\ \hline -89.58 \\ 56.51 \end{array} \right\} \qquad \text{(bb)}$$

or

$$\{\hat{a}^{E}\} = \left\{ \begin{array}{c} F_2^{1k} \\ M_3^{1k} \\ \hline F_2^{2k} \\ M_3^{2k} \\ \hline F_2^{3k} \\ M_3^{3k} \end{array} \right\} = \left\{ \begin{array}{c} 25.57 \\ -89.58 \\ \hline -5.91 \\ 56.51 \\ \hline -15.25 \\ 0 \end{array} \right\} \qquad \text{(cc)}$$

$$\{R\} = \left\{ \begin{array}{c} R_2^{(1)} \\ R_2^{(2)} \\ R_2^{(3)} \\ R_2^{(4)} \end{array} \right\} = \left\{ \begin{array}{c} 13.18 \\ 83.13 \\ 78.94 \\ -15.25 \end{array} \right\} \qquad \text{(dd)}$$

where $\{\hat{a}^{E}\}$ is the matrix of basic nodal actions of the elements of the beam of Fig. d. Moreover, $\{R\}$ is the matrix of the reactions of the beam of Fig. d. It is identical to the matrix of the reactions of the beam of Fig. a.

STEP 7.   We compute the matrix of the basic nodal actions of the elements of the beam of Fig. a. Substituting relations (a) and (cc) into (17.30) we get

$$\{\hat{a}\} = \{\hat{a}^{R}\} + \{\hat{a}^{E}\} = \left\{ \begin{array}{c} 1.25 \\ 35.00 \\ \hline 69.60 \\ -148.00 \\ \hline 0 \\ 0 \end{array} \right\} + \left\{ \begin{array}{c} 25.57 \\ -89.58 \\ \hline -5.91 \\ 56.51 \\ \hline -15.25 \\ 0 \end{array} \right\} = \left\{ \begin{array}{c} 26.82 \\ -54.58 \\ \hline 63.69 \\ -91.49 \\ \hline -15.25 \\ 0 \end{array} \right\} \qquad \text{(ee)}$$

STEP 8.   We compute the components of displacement of the nodes of the beam. Substituting relations (j), (w), (x), and (y) into (17.25) we have

$$\{\hat{\Delta}\} = \{\Delta^{FM}\} = \begin{Bmatrix} \Delta_2^{(1)} \\ \Delta_3^{(1)} \\ \hdashline \Delta_2^{(2)} \\ \Delta_3^{(2)} \\ \hdashline \Delta_2^{(3)} \\ \Delta_3^{(3)} \\ \hdashline \Delta_2^{(4)} \\ \Delta_3^{(4)} \end{Bmatrix} = [[F_{11}] - [F_{12}][F_{22}]^{-1}[F_{21}]][P^{EFM}] = \frac{1}{EI} \begin{Bmatrix} 0 \\ -187.234 \\ \hdashline 0 \\ -85.588 \\ \hdashline 0 \\ 182.648 \\ \hdashline 0 \\ -91.466 \end{Bmatrix}$$

$$\text{(ff)}$$

The matrix $\{\Delta^{FM}\}$ of the model of Fig. e is identical to the matrices $\{\hat{\Delta}\}$ of the beam of Figs. a and d.

Example 3.   Compute the basic nodal actions of the frame loaded as shown in Fig. a and the components of displacements of its nodes. The elements of the frame are made from the same isotropic linearly elastic material and have the same constant cross section with moment of inertia to area ratio $I/A = 0.02 \text{ m}^2$.

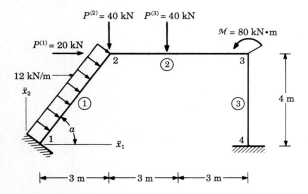

Figure a   Geometry and loading of the frame.

solution

STEP 1.   We analyze the restrained structure (see Fig. bb) and establish the matrix of the basic nodal actions of all its elements $\{\hat{a}^R\}$ as well as the equivalent actions to be placed on the nodes of the structure.

The internal actions at the ends of the elements of the restrained structure

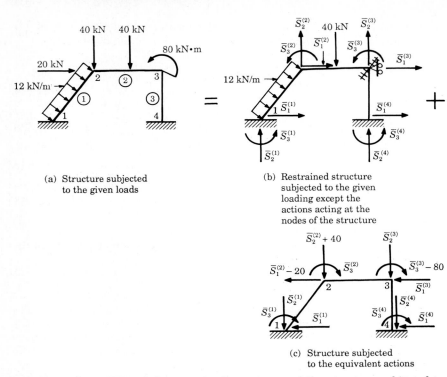

(a) Structure subjected to the given loads

(b) Restrained structure subjected to the given loading except the actions acting at the nodes of the structure

(c) Structure subjected to the equivalent actions

Figure b Superposition of the restrained structure and the structure subjected to the equivalent actions.

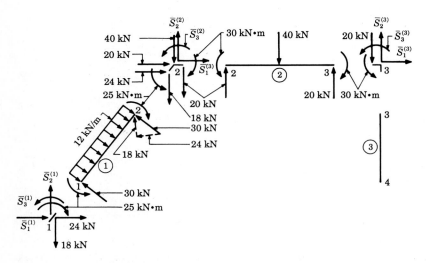

Figure c Free-body diagram of the elements and nodes of the restrained structure.

(fixed-end actions) are obtained by referring to the table on the inside of the back cover and are shown in Fig. c. Referring to Fig. c, the matrix of the basic nodal actions of all the elements of the restrained structure is

$$\{\hat{a}^R\} = \begin{Bmatrix} F_1^{1k} \\ F_2^{1k} \\ M_3^{1k} \\ \hline F_1^{2k} \\ F_2^{2k} \\ M_3^{2k} \\ \hline F_1^{3k} \\ F_2^{3k} \\ M_3^{3k} \end{Bmatrix} = \begin{Bmatrix} 0 \\ 30 \\ -25 \\ \hline 0 \\ 20 \\ -30 \\ \hline 0 \\ 0 \\ 0 \end{Bmatrix} \qquad (a)$$

Moreover, referring to Fig. c from the equilibrium of nodes 2 and 3 of the restrained structure, we obtain

$$\bar{S}_1^{(1)} = -24\text{ kN} \qquad \bar{S}_1^{(2)} = -24\text{ kN} \qquad \bar{S}_1^{(3)} = 0 \qquad\qquad \bar{S}_1^{(4)} = 0$$

$$\bar{S}_2^{(1)} = 18\text{ kN} \qquad \bar{S}_2^{(2)} = 38\text{ kN} \qquad \bar{S}_2^{(3)} = 20\text{ kN} \qquad \bar{S}_2^{(4)} = 0 \qquad (b)$$

$$\bar{S}_3^{(1)} = 25\text{ kN} \qquad \bar{S}_3^{(2)} = 5\text{ kN} \qquad \bar{S}_3^{(3)} = -30\text{ kN} \cdot \text{m} \qquad \bar{S}_3^{(4)} = 0$$

The structure subjected to the equivalent actions is shown in Fig. d.

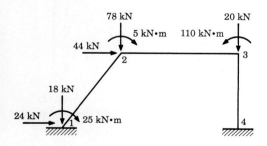

Figure d Structure subjected to the equivalent actions on its nodes.

STEP 2. We make a model of the frame of Fig. d by replacing its supports with reaction elements (see Fig. e). We form the matrix $[B_m^M]$ for the model of

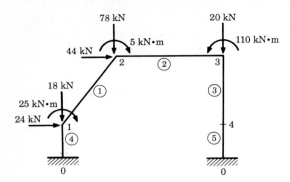

Figure e Model of the frame of Fig. d.

Fig. e following the procedure described in Sec. 15.2. That is,

1. We establish the matrix $[H]$ for each element of the model of the frame shown in Fig. e. Since all the elements of the frame are general planar elements, referring to relations (5.11), (6.48), and (15.15) ($\cos \phi_{11}^1 = \frac{3}{5}$, $\sin \phi_{11}^1 = \frac{4}{5}$, $L_1 = 5$ m), ($\phi_{11}^2 = 0$, $L_2 = 6$ m), ($\phi_{11}^3 = -90°$, $L_3 = 4$ m), ($\phi_{11}^4 = \phi_{11}^5 = 90°$), we obtain

$$[H^1] = \begin{bmatrix} 0.6 & 0.8 & 0 \\ -0.8 & -0.6 & 0 \\ 0 & -5 & -1 \\ \hdashline 0.6 & -0.8 & 0 \\ 0.8 & 0.6 & 0 \\ 0 & 0 & 1 \end{bmatrix} \qquad [H^2] = \begin{bmatrix} -1 & 0 & 0 \\ 0 & -1 & 0 \\ 0 & -6 & -1 \\ \hdashline 1 & 0 & 0 \\ 0 & 1 & 0 \\ 0 & 0 & 1 \end{bmatrix}$$

$$[H^3] = \begin{bmatrix} 0 & -1 & 0 \\ 1 & 0 & 0 \\ 0 & -4 & -1 \\ \hdashline 0 & 1 & 0 \\ -1 & 0 & 0 \\ 0 & 0 & 1 \end{bmatrix} \qquad [H^4] = [H^5] = \begin{bmatrix} 0 & 1 & 0 \\ -1 & 0 & 0 \\ 0 & -L & -1 \\ \hdashline 0 & -1 & 0 \\ 1 & 0 & 0 \\ 0 & 0 & 1 \end{bmatrix}$$

(c)

We form the matrix $[B^M]$ for the model of Fig. e using the matrices $[H^e]$ of its elements. The equations of equilibrium for the nodes of the model are given on p. 583. These equations can be written as

$$\{P^{EFM}\} = [B^M]\{\hat{a}^{EM}\} \qquad (e)$$

STEP 3.   The model of Fig. e is statically indeterminate to the third degree. We choose the basic nodal actions of element 3 as the redundants of the model. Moreover, we interchange the columns of the matrix $[B^M]$ and the rows of the matrix $\{\hat{a}^{EM}\}$ so that in Eqs. (d) the chosen redundants are in the lower part of the matrix $\{\hat{a}^{EM}\}$. Thus, referring to relations (f) and (17.3) we

Element

Node	Equation	Value	F_1^{1k}	F_2^{1k}	M_3^{1k}	F_1^{2k}	F_2^{2k}	M_3^{2k}	F_1^{3k}	F_2^{3k}	M_3^{3k}	$F_1^{(4)}$	$F_2^{(4)}$	$M_3^{(4)}$	$F_1^{(5)}$	$F_2^{(5)}$	$M_3^{(5)}$
1	$\sum \bar{F}_1 = 0$	24	0.6	0.8	0	0	0	0	0	0	0	0	0	1	0	0	0
	$\sum \bar{F}_2 = 0$	−18	−0.8	−0.6	0	0	0	0	0	0	0	−1	0	0	0	0	0
	$\sum \bar{F}_3 = 0$	−25	0	−5	−1	0	0	0	0	0	0	0	1	0	0	0	0
2	$\sum \bar{F}_1 = 0$	44	0.6	−0.8	0	−1	0	0	0	0	0	0	0	0	0	0	0
	$\sum \bar{F}_2 = 0$	−78	0.8	0.6	0	0	−1	0	0	0	0	0	0	0	0	0	0
	$\sum \bar{M}_3 = 0$	−5	0	0	1	0	−6	−1	0	0	0	0	0	0	0	0	0
3	$\sum \bar{F}_1 = 0$	0	0	0	0	1	0	0	−1	0	0	0	0	0	0	0	0
	$\sum \bar{F}_2 = 0$	−20	0	0	0	0	1	0	0	−1	0	0	0	0	0	0	0
	$\sum \bar{M}_3 = 0$	110	0	0	0	0	0	1	0	−4	−1	0	0	0	0	0	0
4	$\sum \bar{F}_1 = 0$	0	0	0	0	0	0	0	1	0	0	0	0	0	0	0	1
	$\sum \bar{F}_2 = 0$	0	0	0	0	0	0	0	0	1	0	0	0	0	0	−1	0
	$\sum \bar{M}_3 = 0$	0	0	0	0	0	0	0	0	0	1	0	0	0	1	0	0

(d)

have

$$[B_0^M] = \begin{bmatrix} 0.6 & 0.8 & 0 & 0 & 0 & 0 & 0 & -1 & 0 & 0 & 0 & 0 \\ -0.8 & -0.6 & 0 & 0 & 0 & 0 & 1 & 0 & 0 & 0 & 0 & 0 \\ 0 & -5.0 & -1 & 0 & 0 & 0 & 0 & 0 & 1 & 0 & 0 & 0 \\ 0.6 & -0.8 & 0 & -1 & 0 & 0 & 0 & 0 & 0 & 0 & 0 & 0 \\ 0.8 & 0.6 & 0 & 0 & -1 & 0 & 0 & 0 & 0 & 0 & 0 & 0 \\ 0 & 0 & 1 & 0 & -6 & -1 & 0 & 0 & 0 & 0 & 0 & 0 \\ 0 & 0 & 0 & 1 & 0 & 0 & 0 & 0 & 0 & 0 & 0 & 0 \\ 0 & 0 & 0 & 0 & 1 & 0 & 0 & 0 & 0 & 0 & 0 & 0 \\ 0 & 0 & 0 & 0 & 0 & 1 & 0 & 0 & 0 & 0 & 0 & 0 \\ 0 & 0 & 0 & 0 & 0 & 0 & 0 & 0 & 0 & 0 & -1 & 0 \\ 0 & 0 & 0 & 0 & 0 & 0 & 0 & 0 & 0 & 1 & 0 & 0 \\ 0 & 0 & 0 & 0 & 0 & 0 & 0 & 0 & 0 & 0 & 0 & 1 \end{bmatrix} \tag{f}$$

$$[B_x^M] = \begin{bmatrix} 0 & 0 & 0 \\ 0 & 0 & 0 \\ 0 & 0 & 0 \\ 0 & 0 & 0 \\ 0 & 0 & 0 \\ 0 & 0 & 0 \\ 0 & -1 & 0 \\ 1 & 0 & 0 \\ 0 & -4 & -1 \\ 0 & 1 & 0 \\ -1 & 0 & 0 \\ 0 & 0 & 1 \end{bmatrix} \tag{g}$$

$$\{a_0^M\} = \begin{Bmatrix} F_1^{1k} \\ F_2^{1k} \\ M_3^{1k} \\ \hline F_1^{2k} \\ F_2^{2k} \\ M_3^{2k} \\ \hline F_1^{4k} \\ F_2^{4k} \\ M_3^{4k} \\ \hline F_1^{5k} \\ F_2^{5k} \\ M_3^{5k} \end{Bmatrix} \tag{h}$$

$$\{a_x^M\} = \left\{ \begin{array}{c} F_1^{3k} \\ F_2^{3k} \\ M_3^{3k} \end{array} \right\} \tag{i}$$

STEP 4.   We assemble the matrix $[\hat{f}^M]$ of all the elements of the model of Fig.
e using relation (16.9). Taking into account that the flexibility of the reaction
elements 4 and 5 is equal to zero, we have

$$[\hat{f}^M] = \begin{bmatrix} 0.1 & 0 & 0 & 0 & 0 & 0 & 0 & 0 & 0 & 0 & 0 & 0 & 0 & 0 & 0 \\ 0 & 41.667 & 12.5 & 0 & 0 & 0 & 0 & 0 & 0 & 0 & 0 & 0 & 0 & 0 & 0 \\ 0 & 12.5 & 5 & 0 & 0 & 0 & 0 & 0 & 0 & 0 & 0 & 0 & 0 & 0 & 0 \\ 0 & 0 & 0 & 0.12 & 0 & 0 & 0 & 0 & 0 & 0 & 0 & 0 & 0 & 0 & 0 \\ 0 & 0 & 0 & 0 & 72 & 18 & 0 & 0 & 0 & 0 & 0 & 0 & 0 & 0 & 0 \\ 0 & 0 & 0 & 0 & 18 & 6 & 0 & 0 & 0 & 0 & 0 & 0 & 0 & 0 & 0 \\ 0 & 0 & 0 & 0 & 0 & 0 & 0.08 & 0 & 0 & 0 & 0 & 0 & 0 & 0 & 0 \\ 0 & 0 & 0 & 0 & 0 & 0 & 0 & 21.333 & 8 & 0 & 0 & 0 & 0 & 0 & 0 \\ 0 & 0 & 0 & 0 & 0 & 0 & 0 & 8 & 4 & 0 & 0 & 0 & 0 & 0 & 0 \\ 0 & 0 & 0 & 0 & 0 & 0 & 0 & 0 & 0 & 0 & 0 & 0 & 0 & 0 & 0 \\ 0 & 0 & 0 & 0 & 0 & 0 & 0 & 0 & 0 & 0 & 0 & 0 & 0 & 0 & 0 \\ 0 & 0 & 0 & 0 & 0 & 0 & 0 & 0 & 0 & 0 & 0 & 0 & 0 & 0 & 0 \\ 0 & 0 & 0 & 0 & 0 & 0 & 0 & 0 & 0 & 0 & 0 & 0 & 0 & 0 & 0 \\ 0 & 0 & 0 & 0 & 0 & 0 & 0 & 0 & 0 & 0 & 0 & 0 & 0 & 0 & 0 \end{bmatrix} \tag{j}$$

We rearrange the rows and columns of the matrix $[\hat{f}^M]$ so that in relation
$\{\hat{d}^M\} = [\hat{f}^M]\{\hat{a}^{EM}\}$ the chosen redundant basic nodal actions and the
corresponding basic deformation parameters are at the lower part of the
matrices $\{\hat{a}^{EM}\}$ and $\{\hat{d}^M\}$, respectively. Moreover, we partition the matrix
$[\hat{f}^M]$ as indicated in relation (17.17). Thus,

$$[f_{00}^M] = \begin{bmatrix} 0.1 & 0 & 0 & 0 & 0 & 0 & 0 & 0 & 0 & 0 & 0 & 0 \\ 0 & 41.667 & 12.5 & 0 & 0 & 0 & 0 & 0 & 0 & 0 & 0 & 0 \\ 0 & 12.5 & 5 & 0 & 0 & 0 & 0 & 0 & 0 & 0 & 0 & 0 \\ 0 & 0 & 0 & 0.12 & 0 & 0 & 0 & 0 & 0 & 0 & 0 & 0 \\ 0 & 0 & 0 & 0 & 72 & 18 & 0 & 0 & 0 & 0 & 0 & 0 \\ 0 & 0 & 0 & 0 & 18 & 6 & 0 & 0 & 0 & 0 & 0 & 0 \\ 0 & 0 & 0 & 0 & 0 & 0 & 0 & 0 & 0 & 0 & 0 & 0 \\ 0 & 0 & 0 & 0 & 0 & 0 & 0 & 0 & 0 & 0 & 0 & 0 \\ 0 & 0 & 0 & 0 & 0 & 0 & 0 & 0 & 0 & 0 & 0 & 0 \\ 0 & 0 & 0 & 0 & 0 & 0 & 0 & 0 & 0 & 0 & 0 & 0 \\ 0 & 0 & 0 & 0 & 0 & 0 & 0 & 0 & 0 & 0 & 0 & 0 \\ 0 & 0 & 0 & 0 & 0 & 0 & 0 & 0 & 0 & 0 & 0 & 0 \end{bmatrix} \tag{k}$$

$$[f_{0x}^M] = \begin{bmatrix} 0 & 0 & 0 \\ 0 & 0 & 0 \\ 0 & 0 & 0 \\ 0 & 0 & 0 \\ 0 & 0 & 0 \\ 0 & 0 & 0 \\ 0 & 0 & 0 \\ 0 & 0 & 0 \\ 0 & 0 & 0 \\ 0 & 0 & 0 \\ 0 & 0 & 0 \\ 0 & 0 & 0 \end{bmatrix} \tag{1}$$

$$[f_{x0}^M] = \begin{bmatrix} 0 & 0 & 0 & 0 & 0 & 0 & 0 & 0 & 0 & 0 & 0 & 0 \\ 0 & 0 & 0 & 0 & 0 & 0 & 0 & 0 & 0 & 0 & 0 & 0 \\ 0 & 0 & 0 & 0 & 0 & 0 & 0 & 0 & 0 & 0 & 0 & 0 \end{bmatrix} \tag{m}$$

$$[f_{xx}^M] = \begin{bmatrix} 0.8 & 0 & 0 \\ 0 & 21.333 & 8 \\ 0 & 8 & 4 \end{bmatrix} \tag{n}$$

STEP 5.   We compute the matrices $[F_0]$ and $[F_x]$ using relations (17.5). That is,

$$[B^M]^{-1} = \left[\begin{array}{cccccccccccc} 0 & 0 & 0 & 0.6 & 0.8 & 0 & 0.6 & 0.8 & 0 & 0 & 0 & 0 \\ 0 & 0 & 0 & -0.8 & 0.6 & 0 & -0.8 & 0.6 & 0 & 0 & 0 & 0 \\ 0 & 0 & 0 & 0 & 0 & 1 & 0 & 6 & 1 & 0 & 0 & 0 \\ 0 & 0 & 0 & 0 & 0 & 0 & 1 & 0 & 0 & 0 & 0 & 0 \\ 0 & 0 & 0 & 0 & 0 & 0 & 0 & 1 & 0 & 0 & 0 & 0 \\ 0 & 0 & 0 & 0 & 0 & 0 & 0 & 0 & 1 & 0 & 0 & 0 \\ \hline 0 & 1 & 0 & 0 & 1 & 0 & 0 & 1 & 0 & 0 & 0 & 0 \\ -1 & 0 & 0 & -0.28 & 0.96 & 0 & -0.28 & 0.96 & 0 & 0 & 0 & 0 \\ 0 & 0 & 1 & -4 & 3 & 1 & -4 & 9 & 1 & 0 & 0 & 0 \\ 0 & 0 & 0 & 0 & 0 & 0 & 0 & 0 & 0 & 0 & 1 & 0 \\ 0 & 0 & 0 & 0 & 0 & 0 & 0 & 0 & 0 & -1 & 0 & 0 \\ 0 & 0 & 0 & 0 & 0 & 0 & 0 & 0 & 0 & 0 & 0 & 1 \end{array}\right]$$

$$\tag{o}$$

$$[F_x] = -[B_0^M]^{-1}[B_x^M] = \begin{bmatrix} -0.8 & 0.6 & 0 \\ -0.6 & -0.8 & 0 \\ -6 & 4 & 1 \\ 0 & 1 & 0 \\ -1 & 0 & 0 \\ 0 & 4 & 1 \\ \hdashline -1 & 0 & 0 \\ -0.96 & -0.28 & 0 \\ -9 & 0 & 1 \\ 1 & 0 & 0 \\ 0 & 1 & 0 \\ 0 & 0 & 1 \end{bmatrix} \tag{p}$$

Furthermore, we compute the matrices $[F_{11}]$, $[F_{12}]$, $[F_{21}]$, and $[F_{22}]$ using relation (17.30). Thus,

$$[F_{11}] = [F_0]^T [f_{00}^M][F_0]$$

$$= \begin{bmatrix} 0 & 0 & 0 & 0 & 0 & 0 & 0 & 0 & 0 & 0 & 0 & 0 \\ 0 & 0 & 0 & 0 & 0 & 0 & 0 & 0 & 0 & 0 & 0 & 0 \\ 0 & 0 & 0 & 0 & 0 & 0 & 0 & 0 & 0 & 0 & 0 & 0 \\ 0 & 0 & 0 & 26.703 & -19.952 & -10 & 26.703 & -79.952 & -10 & 0 & 0 & 0 \\ 0 & 0 & 0 & -19.952 & 15.064 & 7.5 & -19.952 & 60.064 & 7.5 & 0 & 0 & 0 \\ 0 & 0 & 0 & -10 & 7.5 & 5 & -10 & 37.5 & 5 & 0 & 0 & 0 \\ 0 & 0 & 0 & 26.703 & -19.952 & -10 & 26.823 & -79.952 & -10 & 0 & 0 & 0 \\ 0 & 0 & 0 & -79.952 & 60.064 & 37.5 & -79.952 & 357.064 & 55.5 & 0 & 0 & 0 \\ 0 & 0 & 0 & -10 & 7.5 & 5 & -10 & 55.5 & 11 & 0 & 0 & 0 \\ 0 & 0 & 0 & 0 & 0 & 0 & 0 & 0 & 0 & 0 & 0 & 0 \\ 0 & 0 & 0 & 0 & 0 & 0 & 0 & 0 & 0 & 0 & 0 & 0 \\ 0 & 0 & 0 & 0 & 0 & 0 & 0 & 0 & 0 & 0 & 0 & 0 \end{bmatrix} \tag{q}$$

$$[F_{12}] = [F_{21}]^T = [F_0]^T [f_{00}^M][F_x] + [F_0]^T [f_{0x}^M]$$

$$= \begin{bmatrix} 0 & 0 & 0 \\ 0 & 0 & 0 \\ 0 & 0 & 0 \\ -79.952 & 13.297 & 10 \\ 60.064 & -10.048 & -7.5 \\ 37.5 & -10 & -5 \\ -79.952 & 13.177 & 10 \\ 357.064 & -142.048 & -55.5 \\ 55.5 & -34 & -11 \\ 0 & 0 & 0 \\ 0 & 0 & 0 \\ 0 & 0 & 0 \end{bmatrix} \tag{r}$$

$$[F_{22}] = [F_x]^T [f_{00}^M][F_x] + [F_x]^T [f_{0x}^M] + [f_{x0}^M][F_x] + [f_{xx}^M]$$

$$= \begin{bmatrix} 357.864 & -142.048 & -55.5 \\ -142.048 & 144.156 & 42 \\ -55.5 & 42 & 15 \end{bmatrix} \tag{s}$$

STEP 6. We compute the matrix of the basic nodal actions of the elements of the model of Fig. d and the matrix of its reactions. Substituting relations (r) and (t) and $\{P^{EFM}\}$ from relation (d) into (17.23) we obtain

$$\{\hat{a}_x^{EM}\} = -[F_{22}]^{-1}[F_{21}]\{P^{EFM}\} = -\begin{Bmatrix} 33.17 \\ 63.91 \\ -119.59 \end{Bmatrix} \tag{t}$$

Moreover, substituting relations (o), (p), and (t), and $\{P^{EFM}\}$ from relation (d) into (17.24) we get

$$[a_0^{EM}] = [F_0]\{P^{EFM}\} + [F_x]\{a_x^{EM}\} = \begin{bmatrix} -63.81 \\ -22.97 \\ 47.95 \\ -63.91 \\ 13.17 \\ -26.06 \\ -102.83 \\ -32.66 \\ -71.89 \\ -33.17 \\ -63.91 \\ -119.59 \end{bmatrix} \tag{u}$$

$$\{\hat{a}^E\} = \begin{Bmatrix} F_1^{1k} \\ F_2^{2k} \\ M_3^{1k} \\ \hline F_1^{2k} \\ F_2^{2k} \\ M_3^{2k} \\ \hline F_1^{3k} \\ F_2^{3k} \\ M_3^{3k} \end{Bmatrix} = \begin{Bmatrix} -63.81 \\ -22.97 \\ 47.95 \\ \hline -63.91 \\ 13.17 \\ -26.06 \\ \hline -33.17 \\ -63.91 \\ 119.59 \end{Bmatrix} \tag{v}$$

$$\{R\} = \begin{Bmatrix} R_1^{(1)} \\ R_2^{(1)} \\ R_3^{(1)} \\ R_1^{(4)} \\ R_2^{(4)} \\ R_3^{(4)} \end{Bmatrix} = \begin{Bmatrix} 102.83 \\ 32.66 \\ 71.89 \\ 33.17 \\ 63.91 \\ 119.59 \end{Bmatrix} \qquad \text{(w)}$$

where $\{R\}$ is the matrix of the reactions of the frame of Fig. d. It is identical to the matrix of the reactions of the frame of Fig. a.

STEP 7.   We compute the matrix of the basic nodal actions of all the elements of the frame of Fig. a. Substituting relations (a) and (v) into (17.30) we get

$$\{\hat{a}\} = \{\hat{a}^R\} + \{\hat{a}^E\} = \begin{Bmatrix} F_1^{1k} \\ F_2^{1k} \\ M_3^{1k} \\ \hline F_1^{2k} \\ F_2^{2k} \\ M_3^{2k} \\ \hline F_1^{3k} \\ F_2^{3k} \\ M_3^{3k} \end{Bmatrix} = \begin{Bmatrix} 0 \\ 30 \\ -25 \\ \hline 0 \\ 20 \\ -30 \\ \hline 0 \\ 0 \\ 0 \end{Bmatrix} + \begin{Bmatrix} -63.81 \\ -22.97 \\ 47.95 \\ \hline -63.91 \\ 13.17 \\ -26.06 \\ \hline -33.17 \\ -63.91 \\ 119.59 \end{Bmatrix} = \begin{Bmatrix} -63.81 \\ 7.03 \\ 22.95 \\ \hline -63.91 \\ 33.17 \\ 56.06 \\ \hline -33.17 \\ -63.91 \\ 119.59 \end{Bmatrix} \qquad \text{(x)}$$

STEP 8.   We compute the components of displacements of the nodes of the structure. Substituting relations (q), (r), and (s) and $\{P^{EFM}\}$ from relation (d) into (17.25) we have

$$\{\hat{\Delta}\} = \{\Delta^{FM}\} = \begin{Bmatrix} \Delta_1^{(1)} \\ \Delta_2^{(1)} \\ \Delta_3^{(1)} \\ \hline \Delta_1^{(2)} \\ \Delta_2^{(2)} \\ \Delta_3^{(2)} \\ \hline \Delta_1^{(3)} \\ \Delta_2^{(3)} \\ \Delta_3^{(3)} \\ \hline \Delta_1^{(4)} \\ \Delta_2^{(4)} \\ \Delta_3^{(4)} \end{Bmatrix} = [[F_{11}] - [F_{12}][F_{22}]^{-1}[F_{21}]]\{P^{EFM}\} = \frac{1}{EI} \begin{Bmatrix} 0 \\ 0 \\ 0 \\ \hline 278.65 \\ -217.80 \\ -45.05 \\ \hline 271.80 \\ -11.45 \\ 34.74 \\ \hline 0 \\ 0 \\ 0 \end{Bmatrix} \qquad \text{(y)}$$

The matrix $\{\Delta^{FM}\}$ of the model of Fig. e is identical to the matrices $\{\hat{\Delta}\}$ of the frames of Figs. a and d.

17.3   Problems

1 and 2.   Using the flexibility method, compute the internal forces in the members of the truss, loaded as shown in Fig. 17.P1 and the components of displacement of its nodes. The members of the truss are made of the same isotropic linearly elastic material ($E = 210 \text{ kN/mm}^2$). Repeat with the truss of Fig. 17.P2.

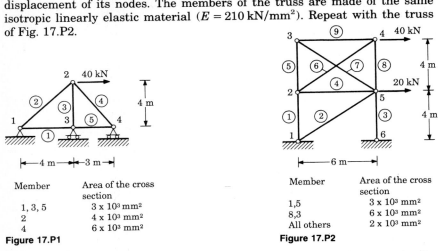

Member	Area of the cross section
1, 3, 5	$3 \times 10^3 \text{ mm}^2$
2	$4 \times 10^3 \text{ mm}^2$
4	$6 \times 10^3 \text{ mm}^2$

Figure 17.P1

Member	Area of the cross section
1,5	$3 \times 10^3 \text{ mm}^2$
8,3	$6 \times 10^3 \text{ mm}^2$
All others	$2 \times 10^3 \text{ mm}^2$

Figure 17.P2

3 to 6.   Using the flexibility method, compute the internal actions in the elements of the beam subjected to the external actions shown in Fig. 17.P3. The elements of the beam have the same constant cross section and are made from the same isotropic linearly elastic material. Choose the reaction at support 2 of the beam as the redundant. Plot the shear and moment diagram of the beam. Repeat with the beams of Figs. 17.P4 to 17.P6.

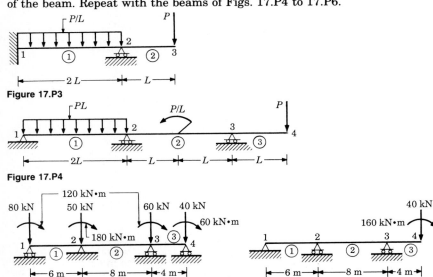

Figure 17.P3

Figure 17.P4

Figure 17.P5

Figure 17.P6

7. Solve Problem 4 choosing the internal moment at point 2 of the beam as the redundant.

8. Using the flexibility method, compute the internal actions in the elements of the structure subjected to the external actions shown in Fig. 17.P8. The elements of the structure have the same constant cross section $(I = 117.7 \times 10^6 \, \text{mm}^4, \ A = 6.24 \times 10^3 \, \text{mm}^2)$ and are made from the same isotropic linearly elastic material.

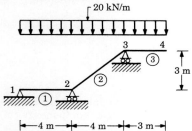

Figure 17.P8

9. Using the flexibility method, compute the internal actions in the elements of the structure resulting from the external force shown in Fig. 17.P9. The area of the steel cable is $A_c = 800 \, \text{mm}^2$. The steel beam has a constant cross section $(I = 369.7 \times 10^6 \, \text{mm}^4, \ A = 13.2 \times 10^3 \, \text{mm}^2)$.

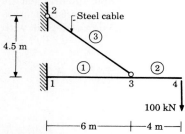

Figure 17.P9

10 to 12. Using the flexibility method, compute the nodal actions in the elements of the frame subjected to the external forces shown in Fig. 17.P10,

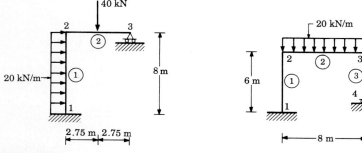

Figure 17.P10

Figure 17.P11

as well as the components of displacement of the nodes of the frame. The elements of the frame are made of the same isotropic linearly elastic material ($E = 210 \text{ kN/mm}^2$) and have the same constant cross section ($I = 117.7 \times 10^6 \text{ mm}^4$, $A = 6.26 \times 10^3 \text{ mm}^2$). Repeat with the frames of Figs. 17.P11 and 17.P12.

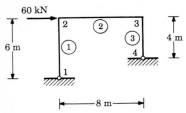

Figure 17.P12

13 to 16. Using the flexibility method, compute the nodal actions in the elements of the frame shown in Fig. 17.P13 and the components of displacements of its nodes for each of the following loading cases.

(a) The external actions shown in the figure.

(b) A settlement of support 1 of 20 mm.

(c) A temperature of the external fibers of $T_e = 35°C$ and of the internal fibers of $T_i = 5°C$. The temperature during construction was $T_0 = 15°C$.

The elements of the frame are made of the same isotropic linearly elastic material ($E = 210 \text{ kN/mm}^2$, $\alpha = 10^{-5}/°C$) and have the same constant cross section ($I = 367.7 \times 10^6 \text{ mm}^4$, $A = 13.2 \times 10^3 \text{ mm}^2$, $h = 425 \text{ mm}$). Repeat with the frames of Figs. 17.P14 to 17.P16.

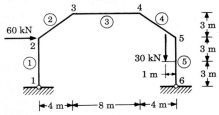

Figure 17.P13

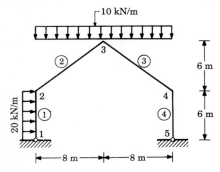

Figure 17.P14

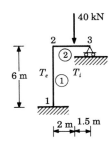

Figure 17.P15

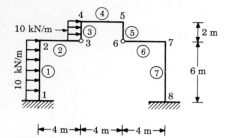

Figure 17.P16

17. Using the flexibility method, compute the nodal actions in the beam loaded as shown in Fig. 17.P17. The beam has a constant cross section, is made from an isotropic linearly elastic material, and is supported on a spring at point 2 whose elastic constant is equal to $k/EI = 0.01$ per cubic meter.

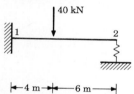

Figure 17.P17

18 and 19. Using the flexibility method, establish the nodal actions of the elements of the frame loaded as shown in Fig. 17.P18 and the components of displacement of its nodes. The elements of the frame are made of the same isotropic linearly elastic material ($E = 210 \text{ kN/mm}^2$) and have the same constant cross section ($I = 83.6 \times 10^6 \text{ mm}^4$, $A = 5.38 \times 10^3 \text{ mm}^2$). Repeat with the frame of Fig. 17.P19.

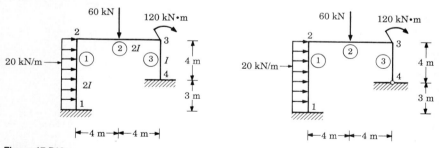

Figure 17.P18

Figure 17.P19

20 to 29. Using the flexibility method, establish the nodal actions of the elements of the frame, loaded as shown in Fig. 17.P20 and the displacements of its nodes. The elements of the frame are made of the same isotropic linearly elastic material ($E = 210 \text{ kN/mm}^2$) and have the same constant cross section ($I = 36.97 \times 10^6 \text{ mm}^4$, $A = 13.2 \times 10^3 \text{ mm}^2$). Repeat with the frames of Figs. 17.P21 to 17.P29.

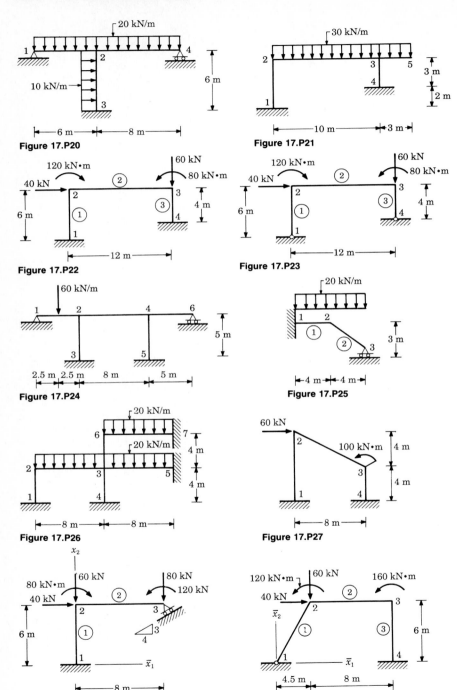

Figure 17.P20

Figure 17.P21

Figure 17.P22

Figure 17.P23

Figure 17.P24

Figure 17.P25

Figure 17.P26

Figure 17.P27

Figure 17.P28

Figure 17.P29

30 to 39. Using the flexibility method, establish the nodal actions of the elements of the frame of Fig. 17.P30 and the components of displacement of its nodes. The elements of the structure are made of the same isotropic linearly elastic material $(E = 210\,\text{kN/mm}^2)$ and have the same constant cross section $(I = 369.7 \times 10^6\,\text{mm}^4,\ A = 13.2 \times 10^3\,\text{mm}^2)$. Repeat with the frames of Figs. 17.P31 to 17.P39.

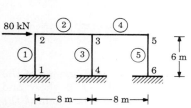

Figure 17.P30

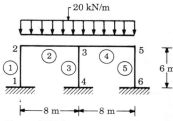

Figure 17.P31

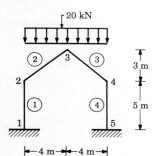

Figure 17.P32

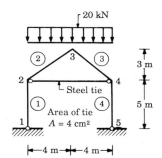

Figure 17.P33

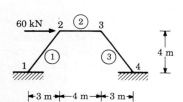

Figure 17.P34

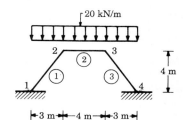

Figure 17.P35

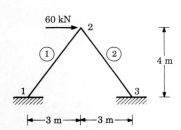

Figure 17.P36

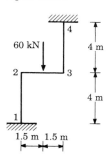

Figure 17.P37

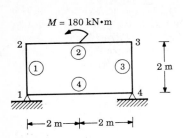

Figure 17.P38

Figure 17.P39

40 and 41. Using the flexibility method, compute the nodal actions of the elements of the grid loaded as shown in Fig. 17.P40, and plot its shear and moment diagrams. Moreover, compute the components of displacement of the nodes of the grid. The elements of the grid are made of the same isotropic linearly elastic material ($E = 210 \, \text{kN/mm}^2$, $v = 0.33$) and have the same constant, tubular cross section ($d_e = 120 \, \text{mm}$, $d_i = 80 \, \text{mm}$). Repeat with the grid of Fig. 17.P41.

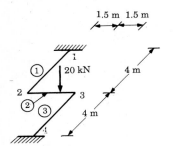

Figure 17.P40

Figure 17.P41

42. Using the flexibility method, compute the nodal forces of the elements of the truss and the components of displacement of its nodes, loaded as shown in Fig. 17.P42. The elements of the truss are made of the same isotropic linearly elastic material ($E = 210 \, \text{kN/mm}^2$) and have the same constant cross section ($A = 8 \times 10^2 \, \text{mm}^2$).

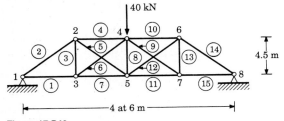

Figure 17.P42

43 and 44. Using the flexibility method, compute the nodal actions of the elements of the space frame of Fig. 17.P43 and the components of displacements of its nodes. The elements of the frame are made of the same isotropic linearly elastic material ($E = 210\,\text{kN/mm}^2$) and have the same constant cross section ($I = 576.8 \times 10^6\,\text{mm}^4$, $A = 19.8 \times 10^3\,\text{mm}^2$). Repeat with the frame of Fig. 17.P44.

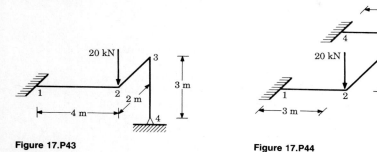

Figure 17.P43   **Figure 17.P44**

45. Using the flexibility method, establish the nodal actions of the elements of the stayed bridge, shown in Fig. 17.P45. The girder of the bridge is subjected to a uniform load of 20 kN/m. Assume that the steel cable slides without friction at the top of the tower. The girder is made of steel, has a constant cross section ($I = 576.80 \times 10^6\,\text{mm}^4$, $A = 19.8 \times 10^3\,\text{mm}^2$), and is continuous at the tower where it is supported on rollers. The cross-sectional area of the cable is $A = 900\,\text{mm}^2$. Disregard the effect of the axial deformation of the girder and the tower ($E = 210\,\text{kN/mm}^2$).

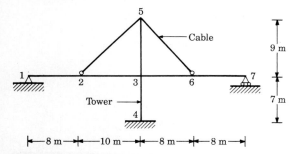

Figure 17.P45

References

1. A. E. Armenàkas, *Classical Structural Analysis: A Modern Approach,* New York, McGraw-Hill, 1988, Sec. 7.2.2.
2. J. Robinson, *Integrated Theory of Finite Element Methods,* New York, John Wiley, 1973.

Photograph and a Brief Description of the Rio Colorado Bridge in Costa Rica

This is a precast, posttensioned concrete suspension bridge with its deck located above rather than below the cables (see Fig. 17.5). It carries the two-lane Pan American highway over a canyon 22 mi (37 km) north of San Jose, Costa Rica. The width of the bridge is 28 ft (8.5 m). Its longitudinal dimensions are given in Fig. 17.6.

It was required that a bridge be designed which would be safe, economical, and aesthetic and whose construction would be compatible with Costa Rica's limited economic development, lack of steel fabrication, lack of capable and skilled construction craftspeople, and difficult site access. Thus a steel structure was not economical because fabricated steel had to be imported. Moreover, a concrete hollow box girder bridge would have been heavy and difficult to erect, while a concrete arch bridge would have required expensive erection falsework because the canyon is deep.

Referring to Fig. 17.7 the bridge was constructed in the following sequence.

Stage 1. The abatements, the piers of the approach spans, and the tunnels for casting the cable anchorages were constructed.

Stage 2. The precast, posttensioned tee beams of the approach spans were put in place and the main 9.14-m (30-ft) piers were cast on hinges. They were kept in the vertical position supported by anchored cables.

Stage 3. The main piers were leaned downward 30° by adjusting the anchored cables, and subsequently their base was fixed by pouring concrete around it. In the inclined position the piers reduce the length of the main span.

Figure 17.5 View of the Rio Colorado Bridge. (*Courtesy T. Y. Lin International Consulting Engineers, San Francisco, Calif.*)

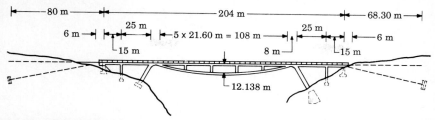

Figure 17.6 Elevation of the Rio Colorado Bridge. (*Courtesy T. Y. Lin International Consulting Engineers, San Francisco, Calif.*)

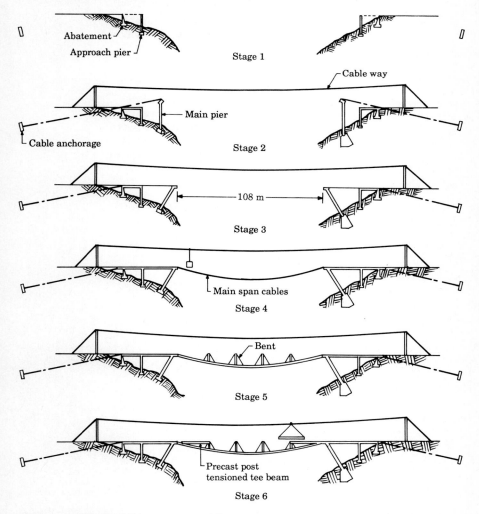

Figure 17.7 Construction sequence. (*Courtesy T. Y. Lin International Consulting Engineers, San Francisco, Calif.*)

Moreover, they exert a stabilizing thrust against the unstable soil of the side slopes. Thus, they increase the safety of the foundations of the abatements.

Stage 4.   The main span cables were draped between the piers, and the topping of the approach spans were completed. Each cable consists of four separate sets of seven tendons and is anchored by means of huge dead anchors at the ends of the bridge.

Stage 5.   Thirty 13×26 ft (4×8 m) precast concrete sections were suspended one at a time from the main span cables with the aid of a cableway situated high above the main span cables (see Fig. 17.7). They form the bottom chord of the bridge which served as a work platform during construction. The concrete sections have two throughs on each side. They were first lowered under the cables with the throughs on their upper side, and then they were raised until each cable of the main span bottomed into one through. The concrete sections were secured in place by 4-in (100-mm) steel pins inserted through holes located on their sides. After all the concrete sections were suspended and secured to the cables, the latter were stressed to 160,000 lb/in^2 ($11{,}034 \times 10^5$ kN/m^2) and then were encased in the throughs with grout. Finally in this stage, the vertical column bents were attached to the cables.

Stage 6.   The precast posttensioned tee beams were lowered into place spanning the distance between the bents.

Stage 7.   A 6-in (150-mm) concrete deck was cast in place.

Derivation and Applications of the Principle of Virtual Work

18

Derivation of the Principle of Virtual Work

18.1 Introduction

In this part of the book we do the following.

1. In Chap. 18 we derive the principle of virtual work.
2. In Chap. 19 we employ the principle of virtual work to establish the response of an element of a framed structure. That is,
   a. We describe the finite-element method and use it in conjunction with the principle of virtual work to establish approximate formulas for the stiffness matrix and the matrix of fixed-end actions of elements of any geometry.
   b. We use the principle of virtual work to obtain exact formulas for the flexibility coefficients for the following types of elements.
      (1) Elements of constant cross section.
      (2) Tapered elements.
      (3) Curved elements of constant cross section.
      (4) Elements for which the effect of shear deformation is included.
   Some of these formulas are used to establish the values of the flexibility coefficients of the previously described elements in many commercially available programs for analyzing groups of framed structures. The stiffness matrix of an element is established from its flexibility matrix using relation (5.93).
3. In Chap. 20 we use the principle of virtual work to analyze framed structures. That is,
   a. We present the dummy load method for computing com-

ponents of displacement of a point of a framed structure. This method is used extensively in classical structural analysis.

b. We use the principle of virtual work for framed structures to establish their stiffness equations. These equations are exact if the elements of a structure have constant cross section and are subjected to such loads along their length that it is easy to establish their exact stiffness matrix and their exact matrix of fixed-end actions. Otherwise, these equations are approximate. A similar approach is employed in classical structural analysis and is known as the *method of virtual displacements*.

18.2 The Boundary-Value Problem for Establishing the Distribution of the Components of Displacement, Strain, and Stress in a Deformable Body

We consider a deformable body initially in a reference stress-free state of mechanical† and thermal‡ equilibrium at a uniform temperature T_0. The body could have been made from a number of pieces connected together. The geometry of one or more of these pieces could have been slightly different than the one required in order that the pieces fit together. In this case in the reference stress-free state we consider the pieces before they are connected, and we regard the body as being in a state of initial strain. We denote the components of initial strain by e_{ij}^I $(i,j = 1, 2, 3)$. They are equal but of opposite sign to those produced by the external actions which must be applied to the pieces to change their geometry to that required, in order to fit and be connected together. Because of the application of specified external disturbances and specified constraints, the body deforms and reaches a second state of mechanical equilibrium. These disturbances and constraints could include the following.

1. Specified components of force applied to every particle of the body due to its presence in a force field such as the gravitational field of the earth. They are called components of *body force* and are given in units of force per unit volume of the body. We denote

---

† When a body is in a state of mechanical equilibrium, its particles do not accelerate. That is, the sum of the forces acting on any portion of the body vanishes and the sum of the moments of these forces about any point vanishes.

‡ When a body is in a state of thermal equilibrium, heat does not flow in or out of it. That is, the temperature of all its particles is the same.

the body force by $\mathbf{B}(x_1, x_2, x_3) = B_1(x_1, x_2, x_3)\mathbf{i}_1 + B_2(x_1, x_2, x_3)\mathbf{i}_2 + B_3(x_1, x_2, x_3)\mathbf{i}_3$.

2. Specified components of force applied on some particles of the surface of the body due to its contact with other bodies. They are called components of *surface traction* and are given in units of force per unit area of the surface of the body. We denote the surface traction by $\mathbf{T}^s(x_1^s, x_2^s, x_3^s) = T_1^s(x_1^s, x_2^s, x_3^s)\mathbf{i}_1 + T_2^s(x_1^s, x_2^s, x_3^s)\mathbf{i}_2 + T_3^s(x_1^s, x_2^s, x_3^s)\mathbf{i}_3$, where x_1^s, x_2^s, x_3^s are the coordinates of the points of the surface of the body.

3. Specified components of displacements of some particles of the surface of the body. The points of the surface of the body where components of displacement are specified are called its *supports*. The forces exerted by the supports of a body on its particles are not known. They are called the *reactions* of the supports of the body.

4. Specified temperature $T(x_1^s, x_2^s, x_3^s)$ of some particles of the surface of the body.

5. Specified temperature gradient $\partial T/\partial x_n$ at some points of the surface of the body, where x_n is the coordinate in the direction normal to the surface of the body at a point.

In general four quantities must be specified at each point of the surface of a body, one from each of the following pairs of quantities:

$$
\begin{array}{ccc}
\hat{u}_1^s & \text{or} & T_1^s \\[4pt]
\hat{u}_2^s & \text{or} & T_2^s \\[4pt]
\hat{u}_3^s & \text{or} & T_3^s \\[4pt]
T & \text{or} & \partial T/\partial x_n
\end{array}
\tag{18.1}
$$

The quantities which are specified at the points of the surface of a body are called its *boundary conditions*. When the components of displacement and the temperature are specified at a point of the surface of a body, the boundary conditions at that point are called *essential*. When the components of traction and the temperature gradient are specified at a point of the surface of a body the boundary conditions at that point are called *natural*.

Notice that the distribution of a component of traction T_i^s on the surface of a body in equilibrium cannot be specified throughout its whole surface independently of the distribution of the component of body force B_i. The distribution of the component of traction T_i^s and the distribution of the component of body force B_i must satisfy the

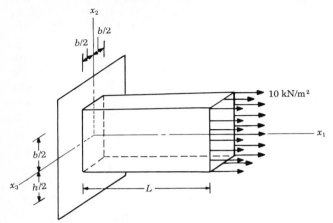

Figure 18.1 Bar subjected to tractions on one face with the opposite face fixed.

conditions imposed by the equilibrium of the body as a whole. This requirement does not arise if the displacement component u_i is specified at some points of the surface of the body. In this case, at the points where the displacement component u_i is specified, the solution of the problem yields components of traction T_i^s satisfying the equilibrium of the body as a whole.

As an example, consider the bar of Fig. 18.1. Its boundary conditions are

$$u_i(0, x_2, x_3) = 0 \qquad i = 1, 2, 3$$

$$T_i^s(x_1, h/2, x_3) = T_i^s(x_1, -h/2, x_3) = 0 \qquad i = 1, 2, 3$$

$$T_i^s(x_1, x_2, b/2) = T_i^s(x_1, x_2, -b/2) = 0 \qquad i = 1, 2, 3$$

$$T_1^s(L, x_2, x_3) = 10 \text{ kN/m}^2 \qquad T_i^s(L, x_2, x_3) = 0 \qquad i = 2, 3$$

If the temperature T is not maintained uniform throughout its volume, the body does not reach a state of thermal equilibrium because heat will be flowing in or out of it. If during a process of deformation the temperature distribution on the surface of the body remains constant $(T = T_0)$, the temperature distribution inside the volume of the body may change slightly and heat may be absorbed from or emitted to the environment. However, when the body reaches the second state of equilibrium, the temperature distribution inside its volume will be uniform and equal to the temperature T_0 of its surface. The amount of heat absorbed from or emitted to the environment during loading is equal to the amount of heat emitted to or absorbed from the environment during unloading (reversible heat). *We assume that the effect of the deformation of the body on the distribution of its temperature is very small and can be neglected.* Thus the distribution of the temperature inside the body can be established independently of its deformation. We assume that it has been established and is known.

The strong form of the boundary-value problem for computing the components of displacement, strain, and stress in a body when it is in the second state of mechanical equilibrium described previously can be formulated† as follows: Establish the components of displacement $\hat{u}_i(x_1, x_2, x_3)$ $(i = 1, 2, 3)$ which have the following attributes.

1. They are bounded, single-valued functions having continuous first derivatives.

2. When substituted into the strain–displacement relations (1.3) they give components of strain which on the basis of the stress–strain relations for the material from which the body is made give components of stress which satisfy the equations of equilibrium at every particle of the body. These equations ensure that the forces acting on each particle of the body are in equilibrium. They can be written as

$$\sum_{j=1}^{3} \frac{\partial \tau_{1j}}{\partial x_j} + B_1 = 0$$

$$\sum_{j=1}^{3} \frac{\partial \tau_{2j}}{\partial x_j} + B_2 = 0 \qquad (18.2)$$

$$\sum_{j=1}^{3} \frac{\partial \tau_{3j}}{\partial x_j} + B_3 = 0$$

3. They satisfy the specified essential boundary conditions at the points of the surface of the body where such conditions are specified. If we assume that the components of displacement of the points of the portion S_1 of the surface S of the body are specified, we have

$$\{\hat{\mathbf{u}}\} = \begin{Bmatrix} \hat{u}_1(x_1, x_2, x_3) \\ \hat{u}_2(x_1, x_2, x_3) \\ \hat{u}_3(x_1, x_2, x_3) \end{Bmatrix} = \begin{Bmatrix} \hat{u}_1^s(x_1^s, x_2^s, x_3^s) \\ \hat{u}_2^s(x_1^s, x_2^s, x_3^s) \\ \hat{u}_3^s(x_1^s, x_2^s, x_3^s) \end{Bmatrix} \quad \text{on } S_1 \quad (18.3)$$

where $\hat{u}_1^s$, $\hat{u}_2^s$, $\hat{u}_3^s$ are the specified components of displacement and x_1^s, x_2^s, x_3^s are the coordinates of a point on the surface of the body.

---

† This formulation of the strong form of the boundary value under consideration is called the *displacement formulation*. Another formulation called the *stress formulation* will not be discussed in this book.

4. When substituted into the strain–displacement relations (1.3) they give components of strain which on the basis of the stress–strain relations for the material from which the body is made give components of stress which satisfy the specified natural boundary conditions at the points of the surface of the body where such conditions are specified. We assume that the components of traction acting at the points of the portion $S - S_1$ of the surface S of the body are specified. Denoting the outward unit vector at a point of the surface of the body by $\mathbf{n} = n_1\mathbf{i}_1 + n_2\mathbf{i}_2 + n_3\mathbf{i}_3$ it can be shown that the components of surface traction acting on a particle located at this point are related to the components of stress acting on this particle by the following relations:

$$\{\mathbf{T}\} = \begin{Bmatrix} \tau_{11}n_1 + \tau_{21}n_2 + \tau_{31}n_3 \\ \tau_{12}n_1 + \tau_{22}n_2 + \tau_{32}n_3 \\ \tau_{13}n_1 + \tau_{23}n_2 + \tau_{33}n_3 \end{Bmatrix} = \begin{Bmatrix} T_1^s(x_1^s, x_2^s, x_3^s) \\ T_2^s(x_1^s, x_2^s, x_3^s) \\ T_3^s(x_1^s, x_2^s, x_3^s) \end{Bmatrix} \quad \text{on } S - S_1$$

(18.4)

where T_1^s, T_2^s, T_3^s are the specified components of traction.

18.3 The Principle of Virtual Work for Deformable Bodies

In this section we establish the principle of virtual work for deformable bodies starting from Bernoulli's principle of virtual displacements for rigid bodies. This principle[1] can be stated as follows: *Consider a rigid body in equilibrium under the influence of a system of body forces, surface actions, and reactions. The work of this system of actions vanishes when the rigid body is subjected to a sufficiently small imaginary displacement, which is called "virtual." The converse is also valid. That is, a rigid body is in equilibrium under the influence of a system of body forces, surface actions, and reactions if the total work performed by this system of actions during every sufficiently small virtual displacement vanishes.* This principle is derived using only the requirements for equilibrium of a rigid body. That is,

$$\sum \mathbf{F} = 0 \qquad \sum \mathbf{M} = 0$$

Before proceeding with the derivation of the principle of virtual work we make the following definitions.

1. We define a statically admissible stress field $\bar{\tau}_{ij}(x_1, x_2, x_3)$ $(i, j = 1, 2, 3)$ in a body in the second state of mechanical equilibrium as a distribution of the components of stress which

satisfies the requirements for equilibrium of its particles. This implies that the components of stress are

a. Differentiable functions with continuous first derivatives at every point inside the volume of the body.

b. They are symmetric ($\tilde{\tau}_{ij} = \tilde{\tau}_{ji}$). This ensures that the sum of the moments of all the forces acting in each particle of the body vanishes.

c. They satisfy the equations of equilibrium at every point inside the volume of the body.

d. They yield the specified components of traction at the points of the surface of the body where components of traction are specified [see relations (18.3)]. Notice that an infinite number of statically admissible distributions of components of stress $\tau_{ij}(x_1, x_2, x_3)$ ($i, j = 1, 2, 3$) exist. Generally, for a given material (given stress–strain relations) the components of strain obtained from a statically admissible stress field may not yield the specified components of displacement at the points of the surface of the body where components of displacement are specified.

2. We define an admissible displacement field $\hat{u}_i(x_1, x_2, x_3)$ ($i = 1, 2, 3$) in the body under consideration as a distribution of the components of displacement which satisfies the following conditions.

a. The components of displacement have continuous first derivatives at every point inside the volume of the body.

b. The magnitude of the components of displacement is in the range of validity of the theory of small deformation.

3. We define a *geometrically admissible displacement field* $\hat{u}_i(x_1, x_2, x_3)$ ($i = 1, 2, 3$) in the body under consideration as one which satisfies the following conditions.

a. It is admissible.

b. It yields the specified components of displacement on the points of the surface of the body where components of displacement are specified.

4. We define an admissible or a geometrically admissible strain field as one which is related to an admissible or geometrically admissible displacement field, respectively, by relations (1.3). A geometrically admissible displacement field $\hat{u}_i(x_1, x_2, x_3)$ does not give components of stress which necessarily satisfy the equations of equilibrium. That is, a geometrically admissible displacement field is not necessarily the actual displacement field. However, if for a particular body the strain field, obtained on the basis of the appropriate strain–stress relations from a statically admissible stress field, yields a geometrically admis-

sible displacement field, then this statically admissible stress field is the actual.

Consider a distribution $\tilde{\tau}_{ij}$ $(i, j = 1, 2, 3)$ of the components of stress in a body which is statically admissible to the specified external forces. Imagine that the body is subjected to an *additional sufficiently small imaginary deformation* specified by the additional admissible displacement field $\hat{u}(x_1, x_2, x_3)$ and corresponding components of strain $\tilde{e}_{ij}(x_1, x_2, x_3)$ $(i, j = 1, 2, 3)$. Generally, this displacement field is unrelated to the loading originally applied to the body. That is, it is produced by another unspecified imaginary loading acting on the body. We call this additional imaginary admissible displacement field "virtual." As a result of the application on the body of the virtual displacement field, any infinitesimal segment (particle) of the body translates and rotates as a rigid body and deforms. Thus, the total work dW_T of the statically admissible components of stress $\tilde{\tau}_{ij}$ $(i, j = 1, 2, 3)$ and of the body force acting on a particle of the body due to the previously described virtual displacement field consists of two parts: the work dW_d due to the deformation of the particle and the work $(dW_T - dW_d)$ due to its displacement as a rigid body. Since the particle under consideration is in equilibrium on the basis of the principle of virtual displacements, the work $(dW_T - dW_d)$ performed by all the forces acting on the particle due to its sufficiently small displacement as a rigid body must vanish. Consequently,

$$dW_T = dW_d \tag{18.5}$$

The total work W_T performed by the statically admissible components of stress acting on all the particles of the body, due to the admissible displacement field is equal to

$$W_T = \iiint_V dW_d \tag{18.6}$$

where V is the volume of the body. However, two adjacent particles have a common boundary, and the components of stress acting on this boundary of the one particle are equal and opposite to those acting on the corresponding boundary of the other particle. Thus, the sum of the work performed by these equal and opposite components of stress vanishes. Consequently, W_T consists of the work of the components of stress acting on the faces of the particles which are part of the surface of the body and of the work of the body forces. Hence, W_T is equal to the work of the known external forces (body forces and surface forces) acting on the body and of the unknown reactions of its supports.

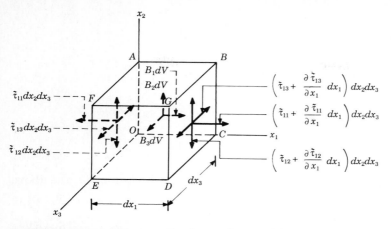

Figure 18.2 Forces acting on the faces of a particle of a body which are normal to the x_1 axis.

On the basis of the previous discussion relation (18.4) becomes

$$W_{\text{ext.act.}} + W_{\text{react.}} = \iiint_V dW_d \qquad (18.7)$$

In what follows we compute the work of deformation dW_d, the statically admissible components of stress $\tilde{\tau}_{ij}$ $(i,j=1,2,3)$ acting on a particle of a body due to its deformation.

Consider an infinitesimal segment (particle) of dimensions dx_1, dx_2, and dx_3 and volume dV of a body subjected to given external disturbances, and denote by $\tilde{\tau}_{ij}$ a distribution of stress in the body which is statically admissible to the given external forces. The forces acting on the faces of this particle which are normal to the x_1 axis and the components of the body force **B** are shown in Fig. 18.2. Moreover, consider an admissible virtual distribution of the components of displacement whose values at the centroid of face *OAFE* of the particle under consideration we denote by $\tilde{\tilde{u}}_1$, $\tilde{\tilde{u}}_2$, and $\tilde{\tilde{u}}_3$. The components of displacement at the centroid of face *BGDC* of this particle may be expressed as

$$\tilde{\tilde{u}}_i\big|_{x_1+dx_1} = \tilde{\tilde{u}}_i\big|_{x_1} + \frac{\partial \tilde{\tilde{u}}_i}{\partial x_1}\bigg|_{x_1} dx_1 \qquad i=1,2,3 \qquad (18.8)$$

The work performed by the statically admissible components of stress acting on the two faces of the particle under consideration which are normal to the x_1 axis due to the previously described admissible displacement field is equal to

$$dW_d^{(1)} = -\tilde{\tau}_{11}\,dx_2\,dx_3\tilde{\tilde{u}}_1 - \tilde{\tau}_{12}\,dx_2\,dx_3\,\tilde{\tilde{u}}_2 - \tilde{\tau}_{13}\,dx_2\,dx_3\,\tilde{\tilde{u}}_3$$

$$+ \left(\tilde{\tau}_{11} + \frac{\partial\tilde{\tau}_{11}}{\partial x_1}\,dx_1\right)dx_2\,dx_3\left(\tilde{\tilde{u}}_1 + \frac{\partial\tilde{\tilde{u}}_1}{\partial x_1}\,dx_1\right)$$

$$+ \left(\tilde{\tau}_{12} + \frac{\partial\tilde{\tau}_{12}}{\partial x_1}\,dx_1\right)dx_2\,dx_3\left(\tilde{\tilde{u}}_2 + \frac{\partial\tilde{\tilde{u}}_2}{\partial x_1}\,dx_1\right)$$

$$+ \left(\tilde{\tau}_{13} + \frac{\partial\tilde{\tau}_{13}}{\partial x_1}\,dx_1\right)dx_2\,dx_3\left(\tilde{\tilde{u}}_3 + \frac{\partial\tilde{\tilde{u}}_3}{\partial x_1}\,dx_1\right)$$

Simplifying and disregarding infinitesimals of higher order, the above relation reduces to

$$dW_d^{(1)} = \left(\tilde{\tau}_{11}\frac{\partial\tilde{\tilde{u}}_1}{\partial x_1} + \tilde{\tau}_{12}\frac{\partial\tilde{\tilde{u}}_2}{\partial x_1} + \tilde{\tau}_{13}\frac{\partial\tilde{\tilde{u}}_3}{\partial x_1} + \frac{\partial\tilde{\tau}_{11}}{\partial x_1}\tilde{\tilde{u}}_1 + \frac{\partial\tilde{\tau}_{12}}{\partial x_1}\tilde{\tilde{u}}_2 + \frac{\partial\tilde{\tau}_{13}}{\partial x_1}\tilde{\tilde{u}}_3\right)dV$$

$$(18.9)$$

This result can be extended to establish the work of the statically admissible components of stress acting on all the faces of the particle under consideration, due to the previously described admissible displacement field. That is,

$$dW_d = \left(\tilde{\tau}_{11}\frac{\partial\tilde{\tilde{u}}_1}{\partial x_1} + \tilde{\tau}_{12}\frac{\partial\tilde{\tilde{u}}_2}{\partial x_1} + \tilde{\tau}_{13}\frac{\partial\tilde{\tilde{u}}_3}{\partial x_1} + \tilde{\tau}_{21}\frac{\partial\tilde{\tilde{u}}_1}{\partial x_2} + \tilde{\tau}_{22}\frac{\partial\tilde{\tilde{u}}_2}{\partial x_2} + \tilde{\tau}_{23}\frac{\partial\tilde{\tilde{u}}_3}{\partial x_2}\right.$$

$$\left.+ \tilde{\tau}_{31}\frac{\partial\tilde{\tilde{u}}_1}{\partial x_3} + \tilde{\tau}_{32}\frac{\partial\tilde{\tilde{u}}_2}{\partial x_3} + \tilde{\tau}_{33}\frac{\partial\tilde{\tilde{u}}_3}{\partial x_3}\right)dV + \left(\frac{\partial\tilde{\tau}_{11}}{\partial x_1}\tilde{\tilde{u}}_1 + \frac{\partial\tilde{\tau}_{12}}{\partial x_1}\tilde{\tilde{u}}_2 + \frac{\partial\tilde{\tau}_{13}}{\partial x_1}\tilde{\tilde{u}}_3\right.$$

$$\left.+ \frac{\partial\tilde{\tau}_{21}}{\partial x_2}\tilde{\tilde{u}}_1 + \frac{\partial\tilde{\tau}_{22}}{\partial x_2}\tilde{\tilde{u}}_2 + \frac{\partial\tilde{\tau}_{23}}{\partial x_2}\tilde{\tilde{u}}_3 + \frac{\partial\tilde{\tau}_{31}}{\partial x_3}\tilde{\tilde{u}}_1 + \frac{\partial\tilde{\tau}_{32}}{\partial x_3}\tilde{\tilde{u}}_2 + \frac{\partial\tilde{\tau}_{33}}{\partial x_3}\tilde{\tilde{u}}_3\right)dV$$

Using the equilibrium equations (18.2) and the strain–displacement relations (1.3) the above relation is simplified to the following.

$$dW_d = (\tilde{\tau}_{11}\tilde{e}_{11} + 2\tilde{\tau}_{12}\tilde{e}_{12} + 2\tilde{\tau}_{13}\tilde{e}_{13} + 2\tilde{\tau}_{23}\tilde{e}_{23} + \tilde{\tau}_{22}\tilde{e}_{22} + \tilde{\tau}_{33}\tilde{e}_{33})\,dV$$

$$(18.10)$$

Relation (18.10) may be rewritten as

$$dW_d = \sum_{p=1}^{3}\sum_{q=1}^{3}\tilde{\tau}_{pq}\tilde{e}_{pq}\,dV \qquad (18.11)$$

Substituting relation (18.11) into (18.5), we obtain

$$W_{\text{ext.act.}} + W_{\text{react.}} = \iiint_V\left(\sum_{p=1}^{3}\sum_{q=1}^{3}\tilde{\tau}_{pq}\tilde{e}_{pq}\right)dV \qquad (18.12)$$

Relation (18.12) may be rewritten as

$$W_{\text{ext.act.}} + W_{\text{react.}} = \iiint_V [\tilde{\tilde{e}}]^T [\tilde{\tau}] \, dV \qquad (18.13)$$

where
$$[\tilde{\tilde{e}}] = \begin{Bmatrix} \tilde{\tilde{e}}_1 \\ \tilde{\tilde{e}}_2 \\ \tilde{\tilde{e}}_3 \\ \tilde{\tilde{e}}_4 \\ \tilde{\tilde{e}}_5 \\ \tilde{\tilde{e}}_6 \end{Bmatrix} = \begin{Bmatrix} \tilde{\tilde{e}}_{11} \\ \tilde{\tilde{e}}_{22} \\ \tilde{\tilde{e}}_{33} \\ \tilde{\tilde{\gamma}}_{12} \\ \tilde{\tilde{\gamma}}_{13} \\ \tilde{\tilde{\gamma}}_{23} \end{Bmatrix} \qquad [\tilde{\tau}] = \begin{Bmatrix} \tilde{\tau}_1 \\ \tilde{\tau}_2 \\ \tilde{\tau}_3 \\ \tilde{\tau}_4 \\ \tilde{\tau}_5 \\ \tilde{\tau}_6 \end{Bmatrix} = \begin{Bmatrix} \tilde{\tau}_{11} \\ \tilde{\tau}_{22} \\ \tilde{\tau}_{33} \\ \tilde{\tau}_{12} \\ \tilde{\tau}_{13} \\ \tilde{\tau}_{23} \end{Bmatrix} \qquad (18.14)$$

$$\tilde{\tilde{\gamma}}_{ij} - 2\tilde{\tilde{e}}_{ij} \qquad i,j = 1, 2, 3 \qquad (18.15)$$

Relation (18.12) or (18.13) is the principle of virtual work for a deformable body. In obtaining relation (18.12) or (18.13) we have not employed the stress–strain relations for the material from which the element is made. Consequently, this relation is valid for elements made of any material (elastic or nonelastic).

We have shown that any stress field $\tilde{\tau}_{ij}(x_1, x_2, x_3)$ $(i,j = 1, 2, 3)$ which is statically admissible to the given external forces acting on a body satisfies the principle of virtual work (18.12) for any admissible distribution of components of displacement $\tilde{\tilde{u}}_i(x_1, x_2, x_3)$ $(i = 1, 2, 3)$ and corresponding components of strain $\tilde{\tilde{e}}_{ij}(x_1, x_2, x_3)$ $(i,j = 1, 2, 3)$. Moreover, it can be shown that a set of functions $\tilde{\tau}_{ij}(x_1, x_2, x_3)$ $(i,j = 1, 2, 3)$, which together with the given external forces acting on a body satisfy the principle of virtual work (18.12) for every admissible distribution of components of displacement $\tilde{\tilde{u}}_i(x_1, x_2, x_3)$ $(i = 1, 2, 3)$ and corresponding components of strain $\tilde{\tilde{e}}_{ij}(x_1, x_2, x_3)$ $(i,j = 1, 2, 3)$, represents a set of components of stress which are statically admissible to the given external forces.

The actual (true) distribution of the components of displacement $\hat{u}_i(x_1, x_2, x_3)$ $(i = 1, 2, 3)$ of a deformable body when substituted into the strain displacement relations (1.3) gives components of strain which when substituted into the appropriate stress–strain relations for the material from which the body is made give the actual (true) components of stress. Since these components represent a statically admissible distribution of stress, they satisfy the principle of virtual work (18.12) for any admissible distribution of components of displacement $\tilde{\tilde{u}}_i(x_1, x_2, x_3)$ $(i = 1, 2, 3)$ and corresponding components of strain $\tilde{\tilde{e}}_{ij}(x_1, x_2, x_3)$ $(i,j = 1, 2, 3)$. Moreover, it can be shown that a geometrically admissible set of functions $\hat{u}_i(x_1, x_2, x_3)$, which on the basis of relations (1.3) gives components of strain $e_{ij}(x_1, x_2, x_3)$ $(i,j = 1, 2, 3)$, which on the basis of the stress–strain relations for the material from which a body is made give components of stress $\tau_{ij}(x_1, x_2, x_3)$, which together with the given set of external forces acting on a body satisfy the principle of virtual work (18.12) for every admissible distribution of the components of displacement $\tilde{\tilde{u}}_i(x_1, x_2, x_3)$ $(i = 1, 2, 3)$ and corresponding distribution of components of strain $\tilde{\tilde{e}}_{ij}(x_1, x_2, x_3)$ $(i,j = 1, 2, 3)$, represents the actual (true) components of displacement of the body.

On the basis of the foregoing discussion the boundary-value problem for computing the components of displacement, strain, and stress of the particles of deformable body can be formulated as follows.

Find the geometrically admissible distribution of components of displacement of the particles of the deformable body which when substituted into the strain–displacement relations (1.3) give components of strain $e_{ij}(x_1, x_2, x_3)$ $(i,j = 1, 2, 3)$ which when substituted into the stress–strain relations for the material from which the body is made give components of stress $\tau_{ij}(x_1, x_2, x_3)$ $(i,j = 1, 2, 3)$ which satisfy the principle of virtual work (18.12) for every admissible distribution of the components of displacement $\tilde{\tilde{u}}_i(x_1, x_2, x_3)$ $(i = 1, 2, 3)$ and corresponding distribution of the components of strain $\tilde{\tilde{e}}_{ij}(x_1, x_2, x_3)$ $(i,j = 1, 2, 3)$.

This formulation of the boundary value under consideration is called *weak* and is equivalent to its strong formulation described in Sec. 18.2.

18.4 Statically Admissible Reactions and Internal Actions—Admissible and Geometrically Admissible Displacements of Framed Structures

In statically determinate framed structures only the actual internal actions are in equilibrium with the given external actions, while in statically indeterminate framed structures, we can find an infinite number of distributions of internal actions which satisfy the equations of equilibrium for any portion of the structure. These distributions of internal actions are referred to as *statically admissible*. In addition to being statically admissible, the actual distribution of internal actions must yield components of displacements which are continuous functions of the space coordinates and satisfy the conditions at the supports of the structure (boundary conditions).

When a framed structure is externally statically indeterminate, an infinite number of sets of reactions can be found which are in equilibrium with the given external actions acting on the structure. These sets of reactions are referred to as *statically admissible*.

The one set of statically admissible reactions which yields components of displacement compatible with the constraints of the structure is the actual set of reactions. For example, two statically admissible sets of reactions for the beam shown in Fig. 18.3*a* are shown in Figs. 18.3*b* and *d*. The corresponding statically admissible distributions of moment are shown in Figs. 18.3*c* and *e*.

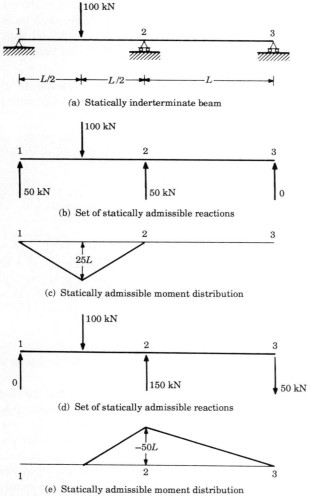

(a) Statically inderterminate beam

(b) Set of statically admissible reactions

(c) Statically admissible moment distribution

(d) Set of statically admissible reactions

(e) Statically admissible moment distribution

Figure 18.3 Statically admissible sets of reactions and corresponding moment distributions.

An admissible displacement field of a framed structure is defined as one that has the following properties.

1. The components of displacement of any element of the structure have continuous derivatives with respect to its axial coordinate of the order required. For example, we may deduce from physical intuition that the transverse component of translation $u_3(x_1)$ and the slope of the elastic curve du_3/dx_1 must be continuous functions. However, the moment and consequently the second derivative of $u_3(x_1)$ has a simple discontinuity or jump at each point of application of a concentrated external moment. Thus it is required that the function $u_3(x_1)$ has continuous first derivatives; that is, it has C^1 continuity. Similarly we can deduce that it is required that the axial component of translation $u_1(x_1)$ is a continuous function; that is, it has C^0 continuity.

2. The global components of nodal displacements of the elements of the structure are equal to the corresponding unknown components of displacement of the nodes of the structure.

3. The magnitude of the components of displacement of the points of a structure is within the range of validity of the theory of small deformation.

A geometrically admissible displacement field of a framed structure is defined as one which has the following properties.

1. It is admissible.

2. The components of nodal displacements of the elements of the structure which are connected to a support are equal to the corresponding specified components of displacement of this support.

The components of strain of a point of a structure obtained from an admissible or a geometrically admissible displacement field on the basis of relations (2.34), (2.48), and (2.62) are called *admissible* and *geometrically admissible,* respectively.

A geometrically admissible displacement field is not the actual displacement field of the structure. Moreover, a statically admissible stress field and a geometrically admissible displacement field of a structure may be independently prescribed. However, if for a particular structure the strain field obtained on the basis of the appropriate stress–strain relations from a statically admissible stress field yields a geometrically admissible displacement field, then this statically admissible stress field is the actual.

18.5   The Principle of Virtual Work for Framed Structures

Consider a framed structure consisting of NE elements originally in a stress-free state of mechanical and thermal equilibrium at a uniform temperature T_0. In this state some elements of the structure may have different lengths and/or curvature than that required to fit in the geometry of the structure. As discussed in Sec. 2.3 these elements may be regarded as being in a state of initial strain. In general in this state the axial component of initial strain in an element could be any specified function of its axial coordinate x_1 and a linear function of its transverse coordinates x_2 and x_3. We specify the axial component of initial strain by its value $e_{11}^{IC}(x_1)$ at the centroid of the cross section of the element and by the initial curvatures $k_2^I(x_1)$ and $k_3^I(x_1)$ of the projection on the x_1x_2 and x_1x_3 planes, respectively, of the axis of the element (see Sec. 2.3).

The structure reaches a second state of mechanical equilibrium due to the application of the following loading.

1. Distributed external forces $\bar{\mathbf{q}}(x_1)$ (including the weight) given per unit length of the elements on which they act.

2. Distributed external moments $\bar{\mathbf{m}}(x_1)$ given per unit length of the elements on which they act.

3. Concentrated external forces $\bar{\mathbf{P}}^{(i)}$ $(i = 1, 2, \ldots, N)$. Some of these forces act on the elements of the structure, while the remaining act on its nodes.

4. Concentrated external moments $\bar{\mathcal{M}}^{(i)}$ $(i = 1, 2, \ldots, M)$. Some of these moments are applied to the elements of the structure, while the remaining are applied to its nodes.

5. Change of temperature.

6. Translation and rotation of the supports of the structure $\bar{\Delta}^{(s)}$ $(s = 1, 2, \ldots, S)$.

For this loading we choose a set of statically admissible reactions of the supports of the structure which we denote by $\tilde{R}^{(s)}$ $(s = 1, 2, \ldots, S)$. Moreover, we denote the corresponding statically admissible internal axial force, shearing forces, torsional moment, and bending moments by $\tilde{N}(x_1)$, $\tilde{Q}_2(x_1)$, $\tilde{Q}_3(x_1)$, $\tilde{M}_1(x_1)$, $\tilde{M}_2(x_2)$, and $\tilde{M}_3(x_1)$, respectively. Referring to relations (2.38) and (2.46) the corresponding statically admissible components of stress are

$$\tilde{\tau}_{11} = \frac{\tilde{N}}{A} + \frac{\tilde{M}_2 x_3}{I_2} - \frac{\tilde{M}_3 x_2}{I_3} \qquad (18.16a)$$

$$\tilde{\bar{\tau}}_{22} = \tilde{\bar{\tau}}_{33} = \tilde{\bar{\tau}}_{23} = 0 \qquad (18.16b)$$

$$\tilde{\bar{\tau}}_{1n} = (\tilde{\tau}_{1n})_Q + (\tilde{\tau}_{1n})_{M_1} \qquad (18.16c)$$

$$\tilde{\bar{\tau}}_{1s} = (\tilde{\tau}_{1s})_Q + (\tilde{\tau}_{1s})_{M_1} \qquad (18.16d)$$

where I_2 and I_3 are the moments of inertia of the cross section of the element about its principal centroidal axes x_2 and x_3, respectively; A is the area of the cross section of the element; $\tilde{\tau}_{in}$ and $\tilde{\tau}_{is}$ are the shearing components of stress acting on a cross section of the element in the directions of the mutually perpendicular unit vectors $\mathbf{n}$ and $\mathbf{s}$, respectively $(\mathbf{n} = n_2\mathbf{i}_2 + n_3\mathbf{i}_3, \quad \mathbf{s} = s_2\mathbf{i}_2 + s_3\mathbf{i}_3) \quad (\mathbf{n} \cdot \mathbf{s} = 0)$. The unit vector $\mathbf{s}$ is in the direction along which the component of shearing stress $\tilde{\tau}_{in}$ does not vary much. Thus the shearing component of stress $\tilde{\tau}_{in}$ can be established from the shearing forces using relation (2.55) (see Sec. 2.4.3). Each shearing component of stress consists of two parts. One part $(\tilde{\tau}_{1n})_Q$ or $(\tilde{\tau}_{1s})_Q$ is due to the shearing forces $\tilde{Q}_2(x_1)$ and $\tilde{Q}_3(x_1)$, while the other part $(\tilde{\tau}_{1n})_{M_1}$ or $(\tilde{\tau}_{1s})_{M_1}$ is due to the torsional moment $\tilde{M}_1(x_1)$.

Suppose that the framed structure under consideration is subjected to an additional unspecified loading resulting in an additional deformation of the structure. We denote the additional translation and rotation vectors of the points of an element of the structure by $\tilde{\tilde{\mathbf{u}}}(x_1)$ and $\tilde{\tilde{\boldsymbol{\theta}}}(x_1)$, respectively. Moreover, we denote by $\tilde{\tilde{\mathbf{u}}}^{(i)}$ the translation vector of the point of application of the concentrated force $\tilde{\mathbf{P}}^{(i)}$, and by $\tilde{\tilde{\boldsymbol{\theta}}}^{(i)}$ the rotation vector of the point of application of the concentrated moment $\tilde{\mathcal{M}}^{(i)}$. Furthermore we denote by $\{\tilde{\Delta}^s\}$ the matrix of the components of displacements of the nodes of the structure corresponding to the second unspecified loading which are inhibited by its supports. The unspecified loading could include additional unspecified movement of the supports of the structure. Thus the additional translation and rotation vectors $\tilde{\tilde{\mathbf{u}}}(x_1)$ and $\tilde{\tilde{\boldsymbol{\theta}}}(x_1)$ of all the elements of the structure constitute an admissible displacement field. Using the notation described we have

Work of distributed actions $\tilde{\mathbf{p}}^{(e)}(x_1)$ and $\tilde{\mathbf{m}}^{(e)}(x_1)$ $(e = 1, 2, \ldots, NE)$ acting on elements of structure due to its additional deformation	$= \sum\limits_{e=1}^{NE} \left[\int_0^L (\tilde{\mathbf{p}}_1 \cdot \tilde{\tilde{\mathbf{u}}} + \tilde{\mathbf{m}} \cdot \tilde{\tilde{\boldsymbol{\theta}}}) \, dx_1 \right]^{(e)}$ (18.17)

Work of concentrated actions acting on structure due to its additional deformation	$= \sum\limits_{i=1}^{N} \tilde{\mathbf{P}}^{(i)} \cdot \tilde{\tilde{\mathbf{u}}}^{(i)} + \sum\limits_{i=1}^{M} \tilde{\mathbf{M}}^{(i)} \cdot \tilde{\tilde{\boldsymbol{\theta}}}^{(i)}$ (18.18)

Work of component of
statically admissible
reactions of structure due $= \{\tilde{\bar{\Delta}}^S\}^T \{\tilde{R}\}$ (18.19)
to additional displacements
of its supports

where N and M represent the number of concentrated forces and moments, respectively, acting on the nodes and the elements of the structure and $\{\tilde{R}\}$ is the matrix of the statically admissible reactions of the structure. It is conjugate to the matrix $\{\tilde{\Delta}\}$. The superscript (e) in relation (18.17) indicates that the quantities of the terms inside the bracket pertain to element e.

Finally we denote by $\tilde{e}_{ij}(x_1)$ $(i,j = 1, 2, 3)$ the additional admissible components of strain of an element of the structure corresponding to the additional admissible displacement vectors $\tilde{\mathbf{u}}(x_1)$ and $\tilde{\boldsymbol{\theta}}(x_1)$. The shearing components of strain $\tilde{e}_{in}$ and $\tilde{e}_{is}$ can be written as

$$\tilde{e}_{1n} = (\tilde{e}_{1n})_Q + (\tilde{e}_{1n})_{M_1} \qquad \tilde{e}_{1s} = (\tilde{e}_{1s})_Q + (\tilde{e}_{1s})_{M_1} \qquad (18.20)$$

where the subscripts Q and M_1 indicate the part of the shearing components of strain due to the shearing forces Q_2 and Q_3 and due to the twisting moment M_1, respectively.

Substituting relations (18.17) and (18.20) into the principle of virtual work (18.12) we obtain

$$\sum_{i=1}^{N} \tilde{\mathbf{P}}^{(i)} \cdot \tilde{\mathbf{u}}^{(i)} + \sum_{i=1}^{M} \tilde{\mathcal{M}}^{(i)} \cdot \tilde{\boldsymbol{\theta}}^{(i)} + \sum_{e=1}^{NE} \left[\int_0^L (\tilde{\mathbf{p}} \cdot \tilde{\mathbf{u}} + \tilde{\mathbf{m}} \cdot \tilde{\boldsymbol{\theta}}) \, dx_1 \right]^{(e)} + \{\tilde{\bar{\Delta}}^S\}^T \{\tilde{R}\}$$

$$= \sum_{e=1}^{NE} \left\{ \int_0^L \iint_A \left[\tilde{\tau}_{11} \tilde{e}_{11} + 2(\tilde{\tau}_{1n})_Q (\tilde{e}_{1n})_Q + 2(\tilde{\tau}_{1n})_{M_1} (\tilde{e}_{1n})_{M_1} \right. \right.$$

$$\left. \left. + 2(\tilde{\tau}_{1s})_{M_1} (\tilde{e}_{1s})_{M_1} \right] dA \, dx_1 \right\}^{(e)} \qquad (18.21)$$

The superscript (e) indicates that the quantities of the terms inside the bracket pertain to element e.

In obtaining relations (18.21) we have taken into account that the effects of the distribution of the shearing forces and torsional moment are not coupled. Thus the work of the shearing components of stress $(\tilde{\tau}_{1n})_{Q_i}$ and $(\tilde{\tau}_{1s})_{Q_i}$ $(i = 2, 3)$ for an element as a result of the additional twisting of the element corresponding to the second unspecified loading vanishes. Moreover, we have taken into account that as discussed in Sec. 2.4.3, the component of stress $(\tilde{\tau}_{1s})_{Q_i}$ $(i = 2, 3)$ is very small compared to $(\tilde{\tau}_{1n})_{Q_i}$ and the component of strain $(\tilde{e}_{1s})_{Q_i}$ is very small compared to $(\tilde{e}_{1n})_{Q_i}$ at most points of an element. Consequently, we have neglected the integral $\iint_A 2(\tilde{\tau}_{1s})_{Q_i}(\tilde{e}_{1s})_{Q_i} \, dA$ $(i = 2, 3)$.

Consider a segment of an element of a structure of length dx_1 subjected to a distribution of shearing components of stress $(\tilde{\tau}_{1n})_{M_1}$ and $(\tilde{\tau}_{1s})_{M_1}$ on its cross section at x_1 and to a distribution of the shearing components of stress $(\tilde{\tau}_{1n})_{M_1} + d(\tilde{\tau}_{1n})_{M_1}$ and $(\tilde{\tau}_{1s})_{M_1} + (d\tilde{\tau}_{1s})_{M_1}$ on its cross section at $x_1 + dx_1$. The moments about the x_1 axis of these shearing components of stress acting on the cross sections at x_1 and $x_1 + dx_1$ are denoted by $\tilde{M}_1$ and $\tilde{M}_1 + d\tilde{M}_1$, respectively (see Fig. 18.4). Suppose that the element is subjected to an additional deformation and denote the shearing components of strain at x_1 by $(\bar{\tilde{e}}_{1n})_{M_1}$ and $(\bar{\tilde{e}}_{1s})_{M_1}$. Because of this deformation, a radial line on the end faces at x_1 and $x_1 + dx_1$ of the segment under consideration rotates by $\bar{\tilde{\theta}}_1$ and $\bar{\tilde{\theta}}_1 + d\bar{\tilde{\theta}}_1$, respectively. The work of the shearing components of stress acting on the end faces of the segment under consideration due to its additional deformation is equal to

$$dW_T = dW_d = \tilde{M}\bar{\tilde{\theta}}_1 - (\tilde{M}_1 + d\tilde{M}_1)(\bar{\tilde{\theta}}_1 + d\bar{\tilde{\theta}}_1) = \tilde{M}_1 \, d\bar{\tilde{\theta}}_1 \quad (18.22)$$

Recalling that the right-hand side of relation (18.21) represents the work of the components of stress $\tilde{\tau}_{ij}(x_1, x_2, x_3)$ $(i, j = 1, 2, 3)$ acting on all the particles of an element due to the additional components of strain $\bar{\tilde{e}}_{ij}(x_1, x_2, x_3)$ $(i, j = 1, 2, 3)$, we conclude that

$$\int_0^{L_e} \iint_A [2(\tilde{\tau}_{1n})_{M_1}(\bar{\tilde{e}}_{1n})_{M_1} + 2(\tilde{\tau}_{1s})_{M_1}(\bar{\tilde{e}}_{1s})_{M_1}] \, dA \, dx_1$$

$$= \int_0^{L_e} \tilde{M}_1 \, d\bar{\tilde{\theta}}_1 = \int_0^{L_e} \tilde{M}_1 \frac{d\bar{\tilde{\theta}}_1}{dx_1} \, dx_1 \quad (18.23)$$

$$\bar{\tilde{e}}_{11} = \frac{d\bar{\tilde{u}}_1}{dx_1} + x_3 \frac{d\bar{\tilde{\theta}}_2}{dx_1} - x_2 \frac{d\bar{\tilde{\theta}}_3}{dx_1} \quad (18.24)$$

Substituting relations (18.16a), (18.23), and (18.24) into (18.21) and taking into account that the axes x_2 and x_3 are principal ($I_{23} = 0$) centroidal axes ($\iint_A x_3 dA = 0$, $\iint_A x_2 \, dA = 0$) we obtain

$$\sum_{i=1}^N \tilde{\mathbf{P}}^{(i)} \cdot \bar{\tilde{\mathbf{u}}}^{(i)} + \sum_{i=1}^M \tilde{\mathcal{M}}^{(i)} \cdot \bar{\tilde{\boldsymbol{\theta}}}^{(i)} + \sum_{e=1}^{NE} \left[\int_0^L (\tilde{\mathbf{p}} \cdot \bar{\tilde{\mathbf{u}}} + \tilde{\mathbf{m}} \cdot \bar{\tilde{\boldsymbol{\theta}}}) \, dx_1 \right]^{(e)} + \{\tilde{\bar{\Delta}}^S\}^T \{\tilde{R}\}$$

$$= \sum_{e=1}^{NE} \left\{ \int_0^L \left(\tilde{N} \frac{d\bar{\tilde{u}}_1}{dx_1} + \tilde{M}_1 \frac{d\bar{\tilde{\theta}}_1}{dx_1} + \tilde{M}_2 \frac{d\bar{\tilde{\theta}}_2}{dx_1} + \tilde{M}_3 \frac{d\bar{\tilde{\theta}}_3}{dx_1} \right) dx_1 \right.$$

$$\left. + \int_0^L \left[\iint_A 2(\tilde{\tau}_{1n})_Q (\bar{\tilde{e}}_{1n})_Q \, dA \right] dx_1 \right\}^{(e)} \quad (18.25)$$

The components of rotation $\bar{\tilde{\theta}}_2$ and $\bar{\tilde{\theta}}_3$ are obtained from the

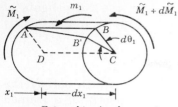

m_1 = External torsional moment
per unit length

Figure 18.4 Segment of an element of a structure subjected to torsion.

components of translation $\tilde{\tilde{u}}_3$ and $\tilde{\tilde{u}}_2$ on the basis of relations (2.47a) and (2.47b), respectively.

Relation (18.25) is the principle of virtual work for framed structures whose elements are made from any material (elastic or nonelastic). The terms on the right-hand side of relation (18.25) represent the sum of the work performed by the internal actions acting on each segment of infinitesimal length of each element of a structure subjected to given loads due to the application on it of the additional displacement field $\tilde{\tilde{u}}_i(x_1, x_2, x_3)$ $(i = 1, 2, 3)$ and $\tilde{\tilde{\theta}}_1(x_1, x_2, x_3)$.

The second integral on the right-hand side of relation (18.25) represents the effect of shear deformation of the elements of the structure. For structures whose elements have lengths which are considerably bigger than their other dimensions the effect of shear deformation is negligible and this integral is disregarded. N and M are the numbers of concentrated forces and moments, respectively, acting on the nodes and elements of the structure.

The first three terms on the left-hand side of relation (18.25) represent the work performed by the given external actions acting on the elements and nodes of a structure due to the application on it of the additional displacement field $\tilde{\tilde{u}}_i$ $(i = 1, 2, 3)$ and $\tilde{\tilde{\theta}}_1$. The last term on the left-hand side of relation (18.25) represents the work performed by the statically admissible reactions of the supports of the structure due to the application on it of the additional displacement field $\tilde{\tilde{u}}_i$ $(i = 1, 2, 3)$ and $\tilde{\tilde{\theta}}_1$. In general it is only required that this field be admissible (see Sec. 18.4). If, however, it is chosen to be geometrically admissible, the last term on the left-hand side of relation (18.25) represents the work of the statically admissible reactions of the supports of the structure due to their specified components of displacement. On the basis of our discussion in Sec. 18.4 the following statements are valid.

1. The actual components of displacements of the elements of a framed structure, when substituted into relations (2.36), (2.49), and (2.64), give components of internal actions $N^{(e)}(x_1)$, $M_1^{(e)}(x_1)$, $M_2^{(e)}(x_1)$, and $M_3^{(e)}(x_1)$ $(e = 1, 2, \ldots, NE)$ which satisfy the principle of virtual

work (18.25) for any admissible distribution of the components of additional virtual displacement $\tilde{\tilde{u}}_1^{(e)}(x_1)$, $\tilde{\tilde{u}}_2^{(e)}(x_1)$, $\tilde{\tilde{u}}_3^{(e)}(x_1)$, and $\tilde{\tilde{\theta}}_1^{(e)}(x_1)$ ($e = 1, 2, \ldots, NE$).

2. The set of continuous functions $u_1^{(e)}(x_1)$ and $\theta_1^{(e)}(x_1)$ ($e = 1, 2, \ldots, NE$) and a set of functions $u_2^{(e)}(x_1)$ and $u_3^{(e)}(x_1)$ ($e = 1, 2, \ldots, NE$) having continuous first derivatives which satisfy the specified components of displacements of the supports of a framed structure and give on the basis of relations (2.36), (2.49), and (2.64) components of internal action for its elements which satisfy the principle of virtual work (18.25) for every admissible displacement field $\tilde{\tilde{u}}_i^e(x_1)$ ($i = 1, 2, 3$) and θ_1^e ($e = 1, 2, \ldots, NE$) are the actual components of displacement of the elements of the structure.

3. Any set of distributions of internal actions $N^{(e)}(x_1)$, $M_1^{(e)}(x_1)$, $M_2^{(e)}(x_1)$, and $M_3^{(e)}(x_1)$ ($e = 1, 2, \ldots, NE$) which are statically admissible to the given external actions acting on a structure satisfy the principle of virtual work (18.25) for any admissible distribution of the components of displacement $\tilde{\tilde{u}}_1^{(e)}(x_1)$, $\tilde{\tilde{u}}_2^{(e)}(x_1)$, $\tilde{\tilde{u}}_3^{(e)}(x_1)$, and $\tilde{\tilde{\theta}}_1^{(e)}(x_1)$ ($e = 1, 2, \ldots, NE$) of the elements of the structure.

4. A set of functions $N^{(e)}(x_1)$, $M_1^{(e)}(x_1)$, $M_2^{(e)}(x_1)$, and $M_3^{(e)}(x_1)$ ($e = 1, 2, \ldots, NE$) which together with the given external actions acting on a framed structure satisfy relation (18.25) for every admissible displacement field $\tilde{\tilde{u}}_1^{(e)}(x_1)$, $\tilde{\tilde{u}}_2^{(e)}(x_1)$, $\tilde{\tilde{u}}_3^{(e)}(x_1)$, and $\tilde{\tilde{\theta}}_1^{(e)}(x_1)$ ($e = 1, 2, \ldots, NE$) of the structure is statically admissible to the given external actions acting on the structure. If it happens that $N^{(e)}(x_1)$, $M_1^{(e)}(x_1)$, $M_2^{(e)}(x_1)$, and $M_3^{(e)}(x_1)$ are obtained from a geometrically admissible displacement field $u_1^{(e)}(x_1)$, $u_2^{(e)}(x_1)$, $u_3^{(e)}(x_1)$, and $\theta_1^{(e)}(x_1)$ ($e = 1, 2, \ldots, NE$) on the basis of relations (2.36), (2.49), and (2.64), then $u_1^{(e)}(x_1)$, $u_2^{(e)}(x_1)$, $u_3^{(e)}(x_1)$, $\theta_1^{(e)}(x_1)$, $N^{(e)}(x_1)$, $M_1^{(e)}(x_1)$, $M_2^{(e)}(x_1)$, and $M_3^{(e)}(x_1)$ ($e = 1, 2, \ldots, NE$) are the actual components of displacement and of internal actions of the elements of the structure.

For a framed structure whose elements are made of one or more isotropic linearly elastic materials, using relations (2.36), (2.44), (2.64), (18.16a), and (18.23), the principle of virtual work (18.25) can be written as:

$$
\sum_{i=1}^{N} \tilde{\mathbf{P}}^{(i)} \cdot \tilde{\mathbf{u}}^{(i)} + \sum_{i=1}^{M} \tilde{\mathcal{M}}^{(i)} \cdot \tilde{\boldsymbol{\theta}}^{(i)} + \sum_{e=1}^{NE} \left[\int_0^{L_e} (\tilde{\mathbf{p}} \cdot \tilde{\mathbf{u}} + \tilde{\mathbf{m}} \cdot \tilde{\boldsymbol{\theta}}) \, dx_1 \right]^{(e)} + \{\tilde{\tilde{\Delta}}^S\}^T \{\tilde{R}\}
$$

$$
= \sum_{e=1}^{NE} \left[\int_0^{L_e} \left(\frac{\tilde{N}\tilde{\tilde{N}}}{EA} + \frac{\tilde{M}_1\tilde{\tilde{M}}_1}{KG} + \frac{\tilde{M}_2\tilde{\tilde{M}}_2}{EI_2} + \frac{\tilde{M}_3\tilde{\tilde{M}}_3}{EI_3} \right) dx_1 \right.
$$

$$
+ \left(K_2 \int_0^{L_e} \frac{\tilde{Q}_2\tilde{\tilde{Q}}_2}{GA} \, dx_1 + K_3 \int_0^{L_e} \frac{\tilde{Q}_3\tilde{\tilde{Q}}_3}{GA} \, dx_1 \right)
$$

$$
\left. + \int_0^{L_e} (\tilde{N}\tilde{\tilde{H}}_1 + \tilde{M}_2\tilde{\tilde{H}}_2 + \tilde{M}_3\tilde{\tilde{H}}_3) \, dx_1 \right]^{(e)} \tag{18.26}
$$

$\tilde{H}_1$, $\tilde{H}_2$, and $\tilde{H}_3$ are defined by relations (2.37), (2.45b), and (2.45a), respectively. N and M are the number of concentrated forces and moments, respectively, acting on the nodes and elements of the structure. The second and third integrals on the right-hand side of relation (18.26) represent an approximation of the effect of the shear deformation. The factors K_2 and K_3 depend on the geometry of the cross section of the element and are referred to as *form factors*.† They are equal to

$$K_2 = \int\int_A \frac{Z_3^2 A}{I_3^2 b_s^2}\, dA \qquad K_3 = \int\int_A \frac{Z_2^2 A}{I_2^2 b_s^2}\, dA \qquad (18.27)$$

where Z_3 and Z_2 are given by relations (2.56) and b_s is defined following relations (2.56). When the lengths of the elements of a framed structure are large compared to their other dimensions, the effect of shear deformation is negligible compared to the effect of bending and can be disregarded. In this case using relations (2.36), (2.47), (2.49), and (2.64) for an element, relation (18.25) can be rewritten as

$$\sum_{e=1}^{NE} \left[\int_0^L [[\mathscr{L}]\{\tilde{u}\}]^T \{\tilde{L}\}\, dx_1 \right]^{(e)} + \{\tilde{\tilde{\Delta}}\}^T [\{\tilde{\hat{P}}^G\} + \{\tilde{\hat{R}}\}]$$

$$= \sum_{e=1}^{NE} \left[\int_0^L \{\widetilde{STR}\}^T \{\widetilde{IA}\}\, dx_1 \right]^{(e)} \qquad (18.28)$$

$\{\tilde{\tilde{\Delta}}\}$ is the matrix of all the components of displacements of the nodes of the structure including those inhibited by its supports. $\{\tilde{\hat{R}}\}$ is the matrix of the reactions of the structure. It is conjugate to the matrix $\{\tilde{\tilde{\Delta}}\}$. The terms of the matrix $\{\tilde{\hat{R}}\}$ which correspond to the components of displacements of the nodes of the structure which are not inhibited by its supports are equal to zero. $\{\hat{P}^G\}$ is the matrix of the given concentrated actions forces and moments acting on all the nodes of the structure, including those directly absorbed by its supports.

$$\{\widetilde{IA}\} = \left\{ \begin{array}{c} \tilde{M}_1(x_1) \\ \tilde{M}_2(x_1) \\ \tilde{M}_3(x_1) \end{array} \right\} \qquad (18.29)$$

$$\{\tilde{u}\} = \left\{ \begin{array}{c} \tilde{\tilde{u}}_1(x_1) \\ \tilde{\tilde{u}}_2(x_1) \\ \tilde{\tilde{u}}_3(x_1) \\ \tilde{\tilde{\theta}}_1(x_1) \end{array} \right\} \qquad (18.30)$$

---

† For a rectangular cross section, $K_2 = K_3 = 1.2$. For a circular cross section, $K_2 = K_3 = 1.185$.

$$\widetilde{\{\overline{STR}\}} = \left\{ \begin{array}{c} \dfrac{d\tilde{\tilde{u}}_1}{dx_1} \\[2mm] \dfrac{d\tilde{\tilde{\theta}}_1}{dx_1} \\[2mm] \dfrac{d\tilde{\tilde{\theta}}_2}{dx_1} \\[2mm] \dfrac{d\tilde{\tilde{\theta}}_3}{dx_1} \end{array} \right\} = [\mathscr{B}]\{\tilde{\tilde{u}}\} \tag{18.31}$$

$$[\mathscr{B}] = \left\{ \begin{array}{cccc} \dfrac{d}{dx_1} & 0 & 0 & 0 \\[3mm] 0 & 0 & 0 & \dfrac{d}{dx_1} \\[3mm] 0 & 0 & -\dfrac{d^2}{dx_1^2} & 0 \\[3mm] 0 & \dfrac{d^2}{dx_2^2} & 0 & 0 \end{array} \right\} \tag{18.32}$$

$$[\mathscr{L}] = \left[\begin{array}{cccc} 1 & 0 & 0 & 0 \\ 0 & 1 & 0 & 0 \\ 0 & 0 & 1 & 0 \\ 0 & 0 & 0 & 1 \\ 0 & 0 & -\dfrac{d}{dx_1} & 0 \\[2mm] 0 & \dfrac{d}{dx_1} & 0 & 0 \end{array} \right] \tag{18.33}$$

$$\{\tilde{L}\} = \left\{ \begin{array}{c} \tilde{P}_1 + \displaystyle\sum_{i=1}^{n_1} \tilde{P}_1^{(i)}\delta(x_1 - a_{1i}) \\ \tilde{P}_2 + \displaystyle\sum_{i=1}^{n_2} \tilde{P}_2^{(i)}\delta(x_1 - a_{2i}) \\ \tilde{P}_3 + \displaystyle\sum_{i=1}^{n_3} \tilde{P}_3^{(i)}\delta(x_1 - a_{3i}) \\ \tilde{m}_1 + \displaystyle\sum_{i=1}^{m_1} \tilde{\mathscr{M}}_1^{(i)}\delta(x_1 - b_{1i}) \\ \tilde{m}_2 + \displaystyle\sum_{i=1}^{m_2} \tilde{\mathscr{M}}_2^{(i)}\delta^1(x_1 - b_{2i}) \\ \tilde{m}_3 + \displaystyle\sum_{i=1}^{m_3} \tilde{\mathscr{M}}_3^{(i)}\delta^1(x_1 - b_{3i}) \end{array} \right\} \tag{18.34}$$

That is, in the matrix $\{L\}$ for an element we include the external actions acting along its length. In relation (18.34) n_i and m_i are the numbers of components of concentrated forces $\tilde{P}_i$ and moments $\tilde{\mathcal{M}}_i$ $(i = 1, 2, 3)$ acting along the length of an element. For an element made from an isotropic linearly elastic material referring to relations (2.36), (2.49), and (2.64) we have

$$\{\widetilde{IA}\} = \begin{Bmatrix} \tilde{N} \\ \tilde{M}_1 \\ \tilde{M}_2 \\ \tilde{M}_3 \end{Bmatrix} = \begin{Bmatrix} EA\left(\dfrac{d\tilde{u}_1}{dx_1} - \tilde{H}_1\right) \\ KG\dfrac{d\tilde{\theta}_1}{dx_1} \\ -EI_2\left(\dfrac{d^2\tilde{u}_3}{dx_1^2} + \tilde{H}_2\right) \\ EI_3\left(\dfrac{d^2\tilde{u}_2}{dx_1^2} - \tilde{H}_3\right) \end{Bmatrix} = [ST][\mathcal{B}]\{\tilde{u}\} - [ST]\{\tilde{H}\}$$

(18.35)

where

$$[ST] = \begin{bmatrix} EA & 0 & 0 & 0 \\ 0 & KG & 0 & 0 \\ 0 & 0 & EI_2 & 0 \\ 0 & 0 & 0 & EI_3 \end{bmatrix}$$

(18.36)

$$\{\tilde{u}\} = \begin{Bmatrix} \tilde{u}_1(x_1) \\ \tilde{u}_2(x_1) \\ \tilde{u}_3(x_1) \\ \theta_1(x_1) \end{Bmatrix}$$

(18.37)

$$\{\tilde{H}\} = \begin{Bmatrix} \tilde{H}_1 \\ 0 \\ \tilde{H}_2 \\ \tilde{H}_3 \end{Bmatrix}$$

(18.38)

Substituting relations (18.31) and (18.35) into relation (18.28) we get the following compact form of the principle of virtual work for a framed structure whose elements are made of isotropic linearly elastic materials:

$$\sum_{e=1}^{NE} \left[\int_0^{L_e} [[\mathcal{B}]\{\tilde{\tilde{u}}\}]^T [ST][\mathcal{B}]\{\tilde{u}\}\, dx_1 \right.$$
$$\left. - \int_0^{L_e} [[\mathcal{B}]\{\tilde{\tilde{u}}\}]^T [ST]\{\tilde{H}\}\, dx_1 - \int_0^{L_e} [[\mathcal{L}]\{\tilde{\tilde{u}}\}]^T \{\tilde{L}\}\, dx_1 \right]^{(e)}$$
$$= \{\tilde{\tilde{\Delta}}\}^T [\{\tilde{P}^G\} + \{\tilde{R}\}]$$
(18.39)

Reference

1. A. E. Armenàkas, *Classical Structural Analysis: A Modern Approach*, McGraw-Hill Book Co., New York, 1988, p. 251.

19

Computation of the Response of an Element Using the Principle of Virtual Work

19.1 The Finite-Element Method

Material bodies are composed of a large number of discrete particles (atoms and molecules) in constant motion. Solids differ from liquids and liquids from gases in the spacing of these particles and in the amplitude of their motion. Therefore, it seems reasonable to expect that the behavior of a body should be inferred by studying the behavior of each molecule or a group of molecules of the body. Consequently, a suitable model for studying the behavior of a body consists of moving particles. The statistical evaluation of the behavior of these particles should lead to conclusions as to the behavior of the body. This model is referred to in the literature as the *corpuscular-statistical model* and it has been used extensively in physics. A second model used in studying the behavior of a body is based on the assumption that the material is distributed in the space which it occupies without leaving gaps or empty spaces. In other words, it is assumed that at every instant of time there is a particle at every point of the space occupied by the body. This model is referred to in the literature as *continuum,* and it is used in all engineering disciplines, including structural analysis.

The mathematical formulation of a physical problem using the continuum model is referred to as a *boundary-value problem*. It involves the determination of one or more independent functions of the space coordinates (called the *state variables*) which satisfy one or more differential equations in a certain region called the *domain of the problem* as well as appropriate specified conditions on the boundary of the domain (*boundary conditions*). In many boundary-value problems there is another set of functions called the *fluxes* which have physical significance and can be established from the state variables. The relations between the state variables and the fluxes are called *constitutive equations* and depend on the material from which the domain is made. For example, the three components of displacement $\hat{u}_i(x_1, x_2, x_3)$ ($i = 1, 2, 3$) of the particles of a body are the state variables for the boundary-value problem for computing the components of displacement, strain, and stress in this body (see Sec. 18.2). The six components of stress $\tau_{ij}(x_1, x_2, x_3)$ ($i, j = 1, 2, 3$) acting on the particles of the body are the fluxes. They are related to the components of strain and consequently to the components of displacement [see relations (1.3)] by the stress-strain relations (constitutive equations) for the material from which the body is made. Moreover, the basic components of displacement $u_1(x_1)$, $u_2(x_1)$, $u_3(x_1)$, and $\theta_1(x_1)$ of the elements of a space framed structure are the state variables for the boundary-value problem for computing the components of displacement and the components of internal action in its elements. The components of internal action are the fluxes. They are related to the components of displacement by relations (2.36), (2.49), (2.50), and (2.64).

Only a few boundary-value problems can be solved exactly with the available mathematical methods. The rest are solved approximately. In order to use a computer in obtaining approximate solutions of boundary-value problems, they must be *discretized*. That is, they must be recast in an algebraic form involving a finite number of unknown parameters. The methods used to discretize boundary-value problems can be classified into two groups: those applied directly to their *strong form* (such as finite differences) and those applied directly to one of their *integral forms*. For example, the principle of virtual work (18.12) is an integral form of the boundary-value problem for computing the components of displacement, strain, and stress of the particles of a body (see Sec. 18.2). Moreover, the principle of virtual work (18.25) is an integral form of the boundary-value problem for computing the components of displacement and the internal actions in the elements of a framed structure.

In this section we describe a method for discretizing boundary-value problems using one of their integral forms. This method has

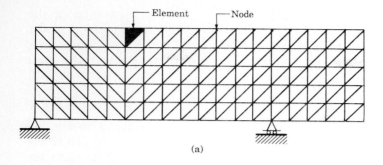

(a)

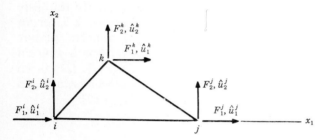

(b) Nodal actions and nodal displacement
of a planar triangular element

Figure 19.1 Plate subdivided into planar triangular finite elements.

been developed in the last 30 years, and it is known as the *finite-element method*. It involves the following steps.

Step 1. The domain of the problem is divided into a finite number of subdomains of simple geometry called *finite elements*. For example, the plate of constant thickness shown in Fig. 19.1 is subdivided into a finite number of planar triangular elements of constant thickness. Framed structures are one-dimensional domains, and as it is done in the direct stiffness or modern flexibility methods they are divided into a finite number of one-dimensional subdomains (line elements).

Step 2. The relations describing the response of each element are established. This is accomplished as follows.

1. Within each finite element, each state variable is approximated by a linear combination of algebraic polynomials of the space coordinates which when evaluated at certain points on the boundary or inside the element becomes equal to the value of the state variable there. These points are called *nodal points* or *nodes*. For

example, we choose as nodal points for the planar triangular element of Fig. 19.1b its three vertices, which we denote i, j, and k. Moreover, we choose as nodal points for a one-dimensional element its two end points.† Within a finite element the state variables are approximated as

$$\{\tilde{u}\} = [\phi]\{\tilde{D}\} \tag{19.1}$$

where $\{\tilde{u}\}$ is an approximation to the matrix of state variables. The state variables for the planar triangular element of Fig. 19.1b are the components of displacement $\hat{u}_1(x_1)$ and $\hat{u}_2(x_1)$ of its particles. Thus, for this element $\{\tilde{u}\}^T = [\tilde{\hat{u}}_1(x_1) \quad \tilde{\hat{u}}_2(x_1)]$. Moreover, the state variables for an element of a framed structure are its basic components of displacement. Thus, as discussed in Sec. 2.5, for an axial deformation element $\{\tilde{u}\} = \tilde{u}_1(x_1)$; for a general planar element in the $x_1 x_3$ plane $\{\tilde{u}\}^T = [\tilde{u}_1(x_1) \quad \tilde{u}_3(x_1)]$; for a general space element $\{u\}^T = [\tilde{u}_1(x_1) \quad \tilde{u}_2(x_1) \quad \tilde{u}_3(x_1) \quad \tilde{\theta}_1(x_1)\}$. $\{\tilde{D}\}$ is an approximation to the local matrix of nodal displacements of an element. For the planar triangular element of Fig. 19.1b

$$\{\tilde{D}\}^T = [\tilde{\hat{u}}_1^i \quad \tilde{\hat{u}}_2^i \quad \tilde{\hat{u}}_1^j \quad \tilde{\hat{u}}_2^j \quad \tilde{\hat{u}}_1^k \quad \tilde{\hat{u}}_2^k].$$

Moreover, referring to relations (4.1c), (4.2c), and (4.3c), for an element of a truss $\{\tilde{D}\}^T = [\tilde{u}_1^j \quad \tilde{u}_1^k]$; for an element of a planar beam or frame in the $x_1 x_3$ plane $\{\tilde{D}\}^T = [\tilde{u}_1^j \quad \tilde{u}_3^j \quad \tilde{\theta}_2^j \quad \tilde{u}_1^k \quad \tilde{u}_3^k \quad \tilde{\theta}_2^k]$; for an element of a space beam or frame

$$\{\tilde{D}\}^T = [\tilde{u}_1^j \quad \tilde{u}_2^j \quad \tilde{u}_3^j \quad \tilde{\theta}_1^j \quad \tilde{\theta}_2^j \quad \tilde{\theta}_3^j \quad \tilde{u}_1^k \quad \tilde{u}_2^k \quad \tilde{u}_3^k \quad \tilde{\theta}_1^k \quad \tilde{\theta}_2^k \quad \tilde{\theta}_3^k].$$

$[\phi]$ is a rectangular matrix whose terms are polynomials of the local coordinates of the element. It is called the *matrix of element shape functions.*

2. The approximations for the state functions (19.1) are substituted into an integral form of the boundary-value problem under consideration. This yields the values of the fluxes at the nodal points of the element as a linear combination of the values of the state functions at the same points. For an element of a framed structure this approach gives its nodal actions as the sum of its corresponding fixed-end actions and its corresponding nodal actions when the element is subjected only to its nodal displacements. That is, it gives relation (4.7) which includes the

---

† In addition to their end points, we could choose as nodal points of one-dimensional elements a finite number of points along their length. This possibility is not explored further in this book.

stiffness equations (4.8) for the element as well as its fixed-end actions.

In general, it is not difficult to establish the exact response of many one-dimensional elements of practical interest (see Sec. 4.4). Moreover, approximate responses of one-dimensional elements can be established using other convenient methods (see Secs. 12.3 and 12.4). For this reason the finite-element method is not used extensively in writing computer programs for analyzing groups of framed structures. However, the exact response of two- and three-dimensional elements cannot be easily established. Moreover, the finite-element method is very well-suited for finding the approximate response of such elements with the aid of a computer. Thus, the finite-element method is of paramount importance for the approximate solution of boundary-value problems involving two- or three-dimensional domains.

Step 3. A set of linear algebraic equations relating the external disturbances acting on the domain and the unknown values of the state variables at its nodes is constructed using the relations obtained for each element of the domain in step 2, part 2. This is accomplished by requiring that at the nodes of the domain the approximations to the state variables and the resulting approximations to the fluxes satisfy the physical laws which govern the boundary-value problem under consideration. The procedures employed in this step, when applied to boundary-value problems involving one-dimensional domains, are those employed for analyzing framed structures by the direct stiffness method (see Chaps. 7 and 8). These procedures are extended to two- or three-dimensional domains. For the boundary-value problem for computing the components of displacements and the internal actions of the elements of a framed structure, in this step we obtain the stiffness equations (8.1) for the structure by requiring that

1. The components of internal action acting on the end of each element connected to a node satisfy the equations of equilibrium for this node.

2. The global components of displacement at the end of each element connected to a node are equal to the corresponding components of displacement of this node.

More specifically, for this problem we use the fixed-end actions of the elements of the structure to compute the equivalent actions to be placed on its nodes (see Chap. 7). Moreover, we assemble the stiffness matrix for the structure from the stiffness matrices for its elements (see Chap. 8).

Step 4. The boundary conditions of the problem are incorporated into the set of algebraic equations established in step 3, which are then solved to establish approximate values of the state variables at the nodes of the domain. For boundary-value problems involving one-dimensional domains, we follow the procedure described in Chap. 9 for computing the components of displacements of the nodes of framed structures using the direct stiffness method. Moreover, these procedures are extended to boundary-value problems involving two- or three-dimensional domains.

Step 5. The approximate values of the state variables at the nodes of the domain established in step 4 are used to obtain approximate values for the fluxes at the nodes of the domain. For one-dimensional domains we follow the procedure described in Chap. 9 for computing the local components of nodal actions of each element of a structure from the global components of displacements of the nodes of the structure to which the element is connected.

Notice that there are differences in the error introduced in step 3 in the solutions of boundary-value problems involving one-dimensional domains and in the solutions of boundary-value problems involving two- and three-dimensional domains. The two nodal points of each one-dimensional element constitute its entire boundary. Thus in step 3 the state variables and the fluxes of boundary-value problems involving one-dimensional domains are required to satisfy the laws which govern their solution at every point of the interelement boundaries of their domain. Consequently, in step 3 no error is introduced in the solutions of boundary-value problems involving one-dimensional domains. However, the nodal points of two- or three-dimensional elements represent only a few points of their boundary (see Fig. 19.1*b*). Thus in step 3 the state variables and the fluxes of boundary-value problems involving two- or three-dimensional domains are required to satisfy the laws which govern their solution only at a few points of the interelement boundaries of their domain. Consequently, in step 3 an error is introduced in the solution of boundary-value problems involving two- and three-dimensional domains. This error is called a *discretization error* and it decreases as the size of the elements decreases. In most finite-element formulations of boundary-value problems involving two- or three-dimensional domains the discretization error is reduced by an appropriate choice of nodal points and shape functions of their elements. That is, they are chosen in a way that some of the laws which govern the solution of a boundary-value problem are automatically satisfied at all points of the boundary of its elements when they are satisfied at their nodal points. For example, in the boundary-value problem for computing the components of displacement, strain, and stress in a body (see Sec. 18.2), the shape functions and the nodes of the elements are chosen in a way that the components of displacement are continuous at all points of the interelement boundaries when they are continuous at the nodes. However, it is difficult to also satisfy the requirement that the components of stress acting on all the particles of the interelement boundaries are in equilibrium.

On the basis of the foregoing discussion, the direct stiffness method for analyzing framed structures presented in this book differs from the finite-element method for one-dimensional domains only in the way the response of each of their elements is established (see step 2). In the next section we establish the approximate response of an element of a framed structure using the finite-element method.

19.2 Computation of the Approximate Response of an Element of a Framed Structure Using the Finite-Element Method in Conjunction with the Principle of Virtual Work

In this section we employ the finite-element method in conjunction with the principle of virtual work to establish approximate stiffness equations and the approximate matrix of fixed-end actions for an element of a framed structure.

Referring to relation (18.39) the principle of virtual work for an element of a framed structure is

$$\int_0^{L_e} [[\mathcal{B}]\{\tilde{\tilde{u}}\}]^T [ST][\mathcal{B}]\{\tilde{u}\} \, dx_1 - \int_0^{L_e} [[\mathcal{B}]\{\tilde{\tilde{u}}\}]^T [ST][\tilde{H}] \, dx_1$$

$$- \int_0^{L_e} [[\mathcal{L}]\{\tilde{\tilde{u}}\}]^T [\tilde{L}] \, dx_1$$

$$= \{\tilde{\tilde{D}}\}^T \{\tilde{A}\} \tag{19.2}$$

where the matrices $[\mathcal{B}]$, $\{\tilde{u}\}$, $[ST]$, $\{\tilde{u}\}$, $\{\tilde{H}\}$, $[\mathcal{L}]$, and $\{\tilde{L}\}$ are defined by relations (18.32), (18.30), (18.36), (18.37), (18.38), (18.33), and (18.34), respectively. Moreover, $\{\tilde{\tilde{D}}\}$ is the local matrix of nodal displacements of the element corresponding to the virtual displacement field $\{\tilde{\tilde{u}}\}$, and $\{\tilde{A}\}$ is an approximation to the local matrix of nodal actions of the element which is statically admissible to the external actions $\{\tilde{L}\}$ applied along the length of the element.

19.2.1   Element shape functions

As discussed in Sec. 19.1, in the finite-element method the basic components of displacements of an element are approximated by relation (19.1). In this section we establish the conditions which must be imposed on the shape functions in order that the approximate solution (19.1) yields the components of nodal displacements when evaluated at the ends of an element.

For an element of a truss, relation (19.1) can be written as

$$\tilde{u}_1(x_1) = [\phi_j(x_1) \quad \phi_k(x_1)]\begin{Bmatrix} u_1^j \\ u_1^k \end{Bmatrix} \tag{19.3}$$

At $x_1 = 0$ relation (19.3) gives

$$\tilde{u}_1(0) = [\phi_j(0) \quad \phi_k(0)]\begin{Bmatrix} u_1^j \\ u_1^k \end{Bmatrix} = u_1^j \tag{19.4}$$

Hence, $\qquad\qquad \phi_j(0) = 1 \qquad \phi_k(0) = 0 \tag{19.5}$

At $x_1 = L_e$, relation (19.3) yields

$$\tilde{u}_1(L_e) = [\phi_j(L_e) \quad \phi_k(L_e)]\begin{Bmatrix} u_1^j \\ u_1^k \end{Bmatrix} = u_1^k \tag{19.6}$$

Consequently,

$$\phi_j(L_e) = 0 \qquad \phi_k(L_e) = 1 \tag{19.7}$$

Referring to Secs. 2.5.1 and 2.5.3 we see that the boundary-value problem for computing the torsional component of rotation $\theta_1(x_1)$ of an element is analogous to the boundary-value problem for computing its axial component of translation $u_1(x_1)$. Thus if we want to have the same order of approximation for both problems, we use the same shape functions for them. Consequently, for an element subjected only to torsional moments relation (19.1) reduces to

$$\tilde{\theta}_1(x_1) = [\phi_j(x_1) \quad \phi_k(x_1)]\begin{Bmatrix} \tilde{\theta}_1^j \\ \tilde{\theta}_1^k \end{Bmatrix} \tag{19.8}$$

It can be easily verified that the shape functions $\phi_j(x_1)$ and $\phi_k(x_1)$ of relation (19.8) satisfy relations (19.5) and (19.7).

For an element subjected only to bending about its x_2 axis relation (19.1) reduces to

$$\tilde{u}_3(x_1) = [\phi_{j3}^u(x_1) \quad \phi_{j3}^\theta(x_1) \quad \phi_{k3}^u(x_1) \quad \phi_{k3}^\theta(x_1)]\begin{Bmatrix} \tilde{u}_3^j \\ \tilde{\theta}_2^j \\ \tilde{u}_3^k \\ \tilde{\theta}_2^k \end{Bmatrix} \tag{19.9}$$

At $x_1 = 0$ relation (19.9) gives

$$\tilde{u}_3(0) = [\phi_{j3}^u(0) \quad \phi_{j3}^\theta(0) \quad \phi_{k3}^u(0) \quad \phi_{k3}^\theta(0)] \begin{Bmatrix} \tilde{u}_3^j \\ \tilde{\theta}_2^j \\ \tilde{u}_3^k \\ \tilde{\theta}_2^k \end{Bmatrix} = \tilde{u}_3^j \quad (19.10)$$

and

$$\frac{d\tilde{u}_3}{dx_1}\bigg|_{x_1=0} = \left[\frac{d\phi_{j3}^u}{dx_1}\bigg|_{x_1=0} \quad \frac{d\phi_{j3}^\theta}{dx_1}\bigg|_{x_1=0} \quad \frac{d\phi_{k3}^u}{dx_1}\bigg|_{x_1=0} \quad \frac{d\phi_{k3}^\theta}{dx_1}\bigg|_{x_1=0}\right] \begin{Bmatrix} \tilde{u}_3^j \\ \tilde{\theta}_2^j \\ u_3^k \\ \tilde{\theta}_2^k \end{Bmatrix}$$

$$= -\tilde{\theta}_2^j \quad (19.11)$$

From relation (19.10) and (19.11) we obtain

$$\phi_{j3}^u(0) = 1 \qquad \frac{d\phi_{j3}^u}{dx_1}\bigg|_{x_1=0} = 0$$

$$\phi_{j3}^\theta(0) = 0 \qquad \frac{d\phi_{j3}^\theta}{dx_1}\bigg|_{x_1=0} = -1$$

$$\phi_{k3}^u(0) = 0 \qquad \frac{d\phi_{k3}^u}{dx_1}\bigg|_{x_1=0} = 0 \qquad (19.12)$$

$$\phi_{k3}^\theta(0) = 0 \qquad \frac{d\phi_{k3}^\theta}{dx_1}\bigg|_{x_1=0} = 0$$

At $x_1 = L_e$ relation (19.9) gives

$$\tilde{u}_3(L_e) = [\phi_{j3}^u(L_e) \quad \phi_{j3}^\theta(L_e) \quad \phi_{k3}^u(L_e) \quad \phi_{k3}^\theta(L_e)] \begin{Bmatrix} \tilde{u}_3^j \\ \tilde{\theta}_2^j \\ \tilde{u}_3^k \\ \tilde{\theta}_2^k \end{Bmatrix} = \tilde{u}_3^k \quad (19.13)$$

and

$$\frac{d\tilde{u}_3}{dx_1}\bigg|_{x_1=L_e} = \left[\frac{d\phi_{j3}^u}{dx_1}\bigg|_{x_1=L_e} \quad \frac{d\phi_{j3}^\theta}{dx_1}\bigg|_{x_1=L_e} \quad \frac{d\phi_{k3}^u}{dx_1}\bigg|_{x_1=L_e} \quad \frac{d\phi_{k3}^\theta}{dx_1}\bigg|_{x_1=L_e}\right] \begin{Bmatrix} \tilde{u}_3^j \\ \tilde{\theta}_2^j \\ u_3^k \\ \tilde{\theta}_2^k \end{Bmatrix}$$

$$= -\tilde{\theta}_2^k \quad (19.14)$$

From relations (19.13) and (19.14) we have

$$\phi_{j3}^u(L_e) = 0 \qquad \frac{d\phi_{j3}^u}{dx_1}\bigg|_{x_1=L_e} = 0$$

$$\phi_{j3}^\theta(L_e) = 0 \qquad \frac{d\phi_{j3}^\theta}{dx_1}\bigg|_{x_1=L_e} = 0$$

$$\phi_{k3}^u(L_e) = 1 \qquad \frac{d\phi_{k3}^u}{dx_1}\bigg|_{x_1=L_e} = 0 \qquad (19.15)$$

$$\phi_{k3}^\theta(L_e) = 0 \qquad \frac{d\phi_{k3}^\theta}{dx_1}\bigg|_{x_1=L_e} = -1$$

For an element subjected only to bending about its x_3 axis, following a procedure analogous to the above, we obtain

$$\phi_{j2}^u(0) = 1 \qquad \frac{d\phi_{j2}^u}{dx_1}\bigg|_{x_1=0} = 0$$

$$\phi_{j2}^\theta(0) = 0 \qquad \frac{d\phi_{j2}^\theta}{dx_1}\bigg|_{x_1=0} = 1$$

$$\phi_{k2}^u(0) = 0 \qquad \frac{d\phi_{k2}^u}{dx_1}\bigg|_{x_1=0} = 0$$

$$\phi_{k2}^\theta(0) = 0 \qquad \frac{d\phi_{k2}^\theta}{dx_1}\bigg|_{x_1=0} = 0 \qquad (19.16)$$

$$\phi_{j2}^u(L_e) = 0 \qquad \frac{d\phi_{j2}^u}{dx_1}\bigg|_{x_1=L_e} = 0$$

$$\phi_{j2}^\theta(L_e) = 0 \qquad \frac{d\phi_{j2}^\theta}{dx_1}\bigg|_{x_1=L_e} = 0$$

$$\phi_{k2}^u(L_e) = 1 \qquad \frac{d\phi_{k2}^u}{dx_1}\bigg|_{x_1=L_e} = 0$$

$$\phi_{k2}^\theta(L_e) = 0 \qquad \frac{d\phi_{k2}^\theta}{dx_1}\bigg|_{x_1=L_e} = 1$$

Comparing relations (19.16) with (19.12) and (19.15) we see that they are satisfied if

$$\phi_{j3}^u = \phi_{j2}^u = \phi_j^u \qquad \phi_{j3}^\theta = -\phi_{j2}^\theta = \phi_j^\theta$$

$$\phi_{k3}^u = \phi_{k2}^u = \phi_k^u \qquad \phi_{k3}^\theta = -\phi_{k2}^\theta = \phi_k^\theta \qquad (19.17)$$

On the basis of the foregoing presentation the matrix of shape functions $[\phi]$ for an element of a truss or for an element subjected only to twisting is

$$[\phi] = [\phi_j \quad \phi_k] \tag{19.18a}$$

The matrix of shape functions $[\phi]$ for a general planar element of a planar beam or frame in the x_1x_3 plane is

$$[\phi] = \begin{bmatrix} \phi_j & 0 & 0 & \phi_k & 0 & 0 \\ 0 & \phi_j^u & \phi_j^\theta & 0 & \phi_k^u & \phi_k^\theta \end{bmatrix} \tag{19.18b}$$

The matrix of shape functions $[\phi]$ for a general planar element of a planar beam or frame in the $\bar{x}_1\bar{x}_2$ plane is

$$[\phi] = \begin{bmatrix} \phi_j & 0 & 0 & \phi_k & 0 & 0 \\ 0 & \phi_j^u & -\phi_j^\theta & 0 & \phi_k^u & -\phi_k^\theta \end{bmatrix} \tag{19.18c}$$

The matrix of shape functions $[\phi]$ for an axial deformation element of a planar beam or frame is obtained from Eqs. (19.18b) and (19.18c) by setting ϕ_j^u, ϕ_j^θ, ϕ_k^u, and ϕ_k^θ equal to zero. The matrix of shape functions for a general space element of a space beam or frame is

$$[\phi] = \begin{bmatrix} \phi_j & 0 & 0 & 0 & 0 & 0 & \phi_k & 0 & 0 & 0 & 0 & 0 \\ 0 & \phi_j^u & 0 & 0 & 0 & -\phi_j^\theta & 0 & \phi_k^u & 0 & 0 & 0 & -\phi_k^\theta \\ 0 & 0 & \phi_j^u & 0 & \phi_j^\theta & 0 & 0 & 0 & \phi_k^u & 0 & \phi_k^\theta & 0 \\ 0 & 0 & 0 & \phi_j & 0 & 0 & 0 & 0 & 0 & \phi_k & 0 & 0 \end{bmatrix} \tag{19.18d}$$

The matrix of shape functions $[\phi]$ for an axial deformation element of a space beam or frame is obtained from Eq. (19.18d) by setting ϕ_j^u, ϕ_j^θ, ϕ_k^u, ϕ_k^θ, and the terms of the last row equal to zero.

On the basis of relations (19.5), (19.7), (19.12), (19.15), and (19.16) the shape functions ϕ_q, ϕ_q^u, and ϕ_q^θ ($q = j$ or k) for an element of a framed structure have the following properties.

1. The shape functions ϕ_q ($q = j$ or k) for an element are equal to unity at its end q and vanish at its other end.
2. The shape functions ϕ_q^u ($q = j$ or k) for an element are equal to unity at its end q and vanish at its other end. Moreover, the derivatives $d\phi_q^u/dx_1$ ($q = j$ or k) vanish at both ends of the element.
3. The shape functions ϕ_q^θ ($q = j$ or k) for an element vanish at both its ends. Moreover the derivative $d\phi_q^\theta/dx_1$ ($q = j$ or k) vanishes at the end p ($p \neq q$) of an element and is equal to 1 or -1 at its end q depending on whether the element is subjected to bending at its x_3 or its x_2 axis, respectively.

4. The shape functions ϕ_q, ϕ_q^u, and ϕ_q^θ ($q = j$ or k) for an element are polynomials of x_1. Their degree is such that when the approximate solution (19.1) is substituted into the internal action-displacement relations (2.36), (2.49), (2.50), and (2.64) give components of internal actions which are different than zero when the corresponding components of nodal displacements of the element do not vanish. Referring to relations (2.36) and (2.64) we see that the component of translation $u_1(x_1)$ and the component of rotation $\theta_1(x_1)$ of an element must be polynomials in x_1 of at least the first degree. Moreover, referring to relations (2.50) we see that the components of translation $u_2(x_1)$ and $u_3(x_1)$ of an element must be polynomial in x_1 of at least the third degree. That is, the simplest shape functions which we can choose have the following form

$$\phi_j(x_1) = A_{j0} + A_{j1}x_1$$
$$\phi_k(x_1) = A_{k0} + A_{k1}x_1 \tag{19.19}$$

and

$$\phi_j^u(x_1) = A_{j0}^u + A_{j1}^u x_1 + A_{j2}^u x_1^2 + A_{j3}^u x_1^3$$
$$\phi_j^\theta(x_1) = A_{j0}^\theta + A_{j1}^\theta x_1 + A_{j2}^\theta x_1^2 + A_{j3}^\theta x_1^3$$
$$\phi_k^u(x_1) = A_{k0}^u + A_{k1}^u x_1 + A_{k2}^u x_1^2 + A_{k3}^u x_1^3 \tag{19.20}$$
$$\phi_k^\theta(x_1) = A_{k0}^\theta + A_{k1}^\theta x_1 + A_{k2}^\theta x_1^2 + A_{k3}^\theta x_1^3$$

The constants A_{qi}^u and A_{qi}^θ ($q = j$ or k) ($i = 0, 1, 2, 3$) are evaluated by requiring that the element shape functions have the properties described previously. That is, referring to relations (19.5), (19.7), and (19.19) we have

$$\phi_j(0) = 1 = A_{j0} \qquad \phi_j(L_e) = 0 = 1 + A_{j1}L_e$$
$$\phi_k(0) = 0 = A_{k0} \qquad \phi_k(L_e) = 1 = A_{k1}L_e \tag{19.21}$$

Thus,

$$A_{j0} = 1 \qquad A_{j1} = -\frac{1}{L_e}$$
$$A_{k0} = 0 \qquad A_{k1} = \frac{1}{L_e} \tag{19.22}$$

Substituting the values of the constants (19.22) into relations (19.19) we get

$$\phi_j = 1 - \frac{x_1}{L_e} \qquad \phi_k = \frac{x_1}{L_e} \tag{19.23}$$

The shape functions ϕ_j and ϕ_k are plotted in Fig. 19.2. Referring to relations (19.12), (19.15), and (19.20), for an element subjected to

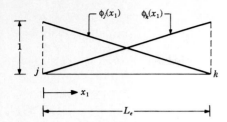

Figure 19.2 Shape functions ϕ_j and ϕ_k for an element.

bending about its x_2 axis we have

$$\phi_j^u(0) = 1 = A_{j0}^u$$

$$\left.\frac{d\phi_j^u}{dx_1}\right|_{x_1=0} = 0 = A_{j1}^u$$

$$\phi_j^u(L_e) = 0 = 1 + L_e^2 A_{j2}^u + L_e^3 A_{j3}^u \tag{19.24}$$

$$\left.\frac{d\phi_j^u}{dx_1}\right|_{x_1=L_e} = 0 = 2L_e A_{j2}^u + 3L_e^2 A_{j3}^u$$

and

$$\phi_j^\theta(0) = 0 = A_{j0}^\theta$$

$$\left.\frac{d\phi_j^\theta}{dx_1}\right|_{x_1=0} = -1 = A_{j1}^\theta \tag{19.25}$$

$$\phi_j^\theta(L_e) = 0 = -L_e + L_e^2 A_{j2}^\theta + L_e^3 A_{j3}^\theta$$

$$\left.\frac{d\phi_j^\theta}{dx_1}\right|_{x_1=L_e} = 0 = -1 + 2L_e A_{j2}^\theta + 3L_e^2 A_{j3}^\theta$$

and

$$\phi_k^u(0) = 0 = A_{k0}^u$$

$$\left.\frac{d\phi_k^u}{dx_1}\right|_{x_1=0} = 0 = A_{k1}^u$$

$$\phi_k^u(L_e) = 1 = L_e^2 A_{k2}^u + L_e^3 A_{k3}^u \tag{19.26}$$

$$\left.\frac{d\phi_k^u}{dx_1}\right|_{x_1=L_e} = 0 = 2L_e^2 A_{k2}^u + 3L_e^3 A_{k3}^u$$

and

$$\phi_k^\theta(0) = 0 = A_{k0}^\theta$$

$$\left.\frac{d\phi_k^\theta}{dx_1}\right|_{x_1=0} = 0 = A_{k1}^\theta \tag{19.27}$$

$$\phi_k^\theta(L_e) = 0 = L_e^2 A_{k2}^\theta + L_e^3 A_{k3}^\theta$$

$$\left.\frac{d\phi_k^\theta}{dx_1}\right|_{x_1=L_e} = -1 = 2L_e A_{k2}^\theta + 3L_e^2 A_{k3}^\theta$$

From relations (19.24) to (19.27) we have

$$A_{j0}^u = 1 \qquad A_{j1}^u = 0 \qquad A_{j2}^u = -\frac{3}{L_e^2} \qquad A_{j3}^u = \frac{2}{L_e^3}$$

$$A_{j0}^\theta = 0 \qquad A_{j1}^\theta = -1 \qquad A_{j2}^\theta = \frac{2}{L_e} \qquad A_{j3}^\theta = -\frac{1}{L_e^2}$$

$$A_{k0}^u = 0 \qquad A_{k1}^u = 0 \qquad A_{k2}^u = \frac{3}{L_e^2} \qquad A_{k3}^u = -\frac{2}{L_e^3}$$

$$A_{k0}^\theta = 0 \qquad A_{k1}^\theta = 0 \qquad A_{k2}^\theta = \frac{1}{L_e} \qquad A_{k3}^\theta = -\frac{1}{L_e^2}$$

(19.28)

Substituting the values of the constants (19.28) into relations (19.20) we get

$$\phi_j^u(x_1) = 1 - \frac{3x_1^2}{L_e^2} + \frac{2x_1^3}{L_e^3}$$

$$\phi_j^\theta(x_1) = -x_1 + \frac{2x_1^2}{L_e} - \frac{x_1^3}{L_e^2}$$

$$\phi_k^u(x_1) = \frac{3x_1^2}{L_e^2} - \frac{2x_1^3}{L_e^3}$$

$$\phi_k^\theta(x_1) = \frac{x_1^2}{L_e} - \frac{x_1^3}{L_e^2}$$

(19.29)

The element shape functions given by relation (19.29) are plotted in Fig. 19.3.

Notice that the element shape functions (19.23) and (19.29) are for two-node elements. We can have elements with more than two nodes (i.e., two at their ends and a number of intermediary nodes). For these elements we must use polynomials of higher order than that of those given by relations (19.23) or (19.29). However, we will not explore this possibility further in this book.

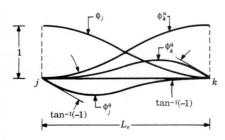

Figure 19.3 Shape functions for an element of a planar beam or frame in the x_1x_3 plane.

Referring to relations (4.45), (4.46), (4.57), (4.58a), (4.65), and (4.67), we see that relation (19.1) in conjunction with (19.17) and (19.23) or (19.29) represents the exact components of displacement of an element of constant cross section subjected only to nodal displacements.

19.2.2 Derivation of approximate stiffness equations for elements of variable cross section

In this section we derive approximate stiffness equations for elements of variable cross section of a framed structure using the finite-element method in conjunction with the principle of virtual work, by adhering to the following steps.

Step 1. We express the actual and the virtual basic components of displacement of an element as follows

$$\{\tilde{u}\} = [\phi]\{\tilde{D}\} \tag{19.30}$$

$$\{\tilde{\tilde{u}}\} = [\phi]\{\tilde{\tilde{D}}\} \tag{19.31}$$

where the matrices $\{\tilde{u}\}$, $\{\tilde{D}\}$, and $[\phi]$ are defined in Sec. 19.2.1. Moreover, $\{\tilde{\tilde{u}}\}$ is the matrix of the virtual basic components of displacement of an element and $\{\tilde{\tilde{D}}\}$ is the matrix of nodal displacements corresponding to the virtual components of displacement $\{\tilde{\tilde{u}}\}$.

Step 2. We substitute relations (19.30) and (19.31) into the principle of virtual work (19.2) to obtain an approximation for the stiffness equations for an element. That is,

$$\{\tilde{\tilde{D}}\}^T\{\tilde{A}\} = \{\tilde{\tilde{D}}\}^T \int_0^{L_e} [[\mathscr{B}][\phi]]^T [ST][\mathscr{B}]\{\phi\}\, dx_1\, \{\tilde{D}\}$$

$$- \{\tilde{\tilde{D}}\}^T \int_0^{L_e} [[\mathscr{B}]\{\phi\}]^T [ST][\tilde{H}]\, dx_1$$

$$- \{\tilde{\tilde{D}}\}^T \int_0^{L_e} [[\mathscr{L}]\{\phi\}]^T \{\tilde{L}\}\, dx_1 \tag{19.32}$$

Since relation (19.32) is valid for any choice of $\{\tilde{\tilde{D}}\}$, we have

$$\{\tilde{A}\} = [\tilde{K}]\{\tilde{D}\} + \{\tilde{A}^R\} \tag{19.33}$$

where
$$[\tilde{K}] = \int_0^L [[\mathscr{B}][\phi]]^T [ST][\mathscr{B}][\phi]\, dx_1 \tag{19.34}$$

is an approximation to the local stiffness matrix for the element, and

$$\{\tilde{A}^R\} = -\int_0^L [[\mathcal{B}][\phi]]^T [ST]\{H\} \, dx_1 - \int_0^L [[\mathcal{L}][\phi]]^T \{L\} \, dx_1 \quad (19.35)$$

is an approximation to the matrix of fixed-end actions of the element.

For elements of constant cross section the integral in relation (19.34) can be easily evaluated. However, for elements of variable cross section it may be necessary to evaluate the integrals in relations (19.34) and (19.35) numerically. Moreover, it may be necessary to evaluate the integrals in relation (19.35) numerically for elements of constant cross section subjected to distributed loads whose variation along the length of the element is complex.

Notice that if relation (19.1) represents the exact basic components of displacement of an element, relation (19.34) represents its exact stiffness matrix. As discussed in Sec. 4.4.1, this occurs for elements of constant cross section.

Example. Using relations (19.34) establish an approximation to the stiffness matrix for the tapered element of a truss shown in Fig. a. The element has constant width.

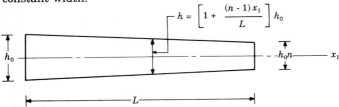

Figure a Geometry of the element.

---

solution For an element of a truss, referring to relations (19.18a), (18.29) to (18.32), (18.35), and (18.36) we have

$$\phi = [\phi_j \quad \phi_k] \tag{a}$$

$$\{\widetilde{IA}\} = N(x_1) \tag{b}$$

$$\{\tilde{u}\} = \tilde{u}_1(x_1) \tag{c}$$

$$\{\widetilde{\widetilde{STR}}\} = \frac{d\tilde{\tilde{u}}_1}{dx_1} = [\mathcal{B}]\{\tilde{\tilde{u}}\} \tag{d}$$

$$[\mathcal{B}] = \frac{d}{dx_1} \tag{e}$$

$$[ST] = EA = EA_j\left[1 + \frac{(n-1)x_1}{L_e}\right] \tag{f}$$

Substituting relations (a), (e), and (f) into (19.34) we obtain

$$[K] = \int_0^{L_e} EA \frac{d[\phi]^T}{dx_1} \frac{d[\phi]}{dx_1} dx_1 = \int_0^{L_e} EA \begin{bmatrix} \phi_j'\phi_j' & \phi_j'\phi_k' \\ \phi_k'\phi_j' & \phi_k'\phi_k' \end{bmatrix} dx_1 \tag{g}$$

where

$$\phi_q' = \frac{d\phi_q}{dx_1} \qquad q = j \text{ or } k \tag{h}$$

Substituting relations (19.23) into (g) we get

$$[K] = \int_0^{L_e} \frac{EA}{L_e^2} \begin{bmatrix} 1 & -1 \\ -1 & 1 \end{bmatrix} dx_1 \tag{i}$$

Referring to Fig. a we have

$$[K] = \int_0^{L_e} \frac{EA_j}{L_e^2} \left[1 + \frac{(n-1)x_1}{L_e} \right] \begin{bmatrix} 1 & -1 \\ -1 & 1 \end{bmatrix} dx_1$$

$$= \frac{EA_j(1+n)}{2L_e} \begin{bmatrix} 1 & -1 \\ -1 & 1 \end{bmatrix} \tag{j}$$

Exact Analysis

The displacement equation of equilibrium (2.39) for the element of Fig. a reduces to

$$\frac{d}{dx_1} \left(EA \frac{du_1}{dx_1} \right) = 0 \tag{k}$$

or

$$\frac{d}{dx_1} \left[\left[1 + \frac{(n-1)x_1}{L_e} \right] \frac{du_1}{dx_1} \right] = 0 \tag{l}$$

Integrating relation (l) twice we get

$$u_1(x_1) = \int_0^{x_1} \frac{B_1}{1 + (n-1)x_1/L_e} dx_1 + B_2 \tag{m}$$

and

$$u_1(x_1) = \frac{L_e}{n-1} \ln \left[1 + \frac{(n-1)x_1}{L_e} \right] B_1 + B_2 \tag{n}$$

The constants B_1 and B_2 in relation (n) are evaluated so that $u_1(0) = u_1^j$ and $u_1(L) = u_1^k$. Thus,

$$u_1 = \left[1 - \frac{\ln [1 + (n-1)x_1/L_e]}{\ln (n)} \right] u_1^j + \left[\frac{\ln [1 + (n-1)x_1/L_e]}{\ln (n)} \right] u_1^k \tag{o}$$

Consequently the exact shape functions for the tapered element of Fig. a are

$$\phi_j = 1 - \frac{\ln [1 + (n-1)x_1/L]}{\ln (n)}$$

$$\phi_k = \frac{\ln [1 + (n-1)x_1/L]}{\ln (n)} \tag{p}$$

Thus

$$\phi'_j = \frac{1 - n}{L_e \ln (n)[1 + (n - 1)x_1/L_e]}$$

$$\phi'_k = \frac{n - 1}{L_e \ln (n)[1 + (n - 1)x_1]/L_e}$$

(q)

Substituting relation (q) into (g) we get

$$[K]_{\text{exact}} = \int_0^{L_e} \frac{EA_j(n - 1)^2}{L_e^2[\ln^2(n)][1 + (n - 1)x_1/L_e]} \begin{bmatrix} 1 & -1 \\ -1 & 1 \end{bmatrix} dx_1$$

$$= \frac{EA_j(n - 1)}{L_e \ln (n)} \begin{bmatrix} 1 & -1 \\ -1 & 1 \end{bmatrix}$$

(r)

The percent error of the approximate analyses for $n = 0.5$ is

$$\text{Percent error} = \frac{[[K]_{\text{approx.}} - [K]_{\text{exact}}]100}{[K]_{\text{exact}}} = 3.82\%$$

(s)

19.3 Exact Stiffness Equations and Matrix of Fixed-End Actions for an Element of a Framed Structure of Constant Cross Section

19.3.1 Element of a truss

Consider an element of a truss of constant cross section subjected to external loads inducing axial deformation. Its axial component of translation $u_1(x_1)$ must satisfy the displacement equation of equilibrium (2.40). That is,

$$EA \frac{d^2 u_1}{dx_1^2} + D_1 = 0$$

(19.36)

where for the loading of Fig. 2.6 we have

$$D_1 = p_1(x_1)\Delta(x_1 - c_1) + \sum_{i=1}^{n_1} P_1^{(i)} \delta(x_1 - a_{1i}) + \frac{d}{dx_1}(EAH_1)$$

(19.36a)

The solution of this equation can be written as

$$u_1(x_1) = u_1^E(x_1) + u_1^R(x_1)$$

(19.37)

where $u_1^E(x_1)$ is the solution of the homogeneous part of Eq. (19.36) which at the ends of the element $x_1 = 0$ and $x_1 = L_e$ gives its axial components of nodal displacements, and $u_1^R(x_1)$ is the particular solution of Eq. (19.36) which vanishes at the ends of the element.

As shown in Sec. 4.4.1, for an element of constant cross section $u_1^E(x_1)$ is equal to

$$u_1^E(x_1) = [\phi]\{D\} \tag{19.38}$$

where

$$\{D\} = \begin{Bmatrix} u_1^j \\ u_1^k \end{Bmatrix} \tag{19.38a}$$

and $[\phi]$ is given by relation $(19.18a)$ with (19.23). Substituting relation (19.38) into (19.37) we get

$$u_1(x_1) = [\phi]\{D\} + u_1^R(x_1) \tag{19.39}$$

If for a given loading the exact particular solution $u_1^R(x_1)$ can be established, relation (19.39) is the exact (within the accuracy of the mechanics of materials theories) expression for the axial component of translation of the element under consideration.

Notice that when the element is subjected to the given loads with its ends fixed, $\{D\}$ is a zero matrix and relation (19.39) reduces to

$$u_1(x_1) = u_1^R(x_1) \tag{19.40}$$

Moreover, notice that when the element is subjected only to the nodal forces required to produce its actual nodal displacements, $u_1^R(x_1)$ vanishes and relation (19.39) reduces to

$$u_1(x_1) = [\phi]\{D\} \tag{19.41}$$

Thus in relation (19.39) we have expressed the axial component of translation of the element under consideration as the sum of its axial component of translation when it is subjected only to the axial component of its nodal translations and when it is subjected to the given loading with its ends fixed (see Fig. 19.4). Finally notice that when an element is subjected on its end k to the axial force required

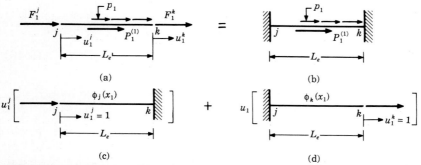

Figure 19.4 Physical significance of the particular solution and of the shape functions in relation (19.39).

to produce an axial component of translation of this end equal to unity ($u_1^k = 1$) while its end j is kept fixed ($u_1^j = 0$), relation (19.39) gives

$$u_1(x_1) = u_1^E(x_1) = \phi_k(x_1) \tag{19.42}$$

Similarly when an element is subjected on its end j to the axial force required to produce an axial component of translation of this end equal to unity ($u_1^j = 1$) while its end k is kept fixed ($u_1^k = 0$), relation (19.39) gives

$$u_1(x_1) = u_1^E(x_1) = \phi_j(x_1) \tag{19.43}$$

Consequently the shape function ϕ_j or ϕ_k for an element of constant cross section represents the distribution of the axial component of translation $u_1(x_1)$ when the element is subjected on its end j or k, respectively, to the axial force required to produce a unit translation of this end while its end k or j, respectively, is kept fixed (see Fig. 19.4).

The actual distribution of the component of translation $u_1(x_1)$ given by relation (19.39) must satisfy the principle of virtual work (19.2) for any admissible virtual displacement field $\tilde{u}_1(x_1)$. For the element under consideration the principle of virtual work (19.2) reduces to

$$\int_0^{L_e} EA \frac{d\tilde{u}_1}{dx_1} \frac{du_1}{dx_1} \, dx_1 - \int_0^{L_e} EA \frac{d\tilde{u}_1}{dx_1} H_1 \, dx_1 - \int_0^{L_e} \tilde{u}_1 D_1 \, dx_1 = \{\tilde{D}\}^T \{A\} \tag{19.44}$$

We assume a virtual component of translation $\tilde{u}_1(x_1)$ of the following form:

$$\tilde{u}_1 = [\phi]\{\tilde{D}\} = \{\tilde{D}\}^T [\phi]^T \tag{19.45}$$

Substituting relations (19.39) and (19.45) into (19.44) we obtain

$$\{\tilde{D}\}^T \left[\left[\int_0^{L_e} \left[EA \frac{d[\phi]^T}{dx_1} \frac{d[\phi]}{dx_1} \right] dx_1 \right] \{D\} + \int_0^{L_e} EA \frac{d[\phi]^T}{dx_1} \frac{du_1^R}{dx_1} \, dx_1 \right.$$
$$\left. - \int_0^{L_e} EA \frac{d[\phi]^T}{dx_1} H_1 \, dx_1 - \int_0^{L_e} [\phi]^T D_1 \, dx_1 \right] = \{\tilde{D}\}^T \{A\} \tag{19.46}$$

Since relation (19.46) is valid for any choice of $\{\tilde{D}\}^T$, we have

$$\left[\int_0^{L_e} \left[EA \frac{d[\phi]^T}{dx_1} \frac{d[\phi]}{dx_1} \right] dx_1 \right] \{D\} + \int_0^{L_e} EA \frac{d[\phi]^T}{dx_1} \frac{du_1^R}{dx_1} \, dx_1$$
$$- \int_0^{L_e} EA \frac{d[\phi]^T}{dx_1} H_1 \, dx_1 - \int_0^{L_e} [\phi]^T D_1 \, dx_1 = \{A\} \tag{19.47}$$

Integrating the second integral of relation (19.47) by parts and taking into account that the element under consideration has a constant cross section we obtain

$$-\int_0^{L_e} EA \frac{d[\phi]^T}{dx_1} \frac{du_1^P}{dx_1} dx_1 = EA \int_0^{L_e} \left(\frac{d^2[\phi]^T}{dx_1^2}\right) u_1^R dx_1 - \left[EA \frac{d[\phi]^T}{dx_1} u_1^R \right]_{x_1=0}^{x_1=L_e}$$

(19.48)

Referring to relation (19.23) we see that

$$\frac{d^2[\phi]^T}{dx_1^2} = [0]$$

(19.49)

Moreover, $u_1^R(x_1)$ vanishes at $x_1 = 0$ and $x_1 = L_e$. Consequently, relation (19.48) reduces to

$$\int_0^{L_e} EA \frac{d[\phi]^T}{dx} \frac{du_1^R}{dx_1} dx_1 = 0$$

(19.50)

Using relation (19.50) relation (19.47) can be written as

$$\{A\} = [K]\{D\} + \{A^R\}$$

(19.51)

where $\{A\}$ = exact matrix of nodal actions of an element of constant cross section

$[K]$ = exact stiffness matrix for an element of constant cross section given by relation (19.34)

$\{A^R\}$ = exact matrix of fixed-end actions of an element of constant cross section given by relation (19.35)

On the basis of the foregoing presentation we may conclude that by expressing the axial component of translation of an element of a framed structure of constant cross section by relation (19.39) we have proven that its exact stiffness matrix is given by relation (19.34), while its exact matrix of fixed-end actions is given by relation (19.35).

19.3.2 Element of a planar beam or frame subjected only to bending about its x_2 axis

Consider an element of constant cross section of a planar beam or frame subjected to given loads inducing bending about its x_2 axis. Its component of translation $u_3(x_1)$ must satisfy the displacement equation of equilibrium (2.53b). That is,

$$-\frac{d^2}{dx_1^2}\left[EI_2\frac{d^2u_3}{dx_1^2}+H_2\right]+D_3=0 \tag{19.52}$$

where for the loading of Fig. 2.7 we have

$$D_3=p_3(x_1)\Delta(x_1-c_3)+\frac{d}{dx_1}[m_2\Delta(x_1-d_2)]$$

$$+\sum_{i=1}^{n_3}P_3^{(i)}\,\delta(x_1-a_{3i})+\sum_{i=1}^{m_3}M_2^{(i)}\,\delta^1(x_1-b_{2i}) \tag{19.52a}$$

The solution of equation (19.52) can be written as

$$u_3(x_1)=u_3^E(x_1)+u_3^R(x_1) \tag{19.53}$$

where $u_3^E(x_1)$ is the solution of the homogeneous part of Eq. (19.52) which at the ends of the element $x_1=0$ and $x_1=L_e$ gives its components of nodal displacements u_3^j and u_3^k, respectively; and u_3^R is the particular solution of Eq. (19.52) which vanishes at the ends of the element.

As shown in Sec. 4.4.1 for an element of constant cross section $u_3^E(x_1)$ is equal to

$$u_3^E(x_1)=[\phi]\{D\} \tag{19.54}$$

where

$$\{D\}=\begin{Bmatrix} u_3^j \\ \theta_2^j \\ u_3^k \\ \theta_2^k \end{Bmatrix} \tag{19.54a}$$

and $[\phi]$ is given by relation (19.18b) with (19.23) and (19.29). Substituting relation (19.54) into (19.53) we get

$$u_3(x_1)=[\phi]\{D\}+u_3^R(x_1) \tag{19.55}$$

If for a given loading the exact particular solution $u_3^R(x_1)$ can be established, relation (19.55) is the exact (within the accuracy of the mechanics of materials theories) expression for the component of translation $u_3(x_1)$ of the element under consideration.

Notice that when the element is subjected to the given loads with its ends fixed, $\{D\}$ is a zero matrix and Eq. (19.55) reduces to

$$u_3(x_1)=u_3^R(x_1) \tag{19.56}$$

Moreover, notice that when the element is subjected only to the nodal actions required to produce its actual nodal displacements, $u_3^R(x_1)$ vanishes and relation (19.55) reduces to

$$u_3(x_1) = u_3^E(x_1) = [\phi]\{D\} \tag{19.57}$$

Thus in relation (19.53) we have expressed the transverse component of translation $u_3(x_1)$ of the element under consideration as the sum of its transverse component of translation when it is subjected only to its nodal displacements and when it is subjected to the given loading with its ends fixed (see Fig. 19.5). Finally notice that when the element with its end j fixed ($u_3^j = 0$, $\theta_2^j = 0$) is subjected on its end k to the actions F_3^k and M_2^k required to translate this end by an amount equal to unity ($u_3^k = 1$) without rotating it, relation (19.55) gives

$$u_3(x_1) = \phi_k^u(x_1) \tag{19.58}$$

Consequently the shape function $\phi_k^u(x_1)$ for an element of constant cross section represents the distribution of its component of translation $u_3(x_1)$ when the element with its end j fixed is subjected on its

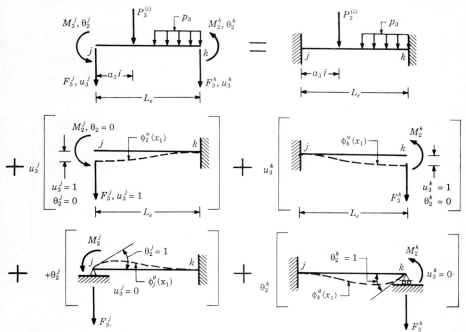

Figure 19.5 Physical significance of the particular solution and of the shape functions in relation (19.55).

end k to the actions required to displace it by $u_3^k = 1$ and $\theta_2^k = 0$ (see Fig. 19.5). Similarly when the element with its end k fixed ($u_3^k = 0$, $\theta_2^k = 0$) is subjected at its end j to the actions F_3^j and M_2^j required to rotate this end by an amount equal to unity ($\theta_2^j = 1$) without translating it ($u_3^j = 0$), relation (19.55) gives

$$u_3(x_1) = \phi_j^\theta(x_1) \tag{19.59}$$

Consequently, the shape function $\phi_j^\theta(x_1)$ for an element of constant cross section represents the distribution of its component of translation $u_3(x_1)$ when the element with its end k fixed is subjected at its end j to the actions required to displace it by $\theta_2^j = 1$ and $u_3^j = 0$ (see Fig. 19.5).

The actual distribution of the component of translation $u_3(x_1)$ given by relation (19.55) must satisfy the principle of virtual work (19.2) for any admissible virtual displacement field $\tilde{u}_3(x_1)$. For the element under consideration the principle of virtual work (19.2) reduces to

$$\int_0^{L_e} EI_2 \frac{d^2\tilde{u}_3}{dx_1^2}\frac{d^2u_3}{dx_1^2}\,dx_1 - \int_0^{L_e} EI_2 \frac{d^2\tilde{u}_3}{dx_1^2}H_3\,dx_1 - \int_0^{L_e} \tilde{u}_3 D_3\,dx_1 = \{\tilde{D}\}^T\{A\} \tag{19.60}$$

We assume a virtual component of translation $\tilde{u}_3(x_1)$ of the following form

$$\tilde{u}_3 = [\phi]\{\tilde{D}\} = \{\tilde{D}\}^T[\phi]^T \tag{19.61}$$

Substituting relations (19.55) and (19.61) into (19.60) we obtain

$$\{\tilde{D}\}^T\left[\left[\int_0^{L_e}\left[EI_2\frac{d^2[\phi]^T}{dx_1^2}\frac{d^2[\phi]}{dx_1^2}\right]dx_1\right]\{D\} + \int_0^{L_e} EI_2\frac{d^2[\phi]^T}{dx_1^2}\frac{d^2u_3^R}{dx_1^2}\,dx_1\right.$$
$$\left.-\int_0^{L_e} EI_2\frac{d^2[\phi]^T}{dx_1^2}H_3\,dx_1 - \int_0^{L_e}[\phi]^T D_3\,dx_1\right] = \{\tilde{D}\}^T\{A\} \quad (19.62)$$

Since relation (19.62) is valid for any choice of $\{\tilde{D}\}^T$, we have

$$\left[\int_0^{L_e}\left[EI_2\frac{d^2[\phi]^T}{dx_1^2}\frac{d^2[\phi]}{dx_1^2}\right]dx_1\right]\{D\} + \int_0^{L_e} EI_2\frac{d^2[\phi]^T}{dx_1^2}\frac{d^2u_3^R}{dx_1^2}\,dx_1$$
$$-\int_0^{L_e} EI_2\frac{d^2[\phi]^T}{dx_1^2}H_3\,dx_1 - \int_0^{L_e}[\phi]^T D_3\,dx_1 = \{A\} \quad (19.63)$$

Integrating the second integral of relations (19.62) by parts and taking into account that the element under consideration has a constant cross section, we obtain

$$\int_0^{L_e} EI_2 \frac{d^2[\phi]^T}{dx_1^2} \frac{d^2 u_3^R}{dx_1^2} dx_1 = EI_2 \int_0^{L_e} \left(\frac{d^4[\phi]^T}{dx_1^4}\right) u_3^R dx_1$$

$$+ \left[EI_2 \frac{d^2[\phi]^T}{dx_1^2} \frac{du_3^R}{dx_1} \right]_{x_1=0}^{x_1=L_e} - \left[EI_2 \frac{d^3[\phi]^T}{dx_1^3} u_3^R \right]_{x_1=0}^{x_1=L_e} \qquad (19.64)$$

Referring to relation (19.29) we see that

$$\frac{d^4[\phi]^T}{dx_1^4} = [0] \qquad (19.65)$$

Moreover, $u_3^R(x_1)$ and du_3^R/dx_1 vanish at $x_1 = 0$ and $x_1 = L_e$. Consequently, relation (19.64) reduces to

$$\int_0^{L_e} EI_2 \frac{d^2[\phi]^T}{dx_1^2} \frac{d^2 u_1^R}{dx_1^2} dx_1 = 0 \qquad (19.66)$$

Using relation (19.66) Eqs. (19.63) can be written as

$$\{A\} = [K]\{D\} + \{A^R\} \qquad (19.67)$$

where $\{A\}$ = exact matrix of nodal actions of an element of constant cross section

$[K]$ = exact stiffness matrix for an element of constant cross section given by relation (19.34)

$\{A^R\}$ = exact matrix of fixed-end actions of an element of constant cross section given by relations (19.35)

On the basis of the foregoing presentation we may conclude that by expressing the component of translation $u_3(x_1)$ of an element of a framed structure of constant cross section by relations (19.55) we have proven that the exact stiffness matrix for this element is given by relation (19.34), while the exact matrix of its fixed-end actions is given by relation (19.35).

In what follows we apply relations (19.34) and (19.35) to establish the stiffness equations for a general planar element of constant cross section of a planar frame subjected to a concentrated force and to uniformly distributed forces over a portion of its length. We consider this element only in order to illustrate the use of relations (19.34) and (19.35). Actually we can establish its local stiffness matrix using relations (4.62) and its fixed-end actions by referring to the table on the inside of the back cover.

Example. Using relations (19.34) and (19.35), establish the local stiffness matrix and the matrix of fixed-end actions $\{A^R\}$ of the element subjected to the external actions shown in Fig. a. The element has a constant cross section and is made of an isotropic linearly elastic material of modulus of elasticity E.

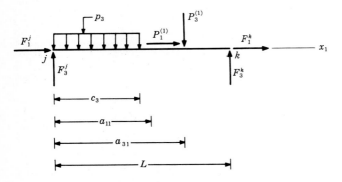

Figure a Geometry and loading of the element.

solution

Computation of the Local Stiffness Matrix for the Element Using Relation (19.34)

For a general planar element in the $x_1 x_3$ plane referring to relations (19.18b) and (18.29) to (18.38) and to Fig. a we have

$$[\phi] = \begin{bmatrix} \phi_j & 0 & 0 & \phi_k & 0 & 0 \\ 0 & \phi_j^u & -\phi_j^\theta & 0 & \phi_k^u & -\phi_k^\theta \end{bmatrix} \tag{a}$$

$$\{\widetilde{IA}\} = \begin{Bmatrix} N(x_1) \\ M_2(x_1) \end{Bmatrix} \tag{b}$$

$$\{\tilde{u}\} = \begin{Bmatrix} \tilde{\tilde{u}}_1(x_1) \\ \tilde{\tilde{u}}_3(x_1) \end{Bmatrix} \tag{c}$$

$$\{\widetilde{\widetilde{STR}}\} = \begin{Bmatrix} \dfrac{d\tilde{\tilde{u}}_1}{dx_1} \\ \dfrac{d\tilde{\tilde{\theta}}_2}{dx_1} \end{Bmatrix} = [\mathscr{B}]\{\tilde{\tilde{u}}\} \tag{d}$$

$$[\mathscr{B}] = \begin{bmatrix} \dfrac{d}{dx_1} & 0 \\ 0 & -\dfrac{d^2}{dx_1^2} \end{bmatrix} \tag{e}$$

$$[\mathcal{L}] = \begin{bmatrix} 1 & 0 \\ 0 & 1 \\ 0 & -\dfrac{d}{dx_1} \end{bmatrix} \tag{f}$$

$$\{L\} = \left\{ \begin{array}{c} P_i^{(1)} \, \delta(x_1 - a_{11}) \\ p_3[1 - \Delta(x_1 - c_3)] + P_3^{(1)} \, \delta(x_1 - a_{31}) \\ 0 \end{array} \right\} \tag{g}$$

$$[ST] = \begin{bmatrix} EA & 0 \\ 0 & EI_2 \end{bmatrix} \tag{h}$$

$$\{H\} = \{0\} \tag{i}$$

Thus,

$$[\mathcal{B}][\phi] = \begin{bmatrix} \dfrac{d\phi_j}{dx_1} & 0 & 0 & \dfrac{d\phi_k}{dx_1} & 0 & 0 \\[2mm] 0 & -\dfrac{d^2\phi_j^u}{dx_1^2} & \dfrac{d^2\phi_j^\theta}{dx_1^2} & 0 & \dfrac{d^2\phi_k^u}{dx_1^2} & -\dfrac{d^2\phi_k^{''}}{dx_1^2} \end{bmatrix} \tag{j}$$

and

$$[ST][\mathcal{B}][\phi]$$

$$= \begin{bmatrix} EA\dfrac{d\phi_j}{dx_1} & 0 & 0 & EA\dfrac{d\phi_k}{dx_1} & 0 & 0 \\[2mm] 0 & -EI_2\dfrac{d^2\phi_j^u}{dx_1^2} & EI_2\dfrac{d^2\phi_j^\theta}{dx_1^2} & 0 & -EI_2\dfrac{d^2\phi_k^u}{dx_1^2} & EI_2\dfrac{d^2\phi_k^\theta}{dx_1^2} \end{bmatrix} \tag{k}$$

Substituting relations (j) and (k) into (19.34) and taking into account that the element has a constant cross section, we obtain

$$[K] = \int_0^{L_e} [[\mathcal{B}][\phi]]^T [ST][\mathcal{B}][\phi] \, dx_1$$

$$= EA \int_0^{L_e} \begin{bmatrix} \phi_j'\phi_j' & 0 & 0 & \phi_j'\phi_k' & 0 & 0 \\ 0 & 0 & 0 & 0 & 0 & 0 \\ 0 & 0 & 0 & 0 & 0 & 0 \\ \phi_k'\phi_j' & 0 & 0 & \phi_k'\phi_k' & 0 & 0 \\ 0 & 0 & 0 & 0 & 0 & 0 \\ 0 & 0 & 0 & 0 & 0 & 0 \end{bmatrix} dx_1$$

$$
+ EI_2 \int_0^{L_e}
\begin{bmatrix}
0 & 0 & 0 & 0 & 0 & 0 \\
0 & (\phi_j^u)''(\phi_j^u)'' & (\phi_j^u)''(\phi_j^\theta)'' & 0 & (\phi_j^u)''(\phi_k^u)'' & (\phi_j^u)''(\phi_k^\theta)'' \\
0 & (\phi_j^\theta)''(\phi_j^u)'' & (\phi_j^\theta)''(\phi_j^\theta)'' & 0 & (\phi_j^\theta)''(\phi_k^u)'' & (\phi_j^\theta)''(\phi_k^\theta)'' \\
0 & (\phi_k^u)''(\phi_j^u)'' & (\phi_k^u)''(\phi_j^\theta)'' & 0 & (\phi_k^u)''(\phi_k^u)'' & (\phi_k^u)''(\phi_k^\theta)'' \\
0 & (\phi_k^\theta)''(\phi_j^u)'' & (\phi_k^\theta)''(\phi_j^\theta)'' & 0 & (\phi_k^\theta)''(\phi_k^u)'' & (\phi_k^\theta)''(\phi_k^\theta)'' \\
0 & 0 & 0 & 0 & 0 & 0
\end{bmatrix}
dx_1
$$

(1)

where
$$
\phi_q' = \frac{d\phi_q}{dx_1} \qquad q = j \text{ or } k \tag{m}
$$

$$
(\phi_q^u)'' = \frac{d^2\phi_q^u}{dx_1^2} \qquad q = j \text{ or } k
$$

$$
(\phi_q^\theta)'' = \frac{d^2\phi_q^\theta}{dx_1^2} \qquad q = j \text{ or } k
$$

(n)

Differentiating relations (19.23) and (19.29) we have

$$
\frac{d\phi_j}{dx_1} = -\frac{1}{L_e}
$$

$$
\frac{d\phi_k}{dx_1} = \frac{1}{L_e}
$$

$$
\frac{d^2\phi_j^u}{dx_1^2} = -\frac{6}{L_e^2}\left(1 - \frac{2x_1}{L_e}\right)
$$

$$
\frac{d^2\phi_j^\theta}{dx_1^2} = \frac{2}{L_e}\left(2 - \frac{3x_1}{L_e}\right)
$$

$$
\frac{d^2\phi_k^u}{dx_1^2} = \frac{6}{L_e^2}\left(1 - \frac{2x_1}{L_e}\right)
$$

$$
\frac{d^2\phi_k^\theta}{dx_1^2} = \frac{2}{L_e}\left(1 - \frac{3x_1}{L_e}\right)
$$

(o)

Substituting relations (o) into (1) we obtain the stiffness matrix $[K]$ for the element [see relation (p) on p. 655]. Integrating the above relation, as expected we get the following stiffness matrix for an element of constant cross section.

$$
[K] = \frac{EI_2}{L_e^3} =
\begin{bmatrix}
\dfrac{AL_e^2}{I_2} & 0 & 0 & -\dfrac{AL_e^2}{I_2} & 0 & 0 \\
0 & 12 & -6L_e & 0 & -12 & -6L_e \\
0 & -6L_e & 4L_e^2 & 0 & 6L_e & 2L_e^2 \\
-\dfrac{AL_e^2}{I_2} & 0 & 0 & \dfrac{AL_e^2}{I_2} & 0 & 0 \\
0 & -12 & 6L_e & 0 & 12 & 6L_e \\
0 & -6L_e & 2L_e^2 & 0 & 6L_e & 4L_e^2
\end{bmatrix}
$$

(q)

$$[K] = EI_2 \left. \int \right|_0^{L_e}
\begin{bmatrix}
\dfrac{A}{I_2 L_e^2} & 0 & 0 & -\dfrac{A}{I_2 L_e^2} & 0 & 0 \\[2ex]
0 & \dfrac{36}{L_e^4}\left(1-\dfrac{2x_1}{L_e}\right)^2 & -\dfrac{12}{L_e^3}\left(1-\dfrac{2x_1}{L_e}\right)\left(2-\dfrac{3x_1}{L_e}\right) & 0 & -\dfrac{36}{L_e^4}\left(1-\dfrac{2x_1}{L_e}\right)^2 & -\dfrac{12}{L_e^3}\left(1-\dfrac{2x_1}{L_e}\right)\left(1-\dfrac{3x_1}{L_e}\right) \\[2ex]
0 & -\dfrac{12}{L_e^3}\left(1-\dfrac{2x_1}{L_e}\right)\left(2-\dfrac{3x_1}{L_e}\right) & \dfrac{4}{L_e^2}\left(2-\dfrac{3x_1}{L_e}\right)^2 & 0 & \dfrac{12}{L_e^3}\left(2-\dfrac{3x_1}{L_e}\right)\left(1-\dfrac{2x_1}{L_e}\right) & \dfrac{4}{L_e^2}\left(2-\dfrac{3x_1}{L_e}\right)\left(1-\dfrac{3x_1}{L_e}\right) \\[2ex]
-\dfrac{A}{I_2 L_e^2} & 0 & 0 & \dfrac{A}{I_2 L_e^2} & 0 & 0 \\[2ex]
0 & -\dfrac{36}{L_e^4}\left(1-\dfrac{2x_1}{L_e}\right)^2 & \dfrac{12}{L_e^3}\left(2-\dfrac{3x_1}{L_e}\right)\left(1-\dfrac{2x_1}{L_e}\right) & 0 & \dfrac{36}{L_e^4}\left(1-\dfrac{2x_1}{L_e}\right)^2 & \dfrac{12}{L_e^3}\left(1-\dfrac{2x_1}{L_e}\right)\left(1-\dfrac{3x_1}{L_e}\right) \\[2ex]
0 & -\dfrac{12}{L_e^3}\left(1-\dfrac{2x_1}{L_e}\right)\left(1-\dfrac{3x_1}{L_e}\right) & \dfrac{4}{L_e^2}\left(2-\dfrac{3x_1}{L_e}\right)\left(1-\dfrac{3x_1}{L_e}\right) & 0 & \dfrac{12}{L_e^3}\left(1-\dfrac{2x_1}{L_e}\right)\left(1-\dfrac{3x_1}{L_e}\right) & \dfrac{4}{L_e^2}\left(1-\dfrac{3x_1}{L_e}\right)^2
\end{bmatrix}
dx_1 \qquad (p)$$

Computation of the Matrix of Fixed-End Actions of the Element Using Relation (19.35)

From relations (a) and (f) we get

$$[\mathcal{L}][\phi] = \begin{bmatrix} \phi_j & 0 & 0 & \phi_k & 0 & 0 \\ 0 & \phi_j^u & \phi_j^\theta & 0 & \phi_k^u & \phi_k^\theta \\ 0 & -\dfrac{d\phi_j^u}{dx_1} & \dfrac{d\phi_j^\theta}{dx_1} & 0 & -\dfrac{d\phi_k^u}{dx_1} & \dfrac{d\phi_k^\theta}{dx_1} \end{bmatrix} \tag{r}$$

Substituting relations (r) and (g) into (19.35) we get

$$\{A^R\} = -\int_0^{L_e} \begin{bmatrix} \phi_j & 0 & 0 \\ 0 & \phi_j^u & -\dfrac{d\phi_j^u}{dx_1} \\ 0 & \phi_j^\theta & \dfrac{d\phi_j^\theta}{dx_1} \\ \phi_k & 0 & 0 \\ 0 & \phi_k^u & -\dfrac{d\phi_k^u}{dx_1} \\ 0 & \phi_k^\theta & \dfrac{d\phi_k^\theta}{dx_1} \end{bmatrix} \left\{ \begin{array}{c} P_1^{(1)}\,\delta(x_1 - a_{11}) \\ p_3[1 - \Delta(x_1 - c_3)] + P_3^{(1)}\,\delta(x_1 - a_{31}) \\ 0 \end{array} \right\} dx_1$$

$$= -\int_0^{L_e} \left\{ \begin{array}{c} P_1^{(1)}\phi_j\,\delta(x_1 - a_{11}) \\ \phi_j^u[p_3[1 - \Delta(x_1 - c_3)] + P_3^{(1)}\,\delta(x_1 - a_{31})] \\ \phi_j^\theta[p_3[1 - \Delta(x_1 - c_3)] + P_3^{(1)}\,\delta(x_1 - a_{31})] \\ P_1^{(1)}\phi_k\,\delta(x_1 - a_{11}) \\ \phi_k^u[p_3[1 - \Delta(x_1 - c_3)] + P_3^{(1)}\,\delta(x_1 - a_{31})] \\ \phi_k^\theta[p_3[1 - \Delta(x_1 - c_3)] + P_3^{(1)}\,\delta(x_1 - a_{31})] \end{array} \right\} dx_1 \tag{s}$$

Substituting relations (19.23) and (19.27) into relation (s) and integrating with the aid of relations (A.3) and (A.11), we obtain

$$\{A^R\} = \left\{ \begin{array}{c} P_1^{(1)}\left(1 - \dfrac{a_{11}}{L_e}\right) \\[2mm] \dfrac{p_3 c_3}{2}\left[2 - 2\left(\dfrac{c_3}{L_e}\right)^2 + \left(\dfrac{c_3}{L_e}\right)^3\right] + P_3^{(1)}\left[1 - 3\left(\dfrac{a_{31}}{L_e}\right)^2 + 2\left(\dfrac{a_{31}}{L_e}\right)^3\right]f \\[2mm] \dfrac{p_3 c_3^2}{12}\left[-6 + 8\left(\dfrac{c_3}{L_e}\right) - 3\left(\dfrac{c_3}{L_e}\right)^2\right] + P_3^{(1)}a_{31}\left[-1 + 2\left(\dfrac{a_{31}}{L_e}\right) - \left(\dfrac{a_{31}}{L_e}\right)^2\right] \\[2mm] \dfrac{P_1^{(1)}a_{11}}{L_e} \\[2mm] \dfrac{p_3 c_3^3}{2L_e^2}\left(2 - \dfrac{c_3}{L_e}\right) + \dfrac{P_3^{(1)}a_{31}^2}{L_e^2}\left(3 - \dfrac{2a_{31}}{L_e}\right) \\[2mm] \dfrac{p_3 c_3^3}{12L_e}\left(4 - \dfrac{3c_3}{L_e}\right) + \dfrac{P_3^{(1)}a_{31}^2}{L_e}\left(1 - \dfrac{a_{31}}{L_e}\right) \end{array} \right\} \tag{t}$$

The terms of the matrix $\{A^R\}$ represent the exact values of the fixed-end actions of an element of constant cross section subjected to the loads shown in Fig. a.

19.3.3 General space element

The results of Secs. 19.3.1 and 19.3.2 can be extended to a general space element of constant cross section subjected to a general loading. The matrix of the exact basic components of displacement of this element can be expressed as

$$\{u\} = [\phi]\{D\} + \{u^R\} \tag{19.68}$$

where the matrix of shape functions $[\phi]$ is given by relation $(19.18d)$ and

$$\{u\}^T = [u_1(x_1) \quad u_2(x_1) \quad u_3(x_1) \quad \theta_1(x_1)] \tag{19.69a}$$

$$\{u^R\}^T = \{u_1^R(x_1) \quad u_2^R(x_1) \quad u_3^R(x_1) \quad \theta_1^R(x_1)\} \tag{19.69b}$$

$$\{D\}^T = [u_1^j \quad u_2^j \quad u_3^j \quad \theta_1^j \quad \theta_2^j \quad \theta_3^j \quad u_1^k \quad u_2^k \quad u_3^k \quad \theta_1^k \quad \theta_2^k \quad \theta_3^k] \tag{19.69c}$$

Following a procedure analogous to the one adhered to for a general planar element in Sec. 19.3.2, it can be shown that the exact stiffness matrix for a general space element is given by relation (19.34), while its exact matrix of fixed-end actions is given by relation (19.35).

19.4 Derivation of the Flexibility Matrix for an Element Using the Principle of Virtual Work

In Sec. 19.4.1 we establish a formula for the flexibility matrix for an element using the principle of virtual work. In Sec. 19.4.2 we specialize this formula to general planar and general space elements of constant cross section including the effect of shear deformation. In Sec. 19.4.3 we specialize this formula to tapered elements, and in Sec. 19.4.4 to planar curved elements of constant cross section whose axis is an arc of a circle.

19.4.1 Derivation of the flexibility matrix for an element

Consider an element of a framed structure made of an isotropic linearly elastic material, subjected only to nodal displacements. The flexibility equations for this element are

$$\{d\} = [f]\{a\} \tag{19.70}$$

where as discussed in Sec. 5.3 the terms of the matrix of basic deformation parameters $\{d\}$ for an element represent the components of displacement $\{D^{Mk}\}$ of the end k of a model, consisting of the element with its end j fixed, when the model is subjected to the basic nodal actions $\{a\}$ (see Fig. 5.12). That is,

$$\{d\} = \{D^{Mk}\} \tag{19.71}$$

We assume that the model is subjected at its end k to an additional virtual set of basic nodal actions $\{\tilde{a}\}$. We denote the resulting additional components of displacement of the end k of the model by $\{\tilde{D}^{Mk}\}$. The work of the basic nodal actions $\{a\}$ acting on the model due to its virtual deformation $\{\tilde{D}^{Mk}\} = \{\tilde{d}\}$ is equal to

$$\text{Work} = \{\tilde{d}\}^T\{a\} \tag{19.72}$$

Substituting the flexibility equations (19.70) into relation (19.72) and taking into account that the flexibility matrix of an element is symmetric, we obtain

$$\text{Work} = \{\tilde{a}\}^T[f]\{a\} \tag{19.73}$$

The principle of virtual work (18.26) for the model of the element under consideration can be rewritten as

$$\{\tilde{a}\}^T[f^i]\{a\} = \int_0^L \left[\frac{\tilde{N}^M N^M}{EA} + \frac{\tilde{M}_1^M M_1^M}{GK} + \frac{\tilde{M}_2^M M_2^M}{EI_2} + \frac{\tilde{M}_3^M M_3^M}{EI_3} \right] dx_1$$
$$+ \left[K_2 \int_0^L \frac{\tilde{Q}_2^M Q_2^M}{GA} dx_1 + K_3 \int_0^L \frac{\tilde{Q}_3^M Q_3^M}{GA} dx_1 \right] \tag{19.74}$$

where $N^M(x_1)$, $Q_n^M(x_1)$ $(n = 2, 3)$ and $M_m^M(x_1)$ $(m = 1, 2, 3)$ are the components of internal action acting on a cross section of the model when it is subjected to basic nodal actions $\{a\}$ at its end k; and $\tilde{N}^M(x_1)$, $\tilde{Q}_n^M(x_1)$ $(n = 2, 3)$ and $\tilde{M}_m^M(x_1)(m = 1, 2, 3)$ are the components of internal action acting on a cross section of the model when it is subjected to basic nodal actions $\{\tilde{a}\}$ at its end k. In Sec. 18.5 we derive the principle of virtual work (18.26) for framed structures having straight elements made from an isotropic linearly elastic material. However, the response of curved elements having a small thickness to radius-of-curvature ratio may be approximated by that of straight elements. Consequently, we can deduce that the principle of virtual work [(18.26) or (19.74)] is also used for curved elements having a small thickness to radius-of-curvature ratio and made from an isotropic linearly elastic material. For such elements the following substitutions must be made in relation (19.74): x_1 becomes s, and x_3 becomes ζ; subscript 1 becomes t, and subscript 3 becomes ζ. The coordinate s is measured along the axis of the element. The coordinate ζ is measured along the normal to the axis of the element and is considered positive in the direction of the center of curvature. L is the total length of the axis of the element.

Relation (19.74) can be rewritten as

$$\{\tilde{a}\}^T[f]\{a\} = \int_0^L \{\widetilde{IA^M}\}^T[FL]\{IA^M\}\, dx_1 \tag{19.75}$$

where $\{IA^M\}$ is the matrix of internal actions acting on a cross section of the model when it is subjected to the basic nodal actions $\{a\}$ at its end k, and $\{\widetilde{IA^M}\}$ is the matrix of internal actions acting on a cross section of the model when it is subjected to the basic nodal actions $\{\tilde{a}\}$ at its end k.

The internal actions acting on a cross section of the model for the element can be expressed in terms of its basic nodal actions by considering the equilibrium of a segment of the model. That is,

$$\{IA^M\} = [Y^M]\{a\} \tag{19.76}$$

$$\{\widetilde{IA^M}\} = [Y^M]\{\tilde{a}\} \tag{19.77}$$

Substituting relations (19.76) and (19.77) into (19.75), we obtain

$$\{\tilde{a}\}^T[f]\{a\} = \int_0^L \{\tilde{a}\}^T[Y^M]^T[FL][Y^M]\{a\}\, dx_1$$

$$= \{\tilde{a}\}^T\left[\int_0^L [Y^M]^T[FL][Y^m]\, dx_1\right]\{a\} \tag{19.78}$$

Hence,

$$[f] = \int_0^L [Y^M]^T[FL][Y^M]\, dx_1 \tag{19.79}$$

In what follows we give the matrices $\{IA^M\}$, $[FL]$, and $[f]$ for the models of the types of the elements which we are considering.

Axial deformation element

$$\{IA^M\} = N^M(x_1) \qquad \{a\} = F_1^k \tag{19.80}$$

$$[FL] = \frac{1}{EA} \tag{19.81}$$

From the equilibrium of a segment of the element we have

$$\{IA^M\} = N^M = F_1^k \tag{19.82}$$

Thus,
$$[Y^M] = 1 \tag{19.83}$$

Substituting relations (19.81) and (19.83) into (19.79), we obtain

$$[f] = \int_0^L \frac{dx_1}{EA} \tag{19.84}$$

General planar element including the effect of shear deformation

$$\{IA^M\} = \begin{Bmatrix} N^M(x_1) \\ Q_2^M(x_1) \\ M_3^M(x_1) \end{Bmatrix} \qquad \{a\} = \begin{Bmatrix} F_1^k \\ F_2^k \\ M_3^k \end{Bmatrix} \tag{19.85}$$

$$[FL] = \begin{bmatrix} \dfrac{1}{EA} & 0 & 0 \\ 0 & \dfrac{K_2}{GA} & 0 \\ 0 & 0 & \dfrac{1}{EI_3} \end{bmatrix} \qquad 0 \tag{19.86}$$

Considering the equilibrium of the segment of the element shown in Fig. 19.6, we have

$$\begin{Bmatrix} N^M(x_1) \\ Q_2^M(x_1) \\ M_3^M(x_1) \end{Bmatrix} = \begin{Bmatrix} F_1^k \\ F_2^k \\ M_3^k + (L - x_1)F_2^k \end{Bmatrix} = \begin{bmatrix} 1 & 0 & 0 \\ 0 & 1 & 0 \\ 0 & L - x_1 & 1 \end{bmatrix} \begin{Bmatrix} F_1^k \\ F_2^k \\ M_3^k \end{Bmatrix} \tag{19.87}$$

Thus,
$$[Y^M] = \begin{bmatrix} 1 & 0 & 0 \\ 0 & 1 & 0 \\ 0 & L - x_1 & 1 \end{bmatrix} \tag{19.88}$$

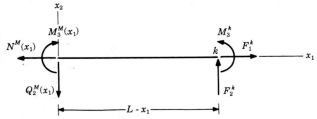

Figure 19.6 Free-body diagram of a segment of the model of a general planar element.

Substituting relations (19.86) and (19.88) into (19.79), we get

$$[f] = \int_0^L \begin{bmatrix} \dfrac{1}{EA} & 0 & 0 \\[2ex] 0 & \left[\dfrac{K_2}{GA} + \dfrac{(L-x_1)^2}{EI_3}\right] & \dfrac{L-x_1}{EI_3} \\[2ex] 0 & \dfrac{(L-x_1)}{EI_3} & \dfrac{I}{EI_3} \end{bmatrix} dx_1 \qquad (19.89)$$

The term K_2/GA represents the effect of shear deformation. The factor K_2 depends on the geometry of the cross section of the element and is defined by relation (18.27). Relation (19.89) can be integrated to yield the flexibility matrix for a general planar element of any geometry (see Secs. 19.4.2 to 19.4.4).

General space element including the effect of shear deformation

$$\{IA^M\} = \begin{Bmatrix} N^M(x_1) \\ Q_2^M(x_1) \\ Q_3^M(x_1) \\ M_1^M(x_1) \\ M_2^M(x_1) \\ M_3^N(x_1) \end{Bmatrix} \qquad \{a\} = \begin{Bmatrix} F_1^k \\ F_2^k \\ F_3^k \\ M_1^k \\ M_2^k \\ M_3^k \end{Bmatrix} \qquad (19.90)$$

$$[FL] = \begin{bmatrix} \dfrac{1}{EA} & 0 & 0 & 0 & 0 & 0 \\[2ex] 0 & \dfrac{K_2}{GA} & 0 & 0 & 0 & 0 \\[2ex] 0 & 0 & \dfrac{K_3}{GA} & 0 & 0 & 0 \\[2ex] 0 & 0 & 0 & \dfrac{1}{KG} & 0 & 0 \\[2ex] 0 & 0 & 0 & 0 & \dfrac{1}{EI_2} & 0 \\[2ex] 0 & 0 & 0 & 0 & 0 & \dfrac{1}{EI_3} \end{bmatrix} \qquad (19.91)$$

Considering the equilibrium of a segment of a general space element we obtain

$$
\{IA^M\} =
\begin{Bmatrix}
N^M(x_1) \\
Q_2^M(x_1) \\
Q_3^M(x_1) \\
M_1^M(x_1) \\
M_2^M(x_1) \\
M_3^M(x_1)
\end{Bmatrix}
=
\begin{Bmatrix}
F_1^k \\
F_2^k \\
F_3^k \\
M_1^k \\
M_2^k - (L - x_1)F_3^k \\
M_3^k + (L - x_1)F_2^k
\end{Bmatrix}
$$

$$
=
\begin{bmatrix}
1 & 0 & 0 & 0 & 0 & 0 \\
0 & 1 & 0 & 0 & 0 & 0 \\
0 & 0 & 1 & 0 & 0 & 0 \\
0 & 0 & 0 & 1 & 0 & 0 \\
0 & 0 & -(L - x_1) & 0 & 1 & 0 \\
0 & (L - x_1) & 0 & 0 & 0 & 1
\end{bmatrix}
\begin{Bmatrix}
F_1^k \\
F_2^k \\
F_3^k \\
M_1^k \\
M_2^k \\
M_3^k
\end{Bmatrix}
\quad (19.92)
$$

Thus,

$$
[Y^M] =
\begin{bmatrix}
1 & 0 & 0 & 0 & 0 & 0 \\
0 & 1 & 0 & 0 & 0 & 0 \\
0 & 0 & 1 & 0 & 0 & 0 \\
0 & 0 & 0 & 1 & 0 & 0 \\
0 & 0 & -(L - x_1) & 0 & 1 & 0 \\
0 & (L - x_1) & 0 & 0 & 0 & 1
\end{bmatrix}
\quad (19.93)
$$

Substituting relation (19.91) and (19.93) into (19.79), we obtain

$$
[f] = \int_0^L
\begin{bmatrix}
\dfrac{1}{EA} & 0 & 0 & 0 & 0 & 0 \\[2ex]
0 & \left[\dfrac{K_2}{GA} + \dfrac{(L - x_1)^2}{EI_3}\right] & 0 & 0 & 0 & \dfrac{L - x_1}{EI_3} \\[2ex]
0 & 0 & \left[\dfrac{K_3}{GA} + \dfrac{(L - x_1)^2}{EI_2}\right] & 0 & -\dfrac{L - x_1}{EI_2} & 0 \\[2ex]
0 & 0 & 0 & \dfrac{1}{KG} & 0 & 0 \\[2ex]
0 & 0 & -\dfrac{L - x_1}{EI_2} & 0 & \dfrac{1}{EI_2} & 0 \\[2ex]
0 & \dfrac{L - x_1}{EI_3} & 0 & 0 & 0 & \dfrac{1}{EI_3}
\end{bmatrix}
dx_1
$$

$$(19.94)$$

The terms K_2/GA and K_3/GA represent the effect of shear deformation. The factors K_2 and K_3 depend on the geometry of the cross section of the element and are defined by relation (18.27). Relation (19.94) can be integrated to yield the flexibility matrix of a general space element of any geometry (see Sec. 19.4.2).

19.4.2 The flexibility matrix of elements of constant cross section including the effect of shear deformation

In this section we establish the flexibility matrix for general planar and general space elements of constant cross section made of an isotropic linearly elastic material, by integrating relations (19.89) and (19.94), respectively.

General planar element

$$[f] = \begin{bmatrix} \dfrac{L}{EA} & 0 & 0 \\[3mm] 0 & \dfrac{K_2 L}{GA} + \dfrac{L^3}{3EI_3} & \dfrac{L^2}{2EI_3} \\[3mm] 0 & \dfrac{L^2}{2EI_3} & \dfrac{L}{EI_3} \end{bmatrix} \tag{19.95}$$

The term $K_2 L/GA$ represents the effect of shear deformation.

General space element

$$[f] = \begin{bmatrix} \dfrac{L}{EA} & 0 & 0 & 0 & 0 & 0 \\[3mm] 0 & \left[\dfrac{K_2 L}{GA} + \dfrac{L^3}{3EI_3}\right] & 0 & 0 & 0 & \dfrac{L^2}{2EI_3} \\[3mm] 0 & 0 & \left[\dfrac{K_3 L}{GA} + \dfrac{L^3}{3EI_2}\right] & 0 & -\dfrac{L^2}{2EI_2} & 0 \\[3mm] 0 & 0 & 0 & \dfrac{L}{KG} & 0 & 0 \\[3mm] 0 & 0 & -\dfrac{L^2}{2EI_2} & 0 & \dfrac{L}{EI_2} & 0 \\[3mm] 0 & \dfrac{L^2}{2EI_3} & 0 & 0 & 0 & \dfrac{L}{EI_3} \end{bmatrix} \tag{19.96}$$

19.4.3 The flexibility matrix for tapered elements

In this section we establish formulas for the flexibility coefficients of the three tapered general planar elements whose geometries are specified in Fig. 19.7. The elements are made of an isotropic linearly elastic material.

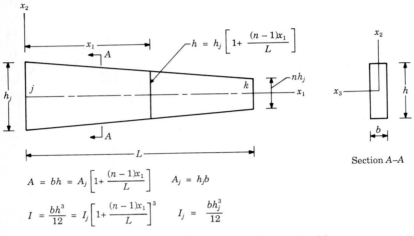

$$A = bh = A_j \left[1 + \frac{(n-1)x_1}{L}\right] \qquad A_j = h_j b$$

$$I = \frac{bh^3}{12} = I_j \left[1 + \frac{(n-1)x_1}{L}\right]^3 \qquad I_j = \frac{bh_j^3}{12}$$

(a) Element of rectangular cross section of constant width

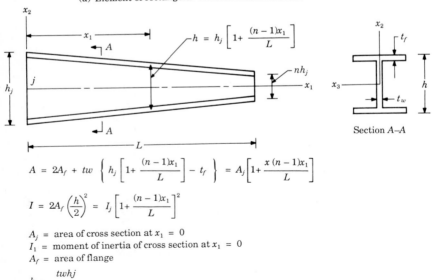

$$A = 2A_f + tw \left\{ h_j \left[1 + \frac{(n-1)x_1}{L}\right] - t_f \right\} = A_j \left[1 + \frac{x(n-1)x_1}{L}\right]$$

$$I = 2A_f \left(\frac{h}{2}\right)^2 = I_j \left[1 + \frac{(n-1)x_1}{L}\right]^2$$

A_j = area of cross section at $x_1 = 0$
I_1 = moment of inertia of cross section at $x_1 = 0$
A_f = area of flange

$$k = \frac{twhj}{A_j}$$

(b) Element of I cross section whose flanges have constant dimensions

Figure 19.7 Description of tapered elements.

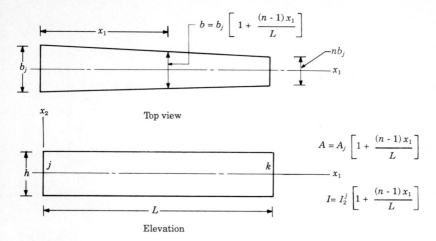

Elevation

(c) Element of rectangular cross section of constant depth and variable width

Figure 19.7 (*Continued*)

The area and moment of inertia I_3 of the elements of Fig. 19.7 can be written as

$$A = A_j + C\xi$$
$$I_3 = I_3^{(j)}[1 + (n-1)\xi]^q \qquad q = 1, 2, 3 \tag{19.97}$$

where

$$\xi = \frac{x_1}{L} \tag{19.98}$$

Using relation (19.98) and noting that $dx_1 = L\,d\xi$, relation (19.89) can be written as

$$[f] = \int_0^1 \begin{bmatrix} \dfrac{1}{EA} & 0 & 0 \\[2ex] 0 & \left[\dfrac{K_2}{GA} + \dfrac{L^2(1-\xi)^2}{EI_3}\right] & \dfrac{L(1-\xi)}{EI_3} \\[2ex] 0 & \dfrac{L(1-\xi)}{EI_3} & \dfrac{1}{EI_3} \end{bmatrix} L\,d\xi \tag{19.99}$$

In order to evaluate the integral in relation (19.99) for the elements whose geometries are specified in Fig. 19.7 we need the following integration formulas.

$$\int_0^1 \frac{d\xi}{A_j + C\xi} = \frac{\ln(1 + C/A_j)}{C} \tag{19.100a}$$

$$\int_0^1 \frac{d\xi}{I_3^{(j)}[1 + (n-1)\xi]} = \frac{\ln(n)}{I_3^{(j)}(n-1)} \tag{19.100b}$$

$$\int_0^1 \frac{d\xi}{I_3^{(j)}[1 + (n-1)\xi]^2} = \frac{1}{I_3^{(j)}n} \tag{19.100c}$$

$$\int_0^1 \frac{d\xi}{I_3^{(j)}[1 + (n-1)\xi]^3} = \frac{(n+1)}{2I_3^{(j)}n^2} \tag{19.100d}$$

$$\int_0^1 \frac{\xi\,d\xi}{I_3^{(j)}[1 + (n-1)\xi]} = \frac{n - 1 - \ln(n)}{I_3^{(j)}(n-1)^2} \tag{19.100e}$$

$$\int_0^1 \frac{\xi\,d\xi}{I_3^{(j)}[1 + (n-1)\xi]^2} = \frac{n\ln(n) - n + 1}{I_3^{(j)}n(n-1)^2} \tag{19.100f}$$

$$\int_0^1 \frac{\xi\,d\xi}{I_3^{(j)}[1 + (n-1)\xi]^3} = \frac{1}{2I_3^{(j)}n^2} \tag{19.100g}$$

$$\int_0^1 \frac{\xi^2\,d\xi}{I_3^{(j)}[1 + (n-1)\xi]} = \frac{3 + n^2 + 2\ln(n) - 4n}{2I_3^{(j)}(n-1)^3} \tag{19.100h}$$

$$\int_0^1 \frac{\xi^2\,d\xi}{I_3^{(j)}[1 + (n-1)\xi]^2} = \frac{n^2 - 1 - 2n\ln(n)}{I_3^{(j)}n(n-1)^3} \tag{19.100i}$$

$$\int_0^1 \frac{\xi^2\,d\xi}{I_3^{(j)}(1 + (n-1)\xi]^3} = \frac{-3n^2 + 4n - 1 + 2n^2\ln(n)}{2I_3^{(j)}n^2(n-1)^3} \tag{19.100j}$$

The flexibility coefficients for the elements of Fig. 19.7 are given in Table 19.1.

19.4.4 The flexibility matrix for a curved element of constant cross section

In this section we establish the flexibility coefficients for the curved element shown in Fig. 19.8. The axis of this element is an arc of a circle. The flexibility equations for the element of Fig. 19.8 can be established by referring to the model of Fig. 19.9. This model is made from the element with its end j fixed and is subjected on its end k to the basic nodal actions F_t^k, F_ξ^k, and M_2^k of the element (see Sec. 5.3).

Consider the free-body diagram of the segment of the element shown in Fig. 19.9b. From the equilibrium of this segment we have

$$\sum F_t = 0 \qquad N = -F_\xi^k \sin\alpha + F_t^k \cos\alpha$$

$$\sum F_\xi = 0 \qquad Q_r = F_\xi^k \cos\alpha + F_t^k \sin\alpha$$

TABLE 19.1 Flexibility Coefficients of Tapered General Planar Elements

Type of element	$[f]$
Tapered element of rectangular cross section of constant width b $A_j = h_j b$ $I_j = \dfrac{b\,h_j^3}{12}$	$\begin{bmatrix} -\dfrac{L\ln(n)}{EA_j(1-n)} & 0 & 0 \\[2ex] 0 & -\dfrac{K_2 L\ln(n)}{GA_j(n-1)} + \dfrac{L^3[n^2-4n+3+2\ln(n)]}{2EI_3^{(j)}(n-1)^3} & \dfrac{L^2}{2nEI_3^{(j)}} \\[2ex] 0 & \dfrac{L^2}{2nEI_3^{(j)}} & \dfrac{L(n+1)}{2n^2EI_3^{(j)}} \end{bmatrix}$
Tapered element of I cross section having a flange of constant dimensions $c = t_w h_j(n-1)$	$\begin{bmatrix} \dfrac{L\ln(1+c/A_j)}{EC} & 0 & 0 \\[2ex] 0 & \dfrac{K_2 L\ln(1+c/A_j)}{Gc} + \dfrac{L^3[n^2-1-2n\ln(n)]}{EI_3^{(j)}(n-1)^3} & \dfrac{L^2[n-1-\ln(n)]}{EI_3^{(j)}(n-1)^2} \\[2ex] 0 & \dfrac{L^2[n-1-\ln(n)]}{EI_3^{(j)}(n-1)^2} & \dfrac{L}{EI_3^{(j)}n} \end{bmatrix}$
Tapered element of rectangular cross section of constant depth and linearly varying width	$\begin{bmatrix} -\dfrac{L\ln(n)}{EA_j(1-n)} & 0 & 0 \\[2ex] 0 & \dfrac{K_2 L\ln(2-n)}{GA_j(1-n)} + \dfrac{L^3[-3n+4n-1+2n^2\ln(n)]}{2EI_3^{(j)}(n-1)^3} & \dfrac{L^2[1-n+n\ln(n)]}{EI_3^{(j)}(n-1)^2} \\[2ex] 0 & \dfrac{L^2[1-n+n\ln(n)]}{EI_3^{(j)}(n-1)^2} & \dfrac{L\ln(n)}{EI_3^{(j)}(n-1)} \end{bmatrix}$

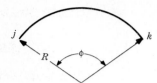

Figure 19.8 Geometry of curved element.

$$\sum M^{(0)} = 0 \qquad M_2 = M_2^k - RF_t^k + RN$$

$$= M_2^k - R(1 - \cos \alpha)F_t^k - R \sin \alpha F_\zeta^k$$

or

$$\{IA^M\} = \begin{Bmatrix} N^M(a) \\ Q_\zeta^M(a) \\ M_2^M(a) \end{Bmatrix} = \begin{bmatrix} \cos \alpha & -\sin \alpha & 0 \\ \sin \alpha & \cos \alpha & 0 \\ -R(1 - \cos \alpha) & -R \sin \alpha & 1 \end{bmatrix} \begin{Bmatrix} F_t^k \\ F_\zeta^k \\ M_2^k \end{Bmatrix}$$

$$(19.101)$$

Thus,

$$[Y^M] = \begin{bmatrix} \cos \alpha & -\sin \alpha & 0 \\ \sin \alpha & \cos \alpha & 0 \\ -R(1 - \cos \alpha) & -R \sin \alpha & 1 \end{bmatrix} \qquad (19.102)$$

Moreover,

$$[FL] = \begin{bmatrix} \dfrac{1}{EA} & 0 & 0 \\ 0 & \dfrac{K_2}{GA} & 0 \\ 0 & 0 & \dfrac{1}{EI_2} \end{bmatrix} \qquad (19.103)$$

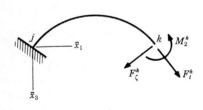

(a) Model of the element

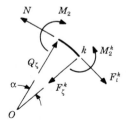

(b) Free-body diagram of a segment of the mode

Figure 19.9 Model for establishing the flexibility coefficients of the curved element of Fig. 19.8.

Substituting relations (19.102) and (19.103) into (19.79) and noting that for curved elements $dx_1 \to ds = R\,d\alpha$, we have

$$[f] =$$

$$
\begin{bmatrix}
\dfrac{\cos^2\alpha}{EA} + \dfrac{\sin^2\alpha K_2}{GA} & -\dfrac{\sin\alpha\cos\alpha}{EA} + \dfrac{K_2\sin\alpha\cos\alpha}{GA} & -\dfrac{R(1-\cos\alpha)}{EI_2} \\
\quad + \dfrac{R^2(1-\cos\alpha)^2}{EI_2} & \quad + \dfrac{R^2(1-\cos\alpha)\sin\alpha}{EI_2} & \\[2ex]
-\dfrac{\sin\alpha\cos\alpha}{EA} & \dfrac{\sin^2\alpha}{EA} + \dfrac{K_2\cos^2\alpha}{GA} & -\dfrac{R\sin\alpha}{EI_2} \\
\quad + \dfrac{K_2\sin\alpha\cos\alpha}{GA} & \quad + \dfrac{R^2\sin^2\alpha}{EI_2} & \\
\quad + \dfrac{R^2(1-\cos\alpha)\sin\alpha}{EI_2} & & \\[2ex]
-\dfrac{R(1-\cos\alpha)}{EI_2} & -\dfrac{R\sin\alpha}{EI_2} & \dfrac{1}{EI_2}
\end{bmatrix}
$$

$$(19.104)$$

Integrating relation (19.104) we obtain

$$[f] =$$

$$
\begin{bmatrix}
\dfrac{2\phi+\sin 2\phi}{4EA} & \dfrac{\cos 2\phi-1}{4EA} & -\dfrac{R}{EI_2}(\phi-\sin\phi) \\
\quad + \dfrac{K_2}{4GA}(2\phi-\sin 2\phi) & \quad -\dfrac{K_2}{4GA}(\cos 2\phi-1) & \\
\quad + \dfrac{R^2}{4EI_2}(6\phi+\sin^2\phi-8\sin\alpha) & \quad + \dfrac{R^2}{4EI_2}(-4\cos\phi+\cos 2\phi+3) & \\[2ex]
\dfrac{\cos 2\phi-1}{4EA} - \dfrac{K_2(\cos 2\phi-1)}{4GA} & \dfrac{2\Phi-\sin^2\phi}{4EA} & \dfrac{R}{EI_2}(\cos\phi-1) \\
\quad + \dfrac{R^2}{4EI_2}(-4\cos\phi+\cos 2\phi+3) & \quad + \dfrac{K_2}{4GA}(2\phi+\sin 2\phi) & \\
 & \quad + \dfrac{R^2}{4EI_2}(2\phi-\sin 2\phi) & \\[2ex]
-\dfrac{R}{EI_2}(\phi-\sin\phi) & \dfrac{R}{EI_2}(\cos\phi-1) & \dfrac{\phi}{EI_2}
\end{bmatrix}
$$

$$(19.105)$$

19.5 The Stiffness Matrix for a General Planar Element of Constant Cross Section Which Includes the Effect of Shear Deformation

We can obtain the stiffness matrix for an element from its flexibility matrix using relation (5.93). Thus substituting the inverse of the flexibility matrix for a general planar element which includes the effect of shear deformation (19.95) and the matrix $[T]$ given in relation (5.11) into (5.93) we get

$$
[K] = \begin{bmatrix} -1 & 0 & 0 \\ 0 & -1 & 0 \\ 0 & -L & -1 \\ 1 & 0 & 0 \\ 0 & 1 & 0 \\ 0 & 0 & 1 \end{bmatrix} \frac{EI_3}{L(L^2/12+\eta)} \begin{bmatrix} \dfrac{A}{I_3}\left(\dfrac{L^2}{12}+\eta\right) & 0 & 0 \\ 0 & 1 & -\dfrac{L}{2} \\ 0 & -\dfrac{L}{2} & \left(\dfrac{L^2}{3}+\eta\right) \end{bmatrix}
$$

$$
\times \begin{bmatrix} -1 & 0 & 0 & 1 & 0 & 0 \\ 0 & -1 & -L & 0 & 1 & 0 \\ 0 & 0 & -1 & 0 & 0 & 1 \end{bmatrix}
$$

$$
= \frac{EI_3}{L(L^2/12+\eta)}
$$

$$
\times \begin{bmatrix} \dfrac{A}{I_3}\left(\dfrac{L^2}{12}+\eta\right) & 0 & 0 & -\dfrac{A}{I_3}\left(\dfrac{L^2}{12}+\eta\right) & 0 & 0 \\[2mm] 0 & 1 & \dfrac{L}{2} & 0 & -1 & \dfrac{L}{2} \\[2mm] 0 & \dfrac{L}{2} & \left(\dfrac{L^2}{3}+\eta\right) & 0 & -\dfrac{L}{2} & \left(\dfrac{L^2}{6}-\eta\right) \\[2mm] -\dfrac{A}{I_3}\left(\dfrac{L^2}{12}+\eta\right) & 0 & 0 & \dfrac{A}{I_3}\left(\dfrac{L^2}{12}+\eta\right) & 0 & 0 \\[2mm] 0 & -1 & -\dfrac{L}{2} & 0 & 1 & -\dfrac{L}{2} \\[2mm] 0 & \dfrac{L}{2} & \left(\dfrac{L^2}{6}-\eta\right) & 0 & -\dfrac{L}{2} & \left(\dfrac{L^2}{3}+\eta\right) \end{bmatrix}
$$

$$\text{(19.106)}$$

where

$$
\eta = \frac{EI_3 K_2}{GA} \qquad \text{(19.107)}
$$

represents the effect of shear deformation.

Application of the Principle of Virtual Work to the Analysis of Framed Structures

20.1 The Dummy Load Method

In this section we describe† the dummy load method, also known as the method of virtual work, and we apply it to an example. This method has been used extensively in classical structural analysis in establishing the following.

1. The component of translation u_m in the direction of the unit vector **m**, of any point A of a framed structure whose elements are made of isotropic linearly elastic materials.

2. The component of rotation θ_m about an axis specified by the unit vector **m** at any point A of a framed structure whose elements are made of isotropic linearly elastic materials.

The component of displacement u_m and the component of rotation θ_m at a point of a structure may be due to external actions, to a change of temperature, to lack of fit of some elements (initial strain), or to specified movement of the supports of the structure. The structure may be statically determinate or statically indeterminate.

In order to establish the component of displacement u_m, or the

---

† For a more detailed description and many examples see A. E. Armenàkas, *Classical Structural Analysis: A Modern Approach,* McGraw-Hill Book Co., New York, 1988, Chap. 5.

component of rotation θ_m at any point A of a structure, we introduce in the principle of virtual work the following.

1. As admissible displacement and strain fields we choose the actual (real) displacement field of the structure subjected to the given loading. With this choice $\{\tilde{\Delta}^S\}$ is the matrix of the specified components of displacements of the supports of the structure.

2. As the loading and corresponding statically admissible distribution of internal actions $\tilde{N}(x_1)$, $\tilde{Q}_2(x_1)$, $\tilde{Q}_3(x_1)$, $\tilde{M}_1(x_1)$, $\tilde{M}_2(x_1)$, $\tilde{M}_3(x_1)$ in the elements of the structure, we choose a virtual loading and a corresponding statically admissible distribution of internal actions in the elements of the structure. If we want to establish the component of translation u_m in the direction of the unit vector $\mathbf{m}$ at point A of the structure, the "virtual" loading consists of a unit force acting at point A in the direction of the unit vector $\mathbf{m}$. If we want to establish the component of rotation θ_m in the direction of the unit vector $\mathbf{m}$ at point A of the structure, the "virtual" loading consists of a unit moment acting at point A in the direction of the unit vector $\mathbf{m}$.

In what follows we limit our attention to structures made of isotropic linearly elastic materials. For the "virtual" loading described in item 2 above, relation (18.23) gives

$$
\begin{aligned}
1xd + \{\Delta^S\}^T\{\tilde{R}\} = \sum_{e=1}^{ME} \Bigg[&\int_0^{L_e} \left(\frac{N\tilde{N}}{EA} + \frac{M_2\tilde{M}_2}{EI_2} + \frac{M_3\tilde{M}_3}{EI_3} + \frac{M_1\tilde{M}_1}{KG} \right) dx_1 \\
&+ K_2 \int_0^{L_e} \frac{Q_2\tilde{Q}_2}{GA} dx_1 + K_3 \int_0^{L_e} \frac{Q_3\tilde{Q}_3}{GA} dx_1 \\
&+ \int_0^{L_e} (H_1\tilde{N} + H_2\tilde{M}_2 + H_3\tilde{M}_3) dx_1 \Bigg]^{(e)}
\end{aligned}
\tag{20.1}
$$

where d is either u_m or θ_m; the superscript (e) indicates that the quantities of the terms inside the brackets pertain to element e; $\{\Delta^S\}$ is the matrix of the specified components of displacements of the supports of the structure; $\{\tilde{R}\}$ is the matrix of statically admissible reactions of the supports of the structure subjected to the "virtual" loading; H_1, H_2, H_3 are specified by relations (2.37), (2.45a), and (2.45b), respectively; and K_2 and K_3 are defined by relations (18.27).

On the basis of the foregoing discussion, we adhere to the following steps in order to compute a component of translation or rotation of a point of a framed structure made of an isotropic linearly elastic material.

Step 1. We establish the actual components of internal actions $N(x_1)$, $Q_2(x_1)$, $Q_3(x_1)$, $M_1(x_1)$, $M_2(x_1)$, $M_3(x_1)$ in the elements of the structure as functions of their axial coordinate. For statically determined structures this can be done by considering the equilibrium of appropriate segments of each element. For statically indeterminate structures this can be done by analyzing the structure using one of the classical methods (a force method or a displacement method).[1]

Step 2. We subject the structure to the "virtual" loading described previously and establish a set of statically admissible reactions $\{\tilde{R}\}$ of the structure and corresponding internal actions $\tilde{N}(x_1)$, $\tilde{Q}_2(x_1)$, $\tilde{Q}_3(x_1)$, $\tilde{M}_1(x_1)$, $\tilde{M}_2(x_1)$, $\tilde{M}_3(x_1)$ of the elements of the structure.

Step 3. We substitute in relation (20.1) the following:

1. The reactions $\{\tilde{R}\}$ established in step 2.
2. The components of internal actions established in steps 1 and 2.
3. The given movement $\{\Delta^S\}$ of the supports of the structure.
4. The given changes of temperature ΔT_c, δT_2, δT_3 of the elements of the structure.
5. The given initial strain e_{11}^{IC}, k_2^I, k_3^I of the elements of the structure.

This step gives the desired component of translation or rotation denoted by d in relation (20.1).

In what follows we illustrate the dummy load method with an example.

Example. Compute the vertical component of translation (deflection) and the rotation at point A of the beam shown in Fig. a due to the following loading.

1. The forces shown in Fig. a.
2. Temperature of the top fibers of 30°C. Temperature of the bottom fibers of -10°C.

The beam is made of an isotropic linearly elastic material ($E = 210 \text{ kN/mm}^2$, $\alpha = 10^{-5}/°\text{C}$) and has a constant cross section ($I = 200 \times 10^6 \text{ mm}^4$, height of cross section = 200 mm).

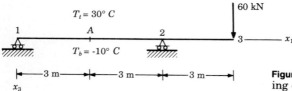

Figure a Geometry and loading of the beam.

solution

STEP 1.   We compute the internal moments in each element of the beam as functions of its axial coordinate. The beam of Fig. a is statically determinate. Its reactions can be computed by referring to its free-body diagram shown in Fig. b and considering its equilibrium. that is,

$$\sum M_2^{(2)} = 0 \qquad 6R_3^{(2)} - 60(9) = 0 \qquad \text{or} \qquad R_3^{(2)} = 90 \text{ kN}$$

$$\sum F_3 = 0 \qquad 60 - 90 - R_3^{(1)} = 0 \qquad \text{or} \qquad R_3^{(1)} = -30 \text{ kN}$$

Referring to Fig. b the moment in the beam is

$$M_2 = \begin{cases} -30x_1 & 0 \leq x_1 \leq 6 \\ -60x_1' & 0 \leq x_1' \leq 3 \end{cases} \tag{a}$$

The difference in temperature between the bottom and top fibers is

$$\delta T = -10 - 30 = -40°C$$

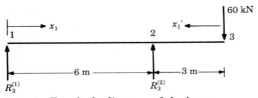

Figure b   Free-body diagram of the beam.

Computation of the Component of Translation $u_3^{(A)}$

STEP 2.   We compute the internal moments in each element of the beam subjected to the "virtual" loading as functions of its axial coordinate. The beam subjected to the "virtual" loading for computing the vertical component of translation of point A is shown in Fig. c. Referring to this figure we have

$$M_2 = \begin{cases} 0.5x_1 & 0 \leq x_1 \leq 3 \\ 0.5x_1 - (x_1 - 3) & 3 \leq x_1 \leq 6 \\ 0 & 0 \leq x_1' \leq 3 \end{cases} \tag{b}$$

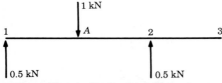

Figure c   Free-body diagram of the beam subjected to the "virtual" loading for computing $u_3^{(A)}$.

STEP 3. We compute the component of translation u_3. Substituting relations (a) and (b) into (20.1) we obtain

$$u_3^{(A)} = \int_0^3 \frac{M_2 \tilde{M}_2}{EI_2} dx_1 + \int_3^6 \frac{M_2 \tilde{M}_2}{EI_2} dx_1 + \int_0^3 \frac{\alpha \tilde{M}_2 \, \delta T}{h} dx_1 + \int_3^6 \frac{\alpha \tilde{M}_2 \, \delta T}{h} dx_1$$

$$= \frac{1}{EI} \left[-\int_0^3 30x_1(0.5x_1) \, dx_1 - \int_3^6 30x_1[0.5x_1 - (x_1 - 3)] \, dx_1 \right.$$

$$\left. + \frac{\alpha \, \delta T}{h} \left[\int_0^3 0.5x_1 \, dx_1 + \int_3^6 [0.5x_1 - (x_1 - 3)] \, dx_1 \right] \right]$$

$$= -\frac{405}{EI} + \frac{\alpha \, \delta T(4.5)}{h} = -\frac{405}{(210)(10^6)(200)(10^{-6})} + \frac{10^{-5}(-40)4.5}{0.2}$$

$$= -0.00964 - 0.00900 = -0.01864 \text{ m} = 18.64 \text{ mm upward}$$

The minus sign indicates that the direction of $u_3^{(A)}$ is opposite to the direction of the "virtual" loading. The deformed configuration (elastic curve) of the beam is shown in Fig. d.

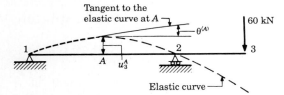

Figure d Deformed configuration of the beam.

Computation of the Rotation of the Beam at Point A

STEP 2. We compute the internal moments in each element of the beam subjected to the "virtual" loading as functions of its axial coordinate. The beam subjected to the "virtual" loading for computing the rotation at point A is shown in Fig. e. Referring to this figure we have

$$M_2 = \begin{cases} \dfrac{x_1}{6} & 0 < x_1 < 3 \\[2mm] \dfrac{x_1}{6} - 1 & 3 < x_1 < 6 \end{cases} \tag{c}$$

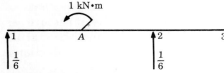

Figure e Free-body diagram of the beam subjected to the virtual loading for computing $\theta_2^{(A)}$.

STEP 3.   We compute the rotation $\theta_2^{(A)}$. Substituting relations (a) and (c) into (20.1) we get

$$\theta_2^{(A)} = \frac{1}{EI}\left[-\int_0^3 30x_1\left(\frac{x_1}{6}\right) dx_1 - \int_3^6 30x_1\left(\frac{x_1}{6}\right) dx_1 \right]$$

$$+ \frac{\alpha\, \delta T}{h}\left[\int_0^3 \frac{x_1}{6} dx_1 + \int_3^6 \left(\frac{x_1}{6} - 1\right) dx_1 \right]$$

$$= \frac{45}{EI} + \frac{\alpha\, \delta T(0)}{h} = \frac{45}{(210)(200)} = 0.00107 \text{ rad counterclockwise}$$

20.2 Derivation of the Stiffness Equations for a Structure Using the Principle of Virtual Work—The Method of Virtual Displacements

In Sec. 8.2 we derived the stiffness equations for a framed structure using directly the following three facts.

1. The nodes of a framed structure are in equilibrium.

2. The components of nodal displacements of each element of a framed structure are compatible with the components of displacements of the nodes to which the ends of this element are connected.

3. The response of each element of a framed structure is specified by its stiffness equations (4.8) and by its fixed-end actions. The response of an element is established using the stress–strain relations for the material from which the element is made and the fact that the element is in equilibrium. We limit our attention to structures whose elements are made from isotropic linearly elastic materials.

In this section we derive the stiffness equations for a structure using the principle of virtual work by adhering to the following steps.

Step 1.   We express the real and the virtual basic components of displacement $\{u^e\}$ and $\{\tilde{u}^e\}$ of each element of the structure as functions of its axial coordinate involving the real and the virtual global components of its nodal displacements using shape functions. That is, referring to relations (19.1) and (6.67) we have

$$\{u^e\} = [\phi^e]\{D^e\} = [\phi^e][\Lambda^e]\{\bar{D}^e\} \tag{20.2a}$$

$$\{\tilde{u}^e\} = [\phi^e]\{\tilde{D}^e\} = [\phi^e][\Lambda^e]\{\bar{\tilde{D}}^e\} \tag{20.2b}$$

For elements of constant cross section subjected only to nodal displacements, relation (20.2a) is exact. For other elements it is an approximation.

Step 2. We substitute relations (20.2) into the principle of virtual work (18.39) to obtain

$$
\sum_{e=1}^{NE} \left[\{\tilde{\bar{D}}\}^T [\Lambda]^T \int_0^{L_e} [[\mathscr{B}][\phi]]^T [ST][\mathscr{B}][\phi]\, dx_1 [\Lambda][\bar{D}] \right.
$$

$$
\left. - \{\tilde{\bar{D}}\}^T [\Lambda]^T \left[\int_0^{L_e} [[\mathscr{B}][\phi]]^T \{H\}\, dx_1 + \int_0^{L_e} [[\mathscr{L}][\phi]]^T \{L\}\, dx_1 \right] \right]^{(e)}
$$

$$
= \{\tilde{\hat{\Delta}}\}^T [\{\hat{P}^G\} + \{\hat{R}\}] \tag{20.3}
$$

where $\{\hat{R}\}$ is the matrix of the reactions of the structure conjugate to the matrix $\{\hat{\Delta}\}$. Its terms corresponding to components of displacements of the nodes of the structure which are not inhibited by its supports are equal to zero. The superscript e of relation (20.3) indicates that the quantities in these brackets pertain to element e. Using relations (6.65), (6.68), (6.74), (19.34), and (19.35), relation (20.3) can be rewritten as

$$
\sum_{e=1}^{NE} [\{\tilde{\bar{D}}\}^T [[\bar{K}]\{\bar{D}\} + \{\bar{A}^R\}]]^{(e)} = \{\tilde{\hat{\Delta}}\}^T [\{\hat{P}^G\} + \{\hat{R}\}] \tag{20.4}
$$

where $[\bar{K}] = [\Lambda]^T [K][\Lambda]$ (20.5)

$[K]$ = local stiffness matrix for an element of structure given by relation (19.34)

$\{\bar{A}^R\} = [\Lambda]^T \{A^R\}$ (20.6)

$[A^R]$ = local matrix of fixed-end actions of an element given by relation (19.35)

Relations (20.4) can be rewritten as

$$
\{\tilde{\bar{D}}\}^T [[\hat{K}]\{\hat{D}\} + \{\hat{A}^R\}] = \{\tilde{\hat{\Delta}}\}^T [\{\hat{P}^G\} + \{\hat{R}\}] \tag{20.7}
$$

where

$$
\{\tilde{\bar{D}}\}^T = [\{\tilde{\bar{D}}^1\}^T, \{\tilde{\bar{D}}^2\}^T, \ldots, \{\tilde{\bar{D}}^{NE}\}^T] \tag{20.8a}
$$

$$
\{\hat{D}\}^T = [\{\bar{D}^1\}^T, \{\bar{D}^2\}^T, \ldots, \{\bar{D}^{NE}\}^T] \tag{20.8b}
$$

$$
[\hat{K}] = \begin{bmatrix} [\bar{K}^1] & & & \\ & [\bar{K}^2] & & \\ & & \ddots & \\ & & & [\bar{K}^{NE}] \end{bmatrix} \tag{20.9}
$$

$$
\{\hat{A}^R\}^T = [\{\bar{A}^{R1}\}^T, \{\bar{A}^{R2}\}^T, \ldots, \{\bar{A}^{RNE}\}^T] \tag{20.10}
$$

Step 3. We express the components of nodal displacements of all the elements of the structure $\{\hat{D}\}$ in terms of the components of displacements of its nodes $\{\hat{\Delta}\}$. That is, referring to relations (8.6) and (8.8) we have

$$\{\hat{D}\} = \{\hat{C}\}\{\hat{\Delta}\} = [\hat{B}]^T\{\hat{\Delta}\}$$
$$\{\tilde{\hat{D}}\} = [\hat{C}]\{\tilde{\hat{\Delta}}\} = [\hat{B}]^T\{\tilde{\hat{\Delta}}\} \tag{20.11}$$

Substituting relation (20.11) into (20.7) we get

$$\{\tilde{\hat{\Delta}}\}^T[[\hat{B}][\hat{K}][\hat{B}]^T\{\hat{\Delta}\} + [\hat{B}]\{\hat{A}^R\}] = \tilde{\hat{\Delta}}\}^T[\{\hat{P}^G\} + \{\hat{R}\}] \tag{20.12}$$

Inasmuch as the above relation is valid for any choice of $\{\tilde{\hat{\Delta}}\}^T$ we have

$$\{\hat{P}^E\} + \{\hat{R}\} = [\hat{S}]\{\hat{\Delta}\} \tag{20.13}$$

where

$$\{\hat{P}^E\} = \{\hat{P}^G\} - [\hat{B}]\{\hat{A}^R\} \tag{20.14}$$

$$[\hat{S}] = [\hat{B}][\hat{K}][\hat{B}]^T \tag{20.15}$$

Equations (20.13) are the stiffness equations for the structure; $[\hat{S}]$ is the stiffness matrix for the structure [see relation (8.15)]; $\{\hat{P}^E\}$ is the matrix of equivalent actions to be applied to the nodes of the structure [see relation (8.13)]. As discussed in Sec. 8.2 the computation of the stiffness matrix for a structure with many nodes using relation (20.15) consumes large computer time and storage, and consequently it has found little practical use. In the direct stiffness method each term of the stiffness matrix for a structure is established directly from the terms of the global stiffness matrices of its elements. This approach is also employed in establishing the stiffness matrix of any domain using the finite-element method.

The stiffness equations (20.13) for a structure are exact (to within the accuracy of the mechanics of materials theories) if relation (20.2a) represents the exact solution of the displacement equations of equilibrium (2.40), (2.53), and (2.67) for each of its elements. This is the case for framed structures whose elements have constant cross sections and are not subjected to external disturbances along their length. If relation (20.2a) does not represent the exact solution of the displacement equations of equilibrium for some elements of a structure, the approach presented in this section gives approximate stiffness equations for this structure. The method for establishing the stiffness equations for framed structures presented in this section has found limited use in classical structural analysis for analyzing statically indeterminate trusses under the label method of virtual displacements. In what follows we employ the method of virtual displacements to analyze a statically indeterminate to the first degree truss.

Example. Using the method of virtual displacements, compute the components of translation of the nodes of the truss loaded as shown in Fig. a. Moreover, compute the internal forces in the elements of the truss. The elements of the truss are made of the same isotropic linearly elastic material and have the same constant cross section ($EA = 200 \times 10^4$ kN).

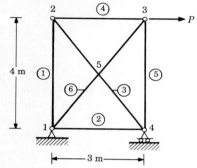

Figure a Geometry and loading of the truss.

---

solution

STEP 1. We express the axial component of translation of each element of the truss as a function of its axial coordinate involving its nodal translations. The elements of the truss have constant cross section. Consequently, the components of their nodal translations have the following form:

$$u_1^{(e)} = u_1^{ej} + \frac{x_1(u_1^{ek} - u_1^{ej})}{L_e} \qquad (e = 1, 2, \ldots, 6) \tag{a}$$

STEP 2. We establish the relations between the nodal translations of the elements of the truss and the components of translation of its nodes. As shown in Fig. b the truss under consideration has five degrees of freedom. In order that the nodal translations of the elements of the truss are compatible

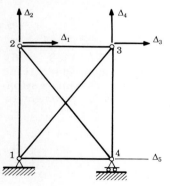

Figure b Components of the translation of the nodes of the truss.

to the components of translation of its nodes and, moreover, satisfy the conditions at its supports, the following relations must be valid:

$$u_1^{1j} = 0 \qquad\qquad u_1^{4j} = \Delta_1$$

$$u_1^{1k} = \Delta_2 \qquad\qquad u_1^{4k} = \Delta_3$$

$$u_1^{2j} = 0 \qquad\qquad u_1^{5j} = -\Delta_4$$

$$u_1^{2k} = \Delta_5 \qquad\qquad u_1^{5k} = 0 \qquad\qquad \text{(b)}$$

$$u_1^{3j} = -\frac{4\Delta_2}{5} + \frac{3\Delta_1}{5} \qquad u_1^{6j} = 0$$

$$u_1^{3k} = \frac{3\Delta_5}{5} \qquad\qquad u_1^{6k} = \frac{3\Delta_3}{5} + \frac{4\Delta_4}{5}$$

STEP 3.   We express the internal force in each element of the truss in terms of the components of translations of its nodes. Substituting relation (a) into (2.36) the internal forces in the elements of the truss are

$$N^{(e)} = \frac{EA(u_1^{ek} - u_1^{ej})}{L_e} \qquad\qquad \text{(c)}$$

STEP 4.   We assume that the virtual components of translation of each element of the truss has the following form

$$\tilde{u}_1^{(e)} = \tilde{u}^{ej} + \frac{x_1(\tilde{u}_1^{ek} - \tilde{u}_1^{ej})}{L_e} \qquad\qquad \text{(d)}$$

Moreover, we denote the corresponding components of displacement of the nodes of the truss by $\tilde{\Delta}_i$ $(i = 1, 2, \ldots, 5)$ (see Fig. b).

STEP 5.   We establish and solve the stiffness equations for the truss. Substituting relations (c) and (d) into the principle of virtual work (18.37) and integrating for the truss under consideration, we obtain

$$P\tilde{\Delta}_3 = \sum_{e=1}^{6} \int_0^{L_e} E_e A_e \left(\frac{u_1^{ek} - u_1^{ej}}{L_e}\right)\left(\frac{\tilde{u}_1^{ek} - \tilde{u}_1^{ej}}{L_e}\right) dx_1$$

$$= \sum_{e=1}^{6} E_e A_e \frac{(u_1^{ek} - u_1^{ej})(\tilde{u}_1^{ek} - \tilde{u}_1^{ej})}{L_e} \qquad\qquad \text{(e)}$$

The actual axial component of nodal translations u_1^{ej} and u_1^{ek} $(e = 1, 2, \ldots, 6)$ of the elements of the truss satisfy equation (e) for any admissible virtual translation field. We specify a virtual field for the truss by choosing a set of values of the components of translations of its nodes. It is convenient to choose the following five virtual translation fields:

$$\hat{\tilde{\Delta}}_1 \neq 0$$

$$\hat{\tilde{\Delta}}_2 = \hat{\tilde{\Delta}}_3 = \hat{\tilde{\Delta}}_4 = \hat{\tilde{\Delta}}_5 = 0 \qquad\qquad \text{(fa)}$$

$$\hat{\tilde{\Delta}}_2 \neq 0$$

$$\hat{\tilde{\Delta}}_1 = \hat{\tilde{\Delta}}_3 = \hat{\tilde{\Delta}}_4 = \hat{\tilde{\Delta}}_5 = 0 \qquad (fb)$$

$$\hat{\tilde{\Delta}}_3 \neq 0$$

$$\hat{\tilde{\Delta}}_1 = \hat{\tilde{\Delta}}_2 = \hat{\tilde{\Delta}}_4 = \hat{\tilde{\Delta}}_5 = 0 \qquad (fc)$$

$$\hat{\tilde{\Delta}}_4 \neq 0$$

$$\hat{\tilde{\Delta}}_1 = \hat{\tilde{\Delta}}_2 = \hat{\tilde{\Delta}}_3 = \hat{\tilde{\Delta}}_5 = 0 \qquad (fd)$$

$$\hat{\tilde{\Delta}}_5 \neq 0$$

$$\hat{\tilde{\Delta}}_1 = \hat{\tilde{\Delta}}_2 = \hat{\tilde{\Delta}}_3 = \hat{\tilde{\Delta}}_4 = 0 \qquad (fe)$$

In order that the chosen virtual translation fields are geometrically admissible the components of the virtual nodal translations of the elements of the truss and the components of the virtual translations of its nodes must be related by relations analogous to (b). Referring to relations (b) and (f) the components of virtual nodal translations are

$$\tilde{u}_1^{1j} = \tilde{u}_1^{1k} = \tilde{u}_1^{2j} = \tilde{u}_1^{2k} = \tilde{u}_1^{3k} = \tilde{u}_1^{4k} = \tilde{u}_1^{5j} = \tilde{u}_1^{5k} = \tilde{u}_1^{6j} = \tilde{u}_1^{6k} = 0$$

$$\tilde{u}_1^{4j} = \tilde{\tilde{\Delta}}_1 \qquad (ga)$$

$$\tilde{u}_1^{3j} = \tfrac{3}{5}\tilde{\tilde{\Delta}}_1$$

$$\tilde{u}_1^{1j} = \tilde{u}_1^{2j} = \tilde{u}_1^{2k} = \tilde{u}_1^{3k} = \tilde{u}_1^{4j} = \tilde{u}_1^{4k} = \tilde{u}_1^{5j} = \tilde{u}_1^{5k} = \tilde{u}_1^{6j} = \tilde{u}_1^{6k} = 0$$

$$\tilde{u}_1^{1k} = \tilde{\tilde{\Delta}}_2 \qquad (gb)$$

$$\tilde{u}_1^{3j} = -\tfrac{4}{5}\tilde{\tilde{\Delta}}_2$$

$$\tilde{u}_1^{1j} = \tilde{u}_1^{1k} = \tilde{u}_1^{2j} = \tilde{u}_1^{2k} = \tilde{u}_1^{3j} = \tilde{u}_1^{3k} = \tilde{u}_1^{4j} = \tilde{u}_1^{5j} = \tilde{u}_1^{5k} = \tilde{u}_1^{6j} = 0$$

$$\tilde{u}_1^{4k} = \tilde{\tilde{\Delta}}_3 \qquad (gc)$$

$$\tilde{u}_1^{6k} = \tfrac{3}{5}\tilde{\tilde{\Delta}}_3$$

$$\tilde{u}_1^{1j} = \tilde{u}_1^{1k} = \tilde{u}_1^{2j} = \tilde{u}_1^{2k} = \tilde{u}_1^{3j} = \tilde{u}_1^{3k} = \tilde{u}_1^{4j} = \tilde{u}_1^{4k} = \tilde{u}_1^{5k} = \tilde{u}_1^{6j} = 0$$

$$\tilde{u}_1^{5j} = -\tilde{\tilde{\Delta}}_4 \qquad (gd)$$

$$\tilde{u}_1^{6k} = \tfrac{4}{5}\tilde{\tilde{\Delta}}_4$$

$$\tilde{u}_1^{1j} = \tilde{u}_1^{1k} = \tilde{u}_1^{2j} = \tilde{u}_1^{3j} = \tilde{u}_1^{4j} = \tilde{u}_1^{4k} = \tilde{u}_1^{5j} = \tilde{u}_1^{5k} = \tilde{u}_1^{6j} = \tilde{u}_1^{6k} = 0$$

$$\tilde{u}_1^{2k} = \tilde{\tilde{\Delta}}_5 \qquad (ge)$$

$$\tilde{u}_1^{3k} = \tfrac{3}{5}\tilde{\tilde{\Delta}}_5$$

Substituting one set of relations (g) at a time into relation (e) we obtain the following set of five equations relating the external force acting on the nodes of the truss to the five unknown components of translation of its nodes.

$$0 = 152\Delta_1 - 36\Delta_2 - 125\Delta_3 - 27\Delta_5 \qquad (ha)$$

$$0 = \tfrac{3}{4}(-48\Delta_1 + 189\Delta_2 + 48\Delta_5) \qquad (hb)$$

$$\frac{375P}{EA} = -125\Delta_1 + 152\Delta_3 + 36\Delta_4 \qquad (hc)$$

$$0 = \tfrac{3}{4}(48\Delta_3 + 189\Delta_4) \qquad (hd)$$

$$0 = -27\Delta_1 + 36\Delta_2 + 152\Delta_5 \qquad (he)$$

These relations may be rewritten as

$$\{P\} = \begin{Bmatrix} 0 \\ 0 \\ P \\ 0 \\ 0 \end{Bmatrix} = \frac{EA}{375} \begin{bmatrix} 152 & -36 & -125 & 0 & -27 \\ -36 & 141.75 & 0 & 0 & 36 \\ -125 & 0 & 152 & 36 & 0 \\ 0 & 0 & 36 & 141.75 & 0 \\ -27 & 36 & 0 & 0 & 152 \end{bmatrix} \begin{Bmatrix} \Delta_1 \\ \Delta_2 \\ \Delta_3 \\ \Delta_4 \\ \Delta_5 \end{Bmatrix} \qquad (i)$$

or

$$\{P^{EF}\} = [S^{FF}]\{\Delta^F\} \qquad (j)$$

where $\{P^{EF}\}$ = matrix of external actions acting on nodes of structure which are not absorbed directly by its supports

$[S^{FF}]$ = basic stiffness matrix of structure

$\{\Delta^F\}$ = matrix of components of displacements of nodes of structure which are not inhibited by its supports

Relations (j) are the basic stiffness equations for the structure.
Solving relation i for $\{\Delta^F\}$, we get

$$\{\Delta^F\} = \begin{Bmatrix} \Delta_1 \\ \Delta_2 \\ \Delta_3 \\ \Delta_4 \\ \Delta_5 \end{Bmatrix} = \frac{P}{AE} \begin{Bmatrix} \dfrac{21}{2} \\ \dfrac{7}{3} \\ \dfrac{189}{16} \\ -3 \\ \dfrac{21}{16} \end{Bmatrix} \qquad (k)$$

STEP 6. We compute the internal forces in the elements of the truss. Substituting the components of translations of the nodes of the truss into relation (b) we obtain the nodal components of translations of the elements of the truss. From these we compute the nodal forces of the elements of the truss using relation (c). Thus,

$$F_1^{1k} = \frac{EA\Delta_2}{L_1} = \frac{7P}{12}$$

$$F_1^{2k} = \frac{EA\Delta_5}{L_2} = \frac{7P}{16}$$

$$F_1^{3k} = EA\left(\frac{\frac{3}{5}\Delta_5 + \frac{4}{5}\Delta_2 - \frac{3}{5}\Delta_1}{5}\right) = -\frac{35P}{48}$$

$$F_1^{4k} = EA\left(\frac{\Delta_3 - \Delta_1}{L_4}\right) = \frac{7P}{16} \tag{1}$$

$$F_1^{5k} = EA\Delta_4 = -\frac{3P}{4}$$

$$F_1^{6k} = EA\left(\frac{\frac{3}{5}\Delta_3 + \frac{4}{5}\Delta_4}{5}\right) = \frac{15P}{16}$$

20.3 Proof of the Contragradient Law Using the Principle of Virtual Work

Consider a structure (statically determinate or indeterminate) subjected to equivalent actions $\{\hat{P}^E\}$ on its nodes. Moreover, consider a set of statically admissible reactions $\{\tilde{R}\}$ of the structure and corresponding statically admissible basic nodal actions of one of its elements $\{\tilde{a}^E\}$. Furthermore, consider an admissible distribution of the components of displacements of the elements and nodes of the structure (see Sec. 18.4). We denote by $\{\tilde{\tilde{\Delta}}\}$ the matrix of the corresponding admissible components of displacements of all the nodes of the structure, including those inhibited by its supports. Moreover, we denote by $\{\tilde{\tilde{\Delta}}^S\}$ the matrix of the corresponding admissible components of displacements of the nodes of the structure which are inhibited by its supports. The matrix $\{\tilde{\tilde{\Delta}}\}$ is conjugate to the matrix $\{\hat{P}^E\}$, while the matrix $\{\tilde{\tilde{\Delta}}^S\}$ is conjugate to the matrix $\{\tilde{R}\}$. Thus, the work of the external actions and the work of the statically admissible reactions of the structure due to the application on it of the admissible distribution of the components of displacements of the elements and nodes of the structure are equal to

$$W_{\text{ext.act.}} = \{\hat{P}^E\}^T\{\tilde{\tilde{\Delta}}\} \tag{20.16}$$

$$W_{\text{react.}} = \{\tilde{R}\}^T\{\tilde{\tilde{\Delta}}^S\} \tag{20.17}$$

As discussed in Sec. 5.3, the deformation of an element of a structure is specified by its basic deformation parameters. The deformation parameters which we have chosen are equal to the components of displacement $\{\tilde{\bar{D}}^{Mk}\}$ of the end k of a model made from the element fixed at its end j and subjected to the basic nodal actions $\{\tilde{a}^E\}$ on its end k. That is,

$$\{\tilde{\bar{d}}\} = \{\tilde{\bar{D}}^{Mk}\} \tag{20.18}$$

Moreover, the work W_d of the statically admissible components of stress acting on the particles of an element due to their deformation is equal to that of the model for the element W_d^M. Thus, applying the principle of virtual work (18.7) to the model of an element we have

$$W_d = W_d^M = W_{\text{ext.act.}}^M = \{\tilde{a}^E\}^T\{\tilde{\bar{D}}^{Mk}\} = \{\tilde{a}^E\}^T\{\tilde{\bar{d}}\} \tag{20.19}$$

Applying the principle of virtual work (18.7) to a structure with NE elements we have

$$W_{\text{ext.act.}} + W_{\text{react.}} = \sum_{e=1}^{NE} W_d^e \tag{20.20}$$

Substituting relations (20.16), (20.17), and (20.19) into (20.20), we obtain

$$\{\hat{P}^E\}^T\{\tilde{\bar{\Delta}}\} + \{\tilde{R}\}^T\{\tilde{\bar{\Delta}}^S\} = \sum_{e=1}^{NE} \{\tilde{a}^{Ee}\}^T\{\tilde{\bar{d}}^e\} = \{\tilde{a}^E\}^T\{\tilde{\bar{d}}\} \tag{20.21}$$

where $\{\tilde{a}^E\}$ is the matrix of the statically admissible basic nodal actions of all the elements of the structure defined by relation (15.18). Moreover, $\{\tilde{\bar{d}}\}$ is the matrix of the admissible deformation parameters of all the elements of the structure defined by relation (15.29a). Relation (20.21) may be rewritten as

$$\{\hat{P}^E\}^T\{\tilde{\bar{\Delta}}\} = \left\{\begin{matrix}\{\tilde{a}^E\}\\-\{R\}\end{matrix}\right\}^T\left\{\begin{matrix}\{\tilde{\bar{d}}\}\\\{\tilde{\bar{\Delta}}^S\}\end{matrix}\right\} \tag{20.22}$$

Any set of statically admissible components of basic nodal actions must satisfy the equations of equilibrium for every node of the structure. Thus, referring to relation (15.3) we have

$$\{\hat{P}^E\} = [\hat{B}]\left\{\begin{matrix}\{\tilde{a}^E\}\\-\{\tilde{R}\}\end{matrix}\right\} \tag{20.23}$$

or

$$\{\hat{P}^E\}^T = \left\{\begin{matrix}\{\tilde{a}^E\}\\-\{\tilde{R}\}\end{matrix}\right\}^T[\hat{B}]^T \tag{20.24}$$

Substituting relations (20.24) into (20.22) we get

$$\left\{ \begin{matrix} \{\tilde{a}^E\} \\ -\{\tilde{R}\} \end{matrix} \right\}^T \left[[\hat{B}]^T\{\tilde{\tilde{\Delta}}\} - \left\{ \begin{matrix} \{\tilde{\tilde{d}}\} \\ \{\tilde{\tilde{\Delta}}^S\} \end{matrix} \right\} \right] = 0 \qquad (20.25)$$

Hence
$$\left\{ \begin{matrix} \{\tilde{\tilde{d}}\} \\ \{\tilde{\tilde{\Delta}}^S\} \end{matrix} \right\} = [\hat{B}]^T\{\tilde{\tilde{\Delta}}\} \qquad (20.26)$$

Thus the following statement is valid for framed structures subjected to equivalent actions on their nodes:

> If the equations of equilibrium for the nodes of a structure subjected to equivalent actions on its nodes are written as indicated by Eqs. (20.23), the compatibility equations for the structure are given by Eqs. (20.26).

This statement is known as the *contragradient law*. It indicates that for any structure, the conditions of equilibrium (20.23) and the conditions of compatibility (20.26) are not independent. Comparing relation (20.26) with (5.29) we have

$$[\hat{C}] = [\hat{B}]^T$$

20.4 Problems

1 and 2. Using the dummy load method compute the vertical component of translation (deflection) and the rotation of node 2 of the beam subjected to the loading shown in Fig. 20.P1. The beam is made of an isotropic linearly elastic material and has a constant cross section. Disregard the effect of shear deformation. Repeat with the beam of Fig. 20.P2.

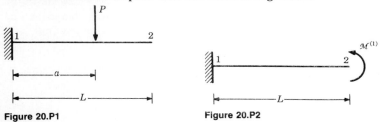

Figure 20.P1 **Figure 20.P2**

3 to 5. Using the dummy load method compute the vertical component of translation (deflection) of node 3 of the beam subjected to the loading shown in Fig. 20.P3. The beam has a constant cross section ($I = 369.70 \times 10^6 \text{ mm}^4$)

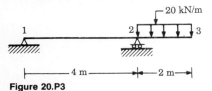

Figure 20.P3

and is made of steel $(E = 210 \, \text{kN/mm}^2)$. Disregard the effect of shear deformation. Repeat with the beams of Figs. 20.P4 and 20.P5.

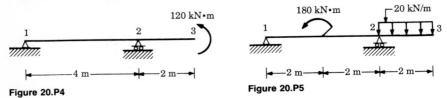

Figure 20.P4   **Figure 20.P5**

6. Using the dummy load method compute the vertical component of translation of node 3 of the structure subjected to the loading shown in Fig. 20.P6. Disregard the effect of shear and axial deformation of the elements of the structure. The elements of the structure are made of steel $(E = 210 \, \text{kN/mm}^2)$ and have the same constant cross section $(I = 369.70 \times 10^6 \, \text{mm}^4)$.

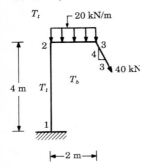

Figure 20.P6

7. Using the dummy load method compute the horizontal component of translation of node 3 of the structure subjected to the following loading:

(*a*) The external forces shown in Fig. 20.P7.
(*b*) A distribution of temperature with $T_t = 25°\text{C}$ and $T_b = -5°\text{C}$ ($\Delta T_c = 0$).

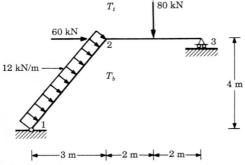

Figure 20.P7

Disregard the effect of shear and axial deformation of the elements of the structure. The elements of the structure are made of steel ($E = 210 \text{ kN/mm}^2$) and have the same constant cross section ($I = 369.7 \times 10^6 \text{ mm}^4$, $\alpha = 10^{-5}/°\text{C}$, $h = 400 \text{ mm}$).

8 to 10. Using the method of virtual displacements compute the components of displacement of the nodes and the internal forces in the elements of the truss subjected to the loads shown in Fig. 20.P8. The elements of the truss are made of steel ($E = 210 \text{ kN/mm}^2$) and have the same constant cross section ($A = 8 \times 10^3 \text{ mm}^2$). Repeat with the trusses of Figs. 20.P9 and 20.P10.

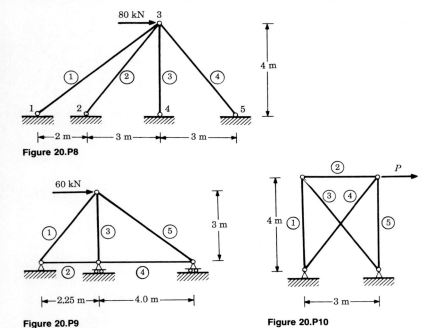

Figure 20.P8

Figure 20.P9

Figure 20.P10

Reference

1. A. E. Armenàkas, *Classical Structural Analysis: A Modern Approach*, McGraw-Hill Book Co., New York, 1988, pp. 476–479.

Photographs and a Brief Description of an Arena With a Domelike Steel Roof

The basketball arena of the University of North Carolina, Chapel Hill (see Fig. 20.1), has an area of 10,591 m^2 (114,000 ft^2) and seats 21,400 people. It is covered by a domelike steel roof (see Figs. 20.2 and 20.3) with a fabric center. The roof is supported on four diagonal, trussed frames (see Fig. 20.4), creating a skew (122 ft × 130 ft) rectangle in the center at the edges of which the fabric is tied down by cables. The fabric is teflon-coated fiberglass and is used as a lantern (an architectural term for a type of skylight). The fabric is supported on steel arches (see Fig. 20.2) pinned at both ends and made of pipes (see Fig. 20.5) so that no sharp edges come into contact with the fabric. The tension tie for the arches is a horizontal, steel, octagonal ring, extending around the perimeter of the roof. As can be seen in Figs. 20.2 and 20.3 the arches were temporarily supported during construction at their four intersection points by trussed columns.

Figure 20.1 The basketball arena of the University of North Carolina in Chapel Hill, N.C., during construction. (*Courtesy Geiger Gossen Hamilton Liao Engineers P.C., New York, N.Y.*)

Figure 20.2 View of the arena's domelike steel roof during construction. (*Courtesy Geiger Gossen Hamilton Liao Engineers P.C., New York, N.Y.*)

Figure 20.3 Another view of the domelike steel roof during construction. (*Courtesy Geiger Gossen Hamilton Liao Engineers P.C., New York, N.Y.*)

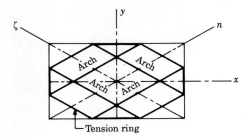

Figure 20.4 Sketch showing the layout of the many arches and the tension ring of the roof. (*Courtesy Geiger Gossen Hamilton Liao Engineers P.C., New York, N.Y.*)

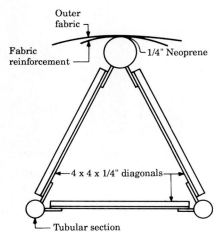

Figure 20.5 Cross section of the arches supporting the fabric. (*Courtesy Geiger Gossen Hamilton Liao Engineers P.C., New York, N.Y.*)

Functions of Discontinuity

A.1   Definition of the Unit Step Function, the Dirac δ-Function, and the Doublet Function

Referring to Fig. A.1, the unit step function $\Delta(x - a)$ is defined as

$$\Delta(x - a) = \begin{cases} 0 & \text{if } x < a \\ 1 & \text{if } x > a \end{cases} \tag{A.1}$$

The unit step function is not defined at $x = a$. In case $a = 0$, the unit step function is denoted as $\Delta(x)$.

In general, for $b < a$ we have

$$\int_b^x f(x)\,\Delta(x - a)\,dx = \int_b^a f(x)\,\Delta(x - a)\,dx + \int_a^x f(x)\,\Delta(x - a)\,dx \tag{A.2}$$

Referring to the definition of the unit step function (A.1) it can be seen that the first integral on the right-hand side of relation (A.2) vanishes. Thus relation (A.2) reduces to

$$\int_b^x f(x)\,\Delta(x - a)\,dx = \Delta(x - a)\int_a^x f(x)\,dx \tag{A.3}$$

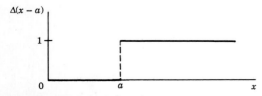

Figure A.1 Unit step function.

From relation (A.3) we obtain

$$\int_b^x \Delta(x-a)\,dx = (x-a)\,\Delta(x-a) \tag{A.4}$$

$$\int_b^x (x-a)^k \Delta(x-a)\,dx = \frac{(x-a)^{k+1}}{k+1}\Delta(x_1-a) \tag{A.5}$$

Referring to Fig. A.2, imagine that there exists a sequence of continuous functions $\psi_n(x-a)$ $(n=1,2,3,\ldots)$ which vanish for $x<a-\varepsilon_n$ and are equal to unity for $x>a+\varepsilon_n$ where ε_n is a small positive number. Moreover, assume that ε_n approaches zero as n approaches infinity. Then, as n increases and ε_n decreases, $\psi_n(x-a)$ approaches $\Delta(x-a)$. Consequently, $\Delta(x-a)$ may be considered as the limit of the sequence of continuous functions $\psi_n(x-a)$ as n approaches infinity. Thus,

$$\Delta(x-a) = \lim_{n\to\infty} \psi_n(x-a) \tag{A.6}$$

Consider the following sequence of functions, one of which is plotted in Fig. A.3:

$$\phi_n(x-a) = \frac{d\psi_n(x-a)}{dx} \tag{A.7}$$

Notice that the cross-hatched area in Fig. A.3 is equal to unity for any value of ε_n. This may be shown as follows:

$$\int_{-\infty}^{\infty} \phi_n(x-a)\,dx = \int_{x=a-\varepsilon_n}^{x=a+\varepsilon_n} \phi_n(x-a)\,dx = \int_{x=a-\varepsilon_n}^{x=a+\varepsilon_n} \frac{d\psi_n(x-a)}{dx}\,dx$$

$$= \psi_n(\varepsilon_n) - \psi_n(-\varepsilon_n) = 1 \tag{A.8}$$

For very small values of ε_n, the cross-hatched area in Fig. A.3 may be

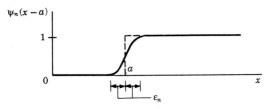

Figure A.2 Approximation of the unit step function by the continuous function $\psi_n(x-a)$.

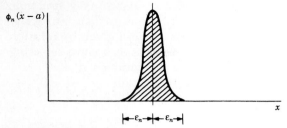

Figure A.3 Derivative of the function $\psi_n(x-a)$.

approximated by $2\varepsilon_n \phi(x-a)$. Thus,

$$\int_{-\infty}^{\infty} \phi_n(x-a)\,dx = 2\varepsilon_n \phi_n(x-a) = 1$$

Thus,

$$\phi_n(x-a) = \frac{d\psi_n(x-a)}{dx} \approx \frac{1}{2\varepsilon_n} \qquad a - \varepsilon_n < x < a + \varepsilon_n \qquad \text{(A.9)}$$

It is apparent that in the limit as $n \to \infty$ the value of $\phi_n(x-a)$ in the region $a - \varepsilon_n < x < a + \varepsilon_n$ increases to infinity (see Fig. A.3). That is, in the limit as $n \to \infty$ the function $\phi_n(x-a)$ does not exist as an ordinary function. We call that limit the Dirac δ-function, that is,

$$\delta(x-a) = \lim_{n \to \infty} \phi_n(x-a) = \lim_{n \to \infty} \frac{d\psi_n(x-a)}{dx}$$

$$= \lim_{\varepsilon_n \to 0} \begin{cases} 0 & \text{if } a + \varepsilon_n \le x \le a - \varepsilon_n \\ \dfrac{1}{2\varepsilon_n} & \text{if } a - \varepsilon_n < x < a + \varepsilon_n \end{cases} \qquad \text{(A.10)}$$

If $a = 0$, the Dirac δ-function is denoted as $\delta(x)$.

We assume that $\delta(x-a)$ has the same properties as ordinary functions when used as an integrand. Furthermore, for any function $f(x)$ which is continuous at $x = a$ the following relation is valid:

$$\int_{-\infty}^{\infty} \delta(x-a)f(x)\,dx = f(a) \qquad \text{(A.11)}$$

Consequently,

$$\int_{-\infty}^{\infty} \delta(x-a)\,dx = 1 \qquad \text{(A.12)}$$

Thus, the δ-function cannot be assigned any value at $x = a$ and vanishes at all other points. However, the integral $\int_{-\infty}^{\infty} \delta(x-a)\,dx$ is equal to unity. Hence, the δ-function is not an ordinary function, having definite values for every value of x, but rather, an entity possessing certain properties as, for instance, those given by relation (A.11).† This relation indicates that $\delta(x - a)$ acts as a sieve selecting from all possible values of $f(x)$ its value at the point $x = a$.

Using relations (A.6), (A.7), and (A.10) we obtain

$$\Delta(x - a) = \lim_{n \to \infty} \psi_n(x - a) = \lim_{n \to \infty} \int_{-\infty}^{x} d\psi_n(x - a)$$

$$= \lim_{n \to \infty} \int_{-\infty}^{x} \phi_n(x - a)\, dx$$

$$= \int_{-\infty}^{x} \delta(x - a)\, dx = \int_{b}^{x} \delta(x - a)\, dx \qquad \text{(A.13)}$$

where $b < a$.

Consider the sequence of functions $d\phi_n(x - a)/dx$, one of which is shown in Fig. A.4. We define the *doublet function* as

$$\delta^1(x - a) = \frac{d\delta(x - a)}{dx} = \lim_{n \to \infty} \frac{d\phi_n(x - a)}{dx} \qquad \text{(A.14)}$$

On the basis of this definition we have

$$\int_{-\infty}^{x} \delta^1(x - a)\, dx = \int_{-\infty}^{x} d\delta(x - a) = \delta(x - a) \qquad \text{(A.15)}$$

---

† For a more detailed discussion see E. Butkov, *Mathematical Physics,* Addison-Wesley Publishing Co., Reading, Mass., 1973, p. 221; and A. Papoulis, *The Fourier Integral and its Applications,* McGraw-Hill Book Co., New York, 1962, p. 269.

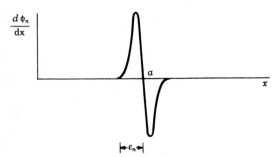

Figure A.4   Derivative of the function $\phi_n(x - a)$.

A.2   Additional Properties of the Dirac δ-Function and of the Doublet Function

1. For any function $f(x)$ which is continuous at $x = a$, using relation (A.14) and integrating by parts we have

$$\int_{-\infty}^{\infty} \delta^1(x-a)f(x)\,dx = \int_{-\infty}^{\infty} f(x)\,d\delta(x-a)$$

$$= f(x)\,\delta(x-a)\Big|_{-\infty}^{\infty} - \int_{-\infty}^{\infty} \delta(x-a)\frac{df}{dx}\,dx$$

$$= -\frac{df}{dx}\Big|_{x=a} \tag{A.16}$$

2. Following a procedure analogous to that employed in proving relation (A.16) it can be shown that

$$\int_{-\infty}^{\infty} \delta^k(x-a)f(x)\,dx = (-1)^k \frac{d^k f(x)}{dx^k}\Big|_{x=a} \tag{A.17}$$

3. For any function $f(x)$ which is continuous at $x = 0$, the following relation is valid

$$f(x+c)\,\delta(x) = f(c)\,\delta(x) \tag{A.18}$$

To prove this, let

$$\xi = x + c$$

Then, on the basis of relation (A.11) we obtain

$$\int_{-\infty}^{\infty} f(x+c)\,\delta(x)\,dx = \int_{-\infty}^{\infty} f(\xi)\,\delta(\xi-c)\,d\xi = f(c)$$

$$= f(c)\int_{-\infty}^{\infty} \delta(x)\,dx \tag{A.19}$$

or $\qquad \displaystyle\int_{-\infty}^{\infty} [f(x+c)\,\delta(x) - f(c)\,\delta(x)]\,dx = 0 \tag{A.20}$

This relation is satisfied only if relation (A.18) is valid.

4. For any constant $c \neq 0$, the following relation is valid

$$\delta(cx) = |c|^{-1}\,\delta(x) \tag{A.21}$$

This may be proven as follows. Let

$$\xi = cx \tag{A.22}$$

Then $\qquad\qquad \displaystyle\frac{d\xi}{c} = dx$

If $c > 0$, using relations (A.11) and (A.22) we obtain

$$\int_{-\infty}^{\infty} f(x)\,\delta(cx)\,dx = \frac{1}{c}\int_{-\infty}^{\infty} f\left(\frac{\xi}{c}\right)\delta(\xi)\,d\xi = \frac{1}{c}f(0) = \frac{1}{c}\int_{-\infty}^{\infty} f(x)\,\delta(x)\,dx$$

(A.23)

Therefore,

$$\int_{-\infty}^{\infty} f(x)[\delta(cx) - c^{-1}\,\delta(x)]\,dx = 0 \qquad (A.24)$$

Inasmuch as this relation is valid for any continuous function $f(x)$, it is apparent that

$$\delta(cx) = c^{-1}\,\delta(x) = |c|^{-1}\,\delta(x) \qquad (A.25)$$

If $c < 0$, using relations (A.11) and (A.22) we get

$$\int_{-\infty}^{\infty} f(x)\,\delta(cx)\,dx = \frac{1}{c}\int_{\infty}^{-\infty} f\left(\frac{\xi}{c}\right)\delta(\xi)\,d\xi$$

$$= -\frac{1}{c}\int_{-\infty}^{\infty} f\left(\frac{\xi}{c}\right)\delta(\xi)\,d\xi$$

$$= -\frac{1}{c}f(0) = -\frac{1}{c}\int_{-\infty}^{\infty} f(x)\,\delta(x)\,dx$$

$$= \frac{1}{|c|}\int_{-\infty}^{\infty} f(x)\,\delta(x)\,dx \qquad (A.26)$$

Therefore,

$$\int_{-\infty}^{\infty} [\delta(cx) - |c|^{-1}\,\delta(x)]\,dx = 0 \qquad (A.27)$$

Hence,

$$\delta(cx) = |c|^{-1}\,\delta(x) \qquad (A.28)$$

5. As a consequence of property (A.28), it is apparent that $\delta(x)$ is an even function, that is,

$$\delta(x) = \delta(-x) \qquad (A.29)$$

6. The following relation is valid

$$(x - a)\,\delta(x - a) = 0 \qquad (A.30)$$

This may be proven by noting that on the basis of relation (A.11) for any function $f(x)$ which is continuous at $x = a$, we have

$$\int_{-\infty}^{\infty} \delta(x - a)(x - a)f(x)\,dx = [(x - a)f(x)]_{x=a} = 0 \qquad (A.31)$$

Inasmuch as this relation holds for any continuous function $f(x)$, it is apparent that relation (A.30) is valid.

7. The following relation may be proven by differentiating relation (A.30) with respect to x:

$$\delta^1(x-a) = -\frac{\delta(x-a)}{(x-a)} \tag{A.32}$$

Similarly,

$$\delta^k(x-a) = (-1)^k k! \frac{\delta(x-a)}{(x-a)^k} \tag{A.33}$$

A.3   Application of the Functions of Discontinuity in Computing the Deflection of Beams

Consider a beam of constant cross section, subjected only to external actions producing bending about its x_2 axis. Referring to relation (2.53b) the component of translation $u_3(x_1)$ (deflection) of the beam satisfies the following differential equation:

$$EI_2 \frac{d^4 u_3(x_1)}{dx_1^4} = p_3(x_1) + \frac{dm_2(x_1)}{dx_1} \tag{A.34}$$

where E = modulus of elasticity of material from which beam is made
  I_2 = moment of inertia of cross section of beam about its x_2 axis
  $p_3(x_1)$ = distribution of external forces per unit length of beam
  $m_2(x_1)$ = distribution of external moments per unit length of beam

Equation (A.34) is not valid at the points of discontinuity of the distribution of the external actions $p_3(x_1)$ and $m_2(x_1)$ acting on the beam. The deflection $u_3(x_1)$ of a simply supported beam of length L must satisfy the following boundary conditions at the ends of the beam.

$$u_3(0) = 0 \tag{A.35a}$$

$$u_3(L) = 0 \tag{A.35b}$$

$$\left. \frac{d^2 u_3}{dx_1^2} \right|_{x_1=0} = 0 \tag{A.35c}$$

$$\left. \frac{d^2 u_3}{dx_1^2} \right|_{x_1=L} = 0 \tag{A.35d}$$

In what follows we establish the deflection of a simply supported beam subjected to different loading cases, using functions of discontinuity.

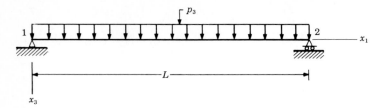

Figure A.5 Simply supported beam subjected to uniformly distributed forces along its entire length.

Case I: Simply supported beam of constant cross section subjected to uniformly distributed forces along its entire length. When the beam is subjected to uniformly distributed forces p_3 along its entire length (see Fig. A.5), Eq. (A.34) is valid along the entire length of the beam and may be easily integrated four times to yield

$$EI_2 u_3(x_1) = \frac{p_3}{24} x_1^4 + Ax_1^3 + Bx_1^2 + Cx_1 + D \qquad (A.36)$$

Substituting relation (A.36) into the boundary conditions (A.35), it can be shown that they are satisfied if

$$A = \frac{p_3 L}{12} \qquad B = 0 \qquad C = \frac{p_3 L^3}{24} \qquad D = 0 \qquad (A.37)$$

Thus, the deflection of the beam is given as

$$u_3(x_1) = \frac{p_3}{24 EI_2} (x_1^4 - 2x_1^3 L + L^3 x_1) \qquad (A.38)$$

Case II: Simply supported beam as constant cross section subjected to uniformly distributed forces along part of its length. Consider the simply supported beam of constant cross section and length L subjected to uniformly distributed forces p_3 as shown in Fig. A.6. In this case the distribution of the external forces has a simple discontinuity (jump) at $x_1 = a$, and consequently Eq. (A.34) is not valid at this point. In order to employ the procedure used in the previous example, the external forces in Eq. (A.34) are represented as follows:

$$p_3(x_1) = p_3 \Delta(x_1 - a) \qquad (A.39)$$

Thus, the differential equation (A.34) becomes

$$EI_2 \frac{d^4 u_3(x_1)}{dx_1^4} = p_3 \Delta(x_1 - \alpha) \qquad (A.40)$$

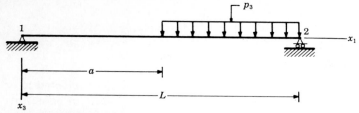

Figure A.6 Simply supported beam subjected to uniformly distributed forces along part of its length.

Integrating Eq. (A.40) 4 times and using relation (A.5), we obtain

$$EI_2 \frac{d^3u_3(x_1)}{dx_1^3} = p_3(x_1 - a)\Delta(x_1 - a) + A$$

$$EI_2 \frac{d^2u_3(x_1)}{dx_1^2} = \frac{p_3(x_1 - a)^2}{2}\Delta(x_1 - a) + Ax_1 + B$$

$$EI_2 \frac{du_3(x_1)}{dx_1} = \frac{p_3(x_1 - a)^3}{6}\Delta(x_1 - a) + \frac{Ax_1^2}{2} + Bx_1 + C \qquad (A.41)$$

$$EI_2 u_3(x_1) = \frac{p_3(x_1 - a^4)}{24}\Delta(x_1 - a) + \frac{Ax_1^3}{6} + \frac{Bx_1^2}{2} + Cx_1 + D$$

The constants A, B, C, and D are evaluated by requiring that the solution satisfies the boundary conditions (A.35). Taking into account that $\Delta(x_1 - a) = 0$ at $x_1 = 0$, and that $\Delta(x_1 - a) = 1$ at $x_1 = L$, we get

$$EI_2 u_3(0) = 0 \qquad D = 0$$

$$EI_2 \frac{d^2u_3}{dx_1^2}\bigg|_{x_1=0} = 0 \qquad B = 0 \qquad (A.42)$$

$$EI_2 \frac{d^2u_3}{dx_1^2}\bigg|_{x_1=L} = 0 \qquad A = \frac{-p_3(L - a)^2}{2L}$$

$$EI_2 u_3(L) = 0 \qquad C = \frac{p_3}{24L}(L - a)^2[2L^2 - (L - a)^2]$$

Substituting the values of the constants (A.42) in the last of relations (A.41), we obtain

$$u_3(x_1) = \frac{p_3}{24EI_2}\left[(x_1 - a)^4\Delta(x_1 - a) - \frac{2(L - a)^2}{L}x_1^3\right.$$

$$\left. + \frac{(L - a)^2}{L}x_1[2L^2 - (L - a)^2]\right] \qquad (A.43)$$

For $a = L/2$, the deflection at $x_1 = L/2$ is

$$u_3\left(\frac{L}{2}\right) = \frac{5p_3 L^4}{768 EI_2} \tag{A.44}$$

Case III: Simply supported beam of constant cross section subjected to a concentrated force. Consider the simply supported beam of constant cross section and length L, subjected to a concentrated force P_3, as shown in Fig. A.7. In this beam the distribution of the external forces has a Dirac delta-type discontinuity at point $x_1 = a$. In order to follow a procedure analogous to the one employed in loading case I, the external forces $p_3(x_1)$ in Eq. (A.34) are represented as follows

$$p_3(x_1) = P_3\,\delta(x_1 - a) \tag{A.45}$$

Referring to relations (A.10) and (A.12) it is apparent that the product $P_3\,\delta(x_1 - a)$ is given in units of force per unit length. Thus, the differential equation (A.33) becomes

$$EI_2 \frac{d^4 u_3(x_1)}{dx_1^4} = P_3\,\delta(x_1 - a) \tag{A.46}$$

Integrating Eq. (A.46) four times, using relations (A.5) and (A.13), we obtain

$$EI_2 \frac{d^3 u_3(x_1)}{dx_1^3} = P_3 \Delta(x_1 - a) + A$$

$$EI_2 \frac{d^2 u_3(x_1)}{dx_1^2} = P_3(x_1 - a)\Delta(x_1 - a) + Ax_1 + B$$

$$EI_2 \frac{du_3(x_1)}{dx_1} = \frac{P_3(x_1 - a)^2}{2}\Delta(x_1 - a) + \frac{Ax_1^2}{2} + Bx_1 + C \tag{A.47}$$

$$EI_2 u_3(x_1) = \frac{P_3(x_1 - a)^3}{6}\Delta(x_1 - a) + \frac{Ax_1^3}{2} + \frac{Bx_1^2}{2} + Cx_1 + D$$

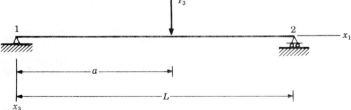

Figure A.7 Simply supported beam subjected to a concentrated force.

The constants A, B, C, and D are evaluated by requiring that the solution (A.47) satisfies the boundary conditions (A.35). Taking into account that $\Delta(x_1 - a) = 0$ at $x_1 = 0$ and that $\Delta(x_1 - a) = 1$ at $x_1 = L$, we get

$$EI_2 u_3(0) = 0 \qquad D = 0$$

$$EI_2 \frac{d^2 u_3}{dx_1^2}\bigg|_{x_1=0} = 0 \qquad B = 0$$

$$EI_2 \frac{d^2 u_3}{dx_1^2}\bigg|_{x_1=L} = 0 \qquad A = -\frac{P_3(L-a)}{L}$$

$$EI_2 u_3(L) = 0 \qquad C = \frac{P_3(L-a)}{6L}(2La - a^2)$$

(A.48)

Substituting the value of the constants in the last of relations (A.47), we get

$$u_3(x_1) = \frac{P_3}{6EI_2L}[L(x_1 - a)^3 \Delta(x_1 - a) - (L-a)x_1^3 + a(L-a)(2L-a)x_1]$$

(A.49)

For $a = L/2$, the deflection at $x = L/2$ is

$$u_3\left(\frac{L}{2}\right) = \frac{P_3 L^3}{48EI_2}$$

(A.50)

Case IV: Simply supported beam of constant cross section subjected to a concentrated moment. Consider the simply supported beam of constant cross section and length L subjected to the loads shown in Fig. A.8. In this beam the distribution of the external moments has a Dirac delta-type discontinuity at $x_1 = b$. In order to follow a procedure analogous to that employed in loading case I, the moment $m_2(x_1)$ in Eq. (A.34) is represented as follows

$$m_2(x_1) = \mathcal{M}_2 \delta(x_1 - b)$$

(A.51)

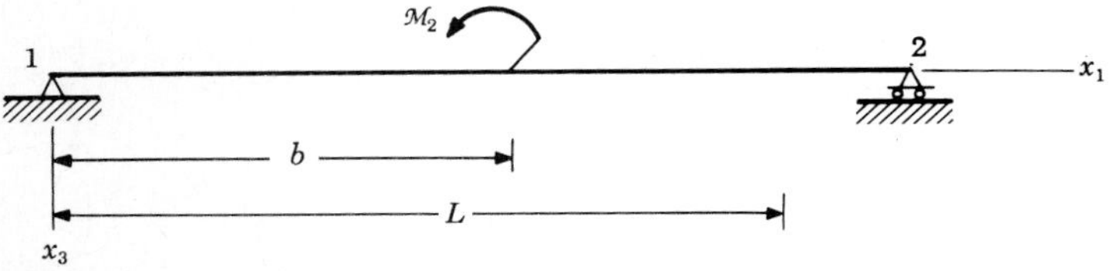

Figure A.8 Simply supported beam subjected to a concentrated moment.

Substituting relation (A.51) into Eq. (A.34) and using relation (A.14) we get

$$EI_2 \frac{d^4 u_3(x_1)}{dx_1^4} = M_2 \, \delta^1(x_1 - b) \tag{A.52}$$

Referring to relations (A.14) or (A.32) it can be seen that the term $M_2 \, \delta^1(x_1 - b)$ is in units of force divided by the square of units of length. Integrating Eq. (A.52) four times, using relations (A.5), (A.13), and (A.15), we obtain

$$EI_2 \frac{d^3 u_3(x_1)}{dx_1^3} = M_2 \, \delta(x_1 - b) + A$$

$$EI_2 \frac{d^2 u_3(x_1)}{dx_1^2} = M_2 \Delta(x_1 - b) + Ax_1 + B$$

$$EI_2 \frac{du_3(x_1)}{dx_1} = M_2(x_1 - b)\Delta(x_1 - b) + \frac{Ax_1^2}{2} + Bx_1 + C \tag{A.53}$$

$$EI_2 u_3(x_1) = \frac{M_2(x_1 - b)^2}{2} \Delta(x_1 - b) + \frac{Ax_1^3}{6} + \frac{Bx_1^2}{2} + Cx_1 + D$$

The constants A, B, C, and D are evaluated by requiring that the solution (A.53) satisfies the boundary conditions (A.35). Taking into account that $\Delta(x_1 - b) = 0$ at $x_1 = 0$, and that $\Delta(x_1 - b) = 1$ at $x_1 = L$, we get

$$A = -\frac{M_2}{L} \qquad B = 0 \qquad C = \frac{M_2}{6}[-2L^2 - 3b^2 + 6bL] \qquad D = 0 \tag{A.54}$$

Substituting the values of the constants in the last of relations (A.53), we obtain

$$EI_2 u_3(x_1) = \frac{M_2}{6} \left[3(x_1 - b)^2 \Delta(x_1 - b) - \frac{x_1^3}{L} + (6bL - 2L^2 - 3b^2)x_1 \right]$$

$$\tag{A.55}$$

Vector Quantities

B.1 Introduction

In order to specify completely certain physical quantities, such as temperature, energy, and mass, it is necessary to give only a real number. These quantities are referred to as *scalars*. In order to specify completely certain other physical quantities, such as, force, moment, velocity, and acceleration, it is necessary to give both their magnitude (a nonnegative number) and their direction. These quantities are referred to as *vectors*. They may be represented in a three-dimensional, euclidian space by directed line segments (arrows) whose length is proportional to the magnitude of the vectors and whose direction is that of the vectors. However, it should be emphasized that not all quantities which can be represented by a directed line segment are vectors. For instance, finite rotations of a rigid body are not vector quantities, notwithstanding the fact that a rotation of a rigid body about an axis may be conveniently represented by an arrow in the direction of this axis.

This may be illustrated by considering the rectangle $OABC$ shown in Fig. B.1, and supposing that it is subjected first to a counterclockwise rotation $\phi_3 = 90°$ about the x_3 axis and then to a counterclockwise rotation $\phi_2 = 90°$ about the x_2 axis, reaching successively the configuration denoted in Fig. B.1, by $OA'B'C'$ and $OA''B''C''$. If finite rotations were vectors, then the final position of the plate would have been independent of the order of application of the aforementioned rotations. That is,

$$\phi = \phi_2 \mathbf{i}_2 + \phi_3 \mathbf{i}_3 = \phi_3 \mathbf{i}_3 + \phi_2 \mathbf{i}_2$$

As illustrated in Fig. B.2, this is not the case when the rectangle is

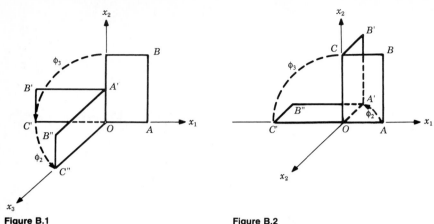

Figure B.1   **Figure B.2**

first subjected to ϕ_2 and then to ϕ_3 reaching successively the configurations denoted by $OA'B'C'$ and $OA''B''C''$.

Two vectors are equal if they have the same magnitude and direction. Consequently, a vector does not change if it moves parallel to itself. However, when we say that two vectors are equal, this does not necessarily indicate that they cause the same effect. For instance, a force accelerates a rigid body differently than an equal force having a different line of action; a force deforms a deformable body differently than an equal force having a different point of application. A vector whose effect depends only on the line of its action is referred to as a *sliding vector,* while a vector whose effect depends on the line of its action and on its point of application is referred to as a *fixed vector.*

Every vector **a** can be expressed as a linear combination of three arbitrary noncoplanar (linearly independent) vectors, **b**, **c**, and **d**, referred to as *base vectors.* That is,

$$\mathbf{a} = m\mathbf{b} + p\mathbf{c} + q\mathbf{d} \tag{B.1}$$

where m, p, and q are real numbers. The vectors $m\mathbf{b}$, $p\mathbf{c}$, and $q\mathbf{d}$ are referred to as the *components* of the vector **a** relative to the base vectors **b**, **c**, and **d**. It can be shown that the components of a vector with respect to a set of base vectors **b**, **c**, and **d** are unique, that is, for any vector **a** there exists only one set of real numbers m, p, and q satisfying relation (B.1). *As base vectors, we choose the three orthogonal unit vectors* $\mathbf{i}_1$, $\mathbf{i}_2$, *and* $\mathbf{i}_3$ *which lie along the positive directions of a set of right-handed rectangular system of axes* (cartesian system of axes) x_1, x_2, and x_3, respectively. Thus we may represent a vector **a** as a linear combination of the three right-handed

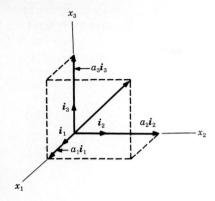

Figure B.3 Components of a vector.

orthogonal unit vectors $\mathbf{i}_1$, $\mathbf{i}_2$, and $\mathbf{i}_3$ as

$$\mathbf{a} = a_1\mathbf{i}_1 + a_2\mathbf{i}_2 + a_3\mathbf{i}_3 \tag{B.2}$$

The vectors $a_1\mathbf{i}_1$, $a_2\mathbf{i}_2$, and $a_3\mathbf{i}_3$ (see Fig. B.3) are the cartesian components of the vector $\mathbf{a}$. However, in the sequel, we will refer to the quantities a_j $(j = 1, 2, 3)$ as the *cartesian components* of the vector $\mathbf{a}$. It is apparent that a vector may be specified by its three cartesian components. Thus, if each of the cartesian components of a vector is equal to the corresponding component of another vector referred to the same rectangular system of axes, the two vectors are equal, that is, their components referred to any rectangular system of axes will be equal.

A vector may be represented as follows.
1. By the symbolic representation employed heretofore, i.e., $\mathbf{a}$, $\mathbf{b}$, $\mathbf{A}$, $\mathbf{B}$, which does not require a choice of a coordinate system.
2. By its three cartesian components with respect to a set of unit orthogonal base vectors $\mathbf{i}_1$, $\mathbf{i}_2$, and $\mathbf{i}_3$ using
   a. Indicial notation, as,
   $$\mathbf{a} \to a_j \qquad (j = 1, 2, 3)$$
   b. Matrix notation, as,

$$\mathbf{a} \to \{a\} = \begin{Bmatrix} a_1 \\ a_2 \\ a_3 \end{Bmatrix} \tag{B.3}$$

From geometric considerations, it can be seen that the magnitude of a vector $\mathbf{a}$ is given by

$$|\mathbf{a}| = (a_1^2 + a_2^2 + a_3^2)^{1/2} = \sqrt{\{a\}^T\{a\}} \tag{B.4}$$

The sum of two or more vectors is the vector whose cartesian components are the sum of the corresponding cartesian components of the added vectors. For instance, if $\mathbf{a} = a_1\mathbf{i}_1 + a_2\mathbf{i}_2 + a_3\mathbf{i}_3$ and $\mathbf{b} = b_1\mathbf{i}_1 + b_2\mathbf{i}_2 + b_3\mathbf{i}_3$, then their sum is a vector $\mathbf{c} = c_1\mathbf{i}_1 + c_2\mathbf{i}_2 + c_3\mathbf{i}_3$ whose components are given as

$$\begin{Bmatrix} c_1 \\ c_2 \\ c_3 \end{Bmatrix} = \begin{Bmatrix} a_1 \\ a_2 \\ a_3 \end{Bmatrix} + \begin{Bmatrix} b_1 \\ b_2 \\ b_3 \end{Bmatrix} = \begin{Bmatrix} a_1 + b_1 \\ a_2 + b_2 \\ a_3 + b_3 \end{Bmatrix} \tag{B.5}$$

B.2 Transformation of the Components of a Vector Upon Rotation of a Rectangular System of Axes

Let us consider two systems of right-handed rectangular cartesian axes x_i and $\bar{x}_j$ ($i, j = 1, 2, 3$) having the same origin at an arbitrary point 0. Referring to Fig. B.4, the direction cosines of the system of axes x_i relative to the system of axes $\bar{x}_j$ are defined as

$$\begin{aligned} \lambda_{11} &= \cos \phi_{11} & \lambda_{21} &= \cos \phi_{21} & \lambda_{31} &= \cos \phi_{31} \\ \lambda_{12} &= \cos \phi_{12} & \lambda_{22} &= \cos \phi_{22} & \lambda_{32} &= \cos \phi_{32} \\ \lambda_{13} &= \cos \phi_{13} & \lambda_{23} &= \cos \phi_{23} & \lambda_{33} &= \cos \phi_{33} \end{aligned} \tag{B.6}$$

or in indicial notation

$$\lambda_{ij} = \cos \phi_{ij} \qquad i, j = 1, 2, 3 \tag{B.7}$$

Denoting by $\mathbf{i}_i$ and $\bar{\mathbf{i}}_j$ the unit vectors acting along the x_i and $\bar{x}_j$ axes,

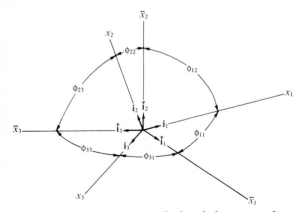

Figure B.4 Rotation of a right-handed rectangular system of axes.

respectively, and referring to Fig. B.4, we have

$$\mathbf{i}_1 = \lambda_{11}\bar{\mathbf{i}}_1 + \lambda_{12}\bar{\mathbf{i}}_2 + \lambda_{13}\bar{\mathbf{i}}_3$$
$$\mathbf{i}_2 = \lambda_{21}\bar{\mathbf{i}}_1 + \lambda_{22}\bar{\mathbf{i}}_2 + \lambda_{23}\bar{\mathbf{i}}_3 \qquad \text{(B.8)}$$
$$\mathbf{i}_3 = \lambda_{31}\bar{\mathbf{i}}_1 + \lambda_{32}\bar{\mathbf{i}}_2 + \lambda_{32}\bar{\mathbf{i}}_3$$

or

$$\mathbf{i}_i = \sum_{j=1}^{3} \lambda_{ij}\bar{\mathbf{i}}_j \qquad i = 1, 2, 3 \qquad \text{(B.9)}$$

Similarly, we obtain

$$\bar{\mathbf{i}}_j = \sum_{i=1}^{3} \lambda_{ij}\mathbf{i}_i \qquad j = 1, 2, 3 \qquad \text{(B.10)}$$

Let us consider the vector **a** whose components relative to the x_i and $\bar{x}_j$ system of axes are a_i and $\bar{a}_j$, respectively, thus,

$$\mathbf{a} = \sum_{i=1}^{3} a_i\mathbf{i}_i = \sum_{j=1}^{3} \bar{a}_j\bar{\mathbf{i}}_j \qquad \text{(B.11)}$$

Substituting relation (B.10) in (B.11), we obtain

$$\sum_{i=1}^{3} a_i\mathbf{i}_i = \sum_{j=1}^{3}\sum_{i=1}^{3} \bar{a}_j\lambda_{ij}\mathbf{i}_i \qquad \text{(B.12)}$$

Consequently,

$$a_i = \sum_{j=1}^{3} \lambda_{ij}\bar{a}_j \qquad \text{(B.13)}$$

Similarly, substituting relation (B.9) into (B.11), we get

$$\bar{a}_i = \sum_{j=1}^{3} \lambda_{ji}a_j \qquad \text{(B.14)}$$

Relations (B.13) and (B.14) can be written in matrix form as

$$\begin{Bmatrix} a_1 \\ a_2 \\ a_3 \end{Bmatrix} = \begin{bmatrix} \lambda_{11} & \lambda_{12} & \lambda_{13} \\ \lambda_{21} & \lambda_{22} & \lambda_{23} \\ \lambda_{31} & \lambda_{32} & \lambda_{33} \end{bmatrix} \begin{Bmatrix} \bar{a}_1 \\ \bar{a}_2 \\ \bar{a}_3 \end{Bmatrix} \qquad \text{(B.15)}$$

and

$$\begin{Bmatrix} \bar{a}_1 \\ \bar{a}_2 \\ \bar{a}_3 \end{Bmatrix} = \begin{bmatrix} \lambda_{11} & \lambda_{21} & \lambda_{31} \\ \lambda_{12} & \lambda_{22} & \lambda_{32} \\ \lambda_{13} & \lambda_{23} & \lambda_{33} \end{bmatrix} \begin{Bmatrix} a_1 \\ a_2 \\ a_3 \end{Bmatrix} \qquad \text{(B.16)}$$

or

$$\{a\} = [\Lambda_S]\{\bar{a}\} \qquad \text{(B.17)}$$

and

$$\{\bar{a}\} = [\Lambda_S]^T\{a\} \qquad \text{(B.18)}$$

$$\text{where} \qquad [\Lambda_S] = \begin{bmatrix} \lambda_{11} & \lambda_{12} & \lambda_{13} \\ \lambda_{21} & \lambda_{22} & \lambda_{23} \\ \lambda_{31} & \lambda_{32} & \lambda_{33} \end{bmatrix} \qquad (B.19)$$

Relations (B.17) and (B.18) represent the transformation relations of the components of a vector referred to two orthogonal systems of axes (x_i and $\bar{x}_j$). The 3×3 matrix $[\Lambda_S]$ is referred to as the *transformation matrix* of the rectangular system of axes x_i relative to the rectangular system of axes $\bar{x}_j$.

Multiplying both sides of relation (B.18) by $[\Lambda_S]$ and using (B.17) we get

$$[\Lambda_S]\{\bar{a}\} = \{a\} = [\Lambda_S][\Lambda_S]^T\{a\} \qquad (B.20)$$

From the above relation, we have

$$[\Lambda_S][\Lambda_S]^T = [I] \qquad (B.21)$$

where $[I]$ is the unit matrix given by

$$[I] = \begin{bmatrix} 1 & 0 & 0 \\ 0 & 1 & 0 \\ 0 & 0 & 1 \end{bmatrix} \qquad (B.22)$$

Relation (B.21) is referred to as the condition of orthogonality and represents the following six equations resulting from the orthogonality of the two systems of axes

$$\mathbf{i}_1 \cdot \mathbf{i}_2 = 0 \rightarrow \sum_{i=1}^{3} \lambda_{1i}\lambda_{2i} = 0 \qquad \mathbf{i}_1 \cdot \mathbf{i}_1 = 1 \rightarrow \sum_{i=1}^{3} \lambda_{1i}\lambda_{1i} = 1$$

$$\mathbf{i}_2 \cdot \mathbf{i}_3 = 0 \rightarrow \sum_{i=1}^{3} \lambda_{2i}\lambda_{3i} = 0 \qquad \mathbf{i}_2 \cdot \mathbf{i}_2 = 1 \rightarrow \sum_{i=1}^{3} \lambda_{2i}\lambda_{2i} = 1 \qquad (B.23)$$

$$\mathbf{i}_1 \cdot \mathbf{i}_3 = 0 \rightarrow \sum_{i=1}^{3} \lambda_{1i}\lambda_{3i} = 0 \qquad \mathbf{i}_3 \cdot \mathbf{i}_3 = 1 \rightarrow \sum_{i=1}^{3} \lambda_{3i}\lambda_{3i} = 1$$

Linear transformations, such as Eqs. (B.17) and (B.18), whose coefficients satisfy relation (B.21) are referred to as *orthogonal transformations*. The orthogonality condition (B.21) implies that the transpose $[\Lambda_S]^T$ of the matrix $[\Lambda_S]$ is equal to its inverse $[\Lambda_S]^{-1}$. That is,

$$[\Lambda_S]^T = [\Lambda_S]^{-1} \qquad (B.24)$$

The position of a point in space may be specified by a position vector $\mathbf{r}$ with respect to a fixed point 0. The components of this vector with respect to an orthogonal system of axes having as its origin point O are the coordinates of the point with respect to that system of axes. For example,

$$\mathbf{r} = x_1\mathbf{i}_1 + x_2\mathbf{i}_2 + x_3\mathbf{i}_3 = \bar{x}_1\bar{\mathbf{i}}_1 + \bar{x}_2\bar{\mathbf{i}}_2 + \bar{x}_3\bar{\mathbf{i}}_3 \qquad (B.25)$$

Referring to relations (B.17) and (B.18) the transformation relations of the coordinates of a point with respect to two orthogonal systems of axes x_i and $\bar{x}_j$ are

$$\{x\} = \begin{Bmatrix} x_1 \\ x_2 \\ x_3 \end{Bmatrix} = \begin{bmatrix} \lambda_{11} & \lambda_{12} & \lambda_{13} \\ \lambda_{21} & \lambda_{22} & \lambda_{23} \\ \lambda_{31} & \lambda_{32} & \lambda_{33} \end{bmatrix} \begin{Bmatrix} \bar{x}_1 \\ \bar{x}_2 \\ \bar{x}_3 \end{Bmatrix} = [\Lambda_S]\{\bar{x}\} \qquad (B.26)$$

and
$$\{\bar{x}\} = [\Lambda_S]^T \{x\} \qquad (B.27)$$

B.3 Transformation of the Components of a Planar Vector Upon Rotation of a Rectangular System of Axes

In the analysis of planar structures we encounter certain vectors such as forces and translations which act in the plane of the structure. In the sequel we establish the relations between the components of such a vector referred to two sets of orthogonal axes laying in the plane of the structure (see Fig. B.5). Consider the vector $\mathbf{F}$ acting in the plane specified by the axes $\bar{x}_1$ and $\bar{x}_2$ and denote its components with respect to these axes by $\bar{F}_1$ and $\bar{F}_2$, respectively. Moreover, consider another set of axes x_1 and x_2 in the plane $\bar{x}_1\bar{x}_2$ and denote the components of vector $\mathbf{F}$ with respect to the axes x_1 and x_2 by F_1 and F_2, respectively. That is,

$$\mathbf{F} = F_1\mathbf{i}_1 + F_2\mathbf{i}_2 = \bar{F}_1\bar{\mathbf{i}}_1 + \bar{F}_2\bar{\mathbf{i}}_2 \qquad (B.28)$$

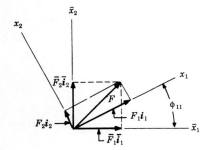

Figure B.5 Components of a planar vector.

Referring to Fig. B.5, we have

$$\begin{Bmatrix} F_1 \\ F_2 \end{Bmatrix} = [\Lambda_P] \begin{Bmatrix} \bar{F}_1 \\ \bar{F}_2 \end{Bmatrix} \tag{B.29}$$

and

$$\begin{Bmatrix} \bar{F}_1 \\ \bar{F}_2 \end{Bmatrix} = [\Lambda_P]^T \begin{Bmatrix} F_1 \\ F_2 \end{Bmatrix} \tag{B.30}$$

where

$$[\Lambda_P] = \begin{bmatrix} \lambda_{11} & \lambda_{12} \\ \lambda_{21} & \lambda_{22} \end{bmatrix} = \begin{bmatrix} \cos\phi_{11} & \sin\phi_{11} \\ -\sin\phi_{11} & \cos\phi_{11} \end{bmatrix} \tag{B.31}$$

From relations (B.29) and (B.30) it can be seen that

$$[\Lambda_P]^{-1} = [\Lambda_P]^T \tag{B.32}$$

Photographs and a Brief Description of a Cable-Fabric Dome

Cable-fabric domes are economical, long-span space structures used as roofs for arenas, stadiums, sport centers, etc., where it is required to cover a large area without interior column support. They are translucent allowing natural light to enter the covered area, creating the impression of an outdoor environment. It is bright enough inside that plants can grow.

The Suncoast Dome at St. Petersburg, Fla. (see Fig. B.6), was completed in 1989. Its diameter is 210 m (690 ft), and it encloses a 43,000-seat stadium primarily for baseball, football, and basketball. Moreover, a utility grid on the floor allows for a 13,940-m^2 (150,000-ft^2) column-free exhibition space, while hydraulically powered movable stands permit different seating arrangements for concerts, shows, and conventions. The Suncoast Dome consists of the following.

1. A circular concrete compression ring of 210 m (690 ft) external diameter which is supported by 24 concrete columns 1.825 m (6 ft) in diameter (see Figs. B.1 and B.7 to B.9). The heights of the columns vary from 53.22 to 30.48 m (175 to 100 ft) giving the compression ring a slope of 6.258° (see Fig. B.7). The 53.22-m height at the one end is required when the Suncoast Dome is used as a baseball stadium; it is reduced to 30.48 m at the other end in order to diminish the volume of the air to be air-conditioned.

2. A steel tension ring at the top of the dome (see Figs. B.7 and B.10) consisting of two horizontal rings connected together by vertical parts (see Fig. B.10).

Figure B.6 The Suncoast Dome of St. Petersburg, Fla. (*Courtesy Geiger Gossen Hamilton Liao Engineers P.C., New York, N.Y.*)

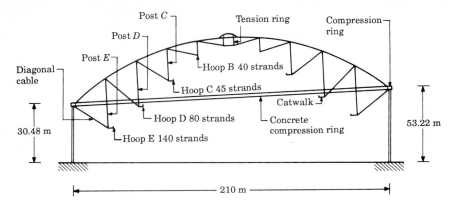

Figure B.7 Section through the Suncoast Dome. (*Courtesy Geiger Gossen Hamilton Liao Engineers P.C., New York, N.Y.*)

3. Four hoops each on a plane parallel to that of the compression ring (see Figs. B.7 to B.9). The largest hoop is made of 140 wire strands, while the smallest is made of 40 wire strands.

4. Diagonal cables whose size varies from 52 wire strands for the outer posts to 4 wire strands at the center tension ring (see Figs. B.7 and B.11).

Figure B.8 View of the Suncoast Dome during construction. (*Courtesy Geiger Gossen Hamilton Liao Engineers P.C., New York, N.Y.*)

Figure B.9 Interior view of the Suncoast Dome during construction. (*Courtesy Geiger Gossen Hamilton Liao Engineers P.C., New York, N.Y.*)

5. Vertical, round, steel posts with castings at both ends. The top casting connects each post to a ridge cable, whereas the bottom casting connects each post to a hoop and to a diagonal (see Figs. B.7 to B.9, B.11, and B.12).

Figure B.10 Steel tension ring. (*Courtesy Geiger Gossen Hamilton Liao Engineers P.C., New York, N.Y.*)

Figure B.11 View of the Suncoast Dome showing the four-strand diagonal cables, the lower part of the tension ring, and vertical posts connecting the two parts of the ring. (*Courtesy Geiger Gossen Hamilton Liao Engineers P.C., New York, N.Y.*)

6. The ridge cable assembly consisting of radial strands, the cable rope (see Figs. B.13 and B.14), and antiponding cables. The latter are used between ridge cables in order to prevent ponding of rain water.

Figure B.12 Detail of connection of diagonal cable and post to hoop. (*Courtesy Geiger Gossen Hamilton Liao Engineers P.C., New York, N.Y.*)

Figure B.13 Detail of part of the ridge cable assembly as the cable net is assembled on the ground. (*Courtesy Geiger Gossen Hamilton Liao Engineers P.C., New York, N.Y.*)

7. The outer fabric and the liner (see Fig. B.14). They are both made of a translucent teflon-coated fiberglass fabric. However, the material of the liner is thinner and of a looser weave. They are assembled from 24 equal panels that span from ridge cable to ridge cable.

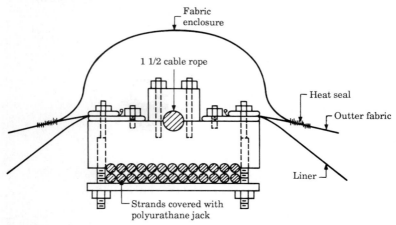

Figure B.14 Detail of the ridge cable assembly in place. (*Courtesy Geiger Gossen Hamilton Liao Engineers P.C., New York, N.Y.*)

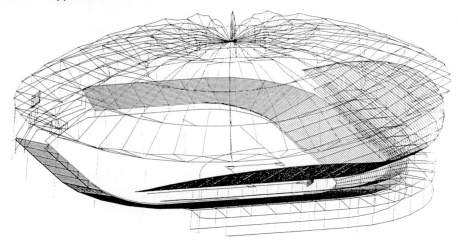

Figure B.15 Line diagram of the Suncoast Dome. (*Courtesy Geiger Gossen Hamilton Liao Engineers P.C., New York, N.Y.*)

The panels have been custom fabricated and were carefully handled in order to prevent damage to the fibers. The liner is used for acoustic purposes.

8. Catwalks supported by the hoop cables. They provide access to the roof and they support equipment (cameras, lights, etc.).

9. A 0.254-m- (10-in) thick concrete masonry block wall covered by metal siding. It closes the spaces between the columns which support the roof.

The strands used in the hoops, the diagonals, and the ridge cable assembly are those used in prestress concrete; they consist of 7 steel wires of 5.08 mm (0.20 in) diameter.

The roof was analyzed as a space framed structure using the model shown in Fig. B.15. In this model the cables are represented by elements which can carry only tensile forces, while the effect of the fabric is disregarded.

Index

ABOUT THE AUTHOR

Anthony E. Armenàkas is a professor of aerospace and mechanical engineering at Polytechnic University, Brooklyn, New York. He has over 35 years' experience teaching undergraduate and graduate courses in structural analysis and applied mechanics, as well as directing and performing research in these areas. He has been associated with the City University of New York, the Cooper Union for the Advancement of Science and Art, the University of Florida, and the National Technical University of Athens. Dr. Armenàkas was a Fulbright lecturer to Greece for two years. He has served as a consultant to many companies and government agencies in the United States and Europe on problems of structural and stress analysis. He has written numerous articles and technical papers on subjects related to structural analysis and is the author of two books: *Free Vibrations of Circular Cylindrical Shells* (Pergamon, 1969) and *Classical Structural Analysis: A Modern Approach* (McGraw-Hill, 1988). He is a fellow of both the American Society of Civil Engineers and the American Society of Mechanical Engineers.

A booklet is available which describes in detail and lists a computer program implementing the procedure for programming the direct stiffness method that is presented in Chapter 11. The program is written in FORTRAN 77 and in modular format. The input data and resulting output for several of this book's problems are included in the booklet. The computer program and the example problems are also available on Macintosh 3½-inch and IBM PC 5¼-inch disks. These items can be purchased from the following:

Computer Program
270 Round Swamp Road
West Hills, NY 11747